# Inorganic Nanoparticles

## Nanomaterials and Their Applications

Series Editor: M. Meyyappan

**Carbon Nanotubes: Reinforced Metal Matrix Composites**
*Arvind Agarwal, Srinivasa Rao Bakshi, Debrupa Lahiri*

**Inorganic Nanoparticles: Synthesis, Applications, and Perspectives**
*Edited by Claudia Altavilla, Enrico Ciliberto*

**Nanorobotics: An Introduction**
*Lixin Dong, Bradley J. Nelson*

**Graphene: Synthesis and Applications**
*Wonbong Choi, Jo-won Lee*

# Inorganic Nanoparticles

## *Synthesis, Applications, and Perspectives*

Edited by
**Claudia Altavilla**
**Enrico Ciliberto**

CRC Press
Taylor & Francis Group
Boca Raton London New York

CRC Press is an imprint of the
Taylor & Francis Group, an **informa** business

CRC Press
Taylor & Francis Group
6000 Broken Sound Parkway NW, Suite 300
Boca Raton, FL 33487-2742



Printed in the United States of America on acid-free paper
10 9 8 7 6 5 4 3 2 1

International Standard Book Number: 978-1-4398-1761-2 (Hardback)



**Visit the Taylor & Francis Web site at**
**http://www.taylorandfrancis.com**

**and the CRC Press Web site at**
**http://www.crcpress.com**

*This book is dedicated*

*to my parents Ida and Nicolò*

*"the guiding lights,"*

*to my husband Giuseppe,*

*"the true love,"*

*and to my daughter Marida*

*"the best reason to become a better person"*

**Claudia Altavilla**

*To my Family and to my Mentors*

**Enrico Ciliberto**

# Contents

# Foreword

Development, characterization, and exploitation of nanophase materials are all fundamental to the anticipated nanotechnology revolution. In the last decade, research activities on carbon nanotubes, inorganic nanowires, quantum dots, and nanoparticles have increased exponentially, as evidenced by the large number of papers in peer-reviewed journals and conference presentations across the world. Among the various nanomaterials, inorganic nanoparticles assume special importance because they are easier and cheaper to synthesize in the laboratory and to mass produce than some other nanomaterials like carbon nanotubes, for example. It is for this reason also that they can be more readily integrated into applications. As synthesis, characterization, and application development using nanoparticles continues strongly, there is a need to capture the fundamentals and the current advances in a textbook for the benefit of researchers, graduate students, and colleagues in various industries. This book by Drs. Claudia Altavilla and Enrico Ciliberto meets the above need admirably. An excellent group of experts have been assembled to discuss the diverse applications of inorganic nanoparticles, which would otherwise have been impossible to cover by just one or two people.

After an overview on material synthesis and general perspectives in Chapter 1, the book delves into myriad applications of nanoparticles. Chapter 2 covers a very interesting and unique application in the conservation of art. Magnetic materials have found their way into magnetic storage media long ago, and Chapter 3 covers the use of nanoparticles in this domain. Oxide thin films, especially tin oxide, have been the conducting media in commercial gas and vapor sensors, and Chapter 4 provides a discussion as to how their performance can be improved using metal and oxide nanoparticles. Solid-state lighting has attracted attention worldwide due to its higher efficiency compared to conventional lighting, but the costs remain very high. Advances in materials, device fabrication, and large-scale production are urgently required to reduce global energy demands. Chapter 5 discusses the advances in semiconductor nanoparticles for light-emitting devices.

Besides lighting, other areas related to the energy sector, such as solar energy and energy storage devices (fuel cells, rechargeable batteries, etc.), can also benefit from the properties of nanomaterials. These are covered in Chapters 7 through 9. Another industrial sector that is likely to feel the impact of nanotechnology is the biomedical field. Several chapters are devoted to quantum dots for bioimaging, nanoparticle-based cancer therapy, drug delivery, antibacterial agents, and others. Last but not the least is the long-standing application in catalysis and the role of nanosized particles in this established field.

I hope the readers find this treatise useful as a textbook and research resource. Nora Konopka of CRC Press deserves praise for initiating the book series on nanomaterials.

**Meya Meyyappan**
*Moffett Field, California*

# *Acknowledgments*

We thank all the contributors to this book for their extra effort in presenting state-of-the-art developments in their areas of expertise. This book would not have been possible without them. Additionally, we would like to acknowledge Dr. Meya Meyyappan for his trust and support in the realization of this project, and Nora Konopka and Kari Budyk of CRC Press for their constant technical support during all the stages of production. Tom Schott, who designed the cover of the book, is also heartily acknowledged. Finally, a special thanks to our families for their endless patience, which allowed us to spend time on the preparation of this book.

# Editors

**Dr. Claudia Altavilla** graduated in chemistry (cum laude) in 2001 from the University of Catania, Italy. She received her PhD in chemistry in 2006 from the same university with a dissertation on the synthesis and characterization of nanostructured materials assembled on inorganic substrates. She worked as a visiting scientist at Ludwig Maximillians Universitat in Munich, Germany, with Professor Wolfgang Parak, and at the University of Florence, Italy, with Professor Dante Gatteschi, where she was involved in the magnetic characterization of nanoparticle monolayers on silicon substrates. Since 2005, she has been a professor of inorganic protective and consolidant methods in cultural heritage at the University of Catania.

Dr. Altavilla's current research includes the chemical synthesis of inorganic nanoparticles of ferrite, chalcogenite, and metals functionalized by different organic coatings for application in magnetic storage media, lubricants, magnetorheological fluids, and biomedicine; and self-assembled monolayers of inorganic and organic nanostructures on different substrates and CVD synthesis of carbon nanotubes on silicon substrates using transition metal oxide nanoparticles as catalyst. She has published several papers and monographs. She is a referee for international journals on material science and nanotechnology such as *ACS Nano*, *Chemistry of Materials*, and the *Journal of Material Chemistry*. Currently she is a research fellow in the Department of Chemical and Food Engineering, University of Salerno, Italy.

**Dr. Enrico Ciliberto** is a full professor of inorganic chemistry at the University of Catania and the president of the Cultural Heritage Technologies Faculty at the University of Syracuse, Italy. His research focuses on the chemistry of materials, including surface science and cultural heritage materials, both from an archaeometric and conservative point of view, and covers Minoan mortars in Crete, Michelangelo's David in Florence, and Saint Mark's Basilica in Venice. His current scientific interest includes the application of nanotechnologies for the conservation of works of art. He has also published over 100 scientific papers.

# Contributors

**Alessandra Aloisi**
National Nanotechnology Laboratory of CNR-INFM
Italian Institute of Technology Research Unit
Lecce, Italy

**Claudia Altavilla**
Department of Chemical and Food Engineering
University of Salerno
Fisciano, Italy

**Guy Applerot**
Department of Chemistry

and

Kanbar Laboratory for Nanomaterials
Bar-Ilan University Center for Advanced Materials and Nanotechnology
Bar-Ilan University
Ramat-Gan, Israel

**Elif Arici**
Linz Institute for Organic Solar Cells
Institute of Physical Chemistry
Johannes Kepler University
Linz, Austria

**Doron Aurbach**
Department of Chemistry
Bar-Ilan University
Ramat Gan, Israel

**Gerd Bacher**
Werkstoffe der Elektrotechnik and Center for Nanointegration Duisburg-Essen
University Duisburg-Essen
Duisburg, Germany

**Piero Baglioni**
Department of Chemistry and Consorzio Interuniversitario per lo Sviluppo dei Sistemi a Grande Interfase
University of Florence
Sesto Fiorentino, Italy

**Maya Bar-Sadan**
Institute of Solid State Research
Ernst Ruska-Centre for Microscopy and Spectroscopy with Electrons
Research Centre Juelich
Juelich, Germany

**Paolo Ciambelli**
Department of Chemical and Food Engineering

and

Centre NANO_MATES
University of Salerno
Fisciano, Italy

**Enrico Ciliberto**
Dipartimento di Scienze Chimiche
Università di of Catania
Catania, Italy

**Bin Fang**
Department of Chemistry
State University of New York, Binghamton
Binghamton, New York

**Natalie A. Frey**
Department of Chemistry
Brown University
Providence, Rhode Island

**Aharon Gedanken**
Department of Chemistry

and

Kanbar Laboratory for Nanomaterials
Bar-Ilan University Center for Advanced Materials and Nanotechnology
Bar-Ilan University
Ramat-Gan, Israel

**Rodorico Giorgi**
Department of Chemistry and Consorzio Interuniversitario per lo Sviluppo dei Sistemi a Grande Interfase
University of Florence
Sesto Fiorentino, Italy

**Lori J. Groven**
Chemical and Biological Engineering Department
South Dakota School of Mines and Technology
Rapid City, South Dakota

**Georg M. Guebitz**
Institute of Environmental Biotechnology
Graz University of Technology
Graz, Austria

**Ortal Haik**
Department of Chemistry
Bar-Ilan University
Ramat Gan, Israel

**Manika Khanuja**
Thin Film Laboratory
Department of Physics
Indian Institute of Technology
New Delhi, India

**Tilmar Kümmell**
Werkstoffe der Elektrotechnik and Center for Nanointegration Duisburg-Essen
University Duisburg-Essen
Duisburg, Germany

**Wilson A. Lee**
Estee Lauder Companies, Inc.
Melville, New York

**Brita Lehmann**
Department of Radiology
School of Medicine
University of California Davis
Sacramento, California

**Joerg Lehmann**
Department of Radiation Oncology
School of Medicine
University of California Davis
Sacramento, California

**Stephanie Lim**
Department of Chemistry
State University of New York, Binghamton
Binghamton, New York

**Rameshwori Loukrakpam**
Department of Chemistry
State University of New York, Binghamton
Binghamton, New York

**Jin Luo**
Department of Chemistry
State University of New York, Binghamton
Binghamton, New York

**B.R. Mehta**
Thin Film Laboratory
Department of Physics
Indian Institute of Technology
New Delhi, India

**Derrick Mott**
Department of Chemistry
State University of New York, Binghamton
Binghamton, New York

**Ekaterina Neshataeva**
Werkstoffe der Elektrotechnik and Center for Nanointegration Duisburg-Essen
University Duisburg-Essen
Duisburg, Germany

**Peter N. Njoki**
Department of Chemistry
State University of New York Binghamton
Binghamton, New York

**Teresa Pellegrino**
National Nanotechnology Laboratory of CNR-INFM
Italian Institute of Technology Research Unit
Lecce, Italy

and

Istituto Italiano di Tecnologia
Genova, Italy

**Ilana Perelshtein**
Department of Chemistry

and

Kanbar Laboratory for Nanomaterials
Bar-Ilan University Center for Advanced Materials and Nanotechnology
Bar-Ilan University
Ramat-Gan, Israel

**Nina Perkas**
Department of Chemistry

and

Kanbar Laboratory for Nanomaterials
Bar-Ilan University Center for Advanced Materials and Nanotechnology
Bar-Ilan University
Ramat-Gan, Israel

**Jan A. Puszynski**
Chemical and Biological Engineering Department
South Dakota School of Mines and Technology
Rapid City, South Dakota

**Miriam Raifailovich**
Material Science and Engineering Department
Stony Brook University
Stony Brook, New York

**Andrea Ragusa**
National Nanotechnology Laboratory of CNR-INFM
Italian Institute of Technology Research Unit
Lecce, Italy

**Diana Sannino**
Department of Chemical and Food Engineering

and

Centre for NANOMAterials and NanoTEchnology
University of Salerno
Fisciano, Italy

**Maria Sarno**
Department of Chemical and Food Engineering

and

Centre for NANOMAterials and NanoTEchnology
University of Salerno
Fisciano, Italy

**Thomas B. Scott**
Interface Analysis Centre
University of Bristol
Bristol, United Kingdom

**V.N. Singh**
Thin Film Laboratory
Department of Physics
Indian Institute of Technology
New Delhi, India

**Shouheng Sun**
Department of Chemistry
Brown University
Providence, Rhode Island

**Reshef Tenne**
Materials and Interfaces Department
Weizmann Institue of Science
Rehovot, Israel

**Naoki Toshima**
Department of Applied Chemistry
Tokyo University of Science, Yamaguchi
Sanyo-Onoda, Japan

and

Core Research for Evolutional Science and Technology (CREST)
Japan Science and Technology Agency
Kawaguchi, Japan

**Bridgid N. Wanjala**
Department of Chemistry
State University of New York, Binghamton
Binghamton, New York

**Eva Wehrschuetz-Sigl**
Institute of Environmental Biotechnology
Graz University of Technology
Graz, Austria

**Jun Yin**
Department of Chemistry
State University of New York, Binghamton
Binghamton, New York

**Antonella Zacheo**
National Nanotechnology Laboratory of CNR-INFM
Italian Institute of Technology Research Unit
Lecce, Italy

**Chuan-Jian Zhong**
Department of Chemistry
State University of New York, Binghamton
Binghamton, New York

# 1

# *Inorganic Nanoparticles: Synthesis, Applications, and Perspectives—An Overview*

**Claudia Altavilla and Enrico Ciliberto**

**CONTENTS**



## 1.1 Introduction

Over the last few years, a variety of inorganic nanomaterials such as nanoparticles, nanowires, and nanotubes have been created or modified in order to obtain superior properties with greater functional versatility. The advent of nanoscale science and technology has stimulated a big effort to develop new strategies for the synthesis of nanomaterials of a controlled size and shape. In particular, nanoparticles due to their size, in the range of 1–100 nm, have been examined for their uses as tools for a new generation of technological devices. Moreover, due to their dimensions and shapes being similar to several biological structures (e.g., membrane cell genes, proteins, and viruses), they have been proposed for investigating biological processes as well as for sensing and treating diseases. Nowadays, the volume of studies dealing with these topics represents one of the most impressive phenomenon in all of scientific history. Even so only one Nobel prize, shared by three scientists, has been awarded for the development of the studies in this field in the last 20 years, in 1996, Robert F. Curl Jr., Sir Harold W. Kroto, and Richard E. Smalley were awarded for their discovery of fullerenes. In Figure 1.1, the number of scientific articles and papers with reference to the themes of nanoparticles from 1996 until 2009 is reported: the exponential trend clearly indicates that the scientific and technological interest is continuing to increase.

Compared with the notable amount of scientific and technological studies in this field, only one Nobel prize could sound quite inadequate. One reason can probably be attributed to the fact that "nanotechnologies" are very old, even though several of the relationships between dimension and properties have only been clarified in the nineteenth century. In fact, very few people know that even in the sixth century BC, nanotechnology was commonly used in the Attic region (Greece). During the Archaic and Classical periods, roughly 620–300 BC, in the region of Attica, dominated by the city of Athens, the production of

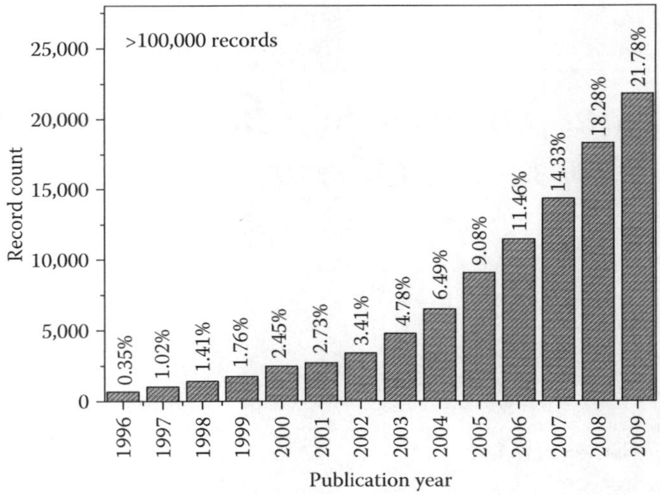


**FIGURE 1.1**
Temporal evolution in the number of scientific papers on nanoparticles published from 1996 to 2009. More than 100,000 records were found and more than 54% of the articles have been published in the last 3 years. (Data from ISI Web of Knowledge.)

decorated vases reached an extraordinary artistic level due to the development of a highly original firing technique that obtained a magnificent black/red dichromatism, the secret of Greek vases (Figure 1.2) (Boardman 1991).

The reason why a deep black color formed on the vase surface was discovered only a few years ago. During the firing process, spinel-like nanoparticles formed inside a glassy layer, which is a few microns thick (Maniatis et al. 1992). In Figure 1.3, a secondary electron microscope (SEM) image of submicron particles inside a glossy layer of a sixth century BC Greek vase is reported. The magnetite particles, looking whitish in the backscattered mode, show different sizes (100–300 nm) and different shapes. A skillful alternation of the

**FIGURE 1.2**
Attic black figure krater, sixth century BC. (Courtesy of Prof. Enrico Ciliberto.)

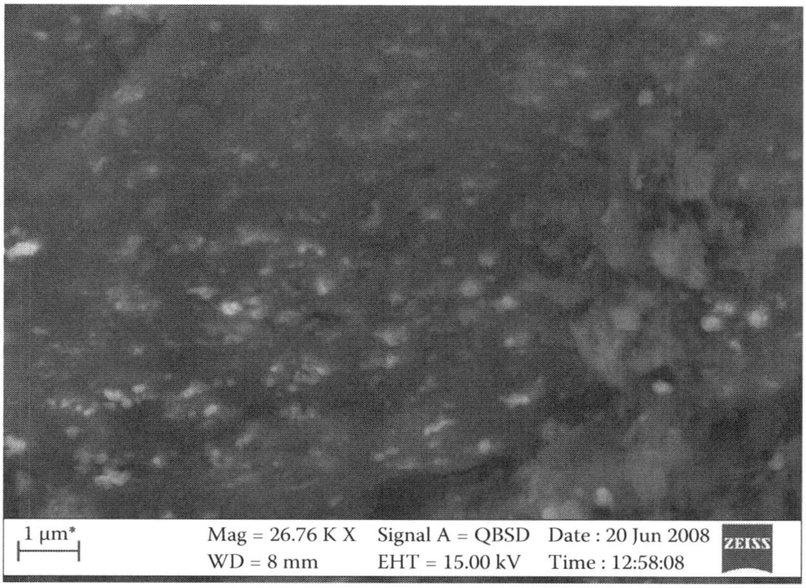


**FIGURE 1.3**
Back scattered electron image of submicron magnetite particles inside the black glossy layer of the vase reported in Figure 1.2. (Courtesy of Prof. Enrico Ciliberto.)

oxidizing and reduction processes, induced by either opening or closing the oven vents, stimulated the formation of black magnetite nanoparticles (Hemelrijk 1991).

In addition, "luster" ceramic decorations have been revealed by transmission electron microscopy (TEM) to have been ancient nanostructured metallic thin films made by man. Considering this type of decoration in the context of cultural heritage, it is a remarkable discovery in the history of technology, because nanocrystal films have been produced empirically since medieval times (Borgia et al. 2002; Padovani et al. 2003). Luster is a type of ceramic decoration, which results in a beautiful metallic shine and colored iridescence on the surface of the ceramic object. The earliest luster was probably made in Iraq in the early ninth century AD on tin-glazed ceramics. However, luster technology spread from the Middle East to Persia, Egypt, Spain, and Italy, and its splendid production continued in the centuries that followed through to the present day. In TEM, luster layers appear with a homogeneous surface microstructure formed by small quasispherical clusters, embedded in an amorphous glassy matrix (Figure 1.4).

The total thickness of this structure is 200–500 nm. An initial outer layer is formed by the biggest clusters, which have a diameter of about 50 nm. The diameter of the next layer, with smaller inner clusters, is 5–20 nm. With respect to the composition of these clusters, transmission electron microscopy (TEM) fitted with energy dispersive x-ray spectroscopy (EDS) analyses indicate that the nanoclusters are particles of pure copper and silver (Perez-Arantegui and Larrea 2003).

In addition, red glasses that are very ancient are colored due to the presence of nanoparticles. In fact, excavations at Qantir, on the Nile Delta, have given insight into the organization and development of an industrial estate in Ramesside, Egypt. In founding the new capital of Egypt, Piramesses, during the nineteenth dynasty, a huge bronze-casting factory was built, accompanied by a range of other, nonmetallic high-temperature industries.

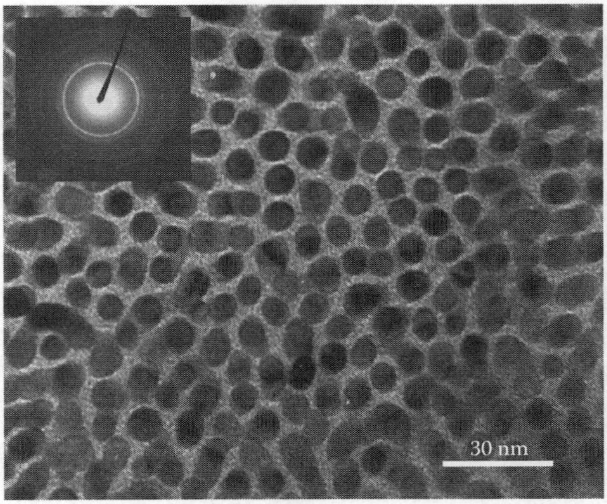


**FIGURE 1.4**
TEM image of a smalt from an Italian Renaissance Luster Majolica. Copper nanoparticles show diameters ranging from 7 to 10 nm. (Courtesy of Prof. Bruno Brunetti.)

Besides, an abundant production of faience implements, coated with copper-colored glazes, and the manufacturing of Egyptian blue, the coloration of large quantities of red glass also played a major role. The production of glass is attested by numerous crucibles, mostly with adhering traces of red glass. While evidence of glass working by artisans is absent, there are indications that the production of both raw glass and glass coloring took place. The nature and complexity of high-temperature industrial debris found at Qantir suggest a highly specialized organization of labor within a framework of shared technologies and skills of closely controlled temperatures and redox conditions. This cross-craft workshop pattern further reveals a significant level of intracraft specialization as well as the spatial separation of glass making, coloring, and finally working in the Late Bronze Age Egypt (Rehren et al. 1998). We now know that the red color is due to metal nanoparticles contained in the glass network. The use of metal nanoparticles dispersed in an optically clear matrix by potters and glassmakers from the Bronze Age up to the present time has been reviewed by Colomban from a solid-state chemistry and material science point of view. The nature of metal (gold, silver, or copper) and the importance of some other elements (Fe, Sn, Sb, and Bi) added to control metal reduction in the glass in relation to the firing atmosphere (combined reducing oxidizing sequences and role of hydrogen and water) are considered in the light of ancient treatises and recent analyses using advanced techniques (TEM, extended x-ray absorption fine structure (EXAFS), etc.) as well as classical methods (optical microscopy, UV–visible absorption). The different types of color production, by absorption/reflection (red and yellow) or diffraction (iridescence), as well as the relationship between nanostructures (metal particle dispersion and layer stacking) and luster color have been also discussed. It has also been shown that Raman scattering is a very useful technique in order to study the local glass structure around the metal particles as well as detect incomplete metal reduction or residues tracing the preparation route; therefore, making it possible to differentiate between genuine artifacts and fakes (Colomban 2009).

In all the aforementioned cases, old technology surpassed the scientific interpretation of the related phenomena and, today, experimental experience remains the basis of modern progress, even if thermodynamic and quantum mechanics have already explained many of the properties of nanoscale materials (Lafait 2006; Cavalcante et al. 2009).

## 1.2 Properties of Nanoparticles

On the nanoscale, materials behave very differently compared to larger scales. In fact, nanoparticles often have unique physical and chemical properties. For example, the electronic, optical, and chemical properties of nanoparticles may be very different from those of each component in the bulk. By increasing the surface area with respect to the volume of a particle, a corresponding increasing of importance of the behavior of the surface atoms can be observed, and a modification of the properties of the particle itself as well as of its interaction with the surrounding environment take place. Moreover, in order to become small enough, a transition from classical physics behavior to a quantum mechanic, one describes the particle that can now be viewed as an artificial atom, an object that possesses discrete electronic states, similar to naturally occurring atoms. An electron in an artificial atom that can be described by a quantum wave-function that is similar to the one used for an electron in a single atom, even though its energy is spread coherently over the lattice of atomic nuclei. As the size of a crystal decreases to the nanometer regime, the size of the particle begins to modify the properties of the crystal. The electronic structure is altered by the continuous electronic bands to discrete or quantized electronic levels. As a result, the continuous optical transitions between the electronic bands become discrete and the properties of the nanomaterial become size-dependent. Therefore, optical, thermal, and electrical properties of the particles become dependent on their sizes and shapes. These properties have been recently reviewed by Burda et al. (2005).

However, some of the properties of the nanoparticles might not be predicted by understanding the increasing influence of surface atoms or quantum effect. For instance, it has been shown that silicon nanoparticles in the range of 20–100 nm are superhard in the 30–50 GPa range after work hardening (Gerberich et al. 2003). The nanosphere hardness falls between the conventional hardness of sapphires and diamonds, which are among the hardest known materials. The extremely small dimensions of nanobuilding blocks have created difficult challenges to many existing instruments, methodologies, and even theories. The methods that have been developed and used for measuring the mechanical properties of isolated individual nanobuilding blocks include uniaxial tensile loading using a nanomanipulation stage, in-situ compression of nanoparticles and nanopillars, mechanical/electric-field-induced resonance, atomic force microscopy (AFM) bending, and nanoindentation (Uchic et al. 2004). These methods certainly represent important instruments that help scientists in designing low-cost superhard materials from nanoscale building blocks.

While nanoparticles display properties that differ from those of bulk samples of the same material, groups of nanoparticles can have collective properties that are different to those displayed by individual nanoparticles and bulk samples. For realizing versatile functions, an assembly of nanoparticles in regular patterns on surfaces and at interfaces is required (Altavilla 2007). Assembling nanoparticles generates new nanostructures, which have unforeseen collective, intrinsic physical properties. These properties can be exploited for multipurpose applications in nanoelectronics, spintronics, sensors, etc. (Nie et al. 2010).

## 1.3 Synthesis Strategies

There are a wide variety of techniques for producing nanoparticles. These essentially fall into three categories: physical methods, chemical syntheses, and mechanical processes such as milling.

Among the physical methods, pulsed laser ablation has been demonstrated to be a powerful and versatile technique for preparing high-purity nanoparticles or nanofilms (Longstreth-Spoor et al. 2008). In general, the targets used for the preparation of nanoparticles or films by laser ablation are bulk sizes, and the lasers are either excimer, pulsed yttrium aluminium garnet (YAG), or femtosecond lasers. The quality and sizes of the nanoparticles prepared by these systems are controlled by optimizing either the laser parameters or ambient-gas pressure. The main advantage of laser ablation is the congruent (stoichiometric) material transport above the threshold fluence, which is used for depositing complex compounds such as high-Tc superconductors (Kang et al. 2006). In addition, high-melting-point materials (e.g., C, W, and refractory ceramics) are easily deposited (Ullmann et al. 2002; Chen et al. 2004). Passivated $\alpha$-Fe nanoparticles can also be prepared at atmospheric pressure by pulsed laser ablation of an Fe wire and a bulk Fe target (Wang et al. 2009). Other physical methods used in preparing nanoparticles belong to the category of vapor condensation. This approach is used to prepare metallic and metal oxide ceramic nanoparticles. It involves the evaporation of a solid metal followed by rapid condensation to form the final nanostructured material. Different methods can be adopted to produce metal vapors. An inert gas is also used to inhibit oxidizing phenomena but in some cases, oxygen atmosphere is used to make metal oxide nanoparticles. The main advantage of this approach is low contamination levels. Final particle size is controlled by the variation of temperature, flux parameters, and gas environment (Swihart 2003).

The most widely used chemical synthesis essentially consists of growing nanoparticles in a liquid medium made up of various reactants. The chemical growth of bulk or nanometer-sized materials inevitably involves the process of precipitation of a solid phase from a solution. For a particular solvent, there is a certain solubility for a solute, whereby addition of any excess solute will result in the precipitation and formation of nanocrystals. Thus, in the case of nanoparticle formation, for nucleation to occur, the solution must be supersaturated either by directly dissolving the solute at higher temperatures and then cooling to low temperatures or by adding the necessary reactants to produce a supersaturated solution during the reaction. The precipitation process then basically consists of a nucleation step followed by particle growth stages (Peng et al. 1998). For a homogeneous nucleation that occurs in the absence of a solid interface, the phenomenon can be described by the overall free energy change ($\Delta G$) because the supersaturated solutions are not stable from a thermodynamic point of view. It has been demonstrated that $\Delta G$ depends on the saturation ratio of the solution as well as the radius of nuclei formed (Burda et al. 2005).

$\Delta G$ shows a maximum critical value of the radius ($r_c$) that corresponds to a critical size of the particle (see Figure 1.5). This maximum free energy is the activation energy for nucleation. Nuclei larger than the critical size will further decrease their free energy for growth and form stable nuclei that grow to form particles.

The growth process of nanocrystals can occur in two different ways, "focusing" and "defocusing," depending on the concentration of the solution. A critical size exists at any given concentration. At a high concentration, the critical size is small so that all the particles grow. In this situation, smaller particles, slightly larger than the critical size, have a high free energy driving force and grow faster than the larger ones. As a result, the size

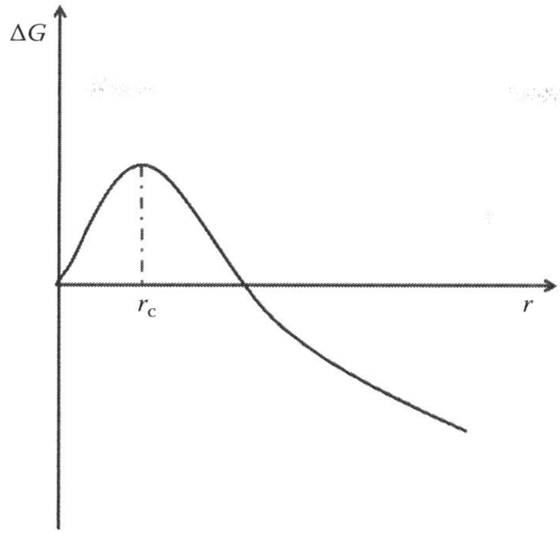


**FIGURE 1.5**
Free energy $\Delta G$ as a function of the radius of particle; $r_c$, critical radius size.

distribution can be focused down to one that is nearly monodisperse. If the monomer concentration is below a critical threshold, small nanocrystals are depleted as larger ones grow and the size distribution broadens, or defocuses (Yin and Alivisatos 2005).

The preparation of nearly monodisperse spherical particles can be achieved by stopping the reaction while it is still in the focusing regime, with a large concentration of monomer still present (Peng et al. 1998). In general, it is desirable for nucleation to be separated in time from the growth step in order to obtain relatively monodisperse samples. This means that nucleation must occur on a short timescale. This may be achieved by rapidly injecting suitable precursors into the solvent at high temperatures to generate transient supersaturation in solutions and induce a nucleation burst. In addition to this kind of growth, where soluble species deposit on the solid surface, particles can grow by aggregation with other particles, and this is called secondary growth. The rate of particle growth by aggregation is much larger than by molecular addition. Finally, the control over size, size distribution, and secondary growth becomes a more challenging problem in such dimensional regimes. In the synthesis of colloidal nanoparticles, the key strategy stands within the use of specific molecules, which act as terminating or stabilizing agents, ensuring a slow growth rate, preventing interparticle agglomeration, and conferring stability as well as further processability to the resulting nanoparticles. These molecules are often chosen among various classes of surfactants. Surfactants are molecules composed of a polar head group and one or more hydrocarbon chains with a hydrophobic nature. The most commonly used in colloidal syntheses include alkyl thiols, amines, carboxylic and phosphonic acids, phosphines, phosphine oxides, phosphates, phosphonates, as well as various coordinating solvents (Cozzoli et al. 2006).

An important step in the generation of colloidal inorganic nanoparticles is the identification of suitable precursor molecules such as metal complexes and organometallic compounds. The precursors need to rapidly decompose or react at the required growth temperature to yield reactive atomic or molecular species (often called monomers), which then cause nanocrystal nucleation and growth (Stuczynski et al. 1989; Steigerwald 1994). In this sense, these chemical methods operating in solutions can be related to the metal organic chemical vapor deposition where volatile precursors in vapor phase react and/or decompose on the substrate surface to produce a desired deposit at much higher growth temperatures (Sun et al. 2004; Apátiga et al. 2007; Creighton et al. 2008). The two approaches share many features including similar basic chemical reactions involved.

Along with the mechanical techniques used to prepare nanoparticles, a method that has received a great deal of interest from the industrial world is bead milling. The particle size achieved from a bead mill is a direct function of the size of the beads used for the grinding process. The average particle size that can be quickly achieved in a bead mill is about 1/1000 the size of the grinding media. The smallest bead size regularly used on a commercial basis is 200–300 μm. The applications of this media are primarily in the pigment manufacturing and ink industry for the fine grinding and dispersion of pigments such as phthalocyanine blue and green as well as carbon black. The uses for these inks are in the ink jet market, textile inks, etc. (Czekai 1996).

Sonochemistry is characterized by both mechanical and chemical properties. In fact, sonochemical methods refer to chemical reactions that are induced by acoustic cavitations. In organic solvents, the high-temperature conditions generated during acoustic cavitations have been used to synthesize metal and other nanomaterials. In water, a variety of primary and secondary radicals are generated during acoustic cavitations that can be used for a series of redox reactions in aqueous solutions. Moreover, it has been demonstrated how the size, size distribution, and, to some extent, the shape of metal nanoparticles may be controlled by the sonochemical preparation method (Muthupandian 2008).

## 1.4 Applications

The goal of this book is to describe the most important applications of nanoparticles.

In Chapter 2, Piero Baglioni and Rodorico Giorgi introduce the use of nanoparticles in the field of cultural heritage conservation. The contribution of science to the conservation of cultural heritage has radically increased over the last years, many thanks to the advancements in the knowledge of the physicochemical composition and properties of the materials constituting the works of art (Ciliberto 2000).

Nanoparticles of calcium hydroxide give a consistent improvement over the classical application of a calcium hydroxide solution. In fact, the use of $Ca(OH)_2$ dispersions overcome the limitation due to the low solubility in water, alcohols are less aggressive than water toward fragile mural paintings, and the quick carbonation of hydroxides gives a strong consolidation effect. Calcium and magnesium hydroxide as a nonaqueous dispersion also give excellent results for the treatment of cellulose-based materials. These preferably require waterless solvents and need an alkaline reserve to protect the object from further degradation due to pollution or internal acid production as a consequence of the natural aging of the materials. Humble particles of calcium or magnesium hydroxide give excellent results and ensure high physicochemical compatibility with the substrates that grant the durability of the treatment and long-lasting protection of the works of art. With illustrative examples on the consolidation of wall paintings and deacidification of books and wood, this contribution also reports on some recent *case studies*, highlighting the improved performances of nanoparticles and nanocontainers (micelles, microemulsion, nanogels, etc.) in respect to traditional conservation methodologies.

The use of magnetic nanoparticles for an information storage application is discussed by Natalie A. Frey and Shouheng Sun in Chapter 3. High-quality monodisperse magnetic nanoparticles with high coercivity can be made from various chemical synthesis routes and provide a way around the superparamagnetic limit that is currently encroaching upon the granular media that is used in hard disk drives today. Considering that synthesized nanoparticles are usually superparamagnetic, some novel approaches have been used to anneal the particles at high temperature, facilitating the face centered cube (fcc) to face centered tetragonal (fct) phase transition in FePt nanoparticles while keeping the particles from sintering and allowing the particles to be dispersed again in organic solvents. In order to increase packing density to maximize areal density for media, the shapes can be controlled and self-assembly can be employed to control interparticle spacing, which in turn provides control over magnetic interactions. Even higher anisotropic rare earth-transition metal nanoparticles are being synthesized, though the challenges associated with their syntheses are significant. In the case of $SmCo_5$, nanoscale powders with high coercivity have been made after reductively annealing core-shell structures. More needs to be done to protect these particles from sintering during the annealing process and the issue of chemical stability needs to be addressed. The results presented paint a promising picture for the future of magnetic nanoparticles in recording technology.

Gas sensor technologies have received a significant boost from nanoparticles. There is a large volume of data on the use of metal oxide nanoparticles and nanoparticle layers for gas sensor applications. Lack of accurate and reliable information about nanoparticle size, size distribution, metal additive, composition, and configuration makes the analysis of this data a challenging task, but B.R. Mehta et al. describe the current state of art of this topic in Chapter 4. Due to the percentage of atoms on the surface increasing with the decrease in particle size as the surface-to-volume ratio is inversely proportional to radius, nanoparticles will offer a large surface area for gas adsorption, which is always the first step in the gas-sensing mechanism. However, for a more detailed and clearer understanding of the dependence of gas sensing properties on nanoparticle size and the nature of the metal additive, it is important to use synthesis methods suitable for yielding well-defined nanoparticle sizes and composite configuration. Some of the current research directions include the use of synthesis methods for well-defined nanoparticle sizes, a reliable and scale electronic characterization of nanoparticles using conducting AFM and scanning tunneling microscopy on gas exposure, as well as the fabrication of nanowire–nanoparticle or decorated nanowire composites.

Nowadays, the demand for low-cost light emitters is high, covering a wide range of different applications in the advertising and giveaway industry, low-cost indicators, and displays for consumer electronics, mobile phones, toys, and many more. In Chapter 5, Ekaterina Neshataeva et al. discuss light-emitting devices (LEDs) based on semiconductor nanoparticles. Versatile implementations of nanocrystals in LEDs are expected to combine the robustness and efficiency of conventional semiconductor LEDs with low-cost processing techniques used for large-area organic LEDs. This fascinating research field not only requires the development of innovative fabrication and processing techniques using nanoparticles but also opens a path toward novel applications and devices. In the chapter, an overview of various device concepts and technical approaches is given focusing on the devices, where nanoparticles are used as active materials. Both direct and alternating current-driven light emitters are also discussed, covering the time span from early to recent developments in the field.

The formation of nanosize aluminum and its applications in condensed phase reaction has been reviewed by Jan A. Puszynski and Lori J. Groven in Chapter 6. They clearly indicate that the use of nanosize reactants in condensed phase exothermic reactions leads to a significant increase in the energy release rate. Such high-energy release rates, not commonly observed between oxidizer and fuel particles, make these nanoenergetic systems suitable candidates for environmentally benign macro- and microinitiators as well as energetic components of microthrusters and other applications requiring fast combustion front velocities. The recent developments in the formation of aluminum nanopowders indicate that high-temperature methods seem to be more suitable for scale-up than low-temperature wet chemistry synthesis routes. The mechanical reduction of aluminum particle size seems to be another promising approach for making larger quantities of reactive aluminum nanopowders.

Fuel cells using hydrogen represent an important form of tomorrow's energy due to it not only being an efficient fuel but also environmentally clean. The auto industry, which relies on oil-fuelled cars, is perhaps the biggest driving force behind the massive investment in fuel cell development. Jin Luo et al. have reviewed this interesting area of research in Chapter 7. In particular, they claim that the molecular encapsulation approach to the synthesis and processing of bimetallic/trimetallic nanoparticles is effective in producing alloy nanoparticles in the 2–5 nm regime with controllable composition and carbon-supported catalysts for fuel cell reactions. This approach differs from other traditional preparation approaches of supported catalysts in the abilities to control the nanoscale size, multimetallic composition, phase properties, and surface properties. As demonstrated by the bimetallic AuPt alloy nanoparticle catalysts, synergistic activity is possible in which Au atoms surrounding Pt provide effective sites for the reaction adsorbates in the electrocatalytic reaction. The fact that this bimetallic nanoparticle system displays a unique single-phase property different from the miscibility gap of its bulk-scale counterpart serves as an important indication of the operation of nanoscale phenomena in the catalysts, which can be further exploited for the design and preparation of the nanostructured bimetallic catalysts for fuel cells. Trimetallic nanoparticle catalysts have displayed enhanced electrocatalytic activity. For carbon-supported ternary PtVFe and PtNiFe nanoparticle catalysts, the size, composition, and loading of the nanoparticles on carbon support have been shown to be controllable, as well as processible by controlled thermal treatment and calcination, which can be optimized in order to achieve the effective shell removal and alloying of the ternary catalysts. The measurements of the intrinsic kinetic activities of the catalysts toward an oxygen reduction reaction have shown high electrocatalytic activities, and the trimetallic PtVFe nanoparticle catalysts prepared by the nanoengineered synthesis and processing methods have exhibited a much better performance in proton exchange membrane (PEM) fuel cell cathode than the commercial Pt catalyst. It also becomes clear that the synthesis and processing approach to the preparation of nanoparticle catalysts is promising for delivering much higher catalyst utilization than those of conventional methods, which has important implications on the improved design of fuel cell cathode catalysts.

Solar cells, devices that convert the energy of sunlight directly into electricity, based on organic–inorganic hybrid blends are discussed by Elif Arici-Bogner in Chapter 8. He describes the current state of art in organic–inorganic hybrid solar cells that use nanocrystalline inorganic materials in two different functions: as anodes and inorganic dyes in dye-sensitized solar cells as well as in bulk heterojunction solar cells. The basic parameters of photovoltaic devices and their characterization, synthesis aspects of inorganic nanoparticles investigated as active materials in solar cells as well as the material characterization methods, and the new developments for integration of inorganic nanoparticles in photovoltaic devices are discussed in this chapter.

The relationship between nanoparticles and rechargeable batteries is described in Chapter 9 by Doron Aurbach and Ortal Haik. This chapter deals with the possible use of nanomaterials in devices for energy storage and conversion, with an emphasis on inorganic species (e.g., alloys transition metal oxides and sulfides and carbon nanotubes). Four types of devices are discussed and classified: batteries (primary and secondary), fuel cells, super electric double-layer capacitors, and photovoltaic cells with the main focus on rechargeable batteries. For fuel cells, the main interest in nanomaterials relates to the catalysts. For low temperatures, hydrogen/oxygen, and alcohol/oxygen (direct) fuel cells, the catalysts are metallic particles comprising mostly platinum and its alloys. The authors mention dye-sensitized photovoltaic cells in which the anode material is semiconducting titanium oxide where the required high surface area is reached through the use of nanoparticles. For super capacitors, whose energy storage mechanism is based on electrostatic interactions, nanostructured carbonaceous materials may provide the necessary high surface area and hence high capacity. For rechargeable batteries, the use of nanomaterials may enable a high capability rate, because of the short length for solid-state diffusion (which is usually the determining step rate for intercalation materials). However, nanomaterials may have high surface reactivity, which can be detrimental for Li ion battery systems, in which there is no thermodynamic stability between most of the relevant electrode materials and the nonaqueous polar electrolyte solutions. There are some cases in which the use of nanomaterials is crucial: $LiMPO_4$ olivine cathode materials, silicon- and tin-based anode materials, as well as anodes based on conversion reactions (e.g., $MO + Li = Li_2O + M$). Nano-alumina and silica may be a desirable component in polymeric electrolytes because of the existence of ionic conductance mechanisms based on the interactions between Li ions and surface oxygens of the nanoparticles. The various battery components are classified and discussed in connection with the possible use of nanomaterials.

Nanobiotechnology, the combination of nanotechnology with biology, allows the use of nanotools and nanodevices to interact with, detect, and alter biological processes at a cellular and molecular level. A. Ragusa et al. in Chapter 10 describe the use of semiconductor quantum dots for biomedical applications. Semiconductor nanocrystals, also known as quantum dots (QDs), represent an emerging class of inorganic fluorescent markers. Due to their inorganic nature, they offer revolutionary fluorescence performance including narrow and symmetrical emission spectra for low interchannel overlap, broad adsorption spectra and extremely bright emitting colors for simple single-excitation multicolor analysis, long-term photostability for live-cell imaging, and dynamics studies. Since the first proof of the concept of the application of QDs as fluorescent probe on living cells in 1998, numerous groups have demonstrated the significant potential of such a tool in biology. In this chapter, the authors provide an overview of the exploitation of QDs in different biological applications ranging from biosensoring to labeling and imaging, both on *in vitro* models and *in vivo* animal studies. They also consider their use in photodynamic therapy and multimodal imaging techniques—fields of research that have only recently been created but are already attracting a lot of attention.

An interesting strategy, with immense potentiality, that can be used to remotely control the delivery of a drug or gene is the use of magnetic nanoparticles manipulated by an external magnetic field. After a brief description of the physical principles underlying some current biomedical applications of nanoparticles (superparamagnetism, hyperthermia, and manipulation of magnetic nanoparticles inside blood vessel), Claudia Altavilla, in Chapter 11, reviews the most important wet chemistry strategies to design, synthesize, protect, and functionalize magnetic nanoparticles and/or multifunctional systems as drug delivery carrier. Some of the most explicative and significant recent studies on the application of these "smart" drug delivery systems *in vivo* and *in vitro* are finally reported.

Thermotherapy, elevation of tissue temperature to above 40°C–41°C, has long been described and researched for cancer therapy. The addition of magnetic nanoparticles was introduced in the hope for a more focused and homogenous distribution in the cancer tissue while sparing healthy tissue. The principle of nanoparticle thermotherapy, discussed by Joerg Lehmann and Brita Lehmann in Chapter 12, is the excitement of magnetic nanoparticles, which have been brought in close proximity to cancer cells through the use of an alternating magnetic field. Heat is produced through the transfer of the energy of the alternating magnetic field via magnetic hysteresis losses and Brownian relaxation losses. There is evidence that the technology is capable of providing a serious blow to cancer, possibly even complete remission. However, in relation to this point, only mice have been cured. The methods of nanoparticle delivery to cancer cells and creating the alternating magnetic field are reviewed in this chapter, as well as the properties of the nanoparticles. Reference is made to animal studies and initial clinical trials. This truly interdisciplinary field involving chemists, biologists, physicians, and physicists is very much under development.

In Chapter 13, Wilson A. Lee and Miriam Raifailovich describe the use of inorganic particles against reactive oxygen species for sun protection products. The chemical grafting of antioxidant molecules and anionic polymer encapsulated in a hydrophobic polymer directly onto $TiO_2$ particle surface is found to mitigate photocatalytic degradation, enabling highly effective filtering against UV radiation. The coating consists of a densely grafted polymer, an anionic polymer, and a free radical scavenger. The addition of the coated particles prevents scission and even possible hydrolysis of the DNA after exposure to UVA, UVB, and even UVC radiation.

Metal oxide nanoparticles can be uniformly deposited onto the surface of different kinds of textiles by a sonochemical method in order to achieve antibacterial properties. The topic is discussed by Nina Perkas et al. in Chapter 14. The coating can be performed by a simple, efficient, one-step procedure using environmentally friendly reagents. The physical and chemical analyses demonstrated that nanocrystals of ~20–30 nm in size are finely dispersed onto fabric surfaces without any significant damage to the structure of the yarn. The mechanism of nanooxide formation and adhesion to the textile is also discussed. It is based on the local melting of the substrate due to the high rate and temperature of the nanoparticles thrown at the solid surface by sonochemical microjets. The strong adhesion of the metal nanooxides to the substrate has been demonstrated in terms of the absence of the leaching of the nanoparticles into the washing solution. The performance of fabrics coated with a low content of nanooxides (<1 wt.%) as an antibacterial has been demonstrated to be very high.

In Chapter 15, Thomas B. Scott explains how engineered nanomaterials can deliver better, faster, and cheaper remediation of contaminated land. An unfortunate by-product of human activity is environmental pollution, the majority of which has occurred since the onset of the industrial revolution in the 1800s. The quantity and type of emissions have changed over the last 50–60 years, with pollution from complex chemicals, dense non-aqueous liquid phases, and radioactive metals adding to those already released into the environment. Only in recent years, their fate, transport, and toxicology have begun to be thoroughly investigated, revealing their potential impact on the environment and ultimately the human population. Many classical remediation methods are regarded as being too costly for extensive deployment in the developing world and well beyond economic feasibility for most rural communities. Nanotechnology seems well placed to restructure the remediation industry, offering significant reductions to the overall cleanup costs for large-scale contaminated sites, while also presenting the prospects of reduced cleanup times and near total removal of contaminants from groundwater.

In Chapter 16, Maya Bar-Sadan and Reshef Tenne discuss the synthesis and applications of inorganic nanotubes and fullerene-like structures. Inorganic fullerene-like (IF) nanoparticles and inorganic nanotubes form a relatively new class of nanomaterials. They are generically produced from layered (2D) materials, which enable the formation of stable, closed, and hollow structures in the nanodomain. These layered structures, with various morphologies, have opened a new field of research in inorganic chemistry. Since then, many more compounds have been synthesized in the cocentric, onion-like structure and more so as nanotubes, with the list increasing. The growth mechanism of some of these materials, like IF-$WS_2$, over the years, is elucidated in the chapter. This progress has permitted the scaling-up of the synthesis of these nanoparticles. Furthermore, it has enabled the understanding of some of these structures to the atomic level, on one side, and correlating these findings with the physical properties of the materials, on the other. Moreover, these developments have created the basis for numerous applications of the IF nanoparticles, some of which are already in commercial use.

Research in nanoscience and nanotechnology is expected to have a great impact on the development of new catalysts, because the detailed understanding of the chemistry of catalytic materials on the nanometer-scale as well as the ability to control their preparation will lead to rational and cost-efficient catalyst design. In Chapter 17, Naoki Toshima critically investigates the relationship between inorganic nanoparticles and catalysts. Metal nanoparticles also exhibit large surface-to-volume ratio as well as increased number of edges, corners, and faces leading to altered catalytic activity and selectivity. The structure of metal nanoparticles, especially bimetallic and trimetallic nanoparticles which is now controlled as well as designed, is one of the topics in this chapter. Not only the recent topics but also the traditional preparation, purification, and characterization methods are briefly reviewed.

Ten years after a scientific debate on the potential impact of nanoscience on catalysis was opened, P. Ciambelli et al. describe in Chapter 18 the enormous progress made in the characterization techniques and synthesizing ways for the construction of nanostructured catalysts. In this chapter, two examples are presented and analyzed. In the case of heterogeneous photocatalysis, the role of electronic properties along with the effect of particle size and morphology are discussed. The singular case of carbon nanotubes growth as affected by the size of catalyst nanoparticles is then analyzed.

## 1.5 Conclusion

Nanoparticle research is currently an area of intense scientific research, due to a wide variety of potential applications in biomedical, environmental, and electronic fields. A total of 100,000 scientific papers on nanoparticles have been published over the last few years. It is an enormous amount that represents a wide and special heritage for the development of scientific research. It is difficult to predict the future of this research, the routes explored, or the applications developed. We can only be sure of one thing that a part of this future has already begun. A new kind of matter has been prepared. Artificial atoms have been used not only due to their intrinsic properties but also to build new materials. Studies from the eighteenth, nineteenth, and twentieth centuries have enabled us to use atoms, ions, and molecules to build not naturally occurring materials. We are now able to synthesize new materials with nanoparticles used as artificial atoms, the building blocks of new generation substances. In this book, we try to discuss the principal applications of nanoparticles and all the authors are researchers that live this kind of research in the front line.

The editors and authors expect that this fundamental knowledge summarized in this book could be useful not only for newcomers to this field but also to specialists of this field to overview and summarize their knowledge.

## References

Altavilla, C. 2007. Innovative methods in nanoparticles assembly. In: *Advanced Materials: Research Trends*, ed. L.A. Basbanes, pp. 217–236. New York: Nova Science Publishers.

Apátiga, L.M., Rivera, E., and Castaño, V.M. 2007. Nucleation and growth of titania nanoparticles prepared by pulsed injection metal organic chemical vapor deposition from a single molecular precursor. *J. Am. Cer. Soc.* 90(3): 932–935.

Boardman, J. 1991. The sixth-century potters and painters of Athens and their public. In: *Looking at Greek Vases*, ed. T. Rasmussen and N. Spivey, pp. 79–102. Cambridge, U.K.: Cambridge University Press.

Borgia, I., Brunetti, B., Mariani, I. et al. 2002. Heterogeneous distribution of metal nanocrystals in glazes of historical pottery. *Appl. Surf. Sci.* 185: 206–216.

Burda, C., Chen, X., Narayanan, R., and El-Sayed, M.A. 2005. Chemistry and properties of nanocrystals of different shapes. *Chem. Rev.* 105: 1025–1102.

Cavalcante, P.M.T., Dondi, M., Guarini, G., Raimondo, M., and Baldi, G. 2009. Colour performance of ceramic nano-pigments. *Dyes Pigments* 80: 226–232.

Chen, G.X., Hong, M.H., Chong, T.C., Elim, H.I., Ma, G.H., and Ji, W. 2004. Preparation of carbon nanoparticles with strong optical limiting properties by laser ablation in water. *J. Appl. Phys.* 95: 1455–1459.

Ciliberto, E. 2000. Analytical methods in art and archaeology. In: *Modern Analytical Methods in Art and Archaeology*, ed. E. Ciliberto and G. Spoto, pp. 1–10. New York: John Wiley & Sons Inc.

Colomban, P. 2009. The use of metal nanoparticles to produce yellow, red and iridescent colour, from Bronze Age to present times in lustre pottery and glass: Solid state chemistry, spectroscopy and nanostructure. *J. Nano Res.* 8:109–132.

Cozzoli, P.D., Pellegrino, T., and Manna, L. 2006. Synthesis, properties and perspectives of hybrid nanocrystal structures. *Chem. Soc. Rev.* 35: 1195–1208.

Creighton, J.R., Coltrin, M.E., and Figiel, J.J. 2008. Observations of gas-phase nanoparticles during InGaN metal-organic chemical vapor deposition. *Appl. Phys. Lett.* 93(17): 171906.1–171906.3.

Czekai, D. 1996. U.S. Patent 5,500,331 (March 19, 1996); Eastman Kodak Company.

Gerberich, W.W., Mook, W.M., Perrey, C.R. et al. 2003. Superhard silicon nanospheres. *J. Mech. Phys. Solids* 51: 979–992.

Hemelrijk, J.M. 1991. A closer look at the potter. In: *Looking at Greek Vases*, ed. T. Rasmussen and N. Spivey, pp. 233–256. Cambridge, U.K.: Cambridge University Press.

Kang, S., Goyal, A., Li, J. et al. 2006. High-performance high-Tc superconducting wires. *Science* 311: 1911–1914.

Lafait, J. 2006. Ruby red-colored glass coloration from metallic nanoparticle dopants. *Verre* 12: 11–21.

Longstreth-Spoor, L., Gibbons, P.C., Kelton, K.F., and Kalyanaraman, R. 2008. Nanostructure of room temperature deposited $TiB_2$ on Si(001) by pulsed laser ablation. *Thin Solid Films* 516: 5981–5984.

Maniatis,Y., Aloupi, E., and Stalios, A.D. 1992. New evidence for the nature of Attic black gloss. *Archaeometry* 35: 23–34.

Muthupandian, A. 2008. Sonochemical synthesis of inorganic nanoparticles. In: *Advanced Wet-Chemical Synthetic Approaches to Inorganic Nanostructures*, ed. P.D. Cozzoli, pp. 107–131. Trivandrum, India: Transworld Research Network.

Nie, Z., Petukhova, A., and Kumacheva, E. 2010. Properties and emerging applications of self-assembled structures made from inorganic nanoparticles. *Nature Nanotechnol*. 5(1): 15–25.

Padovani, S., Sada, C., Mazzoldi, P. et al. 2003. Copper in glazes of renaissance luster pottery: Nanoparticles, ions and local environment *J. Appl. Phys*. 93: 10058–10064.

Peng, X., Wickham, J., and Alivisatos, A.P. 1998. Kinetics of II-VI and III-V colloidal semiconductor nanocrystal growth: "Focusing" of size distributions. *J. Am. Chem. Soc*. 120(1): 5343–5344.

Perez-Arantegui, J. and Larrea, A. 2003. The secret of early nanomaterials is revealed, thanks to transmission electron microscopy. *Trends Anal. Chem*. 22: 327–329.

Rehren, T., Pusch, E.B., and Herold, A. 1998. Glass coloring works within a copper-centered industrial complex in Late Bronze Age Egypt. *Ceram. Civiliz*. 8: 227–250.

Steigerwald, M.L. 1994. Clusters as small solids. *Polyhedron* 13: 1245–1252.

Stuczynski, S.M., Brennan, J.G., and Steigerwald, M.L. 1989. Formation of metal-chalcogen bonds by the reaction of metal-alkyls with silyl chalcogenides. *Inorg. Chem*. 28: 4431–4432.

Sun, Y., Egawa, T., Shao, C., Zhang, L., and Yao, X. 2004. Formation mechanism for high-surface-area Anatase Titania nanoparticles prepared by metalorganic chemical vapor deposition. *Jpn. J. Appl. Phys*. 43: 3544–3547.

Swihart, M.T. 2003. Vapor-phase synthesis of nanoparticles. *Curr. Opin. Colloid Interf. Sci*. 8(1): 127–133.

Uchic, M.D., Dimiduk, D.M., Florando, J.N., and Nix, W.D. 2004 Sample dimensions influence strength and crystal plasticity. *Science*. 305: 986–989.

Ullmann, M., Friedlander, S.K., and Schmidt-Ott, A. 2002. Nanoparticle formation by laser ablation. *J. Nanoparticle Res*. 4: 499–509.

Wang, Z., Zeng, X., Ji, M., and Liu, Y. 2009. The effect of target size on α-Fe nanoparticle preparation by pulsed laser ablation. *Appl. Phys. A* 97: 683–688.

Yin, Y. and Alivisatos, P. 2005. Colloidal nanocrystal synthesis and the organic-inorganic interface. *Nature* 437: 664–670.

# 2

# *Inorganic Nanoparticles for the Conservation of Works of Art*

**Piero Baglioni and Rodorico Giorgi**

**CONTENTS**



## 2.1 Conservation Science: The Contribution of Nanotechnology

For a long time, scientists' work had mainly focused on the improvement of methods for the diagnostic analysis of works of art. Although this is a really important topic, a huge demand for innovative solutions to solve the main issues in restoration was growing from conservators. We pioneered the application of nanoscience to the Conservation of Cultural Heritage, which may offer effective and exhaustive answers to these requests. In the following, we will highlight as case studies some of the most important methodologies developed.

The first application of nanoscience to conservation dated back to the end of 1990 and concerned the consolidation of mural paintings by using $Ca(OH)_2$ nanoparticles to replace the largely used polymer resins. Later, research was focused on the development of soft-matter systems for the cleaning and removal of resins used to consolidate and protect the surfaces of works of art (Baglioni and Giorgi 2006).

For many years, the importance of using materials that could be removed through simple solubilization in the same solvent as that used during application was particularly stressed (Mora et al. 1984, Tate and Tennent 1983). In this sense, the treatment was considered reversible. Reversibility of the treatment was presented as a major feature required for a conservation treatment. Synthetic polymers, being popular in several industrial fields, were considered the panacea for many restoration issues because they were supposed to be completely reversible. The importance of the physicochemical compatibility of the chemicals used with the materials of the works of art was underestimated because of the firm belief that every undesired material could be easily removed when necessary. This is the main reason why several polymer families were used on several different materials such as paintings, stones, glass, bones, wood, paper, ceramic, and bronze (Horie 1987). Inorganic materials, which have largely been in use for a long time (i.e. lime-water, bicarbonate solution, silicate solution), were considered only in a few cases, but as a general convention, the usage was discouraged because the treatment was considered irreversible.

On the contrary, inorganic treatments for wall painting consolidation should be preferred since synthetic polymers degrade faster than the materials to be restored (Feller 1994, Dei et al. 1999) and promote the degradation of the inorganic constituents of the works of art. After degradation, polymers become insoluble and reversibility of the treatment is not achievable anymore. Unfortunately, synthetic polymers have been widely used for decades because of the apparent aesthetically good results associated with some improvements in the mechanical resistance of the paint layer. In the 1970s, they were applied in large amounts over several wall paintings, and today, most of the research in conservation is addressed to the formulation of systems that are able to swell and remove degraded polymers (Carretti et al. 2003, 2004, 2005, 2007). A dramatic illustrative example is in the archaeological site of Mayapan in Yucatan, Mexico. Few decades ago, Maya wall paintings, dating back to the Post-Classic Age (1200–1450), were discovered. Conservators treated the paintings with mowilith resin (vinyl-acetate/*n*-butyl-acrylate 65/35 copolymer) in order to consolidate and protect the paint layer. Recent investigations showed severe degradation of the polymer films and serious changes in the appearance of the paintings. Salts coming from the inner parts of the wall have pushed the paint layer because of the loss of natural "breathing" of the surface due to the presence of the polymers. In Figure 2.1 are reported some pictures of the Maya painting in Mayapan. In particular, the picture on the right shows that a considerable part of the original painting is now missing because of the shrinkage of the polymer coating.

Historically, only very few scientists led by Enzo Ferroni realized in the 1970s that the use of polymers would be extremely deleterious. After the disaster of Florence's flood, Ferroni elaborated a method for the removal of sulfates and the consolidation of frescoes (Ferroni et al. 1969, Ferroni and Baglioni 1986, Ferroni 1990, Baglioni et al. 2003). This method (known as Ferroni–Dini method) employs a barium hydroxide solution to block sulfates in a nonreactive and mobile form and to consolidate paintings through carbonate formation. Few years later the extraordinary restoration performed on the wall paintings by Beato Angelico (in the San Marco convent, Florence), where Ferroni–Dini method was used, Ferroni and Dini realized that good effects from the treatment improved with time. This finding convinced Ferroni and Baglioni that large excess of barium hydroxide used during the application could be responsible for the slow and progressive transformation of calcium carbonate to calcium hydroxide, i.e., for the generation "in situ" of the original painting's binder. The newly formed binder was considered responsible for the excellent results of the treatment, inducing a new lime-setting process (Ferroni and Baglioni 1986).

The logical development of Ferroni's method was the synthesis of the calcium hydroxide binder in a form that could be easily and in large amounts administered to the wall paintings. Very tiny particles (micro and nanoparticles) were for the first time synthesized and dispersed by using organic solvents to obtain a kinetically stable system to be used over the painted surface (Giorgi 1996, Giorgi et al. 2000). Particles were prepared through a simple homogeneous phase reaction in water at 90°C to precipitate $Ca(OH)_2$ (Ambrosi et al. 2001a,b). Nanoparticles from diol solutions of reactants, working up to a temperature of 150°C to favor a higher nucleation rate or from heterogeneous phase reactions (slaking of quicklime, or by using w/o microemulsion as a chemical microreactor) were also investigated with good results (Giorgi et al. 2000, Salvadori and Dei 2001, Nanni and Dei 2003).

**FIGURE 2.1**
**(See color insert following page 302.)** Pictures of mural paintings in (a) Mayapan (Mexico) treated with acrylic/vinyl-acetate copolymer films to protect the surfaces. The pictures show the dramatic effects due to the aging of the polymer film that generates (b) whitening and (c) shrinkage of the paint layer.

Nanoparticle dispersions were successfully tested, for the first time, over the paintings by *Andrea da Firenze* (fifteenth century) in the *Cappellone degli Spagnoli* of the *Santa Maria Novella Basilica* in Florence.

Nowadays, nonaqueous dispersions of calcium and magnesium hydroxide nanoparticles, specifically addressing different conservation issues, are available to restorers (Giorgi et al. 2002a,b, 2005a,b). Recently, nanoparticles of strontium hydroxide have been obtained and proposed as a complementary tool for the consolidation of wall paintings with a high amount of salts, thanks to their high reactivity with sulfate ions (Ciliberto et al. 2008).

In general, the main advantages in the use of nanoparticles are the nanoscale dimensions of the particles, which enable them to penetrate into the porous matrix of the paintings and their reactivity and compatibility with the fresco paintings.

## 2.2 Case Studies: Consolidation of Wall Paintings

In the previous section, we have already mentioned that conservators largely used in the past, and to some extent are still using, synthetic polymers to consolidate and protect wall paintings. These materials, according to the different climate and environmental pollution conditions, degrade with different kinetics. High humidity and temperature, even in unpolluted areas, promote very fast degradation (4–5 years) of the polymer-treated surfaces, i.e. Mesoamerican paintings in the subtropical areas of Mexico. In geographical areas with milder climate, such as Europe, the degradation is slower but unavoidable. The degree of degradation is such that the cleaning of the surfaces treated with acrylic and vinyl polymers is often mandatory and should precede consolidation interventions. Nanotechnology based on microemulsion and micellar solutions proved to be very useful for the cleaning of the painted surfaces (Carretti et al. 2003, 2007). After polymer removal, however, a consolidation treatment is usually necessary. Alcoholic dispersions of calcium hydroxide nanoparticles have been successfully applied to different conservation issues, both in Europe (Ambrosi et al. 1999, 2001a,b, Baglioni et al. 2001, Dei et al. 2001, Croveri et al. 2004) and in tropical settings (Giorgi et al. 2006b). The particles penetrate the first layers of the painted surface, up to an average depth of 200–300 microns, and, once the alcohol evaporates, they turn into calcium carbonate. This way, the original binder for pigments is re-created *in situ*, and consolidation is achieved. The complete physicochemical compatibility with the substrate, moreover, grants a long-term consolidation and avoids drawbacks.

Figure 2.2 reports a particle-size distribution of $Ca(OH)_2$ obtained by analysis of several TEM pictures (more than 300 counts). The log-normal distribution is centered at about 220 nm. The crystallite size of nanoparticles can be estimated by the broadening of peaks in powder XRD pattern. According to Scherrer equation, the average crystallite size of powder grains ($L$) is obtained by the full width at half-maximum of the diffraction peak ($B(2\lambda)$):

$$0.89 \cdot \lambda = L \cdot \left(B(2\theta) \cdot \cos\theta\right)^{-1}$$

$\lambda$ being the wavelength of x-ray (1.54 Å for the Cu K$\alpha$), and $\theta$ being the angle between incident and diffracted beams (Jenkins and Snyder 1996).

Crystallite size of calcium hydroxide nanoparticles obtained through the already mentioned methods (Ambrosi et al. 2001a,b, Giorgi et al. 2005a,b) is around 45 nm; specific surface area obtained by gas porosimetry (Allen 1997) is 8–10 $m^2/g$.

As expected, crystallites are smaller than the average size determined by TEM analysis since the Scherrer equation accounts for the dimension of crystalline domains that do not necessarily match the real size of the particles. The exact determination of particle size by microscopy is rather complicated (see Figure 2.2) due to the tendency of platelet crystal shape to stack perpendicularly to the basal face (Miller index 001). Moreover, Scherrer

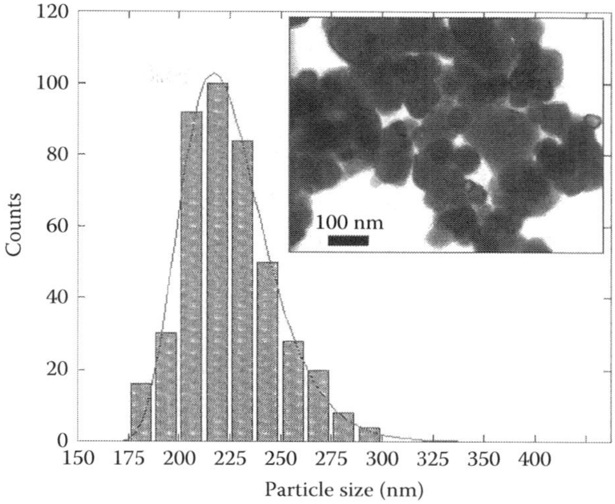


**FIGURE 2.2**
Histogram of nanoparticle sizes obtained by visual analysis of TEM picture performed on more than 300 particles. Inset: TEM picture of calcium hydroxide nanoparticles prepared through a homogeneous phase reaction in water at 90°C.

model presents some limitations in this case, and a nonuniform broadening of lines in the XRD patterns could be due to structural disorder rather than to crystallite size effects (Thomas and Kamath 2006).

We experienced that nanoparticles with dimensions from 150 to 280 nm behave very well for painting consolidation, having a very high kinetic stability due to the tiny dimensions, and are much smaller than the average porosity of plasters and also of several limestones.

In order to obtain more information about the consolidation effects of nanoparticle dispersion on painted plasters, several investigations have been performed first in the laboratory scale. Some plaster model systems, made by using different lime/sand mixtures, have been used. In particular, lime/sand ratios with a very low amount of binder (1:8 and 1:10 lime:sand volume/volume) were selected for testing. Mineral pigments were also applied with a "fresco" technique to reproduce a degraded painted surface. Powdering effect, after complete lime-setting, was obtained, and this made these model systems ideal for experimentation. These samples were treated with nanoparticle dispersions in order to improve the performance of the plasters.

Table 2.1 reports some meaningful parameters to characterize the main properties of the untreated and treated plaster surfaces. Water sorption by capillarity (Normal 11/85 document 1986) and permeability to vapor (Normal 21/85 document 1986) were determined because they are strictly connected to the porosity features of the treated surface. The comparison of the degraded systems with a 1:2 lime/sand ratio plaster, which is typical of a well-preserved fresco, showed that nanoparticle application reduced the capillary sorption of water and the transparency of the plaster, reverting these parameters to those of good quality and unaltered material. Microhardness analysis, carried out according to a procedure elsewhere reported (Giorgi et al. 2000), determined the resistance of the surface to mechanical stress due to indentation and scratching and allowed estimating the hardness of the surface. After the treatment with nanoparticles, the resistance of the treated surface was doubled, confirming the powerful consolidating effect of nanoparticles. Powdering

**TABLE 2.1**

Physicochemical Characterization of Model Plaster Samples Performed Preconsolidation and Postconsolidation by Means of $Ca(OH)_2$ Dispersions

| Lime:Inert | CA Coefficient $[(g/cm^2)s^{-1/2}]$ | VP ($g/m^2 \cdot 24\,h$) | Scratch-Width (μm) | ST Test ($mg/cm^2$) |
|---|---|---|---|---|
| 1:2 | 0.52 | 143 | 0.8 | 0.28 |
| 1:8 untreated | 0.64 | 239 | 3.2 | 5.19 |
| 1:8 treated | 0.35 | 204 | 1.8 | 1.37 |
| 1:10 untreated | 0.72 | 237 | 4.4 | 7.12 |
| 1:10 treated | 0.42 | 172 | 2.4 | 4.12 |

CA, capillary absorption coefficient; VP, vapor permeability.

of the surface was estimated by scotch tape test (STT, Mora and Torraca 1965). Loss of pigments by chemi-mechanical action (due to the glue of the tape and its detaching) was significantly reduced after the consolidation by nanoparticle dispersions. In conclusion, the excellent results obtained with the laboratory tests promoted the first experiments on real cases.

Nowadays, application of nanotechnology is being used in some sub-tropical areas in Mexico, where *in situ* consolidation is a hard challenge due to the severe climatic conditions. One example is Calakmul, located in the southern part of the state of Campeche (Mexico). Calakmul was the most important city in the Classic Maya period (AD 600–800) along with Tikal in Guatemala, and was slowly abandoned until the year AD 900 (Post-Classic period) (Carrasco 1998). Calakmul was discovered by the botanist Cyrus Lundell, in 1931, who gave to the site the actual name that in Maya language means Ca "two," lak "close," mul "artificial hills or mounds."

Excavation work (started in 1993) was recently concentrated in the *Acrópolis Chik Naab*, where well-preserved wall paintings were discovered and are presently being restored (see Figure 2.3). Mural paintings, dated to the Early Classic Period, were found on the walls of the pyramidal structure I that measures 9 m by 9 m at its base and 8 m in height.

(a) (b)

**FIGURE 2.3**

**(See color insert following page 302.)** The Maya ruins of Calakmul: (a) a view of structure I (*Acropolis Chik Naab*), (b) a view of the mural paintings discovered inside the building.

(a) (b) (c)

**FIGURE 2.4**
**(See color insert following page 302.)** Different restoration steps of paintings consolidation by means of $Ca(OH)_2$ nanoparticles. (a) Application of particles dispersion and (b–c) removal of wet cellulose compresses used to maintain surfaces humid, in order to favor a slow calcium carbonate formation.

Exploration and restoration process is still in progress and there are high expectations of finding additional well-preserved paintings.

Natural aging of paintings caused the flaking of the paint layer and the powdering of the painted surface. This was probably due to the action of water and salt solutions soaking the porous matrix of the wall surface. This promoted the slow and progressive degradation and solubilization of the binding materials, with the loss of cohesion between the pigments and the substrate. The consolidation of the pictorial layer of the paintings was performed using nanoparticles dispersed in nonaqueous solvents. This aspect is particularly important since wall paintings usually do not allow the use of binders such as lime water, mainly because water promotes the capillary rise of salts (increasing the degradation), and lime water is too weak a binder due to the very low amount of calcium hydroxide present in solution (about 1.6 g/L) (Brajer and Kalsbeek 1999). Dispersions of nanosized particles in nonaqueous solvents are kinetically stable systems with a very high content of binder ($Ca(OH)_2$) and do not present the drawbacks mentioned above.

Figure 2.4 shows the removal of cellulose compresses used to maintain the paint surfaces wet after consolidation in order to favor a slow carbonation of nanoparticles. They produced in a few days a significant consolidation effect on the treated surfaces. This fact is particularly important in archaeological sites, where the conservation *in situ* usually requires an immediate intervention to protect the painting, avoiding the complete loss of the painted layer (Giorgi et al. 2006a,b).

## 2.3 Case Studies: Deacidification of Paper and Manuscripts

A second example of the use of nanoparticles in conservation deals with the deacidification of cellulose-based materials. This is universally known as a major conservation issue for the extremely large amount of items to be treated (books in libraries, newspapers, textiles, wood, and so on).

It is well known that acid catalyzes the hydrolysis of cellulose with a depolymerization of the cellulose fibers (Barrow 1974, Smith 1987, Roberts 1996). As a consequence, the physical and mechanical properties of the cellulose fibers are compromised, leading to the loss of many ancient books and documents if not properly and timely treated. In the past years, many methods for paper deacidification have been developed, such as Wei T'o (one of the oldest) and Bookkeeper (today used at the Library of Congress in Washington) (Burgess et al. 1992, Buchanan 1994, Kaminska and Burgess 1994, Porck 1996). New methods based on the application of magnesium (or calcium) hydroxide nanoparticles dispersed in a non-aqueous solvent are under investigation in our research group. Nanoparticles, due to their tiny dimensions, penetrate into the network of the cellulose fibers and adhere to them. Some of the hydroxide immediately reacts with the acidity *in situ*, providing deacidification. The hydroxide excess, due to the high reactivity of nanoparticles, is quickly turned into carbonate by carbon dioxide. This way, an alkaline buffer is created in order to withstand the forthcoming acidity that can develop inside the fibers (e.g., due to pollution). Calcium hydroxide nanoparticles provide excellent results, but the smaller size of magnesium hydroxide particles involves a better penetration inside the substrate. This feature is particularly useful when dealing with deacidification of wood cellulose. In that case, penetration of particles must reach depths higher than the 200–300 microns required for wall painting consolidation. For this reason, research efforts are focused on synthetic routes allowing the synthesis of particles with smaller size (Ambrosi et al. 2001a,b, Giorgi et al. 2002a,b). One possible way to minimize the size of the particles is to choose different chemical precursors for magnesium or calcium hydroxide obtained by coprecipitation in homogeneous phase.

Figure 2.5 reports two particle-size distributions of $Mg(OH)_2$ obtained by the image analysis of several TEM pictures (more than 300 particles). The first distribution (Figure 2.5a) refers to particles synthesized starting from $MgCl_2$ and NaOH water solutions, while the second (Figure 2.5b) refers to particles obtained from $MgSO_4$ and NaOH water solutions. We can notice that in the first case, the log-normal distribution is centered at about 75 nm, while in the second case, it is centered at about 50 nm. The reduction in size is due to

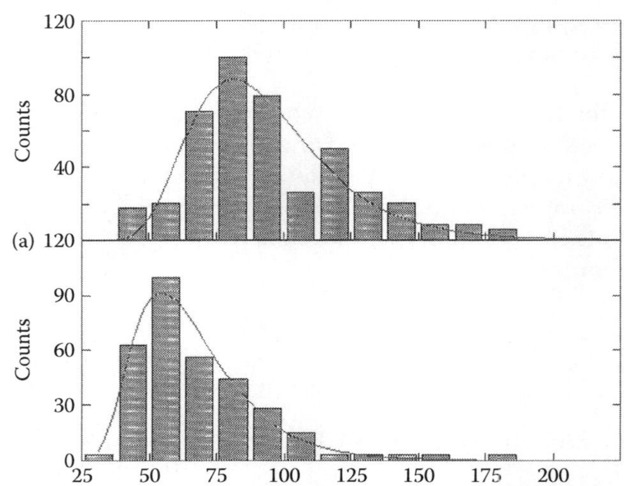


**FIGURE 2.5**
Size distribution of $Mg(OH)_2$ obtained from (a) $MgCl_2$ precursor and (b) $MgSO_4$.

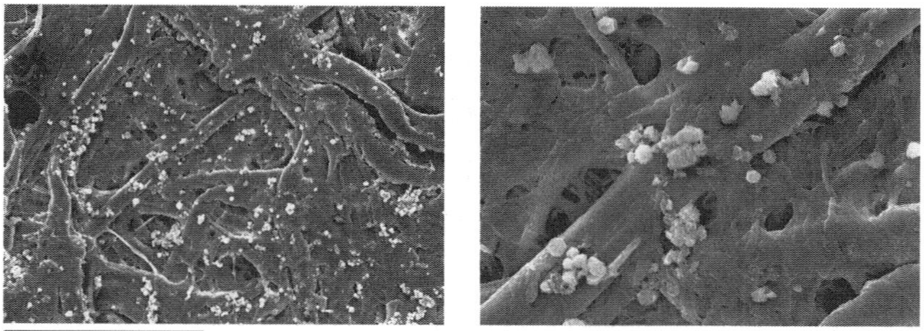

(a) 100 μm (b) 20 μm

**FIGURE 2.6**
(a) SEM picture of magnesium hydroxide nanoparticles adhered to cellulose fibers of Whatman filter paper and (b) detail of the particles.

the counterion effect, with a trend that follows the Hofmeister series (Giorgi et al. 2005a,b). Researches on the effect of the counterions on the particle dimensions are still going on.

The deacidification experimentations on ancient paper samples were very promising (Giorgi et al. 2002a,b). However, in order to obtain clearer evidences about the efficacy of the treatment and a better comprehension of the deacidification mechanisms, standard reference materials were also investigated. Whatman filter paper, composed of pure cellulose, can be considered an ideal material for experimentations. Samples of Whatman no. 1 paper were acidified by soaking them in a sulfuric acid solution (pH 2.5) in order to reproduce the acid degradation of the fibers. Some samples were then deacidified by soaking them in an alcoholic dispersion of $Mg(OH)_2$ nanoparticles.

Figure 2.6 reports a SEM picture showing the fibers of Whatman filter paper after treatment with nanoparticles. The particles are homogeneously distributed onto the surface and adhered to the fibers. In order to quantify the deacidification effect of the treatment, some parameters for degradation should be defined. Apart from pH measurements, the determination of the degree of polymerization (DP) of cellulose was chosen as a quantitative parameter of the conservation *status* of the polymer. DP is determined through viscosity measurements according to international standard protocols (ASTM D 4243 1999). Some samples of acidified paper were artificially aged (by hydrothermal aging; 90°C and 80% RH) and their DP was monitored during 20 days of aging. The same was done for other samples after the deacidification treatment.

Results are reported in Table 2.2. Aging of the acidified samples promoted a strong decrease in the cellulose DP with a loss of about 55%. On the other hand, samples acidified

**TABLE 2.2**
Hydrothermal Aging of Acidified Samples of Whatman Filter Paper before and after Deacidification Treatment with Magnesium Hydroxide Nanoparticles

| DP Polymerization | Hydrothermal Aging | | |
|---|---|---|---|
| **Degree** | **0 h** | **10 Days** | **20 Days** |
| Acid paper | 978 | 438 | 438 |
| Deacidified paper | 978 | 897 | 812 |

and then treated with magnesium hydroxide dispersion resisted the aging very well, with a DP decrease of about 17%. Though a precise conversion of artificial aging time to natural aging is very difficult, these data suggest that treatment with alkaline nanoparticles grant a good protection to paper cellulose.

## 2.4 Case Studies: Deacidification of Waterlogged Wood

The promising results obtained on paper suggested that magnesium hydroxide nanoparticles could be used also for wood deacidification. With respect to paper, wood is a more complex system. In fact, it shows a three-dimensional structure due to the arrangement of cells and also contains lignin and hemicellulose as major components. This implies that small particles must be used in order to get a good penetration into the substrate. In addition, the interaction of the nanoparticles (as well as the carrier solvent) with the different wood components must also be taken into account. The experimentations on wood were done on some samples from the Swedish warship Vasa (Giorgi et al. 2005b, 2006a, Chelazzi 2006, Chelazzi et al. 2006a,b). The Vasa wood shows particular features that made it a unique case in archaeological wood preservation. The warship sank during its maiden voyage in 1628, in the Stockholm port, and was recovered in 1961 (Sandström et al. 2002). This wood showed high acidity due to the oxidation of elemental sulfur contained inside the timber giving sulfuric acid (MacLeod and Kenna 1991, Fors and Sandstrom 2006). The elemental sulfur came from the metabolic action of sulfate-reducing bacteria that dwelled in the wasted water of the Stockholm harbor. The presence of a huge quantity of sulfuric acid is now threatening the preservation of Vasa, so efforts are being made in order to deacidify the wood, and to prevent sulfur oxidation (International Evaluation of the "Preserve the Vasa Project" 2006). Calcium and magnesium nanoparticles were applied by soaking small samples of Vasa wood into alcoholic stable dispersions. Washing of the samples with water, previous to soaking the wood in nanoparticle dispersions, was carried out in order to remove most of the polyethylene glycol (PEG), the wood consolidant used for Vasa wood conservation, whose presence in the wood porous matrix would prevent the alkaline particles to penetrate and adhere to the cellulose fibers. In order to evaluate the efficacy of the deacidification method, pH measurements and thermal analysis were carried out on the treated and untreated samples. Pyrolysis temperature of cellulose, determined through DTG, was monitored, and turned out to be very sensitive to the acidity of cellulose and hemicellulose chains: acidic samples, in fact, showed lower pyrolysis temperature, whereas deacidified ones showed higher values (Giorgi et al. 2005a,b).

Table 2.3 reports the results of the treatment on Vasa oak wood samples: the pH of wood after the nanoparticle treatment turned from highly acidic values (pH = 2.8) to slightly acidic ones (pH = 5.5, very close to the pH of native oak wood), and the pyrolysis temperature reverted back, after the treatment, to the typical values of fresh oak wood. It was shown that magnesium hydroxide nanoparticles penetrated inside the wood matrix up to a depth of 1–2 cm. As for paper applications, nanoparticles neutralized the acidity, and the excess quickly turned to carbonate. This way, possible harmful interactions of the hydroxide excess with residual cellulose and lignin was avoided, and the presence of an alkaline reservoir neutralizes the acidity that develops continuously inside the wood fibers from the sulfur oxidation (Chelazzi 2006).

**TABLE 2.3**

Pyrolysis Temperature and pH of Vasa Oak Wood Samples before Any Treatment, after Washing with Water, and after Treatment with Dispersions of Magnesium Hydroxide Nanoparticle Dispersions

| Vasa Hull Oak Wood | Pyrolysis Temperature of Cellulose (°C) | pH |
|---|---|---|
| Without any treatment | 321 | 2.8 |
| After washing with water | 321 | 3.3 |
| Posttreatment with $Mg(OH)_2$ dispersion in 2-propanol | 345 | 5.5 |

Fluorinated solvents are a possible alternative to the usage of propanol as a carrier for the nanoparticles inside the wood cells. In particular, perfluoropolyethers that are chemically inert, nontoxic, environment-safe, thermally and chemically resistant, and that have low surface tension are good candidates. Several tests showed that fluorinated solvents sorption by capillarity is very good and sufficient to ensure the appropriate soaking of wood in a few tenths of minutes. After a few days, the soaked samples, dried in air, lose more than 90% by solvent weight, reaching 100% in a few weeks. Solvent evaporation from wood can be modulated by selection of the molecular weight of the fluorinated solvent. In order to disperse polar hydroxide nanoparticles in a nonpolar fluorinated solvent, a semi-fluorinated surfactant was used. In particular, carboxyl head linked to perfluorinated chain, after neutralization, was shown to bind hydroxide surface and to produce kinetically stable dispersions. Interestingly, nanoparticles dispersed in fluorinated solvent showed a very good wood penetration even in the presence of a high amount of PEG. Hydrothermal aging (90°C and 85% RH) of the wood treated with nanoparticles showed a good protection toward acidic degradation (Giorgi et al. 2009).

## 2.5 Case Studies: Chemical Gels and Nanomagnetic Sponges for Wall Painting Cleaning

Physical gels (called solvent gels) containing large amount of solvents are today the most commonly used systems for the cleaning of easel and canvas paintings. Physical gels, consisting of a structural polymer network based on relatively weak interaction forces, can be easily applied, for example, by gentle brushing or with a spatula, over the painted surfaces and removed with cotton swabs soaked with white spirit. Solvent-gel method, in spite of its cleaning efficacy, presents some disadvantages. First, neat solvents could penetrate within surface porosity and solubilize the painting binder because they are quite aggressive; second, most of the commonly used solvents are toxic. The removal of the solvent-gel is usually not efficient and after cleaning, gel residues may be found over the surfaces.

Recent works have been focused to avoid the use of free solvents and to replace physical gels with chemical gels. This way, a more efficient cleaning and removal of materials could be achieved. Instead of pure solvents, microemulsion and micellar solutions have been studied to obtain aqueous systems with high performance in the removal of low polarity materials. Microemulsions, as well as micellar solutions, allowed the reduction of the amount of the organic phase to be used during cleaning treatments. This improved the safety of the procedure for the operator and also reduced the environmental impact of the treatment. On the other hand, water is not as well accepted by hydrophilic support as canvas and easel paintings since the paint layers tend to absorb water that induces swelling, deformation, and mechanical stress on the support (canvas) itself. For this reason, in order to maintain all the advantages of aqueous systems and eliminate the disadvantages, microemulsions were loaded into the structure of a very hydrophilic chemical gel to reduce the water penetration. These gels are mainly based on the polymerization of acrylamide and bis-acrylamide monomers to obtain a tridimensional gel network, where molecules are covalently bound to each other. Magnetic cobalt–ferrite nanoparticles ($CoFe_2O_4$) were chemically bonded to the polyacrylamide network to obtain a chemical magnetic-responsive hydrogel. These innovative systems are responsive to magnetic fields and removable by using a permanent magnet. The optimization of the hydrogel properties to obtain an "easy-to-load" reversible system for "oil" phase (i.e. xylene, nitro diluent, or other organic solvents) efficient in the removal of acrylic and vinyl-acetate polymer is giving appreciable results. Hydration tests demonstrated that a microemulsion can be loaded inside the gel up to 92% w/w. After use, a magnetic squeezing allows recovering of microemulsion. These gels have been recently used for the removal of Paraloid® B72 from travertine and marble samples (Bonini et al. 2007). The gels can be shaped as a sheet with dimensions up to some tenths of centimeters and different thicknesses, according to the amount of microemulsion to be soaked and the characteristics of the surface to be cleaned. These systems have been already tested also for cleaning of canvas paintings. In this case, the problem was the removal of glue used in the past for the relining of degraded canvas. Different substances were used with this purpose: mainly animal glue and wax in the past; more recently, synthetic copolymers of acrylate-methacrylate and vinylacetate-butylacrylate. The spreading of water in the textiles constituting the canvas and in the delicate paint layers must be limited as much as possible; for this, the retention capacity of gel determines in a decisive way the performance of the treatment. Although the application times of chemical gels are longer, they are usually a safer treatment as compared to physical gels.

## 2.6 Conclusion

Nanotechnology provides innovative tailored tools for the restoration of works of art. Nanoparticles of calcium hydroxide allowed a consistent improving over the classical application of calcium hydroxide solution. In fact, the usage of $Ca(OH)_2$ dispersions overcomes the limitation of the low solubility in water, alcohols are less aggressive than water toward fragile mural paintings, and quick carbonation of hydroxides provides a strong consolidation effect. Calcium and magnesium hydroxide as nonaqueous dispersions also provide excellent results for the treatment of cellulose-based materials. These preferably require waterless solvents and need an alkaline reserve to protect the object toward further degradation due to pollution or to internal acid production as a consequence of natural aging of the materials. Small particles of calcium or magnesium hydroxide provide excellent results and ensure high physicochemical compatibility with the substrates, which account for the durability of the treatment and long-lasting protection of the works of art.

Inorganic magnetic nanoparticles chemically bonded to a gel network allowed achieving a new class of materials with specific properties to improve the efficacy and limit possible drawbacks of gel cleaning procedures. These nanocomposite materials, based on classical gelators, are a self-explicative example of the huge potential of nanotechnology to open innovative perspectives in replacing and, more often, improving the existing methodology.

## References

Allen, T. 1997. *Particle Size Measurement*, vol. II. London, U.K.: Chapman & Hall.

Ambrosi, M., Baglioni, P., David, P.R., Dei, L., Giorgi, R., Lalli, C., Lanterna, G. et al. 1999. Inorganic consolidants and protectives for architectonic surfaces: Experimental tests on Santa Prisca church apse in Rome. In: *Proceedings of the 2nd International Congress on Science and Technology for the Safeguard of Cultural Heritage in the Mediterranean Basin*, Paris, France, July 5–9, 1999, pp. 873–877. Amsterdam, the Netherlands: Elsevier.

Ambrosi, M., Baglioni, P., Dei, L., Giorgi, R., and Neto, C. 2001a. Stable dispersions of $Ca(OH)_2$ in aliphatic alcohols: Properties and application in cultural heritage conservation. *Progress in Colloid and Polymer Science* 118: 68–72.

Ambrosi, M., Dei, L., Giorgi, R., Neto, C., and Baglioni, P. 2001b. Colloidal particles of $Ca(OH)_2$: Properties and application to restoration of frescoes. *Langmuir* 17: 4251–4255.

ASTM D 4243 1999. *Standard Test Method for Measurement of Average Viscometric Degree of Polymerization of New and Aged Electrical Papers and Boards*. West Conshohocken, PA: ASTM International.

Baglioni, P. and Giorgi, R. 2006. Soft and hard nanomaterials for restoration and conservation of cultural heritage. *Soft Matter* 2: 293–303.

Baglioni, P., Cesari, C., Dei, L., Giorgi, R., Grassi, R., Lorenzetti, M., Mauro, M., Pinzani, D., Ruffo, P., and Schonhaut, G. 2001. Stucco forte veneziano con "tecniche della tradizione": Proprietà chimico-fisiche, degrado in ambiente lagunare e conservazione mediante nanofasi cristalline di $Ca(OH)_2$. In: *Atti del XVII Convegno Internazionale Scienza e Beni Culturali "Lo Stucco-Cultura, Tecnologia, Conoscenza" Bressanone 2001*, pp. 289–298. Venezia, Italy: Edizioni Arcadia Ricerche.

Baglioni, P., Carretti, E., Dei, L., and Giorgi, R. 2003. Nanotechnology in wall painting conservation. In: *Self-Assembly*, ed. B.H. Robinson, pp. 32–41. Amsterdam, the Netherlands: IOS Press.

Barrow, W.J. 1974. *Permanence/Durability of the Book—VII. Physical and Chemical Properties of the Book Papers, 1507–1949*. Richmond, VA: W.J. Barrow Research Laboratory.

Bonini, M., Lenz, S., Giorgi, R., and Baglioni, P. 2007. Nanomagnetic sponges for the cleaning of works of art. *Langmuir* 23: 8681–8685.

Brajer, I. and Kalsbeek, N. 1999. Limewater absorption and calcite crystal formation on a limewater-impregnated secco wall painting. *Studies in Conservation* 44: 145–156.

Buchanan, S. 1994. An evaluation of the bookkeeper mass deacidification process. In: *Technical Evaluation Team Report for the Preservation Directorate, Library of Congress*. Washington, DC: Preservation Directorate, Library of Congress.

Burgess, H.D., Kaminska, E.M., and Boronyak-Szaplonczay, A. 1992. *Evaluation of Commercial Mass Deacidification Processes: Akzo-DEZ, Wei T'o and fmc-mg3. Phase I: Naturally Aged Papers*. Ottawa, Canada: Canadian Conservation Institute.

Carrasco, R. 1998. The metropolis of Calakmul, Campeche. In: *Maya*, eds. P. Schmidt, M. de la Garza, E. Nalda, pp. 372–385. New York: Rizzoli, International Publications Inc.

Carretti, E., Dei, L., and Baglioni, P. 2003. Solubilization of acrylic and vinyl polymers in nanocontainer solutions. Application of microemulsions and micelles to cultural heritage. *Langmuir* 19: 7867–7872.

Carretti, E., Dei, L., and Baglioni, P. 2004. Aqueous polyacrylic acid based gels: Physicochemical properties and applications in cultural heritage conservation. *Progress in Colloid and Polymer Science* 123: 280–283.

Carretti, E., Salvadori, B., Baglioni, P., and Dei, L. 2005. Microemulsions and micellar solutions for cleaning fresco surfaces. *Studies in Conservation* 50: 1–8.

Carretti, E., Giorgi, R., Berti, D., and Baglioni, P. 2007. Oil-in-water nanocontainers as low environmental impact cleaning tools for works of art: Two case studies. *Langmuir* 23: 6396–6403.

Chelazzi, D. 2006. Nanotechnology for artworks conservation, PhD dissertation, University of Florence, Florence, Italy.

Chelazzi, D., Giorgi, R., Baglioni, P. 2006a. Nanotechnologies for the conservation of waterlogged wood: The Vasa *case studies*. In: *Proceedings of the Heritage, Weathering and Conservation Conference—HWC-2006*, eds. R. Fort, M. Alvarez de Buergo, M. Gomez-Heras, and C. Vazquez-Calvo, pp. 797–802. London, U.K.: Taylor & Francis/A.A.Balkema Publishers.

Chelazzi, D., Giorgi, R., and Baglioni, P. 2006b. Nanotechnology for Vasa wood de-acidification. *Macromolecular Symposia* 238: 30–36.

Ciliberto, E., Condorelli, G.G., La Delfa, S., and Viscuso, E., 2008. Nanoparticles of $Sr(OH)_2$: Synthesis in homogeneous phase at low temperature and application for cultural heritage artifacts. *Applied Physics A: Materials Science & Processing* 92: 137–141.

Croveri, P., Dei, L., Giorgi, R., and Salvadori, B. 2004. Consolidation of globigerina limestone (Malta) by means of inorganic treatments: Preliminary results. In: *Proceedings of the 10th International Congress on Deterioration and Conservation of Stone—Stone 2004*, Stockholm, Sweden, June 27–July 2, 2004, eds. D. Kwiatkowski and R. Löfvendahl, pp. 463–470. Stockholm, Sweden: ICOMOS.

Dei, L., Baglioni, P., and Mauro, M. 1999. Materials for wall paintings conservation: Changes of physicochemical properties, aging effects, and reversibility. In: *Preprints of the Conference Reversibility: Does It Exist?* London, U.K., September 8–10, 1999, eds. A. Oddy and S. Carroll, pp. 73–80.

Dei, L., Favaro, M., Giorgi, R., and Portieri, R. 2001. Gli stucchi della Madonna di Lourdes nella chiesa di San Giovanni Evangelista in Venezia: Caratterizzazione del degrado e pre-consolidamento mediante dispersioni alcoliche di calce. In: *Atti del XVII Convegno Internazionale Scienza e Beni Culturali "Lo Stucco- Cultura, Tecnologia, Conoscenza" Bressanone 2001*, pp. 557–566. Venezia, Italy: Edizioni Arcadia Ricerche.

Feller, R.L. 1994. *Accelerated Agings: Photochemical and Thermal Aspects*. Los Angeles, CA: The Getty Conservation Institute.

Ferroni, E. 1990. Riflessioni sui Processi Degradativi dell'Intonaco e sui Metodi di Intervento. In *Atti del Convegno Internazionale di Studi "Piero della Francesca ad Arezzo"—Arezzo*, March 7–10, 1990, ed. G. Centauro and M. Moriondo Lenzini, pp. 219–231. Venezia, Italy: Marsilio.

Ferroni, E. and Baglioni, P. 1986. Experiments on a proposed method for restoration of sulphated frescoes. In: *Proceedings of the Symposium: Scientific Methodologies Applied to Works of Art*, Florence, Italy, May 2–5, 1984, ed. P.L. Parrini, pp. 108–109. Milan, Italy: Montedison Progetto Cultura.

Ferroni, E., Malaguzzi-Valerj, V., and Rovida, G. 1969. Experimental study by diffraction of heterogeneous systems as a preliminary to the proposal of a technique for the restoration of gypsum polluted murals. In: *Reports, Plenary Meeting*, Amsterdam, Netherlands, *1969—ICOM Committee for Conservation*. Paris, France: The International Council of Museums.

Fors, Y. and Sandstrom, M. 2006. Sulfur and iron in shipwrecks cause conservation concerns. *Chemical Society Reviews* 35: 399–415.

Giorgi, R. 1996. Sospensioni stabili di idrossido di calcio in alcoli alifatici a catena corta. Applicazione nell'ambito della conservazione dei beni culturali. Chemistry Graduation thesis, University of Florence, Florence, Italy.

Giorgi, R., Dei, L., and Baglioni, P. 2000. A new method for consolidating wall painting based on dispersions of lime in alcohol. *Studies in Conservation* 45: 154–161.

Giorgi, R., Dei, L., Ceccato, M., Schettino, C.V., and Baglioni, P. 2002a. Nanotechnologies for conservation of cultural heritage: Paper and canvas deacidification. *Langmuir* 18: 8198–8203.

Giorgi, R., Dei, L., Schettino, C.V., and Baglioni, P. 2002b. A new method for paper deacidification based on calcium hydroxide dispersed in nonaqueous media. In: *Preprint of the IIC Congress 2002—Works of Art on Paper, Books, Documents and Photographs: Techniques and Conservation*, ed. V. Daniels, A. Donnithorne, and P. Smith, pp. 69–73. London, U.K.: IIC.

Giorgi, R., Bozzi, C., Dei, L., Gabbiani, C., Ninham, B.W., and Baglioni, P. 2005a. Nanoparticles of $Mg(OH)_2$: Synthesis and application to paper conservation. *Langmuir* 21: 8495–8501.
Giorgi, R., Chelazzi, D., and Baglioni, P., 2005b. Nanoparticles of calcium hydroxide for wood conservation. The deacidification of the Vasa warship. *Langmuir* 21: 10743–10748.
Giorgi, R., Chelazzi, D., and Baglioni, P. 2006a. Conservation of acid waterlogged shipwrecks: Nanotechnologies for de-acidification. *Applied Physics A: Materials Science & Processing* 83: 567–571.
Giorgi, R., Chelazzi, D., Carrasco, R., Colon, M., Desprat, A., and Baglioni, P. 2006b. The Maya site of Calakmul: "in situ" preservation of wall paintings and limestone by using nanotechnologies. In: *Proceedings of the IIC Congress 2006, Munich—The Object in Context: Crossing Conservation Boundaries*, eds. D. Saunders, J.H. Townsend, and S. Woodcock, pp. 162–169. London, U.K.: James & James.
Giorgi, R., Chelazzi, D., and Baglioni, P. 2009. Nanoscience contribution to preservation of acidic shipwrecks. In *Proceedings of the 10th ICOM Group on Wet Organic Archaeological Materials Conference, WOAM 2007—Amsterdam 2007*, eds. K. Straetkvern and D.J. Huisman, pp. 525–538. Amersfoort, the Netherlands: RACM.
Horie, C.V. 1987. *Materials for Conservation*. London, U.K.: Butterworth.
International Evaluation of the "Preserve the Vasa Project" (by McAuley, A., Holmbom, B., Hoffmann, P., and Jensen, P.). Stockholm, Sweden: Statens maritima museer, December 2006 (available at http://www.vasamuseet.se/upload/bevara_vasa_rapport.pdf, last accessed April 16, 2007).
Jenkins, R. and Snyder, R.L. 1996. *Introduction to X-Ray Powder Diffractometry*. New York: John Wiley & Sons.
Kaminska, E.M. and Burgess, H.D. 1994. *Evaluation of Commercial Mass Deacidification Processes: Akzo-DEZ, Wei T'o and fmc-mg3. Phase II: New and Artificially Aged Modern Papers*. Ottawa, Canada: Canadian Conservation Institute.
MacLeod, I.D. and Kenna, C. 1991. Degradation of archaeological timbers by pyrite: Oxidation of iron and sulfur species. In: *Proceedings of the 4th ICOM Group on Wet Organic Archaeological Materials Conference*, Bremerhaven, Germany, 1990, ed. P. Hoffmann, pp. 133–142. Bremerhaven, Germany: ICOM, Committee for Conservation, Working Group on Wet Organic Archaeological Materials.
Mora, P. and Torraca, G. 1965. Fissativi per Pitture Murali. *Bollettino I.C.R.* 109–132.
Mora, P., Mora, L., and Philipot, P. 1984. *Conservation of Wall Painting*. London, U.K.: Butterworths.
Nanni, A. and Dei, L. 2003. $Ca(OH)_2$ nanoparticles from W/O microemulsions. *Langmuir* 19: 933–938.
Normal 11/85 document 1986. *Assorbimento d'acqua per capillarità*. Rome, Italy: C.N.R. & I.C.R.
Normal 21/85 document 1986. *Permeabilità al vapor d'acqua*. Rome, Italy: C.N.R. & I.C.R.
Porck, H.J. 1996. *Mass Deacidification, An Update of Possibilities and Limitations—ECPA reports*. Amsterdam, the Netherlands: ECPA—European Commission on Preservation and Access.
Roberts, J.C. 1996. *The Chemistry of Paper*. London, U.K.: RSC Paperbacks.
Salvadori, B. and Dei, L. 2001. Synthesis of $Ca(OH)_2$ nanoparticles from diols. *Langmuir* 17: 2371–2374.
Sandström, M., Jalilehvand, F., Persson, I., Gelius, U., Frank, P., and Hall-Roth, I. 2002. Deterioration of the seventeenth century warship Vasa by internal formation of sulfuric acid. *Nature* 415: 893–897.
Smith, R.D. 1987. *Conservation of Library and Archive Materials and the Graphic Arts*, ed. G. Petherbridge. London, U.K.: Butterworth.
Tate, J.O. and Tennent, N.H. (eds.) 1983. *Resins in Conservation, Proceedings of the Symposium*, Edinburgh, U.K., May 21–22, 1982. Edinburgh, U.K.: Scottish Society for Conservation and Restoration.
Thomas, G.S. and Kamath, P.V. 2006. Line broadening in the PXRD patterns of layered hydroxides: The relative effects of crystallite size and structural disorder. *Journal of Chemical Sciences* 118: 127–133.

# 3

# *Magnetic Nanoparticle for Information Storage Applications*

**Natalie A. Frey and Shouheng Sun**

**CONTENTS**



## 3.1 Introduction

In 1956, IBM built the RAMAC, the first magnetic hard disk drive, featuring a total storage capacity of 5 MB at a recording density of 2 kbit/in.$^2$ (Moser et al. 2002). The invention of the hard disk drive forever changed information storage, and major research and development has since been concentrated in the area of improving the recording density in hard disk drives. As a result of this mammoth effort, the areal density (number of bits/area) of recording media has increased exponentially for more than 50 years. The annual increase of the storage density has consistently been greater than 25%, at times even outpacing Moore's law with areal density increases of 100% per year in the 1990s (Richter 2009). Even the projected limit of density storage has grown exponentially over time, probably because engineering feats have overcome perceived obstacles in the design of recording media.

Until rather recently, increasing the recording density simply meant down-scaling all of the components in a recording system. But what has become clear over the last 10 years or so is that the design of magnetic media and the continued increase in density storage is fast encroaching upon the fundamental physical limits imposed by the nature of magnetism on the nanoscale (Charap et al. 1997, Bertram et al. 1998, Richter 1999).

In this chapter, we first discuss the basics of magnetism and magnetic nanoparticles. Then, we put the characteristics of magnetic nanoparticles into the context of magnetic recording media and describe the process by which magnetic media is fabricated. Next, we explain how magnetic nanoparticles can address the pressing issues concerning the fundamental changes that must take place in the fabrication of magnetic media in order to continue the shrinkage of magnetic storage components. Then, we go into a detailed description of the synthesis and characterization of some of the most promising magnetic nanoparticles for magnetic media, $L1_0$ phase FePt and CoPt. Then, we describe how the process of self-assembly can be used to implement the fabrication of magnetic nanoparticle-based media, and lastly, we look ahead at some other promising materials for high-density magnetic recording.

## 3.2 Magnetic Materials and Nanomagnetism

A bulk ferromagnetic material displays a distinct nonlinear magnetic response to an applied magnetic field, referred to as a hysteresis loop or *M–H* curve (Figure 3.1). Despite a ferromagnet's characteristic spontaneous internal magnetic moment, without an applied field the magnetization is usually zero, that is, the *M–H* curve starts at the origin. This is because in order to minimize the magnetostatic energy within a bulk ferromagnet, magnetic domains—small regions of magnetic alignment—are formed. When a magnetic field is applied, the energy of the field can overcome the magnetostatic energy associated with the domain formation, and the magnetization of the material increases until all of the moments are aligned with the field. The magnetic alignment with the field occurs in two ways: (1) the walls that make up the domains can shift resulting in the "growth" of domains already aligned with the field at the expense of those not aligned, and (2) the rotation of the magnetization vector away from one crystallographic axis and toward the field. The final magnetization associated with the complete alignment of the electronic

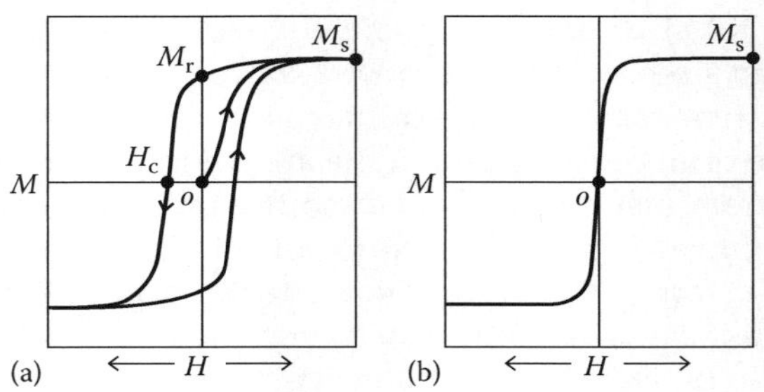


**FIGURE 3.1**
Magnetization versus field (*M–H*) curves for (a) an assembly of ferromagnetic nanoparticles and (b) an assembly of superparamagnetic nanoparticles.

spins is referred to as the saturation magnetization, or $M_s$. Once the magnetization is saturated, if the field is decreased back to zero, the magnetization will not return to zero. The material will maintain some net magnetization, called the remnant magnetization, $M_r$. In order for the magnetization to return to zero, a field must be applied in the opposite direction. The field at which the moments of the domains are again in different directions and the magnetostatic energy is minimized is called the coercivity, $H_c$. The full cycle of the magnetic field can be achieved by saturating the magnetization in the opposite direction, $-M_s$, going through the zero field resulting in a remnant magnetization in the opposite direction, $-M_r$, and increasing the field until $M=0$, corresponding to $-H_c$. When the magnetization is once again saturated in the initial direction the hysteresis loop is complete (Cullity 1972).

Materials used in magnetic information storage are chosen based on their values of the parameters described above. The magnetic components of a disk drive are the read head, the write head, and the media. The media, which is the focus of this chapter, stores data based on the two stable magnetic configurations corresponding to $M_r$ and $-M_r$ on the hysteresis loop. The stability of the media can be evaluated based on the magnetic anisotropy of the material used. The anisotropy energy, $E_A$, describes the amount of energy required to reverse the magnetization from one stable state to the other. This depends on several factors, most importantly the effective anisotropy constant, $K$. The magnetic anisotropy is positively correlated with the coercivity, which is often used to provide a qualitative estimate of the anisotropy of the material, since direct measurements of $K$ can be difficult. In terms of magnetic stability, the more energy that is required to randomize the electronic spins either thermally or via an applied field, the less likely that the information stored in the media will be erased due to stray magnetic fields or excess heat. Therefore, the quest to find better-performing materials to make up hard disk media is centered upon the tailoring of the anisotropy, and thus, the coercivity of the candidate materials.

At odds with the search for high-coercivity materials is the concurrent desire for higher-density data storage. Being able to store data in an ever-decreasing amount of space to continue device miniaturization leads one correctly to conclude that the smaller the magnetic bit, the higher the density of data can be stored within a given amount of space. To that end, it would seem that magnetic nanoparticles would be an obvious choice for the next generation of magnetic data storage. However, analogous to nanoparticles of other functional materials, the size of magnetic nanoparticles encroaches upon a fundamental physical length scale in magnetism, namely the size of a magnetic domain. Like bulk ferromagnets, an array of magnetic nanoparticles containing only one domain (henceforth "single domain") can exhibit hysteresis in the magnetization versus field dependence as well (Stoner and Wohlfarth 1949). The key difference between the magnetic behavior of a bulk magnetic material and a collection of single-domain ferromagnetic nanoparticles arises from the mechanism by which the magnetization is cycled through the hysteresis loop. Since domain wall movement is not possible, only coherent magnetization rotation can be used to overcome the effective anisotropy of the particle. Thus, the maximum coercivity of a given material as a function of particle diameter actually falls in the single-domain range.

The critical diameter for a magnetic particle to reach the single-domain limit is expressed as

$$R_{SD} = \frac{36\sqrt{AK}}{\mu_0 M_s^2}$$

where
$A$ is the exchange constant
$K$ is the effective anisotropy constant
$M_s$ is the saturation magnetization

For most magnetic materials, this diameter is in the range 10–100 nm, though for some high-anisotropy materials the single-domain limit can be several hundred nanometers (Skomski and Coey 1999).

For a single-domain particle, the amount of energy required to reverse the magnetization over the energy barrier from one stable magnetic configuration to the other is proportional to $KV/k_BT$, where $V$ is the particle volume, $k_B$ is Boltzmann's constant, and $T$ is the temperature. If the thermal energy is large enough to overcome the anisotropy energy, the magnetization is no longer stable and the particle is said to be superparamagnetic. That is, an array of nanoparticles each with its own moment can be easily saturated in the presence of a field, but the magnetization returns to zero upon removal of the field as a result of thermal fluctuations (i.e., both $M_r$ and $H_c$ are zero). Figure 3.1 is a schematic of the typical shape of (a) an $M$–$H$ curve for an array of single domain, ferromagnetic nanoparticles (which is qualitatively similar to a bulk ferromagnet) and (b) an array of superparamagnetic nanoparticles. This superparamagnetic behavior is analogous to conventional paramagnets, only, instead of individual electronic spins displaying this fluctuating response, it is the collective moment of the entire particle, and hence the term "superparamagnetism." The temperature at which the thermal energy can overcome the anisotropy energy of a nanoparticle is referred to as the blocking temperature, $T_B$. For an array of nanoparticles with a distribution in volume, $T_B$ represents an average characteristic temperature and can be affected by interparticle interactions as well as the timescale over which the measurement is performed due to the magnetic relaxation of the particles. Keeping these caveats in mind, $T_B$, as the general transition from ferromagnetism to superparamagnetism, is one of the most important quantities that define the magnetic behavior of an assembly of particles.

So, while it is advantageous to try to use single domain, high-anisotropy magnetic nanoparticles for magnetic recording, the crossover in volume at the working temperature from ferromagnetic to superparamagnetic becomes problematic and has been the focus of intense research interest.

## 3.3 Current Approaches to Magnetic Media for Information Storage

The standard design for hard disk drives, throughout the vast majority their existence and until recently, has been the longitudinal recording system (Piramanayagam and Srinivasan 2009). The media in longitudinal recording contains grains whose magnetic moments lie in the plane of the disk. The system also contains a recording head composed of a separate read and write element, which flies in close proximity to a recording medium. The inductive write element records the data in horizontal magnetization patterns (Figure 3.2a). The information is then read back with a giant magnetoresistive (GMR) read element by measuring the stray magnetic field from the transitions between regions of opposite magnetization. Finally, a signal processing unit transforms the analog read back signal into a stream of data bits. The media for these longitudinal systems

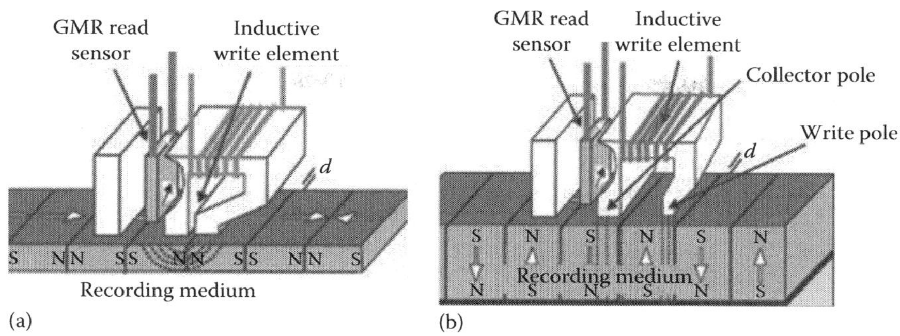


**FIGURE 3.2**
Geometry of a hard disk drive. (a) The traditional longitudinal configuration with the read sensor and write elements passing over a continuous film of magnetic media whose grains have magnetization vectors in the plane of the media. (b) Perpendicular configuration with the magnetic media possessing magnetization vectors, which are orthogonal to the film plane. (Adapted from Moser, A. et al., *J. Phys. D: Appl. Phys.*, 35, R157, 2002. Copyright 2002, the Institute of Physics.)

is fabricated in the form of granular thin films, which are usually deposited by dc magnetron sputtering and consist of a CoPtCrX alloy (X = B, Ta). Along with the recording layer, there are several other layers that make up the media and all are in a polycrystalline state. The extra layers help to nucleate grain growth, control grain size and shape, and help in reducing stray magnetic fields and interactions from the read and write elements.

As a result of the film growth process, the grains of the recording medium usually have random orientations with respect to the film-plane and with respect to the track direction. Because of this randomness, a group of grains are used for each bit or unit of storage. As the signal-to-noise ratio (SNR) from such a medium depends on the number of grains in a bit, optimizing the grain growth to reduce the need for multiple grains per bit is desired. This means reducing the grain size, but more importantly the grain-size distribution, which thus far has been achieved by the use of seed layers or underlayers with small grain sizes or by optimizing the growth of the recording layer itself (Lee et al. 1995, Acharya et al. 2001, Piramanayagam et al. 2002, Mikami et al. 2003). With high-density storage, we also encounter the problem that as each grain is magnetized in one direction, its own magnetic field can influence the surrounding grains, destabilizing them. Therefore, each grain is surrounded by a thin boundary of oxide material to magnetically isolate it from the surrounding grains (Moser et al. 2002).

When a write head passes over the magnetic medium, it records the information by magnetizing the bits in either of the two stable magnetic states. Every disk drive has a clocklike mechanism, which controls the timing at which the write head current is reversed. If the clock interval is $T$, current reversals only occur at $iT$, where $i$ is an integer. This means that the information is recorded as the medium response to a presence or an absence of a write current reversal at times $t = iT$. If $v$ is the linear velocity of the medium at the time of recording, the minimum transition spacing B translates to $B = vT$. It is important to note that no attempt is made to synchronize the rotation speed of the disk with the clock, which means that the actual transition locations are not controlled on a microscopic level (Richter 2009). However, the magnetic media consist of grains that are neither of the same size nor are positioned regularly on the disk. And since the grains are single domain, the transition boundary has to go around the grains. There are a finite number of grains across the width of the track and, hence, the location of the transition (after averaging) will not exactly be at the intended location. Therefore, deviations from the intended positions represent the main source for medium noise in the recording system. This point becomes important once we explore other options for magnetic recording.

As discussed in the previous section, a reduction of grain size eventually leads to thermal instabilities and one has to keep the magnetic energy barriers (KV) of the grains sufficiently high. The mean KV value of current media is typically between 70 and 100 $k_BT$, which ensures a sufficient stability margin for practical operation (Richter 2009). Problematically, a recording medium is composed of grains with a size distribution, and some of the smaller grains are more susceptible to thermal switching.

After over 50 years of grain-size reduction and magnetic-property optimization, longitudinal recording technology is now very close to its superparamagnetic limit. Therefore, perpendicular recording technology was introduced in order to extend the life of hard disk drives. In perpendicular recording, the granular media is grown in such as way that the magnetization vectors of the grains point in the direction perpendicular to the film plane. Figure 3.2 shows a schematic of the geometry of longitudinal versus perpendicular recording. At this time, all companies have introduced perpendicular recording in their hard disk drive products, and longitudinal recording media is being phased out. Iwasaki et al. introduced almost all of the fundamentals of perpendicular magnetic recording in the 1970s, in order to overcome the problem of large demagnetizing fields exhibited by the transitions in longitudinal recording (Iwasaki and Takemura 1975, Iwasaki and Nakamura 1978, Iwasaki et al. 1979). This configuration has allowed grain sizes to shrink even further as the stable magnetic configurations of the grains are in the orientation of the film thickness, allowing each grain to take up less space in the radial direction. However, the fundamental limits associated with the superparamagnetic limit still exist in perpendicular media and it is estimated that as areal densities reach 500–1000 Gb/in.$^2$, thermal instabilities in current CoCrPt:oxide will begin to set in (Piramanayagam and Srinivasan 2009). Therefore, perpendicular media does little more than buy time for researchers to come up with a new design and new recording principles, which do not depend on traditional granular thin films.

An emerging technology that is being explored for increasing areal density is to use patterned media recording, where each bit contains one grain (Ross 2003, Terris and Thomson 2005, Service 2006). In principle, this would allow for larger grain volume than is currently being used, as long as each grain is isolated by a lithographically patterned void or nonmagnetic spacer while still resulting in smaller bit sizes. This approach would result in much higher areal densities, but since each bit is one grain, the requirements for minimizing the distributions of the physical and the magnetic properties of the dots are extremely high. It is estimated that the standard deviations of the magnetic and the structural distributions have to be of the order of 5% to achieve a reasonable system performance (Richter 2009). In addition, there is very little room for error in the intergranular spacing as the reading and writing of the storage is dependent on the timing of the read and write elements over the media and significant errors can occur if the heads are not well aligned with the bits. Consequently, the internal clock of the hard disk drive would have to match precisely with the intended grain, rather than writing to a random collective. The latter problem is more related to the engineering of the servo motor inside the hard disk drive, and while it is beyond the scope of this discussion, it is well worth keeping in mind.

## 3.4 Magnetic Nanoparticles for High-Density Storage

### 3.4.1 Introduction to FePt

One of the most promising candidates for high-density magnetic recording is the face-centered tetragonal (fct) phase of FePt. The FePt alloy of near-equal atomic percentage of Fe and Pt forms at low temperatures in the chemically disordered face-centered cubic (fcc) phase, indicating that any atomic site in the fcc unit cell has an equal chance of being either Fe or Pt (Figure 3.3a). This cubic phase has a very small magnetic anisotropy and displays soft magnetic properties. A high-temperature annealing can result in a phase change to the chemically ordered fct structure, which can be pictured as alternating atomic layers of Fe and Pt stacked along the [001] direction (Figure 3.3b). This phase change from cubic to tetragonal structure results in the breaking of the high symmetry of the cubic structure to the lower symmetry tetragonal structure, causing a preferential magnetic alignment along the *c*-axis. This is manifest as a high magnetic anisotropy with *K* values reaching as high as $4\text{–}10 \times 10^7\,\text{erg/cm}^3$ (Weller and Doerner 2000), among the largest values observed in all hard magnetic materials. Furthermore, the hybridization of the Fe 3*d* states and the Pt 5*d* states renders FePt nanoparticles more chemically stable than many high-moment ferromagnetic nanoparticles composed of Fe and Co, as well as the high-coercivity $SmCo_5$ and $Nd_2Fe_{14}B$ (Staunton et al. 2004) making them useful for myriad applications from biomedicine to catalysis (Sun 2006).

Weller et al. (1992) have calculated that FePt as a recording medium could be thermally stable, even for grain sizes as small as 3 nm. If it is possible to make such small grains or particles and if information could be successfully written onto these materials, the areal density that can be achieved with FePt could easily surpass $1\,\text{Tb/in.}^2$ Currently, the question is whether or how FePt media may be tailored to suit the requirements of recording media. To support $1\,\text{Tb/in.}^2$ and higher densities, the average grain size of FePt has to be less than 4 nm. In addition, the standard deviation in the grain-size distribution has to be less than 0.4 nm (Piramanayagam and Srinivasan 2009).

Because of its high anisotropy and chemical stability, arguments can be made for using FePt in thin-film forms for perpendicular recording, using sputtering techniques similar to those currently in use for longitudinal and perpendicular recording systems. The lowest achieved grain size reported thus far for FePt produced by sputtering is about 6.6 nm (Shen et al. 2005). In this case, an underlayer, RuAl, with small grain size helps to minimize

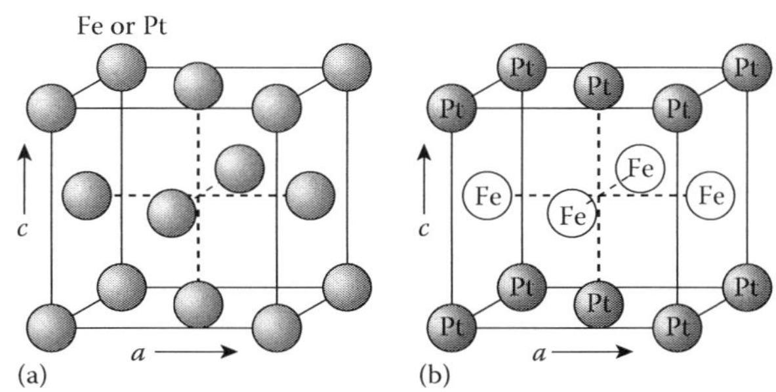


**FIGURE 3.3**
Schematic of the unit cells of (a) chemically disordered fcc FePt and (b) chemically ordered fct FePt.

the grain size in the recording layer. Therefore, reducing the grain size of the FePt media by controlling the underlayer grains is one possible approach, which is similar to current recording media fashioned from a recording layer with several underlayers (Kanbe et al. 2002, Piramanayagam et al. 2002, 2006). In addition, attempts have also been made to dope FePt with additives such as C, Ag, and Cu to reduce the grain size and intergranular distance. Grain sizes of around 4 nm have been obtained by 50% carbon doping by Perumal et al. (2003). However, it should be noted that compositional changes, while resulting in small grain sizes, often lead to changes in magnetic anisotropy and magnetization direction as well as increased intergranular interactions (Chen et al. 2007).

Considering that the future of magnetic recording appears to be moving in the direction of "one grain, one bit," it seems advantageous to thoroughly explore the synthesis and characterization of small, monodisperse FePt nanoparticles. In principle, FePt nanoparticles could be deposited as a thin granular film and used in longitudinal and perpendicular recording, or could be purposefully assembled in such a way as to form patterned media for the next generation of magnetic media.

### 3.4.2 Chemical Synthesis of fcc-FePt Nanoparticles

Until recently, polydisperse FePt nanoparticles had been made by physical, top-down approaches such as vacuum-deposition techniques (Coffey et al. 1995, Li and Lairson 1999, Ristau et al. 1999) and gas-phase evaporation (Liu et al. 2003, Stappert et al. 2003). The as-deposited particles had the chemically disordered fcc structure and could be transformed to fct only after annealing, which resulted in particle aggregation. Subsequent progress in preventing particle aggregation was made by embedding the nanoparticles in a nonmagnetic matrix such as $SiO_2$ or $Al_2O_3$. Though this results in particle isolation and reduction of interparticle interactions, it is difficult to form arrays and superlattices in this manner.

Solution-phase chemical synthesis of FePt nanoparticles has many advantages including a smaller size distribution and the ability to disperse the particles in organic solvents, leading to better self-assembly. A common route to the chemical synthesis of monodisperse FePt nanoparticles is via the simultaneous thermal decomposition of iron pentacarbonyl ($Fe(CO)_5$) and reduction of platinum acetylacetonate ($Pt(acac)_2$) in the presence of 1,2-alkanediol (Sun et al. 2000). $Fe(CO)_5$ is thermally unstable and readily decomposes at high temperature to CO and Fe. It is a common precursor for Fe nanoparticles and Fe-based alloys (Huber 2005). $Pt(acac)_2$ is easily reduced to Pt using a mild reducing agent such as 1,2-alkanediol. During the reaction, a small group of Fe and Pt atoms bond together to form FePt clusters, which act as nuclei for the subsequent nanoparticles. Long-chain carboxylic acids and primary amines (such as oleic acid and oleyl amine) are used to stabilize the nanoparticles and passivate the surface. In a typical synthesis (Sun et al. 2000), $Pt(acac)_2$ (197 mg, 0.5 mmol), 1,2-hexadecanediol (390 mg, 1.5 mmol), and dioctyl ether (20 mL) are first measured into a reaction flask and stirred using a magnetic stirring bar. The mixture is heated to 100°C under a flow of $N_2$ to remove oxygen and moisture. Oleic acid (0.16 mL, 0.5 mmol) and oleyl amine (0.17 mL, 0.5 mmol) are added to the mixture, and after 10 min the $N_2$ outlet is closed and the reaction system is protected under a blanket of $N_2$ for the remainder of the reaction. $Fe(CO)_5$ (0.13 mL, 1 mmol) is then injected into the mixture, which is then further heated to reflux (297°C) for 30 min. The heat source is then removed, and the reaction mixture is allowed to cool to room temperature. The inert gas-protected system is opened to ambient environment at this point. The black product is precipitated by adding ethanol (40 mL) and separated by centrifugation and the yellow-brown supernatant is discarded. The black precipitate is dispersed in hexane (25 mL) in the presence of oleic acid (0.05 mL) and oleyl amine (0.05 mL), and precipitated out by adding ethanol (20 mL) and centrifuging. The product is dispersed in hexane (20 mL), centrifuged to remove any insoluble precipitate (almost no precipitation is usually found at this stage), and precipitated out by adding ethanol (15 mL) and centrifuging. The FePt nanoparticles can be dispersed in any alkane, arene, or chlorinated solvent.

In this reaction, the composition of the FePt particles is controlled by the $Fe(CO)_5$/ $Pt(acac)_2$ ratio, though it is to be noted that a consistent and predictable excess of $Fe(CO)_5$ is required. It has been observed and documented that not all of the $Fe(CO)_5$ contributes to the FePt-alloy formation (Sun et al. 2001). As a result, in this recipe 0.5 mmol of $Fe(CO)_5$ and 0.5 mmol of $Pt(acac)_2$ yield $Fe_{38}Pt_{62}$, while 1.1 mmol of $Fe(CO)_5$ and 0.5 mmol of $Pt(acac)_2$ produce $Fe_{56}Pt_{44}$. However, it has been noted that allowing the solution to reflux for a longer time does yield a higher Fe content, by as much as 15%, in the final composition (Srivastava et al. 2008). Furthermore, the higher the starting ratio of Fe:Pt precursor, the more pronounced the reflux time effect becomes. One hypothesis put forth to explain this observation is the relatively slow decomposition of $Fe(CO)_5$ and/or the slow kinetics of the Fe atoms once released from the carbonyl compound.

It is believed that particle formation ultimately occurs via the following mechanism: (1) Pt-rich nuclei are formed from the simultaneous reduction of $Pt(acac)_2$ and partial decomposition of $Fe(CO)_5$ at temperatures less than 200°C. (2) More Fe atoms will then coat the existing Pt-rich nuclei, forming larger clusters at a higher temperature. (3) Heating the clusters to reflux at 300°C leads to atomic diffusion and formation of fcc-structured FePt nanoparticles. (4) In the presence of excess of $Fe(CO)_5$, the extra Fe will continue to coat the particles, leading to core/shell structured FePt/Fe.

The size of the particles can be tuned by controlling the molar ratio of the stabilizers to $Pt(acac)_2$ and the heating conditions (Momose et al. 2005). A ratio of at least 8:1 is necessary to make FePt nanoparticles larger than 6 nm. At a fixed ratio of 8:1, both heating rate and an interim heating temperature are important in tuning the sizes of the FePt particles. Applying a heating rate of ~15°C/min and an interim heating temperature of 240°C leads to 6 nm diameter FePt, while a slower rate of ~5°C/min and a lower heating temperature of 225°C yields 9 nm diameter FePt. In this particular synthesis, the shape of the particles can be controlled by sequential addition of the surfactants. By first mixing oleic acid, $Fe(CO)_5$, and $Pt(acac)_2$ and heating the mixture at 130°C for about 5 min before oleyl amine is added, cube-like FePt nanoparticles are obtained (Chen et al. 2004). Gaining control over particle shape is an important aspect of nanoparticle synthesis that affects the self-assembly characteristics as well as magnetic properties of an array of magnetic particles. This topic will be revisited in a later section.

Alternatively, to make larger FePt particles, a seed-mediated growth method can be used (Sun et al. 2000). This is carried out by first making monodisperse 3–4 nm seed FePt particles, and then adding more Fe and Pt precursors to enlarge the existing FePt particle seeds to obtain the desired sizes. Still better control of the size of the FePt nanoparticles is obtained via a one-step simultaneous thermal decomposition of $Fe(CO)_5$ and reduction of $Pt(acac)_2$ in the absence of 1,2-alkanediol (Figure 3.4a) (Chen et al. 2004). It is believed that 1,2-alkanediol can lead to the rapid reduction of $Pt(acac)_2$ to Pt, resulting in fast nucleation of FePt and consumption of metal precursors, and, as a result, smaller particles. Exclusion of this reducing agent from the reaction mixture slows down the nucleation rate, allowing more metal precursors to deposit around the nuclei, thus leading to a larger particle size. Figure 3.4 shows transmission electron micrographs (TEM) of FePt nanoparticles of various sizes obtained by different methods.

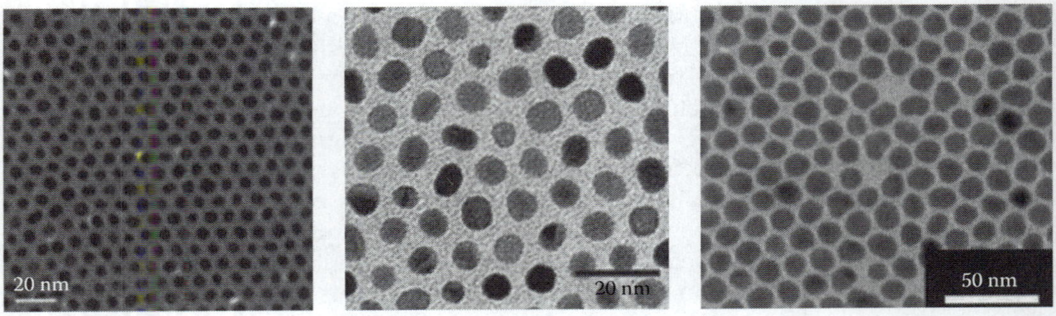


(a) (b) (c)

**FIGURE 3.4**
FePt nanoparticles of various sizes. (a) 6 nm (Reproduced and adapted from Chen, M. et al., *J. Am. Chem. Soc.*, 126, 8394, 2004. Copyright 2004, the American Chemical Society.), (b) 8 nm (Reproduced and adapted with permission from Nandwana, V. et al., *J. Phys. Chem. C*, 111, 4185, 2007. Copyright 2007, the American Chemical Society.), and (c) 11 nm (Reproduced and adapted with permission from Zhao, F. et al., *J. Am. Chem. Soc.*, 131, 5350, 2009. Copyright 2009, the American Chemical Society.)

Large FePt nanoparticles have also been produced by sequential reduction of $Pt(acac)_2$ and thermal decomposition of $Fe(CO)_5$, followed by oxidation and reductive annealing (Teng and Yang 2003, Teng et al. 2003). In this synthesis, Pt nanoparticles are first made via the polyol reduction of $Pt(acac)_2$ in dioctyl ether. A layer of $Fe_2O_3$ is then coated over the Pt nanoparticles via thermal decomposition of $Fe(CO)_5$ and oxidation. The dimensions of the Pt core and the $Fe_2O_3$ shell are controlled by the amount of Pt or Fe precursors, respectively, used in the reaction. To make FePt nanoparticles, the FePt/$Fe_2O_3$ is further annealed under a gas mixture of $H_2$ (5%) and Ar at 550°C or above for 9 h. This high-temperature reductive annealing has two purposes: (1) to reduce $Fe_2O_3$ to Fe and (2) to make Fe and Pt diffuse into FePt alloy particles. After reductive annealing, ~17 nm diameter FePt nanoparticles are obtained. To gain better control over both the stoichiometry and the internal structure of FePt nanoparticles, $Na_2Fe(CO)_4$ has been used to replace $Fe(CO)_5$ in the FePt nanoparticle synthesis (Howard et al. 2005). $Na_2Fe(CO)_4$ acts as both a reducing agent and an Fe source. In this reaction, the anion $Fe^{2-}$ from $Na_2Fe(CO)_4$ transfers two electrons to $Pt^{2+}$. The $Pt^{2+}$ is reduced to metallic Pt while the $Fe^{2-}$ is oxidized to metallic Fe, which then combines with Pt to form FePt. This ensures the 1:1 stoichiometry of the final alloy materials. If the reaction is run in high-boiling tetracosane (boiling point: 389°C), partially ordered fct FePt nanoparticles are obtained. The room temperature coercivity of the particles reaches 1300 Oe.

A slight modification of the decomposition/reduction method by replacement of $Fe(CO)_5$ with $Fe(acac)_2$ or $Fe(acac)_3$ using the polyol method can also lead to monodisperse 2 to 3 nm FePt nanoparticles (Elkins et al. 2003, Liu et al. 2004, Nakaya et al. 2004). The reaction of $Fe(acac)_3$ and $Pt(acac)_2$ in ethylene glycol (Iwaki et al. 2003, Jeyadevan et al. 2003, Harpeness and Gedanken 2005) or tetraethylene glycol (Kitamoto et al. 2005, Minami et al. 2005, Sato et al. 2005) generates FePt nanoparticles that show partially ordered fct structures. Recently, triiron dodecacarbonyl ($Fe_3(CO)_{12}$) has been used as a substitute precursor for $Fe(CO)_5$ because it occurs in a nonvolatile powder form (Kang et al. 2008). It was found that using the polyol method with $Fe_3(CO)_{12}$ and $Pt(acac)_2$ as precursors, ethylene glycol as the solvent/reducing agent, and oleyl amine and oleic acid as surfactants yielded high-quality monodisperse particles. Varying the precursor ratio allows for composition control while an increase in total surfactant amount leads to an increase in particle size up to 8 nm.

Alternatively, the stronger organic reducing agent hydrazine ($N_2H_4$) has been used to reduce metal salts and form FePt nanoparticle in water at low temperature (Gibot et al. 2005). In this synthesis, $H_2PtCl_6{\cdot}H_2O$ and $FeCl_2{\cdot}H_2O$, together with hydrazine and a surfactant, such as sodium dodecyl sulfate (SDS) or cetyltrimethylammonium bromide (CTAB), are mixed in water. The hydrazine reduces the metal cations at 70°C, resulting in fcc-structured FePt nanoparticles. The reduction of $FeCl_2$ and $Pt(acac)_2$ mixtures in diphenyl ether by $LiBEt_3H$ in the presence of oleic acid, oleyl amine, and 1,2-hexadecanediol at 200°C, followed by refluxing at 263°C, has led to ~4nm FePt nanoparticles (Sun et al. 2003a,b). The initial molar ratio of the metal precursors is carried over to the final product, making control over the composition of the FePt much easier than in the decomposition and reduction processes. The strong reducing power of the borohydride also allows the reduction to proceed at room temperature in a predesigned environment.

Reverse micelles have also been used as nanoreactors to make FePt nanoparticles. The micelles are formed using CTAB as a surfactant, 1-butanol as a cosurfactant, and octane as an oil phase (Carpenter et al. 2000). Both an Fe salt and a Pt salt are dissolved in water within the nanoreactors. Adding sodium borohydride leads to the quick reduction of the salts and the formation of FePt nanoparticles. The FePt nanoparticles are also obtained by borohydride reduction of metal salts on a peptide template (Reiss et al. 2004). In general, using water as a solvent allows easy dissolution of various metal salts and application of a well-controlled biological template to synthesize FePt nanoparticles. However, at present, the quality of the particles prepared under these low-temperature conditions has not been as high as those from the high-temperature solution phase syntheses; boron contamination of the final product, a common problem accompanied with borohydride reduction of iron, cobalt, or nickel salts, could also be a concern.

Recently, the use of fatty acids such as a reducing agent as well as an alloy mediator has been reported in the synthesis of FePt particles (Zhao et al. 2009). The air-stable and environmentally friendly iron stearate ($Fe(St)_3$) was used as a precursor along with $Pt(acac)_2$ in the presence of a large excess of stearic acid (HSt:$Fe(St)_3$ of 8:1), octadecylamine, and a high-boiling-point solvent such as tetracosane. While it is believed that $Fe_3O_4$ is formed as an intermediary species after the decomposition of $Fe(St)_3$, the HSt ensures its dissolution at high temperatures and the carboxylate form of HSt acts as a reducing agent for the $Fe^{3+}$ ions. The large excess of HSt also results in the dissolution of any metallic iron formed, leaving FePt as the only stable compound as a final product. The as-synthesized particles are single crystalline with a size up to 17nm (Figure 3.4c), which is tuned by the ratio of $Pt(acac)_2$ to the solvent.

### 3.4.3 Shape Control of FePt Particles

Thus far, we have only discussed the synthesis of small, spherical FePt particles. However, from the standpoint of magnetic media, FePt particles of different physical shapes are advantageous for three reasons: (1) the magnetic anisotropy of nanoparticles has a shape contribution. That is, elongated particles tend to have their easy axis of magnetization lying preferentially along the long axis to reduce the magnetostatic energy of the particle. This added shape anisotropy makes it even more difficult to destabilize the particle and switch the magnetization from one stable configuration to the other. (2) The reduction in symmetry that occurs from a faceted particle as opposed to a spherical particle makes the manipulation of crystallographic axes easier so that when the particles are self-assembled to form patterned media, the magnetic alignment of the [001] axis can be assured. Lastly

(3), in order to maximize the areal density of an array of nanoparticles, a shape that allows for a high packing density should be used. For instance, the packing density of an array of spherical particles is only around 70%, making them less than ideal for high-density applications.

Groups have reported the shape change from spherical to cubic nanoparticles while changing several different parameters of the standard $Fe(CO)_5$ decomposition/$Pt(acac)_2$ reduction recipe (Shukla et al. 2006, 2009, Nandwana et al. 2007). In one report (Shukla et al. 2006), the solvent was changed from dioctyl ether to 1,2-dicholobenzene, and the heating rate was slowed to 4°C/min to a lower reaction temperature of 170°C. However, cubes have also been observed grown in octyl ether under similar conditions (Nandwana et al. 2007). Cubes have also been obtained by the careful sequential addition of surfactants during the growth phase of the nanoparticle synthesis (Figure 3.5a) (Chen et al. 2006). A proposed explanation for this observed phenomenon is that –COOH does not have a strong tendency to bind to Pt. It is known that the surface energy of crystallographic planes of the fcc Pt crystal generally follow the trend of (111) < (100) (Wang 2000). In a kinetic growth process, the Fe-rich species prefer to deposit on the (100) plane, leading to the formation of a cube. If oleyl amine was added first, sphere-like FePt nanoparticles were separated. This indicates that the amine reacts with Pt, forming a stable $Pt\text{-}NH_2-$ complex and hindering the nucleation process. The resulting nuclei may contain more Fe in this case, producing an amorphous spherical structure.

FePt nanorods have also been observed using benzyl ether as a solvent, using the sequential surfactant method (Figure 3.5b) (Nandwana et al. 2007). However, they were not seen when octyl ether was the solvent. Rods were also obtained using dichlorobenzene as a solvent and excess surfactant at high temperature under a pressure of 6 bar (Shukla et al. 2009). Even though the growth mechanism leading to the nanorod morphology is poorly understood, the shape anisotropy associated with the high aspect ratio is especially promising for magnetic recording.

Interestingly, hexagonally-shaped particles can also be obtained using the high pressure method (22 bar) at high temperature (270°C) when toluene is the solvent (Figure 3.5c) (Shukla et al. 2009). It is not entirely clear if it is the solvent itself or the unusual working temperature of the toluene that is responsible for the hexagonal shape (under atmospheric

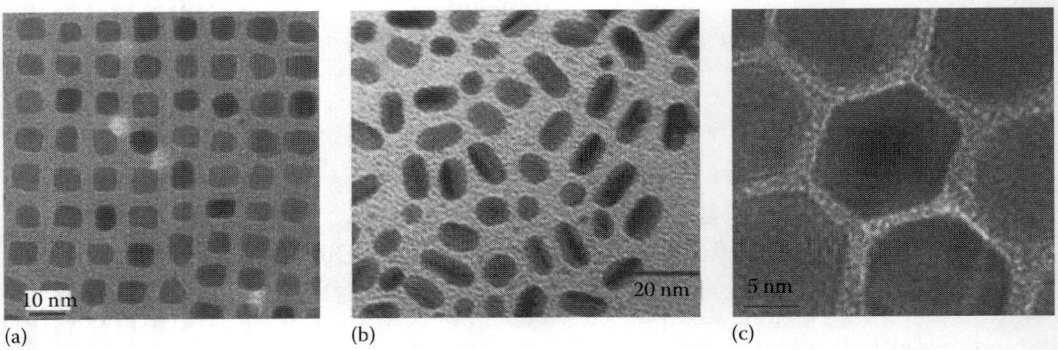


**FIGURE 3.5**
FePt nanoparticles of different shapes. (a) Cubes. (From Chen, M. et al., *J. Am. Chem. Soc.*, 128, 7132, 2006. Copyright 2006, the American Chemical Society.) (b) Rods. (From Nandwana, V. et al., *J. Phys. Chem. C*, 111, 4185, 2007. Copyright 2007, the American Chemical Society.) (c) Hexagons. (From Shukla, N. et al., *Nanotechnology*, 20, 065602-1, 2009. Copyright 2009, the Institute of Physics.)

pressure, toluene boils at 110°C); however, in this case, as well as other shape-controlled syntheses of magnetic nanoparticles, there is much room for further study.

### 3.4.4 $L1_0$ Phase FePt Nanoparticles

The nanoparticle synthesis methods described above almost exclusively result in the soft magnetic fcc phase of FePt or only partially ordered fct phase. In order to fully convert the fcc FePt to magnetically anisotropic fct FePt, an extra annealing step must be incorporated into the process. The change in crystallographic structure upon annealing depends on annealing temperature, time, and Fe and Pt composition. TEM (Dai et al. 2001), x-ray diffraction (XRD) (Sun et al. 2001), and magnetization measurements (Weller et al. 2001) on representative annealed samples of $Fe_{52}Pt_{48}$ particle assemblies show that the onset of this structural transition occurs at around 500°C. Detailed XRD studies on FePt nanoparticle assemblies show that with respect to composition, the $c$-parameter changes mostly in Pt-rich compositions and the $a$-parameter changes mostly in Fe-rich compositions. These results imply that the magnetocrystalline anisotropy in the fct structure will be maximized near the Fe/Pt 50:50 compositions (Klemmer et al. 2002). Because of this known correlation between composition and structure, it follows that the magnetic properties of FePt are composition dependent as well, and are affected by Fe–Pt interactions within the particles (Robach et al. 2003, Ulmeanu et al. 2004, Boyen et al. 2005). The extrapolation of a Gaussian fit of the coercivity and the composition relation yields the best composition to be $Fe_{55}Pt_{45}$ (Sun et al. 2001). This is consistent with what has been observed in physically deposited FePt thin films (Weller et al. 2000). For small nanoparticles sizes, a relatively large $H_C$ (~4000 Oe) at 5 K may also indicate the strong anisotropy contribution from the surface of the nanoparticles (Stahl et al. 2002, 2003). This occurs because the atoms on the surface of the nanoparticles are lacking nearest neighbors, which affects their response to an applied magnetic field. The room-temperature coercivities of the annealed FePt nanoparticle assemblies increase with annealing time and temperature (Sun et al. 2000, 2001, Zeng et al. 2002), reaching a maximum value with the assembly annealed at ~650°C. Even higher annealing temperatures eventually destroy the nanocrystalline features of the particles, leading to the formation of multidomain aggregates and a drop in the coercivity (Zeng et al. 2002).

Although thermal annealing provides FePt nanoparticles with desired magnetic properties, one serious side effect of this annealing is the deterioration of the monodispersity of the particles. Previous in-situ TEM (Dai et al. 2001) and XRD experiments (Sun et al. 2003a,b) have clearly shown the coalescence of the FePt particles annealed at 600°C or above. To prevent this uncontrolled agglomeration, FePt nanoparticles can be embedded in thick organic (Momose et al. 2003) or robust inorganic (Chen et al. 2004, 2005, Zeng et al. 2004a,b, Elkins et al. 2005, Liu et al. 2005) matrices, or assembled on the surface of a robust Si–O network to limit the mobility of the particles at high temperature (Mizuno et al. 2004). An alternative solution to the sintering problem is to dope FePt with another element to lower the structural transition temperature. It is known that the addition of Cu to a FePt film can reduce the transition temperature (Maeda et al. 2002). Both the experimental evidence and the first-principles band calculation indicate that Cu in CuFePt substitutes into the Fe site in the FePt alloy (Kai et al. 2004). The difference in free energy between the ordered and disordered phases is enhanced, and the driving force in the disorder–order transformation increases. In spherical FePt particles, alloying 4 atomic percent Cu into the system has reduced the ordering temperature from 500°C to 400°C

(Takahashi et al. 2002). The same strategy can be applied to chemically made monodisperse FePt nanoparticles. Several classes of monodisperse ternary nanoparticles of FePtCu (Sun et al. 2003a,b), FePtAg (Kang et al. 2003), FePtAu (Kang et al. 2003), and FePtSb (Yan et al. 2005) have been successfully synthesized by a thermal decomposition and reduction method. The fcc-to-fct structural transformation temperature can be decreased to as low as 300°C (Yan et al. 2005).

One promising method for obtaining nonaggregated, dispersible fct-FePt nanoparticles is to follow one of the chemical synthesis procedures outlined in the previous section and perform an extra step to coat the particles individually with an oxide shell during annealing. The particles undergo their phase transformation at the required temperature under the protection of the oxide coating and then another chemical procedure is carried out to remove the shell and redisperse the $L1_0$ FePt nanoparticles back into an organic solvent. The first reports of using an oxide nanoreactor to keep the FePt particles isolated used a $SiO_2$ coating (Yamamoto et al. 2005, Tamada et al. 2007a,b). The effect was to use high-temperature annealing to allow the thermal diffusion of the Fe and Pt atoms during the heat treatment, which can be confined inside the $SiO_2$ nanoreactor. Just as important, the $L1_0$ nanoparticles could be recovered using a solvent-dispersible manner for enabling the ease of handling. In this experiment, precursor 6.5 nm fcc-FePt nanoparticles were prepared using the combination reduction of $Pt(acac)_2$ and decomposition of $Fe(CO)_5$ method, as described in the previous section, to obtain a composition of $Fe_{55}Pt_{45}$. The FePt particles were made water soluble by encapsulation in HTAB using an oil-in-water encapsulation technique outlined by Fan et al. (2004). The surfaces of the particles were coated with $SiO_2$ by addition of tetraethylorthosilicate (TEOS) to the emulsion with subsequent stirring for several hours. The addition of sodium hydroxide (NaOH) and continued stirring caused the particles to precipitate out for subsequent experiments. The $SiO_2$-coated fcc-FePt nanoparticles were annealed at 900°C for 9 h in flowing Ar + 5% $H_2$ gas to convert them to the $L1_0$ structure. Finally, the $SiO_2$ shell was dissolved in a tetramethylammonium (TMA) hydroxide solution (10 wt.% aqueous solution) by simply stirring the suspension at room temperature for 24 h. These particles displayed a coercivity as high as 28 kOe at 300 K. It was also determined that annealing time played an important role as those particles annealed for shorter periods of times yielded a smaller $H_C$ with a kink in the hysteresis loop, indicative of an incomplete fcc–fct transition in the array (Yamamoto et al. 2005).

Since the initial reports of silica-coated FePt, a somewhat simpler method of coating FePt with MgO has been reported (Kim et al. 2008a,b). As mentioned above, FePt particles were synthesized using the simultaneous reduction of $Pt(acac)_2$ and thermal decomposition of $Fe(CO)_5$ method. The MgO coating was prepared by decomposition of $Mg(acac)_2$ in the presence of FePt nanoparticles, 1,2-tetradecanediol, oleic acid, and oleyl amine in benzyl ether, and heated to 298°C. This coats the FePt particles with a flower-like structure of MgO (Figure 3.7a). Thermal annealing of the fcc FePt/MgO nanoparticles was performed under Ar + 5% $H_2$ at various temperatures. The TEM analysis indicated that for particles annealed below 800°C for less than 4 h, there was no obvious FePt morphology change in the FePt/MgO structure. However, upon annealing at 800°C for over 6 h, the FePt/MgO nanoparticles sintered. XRD analyses of the annealed FePt/MgO particles revealed that an fcc-to-fct structural transformation in the FePt nanoparticles is not readily characterized until the annealing temperature reaches above 700°C. This structural transformation temperature is much higher than the 550°C needed for the as-synthesized fcc-FePt nanoparticles, as previously reported (Sun et al. 2000, 2001). XRD peak width measurements confirmed that FePt in the nanoparticles does not experience grain growth during the high-temperature annealing process. Interestingly, the XRD diffraction peaks for the

FePt in the FePt/MgO appeared at lower angles than the corresponding peaks from the FePt nanoparticles without coating, indicating that the nanostructured FePt in FePt/MgO has a slightly larger crystal-lattice spacing than bulk fct-FePt, which is most likely a result of an imperfect fcc-to-fct transformation and/or FePt nanoparticle surface effects. It was hypothesized that the fcc-to-fct transition could be hindered by the limited mobility of the Fe and Pt within the constrained MgO structure.

Magnetic measurements show that the as-synthesized fcc-FePt and fcc-FePt/MgO nanoparticles were superparamagnetic at room temperature. After thermal annealing at 650°C under Ar + 5% $H_2$, the uncoated FePt assembly became ferromagnetic, and the room temperature coercivity reached 20 kOe. Under the same annealing conditions, the FePt/MgO nanoparticles showed only weak ferromagnetism at room temperature. Evidently, the annealing at 650°C for 6 h did not completely convert the fcc-FePt into fct-FePt in the FePt/MgO structure, which is also indicated by the XRD studies. The fct-FePt/MgO was obtained by annealing the fcc-FePt/MgO at 750°C for 6 h. Those fct-FePt/MgO nanoparticles were ferromagnetic, with $H_C$ reaching 10 kOe. By comparing the structural transformation and magnetic property change between the uncoated FePt and the coated FePt/MgO, it was concluded that: (1) MgO in FePt/MgO protects FePt from sintering in a manner similar to $SiO_2$ at annealing temperatures up to 800°C, and (2) in a constrained MgO structure, where atom mobility is limited because of the robust MgO coating, the fcc-to-fct conversion is still possible, but is much more difficult and requires a higher annealing temperature (150°C higher than that for the 7 nm FePt nanoparticles in the same work) and longer annealing time.

A solution proposed to the mobility problem was to slightly alter the Fe and Pt composition of the nanoparticle such that the final nanoparticle product before MgO coating was actually a Pt-rich core and an $Fe_3O_4$ shell (Kim et al. 2008a). The mechanism for the formation of the Pt-rich FePt/$Fe_3O_4$ nanoparticles under the current reaction conditions is similar to what has been proposed for the formation of fcc-FePt/$Fe_3O_4$ (Chen et al. 2004), but the process is controlled so that there is no significant diffusion of Fe into Pt in the reaction condition. The Pt-rich FePt nanoparticles are formed from the familiar simultaneous reduction of $Pt(acac)_2$ and partial decomposition of $Fe(CO)_5$ at temperature <240°C. At a higher reaction temperature, more Fe atoms coat over the existing Pt-rich FePt nanoparticles, forming Pt-rich FePt/Fe nanoparticles that are further oxidized to Pt-rich FePt/$Fe_3O_4$ nanoparticles. The amount of $Fe(CO)_5$ is optimized so that the ratio of Fe/Pt is close to 1:1. MgO is coated over the FePt-$Fe_3O_4$ nanoparticle surface in the condition described above. Hydrogen in the Ar + 5% $H_2$ forming gas reduces $Fe_3O_4$ to Fe, releasing $H_2O$ and causing defects in oxygen sites. Such defects may promote interdiffusion between the Fe shell and Pt-rich FePt core, facilitating the formation of fct-FePt. This easy fct structure formation may be compared with what is observed in the ternary FePtM nanoparticles, discussed in the previous section, in which different Ms are doped into the FePt matrix for decreasing the fcc-to-fct conversion temperature (Kang et al. 2003, Sun et al. 2003a,b, Yan et al. 2005). In this way, well-ordered fct FePt nanoparticles were obtained at 650°C for 6 h (Figure 3.6c), 100°C lower than that for the formation of fct FePt from the fcc-FePt/MgO nanoparticles. In the FePt/$Fe_3O_4$/MgO structure, the $Fe_3O_4$ shell thickness was used to control the final FePt composition. For example, 12 nm fct $Fe_{52}Pt_{48}$ nanoparticles were synthesized by annealing 7 nm/2.5 nm fcc-FePt/$Fe_3O_4$/MgO NPs. Their coercivity reached 32 kOe at 5 K and 20 kOe at 300 K (Figure 3.6b) (Kim et al. 2008a).

The fct-FePt/MgO nanoparticles prepared from the thermal annealing of fcc-FePt/MgO nanoparticles are not easily dispersed in any solvent. Although the MgO in the FePt/MgO structure can be removed by washing with dilute HCl (0.5 M), the bare fct-FePt

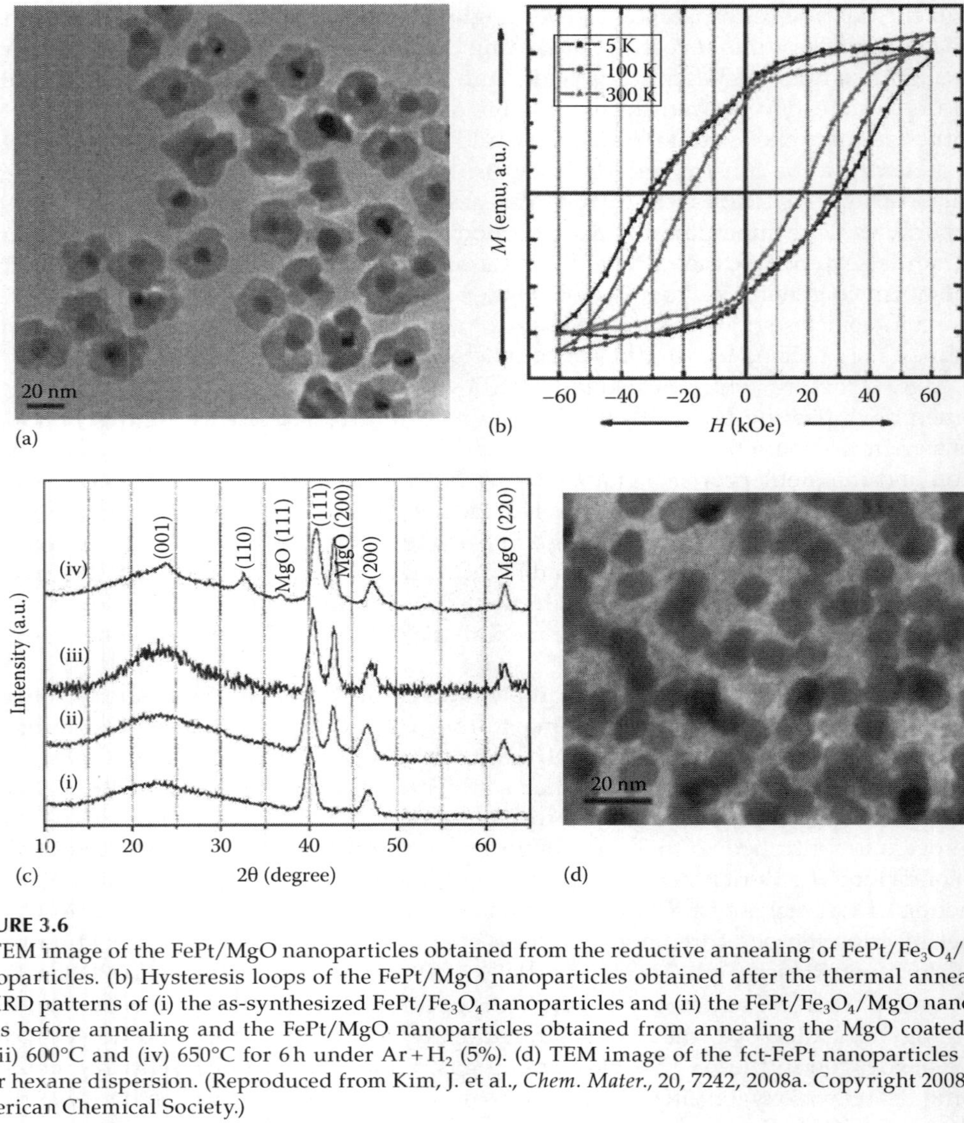


**FIGURE 3.6**
(a) TEM image of the FePt/MgO nanoparticles obtained from the reductive annealing of FePt/$Fe_3O_4$/MgO nanoparticles. (b) Hysteresis loops of the FePt/MgO nanoparticles obtained after the thermal annealing. (c) XRD patterns of (i) the as-synthesized FePt/$Fe_3O_4$ nanoparticles and (ii) the FePt/$Fe_3O_4$/MgO nanoparticles before annealing and the FePt/MgO nanoparticles obtained from annealing the MgO coated NPs at (iii) 600°C and (iv) 650°C for 6h under Ar + $H_2$ (5%). (d) TEM image of the fct-FePt nanoparticles from their hexane dispersion. (Reproduced from Kim, J. et al., *Chem. Mater.*, 20, 7242, 2008a. Copyright 2008, The American Chemical Society.)

nanoparticles quickly aggregate. To protect the fct-FePt nanoparticles from aggregation upon MgO removal, it was necessary to extract them from their aqueous phase into an organic phase by a phase transfer process. Fct-FePt/MgO nanoparticles were added to a mixture of aqueous 0.5M HCl and surfactant-containing hexane, to extract the fct-FePt nanoparticles into hexane with their surface protected by the surfactant. Several different surfactant combinations were tested, which included oleic acid/oleyl amine/hexane, hexadecanethiol (HDT)/hexane, and HDT/oleic acid/hexane. It was found that oleic acid/oleyl amine could not stabilize the fct-FePt nanoparticles under the current extraction conditions. HDT alone offered only temporary stabilization—the nanoparticle dispersion became unstable and aggregated after 2h. The HDT/oleic acid produced the most efficient protection to the fct-FePt nanoparticles. The stabilization difference between HDT and

HDT/oleic acid seems to indicate that thiol (–SH) reacts with Pt, while –COOH binds to Fe, on the surface of the FePt nanoparticle, and the double bond in oleic acid is essential for the stabilization, as the saturated hydrocarbon chain tends to fold on the nanoparticle surface, which leads to strong nanoparticle interactions and aggregation. Figure 3.6d is a TEM image of the particles after annealing and redispersing in hexane.

An in-depth study comparing $SiO_2$ and MgO coatings on FePt particles found that $SiO_2$ coatings resulted in a higher coercivity of the $L1_0$ particles, but the MgO coatings provided better physical and magnetic isolation between particles (Tomou et al. 2007). However, the MgO coated the traditional FePt particles and not the FePt/$Fe_3O_4$ particles, described above, which in general give better results post-annealing.

### 3.4.5 CoPt Nanoparticles and fcc Nanoparticle Synthesis

The cobalt-platinum alloy system (CoPt) has been explored extensively for magnetic recording in thin-film form (Stavroyiannis et al. 1998, Yu et al. 2000, Moser et al. 2002, Piramanayagam and Srinivasan 2009, Richter 2009). It is extremely attractive because, like the $L1_0$ phase of FePt, CoPt also shows an $L1_0$ chemically ordered phase with an anisotropy constant of $K=2–4\times10^7$ erg/cm$^3$ (Bolzoni et al. 1984).

The evolution of CoPt synthesis techniques yielding robust magnetic properties has been noticeably slower than that of $L1_0$ FePt nanoparticles. While coercivity values exceeding 10 kOe have been observed in small-grained CoPt thin films (Stavroyiannis et al. 1998, Yu et al. 2000), coercivity values in analogous, chemically synthesized particles have been, in general, disappointingly low. One reason for this could be that high quality, reproducible CoPt nanoparticles are relatively difficult to synthesize chemically. Since the reduction potentials of Co and Pt are 20.28 and 1.2 V, respectively (Lide 1995), Pt is easily reduced compared to Co, making the simultaneous reduction/decomposition of organometallic precursors prohibitive, and hence one cannot simply modify the well-established FePt recipe to fit CoPt. For example, it has been reported that substitution of dicobalt octacarbonyl ($CO_2(CO)_8$) for $Fe(CO)_5$ results in a broad range of particle compositions manifest magnetically as a mixture of weakly ferromagnetic Co-rich particles and superparamagnetic Pt-rich particles. On the other hand, the use of cobalt tricarbonyl nitrosyl ($Co(CO)_3NO$) instead of $CO_2(CO)_8$ has been used to yield high-quality monodisperse 8 nm $Co_{48}Pt_{52}$ nanoparticles (Chen and Nikles 2002). However, after annealing at 700°C for 3 h, XRD showed that while the fcc-to-fct transition did occur, the particles still did not display an appreciable coercivity ($H_c$ ~ 630 Oe). Compound the large difference in reduction potential with the observation that the coercivity of CoPt seems to be highly sensitive to composition and one has an extremely challenging chemical synthesis problem. For instance, it is believed that a deviation as small as 5% from the 50:50 composition can result in a large fraction of fcc CoPt persisting in particles even after high temperature and long duration of annealing (Weller and Moser 1999, Yu et al. 2000). This soft phase can then be easily converted to the $L1_2$ phase of either $Co_3Pt$ or $CoPt_3$, both of which are highly thermodynamically stable with much lower anisotropy constants (Harp et al. 1993, Rooney et al. 1995, Maret et al. 1997, Albrecht et al. 2002). In this section, we discuss the most promising routes to chemical synthesis of $L1_0$ CoPt particles while pointing out the ample room for optimizing the structural and magnetic properties of this system.

One of the earliest and most compelling reports of CoPt nanoparticle synthesis was via the reaction of organometallic precursors with dihydrogen (Ould Ely et al. 2000). First, three solutions were prepared: 764 mg of poly(vinylpyrrolidone) (PVP) in 35 mL of tetrahydrofuran (THF), 60 mg of cobalt (1,5-cyclooctadiene) (cyclooctadienyl) ($Co(\eta^3\text{-}C_8H_{13})$

($\eta^4$-$C_8H_{12}$)) (0.212 mol of Co) in 5 mL of THF and 110 mg of platinum bis-dibenzylidene acetone ($Pt_2(dba)_3$) (0.201 mol of Pt) in 17 mL of THF. These three solutions were transferred into a bottle that was then pressurized with 3 bar of dihydrogen. After 64 h at room temperature under vigorous stirring, the resulting black precipitate was isolated by filtration, redissolved in 20 mL of methanol, filtered through a 10 μm membrane, and the filtrate precipitated with 40 mL of pentane. The resulting powder was isolated by filtration and evaporated to dryness under vacuum yielding 270 mg of a dark solid. This procedure yielded small (mean diameter 1 nm) monodisperse nanoparticles with a Co:Pt of 1:0.9. The product showed superparamagnetic behavior at room temperature with a very low blocking temperature of around 15 K. The particles were not annealed further to try to facilitate an fcc-to-fct transition, though it is unknown if such small $L1_0$ phase CoPt would be magnetically stable. There have not been many reports utilizing these precursors since the publication of this recipe, probably due to the two-step process involved when one considers the additional reaction reported in order to produce the Co($\eta^3$-$C_8H_{13}$)($\eta^4$-$C_8H_{12}$) and $Pt_2(dba)_3$. However, since the publication of this recipe, both organometallic precursors have become commercially available and this type of reaction is a good candidate for optimization for larger particles, due to the high quality and monodispersity of the product produced.

Due to the differences in reduction potential between Co and Pt, the most common synthesis route for CoPt nanoparticles is the reduction of metal salts in aqueous solutions using very strong reducing agents trying to force the simultaneous coreduction of the two metals. Microemulsion techniques have been used to successfully synthesize CoPt nanoparticles (Kumbhar et al. 2001, Yu et al. 2002, Mandal et al. 2009). Reverse micelles of either CTAB (Kumbhar et al. 2001) or sodium bis(2-ethylhexyl)-sulfosuccinate, Na(AOT) (Yu et al. 2002) were made, using 1-butanol as the cosurfactant and octane as the oil phase. To this solution, an aqueous solution containing the metal ions was added. The molar ratio of water to surfactant governs the size of the reverse micelle, and subsequently the resulting particle diameter, though for the CoPt system mostly small particles (<5 nm), are still observed. The molar ratios of the metallic salts were prepared in the aqueous phase in the same proportions as the desired molar ratio, e.g., 50:50 ratio for CoPt, 3:1 for $Co_3Pt$. The metal and alloy nanoparticles are formed within the reverse micelle by the reduction of the metallic salts using sodium borohydride ($NaBH_4$) as a reducing agent with an excess added to suppress oxidation of cobalt by water. The solution is then mixed and stirred for several hours. For a colloidal suspension, particles can be extracted from the reverse micelles by covalent attachment of oleic acid and oleyl amine, washed and centrifuged with ethanol, and dispersed in hexane. Alternatively, the CoPt powder alone can simply be isolated using magnetic decantation, and the micellar surfactant can be removed by successive washing with chloroform/methanol (1:1) and drying in vacuum.

Room temperature reduction of Co and Pt salts in aqueous media without the use of micelles had also been reported (Sun et al. 2004, Wang et al. 2004, Gibot et al. 2005). In one method (Wang et al. 2004), aqueous and alcohol $NaBH_4$ solution (molarity range 0.006–0.06 M) were added into a mixture solution of $CoCl_2$–$6H_2O$ and $PtCl_4$ (0.002–0.02 M) with equal molarity of oleic acid and oleyl amine added to both solutions. CoPt nanoparticles of varying composition were produced by controlling the density of the precursor solutions. Magnetic decantation was used to collect the particles into mineral oil to stop particle growth. The particles were collected by centrifugation and washed with acetone. They were then dispersed in 2 mL of a 1:1 mixture of hexane and octane, containing 0.025 mL of oleic acid and 0.005 mL of oleyl amine to produce a stable colloid. TEM and HRTEM analyses revealed that particles were formed in two morphologies: small 1–2 nm spheres and 1 × 2 nm rods. Though a full shape control was not pursued with this recipe, the possibility of increasing the aspect ratio to further enhance the effective anisotropy via the shape contribution is intriguing.

Evidence has been found that the reduction of Co and Pt salts using $NaBH_4$ can lead to boron impurities in the particles (Gibot et al. 2005), thus superhydride ($LiBEt_3H$) (Sun et al. 2004) and hydrazine (Gibot et al. 2005) have both been used as alternative reducing agents. To a length of 8 nm rodlike CoPt nanoparticles were synthesized by the superhydride reduction of anhydrous $CoCl_2$ and $Pt(acac)_2$ at 200°C in the presence of oleic acid, oleyl amine, and 1,2-hexadecanediol, followed by refluxing at 260°C. The initial molar ratio of the metal precursors is equal to the final molar composition ratio of the product, and the CoPt composition is easily tuned. As an example, $Co_{50}Pt_{50}$ particles were prepared with 197 mg of $Pt(acac)_2$ (0.5 mmol), 65 mg $CoCl_2$ (0.50 mmol), 520 mg of 1,2-hexadecanediol (2 mmol), and 25 mL of phenyl ether. The reaction was performed under an inert flowing nitrogen atmosphere, where it was heated to 100°C for 10 min. To this, 0.16 mL oleic acid (0.5 mmol) and 0.17 mL oleyl amine (0.5 mmol) were added, and the mixture was continuously heated to 200°C for 20 min. A solution of 1 M $LiBEt_3H$ in THF solution (2.5 mL) was slowly injected into the mixture. The dispersion was then heated to reflux at 260°C for 30 min under flowing nitrogen gas. After the heating source was removed, the reaction mixture was cooled to room temperature. Ethanol was then added, and the particles were precipitated and separated by centrifugation. The final black product, $Co_{50}Pt_{50}$, was dispersed in 10 mL of hexane in the presence of oleic acid (~0.05 mL) and oleyl amine (~0.05 mL).

Despite these advances in CoPt nanoparticle synthesis, handling metal particles in aqueous media can be difficult due to unwanted hydrolysis or oxidation, and using such strong reducing agents can be extremely dangerous. Hence, much effort has been put into developing one-step polyol-derived processes to synthesize CoPt nanoparticles in an organic medium (Chinnasamy et al. 2003, Tzitzios et al. 2005). In the polyol technique, the solvent acts as a reducing and oxidation preventing agent without compromising molecular or atomic level control. The basis for using the polyol technique for CoPt is the observation that the reduction rate of Co could be enhanced by introducing an appropriate amount of OH ions in polyol (Chinnasamy et al. 2003). In one example, cobalt acetylacetonate (0.01 M) and platinum acetylacetonate (0.01 M) were dissolved in trimethylene glycol (TMEG, 200 mL) in a three-neck flask. The appropriate amounts of NaOH were then dissolved in this solution. The solution was placed in an oil bath and heated to 195°C with constant mechanical stirring and allowed to reflux for a maximum period of 3.5 h at this temperature. Intermediate samples were also collected for composition analysis. After refluxing for 3.5 h, the suspension was allowed to cool to room temperature. The precipitated particles were isolated by centrifuging, and were then washed three times with ethanol to remove the by-products.

Even though the individual Co and Pt reduction time was shortened in the presence of OH ions, the reaction was allowed to continue for 3.5 h. Composition analyses of the intermediate samples demonstrated that the release of Co ions increased with reaction time. TEM-EDX analysis of the 1 and 2 h refluxed samples showed the average compositions of $Co_{30}Pt_{70}$ and $Co_{35}Pt_{65}$, respectively. After 3.5 h, the average composition of the nanoparticle was found to be almost equiatomic.

In an alternative polyol approach, CoPt nanoparticles were prepared by the simultaneous reduction of $Pt(acac)_2$ and $Co(ac)_2$ by PEG-200 in diphenyl ether, an organophilic and high-boiling-point solvent (260°C) (Tzitzios et al. 2005). It is reported that the PEG-200 allows for a much quicker reduction of the $Co(ac)_2$ without the need for an alkaline environment. However, because bonding of the organophilic capping molecules to the surface

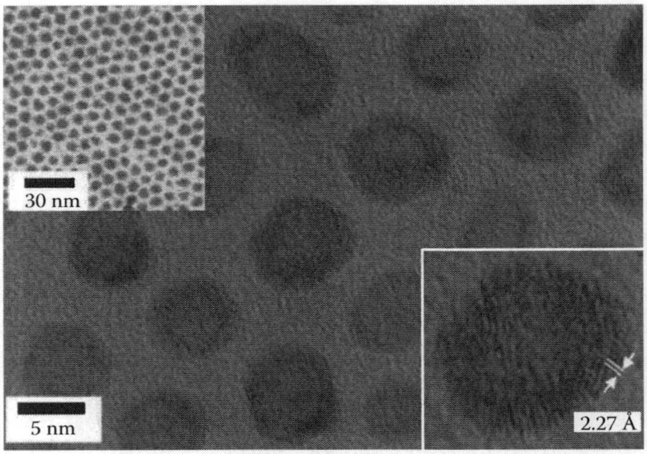


**FIGURE 3.7**
High-resolution TEM images of CoPt core–shell nanoparticles which after annealing become fct-CoPt nanoparticles via interdiffusion. Top inset: TEM image of as-synthesized particles. Bottom inset: Close-up of a particle with Pt lattice spacing indicated. (From Park, J.-I. et al., *J. Am. Chem. Soc.*, 126, 9072, 2004. Copyright 2004, the American Chemical Society.)

of the particles in pure PEG-200 is poor, the reaction is conducted in the organophilic diphenyl ether as a solvent between 200°C and 220°C. The presence of a binary mixture of oleic acid and oleyl amine is also necessary to control and stabilize the growth of the particles. The resulting surface-protected CoPt nanoparticles are air-stable and can be stored under air for months without signs of decomposition.

Lastly, an interesting hybrid approach to the synthesis and annealing of CoPt nanoparticles has been reported (Park et al. 2004). In this synthesis procedure, first Co nanoparticles alone were formed via the thermal decomposition of $Co_2(CO)_8$ in a toluene solution of NaAOT (Figure 3.7). $Co_{core}Pt_{shell}$ nanoparticles were synthesized by transmetalation between platinum hexafluoroacetylacetonate ($Pt(hfac)_2$, 0.375 mmol) and 6.3 nm Co nanoparticles (0.75 mmol) in a nonane solution containing 0.09 mL of dodecyl isocyanide ($C_{12}H_{25}NC$) as a stabilizer. After refluxing for 6 h, the colloidal $Co_{core}Pt_{shell}$ nanoparticles are separated after adding ethanol and centrifugation. Figure 3.7 shows TEM images of the $Co_{core}Pt_{shell}$ particles obtained before annealing. In the main image as well as the bottom inset, the core/shell structure can clearly be seen. During the reaction, the $Pt^{2+}$ of $Pt(hfac)_2$ is reduced to Pt, while the surface Co atom of the Co nanoparticles is oxidized via hfac ligand migration to form $Co(hfac)_2$ as a reaction by-product. The $Co_{core}Pt_{shell}$ nanoparticles were annealed at 700°C for 12 h. The annealed particles had an $H_C$ of 5300 Oe at room temperature.

### 3.4.6 Fcc-to-Fct Transition in CoPt Nanoparticles

Like the FePt system, as-synthesized particles obtained from all of the above-mentioned techniques show superparamagnetism. However, the fcc-to-fct transition in CoPt nanoparticles is not as straightforward, often times yielding frustrating results making high-quality, high-anisotropy CoPt nanoparticles still an elusive goal.

Small (<5 nm) particles synthesized via microemulsion and annealed at only 400°C for 4 h shows only a modest increase in coercivity (up to 500 Oe) (Kumbhar et al. 2001), which indicates that temperature in excess of 400°C are likely needed to make the full fcc-to-fct transition. However, an increased annealing temperature of 550°C allowed particles made with a similar technique (Yu et al. 2002) to show a coercivity of 5500 Oe, a dramatic improvement; however, the same report noted no increase in coercivity with increasing temperature.

A coercivity of 6 kOe was measured after annealing Sb-doped CoPt particles at 650°C for only 1 h (Gibot et al. 2005). The estimated effective anisotropy was $K = 1.7 \times 10^7$ erg/cm$^3$. Originally, the Sb was substituted for Co in order to lower the transition temperature and/or maintain better dispersibility in the organic solvent. While the coercivity was the same with or without the addition of Sb, there is evidence that the magnetic switching was affected, which could be worth pursuing in other CoPt studies.

Still, it seems that even higher annealing temperatures are needed to facilitate the transition to $L1_0$ phase CoPt. Particles synthesized using the polyol method and annealed for 60 min at 700°C exhibit coercivity as high as 7.57 kOe (Chinnasamy et al. 2003). Particles synthesized in a similar fashion and annealed at 700°C for 4 h only displayed a coercivity of 600 kOe, despite the fact that both methods yielded similar particle size. But what is more troubling is that Rietveld analysis on the latter samples showed that a complete transformation from the fcc phase to the fct phase occurred (Tzitzios et al. 2005), which confirms the suspicion that incomplete conversion alone cannot account for the unexpectedly small coercivities measured in some of these experiments. It is likely that Co:Pt in the alloy nanoparticles plays a crucial role, and homogeneity both within the same particle and from particle to particle is more difficult than anticipated.

The best magnetic measurements to date come from annealing the mixture of 1–2 nm spheres and 1 × 2 nm rod-shaped particles obtained by microemulsion (Wang et al. 2004) at 665°C for 30 min. This caused the particles to go from superparamagnetic to having a coercivity of 9 kOe. While there is admittedly particle aggregation after annealing the unprotected particles at this temperature, the coercivity increase cannot be due to aggregation alone. But what is not clear from this study is if the main coercivity contribution comes from all the particles, or predominantly from the high aspect ratio particles. If this question can be answered, it may become clear that gaining control over the shape and aspect ratio of CoPt particles might be the key to optimizing their magnetic properties for data storage applications.

## 3.5 Self-Assembled Nanomagnet Arrays for Magnetic Recording

### 3.5.1 Self-Assembly Techniques

If nanoparticles are to be used to form the bits in patterned magnetic media, fine control has to be gained over the assembly of the nanoparticles onto the substrate, making the understanding of the self-assembly process itself imperative. Self-assembly is a naturally occurring process, generally referring to building blocks forming a structure spontaneously without an additional source of energy (Ulman 1990, Whitesides et al. 1991, Stupp et al. 1997). The structure in either 2D or 3D is usually determined by van der Waals, hydrogen bonding, and electromagnetic dipolar interactions. Self-assembly occurs when the building blocks interact with one another through a balance of attractive and repulsive interactions. The strength of the interactions between the components, therefore, must be comparable to the forces tending to disrupt them. For nanoparticles, the disrupting forces are generated by thermal motion. Self-assembly is normally carried out in solution or at an interface to allow the required motion of the components. Therefore, the interaction of the components with their environment can strongly influence the course of the process.

Magnetic nanoparticles contain hundreds to thousands of single atoms and have extremely large surface energy and magnetic interactions. As a result, they tend to undergo uncontrolled aggregation under common preparation conditions. To stabilize magnetic nanoparticles, repulsive forces must be present to counteract the magnetic and surface-related attractions (Everett 1988). This stabilization can be achieved via electrostatic and steric repulsion. Coating the particles with large molecules, such as polymers or surfactants containing long-chain hydrocarbons, offers efficient stabilization. The presence of a hydrocarbon coating layer will greatly enhance the stability of the particles due to the increased steric interactions when two particles come close. The hydrocarbon coating further facilitates the formation of magnetic nanoparticle dispersions in various hydrocarbon solvents, leading to stable magnetic fluids, known as ferrofluids.

Monodisperse nanoparticles can form close-packed arrays on a variety of substrates as the solvent from the particle dispersion is allowed to evaporate. Preparing particles with narrow size distributions is critical to achieving long-range order in the assemblies: for size distributions $\sigma \sim 7\%$ short to medium range order is usually observed, reducing $\sigma$ to ~5% enables the formation of assemblies with long-range order known as colloidal crystals or superlattices (Murray et al. 2001). However, even a sample with $\sigma \sim 5\%$ can yield disordered particle films if the deposition conditions are not appropriately tailored and the solvent evaporation rate is too high. To form well-ordered nanostructures using the self-assembly method, three factors have to be considered: (1) the interactions between the solvent and the substrate, (2) the interactions between the particles and the substrate, and (3) the interactions between neighboring particles.

A nanoparticle surrounded by hydrocarbon molecules is hydrophobic and can be dispersed only in nonpolar solvents. Such dispersions can spread well over a hydrophobic surface such as carbon or H–Si. By controlling the concentration of the particle dispersion and solvent evaporation rate, large area 2D or 3D nanoparticle superlattices can be readily formed (Figure 3.8). If the substrate (such as glass or silicon oxide) is hydrophilic, then the dewetting nature of the hydrocarbon dispersion of the particles will result in the formation of percolating domains—isolated islands of particles with each island containing either 2D or 3D assemblies.

The interparticle interactions can also affect the particle assembly. During the liquid drying process, attractive molecular interactions between two particles dominate the Brownian motion and gravitational precipitation of the particles, causing particles to form percolating domains. This could be a key reason that local disorder and voids are often present in a compacted monolayer assembly. To maintain a wetting layer on the surface

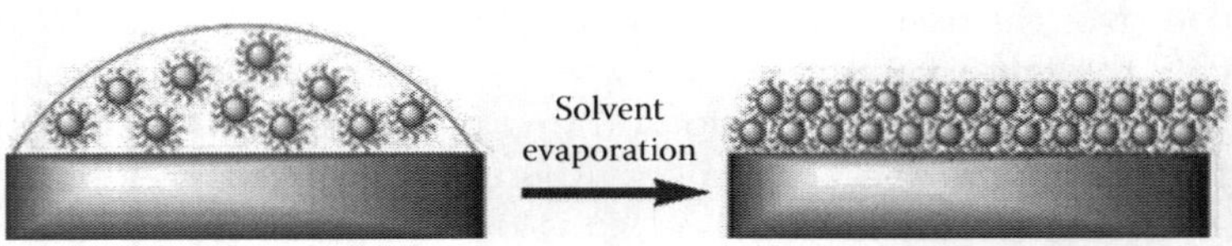


**FIGURE 3.8**
Schematic illustration of self-assembly taking place after the slow evaporation of solvent from a nanoparticle solution.

so that nanoparticles can self-assemble into large periodical structures, the evaporation rate of the solvent must be slowed down, similar to a crystallization process. Controlled solvent evaporation can be performed by covering the substrate with a small container, or by increasing the portion of the high-boiling-point solvent in the dispersion. If the evaporation is performed under a covered environment, the evaporation of hexane under the ambient conditions is slowed down to over 5–10 min. As a result, a well-ordered nanoparticle superlattice structure can be obtained. Depositing nanoparticles from 80% hexane and 20% octane under normal ambient conditions can yield similar nanoparticle superlattices. Here, the higher boiling point octane (boiling point: 126°C) slows the evaporation rate of the dispersion and contributes to the formation of superlattice. The use of a concentrated dispersion of nanoparticles in higher boiling solvents like dodecane (boiling point: 216°C) allows slower evaporation of the solvent at higher temperatures, facilitating the formation of long-range ordered 3D superlattice (Sun and Murray 1999). The added thermal energy permits the particles to diffuse into their lowest energy superlattice sites during the solvent evaporation, producing a well-defined 3D hexagonally closed packed superlattice structure.

Since the nanoparticles are coated with a layer of surfactants, the separation of any two particles in a self-assembled superlattice is dependent on the thickness of the coating layer. However, the interparticle distance is not a simple addition of two thicknesses, as a certain degree of intercalation between the particles is present. Nevertheless, by controlling the length of the surfactant, interparticle spacing can be adjusted. Chemically, the surfactant length can be varied by attaching different surfactant molecules to the particle surface during the synthesis, or by replacing the original surfactant molecules with the new ones via surfactant exchange reaction.

FePt nanoparticles stabilized by oleic acid and oleyl amine have been successfully self-assembled into superlattices using these methods alone (Zeng et al. 2002, Chen et al. 2006, Shukla et al. 2006). For instance, colloidal crystals of monodisperse FePt nanoparticles have been grown by a three-layer technique based on slow diffusion of a nonsolvent (methanol) into the bulk of a concentrated FePt nanoparticle dispersion through a buffer layer of a third component (2-propanol) (Shevchenko et al. 2002). To control the interparticle spacing within the superlattice, surfactant exchange on the surface of the particles is performed before the self-assembly. Figure 3.9 shows two TEM images of 6 nm FePt nanoparticle superlattices assembled on amorphous carbon surfaces (Sun et al. 2000). Figure 3.9a shows

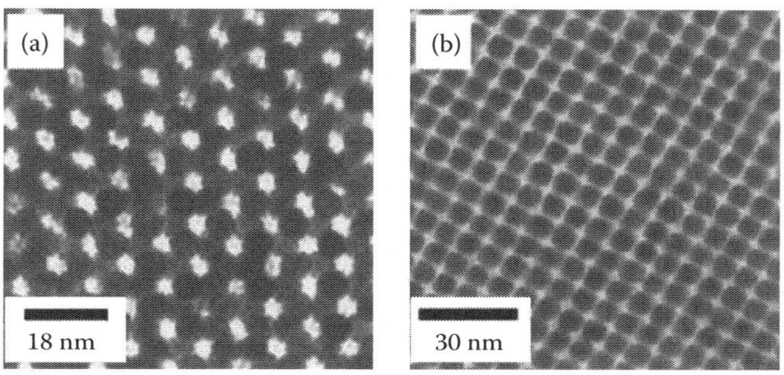


**FIGURE 3.9**
TEM images of (a) oleate/oleylamine coated 6 nm FePt and (b) hexanoate/hexylamine coated 6 nm FePt nanoparticle superlattices. The particles in (b) are obtained by replacing oleate/oleylamine on the surface of the particles in (a) by hexanoate/hexylamine. (Reprinted with permission from Sun, S. et al., *Science*, 287, 1989, 2000. Copyright 2000, the American Association for the Advancement of Science.)

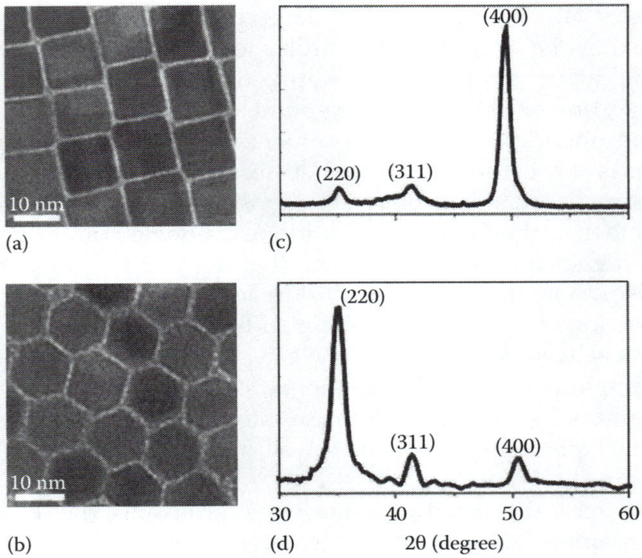


**FIGURE 3.10**
*Left*: TEM images of superlattices of (a) 12 nm cube-like and (b) 12 nm polyhedron-shaped $MnFe_2O_4$ nanoparticles. *Right*: XRD of (c) cube-like and (d) polyhedron-shaped nanoparticle superlattice on Si (100) substrates. (Adapted and reproduced from Zeng, H. et al., *J. Am. Chem. Soc.*, 126, 11458, 2004b. Copyright 2004, the American Chemical Society.)

the assembly of particles coated with oleate and oleyl amine, while Figure 3.9b shows the particles coated with hexanoate/hexyl amine. The interparticle spacing in Figure 3.10a is around 5 nm. That is twice as wide as the distance given by the C18 chain (~2.5 nm) in the oleate and oleyl amine molecules, while in Figure 3.10b the spacing is closer to 1 nm due to the thin-layer coating on the particles.

As the structure of a nanoparticle assembly depends mostly on interparticle interactions, the shape of a particle will alter such interactions and affect the position of the nanoparticles in a superlattice structure. However, if the shape of the particles can be controlled, self-assembly of faceted particles can lead to higher packing densities as well as crystal orientation of each particle in a self-assembled superlattice. For example, $MnFe_2O_4$ nanoparticles have been made in cube-like and polyhedron shapes, as shown in Figure 3.10 (Zeng et al. 2004a,b). Controlled evaporation of the carrier solvent from the hexane dispersion (~2 mg/mL) of the particles led to $MnFe_2O_4$ nanoparticle superlattices. Figure 3.10a shows the superlattice assembly from the cube-like particles, while Figure 3.10b is the assembly from the polyhedron-shaped particles. The fast Fourier transformation (FFT) of these two images reveals that both assemblies have cubic packing. But the different shapes possessed by each group of particles affect the crystal orientation of individual particles within the superlattices. XRD of the self-assembled cube-like particles on Si (100) substrate shows the intensified (400) peak (Figure 3.10c) and that of polyhedron-shaped particles reveals the strong reflections of (220) (Figure 3.10d). These are markedly different from that of a 3D randomly oriented spinel structured $MnFe_2O_4$ nanoparticle assembly, which shows a strong (311) peak. These indicate that each of the cube-like particles in the cubic assembly has preferred crystal orientation with {100} planes parallel to the Si substrate, while for the polyhedron-shaped particle assembly, the {110} planes are parallel to the substrate.

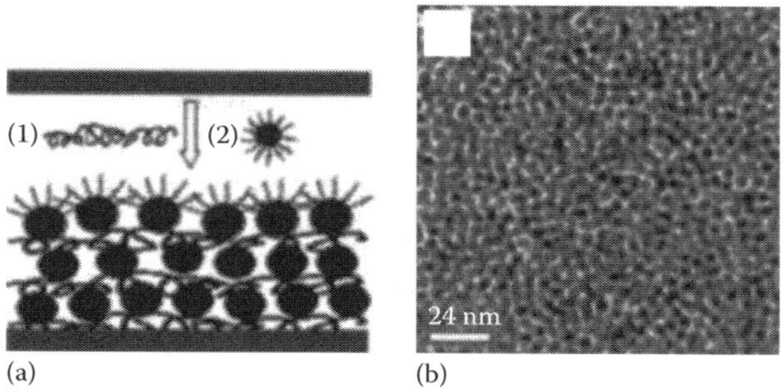


**FIGURE 3.11**
(a) Schematic illustration of PEI-mediated self-assembly of FePt nanoparticles by alternately adsorbing a layer of PEI and a layer of nanoparticles on a solid surface. (b) TEM image of PEI-mediated assembly of 4 nm $Fe_{58}Pt_{42}$ nanoparticles on silicon oxide surface. (Reproduced with permission from Sun, S. et al., *J. Am. Phys. Soc.*, 124, 2884, 2002. Copyright 2002, the American Chemical Society.)

An alternative approach to the control of interparticle spacing is via thermal annealing (Sun et al. 1999). Organic molecules around each nanoparticle are thermally unstable, subject to decomposition and/or evaporation under high-temperature annealing conditions. By controlling the decomposition and evaporation, interparticle spacing can be tuned. It is worth noting that excessive heating can remove all organic coating, making the particles sinter into a larger aggregate.

A substrate functionalized with proper molecules can be used to anchor particles on its surface via surface exchange reaction, leading to controlled assembly of the particles. This self-assembly technique is known as molecule-mediated self-assembly and is commonly used for constructing various composite nanostructures (Decher 1997, Liu et al. 1997, Cassagneau et al. 1998, Hicks et al. 2002). Due to their excellent adhesion capability to various substrates, multifunctional polymers are routinely applied as templates to mediate the assembly of the particles. The assembly is carried out as follows: a substrate is immersed into a polymer solution, and then rinsed, leading to a functionalized substrate. Subsequently, this substrate is dipped into the nanoparticle dispersion and then rinsed, leaving one layer of nanoparticles on the substrate surface. By repeating this simple two-step process in a cyclic fashion, a layer-by-layer assembled polymer/nanoparticle multilayer can be obtained.

One assembly example is polyethylenimine (PEI)-mediated self-assembly of FePt nanoparticles (Sun et al. 2002). PEI is an all-NH-based polymer that can replace oleate/oleyl amine molecules around FePt nanoparticles and attach to hydrophilic glass or silicon oxide surface through ionic interactions (Schmitt et al. 1999). A PEI/FePt assembly is readily fabricated by dipping the substrate alternately into a PEI solution and a FePt nanoparticle dispersion. Figure 3.11 shows (a) the assembly process and (b) TEM images of the 4 nm $Fe_{58}Pt_{42}$ nanoparticle self-assemblies on silicon oxide surfaces. Characterizations of the layered structures with x-ray reflectivity and atomic force microscopy indicate that PEI-mediated FePt assemblies have controlled thickness and the surfaces of the assemblies are smooth with root mean square roughness less than 2 nm.

### 3.5.2 Magnetism in FePt Nanoparticle Assemblies

Interparticle interactions can also influence the magnetic properties of the particles in a self-assembled array. In such an array, each particle is capped with a surfactant layer, and the interparticle spacing separated by the surfactant layer essentially eliminates the exchange interactions, leaving the magnetic dipole interactions a dominant factor. As the magnetic dipole interaction energy is proportional to the $(\text{moment})^2$ and $1/(\text{distance})^3$, the magnetic dipole interaction energy in a close-packed magnetic nanoparticle array with small interparticle distance can have a dramatic effect on the magnetic properties of the array (Vos et al. 1993, Allia et al. 1999, Malkinski et al. 1999, Rubio and Suárez 2000).

The FePt nanoparticle assembly is an ideal system to demonstrate interaction effects on magnetization process of the particles in the assembly (Zeng et al. 2002). Annealed at 550°C, the 4 nm $Fe_{58}Pt_{42}$ nanoparticle assembly consists of well-isolated particles. Its hysteresis loop has a remanence ratio ($M_r/M_s$) of 0.55 (Figure 3.12a), which is close to the value of 0.5 predicted for randomly oriented, noninteracting single-domain particles (Stoner and Wholfarth 1949). The hysteresis loop of the 600°C annealed assembly shows a much higher remanence ratio and steeper slope near the coercivity (Figure 3.12b). This is typically attributed to the cooperative switching behavior, indicating that the dominant interparticle interaction is exchange coupling. For the 800°C annealed assembly, however, the initial curve rises steeply at small fields, tends to saturate at lower fields (Figure 3.12c), indicating that the magnetization reversal is controlled mainly by domain nucleation processes. This further reveals that aggregated particles grow into a large single crystal showing a multidomain behavior.

This example provides excellent evidence that the problem of FePt aggregation cannot be solved simply by annealing arrays of FePT nanoparticles that are self-assembled onto a substrate as a means of creating patterned media for next-generation information storage. While many papers have reported high-temperature annealing of FePt arrays without visible aggregation, the magnetization dynamics of the particles have likely changed leading to arrays of nanoparticles that are no longer magnetically isolated. Furthermore, the fine control over interparticle spacing, which is necessary for patterned media to be integrated into a hard disk design afforded by the techniques mentioned above, are lost due to particle motion during annealing, which is provided by the thermal energy present at high temperatures. Therefore, annealing of the particles in a robust ceramic shell to achieve the $L1_0$ phase and being able to recoat them with surfactant and redisperse them in an organic solvent for self-assembly purposes is the best path for monodisperse, magnetically isolated, organized arrays of nanoparticles for data storage.

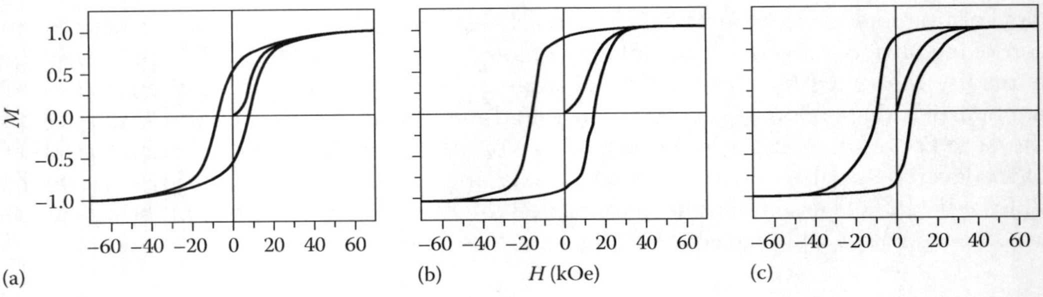


**FIGURE 3.12**
Initial magnetization curves and hysteresis loops for self-assembled $Fe_{58}Pt_{42}$ assemblies annealed at (a) 550°C, (b) 600°C, and (c) 800°C, respectively. (Reproduced from Zeng, H. et al., *Appl. Phys. Lett.*, 80, 2583, 2002. Copyright 2002, American Institute of Physics.)

## 3.6 Future Outlook

### 3.6.1 Future of FePt and CoPt Nanoparticles

It is clear that even after the astounding progress of FePt nanoparticles including synthesis, shape control, oxide-coated protective annealing, and self-assembly, work still needs to be done to make FePt nanoparticles feasible for the next generation of magnetic recording media. While the simultaneous decomposition of $Fe(CO)_5$ and reduction of $Pt(acac)_3$ has led over the years to extremely high-quality particles, by shrinking the size the distribution has even further rendered completely uniform magnetic properties of the particles and facilitated the accompanying technology associated with hard disk drives and proposed components of patterned media. Optimizing the self-assembly properties of the nanoparticles is also necessary. While the simultaneous decomposition of $Fe(CO)_5$ and reduction of $Pt(acac)_3$ has led over the years to extremely high quality particles, shrinking the size distribution even further to render completely uniform the magnetic properties of the particles and facilitate the accompanying technology associated with hard disk drives and proposed components of patterned media. This will require precise control over the kinetics of the particles with respect to the interplay between the steric forces of the stabilizers and the magnetic interactions between particles. Besides being able to assemble them, FePt nanoparticles should also be crystallographically aligned such that the easy and hard axes of magnetization are consistent and predictable.

Progress in making CoPt has been a little slower due to the challenges posed in this synthesis, namely the large difference in reduction potential between Co and Pt. Since the current hard drive media is also composed of Co and Pt alloys, the natural transition would be to their nanoparticle counterparts, which drives advances in synthesis of CoPt nanoparticles. It seems that despite the differences in synthesis necessary to make CoPt nanoparticles, the handling after the synthesis of the particles should, in principle, be similar to FePt. For instance, it seems possible that $SiO_2$ or MgO can be used as a protective coating for annealing CoPt to achieve full conversion to the $L1_0$ phase without sintering. If necessary this step could be facilitated in an analogous reaction to the FePt/$Fe_3O_4$ technique by forming a Pt-rich CoPt core coated with a CoO shell. While progress is admittedly behind in shape control of CoPt, once a polyol process is perfected, shape control seems inevitable.

### 3.6.2 High-Anisotropy Rare Earth Transition Metal Nanoparticles

Even if nanoparticle arrays are implemented into hard disk drives in the next 10 years as a result of the progress made in the area of FePt nanoparticles, scientists will still be looking for materials to form nanostructures of even harder magnets. Alloys composed of rare earth and transition metals (RE-TM) have the highest magnetic anisotropy constants known (Skomski and Coey 1999). In the RE-TM intermetallic compounds, the RE unpaired 4*f* electrons provide the magnetocrystalline anisotropy, whereas the TM 3*d* electrons provide most of the magnetization and determine the Curie temperature. They are currently being explored for use in permanent magnets, though any synthesis of RE-TM nanoparticles for permanent magnets could conceivably have a significant impact on the magnetic recording industry.

$SmCo_5$ is one of the most promising hard magnetic materials being studied for permanent magnets and might also become a contender for magnetic storage. It has a hexagonal close-packed (hcp) structure with Co and Co + Sm present in alternating layers along the *c*-axis. The easy magnetization direction of $SmCo_5$ is aligned along the *c*-direction of the lattice, and the magnetocrystalline anisotropy constant ranges from $1.1–2.0 \times 10^8$ erg/cm$^3$, among the highest known for hard magnetic materials thus giving the lowest thermal limit of grain size, down to 2.2 nm for magnetic recording (Majetich and Kirkpatrick 1997, Gu et al. 2003, Larson et al. 2003, Fidler et al. 2004). Furthermore, this alloy exhibits a very high Curie temperature ($T_c = 1020$ K), giving it excellent magnetic stability at elevated temperatures (Schrefl et al. 2000). Compared to FePt nanoparticles, it is estimated that while $SmCo_5$ nanoparticles have an effective anisotropy that is 1.5 times larger, the cost for preparation is 25 times lower (Ono et al. 2002), making pursuit of $SmCo_5$ nanoparticles even more advantageous. However, as with other rare earth metal materials, metallic $SmCo_5$ nanoparticles are prone to fast oxidation. This chemical instability has made the synthesis of nanostructured $SmCo_5$ extremely difficult, and despite the advances in $SmCo_5$ nanoparticle synthesis, much work will need to be done to stabilize $SmCo_5$ chemically for future applications. Ball milling and melt spinning, the two standard physical methods used for the fabrication of nanostructured $SmCo_5$ magnets, provide only limited control of the sizes of the final magnetic grains (Ding et al. 1995, Chen et al. 1996, Kirkpatrick and Leslie-Pelecky 2000, Zhou et al. 2003, Chakka et al. 2006). For this reason, chemical synthesis of $SmCo_5$ is attractive to get the small size distributions necessary for patterned media.

The most frequently attempted synthesis route for $SmCo_5$ is that inspired by the FePt recipe, that is, by reducing $Sm(acac)_3$ using a 1,2-alkanediol, while decomposing $Co_2(CO)_8$ in the presence of oleic acid and oleyl amine in an organic solvent such as dioctyl ether (Ono et al. 2002, Gu et al. 2003, Hong et al. 2007). In all cases, the as-synthesized particles are superparamagnetic at room temperature, and it seems that chemically synthesized $SmCo_5$ takes on an fcc structure rather than the desirable hcp structure (Ono et al. 2002). For this synthetic strategy, no further reports of attempts at annealing to obtain the correct phase have been made.

In an attempt to stabilize $SmCo_5$ nanoparticles and make a bimagnetic composite nanostructure for permanent magnetic applications, $SmCo_5$ nanoparticles have been synthesized with a $Fe_3O_4$ shell. $SmCo_{5(1-x)}$ ($x \sim 0.1$) nanoparticles were first prepared by the reaction of $Sm(acac)_3$ and $Co_2(CO)_8$ in the presence of 1,2-hexadecandiol (Hong et al. 2007). The resulting nanoparticles were then used as seeds for the magnetite coating via a second polyol process, which involved thermal decomposition of $Fe(CO)_5$. Thickness of the shell was carefully tuned by adjusting the stoichiometric ratio of $SmCo_{5(1-x)}$ nanoparticles to $Fe(CO)_5$. The final core/shell composites were stable in air for several days. The as-prepared $SmCo_{5(1-x)}$ sample shows typical superparamagnetic behavior, but, on the other hand, the *M*(*H*) curve of the $Sm(Co_{1-x}Fe_x)_5/Fe_3O_4$ composite exhibits an enhanced coercivity of 2.5 kOe. While this report clearly demonstrated the feasibility of protecting $SmCo_5$ particles from oxidation using an oxide shell, the $Fe_3O_4$ as the coating material is not optimal for magnetic recording simply because it does not provide magnetic isolation. Similarly concerning is that it is unclear what role the $Fe_3O_4$ plays magnetically in the nanostructure. That the $SmCo_5$ nanoparticles alone are superparamagnetic hints that the phase might not be completely hcp (XRD confirms the presence of some hcp structure), though it is noted that there are consistent vacancies in the Co(2*c*) sites. While there is speculation that the Fe occupies those sites causing the phase to be modified, thus giving better hard magnetic properties, it is also possible that the core/shell coupling gives rise to the onset of coercivity.

Progress was made using a similar recipe to those mentioned above, this time making $Co/Sm_2O_3$ core/shell nanoparticles that were reductively annealed to get the highest coercivity to date of $SmCo_5$ nanoparticles (Hou et al. 2007). Co nanoparticles were first synthesized by the decomposition of $Co_2(CO)_8$ in the presence of oleic acid and dioctylamine

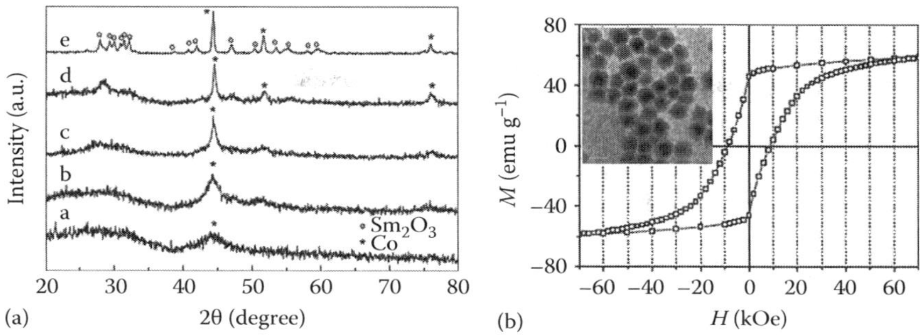


**FIGURE 3.13**
*Left*: XRD patterns of (a) Co/$Sm_2O_3$ nanoparticles and particles annealed under Ar at (b) 450°C, (c) 600°C, (d) 750°C, and (e) 900°C for 2h. The Co and $Sm_2O_3$ phases in (e) match well with the standard fcc pattern of Co (Joint Committee on Powder Diffraction Standards (JCPDS) No. 89-4307) and bcm structure of $Sm_2O_3$ (JCPDS No. 25-0749). *Right*: Hysteresis loop of particles after annealing. Inset shows core–shell particles as-synthesized. (Adapted and reproduced from Hou, Y. et al., *Adv. Mater.*, 19, 3349, 2007. Copyright 2007, Wiley InterScience.)

in 1,2,3,4-tetrahydronaphthalene (tetralin) at 210°C (Puntes et al. 2001, Bao et al. 2005). Monodisperse 8nm Co nanoparticles with a polycrystalline structure are obtained and dispersed in hexane. The core/shell-structured Co/$Sm_2O_3$ nanoparticles were prepared by the decomposition of Sm(acac)$_3$ over the surface of the Co nanoparticles. In this coating process, Sm(acac)$_3$ was first dissolved in oleyl amine and oleic acid at 100°C. Subsequently, a hexane dispersion of Co nanoparticles is added into the Sm(acac)$_3$ solution. The core/shell Co/$Sm_2O_3$ nanoparticles are prepared by heating the mixture to 250°C at a rate of 2°C/min. By adjusting the relative amounts of Sm(acac)$_3$ (1, 0.5, and 0.25mmol) and Co nanoparticles (80mg), Sm/Co molar ratios of 1:4.3, 1:5.2, and 1:7.5, respectively, were obtained for the Co/$Sm_2O_3$ nanoparticles. Co/$Sm_2O_3$ nanoparticles with a Sm/Co molar ratio of 1:4.3 had an average $Sm_2O_3$ shell thickness of 2nm (Figure 3.13).

The core/shell-structured Co/$Sm_2O_3$ nanoparticles served as the starting materials for the synthesis of $SmCo_5$ nanomagnets under reductive annealing conditions. The reductive annealing was performed under Ar+5% $H_2$ at a temperature of 900°C in the presence of metallic Ca. KCl with a melting point of 771°C was used as an inorganic solvent to ensure the complete reduction of $Sm_2O_3$ in the core/shell structure and to promote interface diffusion between Sm and Co. KCl may also act as a physical barrier to prevent the $SmCo_5$ magnets from aggregating too much. After reductive annealing at 900°C for 1.5h, the nanocrystalline $SmCo_5$ became strongly ferromagnetic. The coercivity of the magnets reached 8kOe at room temperature. Figure 3.13a shows the evolution of the XRD peaks of $SmCo_5$ annealed at different temperatures. While there is a broad peak for Co before annealing, the $Sm_2O_3$ is amorphous. Figure 3.13b shows the hysteresis loop obtained after annealing, with the pre-annealed TEM image as an inset. Although the high annealing temperatures used probably still caused sintering and rendered the particles unable to be dispersed in solvent, this is clearly a promising route to eventually obtaining high-anisotropy nanoparticles.

There is still much work needed to be done to get $SmCo_5$ nanoparticles close to being useful for magnetic storage, but the results thus far are promising. Like the FePt system, it seems likely that the correct phase can be obtained by annealing under protection, though chemical stability definitely needs to be controlled to prevent oxidation.

Perhaps the most challenging to synthesize will be the RE-TM ternary compounds such as $Nd_2Fe_{14}B$ and $Sm_2Fe_{17}N_3$. Like $SmCo_5$, they have very high-anisotropy constants

($K = 4.5 \times 10^7$ erg/cm$^3$ for $Nd_2Fe_{14}B$; $K = 8.6 \times 10^7$ erg/cm$^3$ for $Sm_2Fe_{17}N_3$) but display somewhat better chemical stability. Because they are iron-based, the binary compounds $Nd_2Fe_{14}$ and $Sm_2Fe_{17}$ have low Curie temperatures and low in-plane magnetic anisotropy. The addition of the interstitial gas (B or N) causes the lattice distortion that transforms the structures from cubic to rhombohedral and causes the magnetic easy axis to prefer the *c*-axis (Skomski and Coey 1999). However, the small interstitial atoms also pose a great challenge for chemical synthesis methods. Though no reports have been made of the successful synthesis of high-anisotropy $Sm_2Fe_{17}N_3$ nanoparticles, there are hints that chemical synthesis can be used to obtain nanoparticles of $Nd_2Fe_{14}B$ in aqueous solution using $FeCl_2$ and $NdCl_3$ salts and using $NaBH_4$ as a reducing agent (Km et al. 2007), while capitalizing on the known tendency for B-impurities when used as a reducing agent. So far this technique yields only amorphous particles, which after reductive annealing becomes mixed phases of $Nd_2Fe_{14}B$, $\alpha$-Fe, and Fe-B species.

## 3.7 Conclusion

In summary, high-quality monodisperse magnetic nanoparticles with high coercivity can be made from various chemical synthesis routes. These particles provide a way around the superparamagnetic limit that is currently encroaching upon granular media that is used in hard disk drives today. As-synthesized nanoparticles are usually superparamagnetic, but some novel approaches have been used to anneal the particles at high temperature, facilitating the fcc-to-fct transition in FePt nanoparticles while keeping the particles from sintering and allowing the particles to again be dispersed in organic solvents. In order to increase the packing density to maximize areal density for media, the shapes can be controlled and self-assembly can be employed to control interparticle spacing, which in turn provides control over magnetic interactions.

While chemical synthesis and annealing parameters have not yet been perfected for the CoPt system, indications are that the same level of monodispersity, shape control, phase transitions, and self-assembly will be achieved in time.

Even higher anisotropic nanoparticles are being synthesized of the rare earth transition metal species, though the challenges associated with their syntheses are significant. In the case of $SmCo_5$, nanoscale powders with high coercivity have been made after reductively annealing core–shell structures. More needs to be done to protect these particles from sintering during the annealing process and the issue of chemical stability needs to be addressed.

Undoubtedly, further work needs to be done to perfect the processes involved in all syntheses before implementation into hard disk media, but the results presented here paint a promising picture for the future of magnetic nanoparticles in recording technology.

## References

Acharya, B.R., Abarra, E.N., and Okamoto, I. 2001. SNR improvements for advanced longitudinal recording media. *IEEE Transactions on Magnetics* 37: 1475–1477.

Albrecht, M., Maret, M., Maier, A. et al. 2002. Perpendicular magnetic anisotropy in $CoPt_3$(111) films grown on a low energy surface at room temperature. *Journal of Applied Physics* 91: 8153–8155.

Allia, P., Coisson, M., Knobel, M., Tiberto, P., and Vinai, F. 1999. Magnetic hysteresis based on dipolar interactions in granular magnetic systems. *Physical Review B* 60: 12207–12218.

Bao, Y., Beerman, M., Pakhomov, A.B., and Krishnan, K.M. 2005. Controlled crystalline structure and surface stability of cobalt nanocrystals. *Journal of Physical Chemistry B* 109: 7220–7222.

Bertram, H.N., Zhou, H., and Gustafson, R. 1998. Signal to noise ratio scaling and density limit estimates in longitudinal magnetic recording. *IEEE Transactions on Magnetics* 34: 1845–1847.

Bolzoni, F., Leccabue, F., Panizzieri, R., and Pareti, L. 1984. Magnetocrystalline anisotropy and phase transformation in Co-Pt alloy. *IEEE Transactions on Magnetics* 20: 1625–1627.

Boyen, H.-G., Fauth, K., Stahl, B. et al. 2005. Electronic and magnetic properties of ligand-free FePt nanoparticles. *Advanced Materials* 17: 574–578.

Carpenter, E.E., Sims, J.A., Wienmann, J.A., Zhou, W.L., and O'Connor, C.J. 2000. Magnetic properties of iron and iron platinum alloys synthesized via microemulsion techniques. *Journal of Applied Physics* 87: 5615–5617.

Cassagneau, T., Mallouk, T.E., and Fendler, J.H. 1998. Layer-by-layer assembly of thin film Zener diodes from conducting polymers and CdSe nanoparticles. *Journal of the American Chemical Society* 120: 7848–7859.

Chakka, V.M., Altuncevahir, B., Jin, Z.Q., Li, Y., and Liu, J.P. 2006. Magnetic nanoparticles produced by surfactant-assisted ball milling. *Journal of Applied Physics* 99: 08E912-1–08E912-3.

Charap, S.H., Lu, P.L., and He, Y.J. 1997. Thermal stability of recorded information at high densities. *IEEE Transactions on Magnetics* 33: 978–983.

Chen, M. and Nikles, D.E. 2002. Synthesis of spherical FePd and CoPt nanoparticles. *Journal of Applied Physics* 91: 8477–8479.

Chen, S.K., Tsai, J.L., and Chin, T.S. 1996. Nanocomposite $Sm_2Co_{17}$/Co permanent magnets by mechanical alloying. *Journal of Applied Physics* 79: 5964–5966.

Chen, M., Liu, J.P., and Sun, S. 2004. One-step synthesis of FePt nanoparticles with tunable size. *Journal of the American Chemical Society* 126: 8394–8395.

Chen, M.-P., Kuroishi, K., and Kitamoto, Y. 2005. Magnetic properties and microstructure of isolated Fe-Pt nanoparticle-monolayer assembly by protective coating. *IEEE Transactions on Magnetics* 41: 3376–3378.

Chen, M., Kim, J., Liu, J.P., Fan, H., and Sun, S. 2006. Synthesis of FePt nanocubes and their oriented self-assembly. *Journal of the American Chemical Society* 128: 7132–7133.

Chen, J.S., Lim, B.C., Hu, J.F., Lim, Y.K., Liu, B., and Chow, G.M. 2007. High coercivity $L1_0$ FePt films with perpendicular anisotropy deposited on glass substrate at reduced temperature. *Applied Physics Letters* 90:042508-1–042508-3.

Chinnasamy, C.N., Jeyadevan, B., Shinoda, K., and Tohji, K. 2003. Polyol-process-derived CoPt nanoparticles: Structural and magnetic properties. *Journal of Applied Physics* 93: 7583–7585.

Coffey, K.R., Parker, M.A., and Howard, J.K. 1995. High anisotropy $L1_0$ thin films for longitudinal recording. *IEEE Transactions on Magnetics* 31: 2737–2739.

Cullity, B.D. 1972. *Introduction to Magnetic Materials*. Reading, MA: Addison Wesley Publishing Company.

Dai, Z.R., Sun, S., and Wang, Z.L. 2001. Phase transformation, coalescence, and twinning of monodisperse FePt nanocrystals. *Nano Letters* 1:443–447.

Decher, G. 1997. Fuzzy nanoassemblies: Toward layered polymeric multicomposites. *Science* 277: 1232–1237.

Ding, J., McCormick, P.G., and Street, R. 1995. A study of melt-spun $SmCo_5$ ribbons. *Journal of Alloys and Compounds* 228: 102–104.

Elkins, K.E., Vedantam, T.S., Liu, J.P. et al. 2003. Ultrafine FePt nanoparticles prepared by the chemical reduction method. *Nano Letters* 3: 1647–1649.

Elkins, K., Li, D., Poudyal, N. et al. 2005. Monodisperse face-centred tetragonal FePt nanoparticles with giant coercivity. *Journal of Physics D: Applied Physics* 38: 2306–2309.

Everett, D.H. 1988. *Basic Principles of Colloid Science*. Cambridge, U.K.: Royal Society of Chemistry.

Fan, H., Yang, K., Boye, D.M. et al. 2004. Self-assembly of ordered, robust, three-dimensional gold nanocrystal/silica arrays. *Science* 304: 567–571.

Fidler, J., Schrefl, T., Scholz, W., Suess, D., Dittrich, R., and Kirschner, M. 2004. Micromagnetic modelling and magnetization processes. *Journal of Magnetism and Magnetic Materials* 272–276: 641–646.

Gibot, P., Tronc, E., Chanéac, C., Jolivet, J.P., Fiorani, D., and Testa, A.M. 2005. (Co,Fe)Pt nanoparticles by aqueous route; self-assembling, thermal and magnetic properties. *Journal of Magnetism and Magnetic Materials* 290–291: 555–558.

Gu, H., Xu, B., Rao, J. et al. 2003. Chemical synthesis of narrowly dispersed $SmCo_5$ nanoparticles. *Journal of Applied Physics* 93: 7589–7591.

Harp, G.R., Weller, D., Rabedeau, T.A., Farrow, R.F.C., and Toney, M.F. 1993. Magneto-optical Kerr spectroscopy of a new chemically ordered alloy: $Co_3Pt$. *Physical Review Letters* 71: 2493–2496.

Harpeness, R. and Gedanken, A. 2005. The microwave-assisted polyol synthesis of nanosized hard magnetic material, FePt. *Journal of Materials Chemistry* 15: 698–702.

Hicks, J.F., Seok-Shon, Y., and Murray, R.W. 2002. Layer-by-layer growth of polymer/nanoparticle films containing monolayer-protected gold clusters. *Langmuir* 18: 2288–2294.

Hong, J.H., Kim, W.S., Lee, J.I., and Hur, N.H. 2007. Exchange coupled magnetic nanocomposites of $Sm(Co_{1-x}Fe_x)_5/Fe_3O_4$ with core/shell structure. *Solid State Communications* 141: 541–544.

Hou, Y., Xu, Z., Peng, S., Rong, C., Liu, J.P., and Sun, S. 2007. A facile synthesis of $SmCo_5$ magnets from core/shell $Co/Sm_2O_3$ nanoparticles. *Advanced Materials* 19: 3349–3352.

Howard, L.E.M., Nguyen, H.L., Giblin, S.R. et al. 2005. A synthetic route to size-controlled fcc and fct FePt nanoparticles. *Journal of the American Chemical Society* 127: 10140–10141.

Huber, D.L. 2005. Synthesis, properties, and applications of iron nanoparticles. *Small* 1: 482–501.

Iwaki T., Kakihara, Y., Toda, T., Abdullah, M., and Okuyama, K. 2003. Preparation of high coercivity magnetic FePt nanoparticles by liquid process. *Journal of Applied Physics* 94: 6807–6811.

Iwasaki, S. and Nakamura, Y. 1978. The magnetic field distribution of a perpendicular recording head. *IEEE Transactions on Magnetics* 14: 436–438.

Iwasaki, S. and Takemura, K. 1975. An analysis for the circular mode of magnetization in short wavelength recording. *IEEE Transactions on Magnetics* 11: 1173–1175.

Iwasaki, S., Nakamura, Y., and Ouchi, K. 1979. Perpendicular magnetic recording with a composite anisotropy film. *IEEE Transactions on Magnetics* 15: 1456–1458.

Jeyadevan, B., Hobo, A., Urakawa, K., Chinnasamy, C.N., Shinoda, K., and Tohji, K. 2003. Towards direct synthesis of fct-FePt nanoparticles by chemical route. *Journal of Applied Physics* 93: 7574–7576.

Kai, T., Maeda, T., Kikitsu, A., Akiyama, J., Nagase, T., and Kishi, T. 2004. Magnetic and electronic structures of FePtCu ternary ordered alloy. *Journal of Applied Physics* 95: 609–612.

Kanbe, T., Yahisa, Y., Suzuki, H., Kataoka, H., Takahashi, Y., and Hirayama, Y. 2002. Low noise antiferromagnetically coupled media with CrTiB underlayer. *Journal of Applied Physics* 91: 8611–8613.

Kang, S.S., Nikles, D.E., and Harrell, J.W. 2003. Synthesis, chemical ordering, and magnetic properties of self-assembled FePt-Ag nanoparticles. *Journal of Applied Physics* 93: 7178–7180.

Kang, S., Shi, S., Nikles, D.E., and Harrell, J.W. 2008. Easy control of the size and composition of FePt nanoparticles with improved synthesis. *Journal of Applied Physics* 103: 07D503-1–07D503-3.

Kim, J., Rong, C., Lee, Y., Liu, J.P., and Sun, S. 2008a. From core/shell structured $FePt/Fe_3O_4/MgO$ to ferromagnetic FePt nanoparticles. *Chemistry of Materials* 20: 7242–7245.

Kim, J., Rong, C., Liu, J.P., and Sun, S. 2008b. Dispersible ferromagnetic FePt nanoparticles. *Advanced Materials* 20: 1–4.

Kirkpatrick, E.M. and Leslie-Pelecky, D.L. 2000. Structural and magnetic properties of mechanically milled $SmCo_5$:C. *Journal of Applied Physics* 87: 6734–6736.

Kitamoto, Y., Minami, R., Shibata, Y., Chikata, T., and Kato, S. 2005. Magnetocrystalline anisotropy of partially ordered Fe-Pt nanoparticles directly synthesized with microwave-polyol method. *IEEE Transactions on Magnetics* 41: 3880–3882.

Klemmer, T.J., Shukla, N., Liu, C. et al. 2002. Structural studies of $L1_0$ FePt nanoparticles. *Applied Physics Letters* 81: 2220–2222.

Km, C.W., Km, Y.H., Cha, H.G., and Kang, Y.S. 2007. Study on synthesis and magnetic properties of Nd-Fe-B alloy via reduction-diffusion process. *Physica Scripta* T129: 321–325.

Kumbhar, A., Spinu, L., Agnoli, F., Wang, K.-Y., Zhou, W., and O'Connor, C.J. 2001. Magnetic properties of cobalt and cobalt–platinum alloy nanoparticles synthesized via microemulsion technique. *IEEE Transactions on Magnetics* 37: 2216–2218.
Larson, P., Mazin, I.I., and Papaconstantopoulos, D.A. 2003. Calculation of magnetic anisotropy energy in $SmCo_5$. *Physical Review B* 67: 214405-1–214405-6.
Lee L.L., Laughlin D.E., Fang L., and Lambeth, D.N. 1995. Effects of Cr intermediate layers on CoCrPt thin-film media on NiAl underlayers. *IEEE Transactions on Magnetics* 31: 2728–2730.
Li, N. and Lairson, B.M. 1999. Magnetic recording on FePt and FePtB intermetallic compound media. *IEEE Transactions on Magnetics* 35: 1077–1082.
Lide, D.R. 1995. *CRC Handbook of Chemistry and Physics*, 76th edn. Boca Raton, FL: CRC Press, pp. 8–21.
Liu, Y., Wang, A., and Claus, R. 1997. Molecular self-assembly of $TiO_2$/polymer nanocomposite films. *Journal of Physical Chemistry B* 101: 1385–1388.
Liu, C., Klemmer, T.J., Shukla, N. et al. 2003. Oxidation of FePt nanoparticles. *Journal of Magnetism and Magnetic Materials* 266: 96–101.
Liu, C., Wu, X., Klemmer, T. et al. 2004. Polyol process synthesis of monodispersed FePt nanoparticles. *Journal of Physical Chemistry B* 108: 6121–6123.
Liu, C., Wu, X., Klemmer, T. et al. 2005. Reduction of sintering during annealing of FePt nanoparticles coated with iron oxide. *Chemistry of Materials* 17: 620–625.
Maeda, T., Kai, T., Kikitsu, A., Nagase, T., and Akiyama, J. 2002. Reduction of ordering temperature of an FePt-ordered alloy by addition of Cu. *Applied Physics Letters* 80: 2147–2149.
Majetich, S.A. and Kirkpatrick, E.M. 1997. Magnetization reversal in $SmCo_5$ nanoparticles. *IEEE Transactions on Magnetics* 33: 3721–3723.
Malkinski, L.M., Wang, J.-Q., Dai, J., Tang, J., and O'Connor, C.J. 1999. Thickness dependence of magnetic properties of granular thin films with interacting particles. *Applied Physics Letters* 75: 844–846.
Mandal, M., Das, B., and Mandal, K. 2009. Synthesis of $Co_xPt_{1-x}$ alloy nanoparticles of different phase by micellar technique and their properties study. *Journal of Colloid and Interface Science* 335: 40–43.
Maret, M., Cadeville, M.C., Poinsot, R., Herr, A., Beaurepaire, E., and Monier, C. 1997. Structural order related to the magnetic anisotropy in epitaxial (111) $CoPt_3$ alloy films. *Journal of Magnetism and Magnetic Materials* 166: 45–52.
Mikami M., Djayaprawira D.D., Yoshimura S., and Takahashi A. 2003. Effect of oxidation process on grain-size reduction for longitudinal media using Ni-based amorphous seedlayer. *IEEE Transactions on Magnetics* 39: 2258–2260.
Minami, R., Kitamoto, Y., Chikata, T., and Kato, S. 2005. Direct synthesis of $L1_0$ type Fe–Pt nanoparticles using microwave-polyol method. *Electrochimica Acta* 51: 864–866.
Mizuno, M., Sasaki, Y., Yu, A.C.C., and Inoue, M. 2004. Prevention of nanoparticle coalescence under high-temperature annealing. *Langmuir*:20: 11305–11307.
Momose, S., Kodama, H., Ihara, N., Uzumaki, T., and Tanaka, A. 2003. $L1_0$-FePt nanoparticles in a magnetically isolated state. *Japanese Journal of Applied Physics* 42: L1252–L1254.
Momose, S., Kodama, H., Uzumaki, T., and Tanaka, A. 2005. Fine tuning of the sizes of FePt nanoparticles. *Japanese Journal of Applied Physics* 44: 1147–1149.
Moser, A., Takano, K., Margulies, D.T. et al. 2002. Magnetic recording: Advancing into the future. *Journal of Physics D: Applied Physics* 35: R157–R167.
Murray, C.B., Sun, S., Gaschler, W., Doyle, H., Betley, T.A., and Kagan, C.R. 2001. Colloidal synthesis of nanocrystals and nanocrystal superlattices. *IBM Journal of Research and Development* 45: 47–56.
Nakaya, M., Tsuchiya, Y., Ito, K., Oumi, Y., Sano, T., and Teranishi, T. 2004. Novel synthesis of FePt nanoparticles and magnetic properties of their self-assembled superlattices. *Chemistry Letters* 2: 130–131.
Nandwana, V., Elkins, K.E., Poudyal, N., Chaubey, G.S., Yano, K., and Liu, J.P. 2007. Size and shape control of monodisperse FePt nanoparticles. *Journal of Physical Chemistry C* 111: 4185–4189.
Ono, K., Kakefuda, Y., Okuda, R. et al. 2002. Organometallic synthesis and magnetic properties of ferromagnetic Sm-Co nanoclusters. *Journal of Applied Physics* 91: 8480–8482.

Ould Ely, T., Pan, C., Amiens, C. et al. 2000. Nanoscale bimetallic $Co_xPt_{1-x}$ particles dispersed in poly(vinylpyrrolidone): Synthesis from organometallic precursors and characterization. *Journal of Physical Chemistry B* 104: 695–702.

Park, J.-I., Kim, M.G., Jun, Y.-w., Lee, J.S., Lee, W.-r. and Cheon, J. 2004. Characterization of superparamagnetic "core-shell" nanoparticles and monitoring their anisotropic phase transition to ferromagnetic "solid solution" nanoalloys. *Journal of the American Chemical Society* 126: 9072–9078.

Perumal, A., Ko, H.-S., and Shin, S.-C. 2003. Magnetic properties of carbon-doped FePt nanogranular films. *Applied Physics Letters* 83: 3326–3328.

Piramanayagam, S.N. and Srinivasan, K. 2009. Recording media research for future hard disk drives. *Journal of Magnetism and Magnetic Materials* 321: 485–494.

Piramanayagam, S.N., Xu, Y.F., Dai, D.Y., Huang, L., Pang, S.I., and Wang, J.P. 2002. Ultrasmall grain size control in longitudinal recording media for ultrahigh areal densities. *Journal of Applied Physics* 91: 7685–7687.

Piramanayagam, S.N., Pock C.K., Lu, L., Ong, C.Y., Shi, J.Z., and Mah, C.S. 2006. Grain size reduction in $CoCrPt:SiO_2$ perpendicular recording media with oxide-based intermediate layers. *Applied Physics Letters* 89: 162504-1–162504-3.

Puntes, V.F., Krishnan, K.M., and Alivisatos, A.P. 2001. Colloidal nanocrystal shape and size control: The case of cobalt. *Science* 291: 2115–2117.

Reiss, B.D., Mao, C., Solis, D.J., Ryan, K.S., Thomson, T., and Belcher, A.M. 2004. Biological routes to metal alloy ferromagnetic nanostructures. *Nano Letters* 4: 1127–1132.

Richter, H.J. 1999. Longitudinal recording at 10 to 20 Gbit/inch$^2$ and beyond. *IEEE Transactions on Magnetics* 35: 2790–2795.

Richter, H.J. 2009. Density limits imposed by the microstructure of magnetic recording media. *IEEE Transactions on Magnetics* 321: 467–476.

Ristau, R.A., Barmak, K., Lewis, L.H., Coffey, K.R., and Howard, J.K. 1999. On the relationship of high coercivity and $L1_0$ ordered phase in CoPt and FePt thin films. *Journal of Applied Physics* 86: 4527–4533.

Robach, O., Quiros, C., Valvidares, S.M., Walker, C.J., and Ferrer, S. 2003. Structure and Pt magnetism of FePt nanoparticles investigated with X-ray diffraction. *Journal of Magnetism and Magnetic Materials* 263: 202–208.

Rooney, P.W., Shapiro, A.L., Tran, M.Q., and Hellman, F. 1995. Evidence of a surface-mediated magnetically induced miscibility gap in Co-Pt alloy thin films. *Physical Review Letters* 75: 1843–1846.

Ross, C.A. 2003. Patterned magnetic recording media. *Annual Review of Materials Research* 31: 203–235.

Rubio, H. and Suárez, S. 2000. Easy axis distribution in interacting fine-particle systems: Two and three dimensions. *Journal of Applied Physics* 87:7415–7424.

Sato, K., Jeyadevan, B., and Tohji, K. 2005. Preparation and properties of ferromagnetic FePt dispersion. *Journal of Magnetism and Magnetic Materials* 289: 1–4.

Schmitt, J., Mächtle, P., Eck, D., Möhwald, H., and Helm, C.A. 1999. Preparation and optical properties of colloidal gold monolayers. *Langmuir* 15: 3256–3266.

Schrefl, T., Fidler, J., and Scholz, W. 2000. Modeling and limits of advanced HT-magnets. *IEEE Transactions on Magnetics* 36: 3394–3398.

Service, R.F. 2006. Is the terabit within reach? *Science* 314: 1868–1870.

Shen, W.K., Judy, J.H., and Wang, J.-P. 2005. *In situ* epitaxial growth of ordered FePt (001) films with ultra small and uniform grain size using a RuAl underlayer. *Journal of Applied Physics* 97: 10H301-1–10H301-3.

Shevchenko, E., Talapin, D., Kornowski, A. et al. 2002. Colloidal crystals of monodisperse FePt nanoparticles grown by a three-layer technique of controlled oversaturation. *Advanced Materials* 14: 287–290.

Shukla, N., Liu, C., and Roy, A.G. 2006. Oriented self-assembly of cubic FePt nanoparticles. *Materials Letters* 60: 995–998.

Shukla, N., Nigra, M., Nuhfer, T., Bartel, M.A., and Gellman, A.J. 2009. Tailoring the shapes of $Fe_xPt_{100-x}$ nanoparticles. *Nanotechnology* 20: 065602-1–065602-5.

Skomski, R. and Coey, J.M.D. 1999. *Permanent Magnetism*. Bristol, U.K.: Institute of Physics Publishing.
Srivastava, C., Nikles, D.E., and Thompson, G.B. 2008. Compositional evolution during the synthesis of FePt nanoparticles. *Journal of Applied Physics* 104: 064315-1–064315-7.
Stahl, B., Gajbhiye, N.S., Wilde, G. et al. 2002. Electronic and magnetic properties of monodispersed FePt nanoparticles. *Advanced Materials* 14: 24–27.
Stahl, B., Ellrich, J., Theissmann, R. et al. 2003. Electronic properties of 4-nm FePt particles. *Physical Review B* 67: 014422-1–014422-12.
Stappert, S., Rellinghaus, B., Acet, M., and Wassermann, E.F. 2003. Gas-phase preparation of $L1_0$ ordered FePt nanoparticles. *Journal of Crystal Growth* 252: 440–450.
Staunton, J.B., Ostanin, S., Razee, S.S.A. et al. 2004. Temperature dependent magnetic anisotropy in metallic magnets from an *ab initio* electronic structure theory: $L1_0$-ordered FePt. *Physical Review Letters* 93: 257204-1–257204-4.
Stavroyiannis, S., Panagiotopoulos, I., Niarchos, D., Christodoulides, J.A., Zhang, Y., and Hadjipanayis, G.C. 1998. CoPt/Ag nanocomposites for high density recording media. *Applied Physics Letters* 73: 3453–3455.
Stoner, E.C. and Wohlfarth, E.P. 1949. A mechanism of magnetic hysteresis in heterogeneous alloys. *Philosophical Transaction Royal Society—London* 420: 599–642.
Stupp, S.I., LeBonheur, V., Walker, K. et al. 1997. Supramolecular materials: Self-organized nanostructures. *Science* 276: 384–389.
Sun, S. 2006. Recent advances in chemical synthesis, self-assembly, and applications of FePt nanoparticles. *Advanced Materials* 18: 393–403.
Sun, S. and Murray, C.B. 1999. Synthesis of monodisperse cobalt nanocrystals and their assembly into magnetic superlattices. *Journal of Applied Physics* 85: 4325–4330.
Sun, S., Murray, C.B., and Doyle, H. 1999. Controlled assembly of monodisperse cobalt based nanocrystals. *Materials Research Society Symposium Proceedings* 577: 395–398.
Sun, S., Murray, C.B., Weller, D., Folks, L., and Moser, A. 2000. Monodisperse FePt nanoparticles and ferromagnetic FePt nanocrystal superlattices. *Science* 287: 1989–1992.
Sun, S., Fullerton, E.E., Weller, D., and Murray, C.B. 2001. Compositionally controlled FePt nanoparticle materials. *IEEE Transactions on Magnetics* 37: 1239–1243.
Sun, S., Anders, S., Hamann, H.F. et al. 2002. Polymer mediated self-assembly of magnetic nanoparticles. *Journal of the American Physical Society* 124: 2884–2885.
Sun, S., Anders, S., Thomson, T. et al. 2003a. Controlled synthesis and assembly of FePt nanoparticles. *Journal of Physical Chemistry B* 107: 5419–5425.
Sun, X., Kang, S., Harrell, J.W. et al. 2003b. Synthesis, chemical ordering, and magnetic properties of FePtCu nanoparticle films. *Journal of Applied Physics* 93: 7337–7339.
Sun, X., Jia, Z.Y., Huang, Y.H. et al. 2004. Synthesis and magnetic properties of CoPt nanoparticles. *Journal of Applied Physics* 95: 6747–6749.
Takahashi, Y.K., Ohnuma, M., and Hono, K. 2002. Effect of Cu on the structure and magnetic properties of FePt sputtered film. *Journal of Magnetism and Magnetic Materials* 246: 259–265.
Tamada, Y., Morimoto, Y., Yamamoto, S., Takano, M., Nasu, S., and Ono, T. 2007a. Effects of annealing time on structural and magnetic properties of $L1_0$-FePt nanoparticles synthesized by the $SiO_2$-nanoreactor method. *Journal of Magnetism and Magnetic Materials* 310: 2381–2383.
Tamada, Y., Yamamoto, S., Takano, M., Nasu, S., and Ono, T. 2007b. Well-ordered $L1_0$-FePt nanoparticles synthesized by improved $SiO_2$-nanoreactor method. *Applied Physics Letters* 90: 162509-1–162509-3.
Teng, X. and Yang, H. 2003. Synthesis of face-centered tetragonal FePt nanoparticles and granular films from Pt@$Fe_2O_3$ core-shell nanoparticles. *Journal of the American Chemical Society* 125: 14559–14563.
Teng, X., Black, D., Watkins, N.J., Gao, Y., and Yang, H. 2003. Platinum-maghemite core-shell nanoparticles using a sequential synthesis. *Nano Letters* 3: 261–264.
Terris, B.D. and Thomson, T. 2005. Nanofabricated and self-assembled magnetic structures as data storage media. *Journal of Physics D: Applied Physics* 38: R199–R222.

Tomou, A., Panagiotopoulos, I., Gournisa, D., and Kooi, B. 2007. $L_10$ ordering and magnetic interactions in FePt nanoparticles embedded in MgO and $SiO_2$ shell matrices. *Journal of Applied Physics* 102: 023910-1–023910-5.

Tzitzios, V., Niarcho, D., Margariti, G., Fidler, J., and Petridis, D. 2005. Synthesis of CoPt nanoparticles by a modified polyol method: Characterization and magnetic properties. *Nanotechnology* 16: 287–291.

Ulman, A. 1990. Self-assembled monolayers of alkyltrichlorosilanes: Building blocks for future organic materials. *Advanced Materials* 2: 573–582.

Ulmeanu, M., Antoniak, C., Wiedwald, U., Farle, M., Frait, Z., and Sun, S. 2004. Composition-dependent ratio of orbital-to-spin magnetic moment in structurally disordered $Fe_xPt_{1-x}$ nanoparticles. *Physical Review B* 69: 054417-1–054417-5.

Vos, M.J., Brott, R.L., Zhu, J.-G., and Carlson, L.W. 1993. Computed hysteresis behavior and interaction effects in spheroidal particle assemblies. *IEEE Transactions on Magnetics* 29: 3652–3657.

Wang, Z.L. 2000. Transmission electron microscopy of shape-controlled nanocrystals and their assemblies. *Journal of Physical Chemical B* 104: 1153–1175.

Wang, H.L., Zhang, Y., Huang, Y., Zeng, Q., and Hadjipanayis, G.C. 2004. CoPt nanoparticles by chemical reduction. *Journal of Magnetism and Magnetic Materials* 272–276: e1279–e1280.

Weller, D. and Doerner, M.F. 2000. Extremely high-density longitudinal magnetic recording media. *Annual Review of Materials Research* 30: 611–644.

Weller, D. and Moser, A. 1999. Thermal effect limits in ultrahigh-density magnetic recording. *IEEE Transactions on Magnetics* 35: 4423–4439.

Weller, D., Brändle, H., Gorman, G., Lin, C.-J., and Notarys, H. 1992. Magnetic and magneto-optical properties of cobalt-platinum alloys with perpendicular magnetic anisotropy. *Applied Physics Letters* 61: 2726–2728.

Weller, D., Moser, A., Folks, L. et al. 2000. High $K_u$ materials approach to 100 Gbits/in$^2$. *IEEE Transactions on Magnetics* 36: 10–15.

Weller, D., Sun, S., Murray, C., Folks, L., and Moser, A. 2001. MOKE spectra and ultrahigh density data storage perspective of FePt nanomagnet arrays. *IEEE Transactions on Magnetics* 37: 2185–2187.

Whitesides, G.M., Mathias, J.P., and Seto, C.T. 1991. Molecular self-assembly and nanochemistry: A chemical strategy for the synthesis of nanostructures. *Science* 254: 1312–1319.

Yamamoto, S., Morimoto, Y., Ono, T., and Takano, M. 2005. Magnetically superior and easy to handle $L1_0$-FePt nanocrystals. *Applied Physics Letters* 87: 032503-1–032503-3.

Yan, Q., Kim, T., Purkayastha, A., Ganesan, P.G., Shima, M., and Ramanath, G. 2005. Enhanced chemical ordering and coercivity in FePt alloy nanoparticles by Sb-doping. *Advanced Materials* 17: 2233–2237.

Yu, M., Liu, Y., and Sellmyer, D.J. 2000. Nanostructure and magnetic properties of composite CoPt:C films for extremely high-density recording. *Journal of Applied Physics* 87: 6959–6961.

Yu, A.C.C., Mizuno, M., Sasaki, Y., Kondo, H., and Hiraga, K. 2002. Structural characteristics and magnetic properties of chemically synthesized CoPt nanoparticles. *Applied Physics Letters* 81: 3768–3770.

Zeng, H., Sun, S., Vedantam, T.S., Liu, J.P., Dai, Z.-R., and Wang, Z.-L. 2002. Exchange-coupled FePt nanoparticle assembly. *Applied Physics Letters* 80: 2583–2585.

Zeng, H., Li, J., Wang, Z.L., Liu, J.P., and Sun, S. 2004a. Bimagnetic core/shell FePt/$Fe_3O_4$ nanoparticles. *Nano Letters* 4: 187–190.

Zeng, H., Rice, P.M., Wang, S.X., and Sun, S. 2004b. Shape-controlled synthesis and shape-induced texture of $MnFe_2O_4$ nanoparticles. *Journal of the American Chemical Society* 126: 11458–11459.

Zhao, F., Rutherford, M., Grisham, S.Y., and Peng, X. 2009. Formation of monodisperse FePt alloy nanocrystals using air-stable precursors: Fatty acids as alloying mediator and reductant for $Fe^{3+}$ precursors. *Journal of the American Chemical Society* 131: 5350–5358.

Zhou, J., Skomski, R., and Sellmyer, D.J. 2003. Magnetic hysteresis of mechanically alloyed Sm-Co nanocrystalline powders. *Journal of Applied Physics* 93: 6495–6497.

# 4

# *Inorganic Nanoparticles Gas Sensors*

**B.R. Mehta, V.N. Singh, and Manika Khanuja**

**CONTENTS**

## 4.1 Introduction

The importance of maintaining a clean environment and issue related to industrial and home safety makes the development of sensitive, faster, selective gas sensors an important technological task. Probably Seiyama was one of the first few to propose inorganic semiconductor gas in 1962 (Seiyama and Kato 1962). Since then, extensive research has been carried out in this direction. Gas sensors are generally classified on the basis of the type of gas molecule–inorganic material interaction. Gas sensors based on the gas-induced changes in electrical conductivity, permittivity, optical constants, and junction characteristics have been investigated. Some sensors based on the changes in catalytic combustion, field effect transistor, and surface acoustic wave devices have also been reported in literature (Comini et al. 2009). Each type has its own merit and demerits. The choice of a particular type depends on the field of application.

Nanostructured materials are single-phase or multiphase materials that have at least one characteristic dimension less that 100 nm. Due to large surface-to-volume ratio at nanodimensions, a wide variety of size-related effects are observed in these materials. The percentage of atoms on the surface increases with the decrease in the particle size as surface-to-volume ratio is inversely proportional to radius. This can be highly advantageous in term of the gas sensor devices, as nanoparticles will offer large surface area for gas adsorption, which is always the first step in the gas-sensing mechanism. Presence of surface or subsurface sites with higher reactivity can be highly advantageous for enhancement in the gas sensor parameters. In addition, quantum effects due to confinement of carriers to small dimensions can also modify optical, structural, and electronic properties of nanostructured materials.

The central objective of this review is to highlight the important influences of "nano" on science and technology of gas sensor materials and devices. With this objective in view, we have chosen only two types of inorganic sensor materials in the forthcoming discussion. These are resistor gas sensors based on metal oxide and metal nanoparticles. It will be attempted to discuss the effect of nanoparticle character on the sensing mechanism and the performance parameters of these gas sensors.

## 4.2 Sensing Parameters

For resistive gas sensor devices, the following parameters are important.

### 4.2.1 Sensitivity

Sensitivity is proportional to the change in the resistance of the material on exposure to gas. As shown in Figure 4.1, $R_a$ and $R_g$ are the resistance values of the material in air and gas, respectively. Then, the sensitivity $S$ (Watson 1992) can be written as

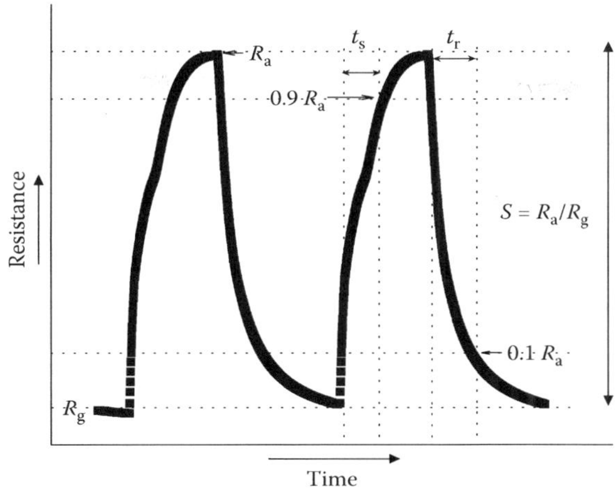


**FIGURE 4.1**
A typical sensing signal showing different sensing parameters: sensitivity ($S$), response time ($t_s$), and recovery time ($t_r$).

$$S = \frac{R_a(\text{air}) - R_g(\text{gas})}{R_g(\text{gas})}$$

### 4.2.2 Selectivity

The sensor material should be able to discriminate between different background gases (Williams 1991). It must detect a particular gas species from a mixture of gases. Resistive gas sensors are mostly sensitive to many gases simultaneously and have poor selectivity.

### 4.2.3 Low Operating Temperature

This is the parameter especially important for resistive gas sensors. As will be discussed later, most of the resistive gas-sensing mechanisms operate at higher temperature. Thus, heating the sensing material is an important part of the sensor device. It is important that the sensor device should require low power.

### 4.2.4 Response and Recovery Time

The response time is the time required for a sensor to respond to the sensing gas. As shown in Figure 4.1, response time can be defined as the time needed to reach 90% of the final signal for a given concentration of gas. The recovery time can be taken to be the time for the signal to fall below 10% of the signal without gas. A good sensor material should respond quickly to any change in the environment, and it should respond and recover immediately.

### 4.2.5 Drift of the Sensing Response

Repeatable and reproducible response is one of the important requirements that must be met before it can be applied in actual field use. In addition to reproducible values of

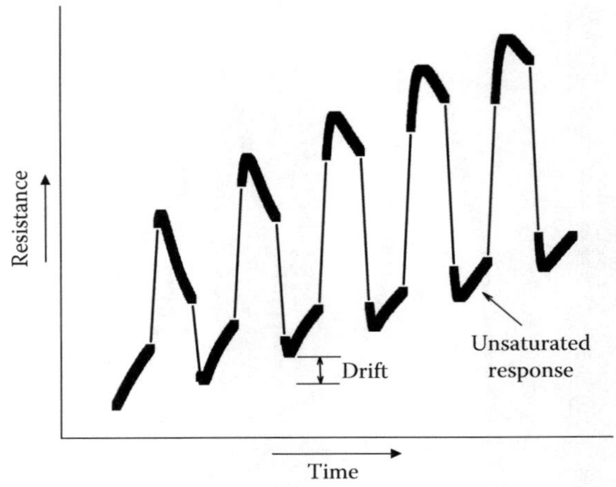


**FIGURE 4.2**
Drift and unsaturated response in a nanoparticle gas sensor. Heat treatments and repeated gas on–off cycles are normally used to stabilize sensor response before making gas-sensing measurements.

sensitivity and response time, it is also important that the value of sensing signal (value of electrical resistance in case of resistive gas sensors) in the on and off stages remains unchanged in each repeated sensing cycle. The change in the value of resistance in a particular sensor state is known as drift in the sensor response (Figure 4.2). Drift is one of the nagging aspects of resistive gas sensors (Eranna et al. 2004, Malyshev and Pislyakov 2008, Pavelko et al. 2008).

### 4.2.6 Device-Related Parameters

The above-mentioned parameters of sensitivity, response and recovery time, and operating temperature are the parameters that are directly related to the sensing mechanism. Parameters like material cost, device fabrication cost, and size are the parameters that are related to the commercial applicability of the device.

## 4.3 Metal Oxide Semiconductors

Various oxide materials have been used for gas sensor applications. The list includes both single-phase oxide materials, e.g., ZnO, $SnO_2$, $WO_3$, $TiO_2$, $In_2O_3$, and $Fe_2O_3$, and also some multicomponent oxide materials (Hoffheins 1996).

### 4.3.1 Synthesis of Nanoparticles for Gas Sensor Devices

A number of techniques have been used for synthesizing oxide semiconductor and metal nanoparticles. General aspects of the deposition techniques used for depositing nanoparticles are summarized below.

#### 4.3.1.1 Thin-Film Deposition Methods

Thin-film deposition methods have been used for preparing a large number of oxide and metal nanocrystallite layers for gas sensor application. Thermal evaporation is one of the oldest methods of producing nanoparticulate layers. It has the advantage of being versatile, maintains high purity due to clean vacuum environment, and can be easily used to produces nanocrystalline films and coatings. The crystallite size can be controlled by varying the substrate temperature and other deposition parameters. Magnetron sputtering is another technique that is especially useful for depositing a wide variety of oxide and composite oxide materials of controllable compositions. Laser ablation is also very useful to deposit oxide nanoparticulate layers of material having low vapor pressure and for maintaining the composition of the layers. Inert gas evaporation method is also used for depositing layers having lower crystallite size. Collision of inert gas atoms and evaporant atoms results in reducing the kinetic energy of adatoms reaching the substrate. Pressure of the inert gas and substrate temperature is varied to prepare films having different nanoparticle size. The above method normally results in a broad nanoparticle size distribution (Chopra 1969).

#### 4.3.1.2 Chemical Techniques

A number of solution-based chemical techniques have been applied to prepare a variety of nanomaterials for sensor applications. In chemical capping method, capping agents block the growth sites at the surface of nanoparticles by passivating dangling bonds. Besides the passivation of the surface states, capping agents prevent the individual nanoparticles from agglomerating and increase the solubility of the nanoparticles in nonpolar solvents. The technique of rheotaxial growth and thermal oxidation (RGTO) for the deposition of metal onto a substrate kept below the melting point followed by oxidation (reactive growth) at a well-defined temperature in an oxidizing atmosphere has also been used (Sberveglieri 1995, Faglia et al. 1999). Low cost and simplicity are the main advantage of these techniques. As a number of precursors, capping agents, and ligands are used to control the particle size, the presence of ligands and precursors on the nanoparticle surface could affect the sensing characteristics (Rao et al. 2007).

#### 4.3.1.3 Sol-Gel Technique

A number of workers have used the sol-gel technique to prepare oxide nanoparticulate layers, and post-synthesis annealing is used to control the nanoparticle size. The process involves the hydrolysis and polymerization of metal alkoxide precursors (Gurlo et al. 2003). The solutions of precursors are reacted to form irreversible gels that shrink to rigid oxide forms. The concept of sol-gel processing is actually quite suitable to the processing of the "nanostructure." First, this process starts with a nanometer-sized molecular unit and goes through the reactions, which also take place on the nanometer scale. The gels also contain "pores" and the nanophase porosity. The particles synthesized through the sol-gel route normally have a relatively better uniformity of size.

#### 4.3.1.4 Gas-Phase Synthesis of Size-Selected Nanoparticles

In this method, a material is evaporated from a ceramic boat and carried by a stream of nitrogen into a tube furnace. The aerosol that is formed by nucleation, condensation, and coagulation is polydisperse. In order to obtain monosized nanoparticles, a differential mobility analyzer (DMA) is used as a size fractionator. The gas containing polydisperse agglomerates is passed through a neutralizer containing a radioactive source, where the particles obtain a known charge distribution. The charged agglomerates, which are mostly singly charged, enter DMA. The DMA selects the size of the nanoparticles on the basis of the electrical mobility of nanoparticles in the carrier gas. For the growth of crystalline and spherical nanoparticles, metal oxide particles are passed through a sintering furnace. Additional oxidation steps can be used for controlling the chemical composition and stoichiometry of the oxide nanoparticles. Syntheses of nanoparticles having well-defined particle size is the main advantage of this technique. But, due to the low yield of this method, nanoparticles can be deposited in a very small deposition area (2–5 mm$^2$) with an extremely low deposition rate (Kennedy et al. 2003a,b).

#### *4.3.1.5 Spray Deposition*

Spray pyrolysis is one of the widely used techniques for the deposition of semiconductor nanoparticles layers. A solution containing chemical compounds (e.g., $SnCl_4$ for depositing $SnO_2$ layers) and carrier liquid (e.g., water and/or ethanol) is sprayed onto heated substrates kept at a temperature suitable for the pyrolysis reaction to take place. The advantage of the method lies in its simplicity and low-cost nature, and can be used to deposit nanoparticulate layers (Gaiduk et al. 2008). The incorporation of unreacted or partially reacted impurities, which can strongly influence the sensing parameters, is one of the major drawbacks of this technique.

#### *4.3.1.6 Special Substrate Requirements for Nanoparticle Deposition for Gas Sensing*

Study of the gas-sensing characteristics of nanoparticles requires that nanoparticles are deposited onto a microhotplate with special electrode configurations. In case of nanoparticles, certain topographical features (roughness, sharp edges, or height of the metal electrodes) can result in nanoparticle agglomeration around these features and can influence sensing characteristics. Although not much work has been done in this direction, some especially designed microhotplate with interdigitated patterned electrodes for nanoparticle deposition have been reported. The schematic of microhotplate having a buried electrode structure is shown in Figure 4.3 (Kennedy et al. 2003a,b). A miniaturized electrode area can help in reducing the effective resistance of the nanoparticle layer. The embedded heating layer can help in reducing the power consumption to reach a particular operating temperature. The buried electrode structure eliminates the influence of electrode walls on the particle deposition.

### 4.3.2 Gas-Sensing Properties of Oxide Semiconductor Nanoparticles

Using the above-mentioned technique, a number of metal oxide materials in nanoparticle form have been deposited. Details of the deposition parameters and the nanoparticle size controlling parameters are summarized in Tables 6.1 through 6.3.

Tin oxide is probably the most widely studied semiconductor material for the gas-sensing applications. Tin oxide has been used for detecting ethanol, methanol, $NH_3$, $H_2$ CO, $CH_4$, and other gases. A summary of the size controlling parameter and the variation in nanoparticle size along with the gas-sensing parameter are given in Table 4.1. In most of the above studies, nanoparticles size has been measured by the XRD technique,

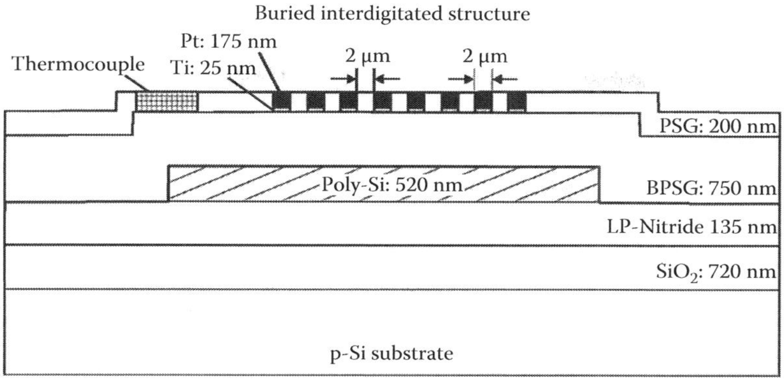


**FIGURE 4.3**
A schematic diagram of the buried electrode structure useful for depositing nanoparticle layers for gas-sensing measurement. (Reprinted with permission from Kennedy, M.K. et al., *J. Appl. Phys.*, 93, 551, 2003a. Copyright 2003, American Institute of Physics; Kennedy, M.K. et al., *Rev. Sci. Instrum.*, 74, 4908, 2003b. Copyright 2003, American Institute of Physics.)

**TABLE 4.1**
Deposition Conditions and Gas-Sensing Parameters of Different $SnO_2$ Nanoparticles

| Synthesis Methods | Size Controlling Parameter | Variation in Size (nm) | $S/R_t$ (–/s) | Gas/Conc./Temp (–/ppm/°C) | References |
|---|---|---|---|---|---|
| Neutralization of Sn salt | Annealing | 5–32 | 180/– | $H_2$/800/400 | Xu et al. (1991) |
| Aerosol | Temperature | 10–35 | 4/10 | $C_2H_5OH$/1000/300 | Kennedy et al. (2003a,b) |
| Spin coat | — | ~10 | 2/– | $CH_3OH$/0.01/350 | Cavicchi et al. (2001) |
| Spray | — | 10–100 | 5/– | $CH_3OH$/1000/330 | Gaiduk et al. (2008) |
| Reverse micelles | — | ~10 | 320/– | CO/200/300 | Yuasa et al. (2009) |
| Wet chemical | Different routes | 3–36 | 120/– | $NO_2$/90/200 | Huang et al. (2005a,b) |
| Screen printing | — | ~40 | | $NH_3$/–/380 | Llobet et al. (2003) |
| Screen printing | — | ~15 | 58/20 | $CH_4$/3500/400 | Hong and Han (2004) |
| Aerosol | Annealing | 5–20 | 20/2 | $C_2H_5OH$/1000/400 | Joshi and Kruis (2006) |
| CVD | — | 20–121 | 2/– | $CH_3OH$/80/400 | Panchapakesan et al. (2006) |
| Laser ablation | — | ~10 | 2/30 | CO/0.1/250 | Dolbec and El Khakani (2007) |
| Sol-gel | Additives | 6–8 | –/– | CO/300/380 | Safonova et al. (2002) |
| Aerosol | Annealing | 10–24 | 1.5/– | $CH_4$/1000/270 | Kennedy et al. (2005) |
| Gas-phase condensations | Annealing | 10–35 | 3.5/38 | $C_2H_5OH$/1000/300 | Kennedy et al. (2003a,b) |
| Solid-state reaction | Additives | 20–80 | 18/– | $H_2$/500/300 | Tan et al. (2008) |

which gives average and approximate value. In many cases, samples were prepared using techniques having no size controlling parameters. In qualitative and general sense, it can be stated that sensitivity for reducing gases like $H_2$, CO, and ethanol increases steeply as D decreases and becomes comparable or lower than $2L$ in the range of 5–15 nm. ($L$ is Debye length, which is estimated to be about 6 nm, in case of tin oxide [Yamazoe 1991]).

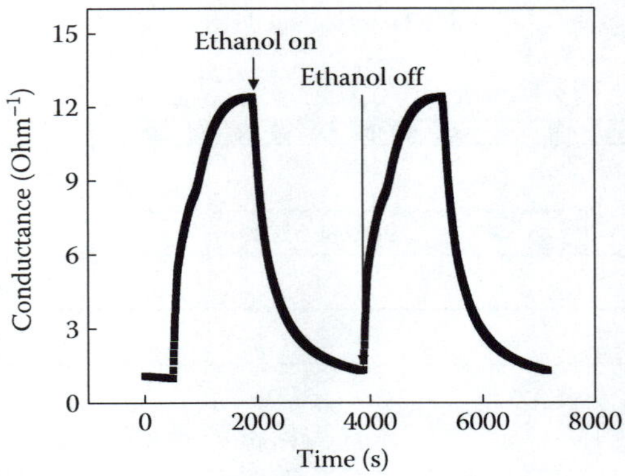


**FIGURE 4.4**
A typical gas-sensing response of nanoparticle layer for ethanol. (Reprinted from Singh, V.N. et al., *Sens. Actuators B*, 125, 482, 2007. Copyright 2007, with permission from Elsevier.)

A typical gas-sensing response curve of $SnO_2$ nanoparticle layers is shown in Figure 4.4. The arrow with "ethanol on" refers to the inclusion of ethanol in the artificial air and "ethanol off" refers to the closing of the ethanol supply. Sensor response (normalized conductance) is the ratio of the conductance of the nanoparticle layer in a sensing gas (ethanol + air) to that in air. Response time has been taken as the time required to reach 90% of the saturated value of resistance in the sensing gas.

In an important study, size-dependent property of $SnO_2$ nanoparticles has been studied. The novelty of this study lies in the use of well-defined nanoparticles. As can be noted from the details given in Tables 4.1 and 4.2, there is a large variation in size of nanoparticles in

**TABLE 4.2**
Summary of Gas-Sensing Studies on Indium Oxide Nanoparticles

| Synthesis Methods | Size Controlling Parameter | Variation in Size (nm) | $S/R_t$ (–/s) | Gas/Conc./Temp (–/ppm/°C) | References |
|---|---|---|---|---|---|
| Magnetron sputtering | Annealing | 11–30 | 33/– | $C_2H_5OH$/26/450 | Papadopoulos et al. (1996) |
| Reactive sputtering | Sputtering power | 20–65 | 104/– | $O_3$/–/– | Kiriadikis (2001) |
| Laser ablation | — | ~12 | 30/45 | — | Starke and Coles (2002) |
| Hydrothermal | — | ~21 | 80/12 | $NH_3$/800/300 | Rout et al. (2007) |
| Sol-gel | Annealing | 50–100 | 5/55 | $NO_2$/1/350 | Gurlo and Ivanovskaya (1997) |
| Sol-gel | — | 6–7 | 16/– | $H_2$/1000/50 | Shukla et al. (2004) |
| Spin coat | Number of layers | 23–27 | 88/12 | $H_2$/2000/350 | Chung et al. (1998) |
| MBE | Substrate temperature | 100–300 | 75/– | $NO_2$/100/30 | Winter et al. (2000) |
| Sol-gel | — | 5–6 | 700/– | $O_3$/0.05/– | Gurlo et al. (2003) |
| Chemical capping | Capping concentration and annealing | 8–29 | 325/8 | $C_2H_5OH$/1000/400 | Singh et al. (2007) |
| Hydrothermal | Co doping | 25–35 | 8/– | $H_2S$/50/150 | Kapse (2008) |

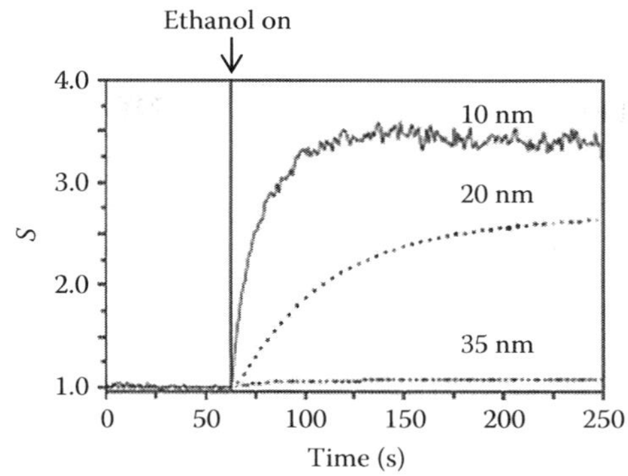


**FIGURE 4.5**
Response transients of $SnO_2$ nanoparticle samples on switching the ethanol flow ($T=300°C$, ethanol concentration 1000 ppm). (Reprinted with permission from Kennedy, M.K. et al., *J. Appl. Phys.*, 93, 551, 2003a. Copyright 2003, American Institute of Physics; Kennedy, M.K. et al., *Rev. Sci. Instrum.*, 74, 4908, 2003b. Copyright 2003, American Institute of Physics.)

most of the reported studies. Figure 4.5 compares the response transients of $SnO_2$ nanoparticles at 300°C with particles sizes 10, 20, and 35 nm in 1000 ppm ethanol (Kennedy et al. 2003a,b). It can be noted that the sensitivity toward ethanol increases as the particle size decreases. The value of response time diminishes with decreasing particle size. For the samples with particle size 10 and 20 nm, the response time was observed to be 10 and 38 s, respectively. Figure 4.6 presents the sensitivity of nanoparticle films as a function of the particle size in the temperature range of 200°C–300°C. The sensitivity increases with decreasing particle size as established in Figure 4.5. The temperature increase leads to an increase in the sensitivity because more oxygen molecules are chemisorbed (Kennedy et al. 2003a,b). Another effect of the rising temperature is that an oxygen species transformation from $O^{2-}$ to $O^-$ takes place at approximately 200°C (Chang 1980); this leads to higher charge transfer rates per oxygen atom.

In comparison to $SnO_2$, there are fewer reports on the application of $In_2O_3$ nanoparticle for gas-sensing applications. Indium oxide has been used for detecting various oxidizing

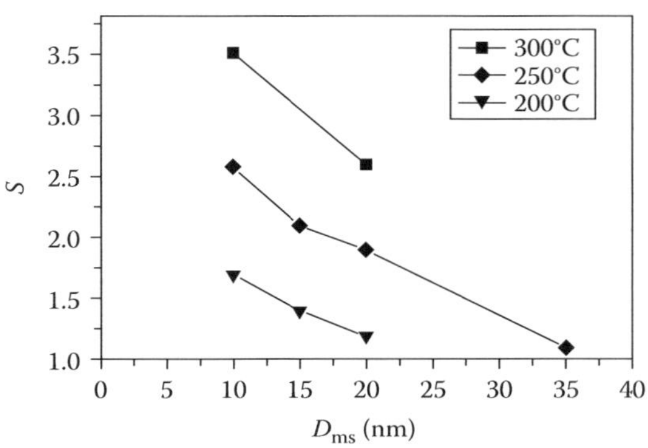


**FIGURE 4.6**
Size dependence of sensitivity at different temperatures for $SnO_2$ nanoparticles in 1000 ppm ethanol. The layer thickness for all the samples is 1.5 μm. (Reprinted with permission from Kennedy, M.K. et al., *J. Appl. Phys.*, 93, 551, 2003a. Copyright 2003, American Institute of Physics; Kennedy, M.K. et al., *Rev. Sci. Instrum.*, 74, 4908, 2003b. Copyright 2003, American Institute of Physics.)

**TABLE 4.3**

Summary of Gas Sensing Studies on Other Oxide Nanoparticles

| Material | Size Controlling Parameter | Variation in Size (nm) | $S/R_t$ (–/s) | Gas/Conc./Temp (–/ppm/°C) | References |
|---|---|---|---|---|---|
| $TiO_2$ | — | ~10 | 40/230 | $H_2$/200/300 | Buso et al. (2008) |
| $TiO_2$ | — | ~15 | 7/5 | $CH_3OH$/50/450 | Benkstein and Semancik (2006) |
| $TiO_2$ | — | ~10 | 11/240 | $C_2H_5OH$/100/350 | Rella et al. (2007) |
| $TiO_2$ | — | ~4 | 80/00 | $H_2$/1000/200 | Du et al. (2002) |
| $WO_3$ | Doping | 50–69 | 82/– | $NH_3$/500/300 | Jiménez et al. (2004) |
| $WO_3$ | — | ~5 | 5000/540 | $H_2S$/10/500 | Hoel et al. (2005) |
| $WO_3$ | — | — | 300/300 | $C_2H_5OH$/50/350 | Ederth et al. (2006) |
| $WO_3$ | — | ~10 | 3900/300 | $NO_2$/10/250 | Reyes et al. (2006) |
| $WO_3$ | — | ~20 | 757/132 | $H_2S$/1000/250 | Rout et al. (2008) |
| $WO_3$ | — | ~5 | 1.05/600 | $C_2H_5OH$/0.2/260 | Ionescu et al. (2005) |
| ZnO | Annealing | 6–20 | 7/35 | $NH_3$/100/300 | Devi et al. (2006) |
| ZnO | — | ~15 nm | 260/48 | $NH_3$/800/300 | Rout et al. (2007) |
| ZnO | — | 10–100 | 26/30 | $H_2S$/50/300 | Li et al. (2009) |
| $V_2O_5$ | — | 30–50 | 10/– | LPG/2000/200 | Rout et al. (2006) |
| $Fe_2O_3$ | — | — | 98/20 | $C_2H_5OH$/200/– | Huo et al. (2000) |
| $Fe_2O_3$ | — | 4–12 | 16/– | $C_2H_5OH$/50/25 | Wu et al. (2008) |
| $Fe_2O_3$ | Doping | 30–50 | 225/35 | $H_2S$/100/160 | Wang et al. (2008) |
| NiO | Annealing | 10–14 | 2/– | $H_2S$/5 ppm/325 | Luyo (2009) |
| NiO | — | — | 6/– | $H_2$/200/– | Cantalini et al. (2005) |
| CdO | Annealing | 10–35 | 341 | LPG/75/450 | Waghulade et al. (2007) |
| CuO | — | ~4 | 130/– | CO/1000/400 | Dutta et al. (2003) |
| $Ga_2O_3$ | — | — | 2.1/205 | $C_2H_5OH$/1500/100 | Yu et al. (2005a,b) |
| $CdSnO_3$ | — | ~32 | 68/28 | $C_2H_5OH$/50/200 | Liu et al. (2004) |

as well as reducing gases. A summary of the size controlling parameter and the variation in nanoparticle size along with the gas-sensing parameter reported in these studies are summarized in Table 4.2. Summary of gas sensing studies on other oxide nanoparticles has been presented in Table 4.3.

Figure 4.7 shows the TEM micrograph of the $In_2O_3$ nanoparticle samples used for gas-sensing applications (Table 4.2). The most probable sizes of the nanoparticles measured using the TEM micrographs are 11 nm. In the inset of Figure 4.7, high resolution transmission electron micrograph (HRTEM) and diffraction pattern of $In_2O_3$ nanoparticles is shown. The lattice spacing measured of nanoparticle is 0.41 nm, which corresponds to (211) plane of cubic indium oxide.

### 4.3.3 Enhancing Gas Sensitivity and Selectivity by Metal Addition

A number of methods have been employed to improve gas sensitivity, response time, and selectivity of gas sensors. Conventional ways to improve the sensor selectivity are (1) optimization of the analysis temperature, (2) introduction of carefully chosen dopants, and (3) use of catalytic materials. As we will see later, the metal additives in nanoparticle form can be highly advantageous. The most commonly used catalysts are expensive noble metals like Pt and Pd. They can either be supported on or doped inside the material. The metal additives

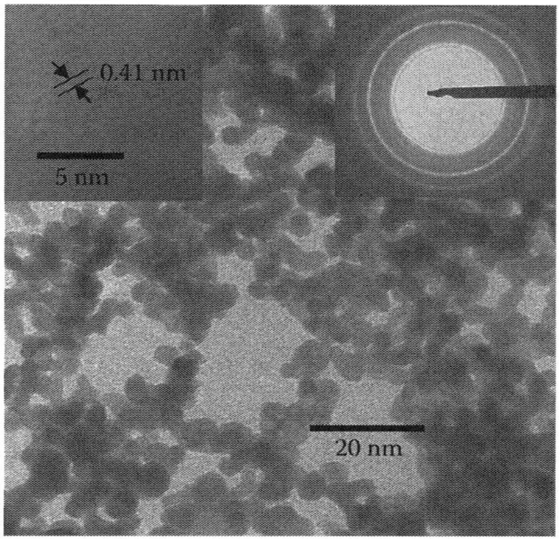


**FIGURE 4.7**
TEM micrograph of indium oxide nanoparticles synthesized by a chemical capping method and used for gas sensing. The HRTEM micrograph and the diffraction pattern are shown in the inset.

may diffuse substitutionally or occupy interstitial positions (Mardare and Hones 1999). The metal additives can also change the optical and electrical properties of the oxide semiconductor nanoparticles. The additive metal can also get atomically dispersed on the semiconductor surface (Reddy and Chandorkar 1999). In some cases, metal addition can lead to the formation of agglomerated metal clusters or nanoparticles on the semiconductor surface. (Cabot et al. 2000, Ivanovskaya et al. 2000, Yamaura et al. 2000). Depending upon the thickness of the metal layer and post-deposition annealing temperature, discontinuous metal clusters or nano- or micron-sized agglomerates can also form upon annealing (Ivanovskaya et al. 2000, Yamaura et al. 2000). The metal addition can also dope the underlying oxide nanoparticles.

A chemical sensor is supposed to perform two functions: (1) the receptor function, to recognize the gas species to be detected; and (2) the transducer function, to transform the chemical signal into an electric signal. Depending upon this, additives can play two different roles for enhancing the gas-sensing response (Yamazoe 1991). Thus, the resulting performance is dependent not only on the type of metal but also on how and where the metal is incorporated. Thus, the configuration of the additive and the base oxide semiconductor is quite important.

Different configurations of metal–metal oxide semiconductors reported in literature for explaining the role of metal additive on gas-sensing properties are schematically described in Figure 4.8. To avoid repletion and give a systematic picture, results from a large volume of data has been summarized in the form of Table 4.4. It is clear from the above discussion and also from the data given in Table 4.4 that in most of the reported work, the morphology

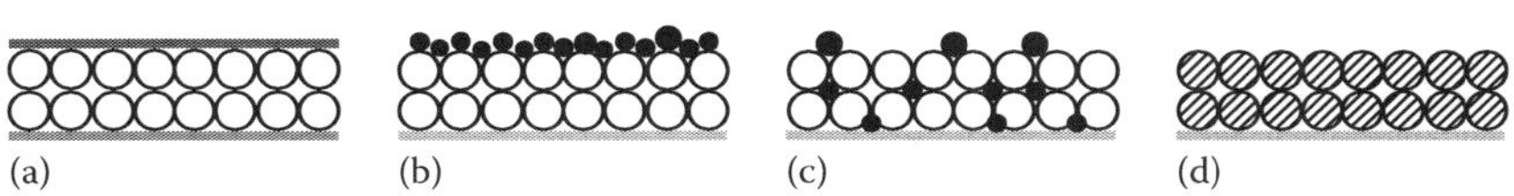


**FIGURE 4.8**
A schematic representation of different ways of metal addition in oxide layer: (a) metal top layer, (b) macro/nano-agglomerated metal, (c) dispersed metal nanoparticle in oxide nanoparticle layer, and (d) metal-doped oxide nanoparticle layer.

**TABLE 4.4**
Summary of Results of Various Studies on Metal Addition to Semiconductor Oxide

| Semiconductor | Method | Metal | Method | Metal/Semiconductor Structure | References |
|---|---|---|---|---|---|
| $In_2O_3$ | Thermal evaporation | $SeO_2$ | Thermal evaporation | Mixed film | Manno et al. (2000) |
| $SnO_2$ and $In_2O_3$ | Magnetron sputtering | $Nb_2O_5$, $TiO_2$ | Co-precursor | Mixed thin films | Comini et al. (2000) |
| $In_2O_3^-$ | Reactive rf sputtering | Au, Ti | Sputtering | Over layer/polycrystalline | Steffes et al. (2001) |
| $In_2O_3$ and $SnO_2$ | Magnetron sputtering | Pt, Ag | Magnetron sputtering | Films/polycrystalline | Galdikas et al. (2003) |
| $SnO_2$, $In_2O_3$, and $WO_3$ | Laser ablation | Pt | Impregnation | Doping/thin film | Starke and Coles (2002) |
| $SnO_2$ or $In_2O_3$ | Magnetron sputtering | Pt, Pd | Electron beam evaporation | Discontinuous cluster/– | Ivanovskaya et al. (2000) |
| $In_2O_3$ | Sol-gel | Ni | Co-precursor | –/Thin film | Zhou (1999) |
| $In_2O_3$ | Sol-gel | Pt- | Sputtered | Thin layer/nanocrystalline | Shukla et al. (2004) |
| $In_2O_3$ | Sol-gel | Ni | Co-precursor | –/Layer | Ivanovskaya and Bogdanov (1998) |
| $In_2O_3$ | Hydrolization-precipitation | $CoO_3$, Au | Impregnation | –/– | Yamaura et al. (2000) |
| $In_2O_3$ | Spray pyrolysis | Pt, Pd, Au | Electron beam evaporation | –/– | Skala et al. (2003) |
| $In_2O_3$ | RGTO | Au | DC sputtering | Film/film | Faglia et al. (1999) |
| $SnO_x$ and $InO_x$ | Reactively sputtered | Pt | Electron beam evaporation | Discontinuous films/– | Papadopoulos et al. (1996) |
| $In_2O_3$ | Sol-gel | $Fe_2O_3$ | Co-precursor | Layer/layer | Ivanovskaya et al. (2003) |

of the metal and semiconductor oxide configuration is not well defined. The topography (nanocluster or nanoparticle, or a thin layer or doping, etc.) of the metal additive is important to pinpoint the mechanism of enhancement.

It is reported that the depth of the space charge layer can be controlled by doping impurities ($Al^{3+}$ or $Sb^{5+}$) inside the lattice of a semiconductor. If trivalent ions were doped (dissolved) partially into the $SnO_2$ lattice, the carrier concentration would decrease and Debye length of $SnO_2$ would increase. On the contrary, pentavalent dopants would bring about a decrease in Debye length. Thus, the sensitivity can be further improved even without changing the particle sizes. Thus, Al-doped $SnO_2$ shows high sensitivity with increasing L even at particles size of 20 nm, while Sb-doped SnO is insensitive in the whole region (Xu et al. 1991).

In a study carried out in author's laboratory, Ag nanoparticles have been dispersed in a layer of $In_2O_3$ nanoparticles. Chemical composition of the metal oxide–metal composite nanoparticle layer can be studied using x-ray photoelectron spectroscopic (XPS) technique. The O 1s, In 3d, and Ag 3d core level spectra for $In_2O_3$: Ag, $In_2O_3$, and Ag samples after sputtering are shown in the Figure 4.9a through g (Singh et al. 2007). The peak O 1s in $In_2O_3$ sample (Figure 4.9d) has been deconvoluted into two peaks at 529.3 and 530.5 eV. The atomic percentage of Ag in $In_2O_3$: Ag sample as obtained from XPS analysis is about 13%, which is quite consistent with the Ag percentage taken in the starting solution (15%). The stoichiometric ratio in $In_xO_y$, $x/y=0.56$ are quite close to the stoichiometric

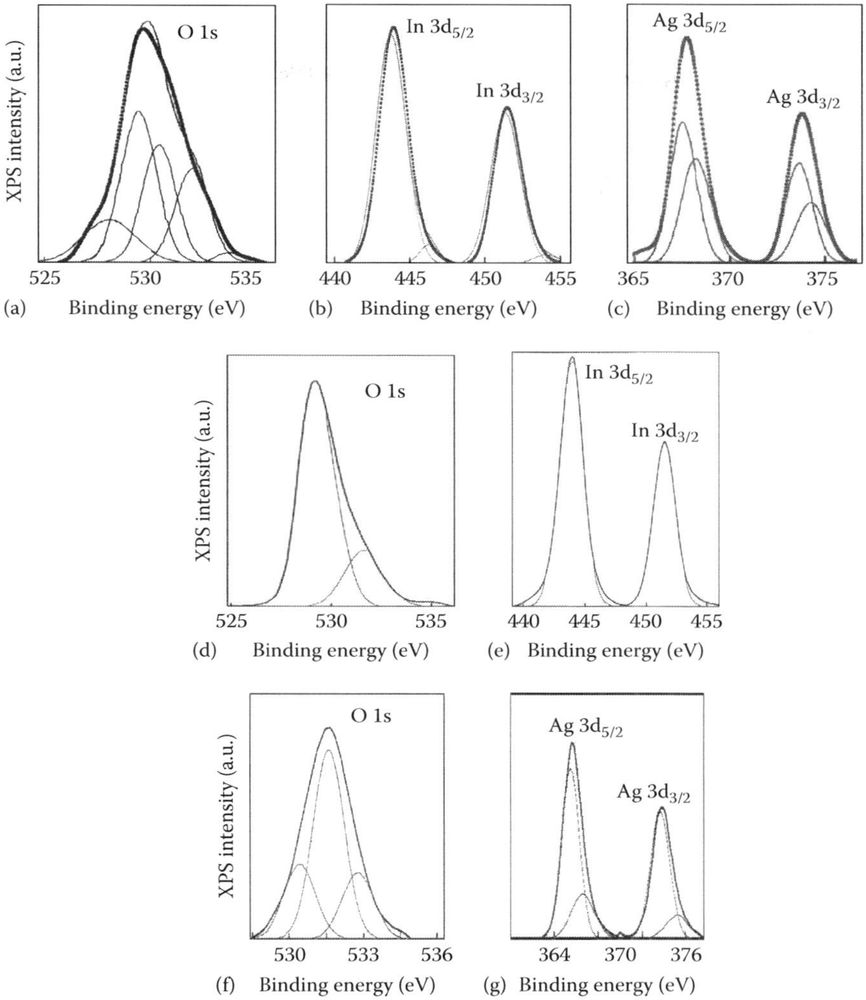


**FIGURE 4.9**
Core levels of (a) O 1s, (b) In 3d, and (c) Ag 3d peaks in the XPS spectra of the $In_2O_3$: Ag (15%) composite nanoparticles, core levels of (d) O 1s and (e) In 3d peaks in the XPS spectra of nanoparticle sample $In_2O_3$, and core levels of (f) O 1s and (g) Ag 3d XPS peaks in nanoparticle sample Ag. (Reprinted from Singh, V.N. et al., *Sens. Actuators B*, 125, 482, 2007. Copyright 2007, with permission from Elsevier.)

value of $In_2O_3$ ($x/y = 0.6$). The Ag $3d_{5/2}$ peak at 368.1 eV corresponding to $Ag_2O$ in Ag nanoparticle sample (Figure 4.9g) is observed to shift to 367.6 eV in the case of composite nanoparticle sample $In_2O_3$: Ag (Figure 4.9c). The In $3d_{5/2}$ peak at 443.7 eV in case of $In_2O_3$ sample (Figure 4.6e) is shifted to 444.0 eV in the composite nanoparticle sample (Figure 4.9b). The positive shift of about 0.3 eV in In $3d_{5/2}$ peak and a negative shift of about 0.6 eV in Ag $3d_{5/2}$ peak in case of the composite nanoparticle sample indicate an electronic interaction between $In_2O_3$ and $Ag_2O$ phases. An increase in the work function of $SnO_2$ of about 0.7 eV has been reported in the presence of silver under oxidizing conditions; the shift disappears completely when the sample is exposed to a reducing atmosphere (Yamazoe et al. 2003). For 1000 ppm ethanol and at 400°C, the values of sensor response

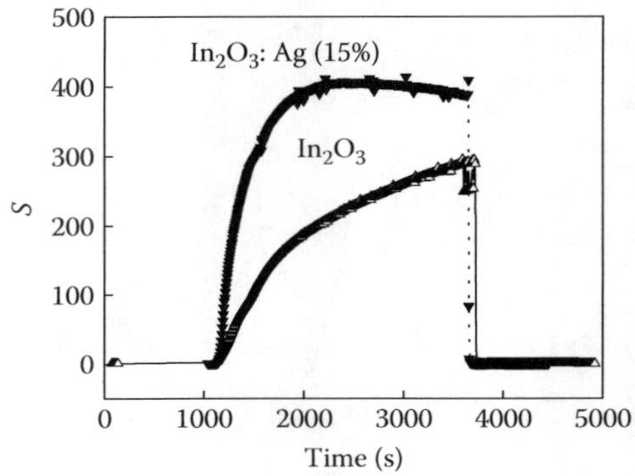


**FIGURE 4.10**
Increase in the sensitivity and the response time upon addition of silver nanoparticles into $In_2O_3$ nanoparticle layer (electronic effect). (Reprinted from Singh, V.N. et al., *Sens. Actuators B*, 125, 482, 2007. Copyright 2007, with permission from Elsevier.)

(response time) for the samples $In_2O_3$ and $In_2O_3$: Ag (15%) are 325 and 436 (8 and 6 s), respectively (Figure 4.10). This enhancement has been explained due to the electronic effect (EE) (Figure 4.10) (Singh et al. 2007).

Figure 4.11 compares the dynamic response of three different sensors toward 50 ppm of CO. Upon exposure to CO, the sensitivity of the 10 nm thick $SnO_2$ sensor is seen to rise up to a value of about 50% with a typical response time in the 30–40 s range. As the sensor thickness is increased to 500 nm, sensitivity increases and reaches a value of ~400%, with a response time very similar to that of 10 nm thick sensors. On the other hand, the deposition of a Pt catalyst nanolayer onto the top of the 500 nm thick $SnO_2$ film is shown to improve markedly the response of the sensors, of which sensitivity is seen to reach a value as high as 1000%. It is worth mentioning that the Pt top layer consists of nanoparticles presenting a mean diameter of about 2 nm. The response time for this sensor was found to be of ~60 s. This enhancement has been explained due to the chemical effect (Dolbec and El Khakani 2007).

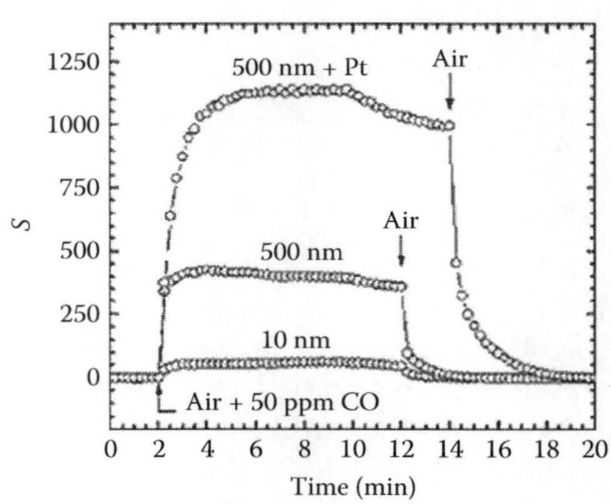


**FIGURE 4.11**
Dynamic response of selected sensors in CO (50 ppm). The sensitivity of $SnO_2$ layer (10–500 nm thick) is compared to that of 500 nm thick $SnO_2$:Pt layer. (From Dolbec, R. and El Khakani, M.A., *Appl. Phys. Lett.*, 90, 173114, 2007. With permission.)

### 4.3.4 Gas-Sensing Mechanism

#### 4.3.4.1 Adsorption/Desorption Mechanism

During the gas molecule–solid-state surface interaction, the gas species initially gets physisorbed and chemisorbed at the semiconductor surface. There is no charge transfer in case of physisorption. A molecule is considered to be chemisorbed when there is an electronic transfer between the gas and the solid, and the conductivity of the material gets affected.

As an illustrative example, the different oxygen species formed as the oxygen molecules interact with a solid-state surface are shown in Figure 4.12 (Kohl 1989, Liu et al. 2007). Oxygen can be adsorbed in the form of $O_2^-$, $O^-$, or $O^{2-}$. Due to energy considerations, near room temperature, the $O^{2-}$ specie dominates the surface coverage, whereas at higher temperatures, $O^-$ is the dominating specie. The adsorption of these species on the surface extracts electrons from the donor levels of the material. The extraction of electrons creates an electron-depleted or positive space charge region near the surface (Figure 4.12). This results in an increase in the resistance of an n-type material.

The adsorbed oxygen ions are present in large concentrations at any semiconductor surface. The gas-sensing mechanism depends on the decrease or increase in the concentration of the adsorbed oxygen ion at the semiconductor surface. It is clear that on interaction with a reducing gas, the number of adsorbed oxygen species will be reduced and a number of electrons will be released back to the semiconductor surface (conduction band of the semiconductor). The $O_2^-$ and $O^-$ species preferentially interact with the C—C bond of the incoming reducing gas. Increase in the concentration of electrons in the conduction band results in an increase in the conductivity of the n-type semiconductor. The large increase in the conductivity of the oxide semiconductor in the presence of a reducing gas is because of a decrease in the adsorbed oxygen species at the oxide surface. It is clear that the effect of an oxidizing gas will be reverse (increase in absorbed oxygen sites and decrease in the conductivity of the semiconductor).

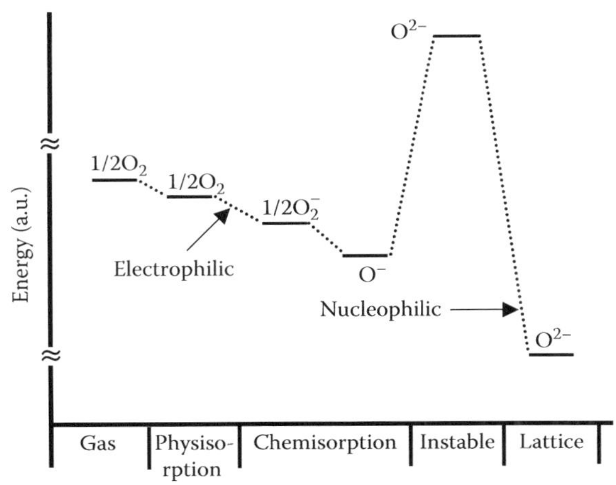


**FIGURE 4.12**
Energy diagram of various oxygen species adsorbed at the surface and within the material. (Reprinted from Kohl, D., *Sens. Actuators*, 18, 71, 1989. Copyright 1989, with permission from Elsevier.)

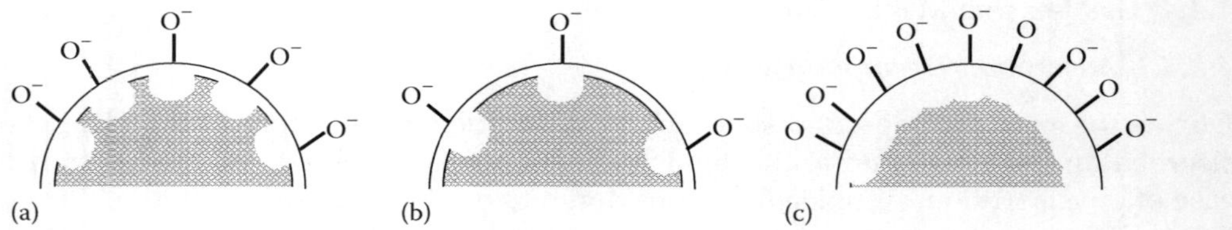


**FIGURE 4.13**
Effect of (a) air, (b) reducing gas, and (c) oxidizing gas on an n-type semiconductor material. Depletion region is shown as white region, and shadowed region shows the conducting regions of the oxide nanoparticles.

The reaction steps (Kohl 1989) are given below:

$$e^- + O_2 \Leftrightarrow O_2^-$$

$$e^- + O_2^- \rightarrow 2O^-$$

$$R_g + O_2^- \rightarrow R_gO_2 + e^-$$

$$R_g + O^- \rightarrow R_gO + e^-$$

Here $R_g$ represents the reducing gas.

If a p-type semiconductor is used in place of an n-type semiconductor, the reducing gas will decrease the adsorbed oxygen sites, take away electrons from the semiconductor surface, and create more holes in the semiconductor. This is also illustrated in Figure 4.13. The conductivity of the p-type semiconductor due to increase in the concentration of holes will thus increase.

#### 4.3.4.2 Nanoparticle Size Dependence of Gas-Sensing Properties

There are two factors that are expected to enhance the gas-sensing response of the nanoparticles: (1) increase in the surface area and (2) nanoparticle size becoming close to Debye length. As mentioned earlier, nanostructured materials for gas-sensing applications are of great interest due to their higher surface-to-volume ratio compared to bulk materials leading to a larger area for gas reaction. For improving the sensing properties, Gopel et al. emphasized the use of oxide nanocrystallites with diameters of the order of the Debye length ($L_D$) (Gopel 1994) (Figure 4.14). The Debye length $L_D$ is defined as

$$L_D = \left(\frac{\varepsilon\varepsilon_0 kT}{e^2 n_b}\right)^{1/2}$$

where $\varepsilon$, $T$, $e$, and $n_b$ are the dielectric constant, the temperature, the electronic charge, and the density of electrons, respectively. How the individual nanoparticles merge into one another for structural and electrical contacts is very important. It is generally considered that for (diameter of the crystallite) $D \gg 2L_D$, the electron conducting channels through necks are too wide to be influenced by the surface effect. The grain boundary contacts share most of the electrical resistance of the chain, and gas sensitivity is controlled by

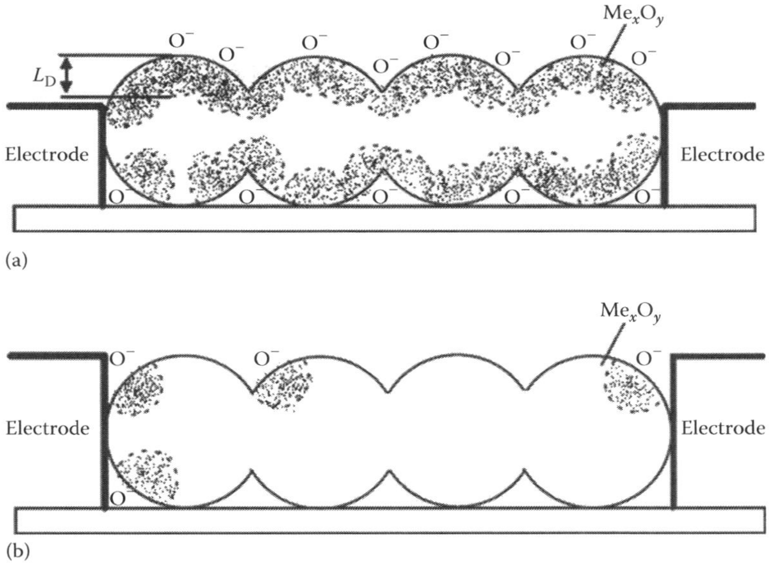


**FIGURE 4.14**
(a) Schematic view of the formation of oxygen adsorbates on the surface of a metal oxide semiconductor, based on monodisperse metal-oxide nanoparticles and (b) sensing layer after exposure to a reducing atmosphere. The oxygen adsorbates are consumed by subsequent reactions so that a lower surface coverage of oxygen adsorbates is obtained.

grain boundary. In this region, the sensitivity is almost independent of $D$. When $D$ decreases to come closer to $2L_D$, the necks become the most resistive part in the chain and thus control the gas sensitivity (neck control) (Gopel 1994). Finally, when $D<2L_D$, each constituent grain is depleted of conduction electrons as a whole. Under this situation, grains share a major part of the resistance of the chain and control the gas sensitivity (grain control). The sensitivity in this region is strongly dependent on $D$. When $D \cong 2L_D$, in the "on state," the complete volume of nanoparticles is depleted of carriers, and there is a very narrow region through which carriers can flow. In the "off state," the complete volume of the nanoparticles has the free carriers. Thus, this will result in maximum sensitivity (Figure 4.15). It may be mentioned that the value of the Debye length of a semiconductor depends upon the effective carrier concentration, density, and porosity. Thus, the

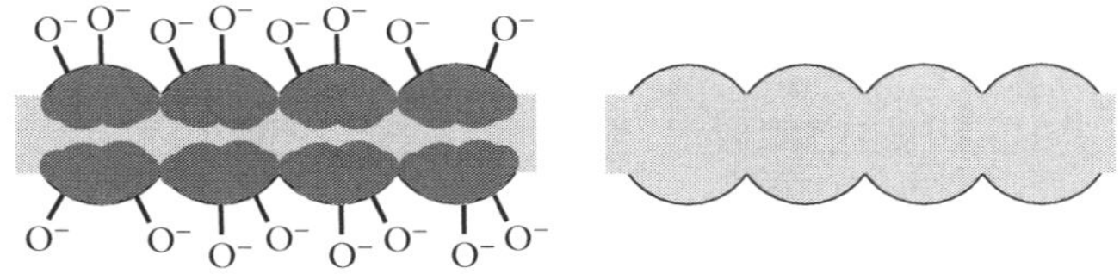


**FIGURE 4.15**
A schematic diagram showing optimal size of nanoparticles for high sensitivity. When the nanoparticle size is equivalent to twice the Debye length ($L_D$), the sensitivity is strongly dependent on particle size ($D$). When $D \sim 2L_D$, (a) complete volume of nanoparticle is depleted and there is a very narrow region through which carriers can flow in the "off state," and (b) the complete volume of the nanoparticles is containing the free carriers in the "on state." This condition will result in maximum sensitivity.

value can significantly vary in materials deposited under different conditions, especially in case of nanoparticles.

#### 4.3.4.3 Effect of Dopants on the Gas-Sensing Properties

It may be mentioned that the value of the Debye length for oxide semiconductor depends upon the carrier concentration, layer porosity, etc. The value can thus vary largely in samples prepared under different conditions. This is especially true in case of nanoparticle samples. Two types of activation mechanisms, chemical (with Pt) and electronic (with Ag and Pd), have been proposed in literature (Yamazoe et al. 2003). In chemical activation mechanism, metal additive increases the gas sensitivity by increasing the rate of the chemical process (a spill-over effect) without affecting the electronic properties of the semiconductor. In electronic activation it occurs through a direct electronic interaction between the metal additives and the semiconductor (Figure 4.16). In a study on $In_2O_3$ nanoparticles layer having Ag nanoparticles, improvements in the sensitivity have been explained on the basis of the energy band diagrams for the $In_2O_3$:$Ag_2O$ nanoparticle interface in air and ethanol ambient, as shown in Figure 4.17. $Ag_2O$ is a p-type semiconductor with a band gap of 1.3 eV and a work function of 5.3 eV (Tjeng et al. 1990). The work function of $Ag_2O$ (5.3 eV) is larger than that of n-type $In_2O_3$ (5.0 eV), which results in the formation of a depletion layer in $In_2O_3$ at the interface (Figure 4.17b) (Tyagi 1984, Jia et al. 2003). Upon exposure to ethanol (a reducing gas), interfacial $Ag_2O$ is reduced to Ag, and the $Ag_2O$–$In_2O_3$ interface transforms to the Ag–$In_2O_3$ interface. Ag has a lower work function (4.3 eV) than that of $In_2O_3$ (Tyagi 1984, Jia et al. 2003). As the Fermi level of Ag is higher than that of $In_2O_3$, electrons from Ag are transported toward $In_2O_3$, resulting in the formation of an accumulation layer at the Ag–$In_2O_3$ nanoparticle interface (Figure 4.17d). The increased gas-sensing response is thus linked with the transformation of a depletion layer at the $Ag_2O$—$In_2O_3$ interface to an accumulation layer in case of Ag—$In_2O_3$ (Bakasybramanian and Subrahmanyam 1991). This explains the high sensor response to ethanol in the presence of Ag. In case of an $In_2O_3$ nanoparticle without Ag attachment, the gas-sensing response depends upon the degree of depletion of electrons from $In_2O_3$ by adsorbed oxygen and the effectiveness of removal of oxygen species by ethanol. In case of $In_2O_3$ having an appropriate amount of Ag, the gas-sensing response depends upon the effectiveness of reduction of $Ag_2O$ to Ag in the presence of a reducing gas (ethanol). Use of Ag nanoparticles is advantageous as it provides a large surface area for interaction with the sensing gas. The fast response time at high temperatures is due

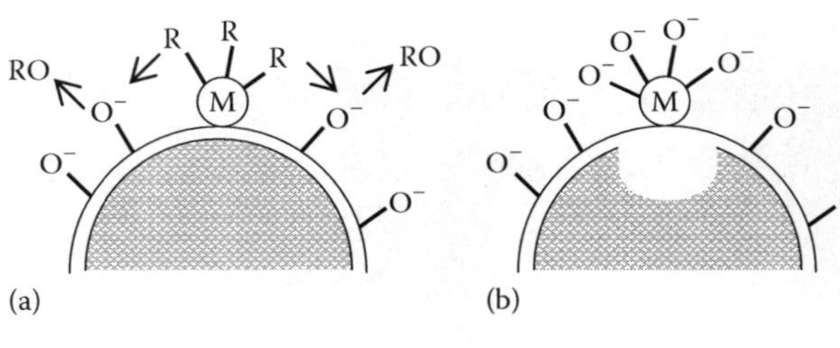


**FIGURE 4.16**
Mechanism of sensitization by metal or metal oxide additive by (a) chemical activation and (b) electronic activation. In chemical activation, the metal particles act like catalyst centers and enhance chemical reaction between the reducing gas molecule and adsorbed oxygen. In electronic activation, metal oxide (additive) particles get converted to metal on reaction with reducing gas. This causes electronic change in the oxide nanoparticles.

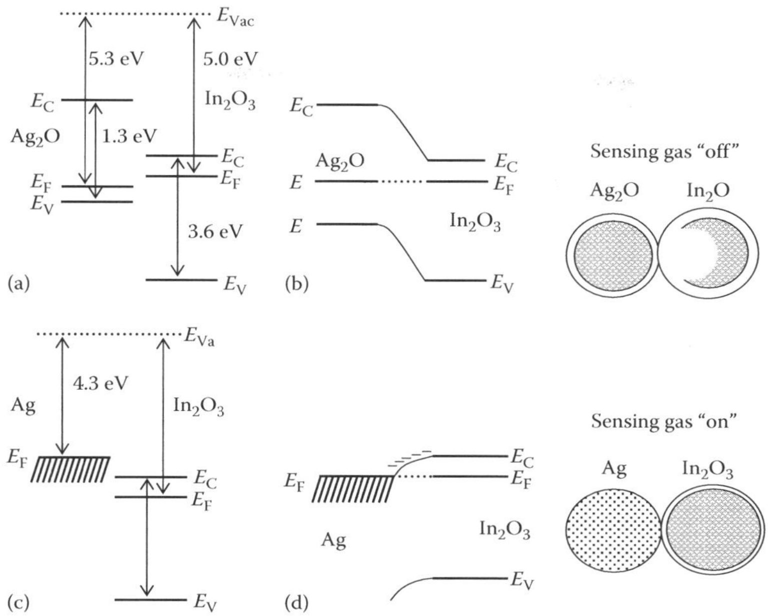


**FIGURE 4.17**
Proposed energy band diagram: (a) $In_2O_3$–$Ag_2O$ (in air) without contact, (b) after contact, (c) $In_2O_3$–Ag (in air + ethanol) without contact, and (d) after contact. $E_c$, $E_v$, and $E_F$ correspond to the conduction band minimum, valence band maximum, and position of Fermi energy. Metal nanoparticle and oxide nanoparticle configuration are also shown. (Reprinted from Singh, V.N. et al., *Sens. Actuators B*, 125, 482, 2007. Copyright 2007, with permission from Elsevier.)

to the fast rate of reaction between ethanol and oxygen at elevated temperatures. The high value of sensor response and the faster response can thus be directly correlated with the electronic interaction and larger surface area due to the nanoparticle nature of both the phases.

#### *4.3.4.4 Some Special Gas Sensor Devices*

Poor selectivity is one of the nagging aspects of oxide nanoparticle sensor. The recent advances in nanomaterial formation have resulted in improved gas sensor response, especially selectivity and drift.

##### *4.3.4.4.1 Sensing Response of a Nanowire (Metal Oxide) Free of Drift*

The possibility of preparing devices based on single nanowires or aligned nanowires opens the prospective to develop sensors free from interfacial region between different nanoparticles, which have been recognized as one of the sources of long-time drift in metal oxide nanoparticulate layers (Comini et al. 2009). Due to the high temperature (usually 200°C–450°C) required to optimize the oxide–gas interaction and transduction mechanisms, oxides usually undergo grain-coarsening phenomena with time. According to XRD and electron microscopy analysis, these phenomena often start from interfacial regions. Nanobelts of semiconducting oxide, with a rectangular cross section in a ribbon-like morphology, are very promising for sensors because their surface-to-volume ratio is very high, the oxide is single crystalline, the faces exposed to the gaseous environment

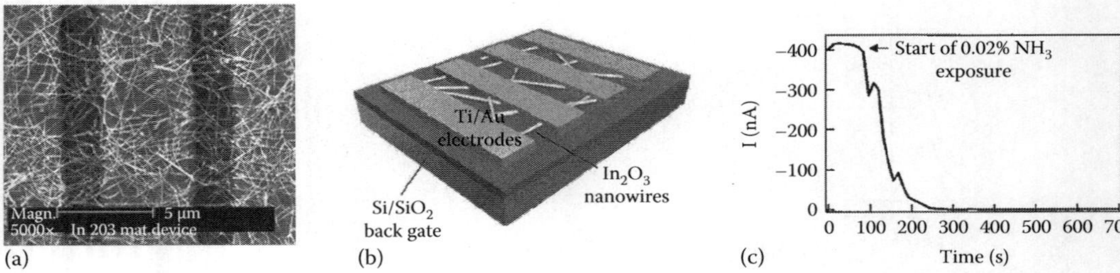

(a) (b) (c)

**FIGURE 4.18**
**(See color insert following page 302.)** (a) SEM image of the $In_2O_3$ nanowires device having electrode structure, (b) multi-nanowire gas sensor device, and (c) response of $In_2O_3$ nanowire sensor to 0.02% $NH_3$. (Reprinted with permission from Zhang, D. et al., *Nano Lett.*, 4, 1919, 2004. Copyright 2004, American Chemical Society; Li, C. et al., *Ann. N. Y. Acad. Sci.*, 1006, 104, 2003. Copyright 2004, American Chemical Society.)

are always the same, and the nano diameter is likely to produce a complete depletion of carriers inside the belt (Wang 2004).

The sensors based on square-shaped $SnO_2$ nanowires do not suffer from long-term drift of conductance. The excellent stability of the sensors could be mainly attributed to the good thermal stability of the materials produced (Qin et al. 2008). The sample was calcined at 600°C for 3 h and investigated using SEM. The SEM observations indicate that the materials have identical morphology before and after being calcined. The conductance of chemical sensors are very stable even after running at 290°C for 8 days (Qin et al. 2008).

#### *4.3.4.4.2 $In_2O_3$ Nanowire Sensors*

Recently, an electronic nose for discriminating hydrogen and carbon monoxide with an array of individual metal oxide nano- and mesowire sensors has been developed on a single chip (Figure 4.18) (Li et al. 2003). In another study, a microarray electronic nose based on percolating $SnO_2$ nanowire sensing elements for discriminating between several reducing gases in air at ppb level has been reported. The discriminating power comes from the temperature gradient across the nanowire layer, density, and morphological inhomogeneities of nanowire mats. The single crystallinity of the nanowire results in reduced aging effects, which are a common drawback for ultrasensitive particulate thin film or nanoparticulate layers. Kolmakov et al. have measured the sensing ability of individual $SnO_2$ nanowires and nanobelts before and after fictionalization with Pd catalyst particles (Kolmakov et al. 2005). Pd-functionalized nanostructures exhibited a dramatic improvement in sensitivity toward oxygen and hydrogen due to the enhanced catalytic dissociation of the molecular adsorbate on the Pd nanoparticle surfaces and the subsequent diffusion of the resultant atomic species to the oxide surface.

#### *4.3.4.4.3 Hybrid Nanowire/Carbon Nanotube Sensor Array*

To enhance the "discrimination power" of important gases, Chen et al. (2009) fabricated a templated structure with four different semiconducting nanostructured materials: $In_2O_3$ nanowires, $SnO_2$ nanowires, ZnO nanowires, and single-walled carbon nanotubes (SWNT). The integration of n-type semiconducting metal oxide nanowires and p-type semiconducting SWNT was carried out to have a large discrimination factor. In addition, the integrated micromachined hot plate enables individual and accurate temperature control for each sensor, thus providing the second discrimination factor. This sensor array was exposed to gases such as hydrogen, ethanol, and nitrogen dioxide at different concentrations and sensing temperatures. Observation of good selectivity has been reported. The data has been used

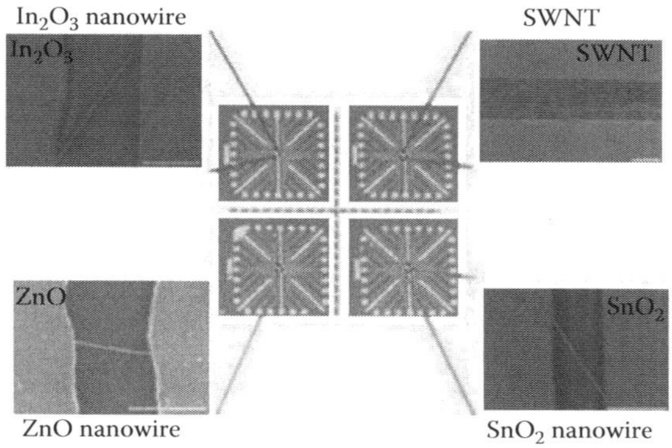


**FIGURE 4.19**
**(See color insert following page 302.)** Hybrid chemical sensor array chip composed of four individual chemical sensors. (Reprinted with permission from Chen, P.-C. et al., *IEEE Trans. Nanotechnol.*, 7, 668, 2008. Copyright 2008, American Institute of Physics.)

to make an interesting "smell-print" data-card for these gases. Selectivity can be further improved by incorporating diverse nanosensor materials into the electronic nose (Zhang et al. 2004, Chen et al. 2008). For instance, using nanowires decorated with Pd nanoparticles may help the discrimination between hydrogen and other gases (Figure 4.19).

## 4.4 Palladium Nanoparticle-Based Hydrogen Sensors

### 4.4.1 Basics of Pd–H Interaction

Pd is the best choice for hydrogen sensing due to its selectivity and sensitivity toward hydrogen. The high diffusion coefficient of hydrogen in Pd is responsible for selectivity of Pd toward H sensing (Lewis 1967). Hydrogen dissolves in palladium and occupies the interstitial sites in fcc Pd lattice. Pd—H interaction takes place in three steps: physisorption, diffusion, and chemisorption (Paál and Menon 1988). On the surface of palladium, hydrogen molecule dissociates into H atoms. H atoms are then physically adsorbed on the Pd surface. On interaction with hydrogen, Pd shows change in electrical, optical, mechanical, and electronic properties. Any of these properties can form the basis of Pd-based hydrogen sensors.

The palladium–hydrogen system exhibits two phases, namely α and β phases. These two phases differ in terms of fraction $x$ in $PdH_x$. Figure 4.20 shows the pressure-lattice constant-temperature plots for the palladium–hydrogen system in case of Pd nanoparticle sample (Khanuja et al. 2009). For alpha phase, $x \leq 0.03$ and lattice constant is 3.889 Å, close to that of bulk Pd. For beta phase, $x \geq 0.6$ and lattice constant is 4.25 Å (Lewis 1967). Switching of $PdH_x$ system from alpha to beta phase depends on the pressure and temperature. The d-band model is quite useful to explain Pd–H interaction. According to d-band model, Pd–H interaction is a strong function of (1) d-band centroid position and (2) degree of orbital overlap between metal (Pd 4d) and adsorbate (H 1s) (Hammer and Nørskov 1995).

Figure 4.21a and b shows the *in situ* x-ray diffractograms of Pd thin film and nanoparticle layer sample at different hydrogen concentrations, $H_c = 2\%$, 5%, and 10%, and substrate temperature, $T_m = 25°C$ (Khanuja et al. 2009a). Pd (111) peak for thin film sample at

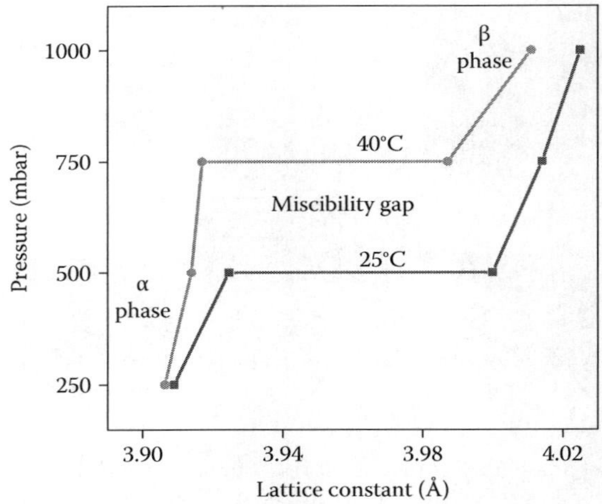


**FIGURE 4.20**
Pressure-lattice constant-temperature plots for the palladium–hydrogen system in Pd nanoparticles at $H_c = 5\%$.

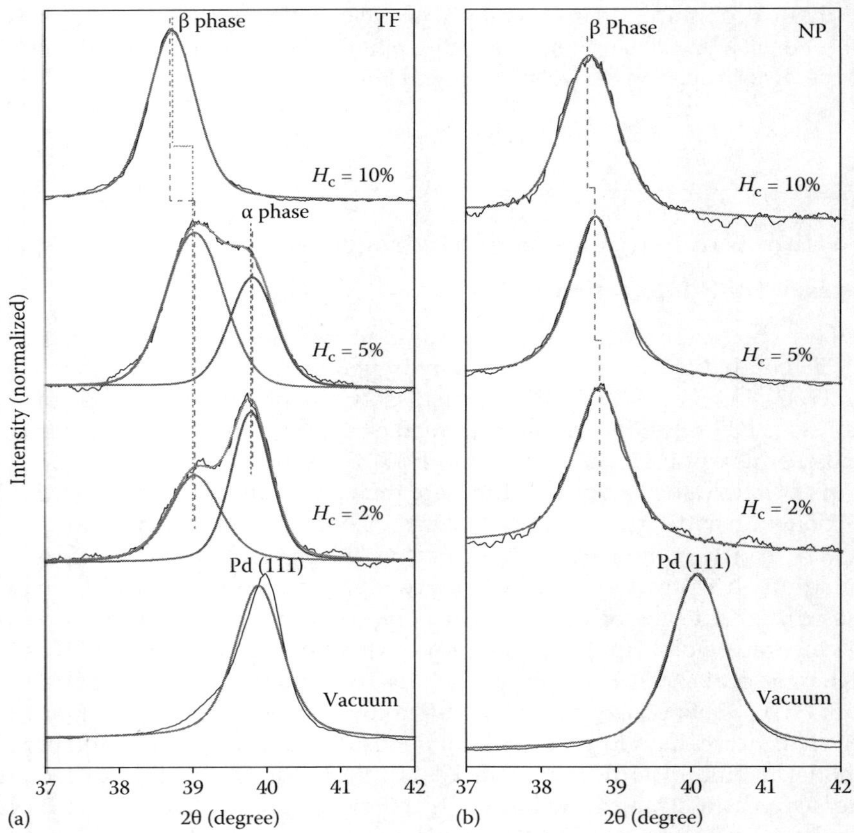


**FIGURE 4.21**
In situ x-ray diffractograms of (a) thin film and (b) nanoparticle samples in vacuum and at hydrogen concentration ($H_c$) = 2%, 5%, and 10% and $T_m = 25°C$. The intensity of these peaks has been normalized. (From Khanuja, M., *J. Appl. Phys.* 106, 093515, 2009a.)

three different $H_2$ concentrations are shown in Figure 4.21a. Hydrogen adsorption causes lattice expansion leading to shift of Pd (111) peak toward lower 2θ value. At $H_c = 2\%$ and 5%, the XRD peak becomes asymmetric, indicating the presence of two hydride phases: α-Pd$H_x$ (α phase) and β-Pd$H_x$ (β phase). At $H_c = 10\%$, only peak corresponding to β phase is observed. Figure 4.21b shows the *in situ* x-ray diffractograms of Pd nanoparticle layer at $H_c = 2\%$, 5%, and 10% and $T_m = 25°C$. In contrast to the thin film sample, only the high hydrogen concentration β phase is observed in nanoparticles in the concentration range of 2%–10%. These results illustrate the enhanced hydrogenation of Pd nanoparticles in comparison to the thin film sample. This is mainly due to the enhanced surface-to-volume ratio; additional surface and subsurface adsorption sites for hydrogen adsorption are present in nanoparticles in addition to the regular interstitial and grain boundary sites (Sachs et al. 2001). It may be noted that microcracks and formation of defects that occur in Pd appear during repeated hydrogen loading and deloading, especially in Pd bulk and thin films.

Pd nanoparticle sensors work on increases in resistivity due to the formation of palladium hydride. Some of the properties of Pd nanoparticles, which make them highly suitable for gas-sensing applications, are described below:

1. A size-induced shift in the position of Pd 4d band centroid has been observed in Pd nanoparticles. According to the d-band model, this should result in the increased interaction between Pd and H (Aruna et al. 2005).
2. Increase in specific surface area with reduction in the nanoparticle dimensions is expected to result in increased solid–gas interaction in general. In case of Pd–H interaction, this is especially advantageous as enhanced physisorption and chemisorptions of hydrogen species is expected to take place via surface and subsurface sites.
3. Nanoparticles have different topography in comparison to thin film or bulk. In contrast to compact topography with crystallites in contact with each other in thin films, nanoparticle layers have slack topography with interparticle voids. The presence of voids is expected to provide space for hydrogen-induced expansion during hydrogen loading and contraction during deloading. This can result in improvement in reproducibility and repeatability of sensor response in Pd nanoparticles in comparison to Pd bulk and thin films or bulk Pd sensors.

### 4.4.2 Pd Nanoparticles-Based H Sensors

A number of H sensors based on different Pd nanostructures have been studied in the literature. The most important examples are (1) Pd nanoparticle layer, (2) size-selected Pd nanoparticles, (3) Pd mesowire array, (4) Pd nanotubes, (5) self-assembled monolayers, (6) functionalized Pd nanostructures, and (7) Pd nanotrench-based sensors.

#### *4.4.2.1 Pulsed Hydrogen-Sensing Response in Pd Nanoparticle Layer*

A comparative study of $H_2$-sensing response of Pd nanoparticle layers and thin film samples has been carried out by studying the electrical resistance change (Khanuja et al. 2007). Thin film samples have been deposited under high vacuum conditions and at substrate temperature of 250°C. Scanning tunneling microscopy (STM) images show well-connected crystallites. Nanoparticle samples are deposited using the inert gas evaporation method in argon ambience at room temperature. STM micrographs of Pd thin film and nanoparticle samples are shown in Figure 4.22.

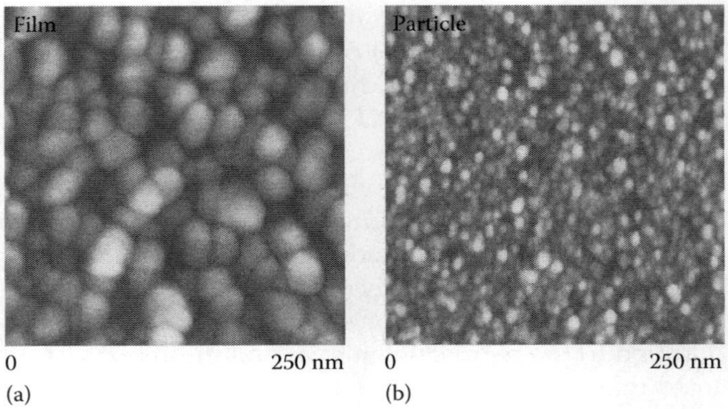


**FIGURE 4.22**
STM (250 × 250 nm²) micrographs showing the surface morphology of Pd samples: (a) thin film and (b) nanoparticle. (Reprinted with permission from Khanuja, M. et al., *Appl. Phys. Lett.*, 91, 253121, 2007. Copyright 2007, American Institute of Physics.)

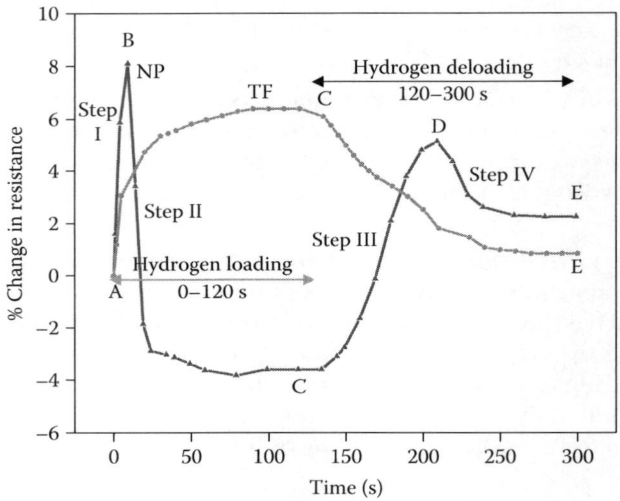


**FIGURE 4.23**
Hydrogen-sensing response of Pd thin film and nanoparticle. (Reprinted with permission from Khanuja, M. et al., *Appl. Phys. Lett.*, 91, 253121, 2007. Copyright 2007, American Institute of Physics.)

Electrical response curves observed during a complete loading and deloading cycle in Pd thin film sample and nanoparticle samples are compared in Figure 4.23. Hydrogen-sensing response has been described in terms of step I and II during $H_2$ loading, and step III and IV during $H_2$ deloading. Step I corresponds to increase in resistance from A → B and step II corresponds to resistance decrease from B → C. Step I arises due to (1) hydrogen acting as scattering centers in Pd lattice and (2) the conversion of Pd to $PdH_x$ (β phase) during hydrogen loading; electrical resistance of $PdH_x$ is higher than that of Pd (Lewis 1967). This is termed as electronic effect (EE). Decrease in resistance on hydrogen loading (step II) can be explained in terms of decrease in the interparticle gaps on lattice expansion due to $PdH_x$ formation. This is termed as geometric effect (GE). This arises as Pd to $PdH_x$ formation is accompanied by 6% lattice expansion. As hydrogen gets removed from $PdH_x$

during hydrogen deloading, inverse of steps II and I take place in the form of steps III and IV, respectively. Resistance increase (step III) is related to the opening of the interparticle gaps as Pd nanoparticles tend to return to their original size. On subsequent deloading, resistance decreases (step IV) as palladium hydride is finally converted back to palladium. Thus, steps I and IV arise due to the reversible $Pd \leftrightarrows PdH_x$ transformation and steps II and III arise because of the closing and opening of the interparticle gaps due to nanoparticle expansion and contraction, respectively. Figure 4.24 shows the schematic diagram of Pd nanoparticles before and after exposure to $H_2$ and the corresponding conduction mechanism during hydrogen loading. In nanoparticles, current transport is through (1) activated electron transport across interparticle barrier and (2) metallic conduction through the Pd nanoparticles (Khanuja et al. 2007). On hydrogen loading, a change in conductivity of Pd nanoparticles due to PdH formation is observed. This is followed by decrease in interparticle gaps.

In case of thin film samples, crystallites are well-connected and thus steps II and III are completely absent or highly suppressed. In the absence of interparticle gaps or voids, the lattice expansion due to hydride formation results in a strained crystal structure in thin film samples. This slows down the hydrogen loading and deloading process during $Pd \leftrightarrows PdH_x$ transformation. In-plane strain in Pd thin films deposited onto a substrate is known to slow down the hydrogenation process due to substrate clamping effects. Hydrogen-induced embrittlement in thin films and bulk Pd is also due to excessive lattice strain (Lewis 1967). The difference in electrical response of Pd nanoparticle layers in comparison to thin film samples has been explained as due to (1) faster reactivity and (2) presence of interparticle gaps in nanoparticle samples (Khanuja et al. 2007). In principle, other physical deposition techniques are also expected to give similar results.

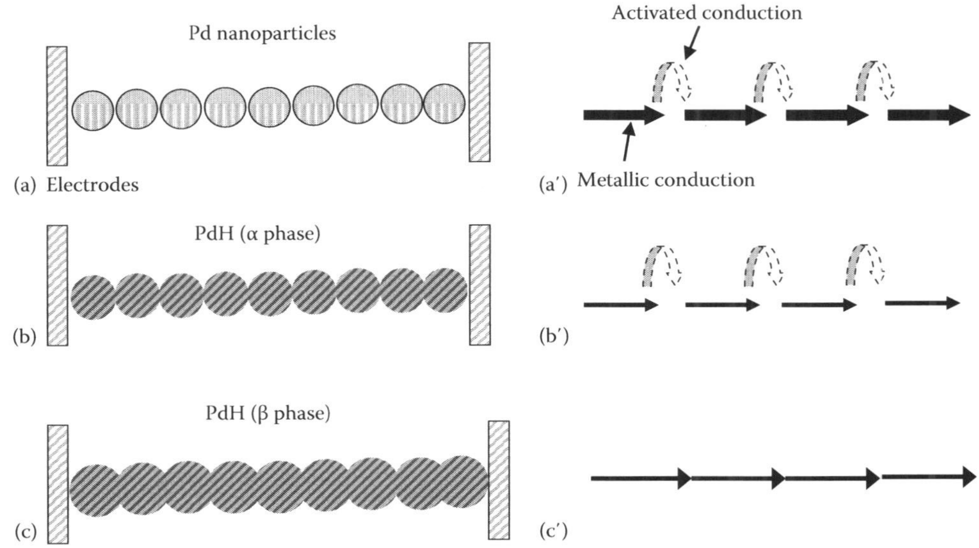


**FIGURE 4.24**
Schematic diagram showing Pd nanoparticles (a) before exposure to hydrogen, (b) initial state of PdH (α phase), and (c) final state of PdH (β phase), and (a′, b′, c′) corresponding conduction mechanisms during hydrogen loading in Pd nanoparticles. Dotted and solid arrows show activated electron transport across interparticle barrier and metallic conduction through the Pd nanoparticles, respectively.

Lith et al. have used a sensor based on electron tunneling between neighboring Pd clusters (Lith et al. 2007). Pd clusters having diameters ranging from 3.5 to 6 nm are deposited by the magnetron sputtering technique between a pair of contacts until a predetermined resistance between the contacts is obtained. It has been demonstrated that the conduction in the cluster film is dominated by electron transport through tunneling gaps. Upon exposure to hydrogen, the clusters expand, reducing the average gaps, and resistance decreases. The sensor response can be modeled by taking into account different step in the Pd–H interaction: (1) change in hydrogen pressure results in a change $\Delta\alpha$ in the lattice constant, $\alpha$; (2) change in lattice constant changes the cluster diameter, $d$; (3) a change in cluster size changes the mean tunneling gap; and (4) change in the mean tunneling gap changes the resistance, $R$. Sensitivity is thus given by

$$\frac{R_0 - R_{(\Delta\alpha)}}{R_0} = 1 - e^{-(\beta p d/\alpha)\Delta\alpha}$$

where

$\beta$ is the system-dependent tunneling constant given by $\beta = \sqrt{8mU_0/\hbar^2}$

$p$ is the average number of clusters aligning a segment of the conduction path within a group with the cluster diameter $d$

### 4.4.2.2 Concentration-Specific Sensing Response in Size-Selected Pd Nanoparticles

Gas phase synthesis method is an important method to prepare well-defined nanoparticles having well-defined sizes and narrow size distribution. In a study carried out by authors, Pd nanoparticles have been synthesized using an integrated process system described earlier (Khanuja et al. 2009).

TEM micrographs of three samples of size 15, 20, and 25 nm prepared by varying DMA voltage clearly show well-formed and spherical Pd nanoparticles, as shown in Figure 4.25. HRTEM image of an individual nanoparticle from the sample P15 is shown in Figure 4.25d. The image clearly shows quasi-spherical and monocrystalline nature of Pd nanoparticles with interplanar spacing of 0.242 nm, which corresponds to the (111) plane of the fcc structure of bulk Pd as marked in Figure 4.25d.

The hydrogen-sensing response of these samples has been studied at different $H_2$ concentrations and measurement temperatures, as shown in Figure 4.26. Two types of sensing behaviors are observed as a function of $H_2$ concentration: (1) a normally observed "saturated" response and (2) a "pulsed" response (Khanuja et al. 2009). As already described, in the slow saturated response, electrical resistance slowly increases on exposure to hydrogen gas and saturated with time. In the pulsed response, the electrical resistance suddenly increase followed by a sharp decrease on exposure to hydrogen gas. The above behavior can be explained in terms of the EE and GE. At 20°C, pulsed response is observed at $H_2$ concentration ≥2.5%; and at $H_2$ concentration ≤2.0%, the saturated response is observed. At low $H_2$ concentrations, small changes in the nanoparticle size are insufficient for causing significant changes in the interparticle gaps. Thus, occurrence of both the EE and GE is observed to be a prerequisite for observing pulse-like sensing response. The variation of sensitivity ($\Delta R/R$%) and response time at 20°C corresponding to the EE and GE as a function of $H_2$ concentration is shown in Figure 4.27. The sensitivity and response time due to GE is larger in comparison to EE at all $H_2$ concentrations. With increase in $H_2$ concentration, the sensitivity due to both the effects increases and response time decreases. The

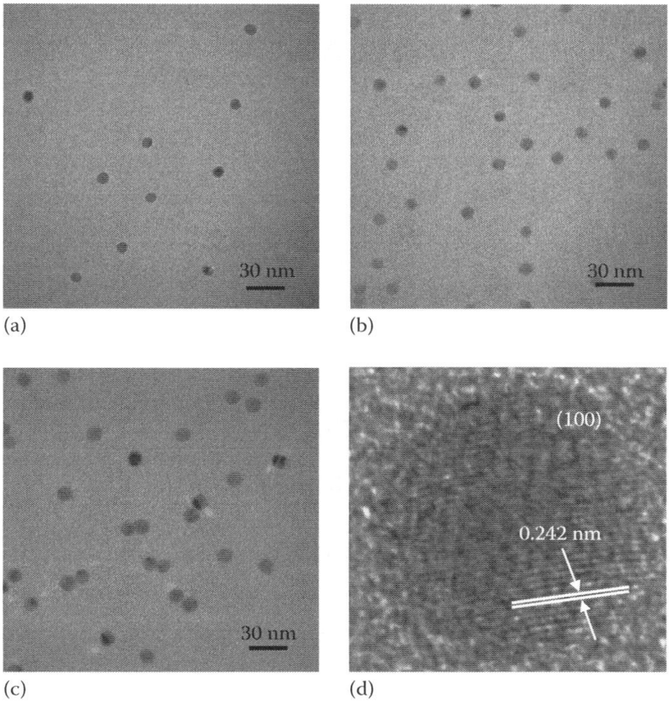


**FIGURE 4.25**
TEM images of samples (a) P15, (b) P20, (c) P25, and (d) shows the HRTEM image of sample P15. (Reprinted from Khanuja, M. et al., *Nanotechnology*, 20, 015502, 2009b. With permission from Institute of Physics.)

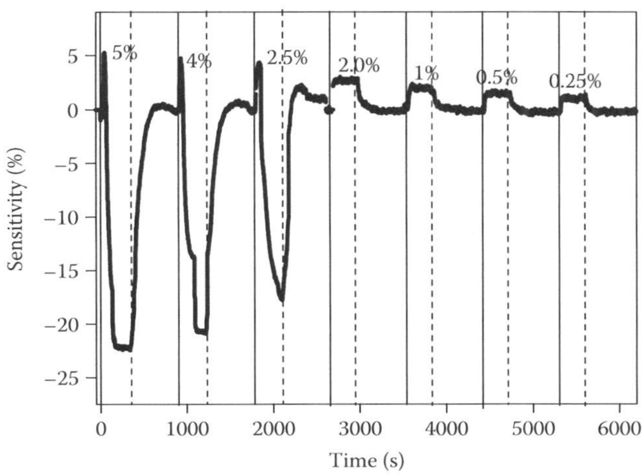


**FIGURE 4.26**
Sensing response for sample P15 at 20°C for different concentrations of $H_2$. Solid and dotted lines represent $H_2$ "on" and "off" states, respectively. (Reprinted from Khanuja, M. et al., *Nanotechnology*, 20, 015502, 2009b. With permission from Institute of Physics.)

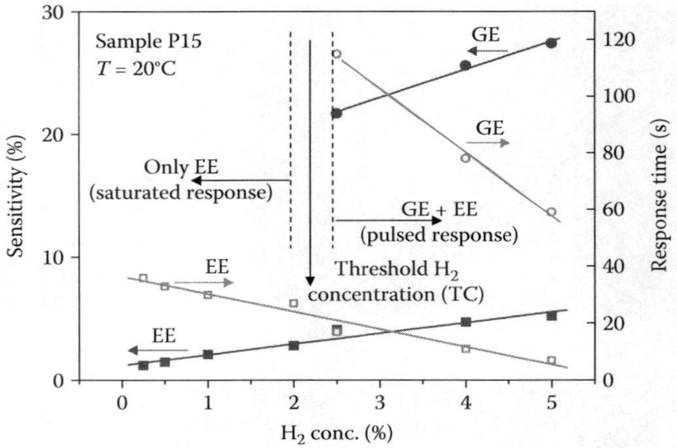


**FIGURE 4.27**
Variation of sensitivity and response time as a function of $H_2$ concentration for sample P15 at 20°C temperature. TC represents threshold $H_2$ concentration required for transition from saturated response to pulsed response. Response time is defined as the time taken to reach 90% of the final value of resistance. (Reprinted from Khanuja, M., et al., *Nanotechnology*, 20, 015502, 2009b. With permission from Institute of Physics.)

electronic phenomenon governing changes in resistance due to incorporation of H scattering centers at the Pd surface is a relatively fast process.

Size-dependent variation of sensitivity and response time is shown in Figure 4.28. With decrease in nanoparticle size, sensitivity due to EE becomes larger and response time becomes smaller (Khanuja et al. 2009).

Hydrogen-sensing response as a function of temperature for different concentrations of $H_2$ is also studied. It was observed that there is a threshold concentration below which the response is saturated and is highly temperature dependent. This is due to a smaller amount of hydrogen getting incorporated into Pd due to reduced physical adsorption at higher temperature (Lewis 1967, Paál and Menon 1988). Table 4.5 shows the percentage

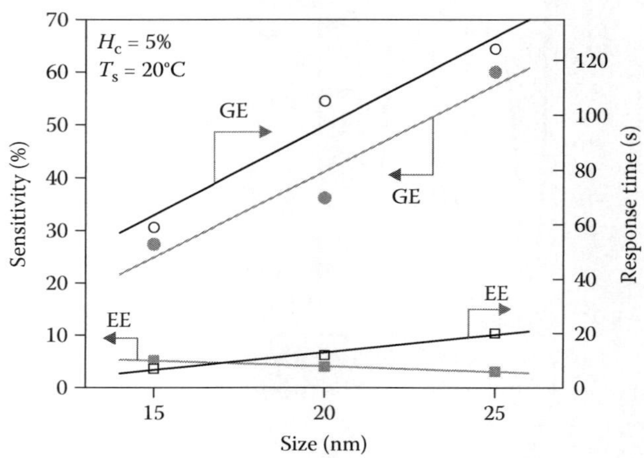


**FIGURE 4.28**
Variation of sensitivity and response time as a function of nanoparticle size at $H_c = 5\%$ and $T_m = 20°C$.

**TABLE 4.5**

Percentage Change in Resistance as a Function of $H_2$ Concentration for Different Temperatures in Monodisperse Pd Nanoparticle Sample

| Hydrogen Concentration ($H_c$) | Resistance Change (%) 20°C EE | 20°C GE | 40°C EE | 40°C GE | 60°C EE | 80°C EE |
|---|---|---|---|---|---|---|
| 5% | 5.2 | 27.4 | 4.5 | 17.8 | 3.6 | 2.6 |
| 4% | 4.7 | 25.6 | 4.3 | 13.9 | 3.3 | 2.2 |
| 2.5% | 4.0 | 21.7 | 3.4 | | 2.2 | 1.7 |
| 2% | 2.8 | | 2.2 | | 1.5 | 1.2 |
| 1% | 2.1 | | 1.4 | | 1.0 | 0.7 |
| 0.5% | 1.5 | | 1.1 | | 0.9 | 0.6 |
| 0.25% | 1.1 | | 1.0 | | 0.7 | 0.5 |

change in resistance as a function of $H_2$ concentration at 20°C, 40°C, 60°C, and 80°C temperatures. In Table 4.1, the rectangular region shows the region where pulse-like sensing response is obtained due to the coexistence of both EE and GE. At 20°C, EE and GE both occur for $H_2$ concentration ≥2.5%; and at 40°C, EE and GE coexist for $H_2$ concentration ≥4.0%. At 60°C and 80°C, GE is absent. It is quite clear that pulse-like response is highly concentration and temperature dependent. At higher temperatures and lower $H_2$ concentration, sensitivity of both EE and GE decreases.

The observed dependence of threshold concentration on temperature is quite useful from the application point of view. A measurement of the temperature, at which the saturated response changes to pulsed response, can give important information about the $H_2$ concentration level. It is important to note that the explosive limit of 4% $H_2$ falls in the concentration range (2%–10%) that can be detected by the proposed methodology. It is important to note that this type of sensing behavior has double advantage of fast response of EE and high sensitivity of GE.

#### *4.4.2.3 Break Junction Phenomenon in Pd Mesowire Array*

In an intensive study, in situ AFM analysis on Pd mesowire arrays has been carried out and results are shown in Figure 4.29. These AFM investigations show that, in air, Pd nanowire has a break junction or open gaps (Figure 4.29a) and in the presence of hydrogen, Pd grains swell leading to closing of gaps (Figure 4.29b). In situ AFM studies provide the direct evidence of "break junction mechanism" or "hydrogen-induced lattice expansion (HILE)," which causes decrease in resistance contrary to increased resistivity due to palladium hydride formation. Figure 4.30 shows the hydrogen-sensing response in Pd mesowire array. It is shown that the initial exposure to hydrogen causes closing of gaps leading to increase in conductance. When hydrogen gas is removed, nano gaps open up again leading to decrease in current. However, there is a threshold concentration ranging from 1% to 2%, which can be detected by the Pd mesowires. This limit can be explained as 1% is the threshold $H_2$ concentration required for transition from alpha to beta phase. Pd to $PdH_x$ formation results in change in lattice expansion of about 4% in beta phase that is responsible for closing of gaps.

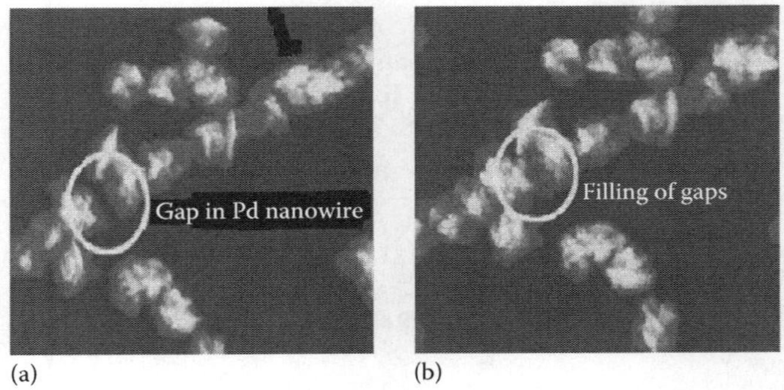


**FIGURE 4.29**
**(See color insert following page 302.)** In situ AFM of Pd mesowire in (a) air and (b) hydrogen. (From Walter, E.C., *Anal. Chem.* 74, 1546, 2002. With permission.)

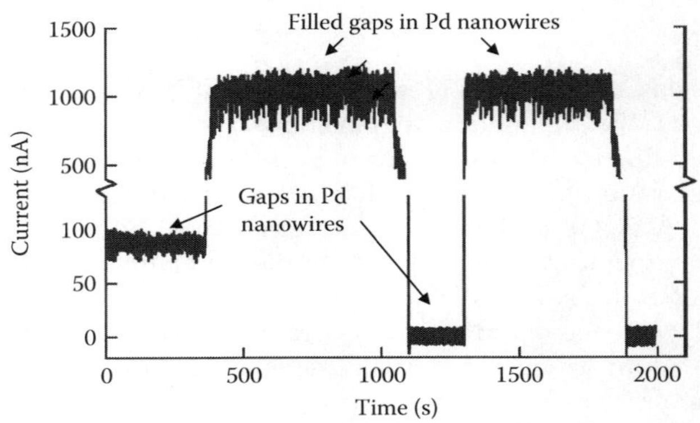


**FIGURE 4.30**
Hydrogen-sensing response in Pd mesowire array. (From Walter, E.C., *Anal. Chem.* 74, 1546, 2002. With permission.)

### 4.4.2.4 Pd Nanotubes

A number of techniques have been used for synthesizing Pd nanotubes. In one study, synthesis was done within the pores or edges of the template by electrochemical reduction of the $Pd^{2+}$ ions. Arrays of palladium nanotubes grown within the template membranes are shown in Figure 4.31. The electrochemical growth of single Pd wire having 1 μm width between lithographically patterned Au contact electrodes has also been reported (Yu et al. 2005a,b).

Hydrogen-sensing response has been studied for different sample geometries. It is observed that given geometry has the highest sensitivity at all values of hydrogen concentration. At 1% $H_2$ concentration, sensitivity is 11.6%; and at 0.5% $H_2$, sensitivity is around 2. This is due to the fact that resistance increase is proportional to the fraction of the surface that is covered by the hydrogen (θ) and can be described as follows:

$$\frac{\Delta\alpha}{R_0} \propto \theta = \frac{Kp}{1+Kp}$$

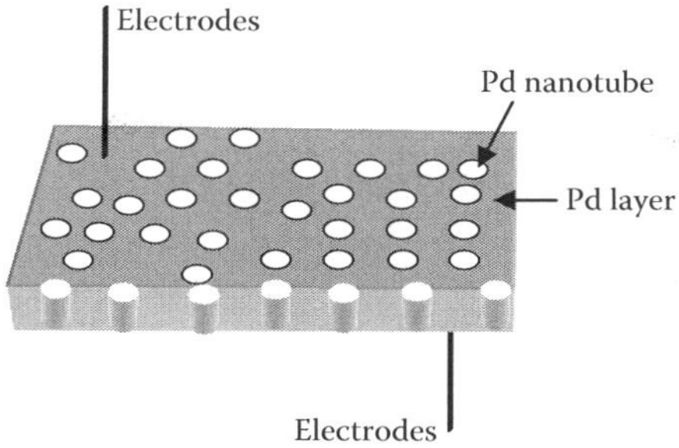


**FIGURE 4.31**
Arrays of Pd nanotube grown on template. (Reprinted with permission from Yu, M.-F. et al., *IEEE Sens. J.*, 5, 20, 2005a. Copyright 2005, American Chemical Society.)

where
- $K$ is the equilibrium constant defined as the adsorption constant ($k_1$) divided by the desorption constant ($k_{-1}$)
- $p$ denotes gas pressure (Paál and Menon 1988)

In this geometry, maximum surface area is exposed for H interaction and hence shows maximum sensitivity.

#### 4.4.2.5 Self-Assembled Pd Monolayers

Xu et al.'s work has demonstrated that a siloxane self-assembled monolayer onto a substrate promotes the formation of small Pd nanoclusters and reduces the stiction between the palladium and the substrate (Xu et al. 2005). Figure 4.32 shows the conductance response of 3.3 nm thick Pd, and inset shows the AFM micrograph of Pd nanoclusters deposited on siloxane-coated glass. The Pd nanocluster film can detect 2% $H_2$ with a rapid response time of 70 ms and is sensitive to 25 ppm hydrogen.

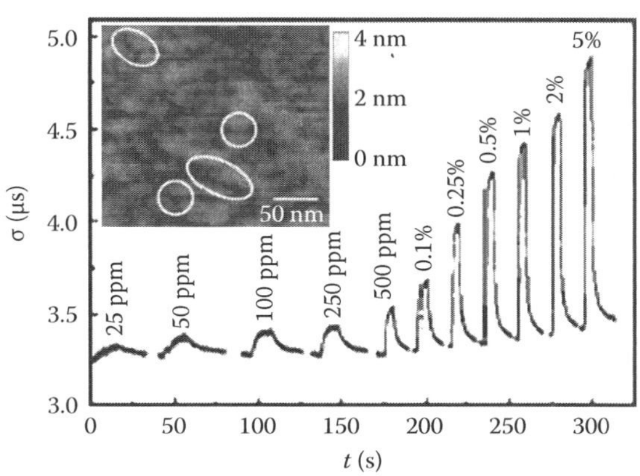


**FIGURE 4.32**
Conductance response of 3.3 nm thick Pd deposited on siloxane coated glass (SCG). (Reprinted with permission from Yu, M.-F. et al., *IEEE Sens. J.*, 5, 20, 2005a. Copyright 2005, American Institute of Physics.)

#### *4.4.2.6 Pd Nanoparticle-Based Composite Hydrogen Sensors*

Pd nanoparticles have been used to functionalize silicon nanowires, carbon nanotubes (CNTs), and tungsten nanowires. The composite structures have been used as hydrogen sensors. One of the advantages of a functionalized sensor based on CNTs is the flexibility. These sensors are of low cost, large area, light weight, and possess mechanical flexibility and mechanical shock resistance. It is expected that these sensors will have better stability as chances of degradation on nanostructures due to repeated loading and deloading are minimum.

In a reported study, a thin layer of Pd nanoparticles has been deposited on SiNWs (Chen et al. 2007). This sensor shows higher and faster sensitivity toward hydrogen than the macroscopic Pd wire. The Pd-functionalized Si NW were dispersed onto a $SiO_2$/Si substrate to form a uniform layer. No signal change is observed in the presence of $NH_3$ or $N_2O$. The current signal of the sensor is observed to increase by about 20 times and response time is 3 s.

In another study, it has been shown that an excellent hydrogen sensor can be fabricated by depositing Pd nanoparticles onto CNTs (Mubeen et al. 2007). The underlying mechanism is the interaction between $H_2$, Pd, and CNT. Adsorption of hydrogen at the Pd surface lowers the work function of Pd. This causes electron transfer from Pd to SWNT, which lowers the hole carriers in p-type CNT. In an interesting study, a sensor based on SWNT decorated with palladium nanoparticles has been developed. It has been reported that this sensor can detect hydrogen as low as 30 ppm in air at room temperature. Deposition of Pd nanoparticles over CNT provides excellent mechanical flexibility.

InN is a suitable hydrogen selector material because of its narrow band gap and good electron transport characteristics. Pd-coated InN nanobelts exhibit higher and faster sensitivity to hydrogen (Wright et al. 2009). Response time in Pd–InN nanobelts is <2 s.

It has been shown that Pd coating on GaN nanowires increases their sensitivity toward hydrogen by 11 times (Lim et al. 2008). A 10 nm thick Pd was deposited by sputtering onto the GaN nanowires. These sensors are working in the $H_2$ concentration range 200–3000 ppm in $N_2$ at 25°C–150°C. It is reported that GaN nanowires are more sensitive to hydrogen than ZnO.

#### *4.4.2.7 Nanotrench*

It has been reported that nanogap-based hydrogen sensors were prepared using top-down approach (Kiefer et al. 2008). Focused ion beam (FIB) technique has been used to create trenches of widths (10, 20, 30, and 40 nm) in palladium microwires, as shown in Figure 4.33. Pd microwires are fabricated by e-beam evaporation followed by a liftoff process. The widths of the trenches were estimated using scanning electron microscopy. The effect of substrate, wire thickness, and milling time on the sensor characteristics has been investigated. Figure 4.34 shows the schematic diagram of illustrating two opposing effects of current on Pd. As already explained in case of size-selected Pd nanoparticles, EE that causes increase in resistance is due to Pd to $PdH_x$ conversion and GE that causes decreases in resistance is due to lattice expansion during Pd to $PdH_x$ formation leading to closing of interparticle gaps.

Palladium-based hydrogen sensors that work on change in optical, mechanical, surface acoustical wave (SAW), piezoelectric, etc., properties have been reported. (Farahi et al. 1987, Huang et al. 2005a,b, Jakubik et al. 2006, Vincenti et al. 2007). In most of these devices, Pd thin films or bulk Pd are used. As a result, these sensors show large response time and poor sensitivity as compared to Pd nanoparticle-based resistive sensors.

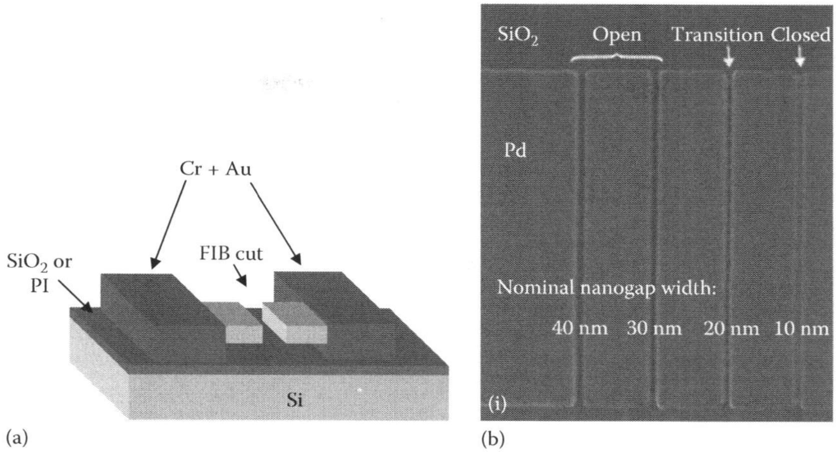


**FIGURE 4.33**
(a) Schematic diagram of sensing device and (b) (i) FIB (30 kV, 1 pA, milling time 104 s cuts trenches of four widths 40, 30, 20, and 10 nm). (From Kiefer, T. et al., *Nanotechnology*, 19, 125502, 2008. With permission from Institute of Physics.)

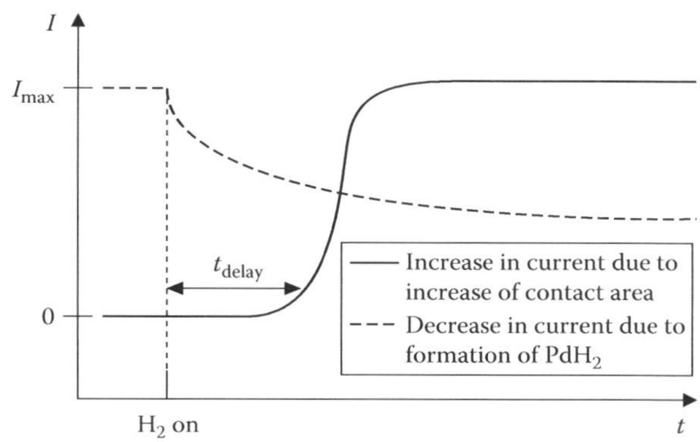


**FIGURE 4.34**
Schematic diagram of opposing effects that contribute to change in current in mesowire array. EE causes decrease in current due to $PdH_x$ formation, and GE caused increase in current due to conducting path formation. (From Kiefer, T. et al., *Nanotechnology*, 19, 125502, 2008. With permission from Institute of Physics.)

### 4.4.2.8 Applications of Hydrogen Sensors

Hydrogen sensors are important for the safe deployment of all hydrogen-based devices like fuel cells, nuclear reactors, space crafts, etc. This is due to the fact that 4% of hydrogen–air mixture causes explosion. The fact that hydrogen is colorless, odorless, and of low mass makes its detection a challenging task. In nuclear reactors, there is a need to frequently measure hydrogen content that can otherwise explode if not properly contained. In nuclear reactor, hydrogen sensor should be capable of measuring hydrogen at high temperatures (in the range of 1650°C) and pressure ranging from less than 1 atmosphere to thousands of atmospheres in the presence of fission products and steam. In fuel cells, two types of hydrogen sensors are required; the first will measure the quality of the hydrogen-fed gas and second, more important, hydrogen sensor for leak detection. These sensors operate from −50°C to 65°C.

## 4.5 Summary

There is a large volume of data on the uses of metal oxide nanoparticles and nanoparticle layers for gas sensor applications. Lack of accurate and reliable information about the nanoparticle size, size distribution, metal additive, composition, and configuration makes the analysis of this data a challenging task. For a clear understanding of the dependence of gas-sensing properties on the nanoparticle size and nature of the metal additive, it is important to use synthesis methods suitable for yielding well-defined nanoparticle sizes and composite configuration. As described above, there have been some efforts in this direction. Difficulty in measuring the electronic properties of nanoparticle layers (in comparison to thin films) makes it difficult to quantify the gas sensing results in terms of semiconductor parameters. Use of metal oxide nanowires seems to reduce the nagging problem of drift in resistive-sensing response. Use of nanowires decorated with nanoparticles may result in improved sensitivity. In case of Pd nanoparticle sensors, use of nanoparticles has "overall" advantages over thin films or bulk sensors. Interparticle gaps and voids provide flexibility to take care of the structural changes during hydrogen loading and deloading. Sensing response of Pd nanoparticles-based sensors is quite well understood in terms of electronic and structural changes. It may be possible to use Pd nanoparticles-based sensor at lower temperature. Some of the current research directions are (1) use of synthesis methods for well-defined nanoparticles sizes, (2) reliable and nanoscale electronic characterization of nanoparticles using conducting atomic force microscopy and STM on gas exposure, (3) fabrication of nanowire-nanoparticle or decorated nanowire composites, and (4) use of nanoparticle deposited in MEMS structures and FIB-fabricated structures.

## References


Aruna, I., Mehta, B.R., and Malhotra, L.K. 2005. Faster H recovery in Pd nanoparticle layer based Gd switchable mirrors, *Appl. Phys. Lett.* 87: 103101.

Bakasybramanian, N. and Subrahmanyam, A. 1991. Studies on evaporated indium tin oxide (ITO)/silicon junctions and an estimation of ITO work function, *J. Electrochem. Soc.* 138: 322–324.

Benkstein, K.D. and Semancik, S. 2006. Mesoporous nanoparticle $TiO_2$ thin films for conductometric gas sensing on microhotplate platforms, *Sens. Actuators B* 113: 445–453.

Buso, D., Post, M., Cantalini, C., Mulvaney, P., and Martucci, A. 2008. Gold nanoparticle-doped $TiO_2$ semiconductor thin films: Gas sensing properties, *Adv. Funct. Mater.* 18: 3843–3849.

Cabot, A., Arbiol, J., Morante, J.R., Weimar, U., Bârsan, N., and Göpel, W. 2000. Analysis of the noble metal catalytic additives introduced by impregnation of as obtained $SnO_2$ sol-gel nanocrystals for gas sensors, *Sens. Actuators B* 70: 87–100.

Cantalini, C., Post, M., Buso, D., Guglielmi, M., and Martucci, A. 2005. Gas sensing properties of nanocrystalline NiO and $Co_3O_4$ in porous silica sol–gel films, *Sens. Actuators B* 108: 184–192.

Cavicchi, R.E., Walton, R.M., Aquino-Class, M., Allen, J.D., and Panchapakesan, B. 2001. Spin-on nanoparticle tin oxide for microhotplate gas sensors, *Sens. Actuators B* 77: 145–154.

Chang, S.C. 1980. Oxygen chemisorption on tin oxide: Correlation between electrical conductivity and EPR measurements, *J. Vac. Sci. Technol.* 17: 366–369.

Chen, Z.H., Jie, J.S., Luo, L.B., Wang, H., Lee, C.S., and Lee, S.T. 2007. Applications of silicon nanowires functionalized with palladium nanoparticles in hydrogen sensors, *Nanotechnology* 18: 345502.

Chen, P.-C., Shen, G., and Zhou, C. 2008. Chemical sensors and electronic noses based on 1-D metal oxide nanostructures, *IEEE T. Nanotechnol.* 7: 668–682.
Chen, P.-C.N., Ishikawa, F., Chang, H.-K., Ryu, K., and Zhou, C. 2009. A nanoelectronic nose: A hybrid nanowire/carbon nanotube sensor array with integrated micromachined hotplates for sensitive gas discrimination, *Nanotechnology* 20: 125503.
Chopra, K.L. 1969. *Thin Film Phenomena*. New York: McGraw-Hill.
Chung, W.-Y., Sakai, G., Shimanoe, K., Miura, N., Lee, D.-D., and Yamazoe, N. 1998. Preparation of indium oxide thin film by spin-coating method and its gas-sensing properties, *Sens. Actuators B* 46: 139–145.
Comini, E., Cristalli, A., Faglia, G., and Sberveglieri, G. 2000. Light enhanced gas sensing properties of indium oxide and tin dioxide sensors, *Sens. Actuators B* 65: 260–263.
Comini, E., Baratto, C., Faglia, G., Ferroni, M., Vomiero, A., and Sberveglieri, G. 2009. Quasi-one dimensional metal oxide semiconductors: Preparation, characterization and application as chemical sensors, *Prog. Mater. Sci.* 54: 1–67.
Devi, G.S., Subrahmanyam, V.B., Gadkari, S.C., and Gupta, S.K. 2006. $NH_3$ gas sensing properties of nanocrystalline ZnO based thick films, *Anal. Chim. Acta* 568: 41–46.
Dolbec, R. and El Khakani, M.A. 2007. Sub-ppm sensitivity towards carbon monoxide by means of pulsed laser deposited $SnO_2$:Pt based sensors, *Appl. Phys. Lett.* 90: 173114.
Du, X., Wang, Y., Mu, Y., Gui, L., Wang, P., and Tang, Y. 2002. A new highly selective $H_2$ sensor based on $TiO_2$/PtO–Pt dual-layer films, *Chem. Mater.* 14: 3953–3957.
Dutta, A., Das, D., Grilli, M.L., Bartolomeo, E.D., Traversa, E., and Chakravorty, D. 2003. Preparation of sol-gel nano-composites containing copper oxide and their gas sensing properties, *J. Sol-Gel Sci. Technol.* 26: 1085–1089.
Ederth, J., Smulko, J.M., Kish, L.B., Heszler, P., and Granqvist, C.G. 2006. Comparison of classical and fluctuation-enhanced gas sensing with $Pd_xWO_3$ nanoparticle films, *Sens. Actuators B* 113: 310–315.
Eranna, G., Joshi, B.C., Runthala, D.P., and Gupta, R.P. 2004. Oxide materials for development of integrated gas sensors—A comprehensive review, *Crit. Rev. Solid State Mater. Sci.* 29: 111–188.
Faglia, G., Allieri, B., Comini, E., Depero, L.E., Sangaletti, L., and Sberveglieri, G. 1999. Electrical and structural properties of RGTO-$In_2O_3$ sensors for ozone detection, *Sens. Actuators B* 57: 188–191.
Farahi, F., Leilabady, P.A., Jones, J.D.C., and Jackson, D.A. 1987. Interferometric fibre-optic hydrogen sensor, *J. Phys. E: Sci. Instrum.* 20: 432–434.
Gaiduk, P.I., Kozjevko, A.N., Prokopjev, S.L., Tsamis, C., and Larsen, A.N. 2008. Structural and sensing properties of nanocrystalline $SnO_2$ films deposited by spray pyrolysis from a $SnCl_2$ precursor, *Appl. Phys. A* 91: 667–670.
Galdikas, A., Kancleris, Ž., Senuliene, D., and Šetkus, A. 2003. Influence of heterogeneous reaction rate on response kinetics of metal oxide gas sensors: Application to the recognition of an odour, *Sens. Actuators B* 95: 244–251.
Gopel, W. 1994. New material and transducers for chemical sensors, *Sens. Actuators B* 18: 1–21.
Gurlo, A. and Ivanovskaya, M. 1997. Grain size control in nanocrystalline $In_2O_3$ semiconductor gas sensors, *Sens. Actuators B* 44: 327–333.
Gurlo, A., Barsan, N., Weimar, U., Ivanovskaya, M., Taurino, A., and Siciliano, P. 2003. Polycrystalline well-shaped blocks of indium oxide obtained by the sol-gel method and their gas-sensing properties, *Chem. Mater.* 15: 4377–4383.
Hammer, B. and Nørskov, J.K. 1995. Why gold is the noblest of all the metals, *Nature* 376: 238–240.
Hoel, A., Reyes, L.F., Saukko, S., Heszler, P., Lantto, V., and Granqvist, C.G. 2005. Gas sensing with films of nanocrystalline $WO_3$ and Pd made by advanced reactive gas deposition, *Sens. Actuators B* 105: 283–289.
Hoffheins, B. 1996. Solid state, resistive gas sensors. In *Handbook of Chemical and Biological Sensors*, eds. R.F. Taylor and J.S. Schultz, Philadelphia, PA: Institute of Physics, pp. 371–398.
Hong, S.-J. and Han, J.-I. 2004. Effect of low temperature composite catalyst loading (LTC2L) on sensing properties of nano gas sensor, *Sens. Actuators A* 112: 80–86.

Huang, J., Matsunaga, N., Shimanoe, K., Yamazoe, N., and Kunitake, T. 2005a. Nanotubular $SnO_2$ templated by cellulose fibers: Synthesis and gas sensing, *Chem. Mater*. 17: 3513–3518.

Huang, X.M.H., Manolidis, M., Jun, S.C., and Hone, J. 2005b. Nanomechanical hydrogen sensing, *Appl. Phys. Lett*. 86: 143104.

Huo, L.H., Li, X.L., Li, W., and Xi, S.Q. 2000. Gas sensitivity of composite Langmuir-Blodgett films of $Fe_2O_3$ nanoparticle-copper phthalocyanine, *Sens. Actuators B* 71: 77–81.

Ionescu, R., Hoel, A., Granqvist, C.G., Llobet, E., and Heszler, P. 2005. Low-level detection of ethanol and $H_2S$ with temperature-modulated $WO_3$ nanoparticle gas sensors, *Sens. Actuators B* 104: 132–139.

Ivanovskaya, M. and Bogdanov, P. 1998. Effect of $Ni^{II}$ ions on the properties of $In_2O_3$-based ceramic sensors, *Sens. Actuators B* 53: 44–53.

Ivanovskaya, M., Bogdanov, P., Faglia, G., and Sberveglieri, G. 2000. The features of thin film and ceramic sensors at the detection of CO and $NO_2$, *Sens. Actuators B* 68: 344–350.

Ivanovskaya, M., Kotsikau, D., Faglia, G., and Nelli, P. 2003. Influence of chemical composition and structural factors of $Fe_2O_3/In_2O_3$ sensors on their selectivity and sensitivity to ethanol, *Sens. Actuators B* 96: 498–503.

Jakubik, W.P., Urbanczyk, M.W., and Maciak, E. 2006. Palladium and metal-free phthalocyanine bilayer structures for hydrogen detection in the SAW sensor system based on interaction speed, *IEEE Sens. J*. 6: 1178–1185.

Jia, H., Zhang, Y., Chen, X., Shu, J., Luo, X., Zhang, Z., and Yu, D. 2003. Efficient field emission from single crystalline indium oxide pyramids, *Appl. Phys. Lett*. 82: 4146–4148.

Jiménez, I., Centeno, M.A., Scotti, R. et al. 2004. $NH_3$ interaction with chromium-doped $WO_3$ nanocrystalline powders for gas sensing applications, *J. Mater. Chem*. 14: 2412–2420.

Joshi, R.K. and Kruis, F.E. 2006. Influence of Ag particle size on ethanol sensing of $SnO_{1.8}$:Ag nanoparticle films: A method to develop parts per billion level gas sensors, *Appl. Phys. Lett*. 153116.

Kapse, V.D., Ghosh, S.A., Chaudhari, G.N., and Raghuwanshi, F.C. 2008. Nanocrystalline $In_2O_3$-based $H_2S$ sensors operable at low temperatures, *Talanta* 76: 610–616.

Kennedy, M.K., Kruis, F.E., Fissan, H., Mehta, B.R., Stappert, S., and Dumpich, G. 2003a. Tailored nanoparticle films from monosized tin oxide nanocrystals: Particle synthesis, film formation, and size-dependent gas-sensing properties, *J. Appl. Phys*. 93: 551–560.

Kennedy, M.K., Kruis, F.E., Fissan, H., and Mehta, B.R. 2003b. Fully automated, gas sensing, and electronic parameter measurement setup for miniaturized nanoparticle gas sensors, *Rev. Sci. Instrum*. 74: 4908–4915.

Kennedy, M.K., Kruis, F.E., Fissan, H., Nienhaus, H., Lorke, A., and Metzger, T.H. 2005. Effect of in-flight annealing and deposition method on gas-sensitive $SnO_x$ films made from size-selected nanoparticles, *Sens. Actuators B* 108: 62–69.

Khanuja, M., Varandani, D., and Mehta, B.R. 2007. Pulse like hydrogen sensing response in Pd nanoparticle layers, *Appl. Phys. Lett*. 91: 253121.

Khanuja, M., Mehta, B.R., Agar, P., Kulriya, P.K., and Avasthi, D.K. 2009a. Hydrogen induced lattice expansion and crystallinity degradation in palladium nanoparticles; effect of hydrogen concentration, pressure and temperature, *J. Appl. Phys*. 106: 093515.

Khanuja, M., Kala, S., Mehta, B.R., and Kruis, F.E. 2009b. Concentration-specific hydrogen sensing behavior in monosized Pd nanoparticle layers, *Nanotechnology* 20: 015502.

Kiefer, T., Favier, F., Mena, O.V., Villanueva, G., and Brugger, J. 2008. A single nanotrench in a palladium microwire for hydrogen detection, *Nanotechnology* 19: 125502.

Kiriadikis, G. 2001. The influence of deposition parameters on room temperature ozone sensing properties of $InO_x$ films, *Sens. Actuators B* 80: 155–161.

Kohl, D. 1989. Surface processes in the detection of reducing gases with $SnO_2$-based devices, *Sens. Actuators* 18: 71–113.

Kolmakov, A., Klenov, D.O., Lilach, Y., Stemmer, S., and Moskovits, M. 2005. Enhanced gas sensing by individual $SnO_2$ nanowires and nanobelts functionalized with Pd catalyst particles, *Nano Lett*. 5: 667–673.

Lewis, F.A. 1967. *The Palladium-Hydrogen System*. New York: Academic Press.
Li, C., Zhang, D., Han, S. et al. 2003. Synthesis, electronic properties and applications of indium oxide nanowires, *Ann. N. Y. Acad. Sci.* 1006: 104–121.
Li, C., Yu, Z., Fang, S. et al. 2009. Preparation and performance of ZnO nanoparticle aggregation with porous morphology, *J. Alloys Compd.* 475: 718–722.
Lim, W., Wright, J.S., Gila, B.P. et al. 2008. Room temperature hydrogen detection using Pd-coated GaN nanowires, *Appl. Phys. Lett.* 93: 072109.
Lith, J.V., Lassesson, A., Brown, S.A., Schulze, M., Patridge, J.G., and Ayesh, A. 2007. A hydrogen sensor based on tunnelling between palladium clusters, *Appl. Phys. Lett.* 91: 181910.
Liu, Y. -Li, Xing, Y., Yang, H.-F. et al. 2004. Ethanol gas sensing properties of nano-crystalline cadmium stannate thick films doped with Pt, *Anal. Chim. Acta* 527: 21–26.
Liu, J., Huang, X., and Meng, F. 2007. The dynamic measurements of $SnO_2$ gas sensors and their applications. In *Science and Technology of Chemiresistor Gas Sensors*, eds. D.K. Aswal and S.K. Gupta, New York: Nova Science Publishers, pp. 177–214.
Llobet, E., Ivanov, P., Vilanova, X. et al. 2003. Screen-printed nanoparticle tin oxide films for high-yield sensor microsystems, *Sens. Actuators B* 96: 94–104.
Luyo, C., Ionescu, R., Reyes, L.F. et al. 2009. Gas sensing response of NiO nanoparticle films made by reactive gas deposition. *Sens. Actuators B* 138: 14–20.
Malyshev, V.V. and Pislyakov, A.V. 2008. Investigation of gas-sensitivity of sensor structures to hydrogen in a wide range of temperature, concentration and humidity of gas medium, *Sens. Actuators B* 134: 913–921.
Manno, D., Micocci, G., Serra, A., Giulio, M.D., and Tepore, A. 2000. Structural and electrical properties of $In_2O_3$–$SeO_2$ mixed oxide thin films for gas sensing applications, *J. Appl. Phys.* 88: 6571–6577.
Mardare, D. and Hones, P. 1999. Optical dispersion analysis of $TiO_2$ thin films based on variable-angle spectroscopic ellipsometry measurements, *Mater. Sci. Eng. B* 68: 42–47.
Mubeen, S., Zhang, T., Yoo, B., Deshusses, M.A., and Myung, N.V. 2007. Palladium nanoparticles decorated single-walled carbon nanotube hydrogen sensor, *J. Phys. Chem. C* 111: 6321–6327.
Paál, Z. and Menon, P.G. 1988. *Hydrogen Effects in Catalysis: Fundamentals and Practical Applications*. New York: Marcel Dekker Inc.
Panchapakesan, B., Cavicchi, R., Semancik, S., and DeVoe, D.L. 2006. Sensitivity, selectivity and stability of tin oxide nanostructures on large area arrays of microhotplates, *Nanotechnology* 17: 415–425.
Papadopoulos, C.A., Vlaehos, D.S., and Avaritsiotis, J.N. 1996. Comparative study of various metal-oxide-based gas-sensor architectures, *Sens. Actuators B* 32: 61–69.
Pavelko, R.G., Vasiliev, A.A., Llobet, E., Vilanova, X., Barrabesb, N., Medinab, F., and Sevastyanov, V.G. 2008. Comparative study of nanocrystalline $SnO_2$ materials for gas sensor application: Thermal stability and catalytic activity, *Sens. Actuators B* 137: 637–643.
Qin, L., Xu, J., Dong, X. et al. 2008. The template-free synthesis of square-shaped $SnO_2$ nanowires: The temperature effect and acetone gas sensors, *Nanotechnology* 19: 185705.
Rao, C.N.R., Thomas, P.J., and Kulkarni, G.U. 2007. *Nanocrystals: Synthesis, Properties and Applications*. New York: Springer-Verlag.
Reddy, M.H.M. and Chandorkar, A.N. 1999. E. Beam deposited $SnO_2$, Pt-$SnO_2$ and Pd-$SnO_2$ thin films for LPG detection, *Thin Solid Films* 349: 260–265.
Rella, R., Spadavecchia, J., Manera, M.G. et al. 2007. Acetone and ethanol solid-state gas sensors based on $TiO_2$ nanoparticles thin film deposited by matrix assisted pulsed laser evaporation, *Sens. Actuators B* 127: 426–431.
Reyes, F., Hoel, A., Saukko, S., Heszler, P., Lantto, V., and Granqvist, C.G. 2006. Gas sensor response of pure and activated $WO_3$ nanoparticle films made by advanced reactive gas deposition, *Sens. Actuators B* 117: 128–134.
Rout, C.S., Govindaraj, A., and Rao, C.N.R. 2006. High-sensitivity hydrocarbon sensors based on tungsten oxide nanowires, *J. Mater. Chem.* 16: 3936–3941.
Rout, C.S., Hegde, M., Govindaraj, A., and Rao, C.N.R. 2007. Ammonia sensors based on metal oxide nanostructures, *Nanotechnology* 18: 205504.

Rout, C.S., Hegde, M., and Rao, C.N.R. 2008. $H_2S$ sensors based on tungsten oxide nanostructures, *Sens. Actuators B* 128: 488–493.

Sachs, C., Pundt, A., Kirchheim, R., Winter, M., Reetz, M.T., and Fritsch, D. 2001. Solubility of hydrogen in single-sized palladium clusters, *Phys. Rev. B* 64: 075408.

Safonova, O., Bezverkhy, I., Fabrichnyi, P., Rumyantseva, M., and Gaskov, A. 2002. Mechanism of sensing CO in nitrogen by nanocrystalline $SnO_2$ and $SnO_2$ (Pd) studied by Mossbauer spectroscopy and conductance measurements, *J. Mater. Chem.* 12: 1174–1178.

Sberveglieri, G. 1995. Recent developments in semiconducting thin-film gas sensors, *Sens. Actuator B* 23: 103–109.

Seiyama, T. and Kato, A. 1962. A new detector for gaseous components using semiconductive thin films, *Anal. Chem.* 34: 1502–1503.

Shukla, S., Seal, S., Ludwig, L., and Parish, C. 2004. Nanocrystalline indium oxide-doped tin oxide thin film as low temperature hydrogen sensor, *Sens. Actuators B* 97: 256–265.

Singh, V.N., Mehta, B.R., Joshi, R.K., Kruis, F.E., and Shivaprasad, S.M. 2007. Enhanced gas sensing properties of $In_2O_3$:Ag composite nanoparticle layers; electronic interaction, size and surface induced effects, *Sens. Actuators B* 125: 482–488.

Skala, T., Veltruska, K., Moroseac, M., Matolinova, I., Korotchenkov, G., and Matolin, V. 2003. Study of Pd-In interaction during Pd deposition on pyrolytically prepared $In_2O_3$, *Appl. Surf. Sci.* 205: 196–205.

Starke, T.K.H. and Coles, G.S.V. 2002. High sensitivity ozone sensors for environmental monitoring produced using laser ablated nanocrystalline metal oxides, *IEEE Sens. J.* 2: 14–19.

Steffes, H., Imawan, C., Solzbacher, F., and Obermeir, E. 2001. Enhancement of $NO_2$ sensing properties of $In_2O_3$-based thin films using an Au or Ti surface modification, *Sens. Actuators B* 78: 106–112.

Tan, E.T.H., Ho, G.W., Wong, A.S.W., Kawi, S., and Wee, A.T.S. 2008. Gas sensing properties of tin oxide nanostructures synthesized via a solid-state reaction method, *Nanotechnology* 19: 255706.

Tjeng, L.H., Meinders, M.B.J., van Elp, J., Ghijsen, J., and Sawatzky, G.A. 1990. Electronic structure of $Ag_2O$, *Phys. Rev. B* 41: 3190–3199.

Tyagi, M.S. 1984. *Metal-Semiconductor Schottky Barrier Junctions and their Applications*, B.L. Sharma, ed. New York: Plenum Press.

Vincenti, M.A., Sario, M.D., Petruzzelli, V. et al. 2007. Fabry-Perot microcavity sensor for $H_2$-breath-test analysis, *J. Appl. Phys.* 102: 074501.

Waghulade, R.B., Patil, P.P., and Pasricha, R. 2007. Synthesis and LPG sensing properties of nano-sized cadmium oxide, *Talanta* 72: 594–599.

Walter, E.C., Favier, F., and Penner, R.M. 2002. *Anal. Chem.* 74: 1546–1553.

Wang, Z.L. 2004. Functional oxide nanobelts: Materials, properties and potential applications in nanosystems and biotechnology, *Annu. Rev. Phys. Chem.* 55: 159–196.

Wang, Y., Wang, Y., and Cao, J. 2008. Low-temperature $H_2S$ sensors based on Ag-doped $\alpha$-$Fe_2O_3$ nanoparticles, *Sens. Actuators B* 131: 183–189.

Watson, J. 1992. A note on the electrical characterization of solid-state gas sensors, *Sens. Actuators B* 8: 173–177.

Williams, D.E. and Moseley, P.T. 1991. Dopant effects on the response of gas-sensitive resistors utilising semiconducting oxides, *J. Mater. Chem.* 1: 809–814.

Winter, R., Scharnagl, K.A., Fuchs, A., Doll, T., and Eisele, I. 2000. Molecular beam evaporation-grown indium oxide and indium aluminium films for low-temperature gas sensors, *Sens. Actuators B* 66: 85–87.

Wright, J.S., Lim, W., Gila, B.P. et al. 2009. Pd-catalyzed hydrogen sensing with InN nanobelts, *J. Vac. Sci. Technol. B* 27: L8–L10.

Wu, Z., Yu, K., Zhang, S., and Xie, Y. 2008. Hematite hollow spheres with a mesoporous shell: Controlled synthesis and applications in gas sensor and lithium ion batteries, *J. Phys. Chem. C* 112: 11307–11313.

Xu, C., Tamaki, J., Miura, N., and Yamazoe, N. 1991. Grain size effects on gas sensitivity of porous $SnO_2$-based elements, *Sens. Actuators B* 3: 147–155.

Xu, T., Zach, M.P., Xiao, Z.L. et al. 2005. Self-assembled monolayer-enhanced hydrogen sensing with ultrathin palladium films, *Appl. Phys. Lett.* 86: 203104.

Yamaura, H., Moriya, K., Miura, N., and Yamazoe, N. 2000. Mechanism of sensitivity promotion in CO sensor using indium oxide and cobalt oxide, *Sens. Actuators B* 65: 39–41.

Yamazoe, N. 1991. New approaches for improving semiconductor gas sensors, *Sens. Actuators B* 5:7–19.

Yamazoe, N., Sakai, G., and Shimanoe, K. 2003. Oxide semiconductor gas sensors, *Catal. Surveys Asia* 7:63–75.

Yu, M.-F., Atashbar, M.Z., and Chen, X. 2005a. Mechanical and electrical characterization of $Ga_2O_3$ nanostructures for sensing applications, *IEEE Sens. J.* 5: 20–25.

Yu, S., Welp, U., Hua, L.Z. et al. 2005b. Fabrication of palladium nanotubes and their application in hydrogen sensing, *Chem. Mater.* 17: 3445–3450.

Yuasa, M., Masaki, T., Kida, T., Shimanoe, K., and Yamazoe, N. 2009. Nano-sized PdO loaded $SnO_2$ nanoparticles by reverse micelle method for highly sensitive CO gas sensor, *Sens. Actuators B* 136: 99–104.

Zhang, D., Liu, Z., Li, C. et al. 2004. Detection of $NO_2$ down to ppb levels using individual and multiple $In_2O_3$ nanowire devices, *Nano Lett.* 4: 1919–1924.

Zhou, H., Cai, W., and Zhang, L. 1999. Synthesis and structure of indium oxide nanoparticles dispersed within pores of mesoporous silica, *Mater. Res. Bull.* 34: 845–849.

# 5

# *Light-Emitting Devices Based on Direct Band Gap Semiconductor Nanoparticles*

**Ekaterina Neshataeva, Tilmar Kümmell, and Gerd Bacher**

**CONTENTS**



## 5.1 Introduction

Nowadays, the demand for low-cost light emitters is high, covering a wide range of different applications in advertisement and give-away industry, low-cost indicators, and displays, for example, consumer electronics, mobile phones, toys, and many more. In addition, cost-efficient large-area emitters are highly desired in automotive industry as well as in architecture, urban development, and interior for innovative design and ambient lightning applications. Additional profitable market segments can be created by introducing emitters on flexible substrates. So far, these demands are mostly covered by organic light-emitting devices (OLEDs) (Müllen 2006; Jain 2007; Li 2007; Held 2009). For both, covering these market segments and developing new ones, nanoparticle-based emitters are regarded as an attractive alternative to OLEDs with the potential of an improvement in lifetime, power consumption, and fabrication costs. Nanoparticle-based light-emitting devices (NP-LEDs) are expected to combine the robustness, long-term stability, and efficiency of traditional semiconductor material systems, known from epitaxially grown LEDs and laser diodes (Burnham 2000; Held 2009; Mottier 2009), with flexible low-cost large-area fabrication techniques known from OLEDs (Müllen 2006; Jain 2007; Li 2007; Held 2009).

The following chapter will focus on the implementation of direct band gap semiconducting nanocrystals in LEDs, where the nanocrystals play the active role in light emission.

Covering other versatile applications like indirect band gap NP-LEDs (Kan et al. 2004; Walters et al. 2005; Walters et al. 2006; Cheong et al. 2008), emission color conversion (Chen et al. 2006; Ahn et al. 2007; Kobayashi et al. 2007; Nizamoglu and Demir 2007; Jang et al. 2008; Yeh et al. 2008; Wood et al. 2009a), and efficiency enhancement (Hussain et al. 2009; Zheng et al. 2009) in conventional emitters and many more would go beyond the scope of the chapter. We therefore would like to concentrate on the different novel light-emitting concepts based on direct band gap semiconducting nanocrystals and the corresponding device fabrication techniques, referring the readers, who are interested in different nanoparticle synthesis routes, to corresponding reviews (Murray et al. 2000; Swihart 2003; Rosenthal et al. 2007).

As other electrically driven LEDs, NP-LEDs can be as well divided into those operated at direct (DC) and at alternating (AC) currents. Typically, the light emission processes differ in these two cases demanding different device architecture types and involved material systems, which are going to be discussed below. The research on these new device concepts and their development started in 1994 (Colvin et al. 1994) and rapidly proceeded since then. First LEDs based on nanoparticles and operated under DC were presented historically earlier than AC devices, mainly pushed by the OLEDs technology and the development of quantum dots (QDs) with high room temperature quantum efficiency (QE) and the availability of processible stable dispersions.

## 5.2 Direct Current-Operated Light-Emitting Devices (DC-LEDs)

The development of OLEDs progressed rapidly in the early 1990s, enabling efficient large-area emitters on flexible substrates. At the beginning, the lifetime of the OLEDs was mainly limited by the high sensitivity of the organic materials toward humidity, oxygen, and the fast degradation under the high operation currents (Jain 2007; Li 2007; Held 2009). An improvement was expected by introducing inorganic semiconducting nanocrystals as the active material in such an OLED structure. The promising idea to combine both robust and highly luminescent inorganic semiconducting nanoparticles with the flexible large-area technology of OLEDs became a reality when the first nanoparticles in stable dispersions, which allow further processing, were available (Steigerwald et al. 1988; Murray et al. 1993; Colvin et al. 1994; Katari et al. 1994).

The typical structure of the NP-LEDs operated under DC was adopted from the OLED technology. Usually, a transparent conductive oxide (TCO)-coated glass is used as a substrate and as a transparent electrode. The most frequently used TCO is indium-doped tin oxide (ITO), utilized in touch screens and liquid crystal displays as a standard. An alternative is flourine-doped tin oxide (FTO), frequently used in solar-energy conversion applications (Bach et al. 1998) because of its stable mechanical and chemical properties (Vossen 1977; Chopra et al. 1983). Recent developments show new substitutes for ITO like aluminum- and gallium-doped zinc oxides (Raniero et al. 2006; Minami 2008a,b) and many others. The other electrode typically consists of metals with a low work function. When using barium (Ba), magnesium (Mg), or calcium (Ca), an additional layer of aluminum (Al) or silver (Ag) on top is needed to passivate the reactive electrode and to protect it from oxidation. Because of its easy handling, Al electrodes are most frequently implemented. The active light-emitting layer consisting of either pure nanoparticles or organic/nanoparticle compounds is sandwiched between both the electrodes. While biased in forward direction, holes are supposed to be injected into the active layer from the TCO and electrons from the top metallic electrode, subsequently recombine and emit light. This is the reason why usually a TCO with a high work function and a cathode metal with a low work function value are preferred to lower the energy barriers for hole and electron injection, respectively.

Historically, the first nanoparticles used for optoelectronic applications were so-called QDs, synthesized chemically in solution-based processes that allow the precise control over the size, composition, and doping. Additionally, different types of ligands can be attached to the surface of the particles in order to passivate, stabilize, and easily disperse them in different solvents. Since the synthesis of these particles is well established and the chemistry is under control, these materials remain up to now the first choice for the nanoparticle LED fabrication. Besides, they also show high QE and pure color-saturated emission. The research on Cd-free, wide band gap semiconducting nanoparticles, their properties, and applications in LEDs is still at the beginning, but nevertheless some first promising results are already demonstrated. The following subsections are reviewing the achievements on QDs and wide band gap semiconductor NP-LEDs.

### 5.2.1 Quantum Dot Light-Emitting Devices (QD-LEDs)

Semiconducting nanoparticles, also called nanocrystals, with sizes comparable or smaller than the bulk exciton diameter are called QDs because of their atomic-like electronic structure with size-dependent optical and electronic properties, evoked by the quantum confinement effect (Efros and Efros 1982; Brus 1991). For instance, the band gap of these particles can increase by several electron volts with respect to the bulk material with decreasing nanoparticle size. This is reflected in a blueshift of the absorption and emission wavelength of some hundreds of nanometers with decreasing particle size, allowing the fine tuning of the desirable emission color/wavelength by synthesizing QDs with the appropriate size (Alivisatos 1996; Dabbousi et al. 1997). Besides, QDs can exhibit high luminous efficiencies of 50% and higher and very narrow emission spectra with a full width at half maximum less than 30 nm at room temperature (Alivisatos 1996; Dabbousi et al. 1997; Reiss et al. 2002; Tan et al. 2007). These adjustable and spectrally narrow emission characteristics make QDs attractive candidates for LEDs. Additionally, the luminescence efficiency and the stability of the QDs can be greatly improved by modifying their surface with, for example, caping agents/ligands, and/or additional crystalline shells. After the development of reliable synthesis routes for the monodisperse QDs in stable and processible dispersions as well as new techniques for the QD assembly, the first QD-LEDs could be realized.

#### *5.2.1.1 Evolution of QD-LEDs*

The first QD material system introduced in electroluminescent devices was CdSe, presented by Alivisatos and coworkers in 1994 (Colvin et al. 1994). A *p*-paraphenylene vinylene (PPV) layer, covered by a multilayer structure with closely packed disordered sheets of CdSe nanocrystals, separated by organic spacers, was sandwiched between ITO and Mg electrodes. Under forward bias, holes were injected into the hole transporting PPV from the transparent ITO and electrons into the nanocrystal layer from the Mg electrode. Electrons and holes recombined on the nanocrystals resulting in a spectral contribution of the characteristic CdSe QD emission to the standard PPV emission, known from the OLEDs. Figure 5.1 shows the dependence of the electroluminescence (EL) on the applied

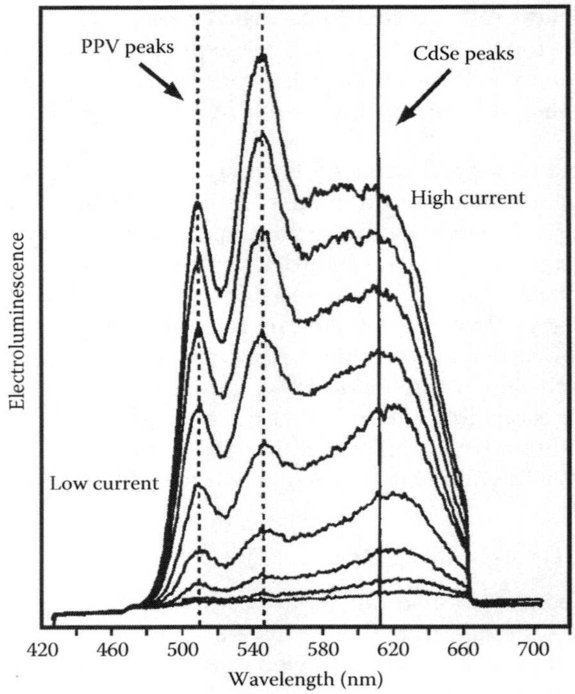


**FIGURE 5.1**
Voltage-dependent color of the CdSe-QDs/PPV-LED. (Reprinted by permission from Colvin, V.L. et al., *Nature*, 370, 354, 1994. Copyright [1994], Macmillan Publishers Ltd.)

voltage. At higher voltages, the contribution of the PPV becomes dominant, changing the overall color of the emission from yellow to green. The change of the emission color from red to yellow was also demonstrated by changing the size of the incorporated CdSe QDs.

A subsequent publication from two groups at MIT (Dabbousi et al. 1995) demonstrated a single-layer CdSe-QD-LED with the nanocrystals incorporated into an organic polymer matrix. In this case, the QDs were blended with a hole and an electron-conducting polymer and spin casted on an ITO substrate in one fabrication step, followed by the top Mg/Ag evaporation. Although these first devices showed quite low external quantum efficiencies (EQEs) of 0.001%–0.01% (Colvin et al. 1994) and 0.0005% (Dabbousi et al. 1995), they demonstrated the principle application potential of QDs and quantum confinement effects in LEDs.

This first device design, as sketched in Figure 5.2a, with a QD/organic polymer blend sandwiched between two electrodes was also further developed by other groups using hole-conducting (Gaponik et al. 2000; Tessler 2002; Zhao et al. 2004; Li et al. 2006), electron-conducting (Mattoussi et al. 1999), isolating (Gao et al. 1998), and semiconducting (Gaponik et al. 1999) organic polymers as a host matrix. The successful implementation of other semiconducting QDs for the applications in the visible (CdTe (Gaponik et al. 1999)) and the near-infrared (NIR) (InAs (Tessler 2002) PbS (Bakueva et al. 2004) and PbSe (Solomeshch et al. 2005)) spectral range using this device concept was demonstrated. Nevertheless, EQEs of these devices remained quite low mainly due to the charge imbalance, caused by inhomogeneous charge transport characteristics, since most QDs exhibit poor conductivity. Additionally, close proximity of the particles to the electrodes can lead to EL quenching (Larkin et al. 2004).

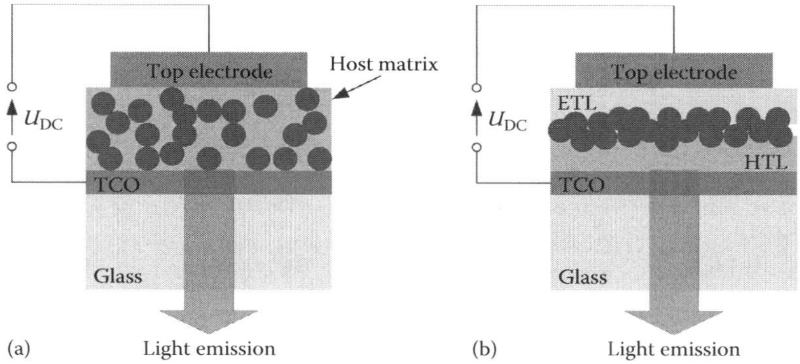


**FIGURE 5.2**
Different device designs of nanoparticle organic/inorganic hybrid LEDs operated under DC with a TCO as an anode, a metallic top electrode as a cathode, and a DC-voltage $U_{DC}$ applied across them. In design (a), nanoparticles are inserted in an organic polymer host matrix. In design (b), nanoparticles are assembled in a tight layer with possible additional organic HTL and electron transport layers (ETLs). The light emission is observed through the transparent TCO-coated glass substrate.

A significant increase of the external efficiency (factor 20) and lifetime (factor 100) with respect to the first devices was achieved in 1997 by introducing core/shell QDs, consisting of a CdSe core epitaxially overcoated by a CdS shell (Schlamp et al. 1997). The band gap of the shell material is higher than that of the core, and the energy levels are aligned in a way that the holes are confined to the core, while the electrons are delocalized throughout the structure (see Figure 5.3). The core/shell QDs therefore exhibited higher photoluminescence

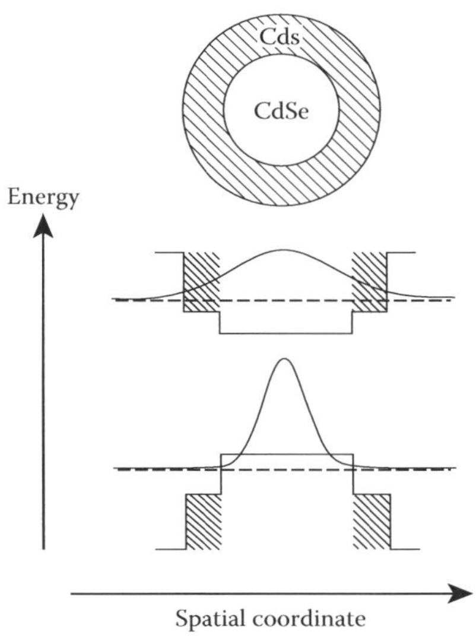


**FIGURE 5.3**
CdSe/CdS core/shell nanocrystal energy diagram. The solid lines represent the bulk potentials and the dashed levels represent the highest occupied molecular orbital (HOMO) and lowest unoccupied molecular orbital (LUMO) in the nanocrystal extrapolated from bulk CdSe. The Gaussians shown are the results of effective mass calculations for electron and hole wavefunctions. Potential offsets between the core and shell valence band (0.5 eV) and conduction band (0.3 eV) are approximated by the differences between the bulk electron affinities and ionization potentials of CdSe and CdS. (Reprinted with permission from Schlamp, M.C. et al., *J. Appl. Phys.*, 82, 5837, 1997. Copyright [1997], American Institute of Physics.)

(PL) quantum yield, photooxidative stability and electronic accessibility, which resulted in higher EQE and improved long-term stability of the QD-LED. In addition, the QDs were assembled in an own layer on top of the organic PPV layer, introducing an alternative design of the QD-LED, where QDs are no longer incorporated into an organic host matrix, but deposited as a self-contained layer from the dispersion. The active luminescent QD layer is sandwiched between the TCO and the top electrode with additional organic layer(s) between the electrode(s) and the QD layer to enable and enhance injection and transport of the charge carriers to the QDs, as schematically shown in Figure 5.2b.

Further 25-fold improvement in luminescence efficiency of QD-LEDs was achieved by Coe et al. (2002) by modifying the deposition technique of the active luminescent QD layer. The developed phase-segregation approach enabled a precise deposition of a single QD monolayer between two organic charge injection/transport layers. The schematic of the device structure is shown in Figure 5.4. ITO and Mg:Ag were used as a transparent anode and a cathode, respectively. Core/shell CdSe/ZnS QDs, passivated with trioctylphosphine oxide caps (inset Figure 5.4), were dispersed in chloroform solution of hole-transporting *N,N′*-diphenyl-*N,N′*-bis(3-methylphenyl)-(1,1′-biphenyl)-4,4′-diamine (TPD) and phase-segregated from it during the spin coating of the mixture on the ITO substrates. The optimized concentration of QDs in the QD/TPD solution led to a complete single monolayer on top of the 35 nm thick TPD film. Subsequently, a 40 nm thick Tris-(8-hydroxyquinoline) aluminum ($Alq_3$) ETL was evaporated on top of the QD monolayer, followed by the top electrode deposition. Figure 5.4a shows the schematic of the device structure I as well as the resulting EL spectrum, where both QDs and $Alq_3$ contribute to the emission. The device structure II, shown in Figure 5.4b, includes an additional 10 nm thick 3-(4-biphenyl)-4-phenyl-5-*tert*-butylphenyl-1,2,4-triazole (TAZ) hole and exciton blocking layer. This layer suppressed the emission from $Alq_3$, as can be seen in the EL spectrum at Figure 5.4b. A peak EQE of 0.52% and a peak brightness of 2000 cd/m$^2$ was achieved by device structure I, without the TAZ layer.

The main improvement of the QD-LEDs was achieved by separating the luminescence function of the QDs from their participation in charge conduction, which was not the case

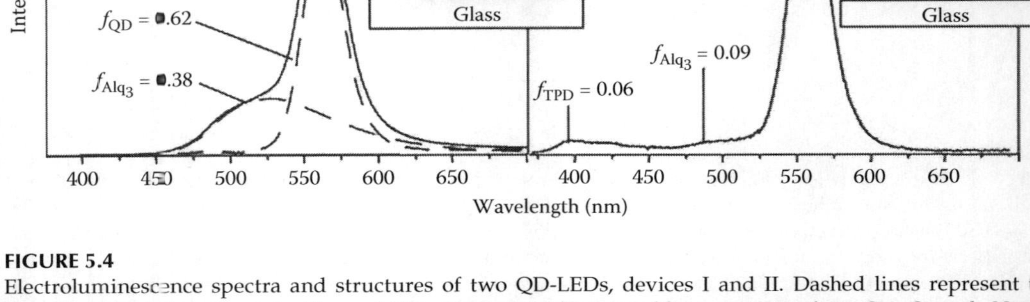


**FIGURE 5.4**
Electroluminescence spectra and structures of two QD-LEDs, devices I and II. Dashed lines represent the decomposition of spectra into $Alq_3$ and QD components. (Reprinted by permission from Coe, S. et al., *Nature*, 420(6917), 800, 2002. Copyright [2002], Macmillan Publishers Ltd.)

in earlier devices. Here, the charge carriers were transported to the vicinity of QDs only by the organic layers. The exciton generation on QDs was attributed to two parallel processes: direct charge injection into the QDs and exciton energy transfer from organic molecules.

One of the negative aspects of using organic charge support layers in QD-LEDs is their contribution to the light emission of the LED, which disturbs the color purity if a saturated monochromatic emission of the QDs is desired. This disturbing emission could be suppressed in some cases (Coe et al. 2002; Li et al. 2006; Zhao et al. 2006; Tan et al. 2007; Kim et al. 2008; Zhu et al. 2008) but, on the other hand, it could also be used to create an efficient white QD-LED, as it was demonstrated in 2005 by Li et al. (2005). A layer of red-emitting CdSe/ZnS core/shell QDs blended with a blue-emitting hole-transporting polymer was sandwiched between a hole injection layer and a green-emitting $Alq_3$ ETL, combining the device designs (a) and (b) in Figure 5.2. An accurate control of the charge transfer and the energy transfer between the different components allowed the fabrication of fairly pure white light emission with Commission Internationale de l'Eclairage (CIE) coordinates of (0.30, 0.33) and a maximum EQE of 0.24%. In the year 2006, the same group demonstrated a 1050 cd/m$^2$ bright QD-LED with a similar device configuration, utilizing 4,4′,*N*,*N*″-diphenylcarbazole (CBP) as a hole-transporting polymer host matrix (Li et al. 2006). The bright, nearly white emission color of the QD-LED (CIE (0.32, 0.45)) resulted from an appropriate mix of different sized CdSe/ZnS core/shell nanocrystals, emitting at 490 (blue), 540 (green), and 615 nm (red). The organic support layers did not show any contribution to the light emission of the device. Figure 5.5 shows the PL spectra of the red, green, and blue QDs incorporated in the device, as well as the EL spectrum of the device, the current density-voltage and the luminance-voltage characteristics of the device, as well as a photograph of the operating device in the inset.

White EL from mixed monolayer of red-, green-, and blue-emitting QDs in a hybrid organic/inorganic QD-LED was reported by Anikeeva et al. (2007) from MIT. The emission color of the device could be precisely tuned by changing the ratio of different color QDs in the active layer without changing the fabrication parameters for the organic support layers and for the whole QD-LED. CdSe/ZnS core/shell QDs emitting at 620 nm were chosen as a red emitter, ZnSe/CdSe alloyed cores overcoated with a shell of ZnS (540 nm) were chosen

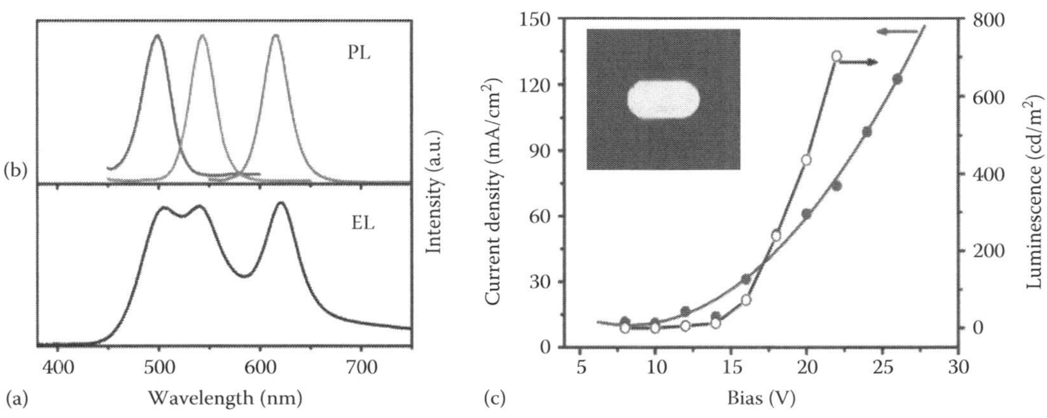


**FIGURE 5.5**
Electroluminescence (a) and photoluminescence (b) spectra, current density–voltage (c, dots) and voltage–luminance (c, circles) characteristics of the white QD-LED. (Reproduced from Li, Y. et al., *J. Appl. Phys.*, 97, 11, 113501, 2005; Li, Y.Q. et al., *Adv. Mater.*, 18, 19, 2545, 2006. Copyright Wiley-VCH Verlag GmbH & Co. KGaA. With permission.)

as a green, and ZnCdS alloyed QDs (440 nm) as a blue emitter within the monolayer. The device showed CIE of (0.35, 0.41) with a corresponding EQE of 0.36% at video brightness.

#### 5.2.1.2 Advanced Fabrication Techniques for QD-LEDs

Not only the QD material system and the device design are important for the device characteristics. The ability to control the quality of the QD layer as well as its easy and reliable fabrication, compatible with the whole QD-LED fabrication process, is crucial for achieving high device performance. A variety of deposition techniques for the QD layer was tested and improved over the last years, such as self-assembly and dip-deposition (Gao et al. 1997, 1998, 1999; Gaponik et al. 1999), electrodeposition (Gaponik et al. 2000), drop-deposition (Colvin et al. 1994; Artemyev et al. 1997; Gaponik et al. 2000; O'Connor et al. 2005), layer-by-layer deposition (Gao et al. 1997, 1999; Bertoni et al. 2007), spray, and mist techniques (Gaponik et al. 2000; Zhu et al. 2008). But the most frequently used technique remains spin coating, which is easy, cost-effective, suitable for large-area devices, and results in homogeneous QD layers (Mattoussi et al. 1998; Hikmet et al. 2003; Caruge et al. 2006; Tan et al. 2007; Kang et al. 2008) as thin as one monolayer by using the phase-separation approach (Coe et al. 2002; Coe-Sullivan et al. 2005). Zhao and coworkers fabricated the active monolayer of CdSe/CdS core/shell QDs in a hybrid QD-LED device without phase-separation technique, only by choosing appropriate parameters for spin coating and QD concentration in the dispersion (Zhao et al. 2006). Nevertheless, spin coating of the QD layers has mainly two limitations: it can not be applied to pattern the QD layer and it places solvent compatibility requirements on the device fabrication process.

As a possible solution to these problems, an elegant contact printing approach for the precise deposition of thin QD layers and monolayers was recently presented by the group of Vladimir Bulović from MIT (Kim et al. 2008) and by the group of Giuseppe Gigli from the University of Salento (Rizzo et al. 2008). Both groups used a poly(dimethylsiloxane) stamp to define the pattern of the designed QD layer. Poly(dimethylsiloxane) is capable of making conformal contact with the surfaces over relatively large areas, at the same time enabling the patterning of micrometer-sized features. After the stamp is formed, it can be inked by spin coating the QDs from the solution. After the solvent is evaporated, the QD layer on the patterned stamp surface can be transferred onto the desired surface (e.g., hole transport layer (HTL) of the QD-LED) by contact printing (Figure 5.6A). When using the QD dispersion directly, the stamp should be coated by a thin film of parylene-C to prevent the stamp from swelling (Kim et al. 2008). A way to avoid the chemical-vapor deposition of the parylene-C is to use a double-transfer approach. The QD layer is first formed on a glass substrate by drop-casting. The patterned stamp becomes covered with the QD layer by the first-step contact with the QD-coated glass. In the second contact printing step, the QD layer is transferred to the desired surface, as shown in Figure 5.7a (Rizzo et al. 2008). In both cases, organic/inorganic hybrid QD-LEDs were fabricated to demonstrate the basic prospects of this technique. Figure 5.7b shows a 1.5 mm × 1.5 mm contact-printed device with corresponding EL and PL spectra (Rizzo et al. 2008), demonstrating the easy large-area patterning. Figure 5.6 shows red- and green-patterned pixels on the same substrate as well as a single pixel, patterned by 25 μm wide stamp features (Kim et al. 2008). This new fabrication technique enables an easy and fast way to create large-area or micrometer-patterned pixels of multilayer multicolor light-emitting structures without any restrictions on the choice of the organic material.

On the other hand, spin coating of the QD layer, also for contact printing, results in loss of more than 94% of the QD solution. In contrast, inkjet printing of the QD layer allows fast

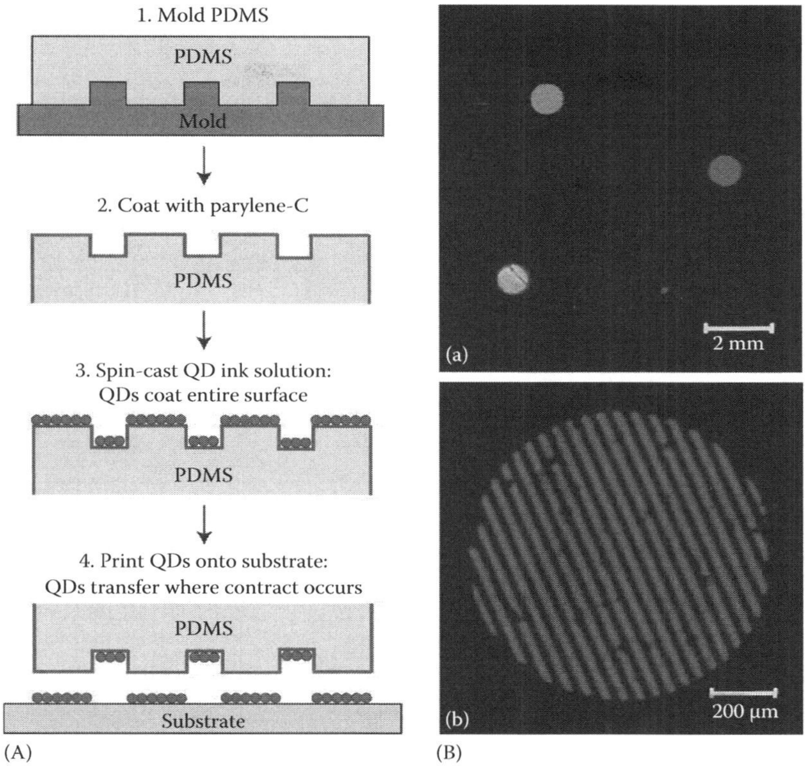


**FIGURE 5.6**
**(See color insert following page 302.)** (A) The schematic of the contact printing process; (B) (a) electroluminescent red and green QD pixels, fabricated on the same substrate. A blue pixel is a result of TPD emission in the area without QDs; (b) QD-LED pixel, patterned with 25 μm wide stamp features. (Reprinted with permission from Kim, L.A. et al., *Nano Lett.*, 8, 4513, 2008. Copyright [2008], American Chemical Society.)

and easy fabrication of patterned structures at low material consumption. Haverinen et al. (2009) presented recently a QD-LED with an inkjet-printed QD layer. The device architecture was a standard HTL, a QD layer, and an ETL sandwiched between ITO and LiF/Al electrodes. The uniformity and close packing of the printed layers were greatly affected by the surface morphology, the size variance of the QDs, and the quality of the underlayer. Figure 5.8 shows a photograph of the emitting device, as well as the corresponding EL spectra with contributions from the QDs and the organic support layers. A peak EQE of 0.19% at 14.2 V and a maximum brightness of 381 cd/m$^2$ at 15.9 V was achieved.

The QD layer fabrication techniques mentioned above are easy and principally allow a roll-to-roll production of large-area patterned emitters on flexible substrates. The vision of flexible QD-LED displays becomes already a reality. Tan and colleagues recently demonstrated mechanically flexible red, green, and blue QD-LEDs on ITO-coated poly(ethylene terephthalate) (PET) substrates (Tan et al. 2009). The device structure and the pictures of the devices under operation as well as the corresponding EL spectra are shown in Figure 5.9. The flexible QD-LEDs showed efficiencies comparable with the devices traditionally fabricated on rigid substrates. Importantly, they exhibited a critical bending radius of ~5 mm, which is sufficient for the applications in large-area conformable flat panel displays that can be manufactured by roll-to-roll processes.

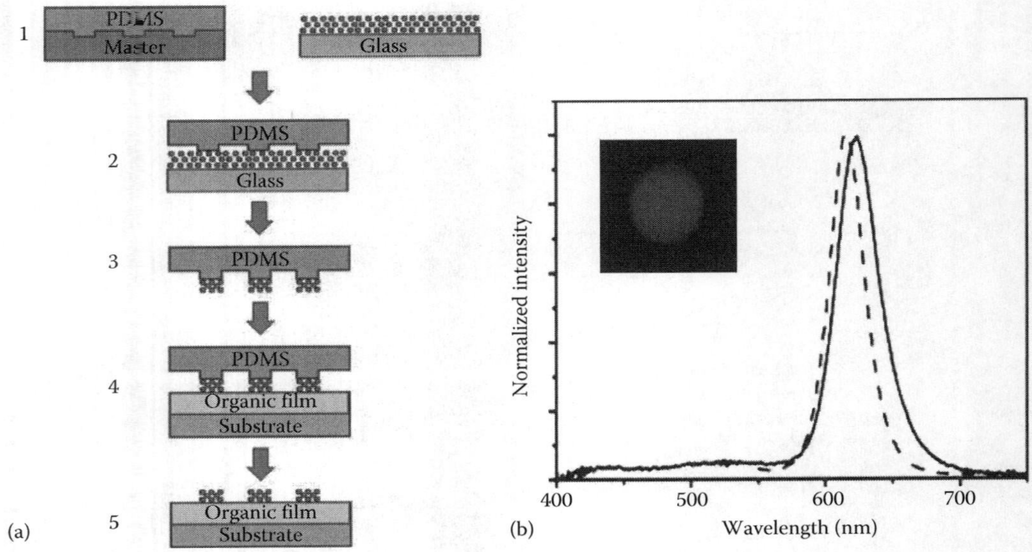


**FIGURE 5.7**
(a) Schematic of the procedure to transfer the QD layer onto thin organic films; (b) EL (solid line) and PL (dashed line) spectra of the QD-LED with a functional area of 1.5 mm × 1.5 mm, shown in the inset picture. (Reproduced from Rizzo, A. et al., *Adv. Mater.*, 20, 1886, 2008. Copyright Wiley-VCH Verlag GmbH & Co. KGaA.)

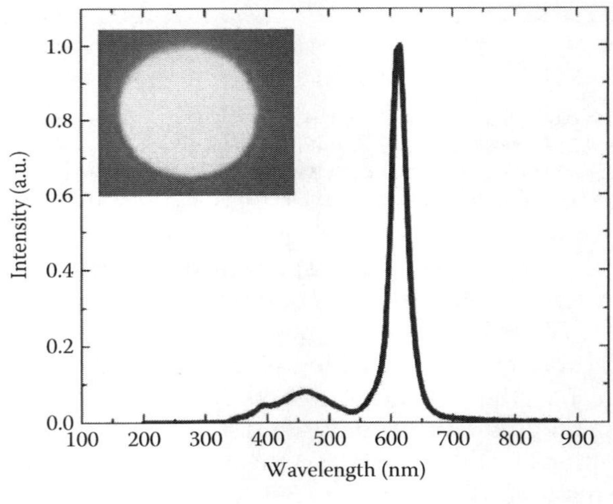


**FIGURE 5.8**
EL spectrum of the light-emitting inkjet-printed QD layer in QD-LED. Inset shows a photograph of the operating device with an area of 0.14 cm². (Reprinted with permission from Haverinen, H.M. et al., *Appl. Phys. Lett.*, 94, 073108, 2009. Copyright [2009], American Institute of Physics.)

### *5.2.1.3 Toward Organic-Free QD-LEDs*

In general, organic/inorganic hybrid QD-LEDs underwent a fast evolution during the last decade, with fabrication techniques adopted from the OLED production or being newly developed and thus becoming more reliable. Established and improved chemical synthesis routes allowed for stable dispersions with nanocrystals of high quantum yield in

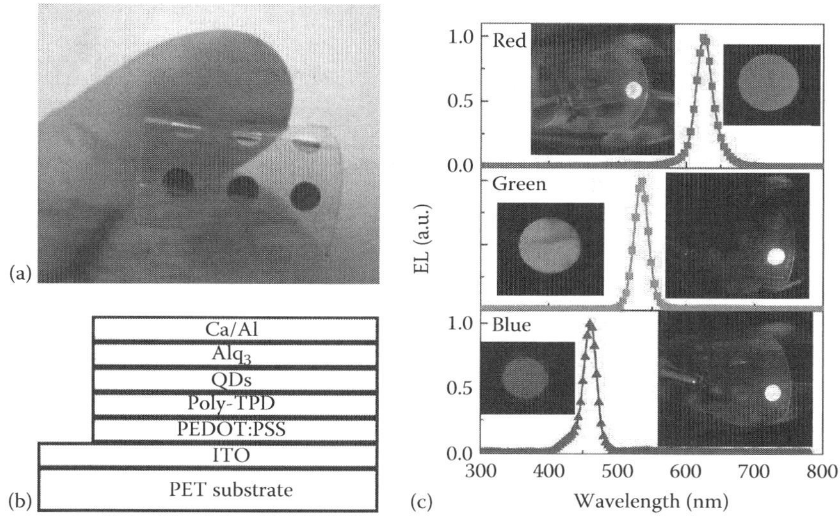


**FIGURE 5.9**
**(See color insert following page 302.)** (a) Photograph, (b) device configuration, and (c) images of the relaxed and bent flexible RGB-emitting QD-LEDs, operated at maximum brightness according to EL spectra. (Reprinted with permission from Tan, Z. et al., *J. Appl. Phys.*, 105, 034312, 2009. Copyright [2009], American Institute of Physics.)

different solvents. Hybrid QD-LEDs in the NIR and the visible spectral range based on different material systems were demonstrated. Nowadays, the EL color can be easily tuned over the whole visible range from red to blue (Tan et al. 2007; Anikeeva et al. 2009). Bright white emission can be achieved by the mixing of red, green, and blue emitting colors (Li et al. 2006; Anikeeva et al. 2007). The so far reported record value for EQE is 2.7% for an orange QD-LED, which is about 1000 times higher than the EQE of the first hybrid QD-LED devices (Anikeeva et al. 2009).

However, the operating lifetime and robustness of the hybrid organic/inorganic QD-LEDs remains still limited due to the presence of the organic charge injection and transport layers within the structure. These layers are quite sensitive toward humidity, oxygen, and high temperatures and are instable under operation at high currents and fields. Thus, the devices need to be packaged for the operation under ambient air conditions. The operation at high current densities, required to achieve population inversion for QD-laser applications, is therefore also challenging. Another mechanism involved in the degradation of the QD-LEDs is the diffusion of reactive metals from the electrodes, like In and Al, into the organic layers during the device operation (Schlatmann et al. 1996; Hirose et al. 1996a,b; Gallardo et al. 2007). Additionally, organic layers often contribute to the overall light emission of the QD-LED, disturbing the desired color purity. Possible solution of these problems is to replace the organic support layers in QD-LEDs by inorganic charge transport layers, which are more robust.

Caruge et al. (2006) demonstrated a successful implementation of p-type NiO in a QD-LED with an additional organic ETL. CdSe/ZnS core/shell QDs exhibited a maximum EQE of 0.18% with a brightness of up to 3000 cd/m$^2$ in these hybrid devices. Stouwdam and Janssen realized an innovative design QD-LED, where the ETL is replaced by inorganic nanocrystals, spin-coated from an isopropanol solution directly on top of the QD layer (Stouwdam and Janssen 2008). Figure 5.10a shows the device architecture and the

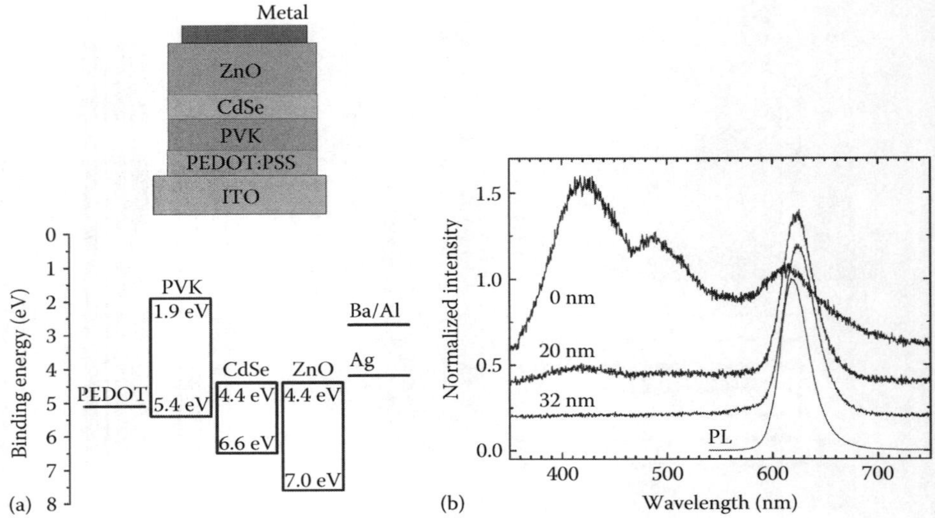


**FIGURE 5.10**
(a) Schematic of the device structure with corresponding energy-level diagram; (b) EL and PL spectra of the device, depending on the ZnO nanoparticle layer thickness. (Reproduced from Stouwdam, J.W. and Janssen, R.A.J., *J. Mater. Chem.*, 18, 1889, 2008. The Royal Society of Chemistry.)

corresponding energy-level diagram. The EL spectra of the devices comprising ZnO nanoparticle layers with different thicknesses are shown in Figure 5.10b. The highest luminous efficiencies achieved in this study were 0.3 $cdA^{-1}$ for red-, 0.65 $cdA^{-1}$ for green-, and 0.03 $cdA^{-1}$ for blue-color emitting devices, lower than the best values reported in literature, but approaching them. More importantly, multilayer QD-LEDs were demonstrated, which opens an efficient way to make inorganic charge transport layers in a cost-effective way. Additionally, the need for a low work function metal cathode that is commonly used in combination with organic ETLs was eliminated, since the conduction band level of ZnO has a quite low energy.

Successful substitution of the organic support layers opened a path for all-inorganic QD-LED. One of the first all-inorganic devices was presented by Artemyev et al. in year 1997, where no additional support layers were used (Artemyev et al. 1997). CdS nanocrystals were drop-deposited in a 200–300 nm thick layer between ITO and Ag electrodes. Spectrally wide EL with a voltage-dependent color was achieved at voltages higher than 20 V. The resulting EL spectrum was attributed to voltage-controlled population of different deep trap states. The overall efficiency and brightness were poor. Nevertheless, EL could be observed by a naked eye. Improved structures with inorganic charge injection/transport support layers were recently reported (Mueller et al. 2005; Caruge et al. 2008; Kang et al. 2008; Gopal et al. 2009). Caruge et al. (2008) demonstrated an efficient all-inorganic QD-LED with sputtered metal-oxide charge transport layers. Figure 5.11 shows a schematic of the QD-LED with NiO hole- and alloyed ZnO and $SnO_2$ ETLs (a) and the corresponding band diagram determined from UV photoemission spectroscopy and optical absorption measurements (b). The EQE and luminosity of the device as a function of the current density are shown in Figure 5.11c with a photograph of the device, operated at 6 V bias as an inset, demonstrating the efficiency and high robustness of the inorganic charge transport layers. Not only NiO (Caruge et al. 2008) but also p-type Si (Kang et al. 2008; Gopal et al. 2009) and p-type GaN (Mueller et al. 2005) were used as hole injection

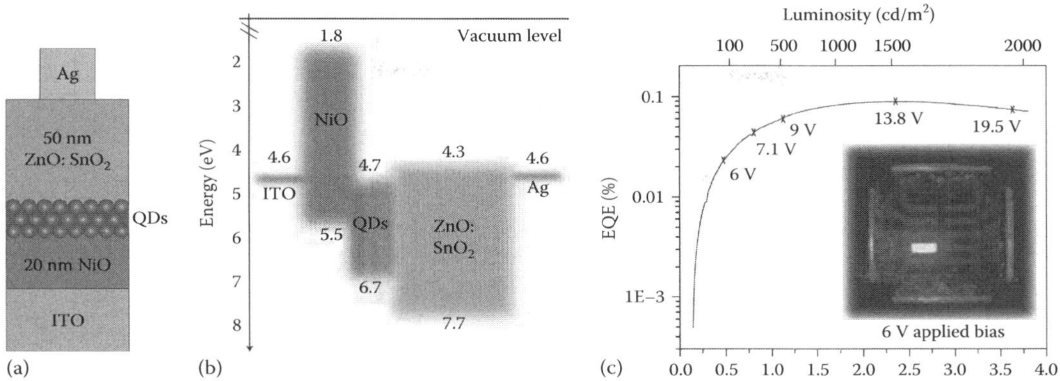


**FIGURE 5.11**
**(See color insert following page 302.)** (a) Schematic of the device structure, (b) band diagram determined from UV photoemission spectroscopy and optical absorption measurements, and (c) QD-LED EQE as a function of the current density J. The inset shows a photograph of the bright and uniform pixel at 6 V applied bias. (Reprinted with permission from Caruge, J.M. et al., *Nat. Photon.*, 2, 247, 2008. Copyright [2008], Macmillan Publishers Ltd.)

and transport layers. On the electron injection side, besides $ZnO:SnO_2$ (Caruge et al. 2008; Gopal et al. 2009), n-type GaN (Mueller et al. 2005) and n-type $TiO_2$ (Kang et al. 2008) have been applied. The active QD layer was deposited by either a Langmuir-Blodgett-method (Mueller et al. 2005), spin coating (Caruge et al. 2008; Kang et al. 2008), or contact printing (Gopal et al. 2009). The deposition of the inorganic support layers was more laborious with various deposition techniques, such as energetic neutral atom beam lithography/epitaxy (Mueller et al. 2005), plasma-enhanced metallorganic chemical vapor deposition (Kang et al. 2008), and sputtering (Caruge et al. 2008; Gopal et al. 2009). These deposition techniques require chambers with controlled pressure, gas flow, and temperatures sometimes up to 500°C (Mueller et al. 2005), which makes the fabrication of large-area emitters on flexible substrates challenging. The simplification of the inorganic layer deposition techniques by utilizing appropriate inorganic nanoparticles might be one key issue for future devices.

### 5.2.2 Wide Band Gap Semiconductor Nanoparticle Light-Emitting Devices

In QD-LEDs, the emission color, as discussed above, is among others strongly controlled by the size of the used nanocrystals due to the confinement effects. Alternatively, EL devices with different colors and even white emission can be realized by using the natural defects in the particles or by an intentional doping with different metals, which act as efficient luminescent centers. In this case, the emission color is not directly linked with the size of the particles, so the monodispersity of the fabricated nanocrystals is no longer coercive. This allows the implementation of crystals with sizes from few nanometer up to 100 nm and higher, which can be easily manufactured in high amounts at low material and synthesis costs (Swihart 2003; Manoharan and Mohammad 2005; Neshataeva et al. 2009b). This is typically the case for material systems like ZnS, ZnO, and GaN. These materials with a bulk direct band gap value higher than 3 eV at room temperature and allow in addition the fabrication of large-area UV emitters. Other advantages of these material systems are their robustness at ambient conditions and the absence of toxic elements like Cd or As, which principally allows environment-friendly devices without additional encapsulation. Certainly, this is one of the driving forces for recent developments, despite the still limited QE of these nanoparticles compared with the well-established CdSe nanocrystal system.

The research on the LEDs based on such wide band gap semiconductors has started years after the first QD-LED demonstration and progressed much more slowly. One of the main challenges is the high-energy barrier for the hole injection due to the wide band gap of the particles. The general development route was still quite similar to QD-LEDs with first devices based on nanocrystals in organic/polymeric matrices (Huang et al. 1997; Que et al. 1998) for better processing, as sketched in Figure 5.2a. Advancements in the host/guest matrix systems resulted in a white light-emitting device with an active luminescent layer of poly(9,9-dioctylfluorene) (PFO) blended with Cu-doped ZnS nanocrystals (Manoharan and Mohammad 2005). The active layer was sandwiched between an ITO anode, covered with hole-conducting poly(3,4-ethylenedioxythiophene):polystyrenesulfonic acid (PEDOT:PSS) and an Al cathode, as shown in Figure 5.12a. The energy-level diagram of the device is shown in Figure 5.12b. The turn on voltage of the device depends on the nanocrystal content in the blend and can be as low as 4 V in case of 1 wt% of ZnS:Cu in PFO. The contribution of PFO and ZnS:Cu emission to the overall EL spectrum depends also on the nanoparticle content, though resulting in a white light emission, as shown in Figure 5.12c. The 1 wt% device demonstrated a brightness of 40 cd/m$^2$ at 9 V. An advanced device concept with nanoparticle layers sandwiched between organic charge injection and transport support layers, as shown in Figure 5.2b, was also successfully implemented in ZnS and ZnO nanoparticle-based systems (Yang et al. 2003; Lee et al. 2006, 2008; Niu et al. 2006). The lifetime and fabrication limitations due to the use of organic support layers are similar to QD-LEDs.

There are only a few all-inorganic device demonstrations so far because of the large energy barrier for the hole injection into the wide band gap nanoparticles, which remains a great challenge. Bright blue/green EL from $Eu^{2+}$-doped GaN/$SiO_2$ nanocomposites was demonstrated by a group from University of Victoria (Canada) (Mahalingam et al. 2007). By synthesizing $Eu^{2+}$-doped GaN nanocrystals on top of a $SiO_2$ nanoparticle, as sketched

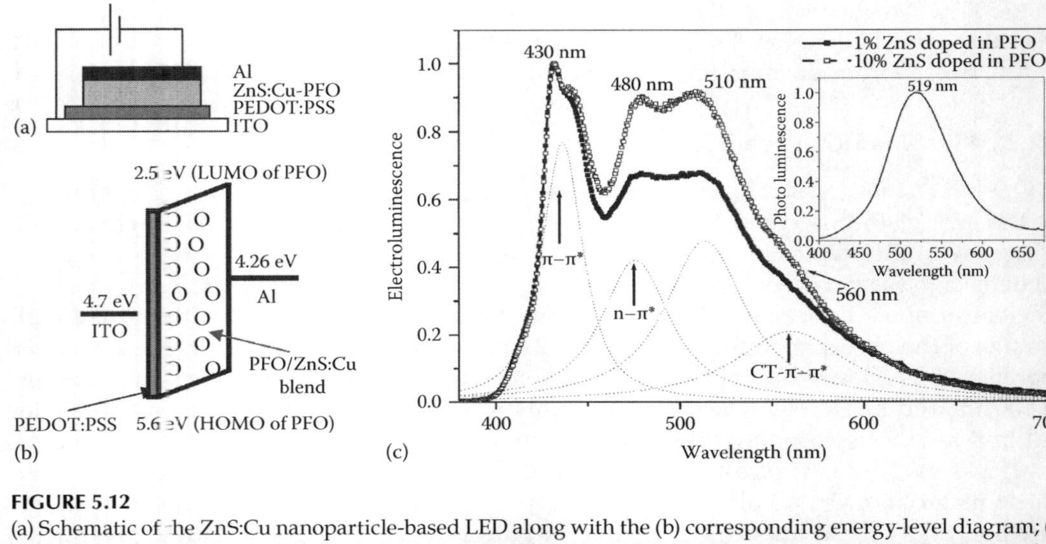


**FIGURE 5.12**
(a) Schematic of the ZnS:Cu nanoparticle-based LED along with the (b) corresponding energy-level diagram; (c) EL spectra of the device with 1 wt% (solid line) and 10 wt% (dashed line). In both cases, three different contributions from PFO and one at 510 nm resulting from the ZnS:Cu nanocrystals are obtained. The inset shows the PL spectrum of ZnS:Cu nanocrystals. (Reproduced from Manoharan, S.S. and Mohammad, Q., *Phys. Status Solidi (a)*, 202, 1124, 2005. Copyright Wiley-VCH Verlag GmbH & Co. KGaA.)

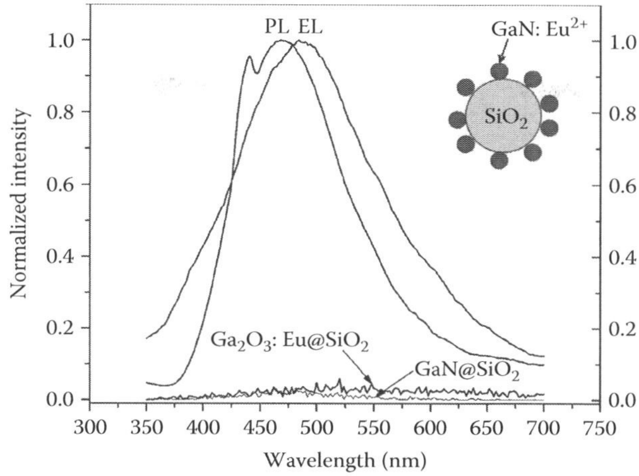


**FIGURE 5.13**
EL spectra collected from an ITO/$Eu^{2+}$-doped GaN-$SiO_2$/Ca/Al EL device and from two control devices such as ITO/GaN-$SiO_2$/Ca/Al and ITO/$Eu^{3+}$-doped $Ga_2O_3$-$SiO_2$/Ca/Al. For comparison, the PL spectrum of $Eu^{2+}$-doped GaN/$SiO_2$ is also shown. The inset shows a schematic of the synthesized $Eu^{2+}$-doped GaN/$SiO_2$ nanocomposite. (Reproduced from Mahalingam, V. et al., *Adv. Funct. Mater.*, 17, 3462, 2007. Copyright Wiley-VCH Verlag GmbH & Co. KGaA.)

in the inset of Figure 5.13, an efficient blue/green luminescent center in GaN was created. Nanocomposites were dispersed in isopropanol by sonication and afterward spin-coated on the ITO substrate and covered by a Ca/Al cathode. The EL of the device under 18 V forward bias is shown in Figure 5.13, compared with the PL spectrum of the nanocomposites. Control devices without Eu-doping or with doping of $Ga_2O_3$ instead of GaN showed no EL, demonstrating the clear origin of the EL in the device from the $Eu^{2+}$-doped GaN/$SiO_2$ nanocomposites.

ZnO nanoparticles prepared by a sol–gel method were used in an all-inorganic light-emitting device by Xiaopeng Zhao and coworkers (Ning et al. 2006; Yan et al. 2006; Li et al. 2008). The devices consisted of a macroscopic, 0.1–0.3 mm thick nanoparticle powder layer, pressed between ITO and an Al electrode. A defect-related green and violet EL in the visible spectral range was retrieved at voltages of 100 V and higher (Ning et al. 2006; Yan et al. 2006; Li et al. 2008). Defect-related EL in the visible spectral range was also achieved by our group at considerably lower operation voltages by spin coating ZnO nanoparticles from the aqueous dispersion onto ITO substrates (Neshataeva et al. 2008). Spin coating resulted in thin and dense homogeneous nanoparticle layers, which were contacted by a top Al electrode through a shadow mask. This allowed defect-related EL in the red/NIR spectral range at voltages as low as 4 V, as shown in Figure 5.14a. At higher voltages, defect states with higher energies can be populated resulting in violet and blue/green emission (Neshataeva et al. 2008). No ZnO band gap-related emission in the UV spectral range could be observed in these devices.

A considerable progress in this respect could be achieved by an improved device design. EL in the visible and the UV range was achieved by us in all-inorganic ZnO nanoparticle-based LEDs (Neshataeva et al. 2009a) by using FTO as a substrate. Commercially available ZnO nanoparticles with an average size of 50 nm were spin-coated onto a FTO-glass substrate resulting in an approximately 350 nm thin and tight ZnO nanoparticle layer. An Al cathode was evaporated on top of the nanoparticle layer. Figure 5.14b shows EL spectra of

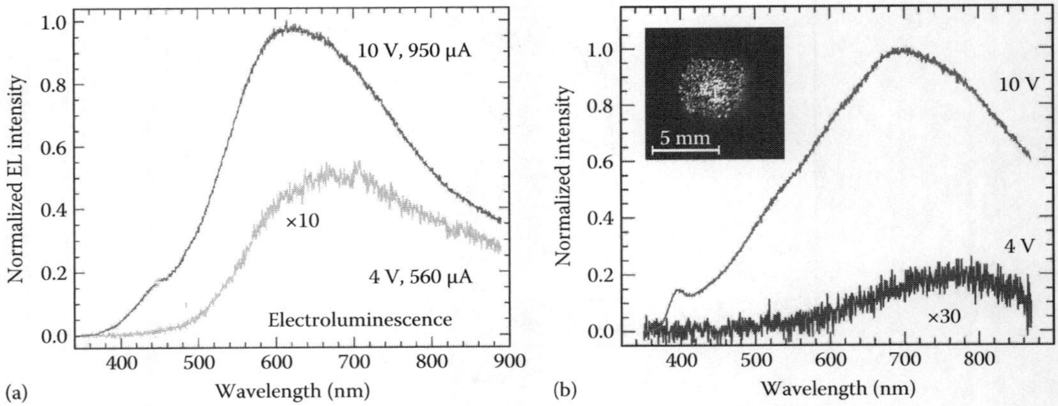


**FIGURE 5.14**
(a) EL spectra of the ZnO nanoparticle LED on ITO substrate at different voltages (From Neshataeva, E. et al., *Electron. Lett.*, 44, 1485, 2008. © 2008 IEEE.); (b) EL spectra of the device based on FTO substrate at different voltages. The inset shows a photograph of the device under 9 V forward bias. (Reprinted with permission from Neshataeva, E. et al., *Appl. Phys. Lett.*, 94, 091115, 2009a. Copyright [2009], American Institute of Physics.)

the device under forward bias. First, ZnO defect-states-related EL could be detected at voltages as low as 4 V, though quite weak. The EL intensity increases with increasing voltage, becoming visible to the naked eye at 6 V. As the voltage increases, defect states with higher energies can be populated and, most important, the band gap emission of ZnO contributes more and more to the EL spectrum. The narrow near-band gap emission can be detected at voltages higher than 9 V. The typical EL spectrum at higher voltages (e.g., at 10 V, shown in Figure 5.14b) exhibits a distinctive narrow peak in the UV range originating from excitonic near-band gap recombination in ZnO, as well as a wide emission in the visible spectral range due to the different native defect states in the particles, resulting in an overall whitish color of the emission. The inset of Figure 5.14b shows a photograph of the device operating at 9 V under forward bias. In contrast to earlier all-inorganic ZnO NP-LEDs (Neshataeva et al. 2008), a large-area emission was successfully achieved but the overall intensity and homogeneity need to be improved. Nevertheless, it was demonstrated that large-area all-inorganic UV emitters based on ZnO nanoparticles are in principle possible.

### 5.2.3 DC Nanoparticle LEDs: Mechanisms Behind

DC light emitters based on nanoparticles is a quite new and fast-developing research field with a lot of different device architectures and fabrication techniques and still many new to come. The understanding of the processes behind the light emission in different devices needs to be engrossed. Charge carrier transport in nanoparticle systems can, in most cases, be described by a space-charge-limited current model assuming hopping of electrons (Colvin et al. 1994; Artemyev et al. 1997; Gao et al. 1997, 1999; Mattoussi et al. 1998; Gaponik et al. 1999; Hikmet et al. 2003; Caruge et al. 2006). The origin of the light emission is still under discussion and can be attributed to different processes. In case of nanocrystals embedded in conductive or semiconducting polymer matrices, as well as devices with organic support layer(s), the EL is most frequently attributed to two processes taking place at the same time: direct charge carrier injection into the particles and energy transfer from the host matrix or from organic support layers to the particles both resulting in excitons, formed on the nanoparticles and their defect states, which recombine and emit light (Li et al. 2005, 2006; Manoharan and Mohammad 2005; Niu et al. 2006; Rizzo et al. 2007; Anikeeva et al. 2008; Lee et al. 2008). In case of nanoparticles imbedded in an isolating host matrix (Huang et al. 1997; Gao et al. 1998; Que et al. 1998; Adachi et al. 2007) or pure inorganic LEDs without additional organic support layers (Artemyev et al. 1997; Mahalingam et al. 2007; Neshataeva et al. 2009a), the exciton formation may not purely be attributed to direct charge carrier injection, but impact ionization has to be considered as well (Artemyev et al. 1997; Adachi et al. 2007).

## 5.3 Alternating Current-Operated Light-Emitting Devices (AC-LEDs)

Alternating current thin-film electroluminescent (AC-TFEL) devices are well known as efficient and robust large-area-emitting devices with low power consumption and are already wide spread on the flat panel market. AC-TFEL devices consist of an emissive phosphor layer, for example, manganese-doped zinc sulfide (ZnS:Mn) sandwiched between insulating layers that are electrically contacted by the electrodes. One of the electrodes is typically metallic, another is a TCO to enable the light emission through it. At high voltages, charge carriers trapped at the interfaces between the layers are injected into the phosphor. They can be trapped within the phosphor and recombine (bipolar field-emission model) or become accelerated by the applied electric field and excite the luminescent centers of the phosphor by impact excitation and ionization mechanisms (Yen 2007). The development of multicolor-displays remains still challenging since most different-colored phosphors comprise different material systems and therefore need different treatment (e.g., annealing at different conditions) and deposition techniques (Ono 1995; Yen 2007).

It is expected that the standard AC-TFEL devices can be improved by replacing the mostly used micropowders by nanoparticles allowing easy fabrication of multicolor displays and lowering the material costs. Manzoor and colleagues demonstrated in 2003 the successful fabrication of nanoparticle-based AC-TFEL devices with blue, green, and orange–red colored light emission out of doped ZnS nanocrystals (Manzoor et al. 2004). Figure 5.15

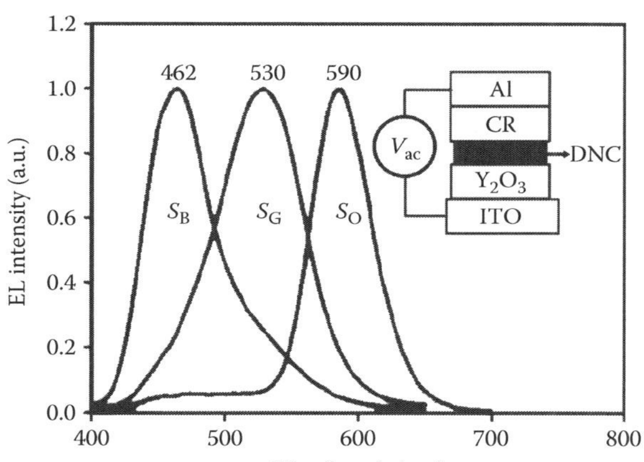


**FIGURE 5.15**
EL spectra of doped ZnS nanocrystals. Inset: schematic diagram of nanocrystals based ac EL device. (Reprinted with permission from Manzoor, K. et al., *Appl. Phys. Lett.*, 84, 284, 2004. Copyright [2004], American Institute of Physics.)

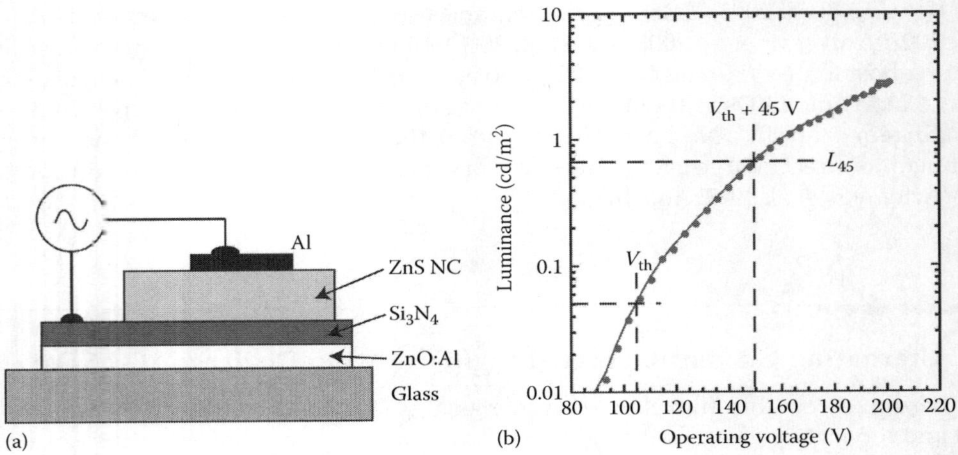


**FIGURE 5.16**
(a) Schematic cross-sectional illustration of an EL device with printed emission layer containing ZnS:Mn nanocrystals. (b) Typical *L–V* characteristic of the ZnS:Mn NC EL device. The definitions of threshold voltage ($V_{th}$) as well as characteristic luminance ($L_{45}$) are shown schematically. (From Toyama, T. et al., *Nanotechnology*, 20, 055203, 2009.)

shows the schematic of the device and the EL spectra. Different colors were achieved by doping the ZnS with Al, Cu, and Mn ions. The doped nanoparticles were spray coated on an yttrium oxide ($Y_2O_3$) insulating layer, afterward covered by another isolating layer of high dielectric cyano resin. Al and transparent ITO were used as the conductive electrodes. The EL emission was observed at low AC voltages of 10 V at a frequency of 100 Hz (Manzoor et al. 2004).

First steps toward large-area printable AC displays were made by Toyama and coworkers. They demonstrated EL of a printed nanocrystal layer of ZnS:Mn with a luminance of almost 3 cd/m² at 200 V AC and 5 kHz sinusoidal voltage (Toyama et al. 2009). Figure 5.16a shows the schematic of the device structure with an aluminum-doped zinc oxide and Al electrodes, $Si_3N_4$ insulating layer, and ZnS:Mn as an active luminescent layer. Figure 5.16b shows the typical luminance behavior of the AC-NP-LED as a function of operating voltage with first increasing light emission followed by a slow saturation at high voltages. The peak luminance of almost 3 cd/m² was measured at 200 V AC. A luminance of 2 cd/m² was demonstrated by Wood et al. (2009b) in an all-transparent light-emitting device, operated at 170 V peak to peak at 30 kHz frequency. These findings present first steps toward printable, transparent multicolor AC-driven displays, based on direct band gap semiconducting nanoparticles.

## 5.4 Outlook

The future development of DC- as well as AC-driven nanoparticle-based devices follows the route toward robust large-area devices on both rigid and flexible substrates. The QDs are more likely to be applied in color display applications, whereas wide band gap semiconducting nanoparticles might allow low-cost ambient lightning applications. All-inorganic light-emitting devices are expected to have longer lifetimes and operate stable at high currents and fields, making even laser applications possible. Another future prospect would be the realization of all-nanoparticle light-emitting devices, which is compatible with printing technology, resulting in customized low-cost emitters on flexible substrates. Recent publications already showed first steps toward multicolor displays and large-area emitters (Tan et al. 2007; Kim et al. 2008; Rizzo et al. 2008; Zhu et al. 2008; Anikeeva et al. 2009; Bae et al. 2009), as well as all-inorganic (Mahalingam et al. 2007; Caruge et al. 2008; Kang et al. 2008; Gopal et al. 2009; Neshataeva et al. 2009a) and even transparent devices (Wood et al. 2009b). New deposition techniques such as inkjet (Haverinen et al. 2009; Toyama et al. 2009) and contact printing (Anikeeva et al. 2008; Kim et al. 2008; Rizzo et al. 2008; Gopal et al. 2009) were implemented. First LEDs, based on nanoparticle multilayers (Stouwdam et al. 2008), and first emitters on flexible substrates (Tan et al. 2009) were successfully demonstrated. The reliability, efficiency, brightness, power consumption, and handling of the nanoparticle LEDs must be further improved. Thus, these findings let us expect a set of exciting future applications like light-emitting wallpapers, lasers on arbitrary substrates, and foldable takeaway displays.

## Acknowledgments

We would like to thank Dr. André Ebbers from Evonik Degussa GmbH for the fruitful discussions. The financial support of the German Research Foundation in the frame of the Graduate Research Group GRK 1240 "Nanotronics" and the support of CeNIDE are gratefully acknowledged.

## References

Adachi, D., T. Hama, T. Toyama, and H. Okamoto. 2007. Electroluminescent properties of chemically synthesized zinc sulfide nanocrystals doped with manganese. *Journal of Materials Science: Materials in Electronics* 20, S1 (11): 130–133.

Ahn, J. H., C. Bertoni, S. Dunn et al. 2007. White organic light-emitting devices incorporating nanoparticles of II–VI semiconductors. *Nanotechnology* 18, 33 (8): 335202.

Alivisatos, A. P. 1996. Semiconductor clusters, nanocrystals, and quantum dots. *Science* 271, 5251 (2): 933–937.

Anikeeva, P. O., J. E. Halpert, M. G. Bawendi, and V. Bulović. 2007. Electroluminescence from a mixed red–green–blue colloidal quantum dot monolayer. *Nano Letters* 7, 8 (8): 2196–2200.

Anikeeva, P. O., C. F. Madigan, J. E. Halpert, M. G. Bawendi, and V. Bulović. 2008. Electronic and excitonic processes in light-emitting devices based on organic materials and colloidal quantum dots. *Physical Review B* 78, 8 (8): 085434–8.

Anikeeva, P. O., J. E. Halpert, M. G. Bawendi, and V. Bulović. 2009. Quantum dot light-emitting devices with electroluminescence tunable over the entire visible spectrum. *Nano Letters* 9, 7 (7): 2532–2536.

Artemyev, M. V., V. Sperling, and U. Woggon. 1997. Electroluminescence in thin solid films of closely packed CdS nanocrystals. *Journal of Applied Physics* 81, 10: 6975.

Bach, U., D. Lupo, P. Comte et al. 1998. Solid-state dye-sensitized mesoporous $TiO_2$ solar cells with high photon-to-electron conversion efficiencies. *Nature* 395, 6702 (8): 583–585.

Bae, W. K., J. Kwak, J. W. Park, K. Char, C. Lee, and S. Lee. 2009. Highly efficient green-light-emitting diodes based on CdSe@ZnS quantum dots with a chemical-composition gradient. *Advanced Materials* 21, 17 (5): 1690–1694.

Bakueva, L., G. Konstantatos, L. Levina, S. Musikhin, and E. H. Sargent. 2004. Luminescence from processible quantum dot-polymer light emitters 1100–1600 nm: Tailoring spectral width and shape. *Applied Physics Letters* 84, 18: 3459.

Bertoni, C., D. Gallardo, S. Dunn, N. Gaponik, and A. Eychmüller. 2007. Fabrication and characterization of red-emitting electroluminescent devices based on thiol-stabilized semiconductor nanocrystals. *Applied Physics Letters* 90, 3: 034107.

Brus, L. 1991. Quantum crystallites and nonlinear optics. *Applied Physics A: Solids and Surfaces* 53, 6 (12): 465–474.

Burnham, G. and Society of Photo-optical Instrumentation Engineers. 2000. *Laser Diodes and LEDs in Industrial, Measurement, Imaging, and Sensors Applications II: Testing, Packaging, and Reliability of Semiconductor Lasers V: 26–26[sic]*, January, 2000. San Jose, CA. Bellingham, WA: SPIE.

Caruge, J. M., J. E. Halpert, V. Bulović, and M. G. Bawendi. 2006. NiO as an inorganic hole-transporting layer in quantum-dot light-emitting devices. *Nano Letters* 6, 12 (12): 2991–2994.

Caruge, J. M., J. E. Halpert, V. Wood, V. Bulović, and M. G. Bawendi. 2008. Colloidal quantum-dot light-emitting diodes with metal-oxide charge transport layers. *Nature Photonics* 2, 4 (3): 247–250.

Chen, H.-S., C.-K. Hsu, and H.-Y. Hong. 2006. InGaN-CdSe-ZnSe quantum dots white LEDs. *IEEE Photonics Technology Letters* 18, 1 (1): 193–195.

Cheong, H. J., A. Tanaka, D. Hippo et al. 2008. Visible electroluminescence from spherical-shaped silicon nanocrystals. *Japanese Journal of Applied Physics* 47, 10 (10): 8137–8140.

Chopra, K., S. Major, and D. Pandya. 1983. Transparent conductors—A status review. *Thin Solid Films* 102, 1 (4): 1–46.

Coe, S., W.-K. Woo, M. Bawendi, and V. Bulović. 2002. Electroluminescence from single monolayers of nanocrystals in molecular organic devices. *Nature* 420, 6917 (12): 800–803.

Coe-Sullivan, S., J. S. Steckel, W.-K. Woo, M. G. Bawendi, and V. Bulović. 2005. Large-area ordered quantum-dot monolayers via phase separation during spin-casting. *Advanced Functional Materials* 15, 7 (7): 1117–1124.

Colvin, V. L., M. C. Schlamp, and A. P. Alivisatos. 1994. Light-emitting diodes made from cadmium selenide nanocrystals and a semiconducting polymer. *Nature* 370, 6488 (8): 354–357.

Dabbousi, B. O., M. G. Bawendi, O. Onitsuka, and M. F. Rubner. 1995. Electroluminescence from CdSe quantum-dot/polymer composites. *Applied Physics Letters* 66, 11: 1316.

Dabbousi, B. O., J. Rodriguez-Viejo, F. V. Mikulec et al. 1997. (CdSe)ZnS core–shell quantum dots: Synthesis and characterization of a size series of highly luminescent nanocrystallites. *The Journal of Physical Chemistry B* 101, 46 (11): 9463–9475.

Efros, Al. L. and A. L. Efros. 1982. Interband absorption of light in a semiconductor sphere. *Soviet Physics—Semiconductors* 16: 772.

Gallardo, D. E., C. Bertoni, S. Dunn, N. Gaponik, and A. Eychmüller. 2007. Cathodic and anodic material diffusion in polymer/semiconductor-nanocrystal composite devices. *Advanced Materials* 19, 20 (10): 3364–3367.

Gao, M., B. Richter, and S. Kirstein. 1997. White-light electroluminescence from self-assembled Q-CdSe/PPV multilayer structures. *Advanced Materials* 9, 10: 802–805.

Gao, M., B. Richter, S. Kirstein, and H. Möhwald. 1998. Electroluminescence studies on self-assembled films of PPV and CdSe nanoparticles. *The Journal of Physical Chemistry B* 102, 21 (5): 4096–4103.

Gao, M., B. Richter, and S. Kirstein. 1999. Electroluminescence and photoluminescence in CdSe/poly (p-phenylene vinylene) composite films. *Synthetic Metals* 102, 1–3 (6): 1213–1214.

Gaponik, N. P., D. V. Talapin, and A. L. Rogach. 1999. A light-emitting device based on a CdTe nanocrystal/polyaniline composite. *Physical Chemistry Chemical Physics* 1, 8: 1787–1789.

Gaponik, N. P., D. V. Talapin, A. L. Rogach, and A. Eychmüller. 2000. Electrochemical synthesis of CdTe nanocrystal/polypyrrole composites for optoelectronic applications. *Journal of Materials Chemistry* 10, 9: 2163–2166.

Gopal, A., K. Hoshino, S. Kim, and X. Zhang. 2009. Multi-color colloidal quantum dot based light emitting diodes micropatterned on silicon hole transporting layers. *Nanotechnology* 20, 23 (6): 235201.

Haverinen, H. M., R. A. Myllylä, and G. E. Jabbour. 2009. Inkjet printing of light emitting quantum dots. *Applied Physics Letters* 94, 7: 073108.

Held, G. 2009. *Introduction to Light Emitting Diode Technology and Applications*. Boca Raton, FL: CRC Press.

Hikmet, R. A. M., D. V. Talapin, and H. Weller. 2003. Study of conduction mechanism and electroluminescence in CdSe/ZnS quantum dot composites. *Journal of Applied Physics* 93, 6: 3509.

Hirose, Y., A. Kahn, V. Aristov, and P. Soukiassian. 1996a. Chemistry, diffusion, and electronic properties of a metal/organic semiconductor contact: In/perylenetetracarboxylic dianhydride. *Applied Physics Letters* 68, 2: 217.

Hirose, Y., A. Kahn, P. Soukiassian, V. Bulovic, and S. R. Forrest. 1996b. Chemistry and electronic properties of metal-organic semiconductor interfaces: Al, Ti, In, Sn, Ag, and Au on PTCDA. *Physical Review B* 54, 19 (11): 13748–13758.

Huang, J., Y. Yang, S. Xue, B. Yang, S. Liu, and J. Shen. 1997. Photoluminescence and electroluminescence of ZnS:Cu nanocrystals in polymeric networks. *Applied Physics Letters* 70, 18: 2335.

Hussain, A. M., B. Neppolian, S. H. Kim et al. 2009. Improved performance of polymer light-emitting diodes with nanocomposites. *Applied Physics Letters* 94, 7: 073306.

Jain, S. 2007. *Conducting Organic Materials and Devices*. Boston, MA: Elsevier/Academic Press.

Jang, H. S., H. Yang, S. W. Kim, J. Y. Han, S.-G. Lee, and D. Y. Jeon. 2008. White light-emitting diodes with excellent color rendering based on organically capped CdSe quantum dots and $Sr_3SiO_5:Ce^{3+},Li^{+}$ phosphors. *Advanced Materials* 20, 14 (7): 2696–2702.

Kan, E. W. H., W. K. Chim, C. H. Lee, W. K. Choi, and T. H. Ng. 2004. Clarifying the origin of near-infrared electroluminescence peaks for nanocrystalline germanium in metal-insulator-silicon structures. *Applied Physics Letters* 85, 12: 2349.

Kang, S.-H., C. K. Kumar, Z. Lee, K.-H. Kim, C. Huh, and E.-T. Kim. 2008. Quantum-dot light-emitting diodes utilizing CdSe/ZnS nanocrystals embedded in $TiO_2$ thin film. *Applied Physics Letters* 93, 19: 191116.

Katari, J. E., V. L. Colvin, and A. P. Alivisatos. 1994. X-ray photoelectron spectroscopy of CdSe nanocrystals with applications to studies of the nanocrystal surface. *The Journal of Physical Chemistry* 98, 15 (4): 4109–4117.

Kim, L. A., P. O. Anikeeva, S. A. Coe-Sullivan, J. S. Steckel, M. G. Bawendi, and V. Bulović. 2008. Contact printing of quantum dot light-emitting devices. *Nano Letters* 8, 12 (12): 4513–4517.

Kobayashi, S., Y. Tani, and H. Kawazoe. 2007. Quantum dot activated all-inorganic electroluminescent device fabricated using solution-synthesized CdSe/ZnS nanocrystals. *Japanese Journal of Applied Physics* 46, 40 (10): 966–969.

Larkin, I. A., M. I. Stockman, M. Achermann, and V. I. Klimov. 2004. Dipolar emitters at nanoscale proximity of metal surfaces: Giant enhancement of relaxation in microscopic theory. *Physical Review B* 69, 12 (3): 121403.

Lee, C.-Y., Y.-T. Haung, W.-F. Su, and C.-F. Lin. 2006. Electroluminescence from ZnO nanoparticles/organic nanocomposites. *Applied Physics Letters* 89, 23: 231116.

Lee, C.-Y., Y.-Y. Hui, W.-F. Su, and C.-F. Lin. 2008. Electroluminescence from monolayer ZnO nanoparticles using dry coating technique. *Applied Physics Letters* 92, 26: 261107.

Li, Z. 2007. *Organic Light-Emitting Materials and Devices*. Boca Raton, FL: CRC/Taylor & Francis.

Li, Y., A. Rizzo, M. Mazzeo et al. 2005. White organic light-emitting devices with CdSe/ZnS quantum dots as a red emitter. *Journal of Applied Physics* 97, 11: 113501.

Li, Y. Q., A. Rizzo, R. Cingolani, and G. Gigli. 2006. Bright white-light-emitting device from ternary nanocrystal composites. *Advanced Materials* 18, 19 (10): 2545–2548.

Li, J., X. Zhao, C. Yan, and L. Yijing. 2008. Electroluminescence of ZnO nanocrystalline particles annealed from mesoporous precursors. *Materials Chemistry and Physics* 107, 1 (1): 177–182.

Mahalingam, V., M. Tan, P. Munusamy, J. B. Gilroy, M. Raudsepp, and F. C. J. M. van Veggel. 2007. Bright blue photo- and electroluminescence from Eu2+-doped GaN/$SiO_2$ nanocomposites. *Advanced Functional Materials* 17, 17 (11): 3462–3469.

Manoharan, S. and Q. Mohammad. 2005. White light electroluminescence in zinc sulfide: Polyflourene blends. *Physica Status Solidi (a)* 202, 6 (5): 1124–1128.

Manzoor, K., S. R. Vadera, N. Kumar, and T. R. N. Kutty. 2004. Multicolor electroluminescent devices using doped ZnS nanocrystals. *Applied Physics Letters* 84, 2: 284.

Mattoussi, H., L. H. Radzilowski, B. O. Dabbousi, E. L. Thomas, M. G. Bawendi, and M. F. Rubner. 1998. Electroluminescence from heterostructures of poly(phenylene vinylene) and inorganic CdSe nanocrystals. *Journal of Applied Physics* 83, 12: 7965.

Mattoussi, H., L. H. Radzilowski, B. O. Dabbousi et al. 1999. Composite thin films of CdSe nanocrystals and a surface passivating/electron transporting block copolymer: Correlations between film microstructure by transmission electron microscopy and electroluminescence. *Journal of Applied Physics* 86, 8: 4390.

Minami, T. 2008a. Substitution of transparent conducting oxide thin films for indium tin oxide transparent electrode applications. *Thin Solid Films* 516, 7 (2): 1314–1321.

Minami, T. 2008b. Present status of transparent conducting oxide thin-film development for indium-tin-oxide (ITO) substitutes. *Thin Solid Films* 516, 17 (7): 5822–5828.

Mottier, P. 2009. *LEDs for Lighting Applications*. London; Hoboken, NJ: Wiley-ISTE.

Mueller, A. H., M. A. Petruska, M. Achermann et al. 2005. Multicolor light-emitting diodes based on semiconductor nanocrystals encapsulated in GaN charge injection layers. *Nano Letters* 5, 6 (6): 1039–1044.

Müllen, K. 2006. *Organic Light-Emitting Devices: Synthesis, Properties, and Applications*. Weinheim, Germany: Wiley-VCH.

Murray, C. B., D. J. Norris, and M. G. Bawendi. 1993. Synthesis and characterization of nearly monodisperse CdE (E = sulfur, selenium, tellurium) semiconductor nanocrystallites. *Journal of the American Chemical Society* 115, 19 (9): 8706–8715.

Murray, C. B., C. R. Kagan, and M. G. Bawendi. 2000. Synthesis and characterization of monodisperse nanocrystals and close-packed nanocrystal assemblies. *Annual Review of Materials Science* 30, 1 (8): 545–610.

Neshataeva, E., T. Kümmell, A. Ebbers, and G. Bacher. 2008. Electrically driven ZnO nanoparticle light emitting device. *Electronics Letters* 44, 25: 1485.

Neshataeva, E., T. Kümmell, A. Ebbers and G. Bacher. 2009a. All-inorganic light emitting device based on ZnO nanoparticles. *Applied Physics Letters* 94, 9: 091115.

Neshataeva, E., T. Kuemmell, A. Ebbers, and G. Bacher. 2009b. Low operation voltage UV-light emitting device based on ZnO nanoparticles. In *Proceedings of SPIE*, pp. 721707–721710, San Jose, CA.

Ning, G., X. Zhao, J. Li, and C. Zhang. 2006. Hugely enhanced electroluminescence from mesoporous ZnO particles. *Optical Materials* 28, 4 (3): 385–390.

Niu, J.-H., R.-N. Hua, W.-L. Li, M.-T. Li, and T.-Z. Yu. 2006. Electroluminescent properties of a device based on terbium-doped ZnS nanocrystals. *Journal of Physics D: Applied Physics* 39, 11 (6): 2357–2360.

Nizamoglu, S. and H. V. Demir. 2007. Hybrid white light sources based on layer-by-layer assembly of nanocrystals on near-UV emitting diodes. *Nanotechnology* 18, 40 (10): 405702.

O'Connor, É., A. O'Riordan, H. Doyle, S. Moynihan, A. Cuddihy, and G. Redmond. 2005. Near-infrared electroluminescent devices based on colloidal HgTe quantum dot arrays. *Applied Physics Letters* 86, 20: 201114.

Ono, Y. 1995. *Electroluminescent Displays*. Singapore; River Edge, NJ: World Scientific.

Que, W., Y. Zhou, Y. L. Lam et al. 1998. Photoluminescence and electroluminescence from copper doped zinc sulphide nanocrystals/polymer composite. *Applied Physics Letters* 73, 19: 2727.

Raniero, L., I. Ferreira, A. Pimentel et al. 2006. Role of hydrogen plasma on electrical and optical properties of ZGO, ITO and IZO transparent and conductive coatings. *Thin Solid Films* 511–512 (7): 295–298.

Reiss, P., J. Bleuse, and A. Pron. 2002. Highly luminescent CdSe/ZnSe core/shell nanocrystals of low size dispersion. *Nano Letters* 2, 7 (7): 781–784.

Rizzo, A., Y. Li, S. Kudera et al. 2007. Blue light emitting diodes based on fluorescent CdSe/ZnS nanocrystals. *Applied Physics Letters* 90, 5: 051106.

Rizzo, A., M. Mazzeo, M. Palumbo et al. 2008. Hybrid light-emitting diodes from microcontact-printing double-transfer of colloidal semiconductor CdSe/ZnS quantum dots onto organic layers. *Advanced Materials* 20, 10 (5): 1886–1891.

Rosenthal, S., J. Mcbride, S. Pennycook, and L. Feldman. 2007. Synthesis, surface studies, composition and structural characterization of CdSe, core/shell and biologically active nanocrystals. *Surface Science Reports* 62, 4 (4): 111–157.

Schlamp, M. C., X. Peng, and A. P. Alivisatos. 1997. Improved efficiencies in light emitting diodes made with CdSe(CdS) core/shell type nanocrystals and a semiconducting polymer. *Journal of Applied Physics* 82, 11: 5837.

Schlatmann, A. R., D. Wilms Floet, A. Hilberer et al. 1996. Indium contamination from the indium–tin–oxide electrode in polymer light-emitting diodes. *Applied Physics Letters* 69, 12: 1764.

Solomeshch, O., A. Kigel, A. Saschiuk et al. 2005. Optoelectronic properties of polymer-nanocrystal composites active at near-infrared wavelengths. *Journal of Applied Physics* 98, 7: 074310.

Steigerwald, M. L., A. P. Alivisatos, J. M. Gibson et al. 1988. Surface derivatization and isolation of semiconductor cluster molecules. *Journal of the American Chemical Society* 110, 10 (5): 3046–3050.

Stouwdam, J. W. and R. A. J. Janssen. 2008. Red, green, and blue quantum dot LEDs with solution processable ZnO nanocrystal electron injection layers. *Journal of Materials Chemistry* 18, 16: 1889.

Swihart, M. 2003. Vapor-phase synthesis of nanoparticles. *Current Opinion in Colloid & Interface Science* 8, 1 (3): 127–133.

Tan, Z., F. Zhang, T. Zhu et al. 2007. Bright and color-saturated emission from blue light-emitting diodes based on solution-processed colloidal nanocrystal quantum dots. *Nano Letters* 7, 12 (12): 3803–3807.

Tan, Z., J. Xu, C. Zhang et al. 2009. Colloidal nanocrystal-based light-emitting diodes fabricated on plastic toward flexible quantum dot optoelectronics. *Journal of Applied Physics* 105, 3: 034312.

Tessler, N. 2002. Efficient near-infrared polymer nanocrystal light-emitting diodes. *Science* 295, 5559 (2): 1506–1508.

Toyama, T., T. Hama, D. Adachi, Y. Nakashizu, and H. Okamoto. 2009. An electroluminescence device for printable electronics using coprecipitated ZnS:Mn nanocrystal ink. *Nanotechnology* 20, 5 (2): 055203.

Vossen, J. L. 1977. Transparent conducting films. *Physics of Thin Films* 9: 1.

Walters, R. J., G. I. Bourianoff, and H. A. Atwater. 2005. Field-effect electroluminescence in silicon nanocrystals. *Nature Materials* 4, 2 (1): 143–146.

Walters, R. J., J. Carreras, T. Feng, L. D. Bell, and H. A. Atwater. 2006. Silicon nanocrystal field-effect light-emitting devices. *IEEE Journal of Selected Topics in Quantum Electronics* 12, 6 (11): 1647–1656.

Wood, V., M. J. Panzer, J. Chen, M. S. Bradley, J. E. Halpert, M. G. Bawendi, and V. Bulović. 2009a. Inkjet-printed quantum dot-polymer composites for full-color AC-driven displays. *Advanced Materials* 21, 21 (6): 2151–2155.

Wood, V., J. E. Halpert, M. J. Panzer, M. G. Bawendi, and V. Bulović. 2009b. Alternating current driven electroluminescence from ZnSe/ZnS:Mn/ZnS nanocrystals. *Nano Letters* 9, 6 (6): 2367–2371.

Yan, C., Z. Chen, and X. Zhao. 2006. Enhanced electroluminescence of ZnO nanocrystalline annealing from mesoporous precursors. *Solid State Communications* 140, 1 (10): 18–22.

Yang, H., P. H. Holloway, and B. B. Ratna. 2003. Photoluminescent and electroluminescent properties of Mn-doped ZnS nanocrystals. *Journal of Applied Physics* 93, 1: 586.

Yeh, D.-M., C.-F. Huang, Y.-C. Lu, and C. C. Yang. 2008. White-light light-emitting device based on surface plasmon-enhanced CdSe/ZnS nanocrystal wavelength conversion on a blue/green two-color light-emitting diode. *Applied Physics Letters* 92, 9: 091112.

Yen, W. 2007. *Phosphor Handbook*. 2. Aufl. Boca Raton, FL: CRC Press/Taylor & Francis.

Zhao, J., J. Zhang, C. Jiang, J. Bohnenberger, T. Basché, and A. Mews. 2004. Electroluminescence from isolated CdSe/ZnS quantum dots in multilayered light-emitting diodes. *Journal of Applied Physics* 96, 6: 3206.

Zhao, J., J. A. Bardecker, A. M. Munro et al. 2006. Efficient CdSe/CdS quantum dot light-emitting diodes using a thermally polymerized hole transport layer. *Nano Letters* 6, 3 (3): 463–467.

Zheng, T., W. C. H. Choy, and Y. Sun. 2009. Nanoparticle-induced resonant tunneling behaviors in small molecule organic light-emitting devices. *Applied Physics Letters* 94, 12: 123303.

Zhu, T., K. Shanmugasundaram, S. C. Price et al. 2008. Mist fabrication of light emitting diodes with colloidal nanocrystal quantum dots. *Applied Physics Letters* 92, 2: 023111.

# 6

# *Formation of Nanosized Aluminum and Its Applications in Condensed Phase Reactions*

**Jan A. Puszynski and Lori J. Groven**

**CONTENTS**



## 6.1 Introduction

During the past decade, a significant research effort has been made to understand the reactivity of elemental nanopowders, such as aluminum, boron, silicon, and several transition metals with different oxidizers. The research effort has also been on investigation of these fuel particles as an energetic enhancement of secondary energetic systems. Energetic materials are a subclass of reactive materials containing both fuel and oxidizer. These materials can be further classified as homogeneous or heterogeneous systems, depending on whether the oxidizer is chemically or physically linked to the fuel. These types of energetic materials are commonly used as propellants, explosives, or pyrotechnics. Homogeneous energetic materials are based on monomolecular compounds, such as TNT, RDX, HMX, and CL-20 (Dreizin 2009). The maximum energy released by these compounds during the combustion process is 50%–500% lower than the energy generated by the combustion of elemental reactants. For example, the oxidation of aluminum or boron generates approximately 30 kJ/g or 58 kJ/g, respectively, compared to 10 kJ/g for HMX energetic material. In order to take advantage of the large energy associated with the oxidation of elemental powders in energetic systems, it is necessary to increase the combustion front velocity by two to three orders of magnitude. Traditionally used micron-sized powders in thermite mixtures are characterized by very low combustion front velocities, only a few meters per second. Therefore, efforts to reduce the average particle size of a fuel reactant are necessary to obtain much faster reaction rates.

The first combustion experiments with ultrafine aluminum and oxidizer powders were performed by researchers from Los Alamos National Laboratory (Martin et al. 1998). These researchers found that the combustion-front velocity in binary systems consisting of nanosized aluminum and various oxidizers, such as $MoO_3$, CuO, and $KMnO_4$, can increase the combustion-front-propagating velocity by a few orders of magnitude. In some cases, these combustion-front velocities may exceed 1000 m/s. Nanothermites, as a subclass of condensed phase reacting systems, have attracted many researchers around the world (Martin et al. 1998, Son et al. 2001, 2002, 2007; Miziolek 2002; Pivkina et al. 2004, 2006; Puszynski et al. 2005, 2006, 2007, 2008, 2009; Hunt and Pantoya 2005; Moore and Pantoya 2006; Kwok et al. 2007; Sullivan et al. 2009). Nanothermites, also called metastable intermolecular composites (MIC), normally consist of a binary mixture of fuel (aluminum, boron, silicon, or transition metals) and oxidizer ($MoO_3$, $WO_3$, CuO, $Bi_2O_3$, $KMnO_4$, or others) in the form of nanopowders (Son et al. 2007). The size of nanopowders is considered to be in the range of 100 nm or below. One of the most important parameters affecting the rate of energy release is the specific surface area of reactants, which gives very high contact area between reacting species.

The main challenge, which limited initial research, was the lack of availability of fuel nanopowders. The first reliable source of aluminum nanopowder, also called ALEX aluminum, was developed in Tomsk in the former Soviet Union (Zelinskii et al. 1984; Yavorovsky et al. 1995; Ivanov et al. 2003). A similar powder was later distributed by the Argonide Company (Tepper 2000). These powders were tested by various military and civilian sectors and have proven to be a very good alternative to micron-sized aluminum, which is commonly used in various propellants (Mench et al. 1998; Pivkina et al. 2002). Unfortunately, the ALEX and Argonide aluminum nanopowders were characterized by wide particle-size distributions, from tens to hundreds of nanometers, and could not be used in applications requiring a significantly narrower distribution. Since that time many other techniques were implemented for the formation of both fuel and oxidizer nanopowders. The paper written by Walter et al. describes the recent progress in the manufacturing of aluminum nanopowders (Walter et al. 2007).

Ultrafine metal powders have also been identified as a very promising fuel additive to improve the performance of both rocket and air-breathing propulsion systems. For example, micron-sized aluminum powder is commonly used in solid rocket propellants to boost specific impulse on the order of 10% over non-aluminized formulations. By using aluminum nanopowders, burn rates of the propellant can be enhanced by 5–10 times with respect to similar formulations using micron-sized aluminum (Dorsett et al. 2001; Dubois et al. 2007).

A completely different approach for preparing reactive nanocomposites was undertaken by Schoenitz et al. (2004). These researchers produced reactive nanocomposites by mechanical alloying, also called arrested reactive milling (ARM). In this technique, the starting materials are typically micron-sized powders of both the oxidizer and fuel. The milling is conducted in such a manner that the exothermic reaction between the fuel and oxidizer particles is not initiated. The ARM process leads to the formation of fully dense, micron-sized composite particles with nanoscaled structural features. Each particle exhibits a three-dimensional composite with intimate contact between the fuel and the oxidizer. The maximum milling time is determined experimentally and can be influenced by the addition of solvent or other additives. According to Dreizin, the ARM is very flexible and versatile (Dreizin 2009). It has been observed that essentially any combination of reactive materials can be processed by ARM to prepare a reactive nanocomposite material.

The use of nanosized aluminum in condensed phase reactions using metallic reactants as oxidizers has been solely focused on the reaction to form nickel aluminides. The combustion synthesis reaction to form NiAl has been well explored and characterized for the use of micron-sized powders; however, only a few published reports detail the use of nanosized aluminum in such reactions (Dong et al. 2002; Hunt et al. 2004, 2005, 2006; Groven and Puszynski 2005). The reported works have focused on the differences in reaction mechanism and ignition characteristics between nanosized aluminum and micron-sized aluminum as a reactant.

Nanoscale energetic materials are very sensitive to many external stimuli, even at ambient conditions, including electrostatic discharge (ESD), friction, impact, shock initiation, etc. Therefore, an understanding of safety and the use of proper equipment is paramount when dealing with these new energetic materials. Therefore, all issues associated with safety, hazards, storage, and the aging of individual nanopowders and nanoenergetic composites must be carefully considered in this type of research. Many nanothermite properties can be tailored by optimizing the particle size of reactants, their morphology, surface functionalization, materials density, and composite composition.

## 6.2 Formation of Aluminum Nanopowders

Over the last 20 years, many new technologies have been developed to produce a variety of nanopowders. The majority of applications were made in the area of the formation of oxide and nonoxide ceramics. During the past decade, a significant effort was also made in the formation of nanosized aluminum. Aluminum nanopowder is considered to be the most common elemental fuel for reactive nanocomposites. Aluminum metal powder is used in many metallurgical, chemical, paint and pigment, and military applications. Micron-sized aluminum powders with $d_{avg} > 2\,\mu m$ are normally produced by air or inert gas atomization. The unit cost of micron-sized aluminum powders fluctuates as world demand changes, but is generally less than \$20/kg. The cost of aluminum powder drastically increases when the particle size of aluminum is in the range of tens of nanometers (Kearns 2004). The current cost of aluminum nanopowders with particle sizes ranging from 80 to 120 nm can be as high as \$10,000/kg for small research quantities. Obviously, this cost can be reduced once an adequate market is established. It should also be noted that the reduction of average particle size makes aluminum, similarly to other metallic powders, pyrophoric in nature. Such pyrophoric nanopowders can be passivated using different techniques, but in each case the reactive content of aluminum in the powder is reduced. The most common passivating method for aluminum nanopowders is the formation of an oxide outer layer. This alumina layer is formed by the exposure of pyrophoric aluminum nanoparticles to environments with low oxygen concentrations (Puszynski 2002, 2007; Li et al. 2006). During the slow oxidation process, a continuous oxide layer is formed on the surface of each nanoparticle. The typical thickness of a passivated alumina layer may vary from 2 to 4 nm depending on the passivation process or storage conditions. As indicated before, the presence of this alumina layer affects the reactive aluminum content in the nanopowder. The reactive aluminum content for a monodispersed spherical aluminum nanopowder passivated with a uniform layer of alumina of 3 nm in thickness significantly decreases with decreasing particle size. Figure 6.1 shows the aluminum reactive content, defined as (mass of unreacted aluminum)/(mass of aluminum nanopowder) × 100% as a function of particle size diameter.

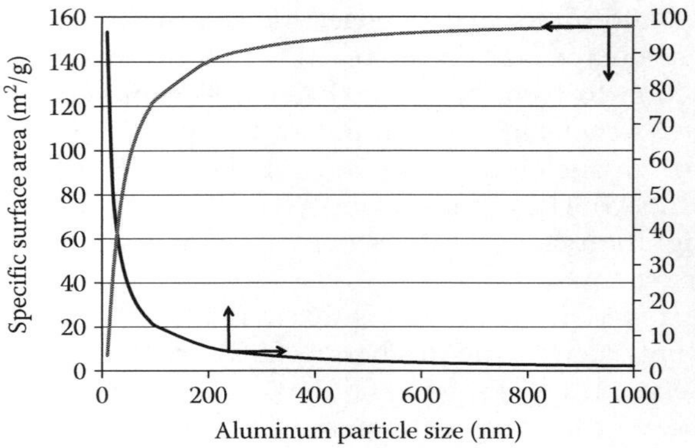


**FIGURE 6.1**
Specific surface area and reactive aluminum content in a monodispersed aluminum nanopowder passivated with 3 nm thick alumina layer as a function of particle size.

In Figure 6.1, the corresponding specific surface area is also plotted. As can be seen, the reduction of the particle size below 50 nm results in aluminum nanopowder having only 59% reactive aluminum content. Therefore, the production of such fine nanoparticles may not be desirable from a practical point of view.

The current manufacturing methods for aluminum nanopowders can be divided into two major categories: (1) high-temperature methods and (2) low-temperature methods.

### 6.2.1 High-Temperature Methods

As indicated in Section 6.1, the first successful production method of aluminum nanopowders was developed in Tomsk in the former Soviet Union (Zelinskii et al. 1984). The well-known nanopowder, called ALEX, was produced by the fast heating of aluminum wire by electric current, which resulted in a wire explosion. It was determined that the required current density should not be lower than $1.0 \times 10^{11}$ A/m$^2$ and the average particle size of ultrafine aluminum depends on the energy released into the wire (Kwon et al. 2001). The typical diameter of the aluminum wire used in this technology varies from tens to hundreds of microns. During the wire explosion process, aluminum vaporizes into an inert atmosphere at the reduced pressure. The effects of inert gas pressure, type of inert gas, and electric discharge characteristics are described elsewhere (Zelinskii et al. 1984; Ivanov et al. 1997; Jiang et al. 1998; Kwon et al. 2001; Sarithi et al. 2007).

The vaporization–condensation process of aluminum in inert gas under reduced pressures was investigated by Granqvist and Buhrman in 1976 (Granqvist and Buhrman 1976). This pioneering research determined that nanosized aluminum, with a relatively narrow particle-size distribution, can be formed by the evaporation of aluminum from an electrically heated boat and subsequent collection on cold surfaces. Figure 6.2 shows the effect of inert gas pressure and its type on the median particle size of aluminum formed on cold surfaces. This approach was later utilized for the pilot-scale production of aluminum nanopowders (Puszynski et al. 2002).

The pilot-scale setup for the vacuum-condensation process to produce larger quantities of aluminum nanopowders (10–15 g/h) is shown in Figure 6.3. High-purity aluminum wire (Al content > 99.95%) was continuously fed into the controlled-vacuum chamber. The aluminum was evaporated from the resistively heated ceramic boat (BN/$TiB_2$ composite) into the flowing helium or argon gases (Puszynski et al. 2002). Prior to the collection

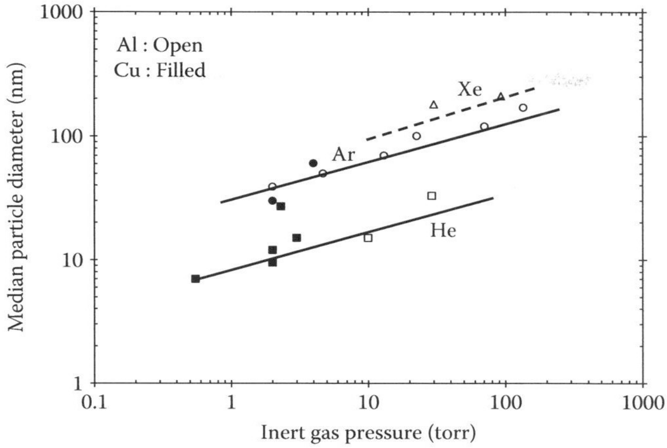


**FIGURE 6.2**
Median diameter versus pressure of He, Ar, or Xe gas for the formed particles of aluminum and copper by evaporation–condensation process. (Reprinted from Granqvist, C.G. and Buhrman, R.A., *J. Appl. Phys.*, 47(5), 2200, 1976. With permission. Copyright [1976], American Institute of Physics.)

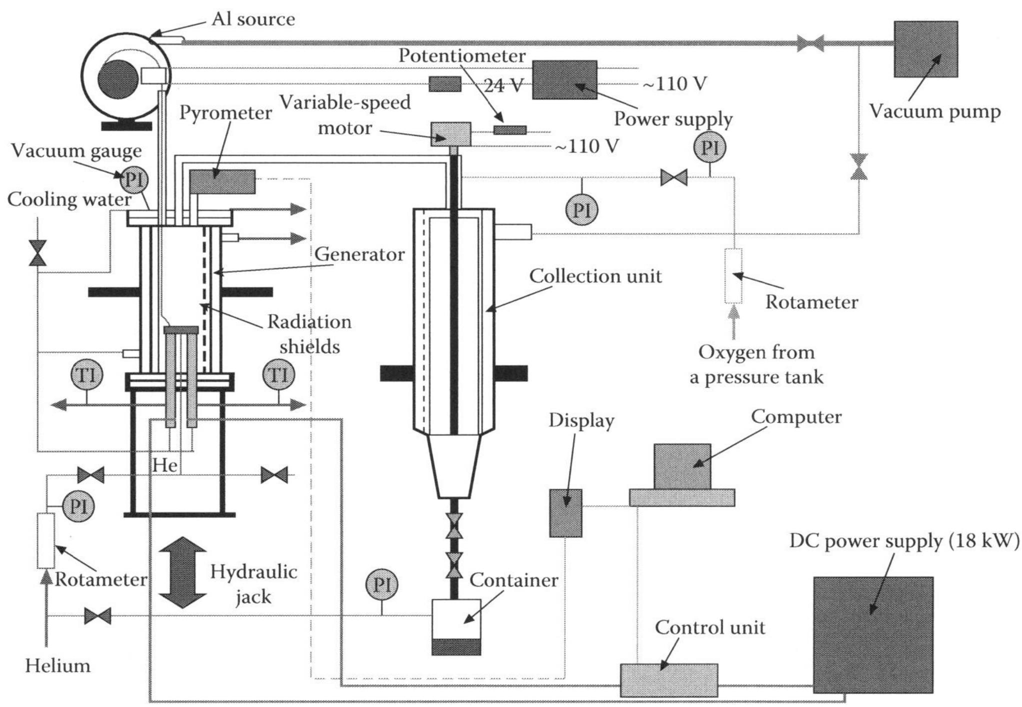


**FIGURE 6.3**
**(See color insert following page 302.)** Schematic of the nanosized aluminum generator and collection systems.

system, generated nanoparticles were passivated with oxygen injected into the system. Powders were collected and further handled in a glove box with controlled atmosphere until they were fully passivated.

The effect of the total inert gas pressure on the specific surface area and reactive aluminum content produced by the vaporization of aluminum at 1620°C is shown in Figure 6.4. It can be clearly seen that the specific surface area decreases with increasing inert gas pressure. It was also found that nanopowders formed in argon atmosphere have significantly lower specific surface area than those generated in helium under identical conditions. These results agree well with those published previously by Granqvist and Buhrman (Granqvist and Buhrman 1976).

The modeling studies of the aluminum particle formation during vapor condensation in an inert gas were conducted by Jayaraman and Puszynski (Jayaraman 2001; Puszynski 2007). They adopted a simple cascade-flow model. In this model, the tubular generator was discretized into *N* cells with various volumes. Each cell was kept at a constant temperature, following a macroscopic temperature distribution in the generator. The nonlinear equations governing the aerosol dynamics were solved for each cell. The main idea in the cascade-flow system is to obtain the output from one cell and to utilize this as input for the next cell, which operates at a different temperature. In this non-isothermal system, nanoparticles of aluminum were generated in several steps, including:

- Vaporization of aluminum into an inert gas from the resistively heated ceramic boat
- Formation of stable particles and subcritical molecular clusters
- Growth of subcritical clusters into stable particles
- Subsequent vapor condensation on the surface of stable particles
- Coalescence of stable particles.

Experimental research clearly indicated that nanosized aluminum with larger particle sizes is obtained when quenching is done in argon instead of helium. To explain that difference, two-dimensional temperature and velocity profiles (Figures 6.5 and 6.6) in the generator under identical operating conditions for helium and argon were analyzed.

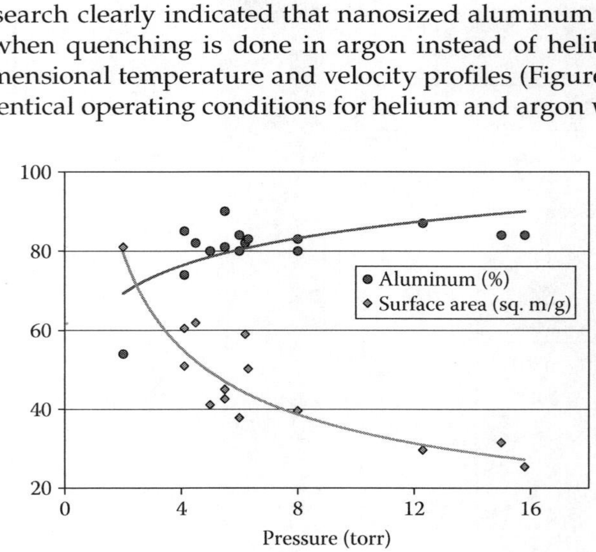


**FIGURE 6.4**
The effect of helium pressure on the specific surface area and the reactive content of aluminum nanopowders generated by vacuum–condensation technique. (From Puszynski, J.A., Formation, passivation, and reactivity of energetic nanopowders, in Kuo, K.K. and Rivera, J.D. (eds.) *Advancements in Energetic Materials and Chemical Propulsion*, Begell House Inc., New York, 3–21, 2007. With permission.)

A comparison of temperature profiles clearly indicates that quenching in helium is more effective than in argon. Therefore, nanosized aluminum powder with smaller average particle size is obtained in helium gas. This prediction agrees well with experimental data. The cascade model was solved for both helium and argon inert gases at different pressures.

The effect of inert gas pressure on the specific surface area of aluminum nanopowders obtained from the proposed model is shown in Figure 6.7. It can be clearly seen that the predicted specific surface area for aluminum nanopowders generated in helium gas under identical operating conditions is higher than in argon gas. The formation of smaller particles in helium is caused by more effective quenching due to the much higher thermal diffusivity of helium gas, $1.64 \times 10^{-4}\,m^2/s$, compared to argon gas, $1.94 \times 10^{-5}\,m^2/s$. In both cases the specific surface area decreases with increasing pressure of the inert gas in the evaporating chamber. Both experimental and modeling results have shown good quantitative agreement.

The vaporization–condensation technique is a very reliable method of producing aluminum nanopowders but has some limitations regarding its scaleability. Pulse-, dc-,

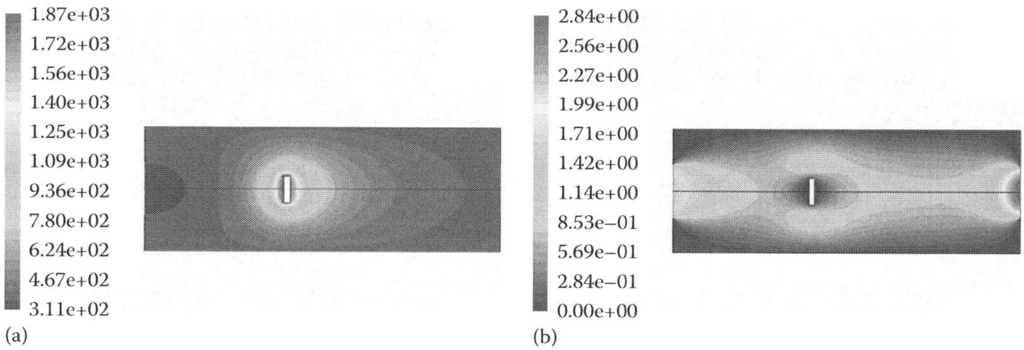


**FIGURE 6.5**
**(See color insert following page 302.)** (a) Temperature (in K) and (b) velocity (in m/s) profiles in the generator filled with helium. (From Puszynski, J.A., Formation, passivation, and reactivity of energetic nanopowders, in Kuo, K.K. and Rivera, J.D. (eds.) *Advancements in Energetic Materials and Chemical Propulsion*, Begell House Inc., New York, 3–21, 2007. With permission.)

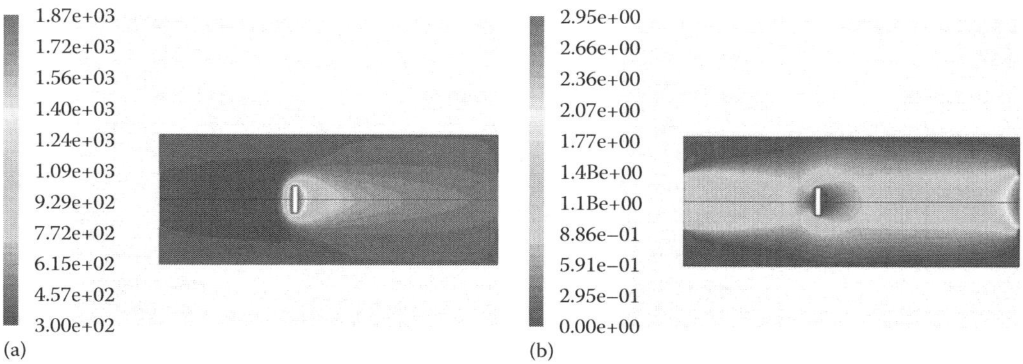


**FIGURE 6.6**
**(See color insert following page 302.)** (a) Temperature (in K) and (b) velocity (in m/s) profiles in the generator filled with argon. (From Puszynski, J.A., Formation, passivation, and reactivity of energetic nanopowders, in Kuo, K.K. and Rivera, J.D. (eds.) *Advancements in Energetic Materials and Chemical Propulsion*, Begell House Inc., New York, 3–21, 2007. With permission.)

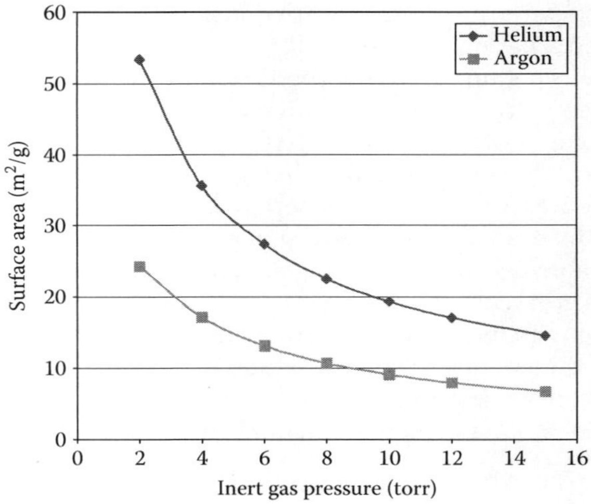


**FIGURE 6.7**
Calculated specific surface area of aluminum nanopowders generated by vapor–condensation technique. (From Puszynski, J.A., Formation, passivation, and reactivity of energetic nanopowders, in Kuo, K.K. and Rivera, J.D. (eds.) *Advancements in Energetic Materials and Chemical Propulsion*, Begell House Inc., New York, 3–21, 2007. With permission.)

or induction-plasma technologies are definitely very important alternatives for the large-scale production of nanopowders, including aluminum. Tetronics company has been successfully producing nanosized aluminum in the range of 50–150 nm (specific surface area: 25–30 $m^2/g$) (Hull 2002). In the Tetronics' dc-plasma process, an aluminum wire is used as the feed, but due to the application of the plasma torch significantly finer particles with a narrower particle-size distribution are formed in comparison to the explosive wire technology. However, the produced aluminum nanopowder contains some impurities, mainly carbon (up to 2.4 wt%). During the last 7 years, the most reliable source of aluminum nanopowders was from Novacentrix (formerly, Nanotechnologies Inc.). This company produced several types of aluminum nanopowders with different average particle sizes ranging from 50 to 140 nm. The reactive aluminum content in these powders varied from 75% to 90%. Aluminum nanopowders were produced using patented radial pulsed arc discharge gun. A general idea of this process is described in U.S. patent (Schroder and Jackson 2004). It should be noted that a radial gun operates in a pulse mode but delivers much higher peak power, $10^8$ W compared to $10^5$ W delivered by transferred arc discharge. Also, the temperature generated using a radial gun is ~50,000 K compared to 10,000 K for the transferred arc discharge, which operates in a continuous mode. Much higher temperature gradients and supersonic expansion quenching mechanism provide better conditions for the growth of nanoparticles from the condensing vapor, resulting in a product with a much narrower particle-size distribution than in the case of transferred arc discharge. Another configuration of high-energy pulsed plasma arc synthesis of nanosized aluminum nanopowder was also described by Kim (2008) and it is presented schematically in Figure 6.8. In this system, when the ignition switch in PFN is closed, the current discharge from the capacitor bank is initiated. Next, the aluminum fuse wire located between the electrodes is rapidly heated by an electric current, and aluminum vaporizes. This vaporization process takes place in approximately 50 ms and creates an initial conducting path for electrical current between the electrodes. As a result, plasma discharge is formed by the arc between the cathode and the anode, and additional aluminum vaporization takes place from the electrode material.

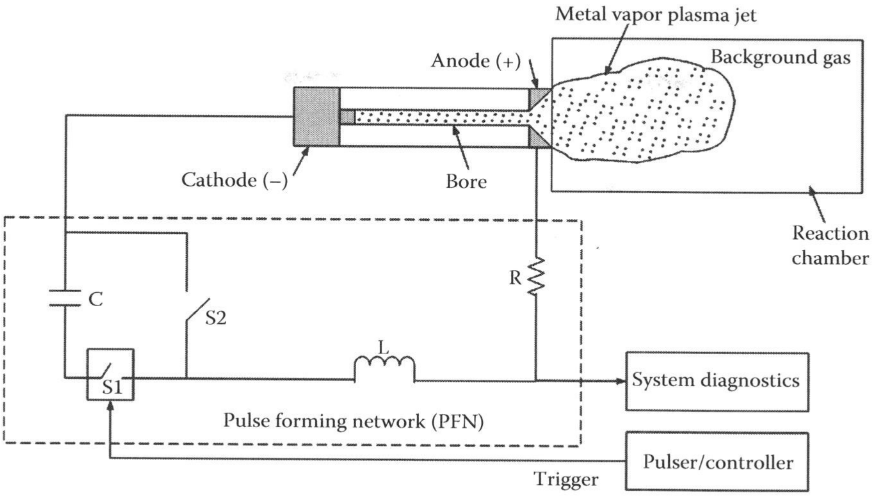


**FIGURE 6.8**
Schematic representation of pulsed arc discharge gun. (Reprinted from Kim, K., *Met. Mater. Int.*, 14(6), 707, 2008. With kind permission of Springer Science and Business Media.)

The formed plasma flows and rapidly accelerates through the bore and expands into the reaction chamber. In contrast, conventional plasma synthesis of nanomaterials, where the plasma discharge is first generated in a dc-plasma torch followed by injection of metal powder or bulk metal, requires much higher arc voltages and electrical currents (Kim 2008).

Recently, the production of aluminum and other nanopowders was demonstrated using a high-power electron accelerator (Korchagin et al. 2008). The production tests were performed on a commercial 100 kW accelerator providing high electron energy (1.4 MeV) and the possibility of extracting the electron beam into the atmosphere. The beam power density can be as high as 5 MW/cm$^2$ and allows evaporation of refractory compounds at atmospheric conditions. Other advantages of this method are high efficiency of direct conversion of electric energy into thermal energy in the heated material, high rates of heating (1000 K/s), and purity of produced materials. That paper describes the production rates up to several kilograms per hour.

### 6.2.2 Low-Temperature Processes

Wet chemistry synthesis methods of aluminum nanopowders have been explored to a lesser extent than high-temperature formation processes. However, the low-temperature methods are attractive because they allow in situ functionalization of aluminum nanoparticle surfaces (Jouet et al. 2005; Chung et al. 2009).

In 1998, a detailed study of aluminum synthesis by low-temperature chemistry methods was published by researchers from Washington University in St. Louis, MO (Haber and Buhro 1998). They synthesized nanocrystalline aluminum by two different techniques. In the first method, they reacted lithium aluminum hydride ($LiAlH_4$) with aluminum chloride ($AlCl_3$) at 164°C in 1,3,5-trimethylbenzene. They obtained aluminum nanoparticles with an average particle size of 160 ± 50 nm. The by-product of this reaction, lithium chloride (LiCl), was removed by washing with methanol at temperatures below 0°C. In the second method, aluminum nanopowder was synthesized by the decomposition of the intermediate compound $H_3Al(NMe_2Et)$ under reflux in 1,3,5-trimethylbenzene at 160°C–164°C, with or without added decomposition catalyst $Ti(O\text{-}i\text{-}Pr)_4$. The intermediate compound needed in this synthesis route was prepared by the method of Frigo and coworkers (Frigo et al. 1994). In this last method, aluminum nanopowder with an average particle size between 40 and 180 nm was obtained. The second method yielded much higher purity aluminum nanopowder than the former, which contained 3–4 wt% each of carbon, oxygen, and chlorine impurities.

A few years ago, researchers from NAVAIR at China Lake, CA demonstrated the wet synthesis of aluminum nanopowder from triethylamine adduct of alane ($AlH_3 \cdot Net_3$) in heptane (Foley et al. 2005). These authors also presented synthesis techniques that led to the formation of palladium-, silver-, gold-, and nickel-coated aluminum nanoparticles. Aluminum nanopowders passivated with palladium, silver, and nickel had specific surface areas of ~23 $m^2/g$.

To date, there have been no literature reports on any large-scale production of aluminum nanopowders using wet chemistry synthesis methods. The lack of scale-up efforts seems to be caused by the complexity of chemical reactions, use of organic solvents, and low stability of synthesized aluminum nanopowders.

Another low-temperature technique of making nanosized aluminum is the attrition milling of micron-sized aluminum powders in organic solvents or under cryogenic conditions. Any milling process of ductile materials tends to result in the formation of flake-like particles, unless the temperature is reduced so the material becomes brittle. In 1993, Eckert et al. applied attrition milling to reduce the particle size of aluminum (Eckert et al. 1993). They accomplished a significant reduction in the size of the starting aluminum particles, from microns to nanometer range. Recently, Innovative Materials and Processes, LLC demonstrated a scalable process of attrition milling of aluminum micron-sized flakes. This very effective size-reduction process has resulted in an increase of the specific surface area of milled aluminum from 0.5 to 26 $m^2/g$ within 4–5 h of milling. The resulting aluminum nanopowder is in the form of nanoflakes (see Figure 6.9).

**FIGURE 6.9**
Al nanoflakes formed by attrition milling in ethanol, using zirconia-grinding media.

## 6.3 Passivation and Characterization of Aluminum Nanopowders

Aluminum metal forms an oxide and/or hydroxide passivation layer at the surface when exposed to oxygen, moisture, or liquid water. This layer effectively slows down the hydration/oxidation reaction of aluminum and it is usually sufficient for the prevention of corrosion of bulk aluminum metal. In the case of aluminum nanopowders, the alumina passivation layer of a few nanometer thickness is also necessary to eliminate the pyrophoricity of the powder. However, the oxide-passivated aluminum nanopowders undergo continuous reaction when exposed to moist air or water (Puszynski 2004; Dubois et al. 2007; Kwon et al. 2007). This unwanted reaction (aging process) continues until a complete oxidation of the metal takes place. An illustration of aging effect on aluminum nanoparticles during exposure to humid air is shown in Figure 6.10.

It was determined experimentally that any aluminum nanopowder reacts with moisture when exposed to humid air. Figure 6.11 shows the effect of humidity and time on the content of reactive aluminum in aluminum nanopowders with an average particle size of 50 nm.

It can be clearly seen that the content of reactive aluminum decreases rapidly with time at higher humidity levels. Even at relatively low humidity levels (43%), the reaction progresses

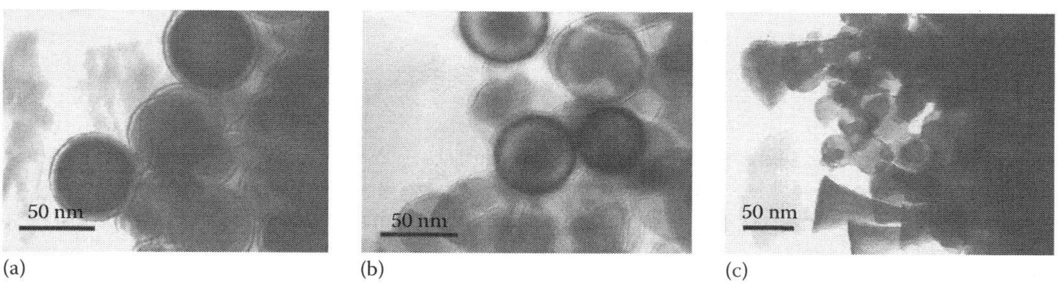


**FIGURE 6.10**
TEM images of aluminum nanoparticles during exposure to humid air (97% RH and 40°C): (a) fresh sample, 74 wt% of reactive Al; (b) after 40 h, 59 wt% of reactive Al; and (c) after 60 h, 17 wt% of reactive Al.

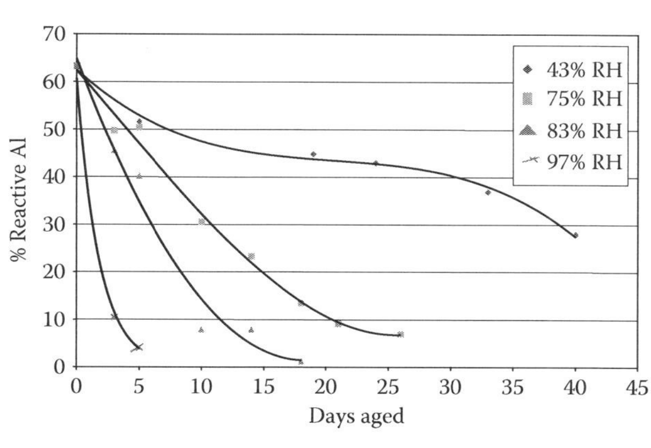


**FIGURE 6.11**
The effect of relative humidity and time on the reactive aluminum content in oxide passivated aluminum nanopowders. (From Puszynski, J.A., *Mater. Res. Soc. Symp. Proc.*, 800, AA4.6.1, 2004. With permission.)

relatively fast. In order to reduce the effect of moisture, the same 50 nm aluminum nanopowder was coated with silanes or fatty acids. In both cases hydrophobic groups have had a significant effect on the reaction with moisture. Figure 6.12 shows the effect of silane Z-6124 from Dow Corning Corp. on the rate of reaction with moisture at 40°C and 97% relative humidity. It was found that samples with a concentration of silane of 5 wt% are well protected over an extensive period of time.

The reaction of aluminum with water initially takes place at the aluminum surface and further progresses following diffusion-controlled mechanism without disturbing the shape of the particles (see Figure 6.10b). At advanced stages of the reaction (less than 20 wt% of reactant left), aluminum hydroxide product forms a distinct, separate phase (see Figure 6.10c).

An effective prevention of the reaction of aluminum nanoparticles in liquid water has recently become of technical importance in colloidal processing. Aluminum nitride powders and aluminum nanopowders with surfaces coated by a protective organic and/or inorganic layer can effectively sustain contact with liquid water for many hours without decreasing the reactive aluminum content (Ferreira et al. 2005; Puszynski et al. 2005).

Many surface modifiers for aluminum nanopowders are known to slow down the reaction of aluminum with water. Long-chain fatty acids reacting with hydroxyl groups on the surface of the oxide-passivated aluminum form a strongly hydrophobic layer, which also effectively reduces the rate of the reaction with water. Other hydrophobic modifiers can be selected from functionalized silanes. TEM images of aluminum nanoparticles coated with phenyltrimethoxysilane reveal an additional layer of silane uniformly surrounding a solid aluminum nanoparticle, as shown in Figure 6.13.

In order to compare the effectiveness of inhibitors, aluminum nanoparticles were treated with the same quantity of various surface modifiers and then re-dispersed in water prior to conducting the reaction in a closed reactor at 48°C. Dynamic pressure profiles in the reactor were recorded and used to calculate the extent of the reaction of aluminum with water for each inhibitor (see Figure 6.14). The lines 2, 3, 4, and 7 in that figure correspond to compounds with hydrophobic functional groups. Although these coatings are very effective in preventing the reaction of aluminum with water (oleic acid being the best), hydrophobic coatings generally inhibit good dispersion of aluminum nanopowder in water.

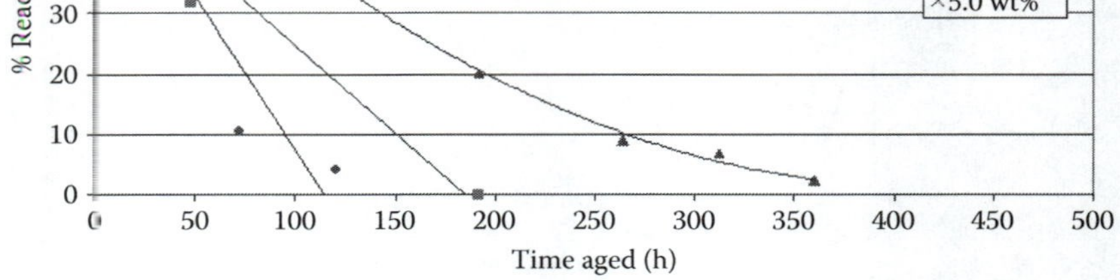


**FIGURE 6.12**
The effect of silane Z-6124 coating and time on the reactive aluminum content of aluminum nanopowders at T = 40°C and RH = 97%. (From Puszynski, J.A., *Mater. Res. Soc. Symp. Proc.*, 800, AA4.6.1, 2004.)

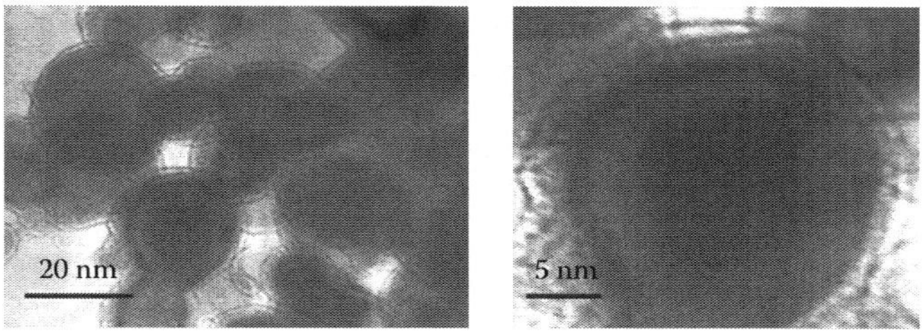


**FIGURE 6.13**
Aluminum nanoparticles coated with phenyltrimethoxysilane. (Reprinted from Puszynski, J.A. et al., *J. Propul. Power*, 23(4), 688, 2007. With permission of the American Institute of Aeronautics and Astronautics, Inc.)

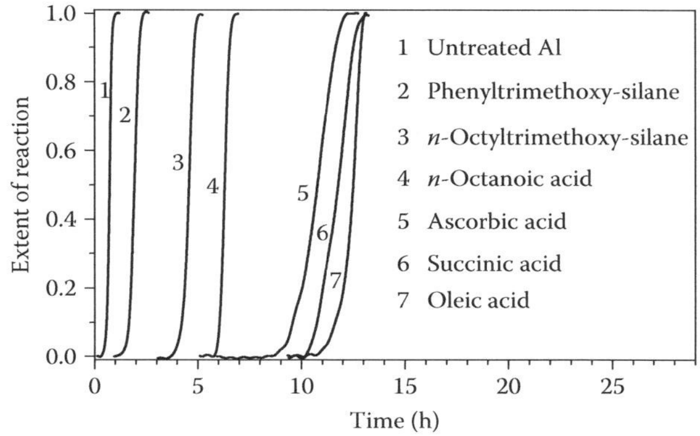


**FIGURE 6.14**
Observed extent of the reaction dependence on time for aluminum nanopowders coated with various organic inhibitors at a temperature of 48°C. (Reprinted from Puszynski, J.A. et al., *J. Propul. Power*, 23(4), 688, 2007. With permission of the American Institute of Aeronautics and Astronautics, Inc.)

Apparently, both dibasic acids protect aluminum very effectively and also allow better dispersion of aluminum nanoparticles in water.

The extent of aluminum nanoparticles' reaction in liquid water can also be monitored by pH change with time. The value of pH increases in a monotonic manner from about pH = 4–7 in experiments conducted at room temperature (~26°C). At or above pH = 7, the reaction accelerates and hydrogen evolves. The solubility and speciation of aluminum hydroxides strongly depend on the pH value (Smith and Martell 1974; Martell and Motekaitis 1992). Accordingly, the oxide/hydroxide suspensions in water may exhibit various pH values depending on the composition of solid phase and concentration of aluminum hydroxide ions (for example, $Al(OH)_2^+$, $AlOH^{+2}$). The minimum solubility of amorphous aluminum hydroxide is in the pH range of approximately 6.1–6.3 at 25°C. In the presence of weak organic acids, the suspension becomes more acidic due to the neutralization of amphoteric aluminum hydroxides. However, the buffering capacity of the system is limited (usually small concentrations of organic acid are used), and pH in water solution will increase as the reaction of aluminum with water continues to produce increasing quantities of aluminum hydroxide. The inhibition of the reaction in slurries consisting of aluminum nanoparticles and succinic acid was monitored by recording the hydrogen release. The reaction induction time increased linearly at low concentrations of acid, as shown in Figure 6.15. Dissolved succinic acid may act as a pH-moderating compound at these low concentrations.

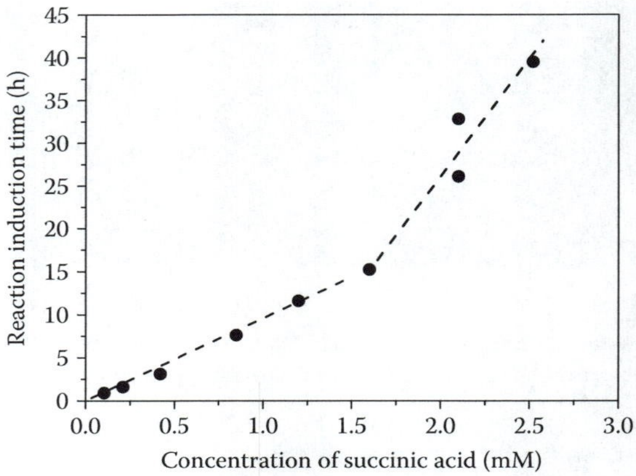


**FIGURE 6.15**
Reaction induction time of aluminum nanopowder in water as a function of succinic acid concentration.

The inhibition effect of succinic acid increases at concentrations higher than ~2 mM. This behavior might be associated with the growth of a protective layer on the aluminum particle. However, once the reaction of aluminum with water begins, it follows the same very steep reaction rate increase. Temperature effect on the reaction rate of aluminum nanopowder in water can be described using the Arrhenius equation for both uncoated and succinic-acid-coated aluminum nanopowders. Effective activation energies derived from the corresponding Arrhenius plots are 77 kJ/mol for succinic-acid-coated aluminum nanopowder and 137 kJ/mol for uncoated aluminum nanopowder. The beneficiary effect of the inhibitor is mainly due to the reduction, by over 10 orders of magnitude, of the pre-exponential factor in the Arrhenius equation.

Another interesting inorganic inhibitor, ammonium dihydrogen phosphate, was successfully used in the processing of aluminum nanopowders in water (Puszynski et al. 2006).

Protecting aluminum nanopowders against the reaction with water allows for the preparation of stable and concentrated aqueous suspensions suitable for colloidal-processing. Water-based processing is particularly important when safety and environmental aspects are of concern. By utilizing colloidal-processing methods, one can achieve a high degree of homogeneity and particle packing—both are key issues in the preparation of reactive heterogeneous mixtures.

## 6.4 Condensed Phase Reactions of Aluminum Nanopowder with Oxides and Metals

### 6.4.1 Reactivity of Nanosized Aluminum with Oxides

Recent advances in the synthesis and processing of aluminum nanopowders have stimulated the exploration and development of a new class of nanoenergetic materials (Mench et al. 1998; Dorsett et al. 2001; Miziolek 2002; Pivkina et al. 2004). Specific examples of such materials are nanothermites, also called metastable interstitial composites (MIC), which consist of a mixture of aluminum and metal oxide nanopowders (Martin et al. 1998;

Son et al. 2001, 2002; Somasundaram et al. 2003; Bulian et al. 2004; Perry et al. 2004; Plantier et al. 2005). Examples of several thermite systems, together with their corresponding heat of reaction, gas generation ability, and adiabatic temperature, calculated at 1 atm argon gas pressure, are listed in Table 6.1 (Fischer and Grubelich 1998).

In the mid-1990s, Aumann et al. and Martin et al. demonstrated that a reduction of solid reactant particle sizes leads to increase in the rate of energy release in thermite systems (Aumann et al. 1995; Martin et al. 1998). The combustion front propagation velocity was increased by two to three orders of magnitude when the particle size of both aluminum and oxidizer was reduced to a nanosize range. Since that time, numerous papers have been published describing combustion front propagation and ignition characteristics in binary-reacting systems consisting of nanosized aluminum and reactive oxide powders, such as $MoO_3$, CuO, $WO_3$, and $Bi_2O_3$ (Son et al. 2001; Somasundaram et al. 2003; Bulian et al. 2004; Perry et al. 2004; Valliappan et al. 2005; Puszynski et al. 2007; Sanders et al. 2007; Walter et al. 2007). A summary of selected experimental data on combustion front velocity in nanothermite systems under unconfined conditions is presented in Figure 6.16.

It should be noted that reported experimental results can be influenced by experimental conditions, including the setup configuration and density of reactants' mixture, which were not identical in each case. As indicated before, the general trend between particle size

**TABLE 6.1**
Thermodynamic Properties of Selected Thermite Reactions

| Thermite Reaction | $Q$ (cal/g) | $Q$ (cal/cm³) | Gas Generation 1 atm, (g gas/g mixture) | $T_{ad}$ (K) |
|---|---|---|---|---|
| $2Al + Fe_2O_3 \rightarrow 2Fe + Al_2O_3$ | 945.4 | 3947 | 0.0784 | 3135 |
| $2Al + Bi_2O_3 \rightarrow 2Bi + Al_2O_3$ | 505.1 | 3638 | 0.894 | 3319[a] |
| $2Al + MoO_3 \rightarrow Mo + Al_2O_3$ | 1124.0 | 4279 | 0.2473 | 3688[a] |
| $2Al + WO_3 \rightarrow W + Al_2O_3$ | 696.4 | 3801 | 0.1463 | 3253 |
| $2Al + 3CuO \rightarrow 3Cu + Al_2O_3$ | 974.1 | 4976 | 0.3431 | 2843 |

[a] Values calculated in this chapter.

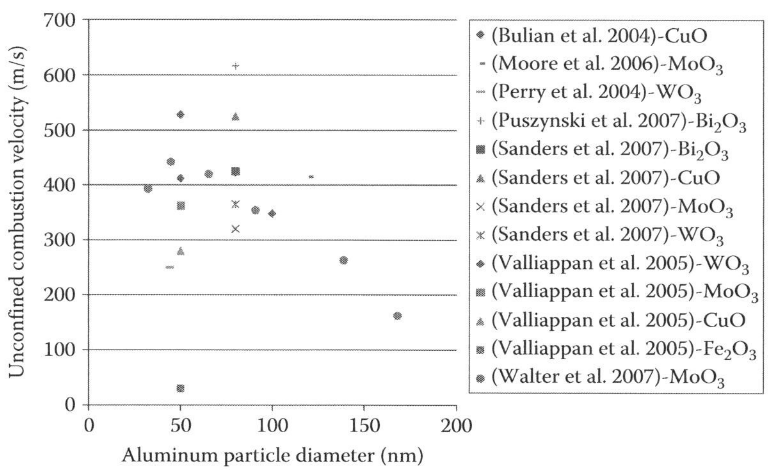


**FIGURE 6.16**
Experimental unconfined combustion velocities for various nanothermite systems.

and reactivity in nanothermite systems is such that as particle size decreases, reaction rate increases. To demonstrate this effect, a detailed study of reaction kinetics in the Al–$Bi_2O_3$ system was conducted (Puszynski et al. 2009). In that paper, three different particle sizes of $Bi_2O_3$ were combined with 80 nm aluminum nanopowder from Novacentrix. The calculated reaction kinetic constants from DSC analysis are shown in Table 6.2. There is a general trend of increased activation energy and pre-exponential factor with increased oxide particle size. A plot of activation energy versus particle size for this system yields a fairly linear relationship of $E = 0.288d_p + 122$, where $d_p$ and $E$ are particle size and activation energy, respectively.

Metal oxide particle size is not the only factor influencing kinetic constants in nanothermite systems. Particle size and the morphology of aluminum particles are also affecting the reaction rate. As shown in Table 6.3, the reduction of aluminum particle size decreases the activation energy as well. The reaction between 80 nm aluminum and 210 nm bismuth trioxide has lower activation energy than the 130 nm aluminum reacting with the identical bismuth trioxide powder.

In DSC experiments, it was observed that aluminum nanopowders react with bismuth trioxide significantly below the melting point of bulk aluminum. In addition, aluminum nanoflakes were used in that study to observe the effect of aluminum particle morphology. The reactive aluminum content and specific surface area of the milled flake aluminum were very similar to that of the 80 nm spherical particles; however, activation energy and pre-exponential factor were higher (see Table 6.3). This can possibly be due to a wider particle-size distribution for the milled aluminum and potentially a lower level of intermixing between aluminum and bismuth trioxide reactants.

Despite high activation energies (>150 kJ/mol), nanoenergetic materials are very susceptible to ignition by ESD, impact, friction, and thermal impulse due to small particle size and high specific surface area. Ultimately, any ignition is caused by a thermal effect, but

**TABLE 6.2**

Kinetic Constants of Al–$Bi_2O_3$ Mixtures Containing 80 nm Aluminum Nanopowder

| Average $Bi_2O_3$ Particle Size (nm) | Activation Energy (kJ/mol) | Pre-Exponential Factor ($min^{-1}$) |
|---|---|---|
| 40 | 60.8 | $3.15 \times 10^4$ |
| 210 | 90.0 | $1.29 \times 10^7$ |
| 410 | 112.9 | $1.46 \times 10^9$ |

*Source:* Puszynski, J.A. et al., *Int. J. Self Prop. High Temp. Syn.*, 2010 (in review).

**TABLE 6.3**

Kinetic Constants of Al–$Bi_2O_3$ Mixtures Containing 210 nm $Bi_2O_3$

| Average Al Particle Size (nm) | Activation Energy (kJ/mol) | Pre-Exponential Factor ($min^{-1}$) |
|---|---|---|
| 80 | 90.0 | $1.29 \times 10^7$ |
| 130 | 81.9 | $4.23 \times 10^6$ |
| Flake | 99.7 | $2.52 \times 10^8$ |

*Source:* Puszynski, J.A. et al., *Int. J. Self Prop. High Temp. Syn.*, 2010 (in review).

the mechanism by which the energetic material reaches its thermal threshold is different for each type of ignition stimulus. Ignition energy thresholds can vary from fractions of a microjoule to several millijoules.

Several nanothermite systems, including Al–$Bi_2O_3$, Al–$Fe_2O_3$, Al–$MoO_3$, and Al–CuO were tested for ESD sensitivity (Puszynski et al. 2008). Aluminum nanopowders with an average particle size of 80 nm were used as the fuel in each nanothermite system. ESD sensitivity tests indicated significant differences in the susceptibilities of the four systems to ignition by that stimulus. Minimum ignition threshold energies were determined to be <1 μJ, 50 μJ, 84.5 μJ, and 1 mJ for the Al–$Bi_2O_3$, Al–$MoO_3$, Al–CuO, and Al–$Fe_2O_3$ systems, respectively (Puszynski et al. 2008).

The effect of oxidizer particle size is another factor that was considered in determining nanothermite's susceptibility to ignition by ESD. Three different sizes of bismuth oxide powders were used to illustrate this effect. The first powder had a specific surface area (SSA) of 0.28 $m^2/g$ and an average particle diameter of 2.4 μm, the second powder had an SSA of 1.62 $m^2/g$ and an average equivalent particle diameter of 416 nm, and the third one, an SSA of 17 $m^2/g$ and an equivalent particle diameter of 40 nm. Each oxide was combined with 80 nm aluminum powder at a fuel-to-oxidizer wt% ratio of 14.5/85.5 and ultrasonically mixed in isopropyl alcohol. It was determined that the increase in the specific surface area of the oxidizer resulted in a decrease in the minimum ignition energy threshold of greater than one order of magnitude. Table 6.4 illustrates the effect of the bismuth oxide average particle size on the minimum ESD ignition energy of the Al–$Bi_2O_3$ nanothermite in a loose-powder form (Puszynski et al. 2008).

The effect of aluminum nanopowder morphology on ESD sensitivity of aluminum and bismuth trioxide mixture was investigated using two aluminum nanopowders with different morphologies but similar specific surface areas. Aluminum nanopowder with an average particle size of 80 nm has a specific surface area of 26.1 $m^2/g$ and consists mostly of spherical particles. The other type of aluminum powder had an SSA of 24.4 $m^2/g$ and was prepared by Innovative Materials and Processes, LLC, as described previously. It was previously indicated that the system comprised of 80 nm aluminum and Accumet bismuth oxide was very sensitive to ignition by ESD, having an ignition threshold energy of only 0.125 μJ. When the same system was tested, but with aluminum nanoflakes used in place of the 80 nm aluminum, the ignition threshold energy was four orders of magnitude higher at ~1.5 mJ while still maintaining its reactivity and impact sensitivity. This indicates that aluminum nanoparticles with larger aspect ratio result in the formation of nanothermites

**TABLE 6.4**

Effect of the Bismuth Oxide Average Particle Size on Minimum ESD Ignition Energy of the Al–$Bi_2O_3$ Nanothermite

| SSA ($m^2/g$) | Average Particle Size (nm) | ESD Sensitivity (μJ) |
|---|---|---|
| 0.28 | 2400 | 2.000 |
| 1.62 | 416 | 0.125 |
| 17.0 | 40 | 0.075 |

*Source:* Puszynski, J.A. et al. Ignition characteristics of nanothermite systems in Kuo, K.K. and Hori, K. (eds.), *Advancements in Energetic Materials and Chemical Propulsion*, Begell House Inc., New York, 73–86, 2008.

that are able to dissipate the absorbed energy better; therefore, resulting in reduced ESD sensitivity. A similar trend was observed in the ignition of nanothermite systems using a laser (Puszynski et al. 2008).

### 6.4.2 Reactivity of Nanosized Aluminum with Other Metallic Reactants

The potential advantage of using nanosized reactants in the synthesis of metal alloys or ceramic composites is that compared to micron-sized materials they are characterized by high chemical activities and reduced enthalpies of fusion. As determined by differential scanning calorimetry at 20°C/min under argon atmosphere, it was shown that the enthalpy of fusion was reduced to 110.8 J/g for 50 nm versus 400 J/g for bulk aluminum (Lide 2005) as listed in Table 6.5. In addition, the use of nanoparticles allows better intermixing of reactants.

Dong et al. published the first experimental work addressing the use of nanosized aluminum, produced via wire electrical explosion, in the combustion synthesis of the intermetallic NiAl (Dong et al. 2002). In that work, micron-sized nickel powders were used (44 μm) with aluminum nanopowder ($d_{avg} = 40$ nm). Typical ignition temperatures reported for nickel and aluminum system consisting of micron-sized powders are in the range of 577–660°C (Plazanet and Nardou 1998; Biswas et al. 2002; Hunt and Pantoya 2005; Groven 2007). In contrast to the reaction between micron-sized aluminum and nickel reactants, these authors observed in differential scanning analysis an endothermic peak at 470°C, which they attributed to the formation of liquid followed by an ignition temperature of 585°C. Hunt et al. have investigated the influence of nanosized aluminum on the ignition characteristics for the reaction to form NiAl using 1 μm nickel powder and 25 nm, 100 nm, and 20 μm sized aluminum powder (Hunt et al. 2004). Measured ignition temperatures were greatly reduced for both 25 and 100 nm reactants to 286°C and 305°C, respectively. The authors suggested that these observations indicate that the reaction is driven by the decrease of the melting point of the aluminum reactant. Theoretical calculations of the aluminum melting point as a function of particle size were done using the relationship developed by Wronski (1967). As shown in Figure 6.17, using this relationship, melting points of 25 and 100 nm aluminum powders correspond to 605°C and 646°C, respectively.

An investigation of the melting behavior of nanosized aluminum was conducted by Trunov et al. for 44, 80, and 121 nm under argon atmosphere and a heating rate of 5°C/min and indicated a reduction in the melting endotherm to 570°C for all three sizes of aluminum powder (Trunov et al. 2006). However, as shown in Figure 6.18, for powders with 50 nm, 80 nm, 100 nm, 2 μm, and 44 μm, no significant shift in the melting endotherm was observed, and therefore no significant reduction in the melting point (<6°C) was recorded. Therefore, the depression of the melting temperature of aluminum nanopowders cannot solely explain the decrease of the ignition temperature for nanosized aluminum–nickel system. The generation of internal stresses within a single nanoparticle of aluminum, coated

**TABLE 6.5**

Latent Heat of Fusion for Aluminum as a Function of Average Particle Size

| Manufacturer | Novacentrix | Novacentrix | Novacentrix | Valimet | JT Baker | |
|---|---|---|---|---|---|---|
| Aluminum particle size | 50 nm | 80 nm | 100 nm | 2 μm | 44 μm | Bulk |
| Latent heat of fusion, J/g | 110.8[a] | 196.1[a] | 174.3[a] | 302.8[a] | 291.9[a] | 395.7[b] |

[a] Groven (2009).
[b] Lide (2005).

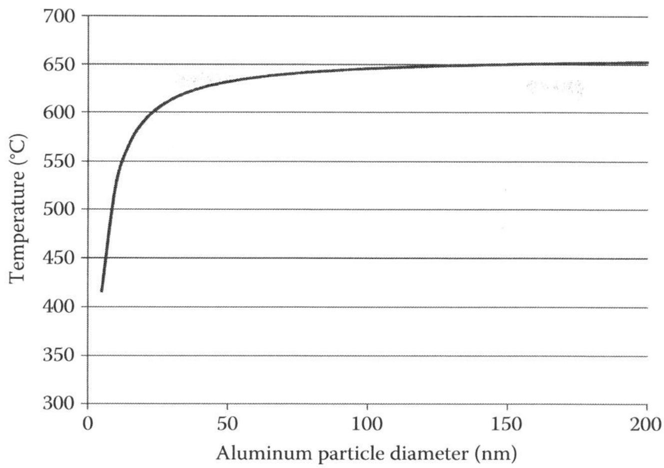


**FIGURE 6.17**
Melting temperature as a function of aluminum particle diameter, using Wronski relationship.

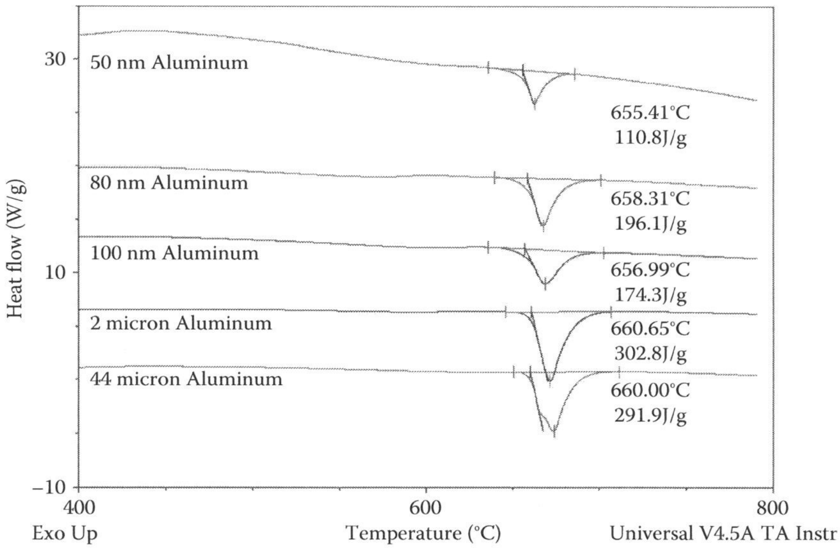


**FIGURE 6.18**
DSC plots at 20 C/min under argon atmosphere for various aluminum powders. The DSC curves are shifted by 6 W/g along the ordinate for clarity.

with a nanolayer of alumina, during the heating process followed by crack formations might be another reason for the ignition to occur at lower temperatures (Levitas et al. 2007).

Further investigation of the ignition temperature and reaction kinetics was also reported by Hunt and Pantoya (Hunt and Pantoya 2005). In their research studies, 40 nm, 80 nm, 4 μm, and 20 μm aluminum particles were tested in reactions with 800 nm nickel and 15 μm nickel particles. Ignition temperatures in the range of 300°C–400°C were reported for the reaction of aluminum nanopowder with nickel. A significant reduction in the activation energy was observed when nanosized aluminum particles were used (17.4 kJ/mol)

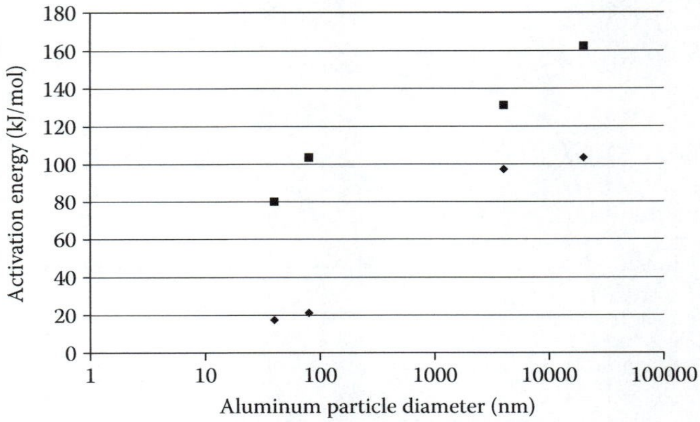


**FIGURE 6.19**
Activation energy as a function of aluminum particle size reacted with 800nm Ni (◆) and 15μm Ni (■). (Reprinted from Hunt, E. and Pantoya, M., *J. Appl. Phys.*, 98, 034909, 2005. With permission. Copyright [2005], American Institute of Physics.)

compared to when micron-sized aluminum was used (162.5kJ/mol). Their results also showed that as the nickel particle size is reduced, the measured activation energy, regardless of aluminum particle size, is also decreased. The results of their study are summarized in Figure 6.19.

Reaction kinetic measurements conducted for 50nm aluminum reacted with 100nm nickel resulted in an average activation energy of 34.5±8.0kJ/mol (Groven 2007). This value closely matches those reported by Hunt and Pantoya and Gennari et al. (Hunt and Pantoya 2005; Gennari et al. 2006). In addition, DSC scans of 80nm aluminum reacted with 100nm nickel confirmed that the use of nanosized reactants leads to the initiation of exothermic reaction between aluminum and nickel at lower temperatures, as shown in Figure 6.20 (Groven 2007).

The use of aluminum and nickel nanopowders in condensed phase reactions is driven by the desire to form nanocrystalline nickel aluminides, with alumina as a reinforcing phase. Alumina is inherently associated with its presence in the starting aluminum nanopowder reactant. To date, there is no significant research conducted on the formation of dense or porous nickel aluminide products from nanosized aluminum. There are a few reports published on in situ densification and combustion using traditional furnace technology (Groven and Puszynski 2005) and spark plasma sintering (Kim et al. 2007), as well as a single report of foam production using passivated aluminum particle replacement (Hunt et al. 2006). Groven and Puszynski ignited nanopowders of aluminum and nickel with subsequent densification of a hot product by means of a uniaxial press at 200MPa. The experiments were conducted in argon atmosphere with a reactant's preheating temperature of 520°C to produce nanostructured composites with bulk densities ranging from 60% to 90% of the theoretical density. Spark plasma sintering of nano-aluminum/nano-nickel mixtures was recently conducted by Kim et al. to produce nickel aluminides with unexpectedly high hardness ranging from 736 to 776HV, depending on reactant mixing method (Kim et al. 2007). These values are in contrast to single-phase intermetallics NiAl and $Ni_3Al$, which in the bulk state show an average of 243HV and 166HV, respectively (Noebe et al. 1991). This significant improvement in hardness is attributed to (1) the small grain size achievable when nanosized powders are used and (2) the presence

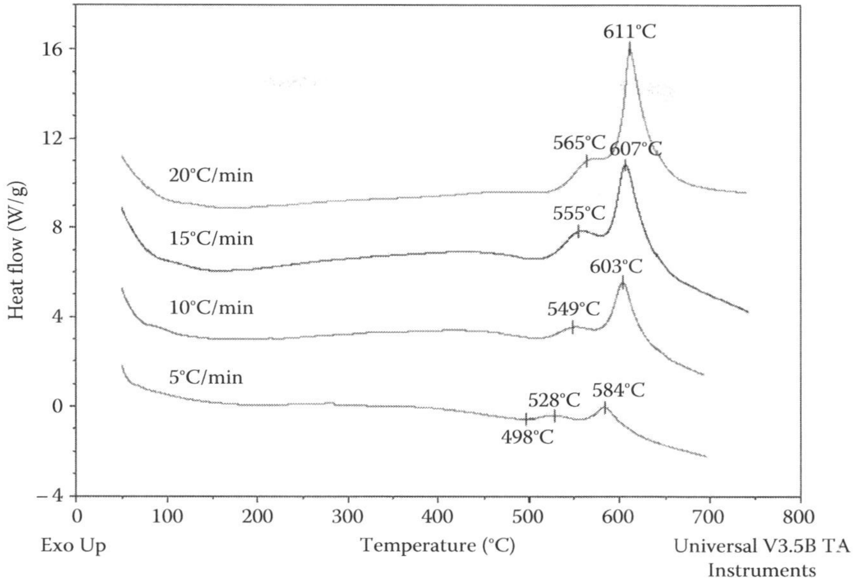


**FIGURE 6.20**
DSC curves for Ni (100 nm—Argonide)–Al (80 nm—Novacentrix) system with a heating rate of 5, 10, 15, and 20°C/min. The curves are shifted along the ordinate for clarity. (From Groven, L.J., Simultaneous combustion synthesis and densification of nickel aluminide/alumina nanocomposites, MS thesis, South Dakota School of Mines and Technology, Rapid City, SD, 2007.)

of oxide on the aluminum reactant, which provided additional reinforcement. Nickel aluminide foams have also been produced with the use of nanosized aluminum particles that were passivated with a gasifying agent, $C_{13}F_{27}COOH$, instead of alumina (Hunt et al. 2006). In that work, a very small quantity of micron-sized aluminum was replaced with passivated nanosized aluminum to produce NiAl foams with 50%–80% porosity.

## 6.5 Summary

This review clearly indicates that the use of nanosized reactants in condensed phase exothermic reactions leads to a significant increase in the rate of energy release. Such high rates of energy release, not commonly observed between oxidizer and fuel particles, make these nanoenergetic systems suitable candidates for environmentally benign macro and micro initiators, as well as energetic components of microthrusters and other applications requiring fast-combustion front velocities.

The recent advances in the formation of aluminum nanopowders indicate that high-temperature methods seem to be more suitable for scale-up than low-temperature wet chemistry synthesis routes. Mechanical reduction of aluminum particle size seems to be another promising approach for making larger quantities of reactive aluminum nanopowders.

This review also addresses some drawbacks associated with the use of aluminum nanopowders, mainly the presence of a passivation layer, which causes a reduction of reactive aluminum content, as well as the necessity of the use of surface modifiers to prevent the reaction of aluminum with moisture.

Definitely aluminum nanopowders and potentially other elemental reactive nanopowders have great potential to be used in both rapid energy release as well as condensed phase reactions, resulting in the formation of new advanced materials in the form of dense or porous structures.

## Acknowledgments

The authors gratefully acknowledge Dr. Jacek Swiatkiewicz and graduate students Chris Bulian, Sundar Jayaraman, and Zac Doorenbos for their contribution and support in the preparation of this chapter.

## References

Aumann, C.E., Skofronick, G.L., and Martin, J.A. 1995. Oxidation behavior of aluminum nanopowders. *J. Vac. Sci. Technol. B*, 13(3): 1178–1183.

Biswas, A., Roy, S.K., Gurumurthy, K.R., Prabhu, N., and Banerjee S. 2002. A study of self-propagating high temperature synthesis of NiAl in thermal explosion mode. *Acta Mater.* 50: 757–773.

Bulian, C.J., Kerr, T.T., and Puszynski, J.A. 2004. Ignition studies of aluminum and metal oxide nanopowders. *Proceedings of the 31st International Pyrotechnics Seminar*, Ft. Collins, CO, pp. 327–338.

Chung, S.W., Guliants, E.A., Bunker, C.E. et al. 2009. Capping and passivation of aluminum nanoparticles using alkyl-substituted epoxides. *Langmuir*: Article ASAP.

Dong, S., Hou, P., and Yang, H. 2002. Synthesis of intermetallic NiAl by SHS reaction using coarse-grained nickel and ultrafine-grained aluminum produced by wire electrical explosion. *Intermetallics* 10: 217–223.

Dorsett, H.E., Brousseau, P., and Cliff, H.E. 2001. The influence of ultrafine aluminum upon explosives detonation. *Proceedings of the 28th International Pyrotechnics Symposium*, Adelaide, Australia, pp. 239–244.

Dreizin, E.L. 2009. Metal-based reactive nanomaterials. *Prog. Energ. Combust.* 35: 141–167.

Dubois, C., Lafleur, P.G., Roy, C., Brousseau, P., and Stowe, R.A. 2007. Polymer-grafted metal nanoparticles for fuel applications. *J. Propul. Power* 23(4): 651–658.

Eckert, J., Holzer, J.C., Ahn, C.C., Fu, Z., and Johnson, W.L. 1993. Melting behavior of nanocrystalline aluminum powders. *Nanostruct. Mater.* 2: 407–413.

Ferreira, J.M.F., Olhero, S.M.H., and Oliveira, M.I.L.L. 2005. Definitive replacement of organic solvents by water in the colloidal processing of aluminium nitride. *Ceram. Trans.* 166: 71–84.

Fischer, S.H. and Grubelich, M.C. 1998. Theoretical energy release of thermites, intermetallics, and combustable metals. *Proceedings of the 24th International Pyrotechnics Seminar*, Monterey, CA, pp. 231–286.

Foley, T.J., Johnson, C.E., and Higa, K.T. 2005. Inhibition of oxide formation on aluminum nanoparticles by transition metal coating. *Chem. Mater.* 17: 4086–4091.

Frigo, D.M., Van Eijdenm, G.J.M., Reuvers, P.J., and Smit, C.J. 1994. *Chem. Mater.* 6: 190–195.

Gennari, S., Tamburini, U., Maglia, F., Spinolo, G., and Munir, Z. 2006. A new approach to the modeling of SHS reactions: Combustion synthesis of transition metal aluminides. *Acta Mater.* 54(9): 2343–2351.

Granqvist, C.G. and Buhrman, R.A. 1976. Ultrafine metal particles. *J. Appl. Phys*. 47(5): 2200–2219.

Groven, L.J. 2007. Simultaneous combustion synthesis and densification of nickel aluminide/alumina nanocomposites. MS thesis, South Dakota School of Mines and Technology, Rapid City, SD.

Groven, L.J. 2009. Combustion synthesis of carbon nanotube reinforced composites, PhD Thesis, South Dakota School of Mines and Technology, Rapid City, SD.

Groven, L.J. and Puszynski, J.A. 2005. Kinetics and product composition of exothermic reacting systems consisting of binary nanopowders. *AICHE Annual Meet. Proc*., Rapid City, SD, No. 602f.

Haber, J.A. and Buhro, W.E. 1998. Kinetic instability of nanocrystalline aluminum prepared by chemical synthesis; facile room-temperature grain growth. *J. Am. Chem. Soc*. 120: 10847–10855.

Hull, M. 2002. Tetronics: Plasma processing holds key to consistent nanopowders. *Powder Metall.* 45(1): 8–9.

Hunt, E. and Pantoya, M. 2005. Ignition dynamics and activation energies of metallic thermites: From nano- to micron-scale particulate composites. *J. Appl. Phys*. 98: 034909.

Hunt, E., Granier, J., Plantier, K., and Pantoya, M. 2004. Nickel aluminide superalloys created by SHS of nano-particle reactants. *Mater. Res. Soc. Symp. Proc*. 800: AA4.6.1.

Hunt, E., Pantoya, M., and Jouet, R.J. 2006. Combustion synthesis of metallic foams from nanocomposite reactants. *Intermetallics* 14: 620–629.

Ivanov, G.V. and Tepper, F. 1997. Activated aluminum as a stored energy source for propellants. In *Challenges in Propellants and Combustion: 100 Years after Nobel*. ed. K.K. Kuo, pp. 636–645. New York: Begell House Inc.

Ivanov, Y.F., Osmonoliev, M.N., Sedoi, V.S. et al. 2003. Productions of ultra-fine powders and their use in high energetic compositions. *Propell. Explos. Pyrot*. 28(6): 319–333.

Jayaraman, S. 2001. Modeling of the formation of nano-particles of aluminum by vapor condensation technique. MS thesis, South Dakota School of Mines and Technology, Rapid City, SD.

Jiang, W. and Yatsui, K. 1998. Pulsed wire discharge for nanosize powder synthesis. *IEEE T. Plasma Sci*. 26(5): 1498–1501.

Jouet, R.J., Warren, A.D., Rosenberg, D.M., Bellitto, V.J., Park, K., and Zachariah, M.R. 2005. Surface passivation of bare aluminum nanoparticles using perfluoroalkyl carboxylic acids. *Chem. Mater*. 17(11): 2987–2996.

Kearns, M. 2004. Development and applications of ultrafine aluminium powders. *Mater. Sci. Eng. A*. 375–377 (1–2 SPEC. ISS.): 120–126.

Kim, K. 2008. High energy pulsed plasma arc synthesis and material characteristics of nanosized aluminum powder. *Met. Mater. Int*. 14(6): 707–711.

Kim, J.S., Choi, H.S., Dudina, D., Lee, J.K., and Kwon, Y.S. 2007. Spark plasma sintering of nanoscale (Ni+Al) powder mixture. *Solid State Phenom*. 119: 35–38.

Korchagin, A.I., Cherepkov, V.V., Fadeev, S.N. et al. 2008. Production of nanopowders using high power electron accelerator. *Proceedings of RuPAC*, Zvenigorod, Russia, pp. 330–332.

Kwok, Q.S.M., Badeen, C., Armstrong, K., Turcotte, R., Jones, D.E.G., and Gertsman, V.Y. 2007. Hazard characterization of uncoated and coated aluminium nanopowder compositions. *J. Propul. Power* 23(4): 659–668.

Kwon, Y.-S., Jung, J.-H., Yavorovsky, N.A., Illyn, A.P., and Kim, J.-S. 2001. Ultra-fine powder by wire explosion method. *Scr. Mater*. 44: 2247–2251.

Kwon, Y.-S., Gromov, A.A., and Strokova, J.I. 2007. Passivation of the surface of aluminum nanopowders by protective coatings of the different chemical origin. *Appl. Surf. Sci*. 253(12): 5558–5564.

Levitas, V.I., Asay, B.W., Son, S.F., and Pantoya, M. 2007. Mechanochemical mechanism for fast reaction of metastable intermolecular composites based on dispersion of liquid metal. *J. Appl. Phys*. 101: 083524-1–083524-20.

Li, Y., Song, W., Xie, C., Zeng, D., Wang, A., and Hu, M. 2006. Influence of humidity on the thermal behavior of aluminum nanopowders. *Mater. Chem. Phys*. 97: 127–131.

Lide, D.R. (Editor-In-Chief). 2005. *CRC Handbook of Chemistry and Physics*. Boca Raton, FL: CRC Press.

Martell, A.E. and Motekaitis, R.J. 1992. *Determination and Use of Stability Constants*. New York: VCH Publishers.

Martin, J.A., Murray, A.S., and Busse, J.R.1998. Metastable interstitial composites. *Warhead Technology*, pp. 179–191.

Mench, M.M., Kuo, K.K., Yeh C.L., and Lu, Y.C. 1998. Comparison of thermal behavior of regular and ultra-fine aluminum powders (ALEX) made from plasma explosion process. *Combust. Sci. Technol*. 135(1–6): 269–292.

Miziolek, A.W. 2002. Nanoenergetics: An emerging technology area of national importance. *Amptiac Quart*. 6(1): 43–48.

Moore, K. and Pantoya, M.L. 2006. Combustion of environmentally altered molybdenum trioxide nanocomposites. *Propell. Explos. Pyrot*. 31(3): 182–187.

Noebe, R.D., Misra, A., and Gibala, R. 1991. Plastic flow and fracture of B2 NiAl-based intermetallic alloys containing a ductile second phase. *ISIJ Int*. 31(10): 1172–1185.

Perry, W.L., Smith, B.L., Bulian, C.J., Busse, J.R., Macomber, C.S., Dye, R.C., and Son, S.F. 2004. Nano-scale tungsten oxides for metastable intermolecular composites. *Propell. Explos. Pyrot*. 29(2): 106–111.

Pivkina, A.N., Zavyalov, S.A., Ulyanova, P., and Frolov, Y.V. 2002. Heterogeneous combustion with nanomaterials. *Proceedings of the 29th International Pyrotechnics Seminar*, Westminster, Colorado, pp. 97–116.

Pivkina, A., Frolov, Y., Zavyalov, S., Ivanov, D., Schoonman, J., Streletskii, A., and Butyagin, P. 2004. Ultra-fine and nanosized oxidizers and metals: Thermal ignition properties and use in high energetic compositions. *Proceedings of the 31st International Pyrotechnics Seminar*, Fort Collins, Co, pp. 285–298.

Pivkina, A., Ivanov, D., Frolov, Y., Mudretsova, S., Nickolskaya, A., and Schoonman, J. 2006. Plasma synthesized nano-aluminum powders: Structure, thermal properties and combustion behavior. *J. Therm. Anal. Calorim*. 86(3): 733–738.

Plantier, K.B., Pantoya, M.L., and Gash, A.E. 2005. Combustion wave speeds of nanocomposites Al/$Fe_2O_3$: The effects of $Fe_2O_3$ particle synthesis technique. *Combust. Flame* 140: 299–309.

Plazanet, L. and Nardou, F. 1998. Reaction process during relative sintering of NiAl. *J. Mater. Sci*. 33: 2129–2136.

Puszynski, J.A. 2004. Reactivity of nanosize aluminum with metal oxides and water vapor. *Mater. Res. Soc. Symp. Proc*. 800: AA4.6.1.

Puszynski, J.A. 2007. Formation, passivation, and reactivity of energetic nanopowders. In *Advancements in Energetic Materials and Chemical Propulsion*. ed. K.K. Kuo and J.D. Rivera, pp. 3–21. New York: Begell House Inc.

Puszynski, J.A., Jayaraman, S., Carpenter, G., Carpenter, P., and Bichay, M. 2002. Formation and reactivity of nanosized aluminum powders. *Proceedings of the World Congress on Particle Technology*, Sydney, Australia, 4: P164.

Puszynski, J.A., Bulian, C.J., and Swiatkiewicz, J.J. 2005. Combustion characteristics in condensed phase systems consisting of nanosized reactants. *Proceedings of the 7th World Congress of Chemical Engineering*, Glasgow, Scotland, P15–32, 1–9.

Puszynski, J.A., Bulian, C.J., Swiatkiewicz, J.J., and Bichay, M.M. 2006. Challenges in processing of aluminum and metal oxide nanopowders and water. *Proceedings of the 5th World Congress on Particle Technology*, Orlando, FL, P247D.

Puszynski, J.A., Bulian, C.J., and Swiatkiewicz, J.J. 2007. Processing and ignition characteristics of aluminum-bismuth trioxide nanothermite system. *J. Propul. Power* 23(4): 688–706.

Puszynski, J.A., Bulian, C.J., and Swiatkiewicz, J.J. 2008. Ignition characteristics of nanothermite systems. In *Advancements in Energetic Materials and Chemical Propulsion*. ed. K.K. Kuo and K. Hori, pp. 73–86. New York: Begell House Inc.

Puszynski, J.A., Bulian, C.J., Doorenbos, Z., and Swiatkiewicz, J.J. 2010. Combustion mechanisms and kinetics of nanothermite reactions. *Int. J. Self Prop. High Temp. Syn*. (in review).

Sanders, V.E., Asay, B.W., Foley, T.J., Tappan, B.C., Pacheco A.N., and Son, S.F. 2007. Reaction propagation of four nanoscale energetic composites (Al/$MoO_3$, Al/$WO_3$, Al/CuO, and $Bi_2O_3$). *J. Propul. Power* 23(4): 707–714.

Sarithi, R., Sindhu, T.K., and Chakravarthy, S.R. 2007. Generation of nano aluminium powder through wire explosion process and its characterization. *Mater Charact*. 58: 148–155.
Schoenitz, M., Ward, T., and Dreizin, E.L. 2004. Preparation of energetic metastable nanocomposite materials by arrested reactive milling. *Mater. Res. Soc. Symp. Proc.* 800: AA2.6.1–AA2.6.6.
Schroder, K.A. and Jackson, D.K. 2004. Radial pulsed arc discharge gun for synthesizing nanopowders. *U.S. Patent No. 6,777,639 B2.*
Smith, R.M. and Arthur E. Martell. 1974. *Critical Stability Constants*, Vol. 4. New York: Plenum.
Somasundaram, V., Bulian, C.J., and Puszynski, J.A. 2003. Investigation of Al-CuO nanocomposite reacting systems. *Proceedings of EUROPYRO 2003*, Saint Malo, France, vol. 2, pp. 536–541.
Son, S.F., Hiskey, H.L., Asay, B.W., Busse, J.R., Jorgensen, B.S., Bockmon, B., and Pantoya, M. 2001. Reaction propagation physics of $Al/MoO_3$ nanocomposite thermites. *Proceedings of the 28th International Pyrotechnics Seminar*, Adelaide, Australia, pp. 833–842.
Son, S.F., Busse, J.R., Asay, B.W., Peterson, P.D., Mang, J.T., Bockmon, B., and Pantoya, M.L. 2002. Propagation studies of metastable intermolecular composites (MIC). *Proceedings of the 29th International Pyrotechnics Seminar*, Westminster, Colorado, pp. 203–212.
Son, S.F., Yetter, R.A., and Yang, V. 2007. Introduction: Nanoscale composite energetic materials. *J. Propul. Power* 23(4): 643–644.
Sullivan, K., Young, G., and Zachariah, M.R. 2009. Enhanced reactivity of nano-B/Al/CuO MIC's. *Combust. Flame* 156(2): 302–309.
Tepper, F. 2000. Nanosize powders produced by electro-explosion of wire and their potential applications. *Powder Metall.* 43(4): 320–322.
Trunov, M.A., Umbrajkar, S., Schoenitz, M., Mang, J.T., and Dreizin, E.L. 2006. Oxidation and melting of aluminum nanopowders. *J. Phys. Chem. B.* 110(26): 13094–13099.
Valliappan, S., Swiatkiewicz, J.J., and Puszynski, J.A. 2005. Reactivity of aluminum nanopowders with metal oxides. *Powder Technol.* 156(2–3): 164–169.
Walter, K.C., Pesiri, D.R., and Wilson, D.E. 2007. Manufacturing and performance of nanometric $Al/MoO_3$ energetic materials. *J. Propul. Power* 23(4): 645–650.
Wronski, C.R.M. 1967. The size dependence of the melting point of small particles of tin. *Brit. J. Appl. Phys.* 18: 1731–1737.
Yavorovsky, N.A., Davydovich, V.I., and Bil, B.A. 1995. Apparatus for production of highly dispersed powders of inorganic materials by electric explosion and reactor for explosion of metallic component. *Patent of Russian Federation No. 2048278.*
Zelinskii, V.Y., Yavorovskii, N.A., Proskurovskaya, L.T., and Davydovich, V.I. 1984. Structural state of aluminum particles prepared by electric explosion. *Fizika i Khimiya Obrabotki Materialov* (In Russian) 1: 57–59.

# 7

# *Nanoparticles for Fuel Cell Applications*

**Jin Luo, Bin Fang, Bridgid N. Wanjala, Peter N. Njoki, Rameshwori Loukrakpam, Jun Yin, Derrick Mott, Stephanie Lim, and Chuan-Jian Zhong**

**CONTENTS**



## 7.1 Introduction

Fuel cells utilizing hydrogen as fuels represent an important form of tomorrow's energy because hydrogen is an efficient fuel and it is environmentally clean. Auto industry, which relies on oil-fueled cars, is perhaps the biggest driving force behind the massive investment in fuel cell development (Zhong et al. 2008). Harmful emissions of $CO_2$, CO, $SO_2$, $NO_x$, and volatile organic compounds into the atmosphere cause serious environmental damage, increase respiratory problems in humans, and produce "greenhouse gas" that contributes to global warming. With these problems, fuel cell technology is inevitably seen as a viable alternative. Energy sources of the future will have to be cleaner and more efficient than current sources—fuel cells fulfill these requirements. "Hydrogen Economy" offers an energy system based upon hydrogen for energy generation, storage, distribution, and utilization. There are three main areas of challenges for the realization of hydrogen energy: hydrogen production, hydrogen storage, and hydrogen utilization. Fuel cells based on hydrogen fuel represent one of the most effective ways of hydrogen utilization. Proton exchange membrane fuel cell (PEMFC) (Figure 7.1) has become attractive because of high conversion efficiency, low pollution, lightweight, high power density, and a wide range of applications from power sources in automobiles and space shuttles to power grids for buildings and factories. Fuel cells are essentially electrochemical cells and operate by the same basic mechanism as regular batteries. Hydrogen fuel cells convert flows of hydrogen and oxygen into water and produce electricity. At the anode, hydrogen is forced through a catalyst where it is ionized. At the cathode, oxygen reacts with the products from the anode (the protons and electrons) to produce water. The close circuit of the two electrodes produces electricity and heat, and water as the only product.

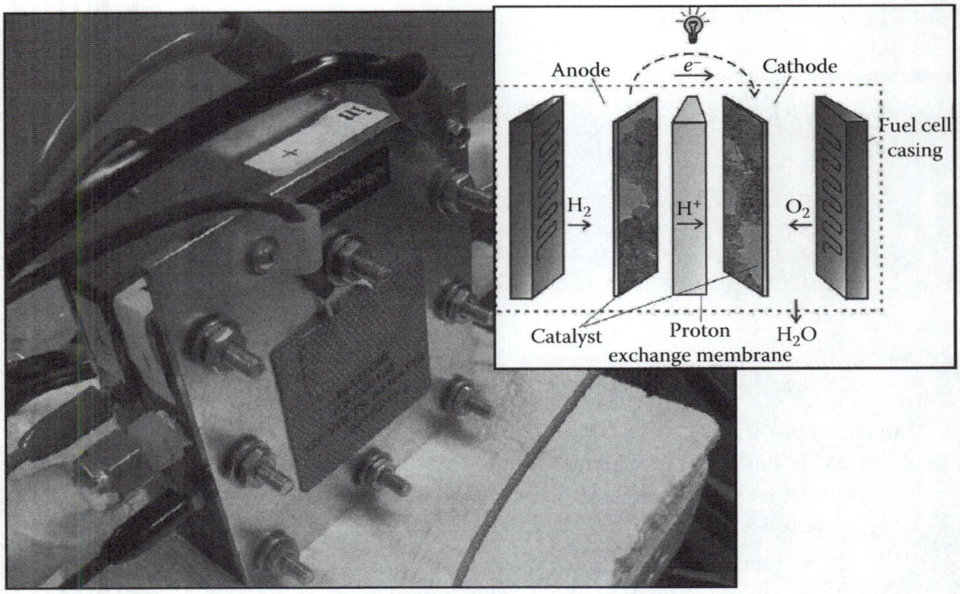


**FIGURE 7.1**
**(See color insert following page 302.)** The picture of a PEMFC. Inset: a simplified and dissected scheme of the basic components of PEMFC.

In PEMFC, electrochemical reactions occur at the surface of the catalyst at the interface between the electrolyte and the membrane. Hydrogen fed on the anode side of the membrane splits into protons and electrons. Protons travel through the membrane, while the electrons travel through the outside circuit where they perform useful work and return to the cathode side of the membrane. At the catalyst sites of the cathode oxygen is reduced, which combines with the protons, forming water. The net result of these simultaneous reactions is the current of electrons through an external circuit—direct electrical current. There are many types of fuel cells. In addition to the most popular hydrogen/air fuel cells, another type of fuel cells such as direct methanol fuel cell (DMFC) has also become attractive because of high conversion efficiency, low pollution, lightweight, high power density, and applications from small power supplies for electronic devices such as PCs, notebooks, and cellular phones.

The large overpotential for oxygen reduction at the cathode represents a loss of about 20% from the theoretical maximum efficiency for the hydrogen/air fuel cells. The situation is even worse with the DMFCs. The thermodynamic potential for a DMFC is 1.21 V, which is only 20 mV less than that for the PEMFC. Both methanol oxidation and oxygen-reduction reactions (ORRs) are highly irreversible, and thus there is a loss of about 0.2 V at the anode for DMFC under open-circuit conditions, and an enhanced loss of about 0.1 V at the oxygen electrode because of the crossover of methanol from the anode to the cathode (Shukla and Raman 2003).

There are many basic components in a fuel cell, including gas diffusion electrode, polymer membrane, and catalysts. Catalyst is one of the key components. According to the cost breakdown of fuel cell components (Stone 2005), the cost of catalysts in manufacturing fuel cells is the highest (~30%) for small production volume, and remains very high with increasing production volume. Currently, low activity, poor durability and high cost of the platinum-based anode and cathode catalysts in PEMFCs and DMFCs constitute some of the major barriers to the commercialization of fuel cells. In addition, the durability of Pt-based catalysts can be compromised by the sintering and dissolution of the catalysts in fuel cells. There is a major gap in the development of catalyst technology between the laboratory test and the practical application, which largely lies at the lack of abilities to engineer the size, composition and stability. A key challenge to the ultimate commercialization of fuel cells is the development of active, robust, and low-cost catalysts (Gasteiger et al. 2005, Mallouk and Smotkin 2003). In contrast to most existing catalyst preparations based on traditional co-precipitation or impregnation methods, the foundation of the nanoengineering of multimetallic alloy catalysts (Bonnemann and Nagabhushana 2004, He et al. 2008, Luo et al. 2006a,b,c,d, Wang et al. 2010, Zhong et al. 2006, 2007, 2010) has many important attributes to bridge the existing gap and lead to new opportunities in creating synergistic balance of activity and stability of the catalysts. Selected examples from recent studies of the preparation and characterization of multimetallic nanoparticles and fuel cell catalysts will be discussed.

## 7.2 Catalyst Preparation

One of the important challenges for fuel cell commercialization is the preparation of active, robust, and low-cost electrocatalysts. However, the lack of abilities in the controlled preparation of nanoscale size and composition, and the prevention of the intrinsic propensity of aggregation of nanoscale materials constitute an obstacle to exploiting the nanometer-sized catalytic properties. The ability to harness the large surface area-to-volume ratios and the unique binding sites of nanoparticles constitutes a major driving force in both fundamental research and practical applications of nanotechnology in catalysis. It is the nanoscale size range over which metal particles undergo a transition from atomic to metallic properties, leading to new electronic and catalytic properties. The synthesis of molecularly capped metal nanoparticles as building blocks for engineering the nanoscale catalytic materials takes advantage of diverse attributes including monodispersity, processability, solubility, stability, and self-assembly capability in terms of size, shape, composition, and surface properties.

Some of the serious problems for fuel cells operated by electrochemical reduction of oxygen at the cathode include the poor activity of the anode and cathode catalysts, and the "methanol cross-over" to the cathode electrode (Adler 2004, Litster and McLean 2004, Mehta and Cooper 2003, Ren et al. 2000, Russell and Rose 2004, Spendelow and Wieckowski 2004), leading to a loss of about one-third of the available energy at the cathode. The propensity of poisoning of Pt by CO species (including trace level CO in reformed hydrogen from methanol, natural gas, or gasoline (Acres et al. 1997, Roucoux et al. 2002, Wasmus and Kuever 1999, Wieckowski and Lu 2000)) is another problem for the extensively studied Pt catalysts. The binary Pt–Ru on carbon support has been studied for decades because of the bifunctional catalytic capability (Ley et al. 1997, Long et al. 2000, Roucoux et al. 2002, Wieckowski and Lu 2000). The kinetic limitation of the oxygen reduction at cathode catalysts is another problem for fuel cells operating at low temperature (<100°C) because the rate of breaking O=O bond to form water strongly depends on the degree of its interaction with adsorption sites of the catalyst, and the competition with other species in the electrolyte (e.g., $CH_3OH$). Metalloporphyrins (Chu and Jiang 2002, Collman et al. 1983) and PGM alloys (Paulus et al. 2002) have been used to design catalyst for $4e^-$ reduction of $O_2$ to water.

Many studies focused on understanding the mechanism of oxygen reduction on Pt–Fe, Pt–Ni, and Pt–Co (Brandon et al. 2003), including CO- or methanol-tolerant (Adler 2004, Shukla and Raman 2003). Bulk-melted PtBi, PtIn, and PtPb intermetallic phases (Casado-Rivera et al. 2004) and Ru nanoparticles modified with Pt (Sasaki et al. 2004, Zhang et al. 2005) showed some promises for fuel cell applications. The application of high throughput, combinatorial screening methods to screen a large number of catalysts using an optical fluorescence technique by detecting proton production on an array of catalyst inks with different compositions has been successfully demonstrated (Chan et al. 2005, Chen et al. 2001, Mallouk and Smotkin 2003, Sun et al. 2001). Individually addressable array electrodes have also been investigated for rapid screening (Guerin et al. 2004, Liu and Smotkin 2002, Strasser et al. 2003). Recently, simple thermodynamic principles are proposed as guidelines that assume that one metal breaks the oxygen–oxygen bond of $O_2$ and the other metal acts to reduce the resulting adsorbed atomic oxygen (Fernandez et al. 2005). The high-throughput combinatorial screening of catalysts is very useful for rapid screening, including our latest work (He et al. 2006). Theoretical work also begins to demonstrate the importance of binary or ternary alloy catalysts, including computational chemistry (e.g., density functional theory [DFT]) (Pedersen et al. 1999) to determine optimal structures and adsorption energies and to predict synergetic effects. Because the preparation of fuel cell catalysts has been mostly based on traditional methods such as co-precipitation or impregnation (Bond and Thomposon 1999, Bond 2002, Campbell 2004, Chen and Goodman 2004, Davis 2003, Haruta and Date 2001, Haruta 2004, Lai and Goodman 1998, Rolison 2003), little is known about how to optimize size, composition, and morphology of multimetallic catalysts. The knowledge is very important since the nanoscale synergistic effects are very different from their macroscopic counterparts, as supported by the examples of bimetallic nanoparticles (PtRu, AuPt, etc.), including our recent work (Kariuki et al. 2004, Luo et al. 2005a, Njoki et al. 2005).

To achieve durable and active catalysts with a low cost, new concepts and strategies must be developed for the creation of size-, composition-, and morphology-controlled multimetallic nanoparticles and catalysts. Nanoscale phenomena differ from bulk counterparts in many significant ways, including atomic–metallic transition, possible phase reconstitution, different melting points due to size or alloying effects, and synergistic effects due to modified electronic band structure. The understanding of whether the formation of alloy or phase segregation in multimetallic nanoparticles is different from bulk-scale materials, and how the catalytic activity and stability are influenced by size, composition, and morphology could hold keys to the engineering of durable and active catalysts.

The recent studies of nanogold catalysts serve as best examples to illustrate the unique properties displayed by nanoparticles. Despite the intensive research into the catalytic activity of gold in a restricted nanoscale size range (Haruta 2005), the catalytic origin of nanosized gold and gold-based bimetallic catalysts remains elusive. One of main problems is the lack in understanding of the nanoscale core-surface property correlation. Gold–platinum nanoparticles of 2–5 nm diameter present an intriguing system for delineating the correlation in view of recent ability in synthesizing AuPt nanoparticles in a wide range of bimetallic composition (Luo et al. 2005a, Mott et al. 2007a,b). Whether the AuPt nanocrystal core is alloyed or phase segregated and how the surface-binding properties are correlated with the nanoscale bimetallic properties are important questions for the exploitation of catalytic activity of the nanoscale bimetallic catalysts. Our x-ray diffraction (XRD) studies (Luo et al. 2005a, Mott et al. 2007a,b) revealed alloy properties for the nanocrystal core, which is in contrast to the miscibility gap known for the bulk counterparts (Ponec and Bond 1995). There are also infrared spectroscopic studies of the adsorption of CO on the nanoparticles to address the surface-binding properties (Chen et al. 2005, Kim and Korzeniewski 1997, Lang et al. 2004, Mihut et al. 2002), including our own recent work (Mott et al. 2007a,b). AuPt nanoparticles could provide a synergistic catalytic effect that involves the suppression of adsorbed poisonous species and a change in electronic band structure to modify the strength of the surface adsorption. The decrease in activation energy to facilitate oxidative desorption or suppress CO adsorption was previously considered to lead to sufficiently high adsorptivity to support catalytic oxidation in alkaline electrolytes (Anderson et al. 1996, Burke et al. 2000, Morita et al. 1991, Nishimura et al. 1989). Since the alloy properties of the bimetallic AuPt nanoparticles (Luo et al. 2005a, Mott et al. 2007a,b) are in sharp contrast to the bimetallic miscibility gap known for the bulk counterparts in a wide composition range (10%–80% Au), the understanding of how the bimetallic nanocrystal and surface alloy properties are related to the surface binding and catalytic activities is very important. The recent report (Zhang et al. 2007) on the stabilization of platinum oxygen-reduction electrocatalysts using gold clusters demonstrated that the Pt catalysts can be stabilized against dissolution under potential cycling regimes by modifying Pt nanoparticles with Au clusters.

Arrays of various binary combinations were studied using scanning electrochemical microscopy, and Pd–Co (10%–30% Co) was found to exhibit activity close to that of Pt. In view of the strong adsorption of OH-forming Pt–OH, which causes inhibition of the $O_2$ reduction, there are several important aspects of the recent progress in understanding the synergistic properties of PtM catalysts in ORR in terms of the role of M (metal) in the adsorption of oxygenated species (e.g., O, OH), including near-surface alloys (Zhang et al. 2005, 2007) such as $M_xPt_{1-x}$/Pd(111) (Zhang et al. 2005) in which M has the ability to adsorb OH or O, and Pt/M(111) (Zhang et al. 2005) with which a volcano-type dependence of the activity on d-band centers was shown due to an optimum compromise between a higher lying d-band center for O=O bond breaking and a lower lying one for H–O bond formation.

Gold is a special metal of increasing interest in view of its inertness in the bulk state and its high catalytic activity at the nanoscale (Bond and Thompson 1999, Bond 2002, Campbell 2004, Chen and Goodman 2004, Chen et al. 2005, Davis 2003, Haruta 2005, Hughes et al. 2005, Lai and Goodman 1998, Rolison 2003). The detection of $AuO^-$, $AuO_2^-$, and AuOH ion clusters in the latest time-of-flight secondary ion mass spectroscopic study of $\gamma$-$Al_2O_3$- and $TiO_2$-supported Au catalysts provided a direct evidence for oxidized gold on the supported gold catalysts (Fu et al. 2005). This finding substantiates the favorable formation of surface oxygen–species on gold. The formation for surface oxygen–species has been implied in studies of gold nanoparticles (Luo et al. 2001, Raj et al. 2005), gold single-crystal electrodes (Blizanac et al. 2004, Strabac and Adzic 1996) in alkaline electrolytes, and Au adlayers on Pt or Pt adlayers on Au electrode (El-Deab and Ohsaka 2003, Van Brussel et al. 2002). Using a pyramid model for Au, Pt, and AuPt (organized alloying) (El-Deab and Ohsaka 2002), the adsorption of $O_2$ on a pure Au is indeed found to be exothermic by 0.77 eV. While the interaction energy does not vary considerably on different types of Pt surfaces, significant changes in the adsorption energy were observed for Au with different geometries.

In view of the unique catalytic properties of nanosized gold (El-Deab and Ohsaka 2002, 2003, Kim et al. 2003, Van Brussel et al. 2002, World Gold Council 2003, Xu and Mavrikakis 2003) and the high catalytic hydrogenation activity of platinum, bimetallic AuPt nanoparticles of controllable composition may serve as a synergistic catalyst system. For example, in alkaline medium, the presence of Au in Pt catalysts could reduce the strength of the Pt–OH formation (El-Deab and Ohsaka 2002, 2003, Van Brussel et al. 2002), while providing the needed adsorption sites for –OH species. While the presence of Au in Pt increases the lattice distance of Pt, the higher electronegativity of Au than Pt could cause an increase of the amount of charge being transferred from Pt to Au, which was in fact supported by high-resolution x-ray photoelectron spectroscopy (XPS) data showing Au $4f_{7/2}$ binding energy 83.32 eV for Au/Pt and 83.87 eV for bulk-like Au atoms (Berg et al. 1998), and consequently an increase of the d-orbital vacancy in the PtAu.

The preparation of Pt-group catalysts, especially Pt alloyed with other transition metals, has been extensively studied for fuel cell catalytic reactions (Adler 2004, Antolini 2003, Litster and Mclean 2004, Rao and Trivedi 2005, Ressell and Rose 2004, Spendelow and Wieckowski 2004, Yang et al. 2004). Traditional approaches to preparing supported nanoparticle catalysts involve co-precipitation, deposition precipitation, ion exchange, impregnation, successive reduction, and calcination, etc. These methods have been widely used for preparing noble metal catalysts on support materials (Klabunde 2001). While a variety of supported Pt-group binary or ternary catalysts have been prepared by traditional methods (Dickinson et al. 2002, Feldheim and Foss 2002, Jones et al. 2002, Klabunde 2001, Park et al. 2002, Paulus et al. 2002, Raja et al. 2001, Schmidt et al. 1999, Waszczuk et al. 2002, Yang et al. 2004), the ability to control the size and composition is limited due to the propensity of aggregation of metals at the nanoscale. Among many emerging approaches to the preparation of nanoparticles or nanostructures, one particular class of nanoparticles with core–shell-type structures is beginning to attract interest for addressing some of the challenges in nanoscale catalyst preparation (Brust et al. 1994, Paulus et al. 2000, Schmid et al. 1996, Templeton et al. 2000). The core–shell-type nanomaterials can be broadly defined as core and shell of different matters in close interaction, including inorganic/organic and inorganic/inorganic combinations (Brust et al. 1994, El-Sayed 2001, Kiely et al. 2000, Paulus et al. 2000, Schmid et al. 1996, Templeton et al. 2000, Whetten et al. 1996, Zhong and Maye 2001). In recent years, there has been an increasing volume of studies aimed at synthesizing metal nanoparticles in the presence of organic capping agents (Caruso 2001, Chen and Nikles 2002, Crooks et al. 2001, El-Sayed 2001, Kiely et al. 2000, Mbindyo et al. 2001, Paulus et al. 2000, Schärtl, 2000, Schneider 2001, Storhoff and Mirkin 1999, Sun et al. 2000, Zhong and Maye 2001). While the synthesis of monometallic nanoparticles such as gold, silver, platinum, and palladium have extensively been studied using molecular encapsulation–based synthesis methods (e.g., two-phase protocol), relatively limited studies of bi- or trimetallic nanoparticles have been reported using such synthetic methods (Chen and Nikles 2002, Hostetler et al. 1998, Paulus et al. 2000, Sun et al. 2000).

In addition to controllable nanoscale dimensions, the prevention of the intrinsic propensity of aggregation of nanoscale materials is another challenging area. Aggregation of nanoparticles leads to eventual loss of the nanoscale catalytic activity in practical applications. The use of naked metal nanoparticle catalysts on supporting materials based on traditional preparative methods has been well demonstrated for different catalytic reactions (Aiken and Finke 1999, Crown et al. 2000). Recently, nanoparticles capped in monolayers, polymers, or dendrimers are rapidly emerging, demonstrating remarkable parallels to catalytic activities for supported nanoparticles. The catalysis includes those utilizing functional groups at the capping shell (Ingram and Murray 1998, Li et al. 1999, Oldenburg et al. 1998, Peng et al. 1997) and those exploiting surface sites on the nanocrystals (Aiken and Finke 1999, Crown et al. 2000, Zhong and Maye 2001). Polymer-mediated self-assembly of monolayer-functionalized Pd nanoparticles and $SiO_2$ particles through "bottom-up" approach and thermal treatment have recently been demonstrated as highly reactive, recyclable heterogeneous catalysts for both hydrogenation and carbon–carbon bond formation reactions (Galow et al. 2002). The nanotechnology-guided design and fabrication approach for enhancing the catalytic activity and reducing the cost of catalysts will have enormous impacts to better catalyst preparation for fuel cell application (Mehta and Cooper 2003, Ren et al. 2000, Zhong et al. 2004). Importantly, deliberate tailoring of nanoparticle size, shape, and composition ranging down to a few nm could lead to improved or new catalytic properties.

## 7.3 Multimetallic Nanoparticles and Catalysts

In contrast to traditional approaches of preparing supported catalysts, the molecular encapsulation–based synthesis and processing strategy of multimetallic catalyst preparation involves a sequence of three steps: solution-phase synthesis of the nanocrystal cores with molecular encapsulation, assembly of the encapsulated nanoparticles on support materials, and thermal treatment of the supported nanoparticles. Two examples, including bimetallic gold–platinum (AuPt) alloy and core–shell nanoparticles and trimetallic platinum–vanadium–iron (PtVFe) and platinum–nickel–iron (PtNiFe) catalysts, are described here to illustrate how such nanoscale-engineered materials can advance the development of fuel cell catalysts.

### 7.3.1 Bimetallic Catalysts—AuPt Alloy and Core–Shell Nanoparticles

Using a modified two-phase method (Hostetler et al. 1998, Luo et al. 2005a,c, 2006a), AuPt alloy nanoparticles of 2 nm core size and different compositions encapsulated with organic shells were synthesized by controlling the feed ratios of the two metal precursors ($AuCl_4^-$ and $PtCl_6^{2-}$) (Luo et al. 2006a,b). Bimetallic nanoparticles in which the nanocrystal core consists of one metal core and another metal shell (core@shell) such as Au@Pt and Pt@Au were also synthesized by seeded growth method (Luo et al. 2008). Such nanoparticles were assembled onto carbon black materials through interactions between the capping shells and the carbon surface. The carbon-supported nanoparticles were then subjected to thermal treatment under controlled atmosphere, which involves the removal of organic shells and calcination of the alloy nanoparticles.

The as-synthesized $Au_mPt_{100-m}$ nanoparticles with different compositions are capped with thoil/amine monolayer shells (Figure 7.2a). Varying the feeding ratio of the metal precursors used in the synthesis controls the bimetallic composition and sizes of the nanoparticles. The high-resolution transmission electron microscopy (TEM) image indicates that the thermally treated nanoparticles exhibit a highly crystalline morphology (Figure 7.2b). The effective removal of the capping monolayers from the nanoparticles by the thermal-treatment process (Luo et al. 2004) was supported by the absence of the vibrational bands characteristic of the capping molecules in the C–H stretching region detected by Fourier transform infrared (FTIR), and the absence of the bands associated with sulfur species after the thermal treatment detected XPS analysis. The average size of the particles in $Au_{72}Pt_{28}/C$ catalyst treated at 400°C from the TEM data (4.6 nm) is very close to the value determined from XRD data (Luo et al. 2005a) (4.8 nm).

As shown in Figure 7.3, in contrast to the bulk Au–Pt counterparts, which display a miscibility gap at 20%–90% Au (Ponec and Bond 1995), the lattice parameters of the bimetallic nanoparticles were found to scale linearly with Pt%. Such a relationship follows a Vegard-type law typically observed with binary metallic alloys, demonstrating the alloy

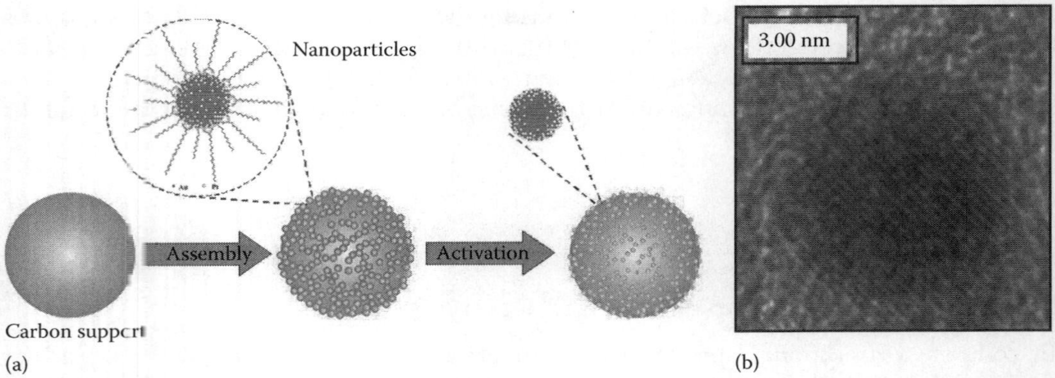


**FIGURE 7.2**
(a) A schematic illustration of AuPt nanoparticle core encapsulated by a monolayer shell of thiol/amine, and shell remover (on carbon support). (b) High-resolution transmission electron microscopy (HRTEM) of thermally treated AuPt/C catalysts (e.g., $Au_{72}Pt_{28}/C$). (Reproduced from Zhong, C.J. et al., *Energy Environ. Sci.*, 1, 454, 2008. With permission from the Royal Society of Chemistry.)

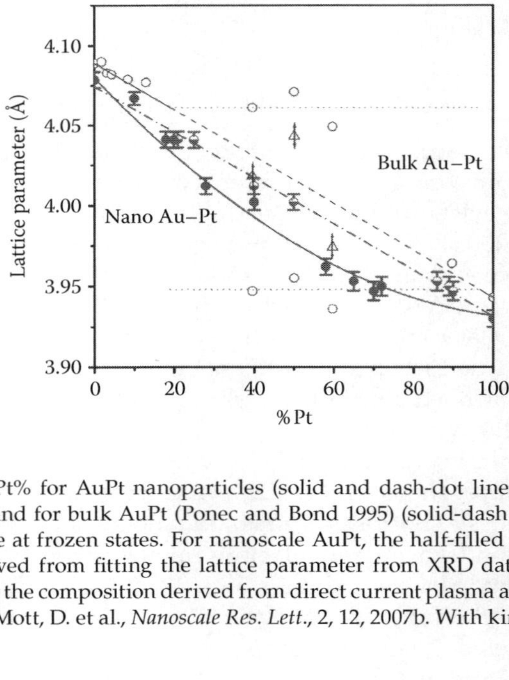


**FIGURE 7.3**
The lattice parameters vs. Pt% for AuPt nanoparticles (solid and dash-dot lines), part of the data reported recently (Luo et al. 2005a), and for bulk AuPt (Ponec and Bond 1995) (solid-dash line). For bulk AuPt, the triangle points represent those at frozen states. For nanoscale AuPt, the half-filled circle points represent those using the composition derived from fitting the lattice parameter from XRD data, whereas the filled circled points represent those using the composition derived from direct current plasma atomic emission spectrometry (DCP-AES) analysis. (From Mott, D. et al., *Nanoscale Res. Lett.*, 2, 12, 2007b. With kind permission from Springer Science+Business Media.)

properties for the bimetallic AuPt nanoparticles (Luo et al. 2005a). The XRD pattern for $Au_{52}Pt_{48}$ synthesized using the different protocol also showed alloy feature (Kariuki et al. 2004, Njoki et al. 2005). While the diffraction patterns are characteristic of the fcc-type lattice, there are subtle differences in peak shape, width, and position. The lattice parameters were determined for each AuPt sample by carefully determining the positions of the Bragg peaks in the diffraction patterns. In addition, the values for the lattice parameter of the nanoscale AuPt are all smaller than those for the bulk AuPt. This intriguing phenomenon suggests that nanoparticles have smaller interatomic distances than those for the bulk counterparts (Mott et al. 2007a,b). To our knowledge, this is the first example demonstrating that the nanoscale AuPt nanoparticles could exhibit single-phase character and small interatomic distances in the entire bimetallic composition range, both which are in sharp contrast to those known for their bulk counterparts. This finding is in fact supported by a recent theoretical modeling study (Xiao et al. 2006), which showed that the nanoscale alloying is thermodynamically favored for AuPt nanoparticles smaller than ~6 nm because the heat of formation is negative.

A comparison of the FTIR spectra for CO adsorption on AuPt nanoparticles over a wide range of bimetallic composition (Figure 7.4) provides important information for assessing the surface-binding properties of these bimetallic nanomaterials. By comparing CO spectra for $Au/SiO_2$, $Pt/SiO_2$, physical mixtures of $Au/SiO_2$ and $Pt/SiO_2$, and an $Au_{72}Pt_{28}/SiO_2$ alloy, the CO bands for the bimetallic alloy catalyst are detected at 2115 and 2066 $cm^{-1}$, which are distinctively different from the single band feature at 2115 $cm^{-1}$ for CO linearly adsorbed on atop sites of Au (Bailie and Hutchings 1999, Hsu and Lai 2006, Kim and Korzeniewski 1997, Meier and Goodman 2004), and the single-band feature at 2096 $cm^{-1}$ for CO on atop sites for Pt (Kim and Korzeniewski 1997, Meyer et al. 2004). From FTIR spectra comparing CO adsorption on $AuPt/SiO_2$ with a wide range of bimetallic compositions, two important features can be observed from the spectral evolution as a function of bimetallic composition. First, the 2115 $cm^{-1}$ band observed for $Au/SiO_2$ (a) displays a clear trend of diminishing absorbance as Pt concentration increases in the bimetallic catalysts. It is very interesting that this band becomes insignificant or even absent at >~45% Pt. Second, the lower-wavenumber CO band (~2050 $cm^{-1}$) shows a clear trend of shift toward that for the Pt-atop CO band observed for $Pt/SiO_2$ (i) as Pt concentration increases. This trend is shown in Figure 7.4. For higher concentrations of Au, this band is strong and broad. Such a dependence of the CO bands on the bimetallic concentration is remarkable, and is to our

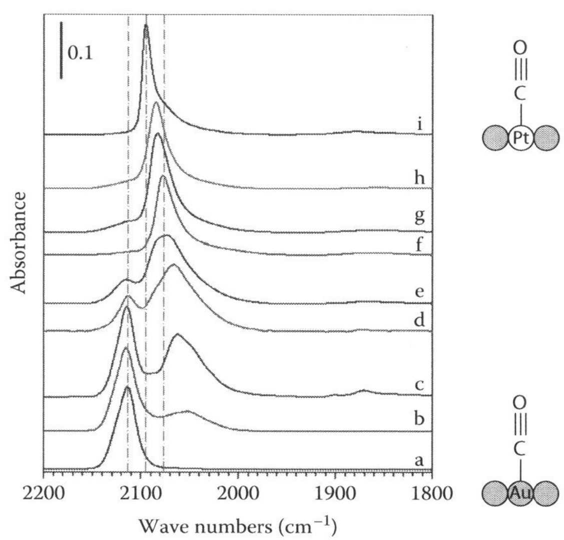


**FIGURE 7.4**
Comparison of FTIR spectra of CO adsorption: (a) $Au/SiO_2$, (b) $Au_{96}Pt_4/SiO_2$, (c) $Au_{82}Pt_{18}/SiO_2$, (d) $Au_{72}Pt_{28}/SiO_2$, (e) $Au_{65}Pt_{35}/SiO_2$, (f) $Au_{56}Pt_{44}/SiO_2$, (g) $Au_{43}Pt_{57}/SiO_2$, (h) $Au_{35}Pt_{65}/SiO_2$, and (i) $Pt/SiO_2$. The schemes illustrate the band assignment for CO adsorbed to Au-atop site surrounded by Au atoms, and Pt-atop site surrounded by Pt atoms for an AuPt alloy surface. (From Mott, D. et al., *Nanoscale Res. Lett.*, 2, 12, 2007b. With kind permission from Springer Science+Business Media.)

knowledge observed for the first time. The higher-wavenumber band (2115 $cm^{-1}$) is attributed to CO adsorption on Au-atop sites in an Au-rich surface environment, whereas the lower-wavenumber band and its composition-dependent shift reflect an electronic effect of the surface Pt-atop sites alloyed in the bimetallic nanocrystal. The fact that the disappearance of the Au-atop CO band at >~45% Pt is accompanied by a gradual shift of the Pt-atop CO band is indicative of a unique synergistic surface property in which the Pt-atop CO adsorption is greatly favored over the Au-atop CO adsorption.

The understanding of the electronic effect is based on the correlation between the spectral features and findings from a previous DFT calculation on the d-band of Pt atoms in bimetallic AuPt surfaces (Pedersen et al. 1999). The DFT calculation showed that the d-band center of Pt atoms increases with Au concentration in the AuPt alloy on a Au(111) or Pt(111) substrate. For an AuPt alloy on Au(111), the d-band center of Pt atoms was found to show an increase from 0% to 65%–70% Au, after which a slight decrease was observed. For an AuPt alloy on Pt(111), the d-band center of Pt atoms is found to increase almost linearly with the concentration of Au. Both findings were supported by experimental data in which the adsorption of CO showed an increased binding energy in comparison with Pt(111), due to the larger lattice constant of Au, leading to an expansion of Pt (Pedersen et al. 1999). Since the theoretical data for nanocrystal alloys are not yet available, the average d-band shift for Pt atoms from these two sets of DFT calculation results is used to illustrate the general trend (Mott et al. 2007a,b). Interestingly, a subtle transition for the lower-wavenumber band, i.e., from a relatively broadband feature to a narrow band feature that resembles that of the Pt-atop CO band is observed to occur at ~65% Au, below which the Au-atop CO band basically disappeared. There exists a stronger electron donation to the CO band by a Pt-atop site surrounded by Au atoms in the bimetallic alloy surface than that from the monometallic Pt surface as a consequence of the upshift in d-band center of Pt atoms surrounded by Au atoms, which explains the preference of Pt-atop CO over the Au-atop CO adsorption. The observed decrease of the Pt-atop CO band frequency with increasing Au concentration is clearly in agreement with the d-band theory for the bimetallic system (Pedersen et al. 1999). Note that the observed wavelength region of 2050–2080 $cm^{-1}$ is quite close to those found recently based on DFT calculations of CO adsorption on AuPt clusters (2030 and 2070 $cm^{-1}$) depending on the binding site (Pt or Au) (Meyer et al. 2004).

The complete disappearance of the Au–CO band for samples with a concentration below 65% Au does not necessarily imply the absence of Au on the surface of the nanoparticles; but rather implies the preferential Pt-atop CO adsorption over Au-atop CO adsorption, which is supported by the DFT calculation results (Pedersen et al. 1999). The observation of the maximum mass activity for electrocatalytic methanol oxidation reaction (MOR) under alkaline conditions around the composition of 65%–85% Au coincides remarkably with the finding of the composition of ~65% Au for the transition of the band features for CO adsorption, suggesting a synergistic effect of the surface reactivity for Pt atoms surrounded by Au atoms. AuPt catalysts could display synergistic bifunctional properties by the reduction of the strength of Pt–OH formation, the added adsorption sites for –OH species, the increase of the lattice distance of Pt, the increase of electronegativity of Pt, or increase of the d-orbital vacancy by the presence of Au in Pt catalysts.

Based on the kinetic current extracted from the rotating disk electrode (RDE) data for ORR (Figure 7.5), the mass activities were found to be strongly dependent not only on the bimetallic composition ($Au_mPt_{100-m}$), but also on the nature of the electrolyte. The strong dependence of the mass activities on the bimetallic composition in the alkaline electrolyte is evident by the exhibition of a maximum in the composition region of 60%–80% Au, which is higher than those for Pt/C and Au/C by a factor of 2–3. This finding is in contrast

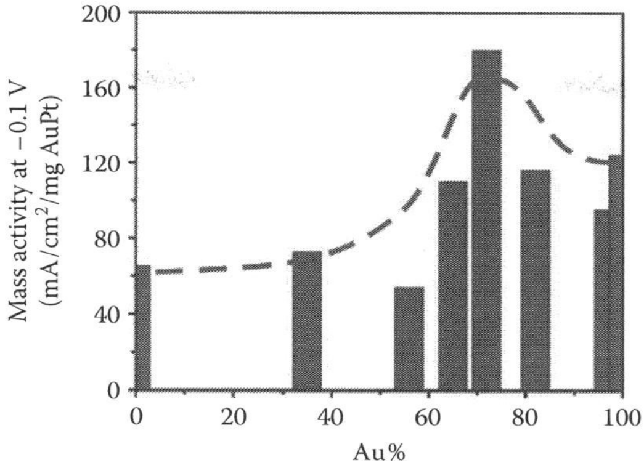


**FIGURE 7.5**
The composition dependence of mass activities (from RDE curves at 1600 rpm) determined for different $Au_mPt_{100\text{-}m}$/C catalysts in 0.5 M KOH electrolytes saturated with $O_2$. Dash line (not fitting) illustrates the general trend. (Reprinted from Luo, J. et al., *Electrochem. Commun.*, 8, 581, 2006a. Copyright [2006], with permission from Elsevier.)

to the gradual increase of mass activity in the acidic electrolyte displaying a relatively smooth transition from the low activity of Au to the high activity of Pt. In acidic solution, Au is not capable of proving adsorption sites for –OH and the electrocatalytic activity is thus rather low.

While the mass activity in the acidic electrolyte could reflect a collective effect of the activities from both Au and Pt, the concurrence of a maximized activity in the 60%–80% Au region in the alkaline electrolyte suggests the operation of a remarkable synergistic effect. The possibility of an optimal fraction of Au atoms surrounding Pt could have played an important role in the observed activity maximum. These Au atoms could thus function as the sites for chemisorbed $OH^-_{ad}$ for the dissociative adsorption of $O_2$ via interaction with $OH^-_{ad}$ or for chemisorption of the reaction intermediate $HO^-_{2\,ad}$ (Strbac and Adzic 1996). In addition to reducing –OH on Pt by alloying Au in the Pt catalyst, similarly to those recently revealed for other metals in $M_xPt_{1-x}$/Pd(111) (Zhang et al. 2005) and Pt/M(111) (Zhang et al. 2005), the favorable chemisorption of oxygen on gold nanoparticles, as evidenced by the detection of $AuO^-$, $AuO_2^-$, and AuOH on oxide-supported Au nanoparticles (Fu et al. 2005) and $OH^-_{ad}$ on Au(110) in alkaline electrolyte (Strbac and Adzic 1996), must also have played an important role in the synergistic activity. Based on the recent DFT calculation using "organized alloy" pyramid model (Tielens et al. 2005), there seems to be a minimum in the transition of $O_2$-adsorption energy at 55% Au (–0.5 eV) for the change of the adsorption energy from ~–1.5 eV in the 55%–97%Pt region to ~–0.7 eV in the 0%–45%Pt region. This finding seems to suggest some correlation in comparison with the finding in an earlier calculation (Pedersen et al. 1999) that showed small increase of the d-band as a function of Au concentration with a maximum at ~70% Au for PtAu alloy on Au(111). The observed maximum likely represents a compromised balance between the dissociation at the Pt-site activation and the promotion at the Au-site adsorption (–OH). Some of the mechanistic aspects implied from the above discussion for a bifunctional catalytic activity are illustrated in Scheme 7.1. The formation of Au-OH is operative for the alloy catalyst with 60%–80% Au atoms surrounding Pt in the chemisorption of the reaction intermediate from $O_2$ dissociation at Pt via forming $HO^-_{2\,ad}$, and such species should have sufficient binding strength with Au on further 2e reduction to $H_2O$. It is important to note that the key emphasis in Scheme 7.1 is the adsorption of –OH on Au atoms surrounding Pt atom,

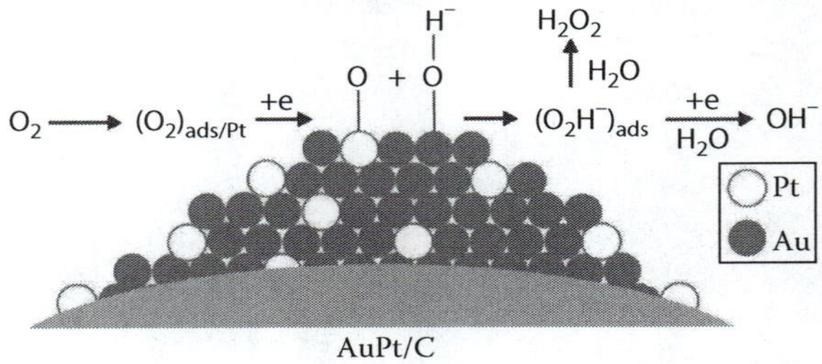


**SCHEME 7.1**
A schematic illustration of the synergistic ORR activity of AuPt/C catalyst with a relatively high concentration of Au in the alloy. (Reprinted from Luo, J. et al., *Electrochem. Commun.*, 8, 581, 2006. Copyright [2006], with permission from Elsevier.)

providing sites for further reduction reaction. The other mechanistic aspects are already known in previous studies (Strbac and Adzic 1996).

Similar to the oxygen reduction, the bifunctional electrocatalytic property of AuPt catalyst was also found to be operative for MOR. For example, in the alkaline electrolyte, the mass activity for the bimetallic catalyst with 65%–85% Au was found to be similar to those for pure Pt catalyst, which is suggestive of the participation of Au in the catalytic reaction of Pt (Luo et al. 2006b, Mott et al. 2007a). A Pt atom for an AuPt alloy with ~75% Au would be practically surrounded by Au atoms. Scheme 7.2 illustrates a proposed reaction pathway for the conversion of methanol to carbonate ion mediated by an AuPt catalyst.

The examination of the electrocatalytic activity for carbon-supported core@shell nanoparticles such as Au@Pt and Au@Pt further substantiated the importance of the nanoscale surface properties (Luo et al. 2008). As revealed by the order for the increase of kinetic current and the positive-shift of the reduction potential, Au < Pt@Au < Au@Pt < Pt., the increased activity for Au@Pt catalyst is consistent with the presence of Pt on the surface.

The recent study of CO adsorption on core@shell nanoparticles (Au@Pt and Pt@Au) (Luo et al. 2008) has also revealed insights into the surface-binding difference of bimetallic nanoparticles. The resemblance of the observed CO bands for Au@Pt and Pt@Au nanoparticles Pt and Au nanoparticles, respectively, is consistent with the core@shell nanostructures by design. The importance of the relative surface alloying or layering

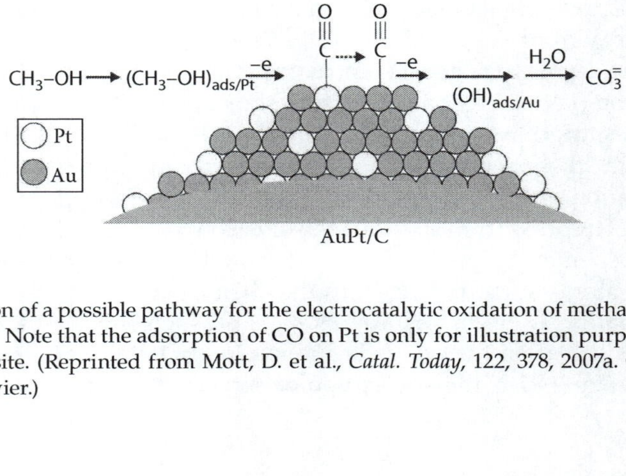


**SCHEME 7.2**
A schematic illustration of a possible pathway for the electrocatalytic oxidation of methanol on AuPt/C catalyst in alkaline electrolyte. Note that the adsorption of CO on Pt is only for illustration purpose, not necessarily for the exact adsorption site. (Reprinted from Mott, D. et al., *Catal. Today*, 122, 378, 2007a. Copyright [2007], with permission from Elsevier.)

arrangements of metals on single-crystal substrates has been recognized, including the "near surface alloy" model of metal adlayer on metal substrate for ORR (Zhang et al. 2005). Despite the extensive research in metal or oxide core@shell nanoparticles (Damle et al. 2001, Garcia-Gutierrez et al. 2005, Park et al. 2004, Teng et al. 2003, Toshima et al. 2001, 2005, Xu et al. 2007, Zeng et al. 2004), the correlation between the synergistic catalytic properties and the composition and spatial arrangement for metal–metal and metal-oxide core@shell nanoparticles remain elusive. Built upon our recent demonstration of the alloy character of bimetallic AuPt nanoparticle catalysts (Luo et al. 2005a, Njoki et al. 2005), the core–shell character of $Fe_3O_4$@Au nanoparticles (Park et al. 2007, Wang et al. 2003, 2005) and some recent theoretical insights (Ge et al. 2006, Song et al. 2005, Xiao et al. 2006), core@shell nanoparticles (Au@Pt, Pt@Au, $Fe_3O_4$@Au@Pt) were also studied for assessing their nanostructural correlation of the electrocatalytic properties for MOR and ORR. The presence of organic capping shells in each step is important for controlling the size and monodispersity of the M1@M2 nanoparticles (Luo et al. 2008).

Figure 7.6a shows a representative set of RDE curves for ORR in acidic electrolyte to assess the electrocatalytic activities of the carbon-supported core@shell nanoparticle catalysts. The relative changes in the RDE characteristics were compared for several different monometallic and core–shell nanoparticle catalysts. The increase of kinetic current and the positive shift of the reduction potential followed the order of Au < Pt@Au < Au@Pt < $Fe_3O_4$@Au@Pt < Pt. The increased activity for Au@Pt and $Fe_3O_4$@Au@Pt catalysts is consistent with the presence of Pt on the surface. This is also reflected by the shift of the reduction potential of Au@Pt/C from Au to Pt, and the values obtained from Levich plots for the electron transfer number (2.2 for Au, 3.9 for Pt and 3.9 for Au@Pt). The reduction overpotential can be dramatically shifted from high (for Au) to low (for Pt) depending on the relative Au–Pt-oxide spatial configuration.

Figure 7.6b shows a representative set of CV data of different catalysts for MOR. In contrast to the absence of activities for Au and Au-coated nanoparticles in acidic solution,

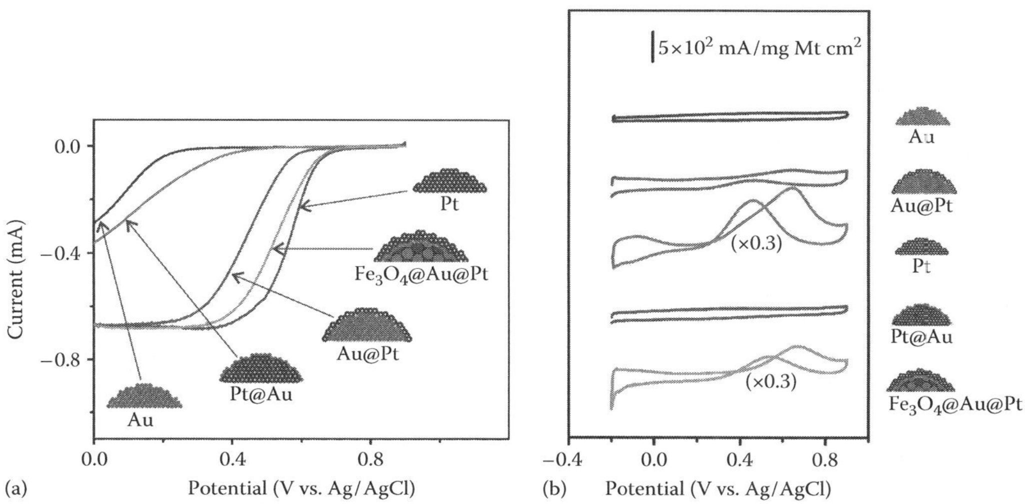


**FIGURE 7.6**
(a) RDE curves of different catalysts in 0.5 M $H_2SO_4$ saturated with $O_2$ (metal loading 10%–30%) (10 mV/s, 1600 rpm, surface area: 0.2 $cm^2$). The current for the curves is not normalized by the metal loading and the surface area. (b) CV curves for MOR for different catalysts in 0.5 M $H_2SO_4$ with 0.5 M MeOH (50 mV/s). The inserted schemes illustrate the corresponding nanoparticle surface structures by design.

the increased activities for Pt-coated catalysts is characteristic of the electrocatalytic properties of Pt component in Au@Pt and $Fe_3O_4$@Au@Pt catalysts. This is supported by the analysis of the methanol oxidation peak currents normalized with total metals. The catalyst treatment temperature was also found to exhibit an effect on the observed activity, which reflects the relative enrichment of metals in the core–shell nanostructure.

A remarkable finding is the high activity for $Fe_3O_4$@Au@Pt catalyst, as evidenced by the greater overpotential shift and mass activity increase. The fact that the mass activity for $Fe_3O_4$@Au@Pt is close to or higher than Pt demonstrates the feasibility of producing the synergistic catalytic effect by the metal-oxide core–shell nanostructures. Preliminary XRD examination of the core–shell structure revealed phase segregation of Pt and Au, as evidenced by the assymmetric peak in the XRD for $Fe_3O_4$@Au@Pt/C (thermal activation at <300°C). This observation is in sharp contrast to the single-phase character found for AuPt alloy nanoparticles (Luo et al. 2005a, Njoki et al. 2005). The hydrogen adsorption/evolution currents at −0.2 V and the shift of its reduction potential positively to a potential almost comparable with Pt catalysts agrees with the presence of Pt-shell. The thermal treatment temperatures were also found to influence the relative surface distribution or alloying properties. The mass activity for $Fe_3O_4$@Au@Pt was higher than the other Au–Pt combinations (Au@Pt or AuPt), suggesting the important role played by the oxide core. While more quantitative study is yet to be carried out for a detailed correlation of the catalytic activity with various parameters in controlling the relative core–shell composition and structure, the electrocatalytic activities were shown to depend on the nanoscale spatial arrangement of the metals. The relative changes in the catalytic activity should correlate with the core size, shell thickness, composition, and spatial properties.

### 7.3.2 Trimetallic Catalysts—PtVFe and PtNiFe Nanoparticles

The general synthesis of PtVFe nanoparticles involved the use of three metal precursors, $Pt^{II}(acac)_2$, $V^{V}O(acac)_2$, and $Fe^0(CO)_5$, in controlled molar ratios (Luo et al. 2005b). The general reaction for the synthesis of the (oleylamine/oleic acid)-capped PtVFe nanoparticles involves a combination of thermal decomposition and reduction reactions. The composition of the $Pt^0_{n1}$ $V^0_{n2}$ $Fe^0_{n3}$ nanoparticles, where $n_1$, $n_2$ and $n_3$ represent the atomic percentages of each metal, is controlled by the feeding ratio of the metal precursors, which is expressed in molar percentages ($m_1$, $m_2$, and $m_3$). For the synthesis of (oleylamine/oleic acid)-capped PtNiFe nanoparticles, a similar method was used involving three metal precursors, $Pt^{II}(acac)_2$, $Ni^{II}(acac)_2$, and $Fe^{II}(acac)_2$ in controlled molar ratios. By controlling the relative concentrations of metal precursors, and capping agents such as oleylamine and oleic acid, PtNiFe nanoparticles of different compositions, $Pt^0_{n1}$ $Ni^0_{n2}$ $Fe^0_{n3}$ nanoparticles ($n_1$, $n_2$ and $n_3$ represent the atomic percentages of each metal), were synthesized. A series of techniques including TEM, direct current plasma atomic emission spectrometry (DCP-AES), FTIR, and thermogravimetric analysis (TGA) have been used to characterize the nanostructure in terms of the average size and monodispersity of the nanoparticles, the composition of metals in the ternary nanocrystal core, and the structure of the organic monolayer shell.

The basic morphology of PtVFe nanoparticles (Luo et al. 2005b, 2006c) is largely characterized by the highly faceted nanocrystal feature, which is observable by a close examination of the shapes of the individual nanocrystals. The particles are highly monodispersed, with an average size of 1.9 ± 0.3 nm. The fact that the nanoparticles have well-defined interparticle spacing and display domains of hexagonal ordering is indicative of the encapsulation of the nanocrystal cores by organic monolayers. While the sizes of the ternary nanoparticles varied slightly depending on the actual composition, the data demonstrated the controllability over size monodispersity. Nanoparticles with average diameters ranging from 1.4 to 3.2 nm have been obtained. The size monodispersity in most cases was very high, ranging from ±0.2 to ±0.6 nm.

The trend for the variation of the metal composition in nanoparticles is very similar to the trend for the variation of metal precursor rations. In general, as the relative Pt concentration used in the synthesis increases, the relative Pt nanoparticles increase. The relative variation of V and Fe also followed similar trends. Note that the relative composition found in the nanoparticle products were not exactly the same as the feed ratios of the metal precursors used for the synthesis of nanoparticles. The difference likely reflects the thermodynamic differences between the reduction of $Pt^{II}$ and $V^{IV}$ to $Pt^0$ and $V^0$ and the decomposition of $Fe^0(CO)_5$ into $Fe^0$, and their relative driving force for atomic aggregation in forming the ternary nanoparticles. The detailed comparison between the compositions obtained for the nanoparticle products ($n_1$:$n_2$:$n_3$) and the composition feeding ($m_1$:$m_2$:$m_3$) in the synthetic reaction provides information about the relationship of the relative composition between the nanoparticle product and the synthetic feeding ratio. First, the increase of $m_1$(Pt)% leads to an increase in $n_1$(Pt)% in the nanoparticles. Second, $n_2$(V)% found in the nanoparticles was less than those used in the synthetic feeding $m_2$(V)% in most cases. Finally, $n_3$(Fe)% found in the nanoparticles was generally quite close to those used in the synthetic feeding $m_3$(Fe)%. The thermodynamic competition for the formation of each metal in the ternary nanoparticles is different. The reduction of $Pt(acac)_2$ seemed to be more favorable than that for $VO(acac)_2$.

The XRD data revealed a typical fcc pattern with some insignificant features indicative of chemically disordered structure. The calcinations treatment of the nanoparticles led to the rearrangement of Pt, V, and Fe atoms in the nanoparticles into long-range chemically ordered fcc structure. The formation of the alloyed nanocrystalline cores is supported by the fact that the diffraction peaks of metallic platinum shift to higher angles due to lattice shrinking resulting from the doping of smaller vanadium and iron atoms. Indeed, the diffraction peak position falls in between those for the monometallic Pt and those for monometallic V and Fe. For example, the strongest peak for Pt appears at $2\theta = 40.5$, slightly higher than Pt(111) peak ($2\theta = 39.8$) (Luo et al. 2005b). Diffraction peaks corresponding to V(110) ($2\theta = 44.7$) and Fe (110) ($2\theta = 42.2$) were not detected. The (111) peak for PtFe alloy nanoparticles ($2\theta = 41.2$ (Sun et al. 2000)) was also not detected, suggesting the absence of PtFe nanoparticles in the PtVFe nanoparticles.

The PtVFe nanoparticles can be easily assembled on carbon support materials with controllable dispersion and mass loading. In our recent work, after calcination treatment, the carbon-supported alloy nanoparticles were found to display high electrocatalytic activities for oxygen reduction. The particle sizes were basically unaffected after their assembly onto carbon materials. The thermal treatments of PtVFe/C catalysts involved the removal of organic shells and calcination of the ternary nanoparticles. A representative transmission electron microscopy-energy dispersive x-ray (TEM-EDX) image for a sample of the thermally treated $Pt_{32}V_{14}Fe_{54}/C$ catalyst is shown in Figure 7.7. The average particle sizes after the calcination treatment were found to show a slight increase (~0.5 nm) in comparison with that before the treatment. The subtle increase in size was found to be dependent on the calcination temperature. The particle sizes displayed a certain degree of increase (in avg., by ~0.5 nm) after the thermal treatment, especially at higher temperatures. The metal-loading data will be discussed in the next section.

From XRD spectra for $Pt_{32}V_{14}Fe_{54}/C$ treated under 550°C, the broad peak at low angles is from carbon support materials. The diffraction peak positions for PtVFe fall between those

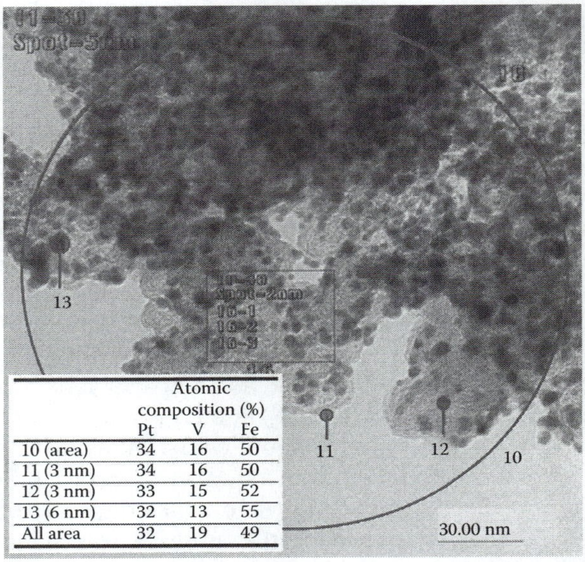


| | Atomic composition (%) | | |
|---|---|---|---|
| | Pt | V | Fe |
| 10 (area) | 34 | 16 | 50 |
| 11 (3 nm) | 34 | 16 | 50 |
| 12 (3 nm) | 33 | 15 | 52 |
| 13 (6 nm) | 32 | 13 | 55 |
| All area | 32 | 19 | 49 |

**FIGURE 7.7**
High-resolution TEM-EDX data for the ternary nanoparticle catalyst ($Pt_{33}V_{14}Fe_{53}$/C).

for the monometallic Pt, V, and Fe (Luo et al. 2005b), which indicates that these particles are largely alloyed. No secondary phase was detected. The diffraction peaks of metallic platinum shift to higher angles due to lattice shrinking resulting from the doping of smaller vanadium and iron atoms. The as-synthesized PtVFe nanoparticles were shown to be of fcc structure with little chemically disordered structure (Luo et al. 2005b). After thermal treatment, there were some indications of rearrangement of Pt, V, and Fe atoms in the PtVFe nanoparticles, leading into long-range chemically ordered fcc structure. The average sizes of the nanoparticles estimated by Scherrer correlation are slightly larger than the value determined from TEM data. The XRD data support that the ternary nanoparticles are alloy in character. The XRD data for nanoparticles treated at different temperatures (400°C and 450°C) were also compared. The data reveal subtle difference in peak width, which can be related to difference in size change.

The question whether the nanoparticles are multimetallic in individual nanoparticles or in an ensemble of the nanoparticles was addressed using TEM-EDX (He et al. 2006). TEM-EDX data (nano-composition) are also compared with DCP-AES analysis (macro-composition). By controlling the electron beam diameter and current, this measurement will yield reproducible compositions of individually isolated nanoparticles. For example, the atomic compositions for PtVFe nanoparticles (Figure 7.7) are found to be almost identical, independent of the actual sizes, in contrast to the results observed from the traditional synthesis where large-sized particles are usually base metal rich and small particles are Pt rich. Similar compositions were found for as-synthesized PtVFe nanoparticles, indicating the composition uniformity in the individual nanoparticles.

FTIR technique was used to examine the removal of shell components and the presence of possible surface reactivities after the shell removal. The disappearance of C–H stretching bands (2800–3000 $cm^{-1}$) in the high-frequency region indicates the effective removal of alkyl chains of the organic capping shells. The removal of OAC and OAM is evidenced by the disappearance of diagnostic bands C=O (in $CO_2H$) stretching (~1700 $cm^{-1}$), and N–H bending (1600 $cm^{-1}$) in the low-frequency region. The relative core–shell mass percentages, loading of particles on carbon, and interfacial reactivities associated with the various treatments were further studied using TGA. From the mass change, the metal loading was determined to be 37% without oxide correction.

Similar to PtVFe nanoparticles, the PtNiFe nanoparticles consist of ternary nanocrystal cores and the monolayer organic capping shells. A close examination of the shapes of the individual nanocrystals indicates that the nanoparticles are highly faceted nanocrystals. Using the synthetic protocols as described in the experimental section, PtNiFe nanoparticles with average diameters ranging from 1.4 to 1.8 nm have been obtained. The size monodispersity in most cases was very high, ranging from ±0.2 to ±0.4 nm. On the basis of the DCP-AES analysis of the composition of PtNiFe nanoparticles, the efficiencies for the synthetic conversion were found to be 70%–80% for Pt, ~90% for Ni, and ~60% for Fe, depending on the reaction temperature, precursor concentrations, and reaction time. The trimetallic alloy composition of the nanoparticles could be controlled by the feed ratio of the three metal precursors in the synthetic solution. The average particle sizes after the thermal treatment were found slightly increased in comparison with that before the thermal treatment. For $Pt_{30}Ni_{29}Fe_{41}/C$, the size increased by ~0.2 nm, and for $Pt_{30}Ni_{37}Fe_{33}/C$ it increased by ~1.1 nm. From the mass change in TGA data, the metal loading was determined to be 35% (without oxide correction). In comparison with the metal loading used in the synthetic feeding ~33% for the assembly of the ternary nanoparticles on carbon support, the TGA-determined metal loading is higher by ~6%. The alloy properties of the catalysts were examined using XRD technique. The XRD data revealed a typical fcc pattern. The fact that the diffraction peaks of metallic platinum shift to higher angles due to lattice shrinking resulting from the doping of smaller nickel and iron atoms supports the formation of the alloyed nanocrystalline cores. The position of the diffraction peaks was found to fall in between those for the monometallic Pt and those for monometallic Ni and Fe (Luo et al. 2006d). The XRD data for PtNiFe/C before thermal treatment showed some features indicative of chemically disordered structure, but a tetragonal-type structure of the PtFe type after the thermal treatment.

The carbon-supported trimetallic (e.g., PtVFe/C and PtNiFe/C) alloy nanoparticle catalysts are expected to be electrocatalytically active for ORR based on combinatorial studies of a series of Pt-based bimetallic alloy thin film catalysts, which reveal significant increase of activities for ORR (e.g., PtFe, PtNi, and PtV thin films exhibit high activities for ORR) (He et al. 2006). The introduction of a third metal to the alloy is expected to produce a combination of effects such as the reduction of the lattice distance, the addition of surface sites for the formation of metal–oxygen bond and, the adsorption of $OH^-$, and the modification of the d-band center. In the next two subsections, the results for PtVFe/C and PtNiFe/C alloy nanoparticle catalysts are described as two examples of the study. The RDE data for ORR at PtVFe/C and PtNiFe/C showed that the electron transfer number ($n$) is close to 4, as expected for ORR at this type of catalysts (Luo et al. 2006c,d).

The electrocatalytic activity data were compared for the $Pt_{32}V_{14}Fe_{54}/C$ catalysts in terms of relative mass activities at 0.8 V. The relative mass-specific activities for each of the seven samples are compared with that for standard commercial Pt/C catalyst (36.4% Pt loading). All these ternary catalysts showed increased electrocatalytic activities in comparison with the polycrystalline Pt/C catalysts.

Figure 7.8 shows the electrochemical activities toward ORR of various Pt-based materials made by the core–shell method described here in comparison with a commercially available standard Pt/C catalyst (TKK). PtFe and PtVFe have relative activities of almost 2 and 4 times of the commercially available Pt catalyst. Note that carbon-supported platinum

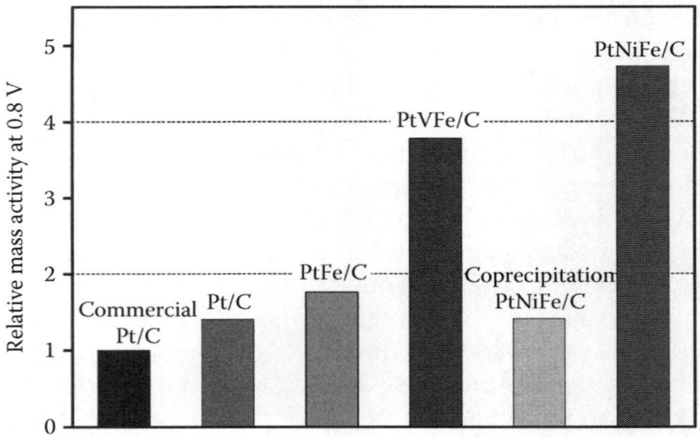


**FIGURE 7.8**
Comparison of the relative electrocatalytic activities for carbon-supported monometallic, binary and ternary catalysts.

nanoparticles prepared in our laboratory show a 30% increase in molecular oxygen reduction in comparison with the commercial Pt/C catalyst under the similar loading and test condition. The activity displays the order of PtVFe/C > Pt/C, demonstrating the effectiveness of the alloy composition in enhancing the electrocatalytic performance. This finding can be understood by the fact that the alloy nanoparticles are all located on the surface of the supporting carbon, whereas the traditional synthesis method cannot avoid burying some nanoparticles inside of the micropores of the carbon, which are not accessible by the molecular oxygen. The utilization of the catalytic nanoparticles was much higher on the surface than those inside the micropores.

Similar results were observed for PtNiFe/C catalyst. The catalyst prepared by the coprecipitation method had a composition of $Pt_{34}Ni_{46}Fe_{20}$, an average size of 3.0 nm. The PtNiFe/C catalyst prepared by the core–shell based synthesis method displayed a relative activity of almost 5 times larger than that of the commercially available Pt/C catalyst. It is also found that the PtNiFe/C catalysts prepared by the core–shell synthesis method display much higher electrocatalytic activity than PtNiFe/C prepared by the traditional co-precipitation method.

The ternary PtVFe/C catalyst was evaluated for determining the fuel cell performance (Fang et al. 2009). Membrane electrode assemblies (MEAs) (5 cm² active area) used in this study were prepared by traditional catalyst-coated substrate (CCS) method. In-house gas diffusion electrodes (GDEs) were fabricated using electrocatalyst-Nafion ink painted on a wet-proofed carbon paper (Toray™ EC-TP1-060T). $Pt_{42}V_{19}Fe_{39}/C$ catalyst (21% metal loading, 0.4 mg Pt/cm²) was used for the cathode and Pt/C catalyst (20% Pt/C, E-tek, 0.4 mg Pt/cm²) for the anode. A reference MEA was fabricated with Pt/C (20% Pt/C, E-tek, 0.4 mg Pt/cm²) as electrocatalyst for both anode and cathode. The MEAs were prepared by hot pressing the sandwich-structured Nafion 212 membrane (DuPont) and catalyst-coated electrodes at 120°C. The MEAs were tested in an Electrochem Inc. single fuel cell test station. The testing conditions included 100% humidified $H_2$ and $O_2$ at a flow rate of 100 mL/min. Backpressure for both electrodes was kept at 30 psi. The operating temperatures for the fuel cell testing was 75°C.

Figure 7.9 shows a representative set of fuel cell performance data for $Pt_{42}V_{19}Fe_{39}/C$ and Pt/C catalysts under the loading condition of 0.4 $mg_{Pt}/cm^2$. As a validation of the quality

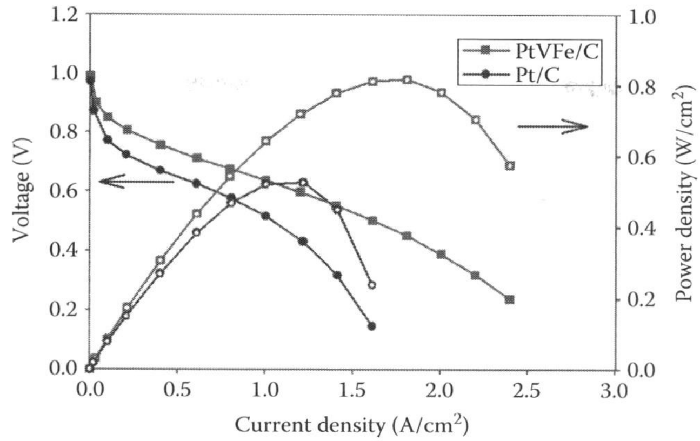


**FIGURE 7.9**
Polarization (close symbols) and power density (open symbols) curves of homemade MEA with $Pt_{42}V_{19}Fe_{39}/C$ (square) or Pt/C (circle) as cathode in PEMFC at 75°C (Fang et al. 2009). Pt loading in both anode and cathode was 0.4 mg $Pt/cm^2$ for the PVF/C and Pt/C MEA. (Reprinted from Fang, B. et al., *Electrochem. Commun.* 11, 1139, 2009. Copyright [2009], with permission from Elsevier.)

of the MEAs prepared in our laboratory, the data between our MEAs and those using the same commercial Pt/C catalyst were compared. The MEA with Pt/C catalyst exhibited a value of 0.52 V at 1.0 $A/cm^2$. This value is largely comparable to those reported under similar operation conditions using CCS MEA fabrication method (Seo et al. 2006), thus validating the quality and effectiveness of our MEA preparation for the comparison of the fuel cell performance data comparing $Pt_{42}V_{19}Fe_{39}/C$ and Pt/C catalysts (Figure 7.9).

It is evident that both the cell voltage and the power density for the fuel cell with PtVFe/C catalyst in the cathode are higher than those for the Pt/C catalyst under the same test conditions. By comparing the FC performance data in terms of polarization curve and power density between our $Pt_{42}V_{19}Fe_{39}/C$ catalyst and the commercial Pt/C catalyst, the most important observation is that both the cell voltage and the power density for the fuel cell with PtVFe/C catalyst in the cathode are overall higher than those for the cell with the Pt/C catalyst. The peak power density of Pt/C was found at 1.23 $A/cm^2$ and 0.53 $W/cm^2$ at 0.43 V, whereas values of 1.81 $A/cm^2$ and 0.82$W/cm^2$ at 0.45 V were found for the PtVFe/C catalyst. The power densities for both Pt/ and PtVFe/C catalysts reached the maximum at the similar cell voltage (0.4–0.5 V). The fuel cell with PtVFe/C catalyst showed a 50% increase in peak power density in comparison with that of Pt/C at both testing temperatures. These results demonstrate that $Pt_{42}V_{19}Fe_{39}/C$ catalyst has a much better fuel cell performance than Pt/C catalyst. The finding is consistent with the electrocatalytic activity trend revealed by the RDE data described in the previous subsection.

## 7.4 Summary

In summary, the molecular-encapsulation approach to the synthesis and processing of bimetallic/trimetallic nanoparticles is effective in producing alloy nanoparticles in the 2–5 nm regime with controllable composition and carbon-supported catalysts for fuel cell reactions. This approach differs from other traditional approaches to the preparation of supported catalysts in the abilities to control the nanoscale size, multimetallic composition, phase properties, and surface properties. As demonstrated by the bimetallic AuPt alloy nanoparticle catalysts, synergistic activity is possible in which Au atoms surrounding Pt provide effective sites for the reaction adsorbates in the electrocatalytic reaction. The fact that this bimetallic nanoparticle system displays the unique single-phase property different from the miscibility gap of its bulk-scale counterpart serves as an important indication of the operation of nanoscale phenomena in the catalysts, which can be further exploited for the design and preparation of the nanostructured bimetallic catalysts for fuel cells (Zhong et al. 2007). The trimetallic nanoparticle catalysts have displayed enhanced electrocatalytic ORR activity. For carbon-supported ternary PtVFe and PtNiFe nanoparticle catalysts, the size, composition, and loading of the nanoparticles on carbon support were shown to be controllable, and also processable by controlled thermal treatment and calcination, which can be optimized for achieving effective shell removal and alloying of the ternary catalysts. The measurements of the intrinsic kinetic activities of the catalysts toward ORR have shown high electrocatalytic activities, and the trimetallic PtVFe nanoparticles catalysts prepared by the nanoengineered synthesis and processing methods have exhibited a much better performance in PEM fuel cell cathode than that for the commercial Pt catalyst. It also becomes clear that the synthesis and processing approach to the preparation of nanoparticle catalysts is promising for delivering much higher catalyst utilization than those of conventional methods, which has important implications for the better design of fuel cell cathode catalysts.

## Acknowledgments

The work described in this article was supported in part by NSF (CBET-0709113 and CHE-0848701), NYSTAR, and Honda.

## References

Acres, G. J. K., Frost, J. C., Hards, G. A. et al. 1997. Electrocatalysts for fuel cells. *Catal. Today*, 38: 393–400.

Adler, S. B. 2004. Factors governing oxygen reduction in solid oxide fuel cell cathodes. *Chem. Rev.*, 104: 4791–4843.

Aiken III, J. D. and Finke, R. G. 1999. A review of modern transition-metal nanoclusters: Their synthesis, characterization, and applications in catalysis. *J. Mol. Catal. A*, 145: 1–44.

Anderson, A. B., Grantscharova, E., and Seong, S. 1996. Systematic theoretical study of alloys of platinum for enhanced methanol fuel cell performance. *J. Electrochem. Soc.*, 143: 2075–2082.

Antolini, E. 2003. Formation of carbon-supported PtM alloys for low temperature fuel cells: A review. *Mater. Chem. Phys.*, 78: 563–573.

Bailie, J. E. and Hutchings, G. J. 1999. Promotion by sulfur of gold catalysts for crotyl alcohol formation from crotonaldehyde hydrogenation. *Chem. Commun.*, 21: 2151–2152.

Berg, C., Venvik, H. J., Strisland, F. et al. 1998. Nucleation and growth of Au overlayers on Pt(100)-hex-R0.7 degrees studied by STM and photoelectron spectroscopy. *Surf. Sci.*, 409: 1–15.

Blizanac, B. B., Lucas, C. A., Gallagher, M. E. et al. 2004. Anion adsorption, CO oxidation, and oxygen reduction reaction on a Au(100) surface: The pH effect. *J. Phys. Chem. B*, 108: 625–634.

Bond, G. C. 2002. Gold: A relatively new catalyst. *Catal. Today*, 72: 5–9.

Bond, G. C. and Thompson, D. T. 1999. Catalysis by gold. *Catal. Rev.*, 41: 319–388.

Bonnemann, H. and Nagabhushana, K. S. 2004. Advantageous fuel cell catalysts from colloidal nanometals. *J. New Mater. Electrochem. Syst.*, 7: 93–108.

Brandon, N. P., Skinner, S., and Steele, B.C.H. 2003. Recent advances in materials for fuel cells. *Annu. Rev. Mater. Res.*, 33: 183R–213R.

Brust, M., Walker, M., Bethell, D. et al. 1994. Synthesis of thiol-derivatized gold nanoparticles in a 2-phase liquid-liquid system. *J. Chem. Soc., Chem. Comm.*, 7: 801–802.

Burke, L. D., Collins, J. A., Horgan, M. A. et al. 2000. The importance of the active states of surface atoms with regard to the electrocatalytic behaviour of metal electrodes in aqueous media. *Electrochim. Acta*, 45: 4127–4134.

Campbell, C. T. 2004. The active site in nanopaticle gold catalysis. *Science*, 306: 234–235.

Caruso, F. 2001. Nanoengineering of particle surfaces. *Adv. Mater.*, 13: 11–22, and references therein.

Casado-Rivera, E., Volpe, D. J., Alden, L. et al. 2004. Electrocatalytic activity of ordered intermetallic phases for fuel cell applications. *J. Am. Chem. Soc.*, 126: 4043–4049.

Chan, B. C., Liu, R. X., Jambunathan, K. et al. 2005. Comparison of high-throughput electrochemical methods for testing direct methanol fuel cell anode electrocatalysts. *J. Electrochem. Soc.*, 152: A594–A600.

Chen, G., Delafuente, D. A., Sarangapani, S. et al. 2001. Combinatorial discovery of bifunctional oxygen reduction—water oxidation electrocatalysts for regenerative fuel cells. *Catal. Today*, 67: 341–355.

Chen, M. and Nikles, D. E. 2002. Synthesis, self-assembly, and magnetic properties of FexCoyPt100-x-y nanoparticles. *Nano Lett.*, 2: 211–214.

Chen, M. S. and Goodman, D. W. 2004. The structure of catalytically active gold on titania. *Science*, 306: 252–255.

Chen, M. S., Kumar, D., Yi, C.-W. et al. 2005. The promotional effect of gold in catalysis by palladium-gold. *Science*, 310: 291–293.

Chu, D. and Jiang, R. 2002. Novel electrocatalysts for direct methanol fuel cells. *Solid State Ionics*, 148: 591–599.

Collman, J. P., Bencosme, C. S., Durand, R. R. et al. 1983. Mixed-metal face-to-face porphyrin dimers. *J. Am. Chem. Soc.*, 105: 2699–2703.

Crooks, R. M., Zhao, M. Q., Sun, L. et al. 2001. Dendrimer-encapsulated metal nanoparticles: Synthesis, characterization, and applications to catalysis. *Acc. Chem. Res.*, 34: 181–190.

Crown, A., Kim, H., Lu, G.Q. et al. 2000. Research toward designing high activity catalysts for fuel cells: structure and reactivity. *J. New Mater. Electrochem. Syst.*, 3: 275–284.

Damle, C., Biswas, K., and Sastry, M. 2001. Synthesis of Au-core/Pt-shell nanoparticles within thermally evaporated fatty amine films and their low-temperature alloying. *Langmuir*, 17: 7156–7159.

Davis, R. J. 2003. All that glitters is not Au-0. *Science*, 301: 926–927.

Dickinson, A. J., Carrette, L. P. L., Collins, J. A. et al. 2002. Preparation of a Pt-Ru/C catalyst from carbonyl complexes for fuel cell applications *Electrochim. Acta*, 47: 3733–3739.

El-Deab, M. S. and Ohsaka, T. 2002. Hydrodynamic voltammetric studies of the oxygen reduction at gold nanoparticles-electrodeposited gold electrodes. *Electrochim. Acta*, 47: 4255–4261.

El-Deab, M. S. and Ohsaka, T. 2003. Electrocatalysis by nanoparticles: Oxygen reduction on gold nanoparticles-electrodeposited platinum electrodes. *J. Electroanal. Chem.*, 553: 107–115.

El-Sayed, M. A. 2001. Some interesting properties of metals confined in time and nanometer space of different shapes. *Acc. Chem. Res.*, 34: 257–264.

Fang, B., Luo, J., Njoki, P. N. et al. 2009. Nanostructured PtVFe catalysts: Electrocatalytic performance in proton exchange membrane fuel cells. *Electrochem. Commun.*, 11: 1139–1141.

Feldheim, D. L. and Foss Jr., C. A. 2002. *Metal Nanoparticles: Synthesis, Characterization, and Applications*, New York: Marcel Dekker, Inc.

Fernandez, J. L., Walsh, D. A., and Bard, A. J. 2005. Thermodynamic guidelines for the design of bimetallic catalysts for oxygen electroreduction and rapid screening by scanning electrochemical microscopy M-Co (M: Pd, Ag, Au). *J. Am. Chem. Soc.*, 127: 357–365.

Fu, L., Wu, N.Q., Yang, J. H. et al. 2005. Direct evidence of oxidized gold on supported gold catalysts. *J. Phys. Chem. B*, 109: 3704–3706.

Galow, T. H., Drechsler, U., Hanson, J. A. et al. 2002. Highly reactive heterogeneous Heck and hydrogenation catalysts constructed through 'bottom-up' nanoparticle self-assembly. *Chem. Commun.*, 10: 1076–1077.

Garcia-Gutierrez, D. I., Gutierrez-Wing, C. E., Giovanetti, L. et al. 2005. Temperature effect on the synthesis of Au-Pt bimetallic nanoparticles. *J. Phys. Chem. B*, 109: 3813–3821, and references therein.

Gasteiger, H. A., Kocha, S. S., Sompalli, B. et al. 2005. Activity benchmarks and requirements for Pt, Pt-alloy, and non-Pt oxygen reduction catalysts for PEMFCs. *Appl. Catal. B*, 56: 9–35.

Ge, Q., Song, C., and Wang, L. 2006. A density functional theory study of CO adsorption on Pt-Au nanoparticles. *Comput. Mater. Sci.*, 35: 247–253.

Guerin, S., Hayden, B. E., Lee, C. E. et al. 2004. Combinatorial electrochemical screening of fuel cell electrocatalysts. *J. Comb. Chem.* 6: 149–158.

Haruta, M. 2004. Gold as a novel catalyst in the 21st century: Preparation, working mechanism and applications. *Gold Bull.*, 37: 27–36.

Haruta, M. 2005. Catalysis—Gold rush. *Nature*, 437: 1098–1099.

Haruta, M. and Date, M. 2001. Advances in the catalysis of Au nanoparticles. *Appl. Catal. A*, 222: 427–437

He, T., Kreidler, E., Xiong, L. et al. 2006. Alloy electrocatalysts—Combinatorial discovery and nanosynthesis. *J. Electrochem. Soc.*, 153: A1637–A1643.

He, T., Zhong, C. J., Luo, J. et al. 2008. Metal and alloy nanoparticles and synthesis methods thereof. US Patent No 7,335,245.

Hostetler, M. J., Zhong, C. J., Yen, B. K. H. et al. 1998. Stable, monolayer-protected metal alloy clusters *J. Am. Chem. Soc.*, 120: 9396–9397.

Hsu, P. J. and Lai, S. K. 2006. Structures of bimetallic clusters. *J. Chem. Phys.*, 124: 44711.

Hughes, M. D., Xu, Y. J., Jenkins, P. et al. 2005. Tunable gold catalysts for selective hydrocarbon oxidation under mild conditions. *Nature*, 437: 1132–1135.

Ingram, R. S. and Murray, R. W. 1998. Electroactive three-dimensional monolayers: Anthraquinone omega-functionalized alkanethiolate-stabilized gold clusters. *Langmuir*, 14: 4115–4118.

Jones, F. E., Milne, S. B., Gurau, B. et al. 2002. Synthesis and characterization of PtSn/carbon and Pt3Sn/carbon nanocomposites as methanol electrooxidation catalysts. *J. Nanosci. Nanotechnol.*, 2: 81–87.

Kariuki, N. N., Luo, J., Maye, M. M. et al. 2004. Composition-controlled synthesis of bimetallic gold-silver nanoparticles. *Langmuir*, 20: 11240–11246.

Kiely, C. J., Fink, J., Zheng, J. G. et al. 2000. Ordered colloidal nanoalloys. *Adv. Mater.*, 12: 640–643.

Kim, C. S. and Korzeniewski, C. 1997. Vibrational coupling as a probe of adsorption at different structural sites on a stepped single-crystal electrode. *Anal. Chem.*, 69: 2349–2353.

Kim, Y. D., Fischer, M., and Gantefor, G. 2003. Origin of unusual catalytic activities of Au-based catalysts. *Chem. Phys. Lett.*, 377: 170–176.

Klabunde, K. J. 2001. *Nanoscale Materials in Chemistry*, New York: John Wiley & Sons, Inc.

Lai, M. X. and Goodman, D. W. 1998. Onset of catalytic activity of gold clusters on titania with the appearance of nonmetallic properties. *Science*, 281: 1647–1650.

Lang, H., Maldonado, S., Stevenson, K. J. et al. 2004. Synthesis and characterization of dendrimer templated supported bimetallic Pt-Au nanoparticles. *J. Am. Chem. Soc.*, 126: 12949–12956.

Ley, K. L., Liu, R. X., Pu, C. et al. 1997. Methanol oxidation on single-phase Pt-Ru-Os ternary alloys. *J. Electrochem. Soc.*, 144: 1543–1548.

Li, H., Luk, Y.-Y., and Mrksich, M. 1999. Catalytic asymmetric dihydroxylation by gold colloids functionalized with self-assembled monolayers. *Langmuir*, 15: 4957–4959.

Litster, S. and McLean, G. 2004. PEM fuel cell electrodes. *J. Power Sources*, 130: 61–76.

Liu, R. and Smotkin, E. S. 2002. Array membrane electrode assemblies for high throughput screening of direct methanol fuel cell anode catalysts. *J. Electroanal. Chem.* 535: 49–55.
Long, J. W., Stroud, R. M., Swider-Lyons, K. E. et al. 2000. How to make electrocatalysts more active for direct methanol oxidation—Avoid PtRu bimetallic alloys! *J. Phys. Chem. B*, 104: 9772–9776.
Luo, J., Lou, Y. B., Zhong, C. J. et al. 2001. An EQCN assessment of electrocatalytic oxidation of methanol at nanostructured Au-Pt alloy nanoparticles. *Electrochem. Commun.*, 3:172–176.
Luo, J., Maye, M. M., Han, L. et al. 2004. Spectroscopic characterizations of molecularly linked gold nanoparticle assemblies upon thermal treatment. *Langmuir*, 20: 4254–4260.
Luo, J., Maye, M. M., Petkov, V. et al. 2005a. Phase properties of carbon-supported gold-platinum nanoparticles with different bimetallic compositions. *Chem. Mater.*, 17: 3086–3091.
Luo, J., Han, L., Kariuki, N. N. et al. 2005b. Synthesis and characterization of monolayer-capped PtVFe nanoparticles with controllable sizes and composition. *Chem. Mater.*, 17: 5282–5290.
Luo, J., Maye, M. M., Kariuki, N. N. et al. 2005c. Electrocatalytic oxidation of methanol: Carbon-supported gold-platinum nanoparticle catalysts prepared by two-phase protocol. *Catal. Today*, 99: 291–297.
Luo, J., Njoki, P., Lin, Y. et al. 2006a. Activity-composition correlation of AuPt alloy nanoparticle catalysts in electrocatalytic reduction of oxygen. *Electrochem. Commun.*, 8: 581–587.
Luo, J., Njoki, P., Lin, Y. et al. 2006b. Characterization of carbon-supported AuPt nanoparticles for electrocatalytic methanol oxidation reaction. *Langmuir*, 22: 2892–2898.
Luo, J., Kariuki, N., Han, L. et al. 2006c. Preparation and characterization of carbon-supported PtVFe electrocatalysts. *Electrochim. Acta*, 51: 4821–4827.
Luo, J., Wang, L., Mott, D. et al. 2006d. Ternary alloy nanoparticles with controllable sizes and composition and electrocatalytic activity. *J. Mater. Chem.*, 16: 1665–1673.
Luo, J., Wang, L., Mott, D. et al. 2008. Core@shell nanoparticles as electrocatalysts for fuel cell reactions. *Adv. Mater.*, 20: 4342–4347.
Mallouk, T. E. and Smotkin, E. S. 2003. Combinatorial catalyst development methods. In *Handbook of Fuel Cells: Fundamentals, Technology and Application*, ed. W. Vielstich, A. Lamm, and H. A. Gasteiger, John Wiley & Sons, Chichester, U.K.
Mbindyo, J. K. N., Reiss, B. D., Martin, B. R. et al. 2001. DNA-directed assembly of gold nanowires on complementary surfaces. *Adv. Mater.*, 13: 249–254.
Mehta, V. and Cooper, J. S. 2003. Review and analysis of PEM fuel cell design and manufacturing. *J. Power Sources*, 114: 32–53.
Meier, D. C. and Goodman, D. W. 2004. The influence of metal cluster size on adsorption energies: CO adsorbed on Au clusters supported on $TiO_2$. *J. Am. Chem. Soc.*, 126: 1892–1899.
Meyer, R., Lemire, C., Shaikhutdinov, Sh. K. et al. 2004. Surface chemistry of catalysis by gold. *Gold Bull.*, 37: 72–124.
Mihut, C., Descorme, C., Duprez, D. et al. 2002. Kinetic and spectroscopic characterization of cluster-derived supported Pt–Au catalysts. *J. Catal.*, 212: 125–135.
Morita, M., Iwanaga, Y., and Matsuda, Y. 1991. Anodic-oxidation of methanol at a gold-modified platinum electrocatalyst prepared by re-sputtering on a glassy-carbon support. *Electrichim. Acta*, 36: 947–951.
Mott, D., Luo, J., Njoki, P. N. et al. 2007a. Synergistic activity of gold-platinum alloy nanoparticle catalysts. *Catal. Today*, 122: 378–385.
Mott, D., Luo, J., Smith, A. et al. 2007b. Nanocrystal and surface alloy properties of bimetallic gold-platinum nanoparticles. *Nanoscale Res. Lett.*, 2: 12–16.
Nishimura, K., Kunimatsu, K., and Enyo, M. 1989. Electrocatalysis on Pd+Au alloy electrodes 3: IR spectroscopic studies on the surface species derived from CO and $CH_3OH$ in nano solution. *J. Electroanal. Chem.*, 260: 167–179.
Njoki, P. N., Luo, J., Wang, L. et al. 2005. Platinum-catalyzed synthesis of water-soluble AuPt nanoparticles. *Langmuir*, 21:1623–1628.
Oldenburg, S. J., Averitt, R. D., Westcott, S. L. et al. 1998. Nanoengineering of optical resonances. *Chem. Phys. Lett.*, 288: 243–247.
Park, K. W., Choi, J. H., Kwon, B. K. et al. 2002. Chemical and electronic effects of Ni in Pt/Ni and Pt/Ru/Ni alloy nanoparticles in methanol electrooxidation. *J. Phys. Chem. B*, 106: 1869–1877.

Park, J. I., Kim, M. G., Jun, Y. W. et al. 2004. Characterization of superparamagnetic "core-shell" nanoparticles and monitoring their anisotropic phase transition to ferromagnetic "solid solution" nanoalloys. *J. Am. Chem. Soc.*, 126: 9072–9078.

Park, H. Y., Schadt, M. J., Wang, L. et al. 2007. Fabrication of magnetic core @ shell Fe oxide @ Au nanoparticles for interfacial bioactivity and bio-separation. *Langmuir*, 23: 9050–9056.

Paulus, U. A., Endruschat, U., Feldmeyer, G. J. et al. 2000. New PtRu alloy colloids as precursors for fuel cell catalysts. *J. Catal.*, 195: 383–393.

Paulus, U. A., Wokaun, A., Scherer, G. G. et al. 2002. Oxygen reduction on carbon-supported Pt-Ni and Pt-Co alloy catalysts. *J. Phys. Chem. B*, 106: 4181–4191.

Pedersen, M. Ø., Helveg, S., Ruban, A. et al. 1999. How a gold substrate can increase the reactivity of a Pt overlayer. *Surf. Sci.*, 426: 395–409.

Peng, X., Schlamp, M. C., Kadavanich, A. V. et al. 1997. Epitaxial growth of highly luminescent CdSe/CdS core/shell nanocrystals with photostability and electronic accessibility. *J. Am. Chem. Soc.*, 119: 7019–7029.

Ponec, V. and Bond, G. C. 1995. Catalysis by metals and alloys—Prologue. *Catalysis by Metals and Alloys*, Elsevier, Amsterdam, the Netherlands, Vol. 95, pp. 1–5.

Raj, C. R., Abdelrahman, A. I., and Ohsaka, T. 2005. Gold nanoparticle-assisted electroreduction of oxygen. *Electrochem. Commun.*, 7: 888–893.

Raja, R., Khimyak, T., Thomas, J. M. et al. 2001. Single-step, highly active, and highly selective nanoparticle catalysts for the hydrogenation of key organic compounds. *Angew. Chem. Int. Ed.*, 40: 4638–4642.

Rao, C. R. K. and Trivedi, D. C. 2005. Chemical and electrochemical depositions of platinum group metals and their applications. *Coord. Chem. Rev.*, 249: 613–631.

Ren, X. M., Zelenay, P., Thomas, S. et al. 2000. Recent advances in direct methanol fuel cells at Los Alamos National Laboratory. *J. Power Sources*, 86: 111–116.

Rolison, D. R. 2003. Catalytic nanoarchitectures—The importance of nothing and the unimportance of periodicity. *Science*, 299: 1698–1701.

Roucoux, A., Schulz, J., and Patin, H. 2002. Reduced transition metal colloids: A novel family of reusable catalysts?. *Chem. Rev.*, 102: 3757–3889.

Russell, A. E. and Rose, A. 2004. X-ray absorption Spectroscopy of low temperature fuel cell catalysts. *Chem. Rev.*, 104: 4613–4635.

Sasaki, K., Wang, J. X., Balasubramanian, M. et al. 2004. Ultra-low platinum content fuel cell anode electrocatalyst with a long-term performance stability. *Electrochim. Acta*, 49: 3873–3877.

Schärtl, W. 2000. Crosslinked spherical nanoparticles with core-shell topology. *Adv. Mater.*, 12: 1899–1908.

Schmid, G., Maihack, V., Lantermann, F. et al. 1996. Ligand-stabilized metal clusters and colloids: Properties and applications. *J. Chem. Soc., Dalton Trans.*, 5: 589–595.

Schmidt, T. J., Gasteiger, H. A., and Behm, R. J. 1999. Methanol electrooxidation on a colloidal PtRu-alloy fuel-cell catalyst. *Electrochem. Commun.*, 1: 1–4.

Schneider, J. J. 2001. Magnetic core/shell and quantum-confined semiconductor nanoparticles via chimie douce organometallic synthesis. *Adv. Mater.*, 13: 529–533.

Seo, A., Lee, J., Han, K. et al. 2006. Performance and stability of Pt-based ternary alloy catalysts for PEMFC. *Electrochim. Acta*, 52: 1603–1611.

Shukla, A. K. and Raman, R. K. 2003. Methanol-resistant oxygen-reduction catalysts for direct methanol fuel cells. *Annu. Rev. Mater. Res.*, 33: 155–168.

Song, C., Ge, Q., and Wang, L. 2005. DFT studies of Pt/Au bimetallic clusters and their interactions with the CO molecule. *J. Phys. Chem. B*, 109: 22341–22350.

Spendelow, J. S. and Wieckowski, A. 2004. Noble metal decoration of single crystal platinum surfaces to create well-defined bimetallic electrocatalysts. *Phys. Chem. Chem. Phys.*, 6: 5094–5118.

Stone, H. 2005. *Economic Analysis of Stationary PEMFC Systems*. DOE Hydrogen Program FY 2005 Progress Report 961.

Storhoff, J. J. and Mirkin, C. 1999. Programmed materials synthesis with DNA. *Chem. Rev.*, 99: 1849–1862.

Strasser, P., Fan, Q., Devenney, M. et al. 2003. High throughput experimental and theoretical predictive screening of materials—A comparative study of search strategies for new fuel cell anode catalysts. *J. Phys. Chem. B*, 107: 11013–11021.

Strbac, S. and Adzic, R.R. 1996. The influence of OH-chemisorption on the catalytic properties of gold single crystal surfaces for oxygen reduction in alkaline solutions. *J. Electroanal. Chem.*, 403: 169–181.

Sun, S. H., Murray, C. B., Weller, D. et al. 2000. Monodisperse FePt nanoparticles and ferromagnetic FePt nanocrystal superlattice. *Science*, 287: 1989–1992.

Sun, Y., Buck, H., and Mallouk, T. E. 2001. Monodisperse FePt nanoparticles and ferromagnetic FePt nanocrystal superlattices. *Anal. Chem.*, 73: 1599–1992.

Templeton, A. C., Wuelfing, W. P., and Murray, R. W. 2000. Monolayer protected cluster molecules. *Acc. Chem. Res.*, 33: 27–36 and references therein.

Teng, X. W., Black, D., Watkins, N. J. et al. 2003. Platinum-maghemite core-shell nanoparticles using a sequential synthesis. *Nano Lett.*, 3: 261–264.

Tielens, F., Andres, J., Van Brussel, M. et al. 2005. DFT study of oxygen adsorption on modified nanostructured gold pyramids. *J. Phys. Chem. B*, 109: 7624–7630.

Toshima, N., Shiraishi, Y., Shiotsuki, A. et al. 2001. Novel synthesis, structure and catalysis of inverted core/shell structured Pd/Pt bimetallic nanoclusters. *Eur. Phys. J. D*, 16: 209–212.

Toshima, N., Kanemaru, M., Shiraishi, Y. et al. 2005. Spontaneous formation of core/shell bimetallic nanoparticles: A calorimetric study. *J. Phys. Chem. B*, 109: 16326–16331.

Van Brussel, M., Kokkinidis, G., Vandendael, I. et al. 2002. High performance gold-supported platinum electrocatalyst for oxygen reduction. *Electrochem. Commun.*, 4: 808–813.

Wang, L., Luo, J., Fan, Q. et al. 2005a. Monodispersed core-shell $Fe_3O_4$@Au nanoparticles. *J. Phys. Chem. B*, 109: 21593–21601.

Wang, L., Luo, J., Maye, M. M. et al. 2005b. Iron oxide-gold core-shell nanoparticles and thin film assembly. *J. Mater. Chem.*, 15: 1821–1832.

Wang, L., Luo, J., Schadt, M. J. et al. 2010. Thin film assemblies of molecularly-linked metal nanoparticles and multifunctional properties. *Langmuir*, 26: 618–632.

Wasmus, S. and Kuever, A. 1999. Methanol oxidation and direct methanol fuel cells: A selective review. *J. Electroanal. Chem.*, 461: 14–31.

Waszczuk, P., Lu, G. Q., Wieckowski, A. et al. 2002. UHV and electrochemical studies of CO and methanol adsorbed at platinum/ruthenium surfaces, and reference to fuel cell catalysis. *Electrochim. Acta*, 47: 3637–3652.

Whetten, R. L., Khoury, J. T., Alvarez, M. M. et al. 1996. Nanocrystal gold molecules. *Adv. Mater.*, 8: 428–433.

Wieckowski, G. Q. and Lu, A. 2000. Heterogeneous electrocatalysis: A core field of interfacial science. *Curr. Opin. Colloid Interf. Sci.*, 5: 95–100.

World Gold Council, 2003. *Gold 2003: New Industrial Applications for Gold*, Gold Bull., 36: 144–151.

Xiao, S., Hu, W., Luo, W. et al. 2006. Size effect on alloying ability and phase stability of immiscible bimetallic nanoparticles. *Eur. Phys. J. B*, 54: 479–484.

Xu, Y. and Mavrikakis, M. 2003. Adsorption and dissociation of O-2 on gold surfaces: Effect of steps and strain. *J. Phys. Chem. B*, 107: 9298–9307.

Xu, Z., Hou, Y., and Sun, S. 2007. Magnetic core/shell $Fe_3O_4$/Au and $Fe_3O_4$/Au/Ag nanoparticles with tunable plasmonic properties. *J. Am. Chem. Soc.*, 129: 8698–8699.

Yang, H., Vogel, W., Lamy, C. et al. 2004. Structure and electrocatalytic activity of carbon-supported Pt-Ni alloy nanoparticles toward the oxygen reduction reaction. *J. Phys. Chem. B*, 108: 11024–11034.

Zeng, H., Li, J., Wang, Z. L. et al. 2004. Bimagnetic core/shell FePt/$Fe_3O_4$ nanoparticles. *Nano Lett.*, 4: 187–190.

Zhang, J., Vukmirovic, M. B., Sasaki, K. et al. 2005a. Mixed-metal Pt monolayer electrocatalysts for enhanced oxygen reduction kinetics. *J. Am. Chem. Soc.*, 127: 12480–12481.

Zhang, J. L., Vukmirovic, M. B., Xu, Y. et al. 2005b. Controlling the catalytic activity of platinum-monolayer electrocatalysts for oxygen reduction with different substrates. *Angew. Chem. Int. Ed.*, 44: 2132–2135.

Zhang, J., Sasaki, K., Sutter, E. et al. 2007. Stabilization of platinum oxygen-reduction electrocatalysts using gold clusters. *Science*, 315: 220–222.

Zhong, C. J. and Maye, M. M. 2001. Core-shell assembled nanoparticles as catalysts. *Adv. Mater.*, 13: 1507–1511.

Zhong, C. J., Luo, J., Maye, M. M. et al. 2004. Nanostructured gold and alloy electrocatalysts. In *Nanotechnology in Catalysis*, eds. B. Zhou, S. Hermans, and G.A. Somorjai, Vol. 1., Chapter 11, pp. 222–248. Kluwer Academic/Plenum Publishers, New York.

Zhong, C. J., Luo, J., Maye, M. M. et al. 2006. Core-shell synthesis of C-supported alloy nanoparticle catalysts. US Patent No 7,053,021.

Zhong, C. J., Luo, J., Maye, M. M. et al. 2007. Gold-based alloy catalysts for use in fuel cell catalysts. US Patent No 7,208,439.

Zhong, C. J., Luo, J., Njoki, P. N. et al. 2008. Fuel cell technology: Nano-engineered multimetallic catalysts. *Energy Environ. Sci.*, 1: 454–466.

Zhong, C. J., Luo, J., Fang, B. et al. 2010. Nanostructured catalysts in fuel cells. *Nanotechnology*, 21: 062001.

# 8

# *Inorganic Nanoparticles for Photovoltaic Applications*

**Elif Arici**

**CONTENTS**



## 8.1 Introduction

A photovoltaic cell harvests the energy from sunlight and converts it directly into electrical power. Therefore, photovoltaic technology is one of the most promising renewable energy technologies, such as wind, biomass, and water. The idea of this chapter is to first describe the basic organic solar cell configurations, and then to collect arguments, where an active part of a photovoltaic device could be replaced by inorganic nanoparticles modified with organic surfactants. Inorganic photoactive materials, which we are interested in, are surface-modified inorganic clusters of different shapes, in dimensions of a few nanometers. Similar to purely organic materials for photovoltaic applications, their thin film preparation is possible with solutions using low-cost fabrication technologies, such as spin cast, doctor blade, and ink-jet printing on various substrates. The dispersion properties of the inorganic nanoparticles in common organic solvents are determined by the chosen organic surfactants shielding the nanoparticle surface.

We start with a brief description of the history of photovoltaics and figure out the basic working principles of the devices as well as their most important solar cell parameters.

New methods to prepare hybrid solar cells and their performances will be discussed next. This chapter is based on recent publications (2004–2008) on hybrid solar cells prepared from solutions. It does not cover all of the interesting work done for the development of the purely organic, organic/inorganic hybrid, and dye-sensitized solar cells (DSSCs). Some recent review papers are given in references (Spanggaard and Krebs 2004, Beek et al. 2005a, Xue et al. 2005, Lloyd et al. 2007, Kamat 2008, Saunders and Turner 2008) to cover the literature in detail. The scientific publications cited in this chapter are chosen among the many others because of the new concepts introduced using similar materials to enable a rough comparison of the studies.

### 8.1.1 A Brief Historical Background of Solar Cell Developments

Monocrystalline silicon cells underwent major development because they were the first photovoltaic structures responsible for converting light into electricity with a reasonable efficiency of 4.5%, as published by Bell Labs researchers Chapin, Fuller, and Pearson (Chapin et al. 1954) in 1954. Solar cells based on the monocrystalline silicon p–n junctions are called *first generation*. In the early 1970s, interest in terrestrial applications developed, leading to an interest of an increase in the production volume accompanied by a significant reduction in solar cell costs. In 1976, David Carlson and Christopher Wronski in RCA (Radio Corporation of America) Labs demonstrated the first solar cell based on amorphous silicon with an efficiency of 1.1%. This was the first example of *second-generation* solar cells based on thin film technology. The so-called second-generation solar cells aim to reduce solar cell production costs by using thin films of silicon and other inorganic semiconductors. The most successful second-generation materials are cadmium telluride (CdTe), copper indium selenide (CIS), copper indium gallium selenide (CIGS), and amorphous silicon. These materials are formed into a thin film on a supporting substrate, reducing material mass. But while much cheaper than monocrystalline silicon cells, these second-generation devices suffer from crystal defects that make them less efficient than their single-crystal counterparts.

Research on *third-generation* solar cells aims to increase the efficiencies two to three times of those of the currently existing solar cells by using novel technologies and materials (Trupke et al. 2002, Nazeeruddin et al. 2005, Green 2008). The third generation is somewhat ambiguous in the technologies that it encompasses: tandem/multi-junction/hot-carrier cells, organic/polymer or quantum dot (QD)-based cells, and other technologies such as up-conversion. The ultimate argument for third-generation solar cell technology lies in the ability to manufacture products using very-low-cost processing methods for roll-to-roll manufacturing.

The two main classes of third-generation photovoltaics having at least one key function for energy conversion by organic materials are DSSCs and bulk heterojunction (BHJ) solar cells. Because of the close connection between hybrid concepts and the two classes of solution-processed organic photovoltaic devices, a detailed description of the development is given next.

#### *8.1.1.1 Dye-Sensitized Solar Cells*

In the early 1980s, there was a reawakening of the concept of photoelectrode sensitization discovered by Moser in 1887. The discovery of Moser was that silver halogenides produce a higher photocurrent by inserting them in a dye solution such as erythrosine (Moser 1887). The dye-sensitization process and its theoretical understanding steadily improved since its first discovery, mainly due to its importance in photography. Titanium dioxide (titania, $TiO_2$) is low cost, widely available, and nontoxic. Therefore, $TiO_2$ became the semiconductor of choice as the transparent nanocrystalline metal oxide electrode. An alternative material as a transparent electrode is zinc oxide.

Unlike conventional p–n junction solar cells, a DSSC employs a dye on the top of nanocrystalline porous titania permeated by a hole transport material and a counter electrode. The Lewis acidity of the electrode surface allows efficient attachment of the dyes having a carbonyl group as an anchor. Light absorption takes place in the sensitizing dye molecules, followed by the injection of an electron from the dye's excited states into the conduction band of $TiO_2$.

Electron transport to the anode occurs via a diffusion of the electrons through the disordered network of $TiO_2$ nanoparticles. A redox mediated as the hole-conducting material regenerates the oxidized dye, while is itself reproduced at the counter electrode (usually Au or Pt) by electrons passed though the external circuit load. A typical design of DSSCs is shown in Figure 8.1.

An advantage of a classical DSSC is the insensitivity of the liquid hole transport material to the porosity of the electrode for an effective interpenetration. Oxidized dyes upon transferring electrons to the titania electrode can be regenerated by the electrolyte. Intermediate ionic species formed in the electrolyte after the dye regeneration leads to a retardation of the charge recombination at the interface of $TiO_2$/liquid electrode (Pelet et al. 2000). The intermediate ionic species has to form triiodide first, to be able to accept further electrons from the titania (Figure 8.2).

So far, using ruthenium derivatives as dyes on a nanoporous $TiO_2$ electrode and $I^{-1}/I^{-3}$ as a redox couple, a maximum conversion efficiency of over 11% under standard global AM 1.5 solar light conditions was obtained (Nazeeruddin et al. 2005, Chiba et al. 2006).

Although the liquid DSSCs display high power conversion efficiencies, their commercial applications are still limited due to stability problems as well as technological aspects of the large module production. The presence of a liquid electrolyte makes the manufacturing process difficult. Therefore, recent efforts in DSSC research are focusing more on replacing the liquid electrolyte with a thin film of a p-type organic hole transport material to eliminate practical problems with sealing (Tennakone et al. 1988, Gebeyehu et al.

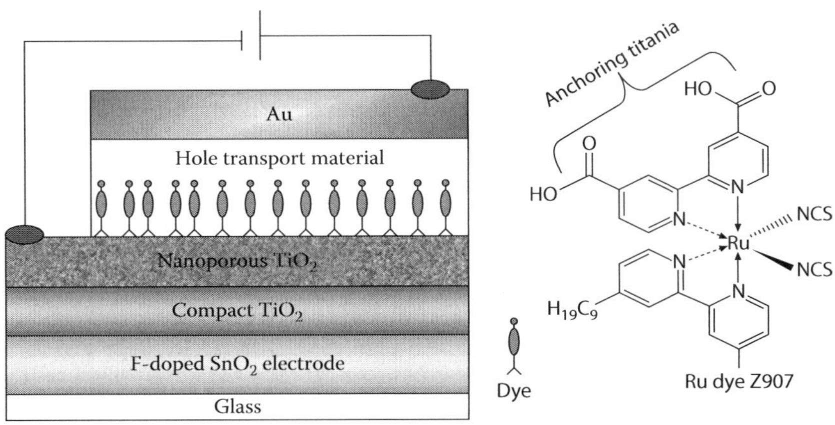


**FIGURE 8.1**
Typical structure of a DSSC.

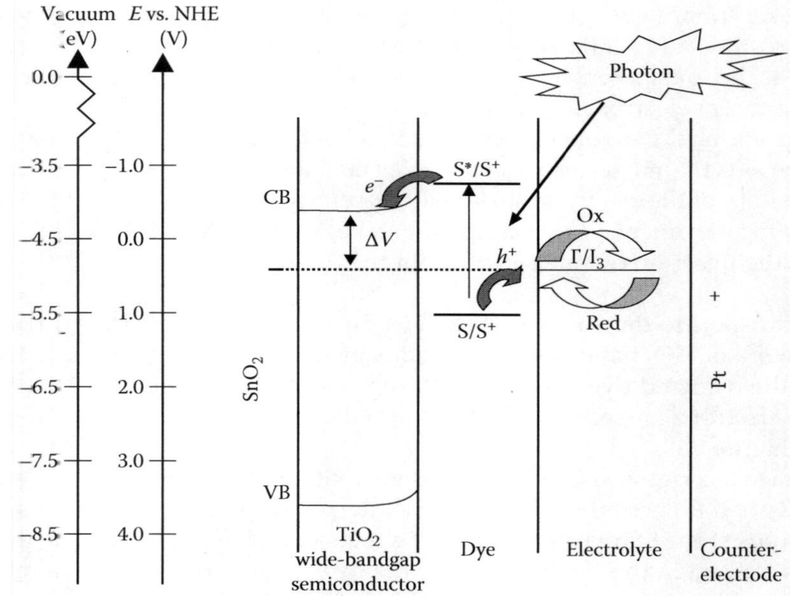


**FIGURE 8.2**
Schematic representation of charge transfer processes in liquid-state DSSCs.

2002, Krüger et al. 2003, Peng et al. 2004, Fujishima et al. 2005, Schmidt-Mende et al. 2005, Lancelle-Beltran et al. 2006, Li et al. 2006, Xia et al. 2008).

The major requirements to be fulfilled for efficient DSSCs are

1. Continuous, dense, and pinhole-free $TiO_2$ films for a good rectifying behavior of the diode
2. High surface area of $TiO_2$ for dye adsorption and charge generation
3. Monolayer of dyes with a broad absorbance range and a high extinction coefficient
4. Sufficient transport of the charges to the electrodes

Today, chemists are highly motivated to discover new methods for the synthesis of highly crystalline, elongated nanoparticles for transparent electrodes with improved electron-collecting properties, together with a high surface area for dye adsorption.

Semiconducting QDs with extinction coefficients as high as $10^5$ can absorb nearly all of the incident solar radiation for wavelengths above their absorption onset, so that they can easily replace the organic dyes. We discuss new ideas for DSSCs using inorganic nanoparticles and nanostructures in Section 8.2.1.

#### *8.1.1.2 Bulk Heterojunction Solar Cells*

Solid-state BHJ devices are one of the promising recent technologies for third-generation solar cells. Figure 8.3 shows an illustration of the general structure of a BHJ device. A photoactive component consisting of an electron donor, usually a p-type polymer, and an electron acceptor, usually n-type organic material, can be deposited from the solution. The anode is often a transparent indium tin oxide (ITO)-coated glass. A thin layer

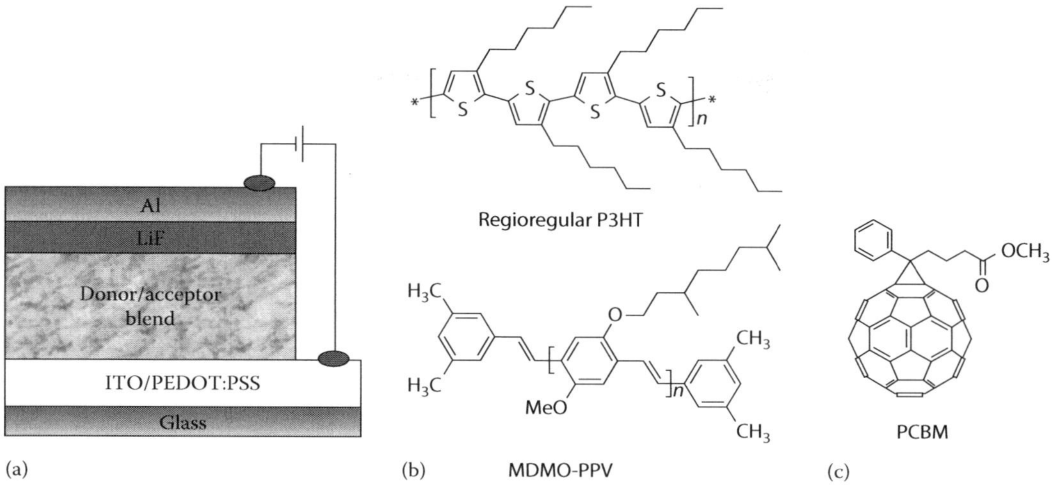


**FIGURE 8.3**
(a) Typical structure of a BHJ device, (b) p-type polymers, and (c) electron acceptors mostly investigated in BHJ devices.

(ca. 80 nm) of a highly p-doped polymer mixture, poly(3,4-ethylendioxythiophene):poly(styrenesulfonate) (PEDOT:PSS), is used to reduce the roughness of the ITO, for a good interfacial contact between the photoactive layer and the anode. The cathode is usually a bilayer of lithium fluoride, LiF (0.3–05 nm), and aluminum, Al (100 nm), but other metals such as Ag have also been used.

The basic working principle of BHJs is the dissociation of photogenerated excitations at the interface between electron-donor and electron-acceptor phases by a photoinduced charge transfer process (Sariciftci et al. 1992). As the excitons in organic materials—with diffusion lengths in the range of 6.5–40 nm—are usually not very mobile (Kroeze et al. 2003, Peumans et al. 2003), only a small volume around the interface between the donor and acceptor materials is active in the charge generation. Therefore, the BHJ concept is a very good method to extend the charge carrier generation volume by the distribution of the donor/acceptor interface over the whole active layer of the solar cell. This can be done rather easily with organic materials by casting the two materials from a common solution (Yu et al. 1995). The blends are phase separated. The morphological investigations of the blends involve chain-packing of the semiconducting polymer and the solubility parameter of the donor/acceptor materials in different solvent mixtures. X-ray investigations and differential scanning calorimetric measurements of blends depending on the preparation history give information about the crystallinity and arrangement of the polymer chains depending on the sample preparation. Atomic force microscopy (AFM) is used for topological characterizations.

In an optimized blend mixture, almost all excited states generated by absorbed photons undergo a photoinduced charge transfer and lead to separated charges with the holes in the donor phase and the electrons in the acceptor phase. Subsequently, the charges have to travel to their respective electrodes, where they can be extracted as photocurrent. The morphology, which is determined by many parameters, such as solubility in the common solvent, composition of the solvent mixture, and temperature treatment, strongly influences the device performance.

Successful BHJ devices have been fabricated by the solution deposition of mixtures of a soluble conjugated polymer and the fullerene derivative [6,6]-phenyl C61 butyric acid methyl ester (PCBM). Commonly used soluble conjugated polymers are poly(phenylene vinylene) (PPV) (Shaheen et al. 2001, Munters et al. 2002), polyfluorene (Svennson et al. 2003, Gadisa et al. 2007), and polythiophene (Dennler et al. 2006, Peet et al. 2007) derivatives.

Though several different BHJ approaches are investigated, those employing p-type polyalkylthiophene as the donor and PCBM as the acceptor are among the most interesting to date. By controlling the BHJ nanomorphology, power conversion efficiencies up to 4%–5% have been achieved using regioregular poly(3-hexylthiophene) (RR-P3HT)/PCBM mixtures (Dennler et al. 2006, Peet et al. 2007). The high efficiency of these devices is proposed to be due to a microcrystalline lamellar stacking in the solid-state packing, resulting in a high hole mobility of the polymer RR-P3HT and, consequently, reduced recombination of the charges. Moreover, interchain interactions cause a redshift of the optical absorption of RR-P3HT due to this stacking. The key idea for the profitable morphology changes was to incorporate a few volume percent of alkane dithiols in chlorobenzene and use the solution to spin-cast the blends.

The major requirements for BHJs are

1. A thin light-absorbing layer (~80 nm) with a broad absorbance range and a high extinction coefficient
2. An interpenetrated network of donor/acceptor phases in the light-absorbing layer for an effective charge generation using a high donor/acceptor interface (Yu et al. 1995)
3. Carrier transport properties of the absorber materials involved (Hoppe et al. 2004)
4. Sufficient transport of the charges to the electrodes by a profitable morphology of the donor/acceptor phases (Zerza et al. 2001)

There are very promising ideas and fantastic developments in the BHJ concept using only organic materials, such as low-bandgap polymers and better absorbing derivatives of fullerene, which are not within the scope of this chapter (Dennler et al. 2006).

Nowadays, semiconductor QDs exhibit a nearly monodispersed size distribution and controlled optical properties as a function of their size. Consequently, a judicious choice of the QD material and size allows to access different regions of the visible spectrum and far beyond. Colloidal nanoparticles, such as CdSe (Murray et al. 1993, Katari et al. 1994), $CuInS_2$ (Arici et al. 2003a), PbSe (Cui et al. 2006), $TiO_2$ (Van Hal et al. 2003, Jiu et al. 2006, Kashiwa et al. 2008), and ZnO (Beek et al. 2005b), have been proposed as a cost-effective alternative for developing BHJ hybrid solar cells because of their size-dependent optical properties together with the possibility to fabricate them into thin layers from the solution. These characteristics can be merged with processing advantages of the polymers to give a novel class of hybrid systems for several functions, such as light absorbing (Huynh et al. 2002, Arici et al. 2003, Rogach et al. 2007b), light emitting (Colvin et al. 1994, Dabbusi et al. 1995), and light coupling within the device (Trupke et al. 2002). A detailed discussion on hybrid BHJs is given in Section 8.2.2. The highest occupied molecular orbital (HOMO)—lowest unoccupied molecular orbital (LUMO) levels of typical electron acceptors and electron donors are given in Table 8.1.

### 8.1.2 Characterization of Solar Cells

The performance of solar cells is described by their current versus voltage characteristics. The voltage across the cell is varied using an applied load and the current measured.

**TABLE 8.1**

HOMO-LUMO Levels of Some Electron Donors and Electron Acceptors

| Material | HOMO [eV] | LUMO [eV] | $E_g$ [eV] | Reference |
|---|---|---|---|---|
| *Electron donors* | | | | |
| MEH-PPV | −2.9 | −5.3 | 2.4 | Dennler et al. (2006) |
| P3HT | −3.0 | −5.1 | 2.1 | Scharber et al. (2006) |
| CdTe | −4.0 (bulk) | −5.5 (bulk) | 1.5 | Rogach et al. (2007b) |
| *Electron acceptors* | | | | |
| PCBM | −4.3 | −6.0 | 1.7 | Scharber et al. (2006) |
| CdSe | −3.7 (bulk) | −5.8 (bulk) | 2.1 | Talapin et al. (2001) |
| PbSe | −4.2 (bulk) | −5.0 (bulk) | 0.8 | Cui et al. (2006) |
| ZnO | −4.4 | −7.6 | 3.2 | Beek et al. (2005a) |

The diode behavior of solar cells is characterized in the dark and under illumination—usually in inert atmosphere—with a source measure unit. For the sake of comparisons, the illumination conditions should be the same in all laboratories. A standard is defined by sun illumination in a clear atmosphere with certain geometry. The radiation properties of the sun are simulated with a white light source having an integrated power density of 100 mW/cm$^2$. The angle of irradiation on the solar cell is defined as a function of the parameter "air mass" (AM). The parameter AM is the relative path length of solar beam radiation through the atmosphere. This is an important parameter, because the intensity of the illumination light chances via absorption and reflection through the atmosphere. $AM_0$ is a vertical vector to the earth. $AM_{1.5}$—the standard value for solar cell characterization—is a vector with an angle of 48° from $AM_0$.

The key parameters determined from current/voltage measurements are short-circuit current ($I_{sc}$), open-circuit voltage ($V_{oc}$), and maximum power point (MPP). Figure 8.4 displays the basic parameters of the *I–V* characteristics of solar cells.

While $I_{sc}$ is determined by the creation and subsequent dissociation of excitons at the donor/acceptor interface followed by transport of free charge carriers toward the collecting electrodes (Kroon et al. 2008), $V_{oc}$ is primarily determined by the effective bandgap of the active layers. The ionization potential of the donor (HOMO) and electron affinity of the acceptor (LUMO) directly affect the $V_{oc}$ (Gadisa et al. 2004, Scharber et al. 2006). MPP is the product of maximal current ($I_{max}$) and maximal voltage ($V_{max}$) within the area defined by the coordinates of $I_{sc}$ and $V_{oc}$ (Figure 8.4):

$$\text{MPP} = I_{max} \times V_{max}$$

The fill factor (FF) is defined as the ratio of MPP to the product of $I_{sc}$ and $V_{oc}$:

$$\text{FF} = \frac{\text{MPP}}{I_{sc} \times V_{oc}}$$

Understanding the factors that limit photovoltaic parameters such as $I_{sc}$ and $V_{oc}$ helps to optimize material and device structures for higher efficiencies. Important limiting factors for device performance are the carrier recombination and the low charge mobility of photovoltaic materials, both decreasing the FF and the overall photon harvesting by reducing

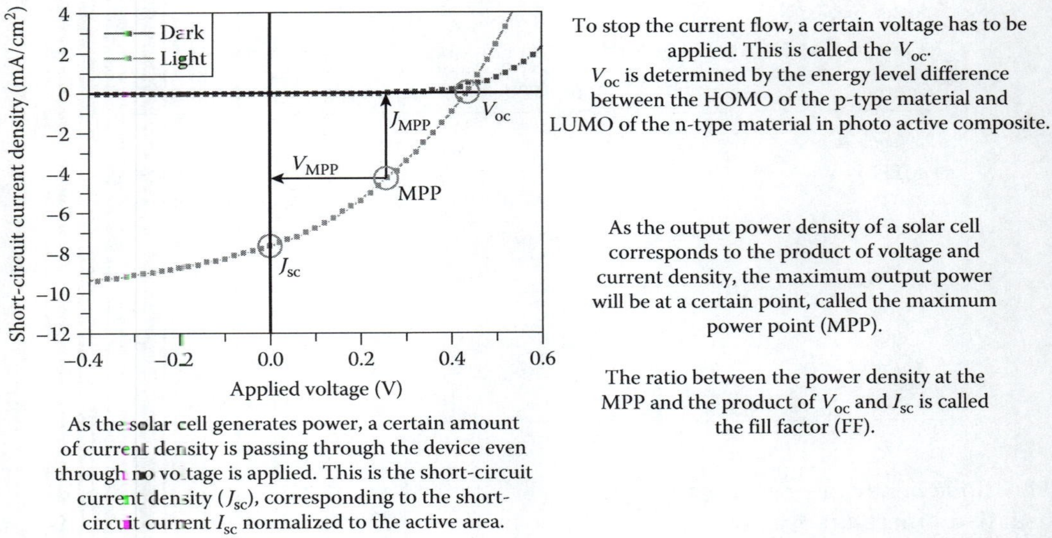


**FIGURE 8.4**
Basic parameters of a solar cell shown in *I–V* characteristics in the dark and under illumination.

the optimum active-layer thickness. A good interfacial contact between the photoactive material and electrodes is essential to minimize the serial resistance across the cell.

Higher values of these three parameters yield larger light-to-electricity power conversion efficiency ($\eta_e$):

$$\eta_e = I_{sc} \times V_{oc} \times \mathrm{FF}$$

Incident photon-to-current efficiency (IPCE) is measured by illuminating the solar cell with monochromatic light and recording the short-circuit current, $I_{sc}$, at different wavelengths. For the wavelength-dependent photocurrent measurement, a 900 W xenon lamp combined with a monochromator ($\Delta\lambda < 2\,\mathrm{nm}$ resolution) can be used as the light source. The IPCE can be calculated as follows:

$$\mathrm{IPCE}(\%) = \left(\frac{\text{number of collected electrons}}{\text{number of incident photons}}\right) \times 100$$

In optimized organic solar cells, the IPCE can reach very high values of above 80%, so that almost all photons absorbed in the active layer generate charges that can be collected at the electrodes. If there are not any light-filtering effects in the photovoltaic active area, the absorbance spectra and the photocurrent spectra are similar in their spectral form.

### 8.1.3 New Materials for Organic Photovoltaics: Colloidal Nanoparticles

#### *8.1.3.1 Synthetic Aspects*

Nanoparticles can be synthesized from a variety of different materials in several surroundings (Figure 8.5).

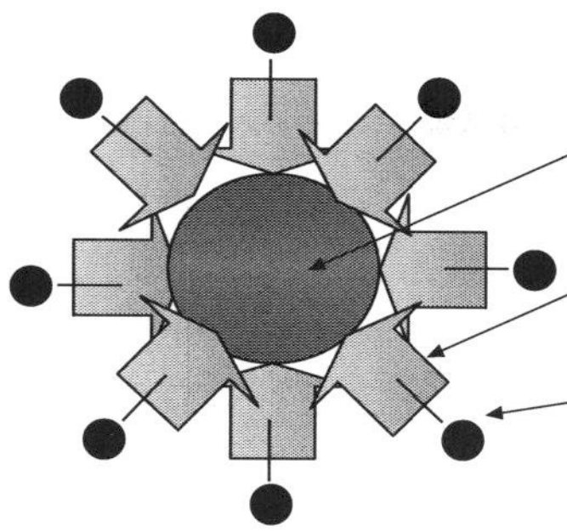


**FIGURE 8.5**
Anatomy of a QD and surfactants.

We can classify the synthesis procedures of solution-processable nanoparticles for solar cell applications briefly in three groups:

1. Reactive methods in high-boiling-point solvents (CdSe (Murray et al. 1993, Katari et al. 1994))
2. Precipitative methods using polar solvents (CdTe (Murphy et al. 2002, Rogach et al. 2007a) and $CuInS_2$ (Czekelius et al. 1999))
3. Reactions involving templates, micelles, or confined solids (Jiu et al. 2006, Kashiwa et al. 2008)

The first procedure involves the decomposition of molecular precursors at relatively high temperatures. A breakthrough in nanoparticle synthesis was developed 15 years ago by Murray et al. by using n-trioctylphosphine oxide (TOPO) to stabilize the particles. Selenium and tellurium salts were dissolved in n-octylphosphine mixed with a solution of dimethylcadmium, and then were injected into hot TOPO at 300°C (Murray et al. 1993). Immediately, color changes from transparent to yellow/orange/red occurred in the solution indicating the particle growth. Usually, the use of organic solvents has the advantage of tuning the reaction temperature over a wide range, enabling highly crystalline nanoparticles. By varying the temperature of the reaction, particle nucleation and growth can be separated, leading to nearly monodispersed size distribution of the quantum dots in solution (Murray et al. 1993, Huynh et al. 2002). Most of the polymers used in BHJs are soluble in an organic medium, such as toluene and chlorobenzene. Therefore, nanoparticles synthesized in an organic medium can be integrated in BHJ solar cells easily.

The choice of the surfactant, which exhibits a polar head group and a nonpolar part, plays a very critical role for controlled particle growth. The binding strength of the polar group influences the dynamics of adsorbing and deadsorbing of the surfactant on the surface of the growing nanocrystals. The nonpolar group usually consists of an alkyl chain. The shape of the nonpolar group affects the diffusion properties of the monomer units during crystal growth. The nature of the polar as well as the nonpolar group of the surfactant may influence also the shape of the nanoparticles. Examples of surfactants and their boiling points ($B_p$) are shown in Figure 8.6.

The main developments for the integration of inorganic nanoparticles in optoelectronically active blends were replacing the toxic precursors with less toxic ones, controlling

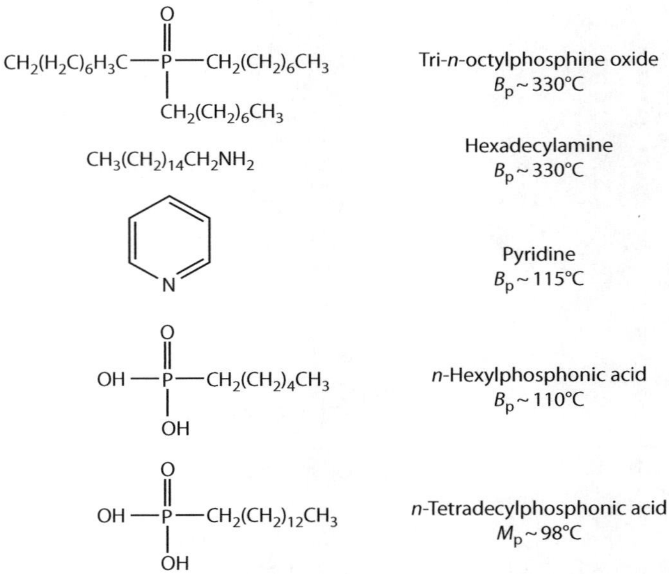


**FIGURE 8.6**
Examples of surfactants used by the QD synthesis.

the shape and size of the nanoparticles, and finding thin film preparation methods with profitable optical, electrical, and morphological properties.

Generally, the shape of a crystal depends on the relative speeds of the individual facet growth, which is determined by the surface tensions of the crystal facets. The crystal facets can have different surface tensions due to the variations in atom densities and the changes in the number of unsaturated bonds of each facet. Therefore, the growth of spherical crystals is more an exception than a rule. By using different surfactants during the synthesis, chemists are able to manipulate the growth kinetics. If a surfactant binds stronger to one facet than to its neighboring facets, new monomers are more likely to be incorporated into these neighboring facets. CdE (E: S, Se, Te) nanocrystals with a wurtzite structure, for example, have a high-symmetry axis distinguished from all the other axes. This high-symmetry axis serves as the directional axis for the asymmetric growth for the synthesis of nanorods, which is proven to be an advantage for solar cell applications (Huynh et al. 2002). With the increasing aspect ratio of the nanorods, only a slight shift of the bandgap is observed, which saturates at aspect ratios of the order of 10 (Kan et al. 2003). Therefore, it is possible to take advantage of size quantization together with improved charge transport properties of nanorods.

A promising idea was to use nanoparticles with four rod-shaped arms that branch out at tetrahedral angles from a central region. The so-called tetrapods exhibit a large surface area for charge separation and still provide a pathway for the transport of the charges due to their geometry (Yin et al. 2005).

The core and the four nanorods grown on the core have an identical crystallite face each, but different crystallite structures. For CdSe tetrapods, the CdSe core is a zinc blende structure with a tetrahedral geometry, while the CdSe arms exhibit the wurtzite structure. Both CdSe structures, zinc blende and wurtzite, are enantiotropic systems at a certain temperature. Experiments showed that the tetrapod-shaped nanocrystals grow only at high

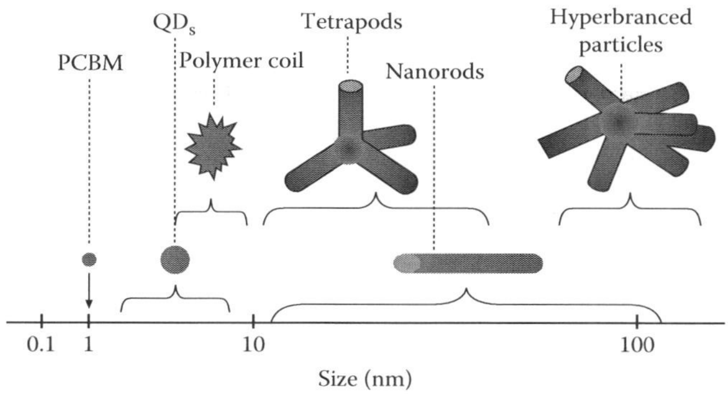


**FIGURE 8.7**
Approximate dimensions of different nanoparticles used in BHJ hybrid solar cells.

monomer concentrations, whereas the system is not thermodynamically stable. In thermodynamically stable conditions, nanoparticles of spherical shapes occur. Details on the mechanism of the formation of tetrapod structures are provided in Yin et al. (2005). Their photovoltaic performance in organic/inorganic BHJ solar cells is discussed in Section 8.2.2. In Figure 8.7, approximate dimensions for different nanoparticles tested in hybrid solar cells are displayed.

Precipitative methods at low temperatures (70°C–100°C) (in the second procedure) have been used for the synthesis of water-soluble inorganic nanoparticles, such as CdTe and CdHgTe (Rogach et al. 2007a). Similar to the chemical structure of the ruthenium dye, surfactants having a carboxylate or phosphonate group can be used to attach QDs to the $TiO_2$ surface. Therefore, QDs for DSSC applications are preferable surrounded with a surfactant such as 2-mercaptopropionic acid, thiolactic acid, or mercaptobenzoic acid (Figure 8.8).

Improvement of the synthesis conditions requires an empirical adjustment of the acidity of the reaction medium, sharply defined temperature for the nucleation and growth of the nanoparticles, as well as the right choice of the ratio of surfactants to inorganic precursors.

Nanoparticles using the low-temperature procedure have, in general, a spherical form. The growth of the nanoparticles can be controlled by choosing the right surfactants. Using more rigid surfactants, the diffusion of the monomer units can be hindered, resulting in very small nanoparticles (less than 4 nm in diameter).

Another important material for the purpose of this chapter is titania, a high-bandgap material used as nanostructured electrodes in hybrid solar cells. Nanostructured titania can be synthesized using reaction routes involving templates, micelles, or confined solids.

It is well known that a $TiO_2$ crystal in an anatase structure has a tetragonal geometry (the *c*-axis being 2.7 times larger than the *a*-axis) and nucleates as truncated octagonal bipyramidal seeds, exposing eight equivalent {101} faces and two equivalent {001} faces. Using two surfactants that bind selectively to the respective surface planes is the key method to control the crystal growth for elongated titania nanorods (Puntes et al. 2001). The diameter and the length of titania nanorods are controlled by the amount of surfactants, such as cetyltrimethylammonium bromide (CTAB). To avoid the aggregation of a titania nanoparticle, a triblock copolymer of poly(ethylene oxide)$_x$–poly(propylene oxide)$_y$–poly(ethylene oxide)$_x$ has been used as a template, diluting the reaction media and dispersing titania nanorods inside the polymer matrices (Jiu et al. 2006). Reaction mixtures without the polymer template lead to branched titania rods. Existing theories hold that the hydrated hydrophilic PEO of a copolymer anchors the titania surface, while the hydrated hydrophobic

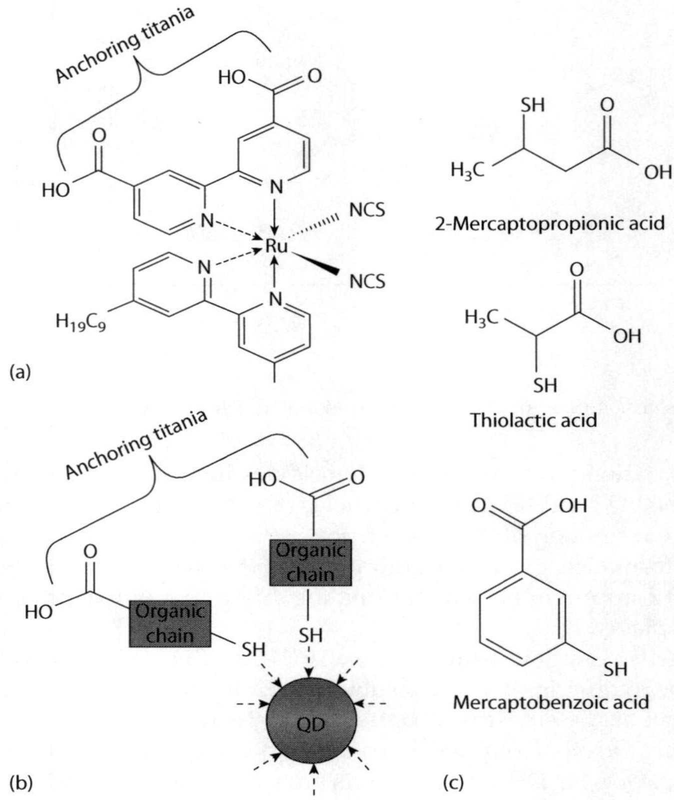


**FIGURE 8.8**
(a) Chemical structure of a typical ruthenium dye anchoring titania, (b) scheme of a QD surrounded with organic surfactants anchoring titania, and (c) chemical formula of some suitable surfactants attached on QDs for DSSC applications.

PPO chains extend out into the solution, forming a "brush layer" that inhibits the aggregation of the nanoparticle by steric repulsion. After sintering the titania at high calcination temperatures of 450°C for 1 h, the rod shape of the nanoparticle was still evident in scanning electron microscopic (SEM) images, whereas the copolymer template was removed. $TiO_2$ nanorods having a diameter of ~20–30 nm and a length of >100 nm have been characterized using this procedure.

A new concept to create vertically ordered $TiO_2$ electrodes on transparent bulk titania was introduced by Kashiwa et al. (2008). ZnO tetrapods were used as templates. Blends of ZnO tetrapods and nanocrystalline titania were deposited on bulk titania using an electro-spray apparatus. A voltage of 17 kV was applied between the needle and the electrode surface to assemble the ZnO tetrapods first (Kroon et al. 2007). The annealing procedure for the titania electrode was performed at 500°C for 0.5 h. After removing the ZnO nanoparticle, simply by washing in an HCl solution, straight holes in the sputtered titania layer were formed for an efficient charge transport.

Alternatively, the electrochemical etching of a titanium foil in a fluoride media produces an ordered array of hollow $TiO_2$ tubes. The details on the mechanism of the formation of a $TiO_2$ tubular array structure on a titanium substrate are provided in Kuo et al. (2008) and

Paulose et al. (2006). The nanotubes were approximately 80–90 nm in diameter and 8 μm in length. An advantage of this method is the hollow nature of these nanotubes, which makes both inner and outer surface areas accessible for dye adsorption for DSCCs. A comparison of the photovoltaic performance of DSCCs using different concepts to structure titania is given in Section 8.2.2.

#### 8.1.3.2 Characterization Methods for Inorganic Nanoparticles

##### *8.1.3.2.1 Absorption and Photoluminescence*

In a bulk semiconductor, an electron can be excited from the valence band to the conduction band by the absorption of a photon with an appropriate energy, leaving a hole in the valence band. The formed electron–hole pair is called an exciton. Its lowest energy state is slightly below the lower edge of the conduction band because of the Coulomb interaction between the electron–hole couple. For CdSe, for example, the exciton radius is approximately 5 nm. The reduction of CdSe nanoparticle sizes to values less than 5 nm creates an unusual situation—that the exciton radius can theoretically exceed the particle dimensions. The kinetic energies of the charge carriers increase to overcome the situation, leading to an increasing bandgap and the quantization of the energy levels to discrete values depending on the particle size. This phenomenon is called "quantum confinement effect" and is theoretically studied using a quantum mechanical particle in a box model (Bawendi et al. 1990a,b). Because of the quantum confinement effect, decreasing the particle size results in a hypsochromic shift of the absorption onset. A relatively sharp absorption feature near the absorption onset corresponds to the excitonic peak, that is, the lowest excited state exhibiting large oscillator strength. While its position depends on the bandgap and, consequently, on the particle size, its form and width is strongly influenced by the particle size distribution and also by the stoichiometry of the nanoparticles.

##### *8.1.3.2.2 Electron Microscopy*

Transmission electron microscopy (TEM) is a straightforward technique to image the shape of the individual crystallites and the statistical distribution of their size. It can also reveal information on the composition and the internal structure of nanoparticles, for example, by detecting the characteristic x-rays, which are produced by the interaction of electrons with the material. In TEM, electrons are accelerated at high voltages (~300 kV) to high velocities. Condenser lenses focus the electron beam on the sample. An objective lens forms diffraction in the back focal plane and the image of the sample. Some intermediate lenses magnify the diffraction pattern and the image on the screen. The resolution is in the order of 0.1 nm. To obtain a high contrast between crystallites in a blend, where the nanoparticles are not ordered and can overlap, a bright field-imaging mode is usually applied. Hereby, an objective diaphragm is inserted in the back focal plane to select the transmitted beam. Only the crystalline parts in the Bragg orientation appear dark and the amorphous parts appear bright. The selected area diffraction mode permits to obtain the symmetry of the monocrystallite lattice and calculates its interplanar distances using the Bragg law (Flegler et al. 1995).

##### *8.1.3.2.3 X-Ray Analysis*

The wide part of x-ray diffraction (WAXS) pattern corresponds to the diffraction of x-rays on atoms the nanocrystals consist of and allows the estimation of the average size of crystalline domains within each nanocrystal. A wide-angle XRD (x-ray diffraction) of nanoparticles reveals the internal structure of the average nanocrystal core and permits an approximate assumption of nanoparticle dimensions using the Debye–Scherrer approximation, with $\lambda = 0.15418\,\text{nm}$ for Cu K$\alpha$ radiation, $\Delta 2\theta$ for line width in radians, and $\cos\theta$ for half of the XRD angle (Borchert et al. 2005):

$$D = \frac{1.2\lambda}{\left[(\Delta 2\theta)(\cos\theta)\right]}$$

Broadening of the WAXS pattern is associated with decreasing of the particle size, but only when the nanocrystals are defect free and spherical in shape. Therefore, nanoparticle dimensions should be approved using additional methods.

*8.1.3.2.4 Elemental Analysis*

The chemical analysis of a material under investigation can be performed by analytical techniques, such as energy-dispersive x-ray spectroscopy (EDS) and x-ray photoelectron spectroscopy (XPS). The principle of analysis depends on the inelastic scattering between electrons and the material, giving different kinds of signals: secondary electrons, Auger electrons, and x-rays. The x-ray energy corresponds to a difference between the energy levels of the electron orbitals. Since these levels are quantified, the x-ray energy spectrum represents the signature of the atoms present in the material.

## 8.2 Applications of Nanoparticles in Photovoltaic Devices

### 8.2.1 Nanoparticles for Novel Transparent Electrodes

To evaluate the performance of the nanocrystalline network as the electron-collecting electrode, different electrode structures and the corresponding solar cell characteristics are compared and discussed in this section. In all examples compared here, a monolayer of a ruthenium dye is used as the sensitizer. The redox electrode as the hole transport medium is composed of $I^{-1}/I^{-3}$ salt in an acetonitrile solution. The anodes of DSSCs are typically constructed using $TiO_2$ or less often using ZnO nanoparticles, deposited as a paste and sintered to produce electrical continuity. Au has been chosen as the counter electrode.

A well-established application of nanoparticles as electrode materials is the replacement of a flat electrode surface by a mesoporous titania delivering a 100-fold enhancement in the surface area per micrometer thickness, when compared with a flat film. The increased active area for dye adsorption leads to a high-density packed monolayer of the dye for a better light harvesting. The most broadly researched "liquid-electrolyte-based" DSSC is composed of a mesoporous titania fabricated from solgel-processed sintered nanoparticles coated upon a bulk titania on a transparent conducting glass ($SnO_2$:F). The typical $TiO_2$ film thickness for liquid-state DSSCs with the highest light conversion efficiency ranges from 8 to 12 μm. The grain size of the $TiO_2$ film (Figure 8.9a) can range from 10 up to 80 nm, depending on the processing technique (Ravirajan et al. 2006). The porosity is approximately 50%. The interpenetration of the liquid hole transport material is expected not to be sensitive to the porosity of the electrode, so that the oxidized dyes upon transferring electrons to the titania electrode can be regenerated by the electrolyte.

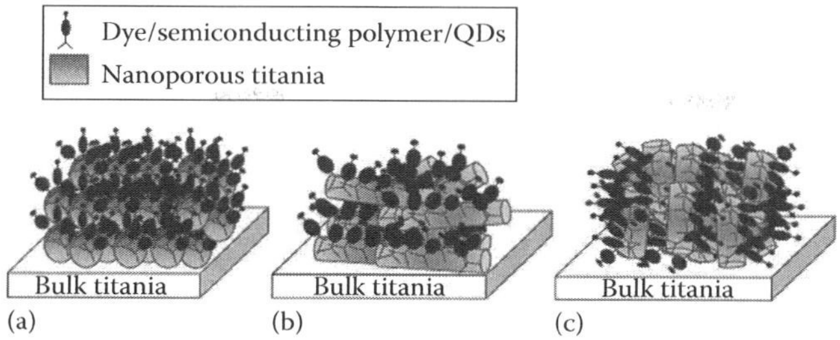


**FIGURE 8.9**
Different concepts for using nanoparticles as transparent electrodes in solar cells: (a) randomly aligned spherical nanoparticles, (b) horizontally aligned nanorods, and (c) vertically aligned nanorods.

The nature of electron transport in titania layers is fairly well understood. Drift transport, a vital mechanism in most photovoltaic cells, is prevented in DSSCs by ions in the electrolyte screening the electric field within the device (Law et al. 2005). Time-resolved photocurrent and modeling studies indicate that electron transport proceeds by a trap-limited diffusion process, in which photogenerated electrons repeatedly interact with a distribution of traps as they undertake a random walk through the nanoparticle network. Under sunlight, an average injected electron may experience a million trapping events before either percolating into the collecting electrode or recombining with an oxidizing species, predominantly $I_3^-$ in the electrolyte.

New ideas to optimize the electron transport properties in the DSSCs include the replacement of the traditional nanoporous titania layer by a dense array of oriented, highly crystalline nanowires (Figure 8.9b and c). The direct electrical pathways provided by the nanowires ensure the rapid collection of carriers generated throughout the device. Because of the reduced number of crystallite boundaries, electron transport in crystalline wires is expected to be several orders of magnitude faster than percolation through a random polycrystalline network.

A comparison of the electrical properties of spherical and elongated nanoparticles has been studied using thin films of zinc oxide as transparent electrodes (Law et al. 2005): Two-point electrical measurements of nanowires on indium tin oxide (FTO) substrates gave a linear current–voltage (*I–V*) trace, indicating barrier-free contacts between the nanowire and the substrate. Individual nanowires were extracted from the arrays; fashioned into field-effect transistors using standard electron-beam lithography procedures; and studied to determine their resistivity, carrier concentration, and mobility. Measured resistivity values ranged from 0.3 to 2.0 Ω cm, with an electron concentration of $(1–5) \times 10^{18}\,cm^{-3}$ and a mobility of $1^{-5}\,cm^2/V\,s$. Using the Einstein relation, $D = kBT\mu/e$, the electron diffusivity has been estimated to be $D_n = 0.05 - 0.5\,cm^2/s$ for single dry nanowires. This value is several hundred times larger than the highest reported diffusivity for $TiO_2$ or ZnO nanoparticle films in operating cells.

A switch from particles to wires may also affect the kinetics of charge transfer at the dye–semiconductor interface, as particle and wire films have dissimilar surfaces onto which the sensitizing dye adsorbs. Whereas nanoporous electrodes present an ensemble of surfaces having various bonding interactions with the dye, nanowire arrays as electrodes are dominated by a single crystal plane. Femtosecond transient absorption spectroscopy has been utilized to measure the rate of electron injection from photoexcited ruthenium dyes into nanowire and nanoparticle films. Injection in wires was characterized by bi-exponential kinetics with time constants of less than 250 fs and around 3 ps, whereas the particle response was tri-exponential and significantly slower (time constants: <250 fs, 20 ps, 200 ps) (Law et al. 2005). On the other hand, a high-performance nanowire photoanode must foremost have a large surface area for dye adsorption, comparable to that of a nanoparticle film.

$TiO_2$ nanorods having a diameter of 20–30 nm and a length of >100 nm have been characterized in their photovoltaic performance in liquid-state DSSCs by the group of Adachi et al. (Adachi et al. 2004, Jiu et al. 2006). Titania nanorods have been synthesized by reactions involving polymer templates. A power conversion efficiency of 7.1%, an $I_{sc}$ of 13.4 mA/cm², and an open-circuit voltage of 0.75 V were obtained using these titania layers with a thickness of 16 μm in liquid DSSCs. The FF was 0.73, respectively. The extent of dye adsorption was relatively low in comparison to flat layers; therefore, the maximum IPCE performance was only slightly better than flat layers of titania increasing from 42% at 550 nm to 58% at the same wavelength.

Recently, in a similar concept, Sung et al. compared the efficiencies of liquid-state DSSCs using titania nanorods and nanodots (Kang et al. 2008). A power conversion efficiency of 4.95%, an $I_c$ of 11.70 mA/cm², and an open-circuit voltage of 0.70 V were obtained using these titania layers consisting of nanorods with a layer thickness of 16 μm (Figure 8.10b). The FF was 0.60, respectively. On the other hand, DSSCs using titania nanodots as an electrode (Figure 8.10a) had a power conversion efficiency of 3.36%, an $I_{sc}$ of 6.90 mA/cm², and an open-circuit voltage of 0.68. The FF was 0.71, respectively.

The significant improvement of the short-circuit current by 70% cannot be explained just by the increased specific surface area. Improved electron transport properties of titania nanorods might also have an influence. The reduced number of grain boundaries and, consequently, less charge-trapping sides are the arguments for a better charge transport in elongated nanoparticles. Additionally, aligned nanotube arrays of titania should permit electron transport along the length of the nanotube toward the electrode, avoiding random charge hopping. The FF is a parameter, which is determined by the resistance of the

**FIGURE 8.10**
Various types of $TiO_2$ layers used in DSSCs: (a) nanoporous titania for classical DSSCs, (b) nanorods in plane with lengths of 100–300 nm and diameters of 10–30 nm, (c) tubular hollow nanotubes of $TiO_2$ with a length of 8 μm and a diameter of 80 nm, and (d) nanoembossed hollow spherical $TiO_2$ with a diameter of 1–3 μm and a wall thickness of ca 0.25 μm.

solar cell. A decrease of the FF using nanorods, instead of spherical titania nanoparticles for DSSCs, is an indicator of higher series resistance within the cell, which might be optimized by additional purification techniques to remove the polymer template completely.

Nanorod titania films have been also prepared by using the electro-spinning method. Solid-state DSSCs yielded a total efficiency of 0.29%. Unfortunately, the device performance in liquid-state DSSCs was not reported for comparison (Zhu et al. 2008).

Alternatively, the electrochemical etching of a titanium foil in a fluoride media has been used to produce an ordered array of hollow $TiO_2$ tubes (Paulose et al. 2006, Kongkanand et al. 2008, Kuo et al. 2008). The hollow nature of these nanotubes makes both inner and outer surface areas accessible for dye adsorption (Figure 8.10c). Although the absorption spectra indicate that dye-attached nanotube $TiO_2$ films generally absorb more light than nanoparticle $TiO_2$ electrodes in the same conditions, this difference accounts for no more than a 5% increase in the overall photons absorbed. Comparing this with a 10% improvement in the IPCE of the nanotube film over the nanoparticle film demonstrates the measurable advantage of the nanotube architecture not only because of the increased area for dye adsorption, but also because of the optimized electron transport properties. This statement was experimentally proven by electron lifetime and electron diffusion coefficient measurements (Kongkanand et al. 2008). The electrons in the particulate $TiO_2$ films might be more susceptible to loss processes at grain boundaries than those in nanotube $TiO_2$ films.

Using ZnO tetrapods as templates, vertically ordered $TiO_2$ electrodes on transparent bulk titania have been fabricated by Hayase et al. (Kashiwa et al. 2008) (Figure 8.9c). A high power conversion efficiency of 7.43%, an $I_{sc}$ of 16.16 mA/cm$^2$, and an open-circuit voltage of 0.73 V were obtained using these titania layers in liquid DSSCs. The FF was 0.69, respectively.

### 8.2.2 Nanoparticles to Optimize the Absorption Behavior

#### *8.2.2.1 Increasing the Light-Scattering Properties*

Large titania particles with spherical shapes are generally used as scattering particles, confining the incident light within an optoelectronically active layer, and thereby enhancing the photocurrent density (Ito et al. 2006). Combining the light-scattering function with the idea of high-surface-area electrodes, new multifunctional materials have been developed, such as nanoembossed hollow spherical $TiO_2$ (Toyoda et al. 2008). The diameter of the spheres was in the range of 1–3 μm, with a wall thickness of ca. 0.25 μm to enable light scattering (Figure 8.10d). The average pore size within the spheres was about 10 nm, leading to an additional surface area for dye adsorption. After calcinations at 450°C in air for 2 h, the spherical structures have not been greatly deformed or damaged, and there were no apparent pinholes or cracks on their surfaces. In comparison to devices with scattering but nonporous titania, a photocurrent increase of 21% has been observed. The solar cell parameters observed were an $I_{sc}$ of 15.80 mA/cm$^2$, a $V_{oc}$ of 0.84 V, and a FF of 0.71.

#### *8.2.2.2 Enhancing Light Absorption*

Short-bandgap materials, such as CdS, CdSe, PbS, InP, and HgTe, can also serve as sensitizers for DSSCs, because they can transfer electrons to large-bandgap-semiconductor $TiO_2$ under visible light excitation, whereas the holes are transported via a semiconducting polymer to the counter electrode (Plass et al. 2002, Fritz et al. 2008).

Studies demonstrated that chemically and electrochemically deposited CdS nanocrystallites on $TiO_2$ surfaces are capable of injecting excited electrons and generating photocurrent under visible light irradiation (Lin et al. 2007, Shen et al. 2008). A bifunctional surface modifier, such as 3-mercaptopropyl trimethyoxysilane, has been used as a linker molecule to anchor a CdS-QD monolayer on the $TiO_2$ surface (Lin et al. 2007). The self-assembled CdS-QD monolayer on titania had an absorbance edge at 520 nm, revealing the optical characteristics of a quantum-sized CdS. When the number of self-assembly cycles was increased, the resulting absorbance spectra showed an increase in the intensity with the increasing number of the self-assembly cycles. Repeating the self-assembly cycles for five times, the monolayer of CdS QDs on the top of titania were complete under the given preparation conditions; a further increase in the number of self-assembly cycles resulted in a decrease of cell performance (Table 8.2). The onset of the absorption and the IPCE spectra shifts from 520 to 550 nm, an indication for the growth of self-assembled nanoparticles during the preparation. A maximum IPCE value as high as 30% has been obtained at 520 nm.

Other preparative factors important for QD-sensitized solar cells are:

1. Substrate temperature (Toyoda et al. 2008)
2. Time for self-assembly (Huynh et al. 2002, Toyoda et al. 2008)
3. Chemistry of the surfactants (Zayats et al. 2003, Lin et al. 2007)
4. Effective filling of nanoporous titania (Kang et al. 2008)

There are also some examples (Kang et al. 2008) where different-sized CdSe QDs have been used to co-sensitize titania (Table 8.3). Devices using CdSe QD nanoparticles of different sizes had incident photon-to-current efficiencies up to 47% at around 550 nm monochromatic irradiation, short-circuit current densities of 3.3 mA/cm$^2$, and an open-circuit

**TABLE 8.2**

Solar Cell Characteristics of CdS-Sensitized Solar Cells Depending on the Number of CdS Deposition Cycles

| Number of Self-Assembly Cycles | $I_{sc}$ [mA/cm$^2$] | $V_{oc}$ [mV] | Fill Factor | Efficiency |
|---|---|---|---|---|
| 1 | 0.89 | 602 | 0.53 | 0.28 |
| 3 | 1.93 | 634 | 0.56 | 0.68 |
| 5 | 3.44 | 657 | 0.60 | 1.35 |
| 7 | 2.52 | 615 | 0.52 | 0.81 |

**TABLE 8.3**

Performances of CdSe-Sensitized Solar Cells

| Diameter of CdSe QD | $I_{sc}$ [mA/cm$^2$] | $V_{oc}$ [mV] | IPCE ($\lambda_{max}$) |
|---|---|---|---|
| 2.6 nm | 2.5 | 620 | 35 (520 nm) |
| 3.0 nm | 2.2 | 620 | 30 (550 nm) |
| Co-sensitization | 3.3 | 620 | 45 (520 nm)<br>47 (550 nm) |

voltage of 620 mV under simulated AM 1.5 (100 mW/cm$^2$) illumination. The power conversion efficiency was 1.20%. The co-sensitization of titania resulted in a higher solar cell performance than the single-size QD sensitization, by broadening the absorption range and by effective filling of titania.

Most of the published work to date are dedicated to cadmium chalcogenides; however, there are also some studies on *proof of concepts*, employing broad-range absorbers, such as HgTe and PbS QDs, in solid-state DSSCs.

An inherent advantage of using PbS and PbSe QDs is the increased absorption capability in the NIR (near-infrared region) of sunlight. These particles have also gained considerable interest since reports on efficient multiple-exciton generation upon excitation with photons exceeding 2.5 $E_g$ have been published (Liu et al. 2004). A high-surface-area p–n heterojunction between $TiO_2$ and an organic p-type charge transport material (spiro-OMeTAD) have been sensitized to visible light using lead sulfide (PbS) QDs by Grätzel et al. PbS QDs have been formed in situ on a nanocrystalline $TiO_2$ electrode using chemical bath deposition techniques. The sensitized heterojunction showed (Jiu et al. 2006) incident photon-to-electron conversion efficiencies up to 45% and energy conversion efficiencies of 0.49% under simulated white light having low light intensities (10 mW/cm$^2$).

It has been demonstrated that the HgTe nanocrystals improve the photon-harvesting efficiency of hybrid solar cells over a broad spectral region between 350 and 1500 nm (Guenes et al. 2006b). In this study, solar cells are prepared both using HgTe nanocrystals deposited on nanoporous $TiO_2$ electrodes as well as by blending them into a hole-transporting polymer (Figure 8.11).

The fabrication of such solar cells requires two types of QDs: The first type (shortened as aqueous solution, AS), which is deposited on the $TiO_2$ electrode, contains carboxyl groups to anchor the titania surface and is preferably soluble in water to avoid dissolving effects during the deposition of the second layer. The second type (shortened as organic solution, OS) should be dispersive in polymer, and thus has to be soluble in organic solvents. Poly(3-hexylthiophene) (P3HT) has been used as the hole-transporting material.

P3HT has a broad absorption peak with a maximum at 553 nm, and is almost transparent to wavelengths longer than 670 nm. The reference sample, a thin film of P3HT

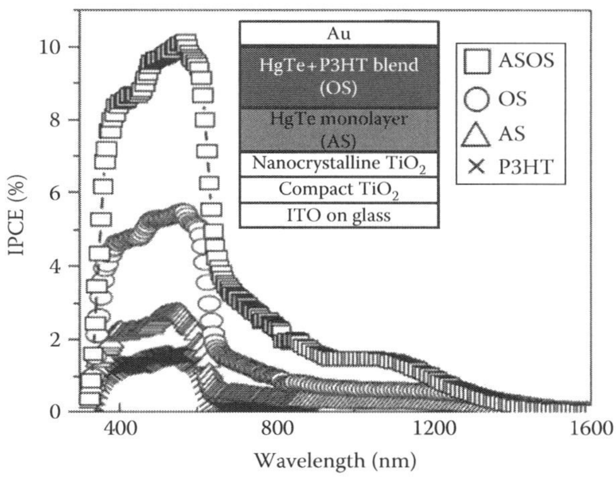


**FIGURE 8.11**
Device configuration and photoresponses of references P3HT, AS, OS, and the ASOS devices in solid-state DSSCs.

in between titania and an Au electrode showed a photoresponse only at wavelengths shorter than 650 nm and with a maximum IPCE of 1.6%. The HgTe nanocrystals, on the other hand, have shown an absorption peak at long wavelengths close to 1400 nm, as measured for the HgTe-OS nanocrystals (not shown here). This absorption corresponds to the fundamental optical transition across the bandgap of the HgTe nanocrystals with sizes in the range of 3–6 nm. Incorporating HgTe nanocrystals into the P3HT lead to an increase of the IPCE values and an extension of the spectral range of the photocurrent action spectra.

Solar cells using blends of P3HT and HgTe QDs (reference OS) had a maximum IPCE of 5.7% at wavelengths around 550 nm, and were sensitive to the spectral region up to 1500 nm. Similarly, solar cells using just sandwich structures of HgTe QD monolayers and P3HT (reference AS) had a broad IPCE with a maximum response of 2.5% at 550 nm. Combining both concepts, the device structure shortened as ASOS (aqueous solution–organic solution), the photoresponse is further increased, resulting in a maximum IPCE of 10.0% at wavelengths around 550 nm under simulated sunlight AM 1.5.

In the well-known blends of P3HT/PCBM for a BHJ device, PCBM with a higher electron affinity is the electron acceptor, while the hole remains within P3HT. PCBM absorbs quite weakly in the visible range. An alternative to PCBM to achieve effective harvesting of sunlight are broad-band-absorbing inorganic semiconducting nanoparticles. The main challenge in hybrid BHJ devices is related to the morphology of the blends together with the charge transport properties. Polymer domains must be limited in dimension to twice the exciton diffusion length. In addition, both the donor and acceptor phases must form high-quality percolation networks spanning the thickness of the devices to ensure efficient charge collection.

Typically, hybrid blends are co-deposited via spin casting from a solution containing solvent mixtures with different solubility parameters, such as pyridine and chloroform. An optimal mixture of these two solvents can stabilize both the organic and inorganic phases. A perfect solvent mixture has to be found out for every new material combination depending on the synthesis, especially the chemistry of the surfactants, as well as processing techniques. The random nature of the blend film limits the alignment of domains. Blend defects such as islands are common, negatively influencing the device behavior.

The first example of organic/inorganic BHJs has been presented by the Alivisatos group using CdSe nanoparticles in P3HT (Huynh et al. 2002). The spectral response of pure P3HT displays an absorption edge of 660 nm, beyond which the polymer is transparent to incident radiation. Any response from blend devices at wavelengths greater than 660 nm must therefore be the direct result of photocurrent generation in the nanocrystalline phase. Thus, the relative current contribution from this low-energy portion of a given spectrum directly reflects the degree to which carriers created in the CdSe are extracted from the device. Using CdSe nanoparticles, the light harvesting could be extended only up to 700 nm. The main output of the study was the relation between the nanoparticle shape and the corresponding cell performance. Changing the CdSe nanoparticle shape from spherical, with a diameter of 7 nm, to elongated, with aspect ratios of 7/30 and 7/60, a systematic development of device performance could be demonstrated. The best device performances with an external quantum efficiency of 1.7% have been observed for CdSe nanorods with an aspect ratio of 7/60.

A successful strategy in the next step was to use branched nanoparticles, allowing improved charge transport for an enhanced solar cell performance. The creation of 3D dentritic inorganic nanocrystals enables percolation within the device, which is less sensitive to thin film preparation parameters, because each hyperbranched particle contains a preformed percolation path. Blends of hyperbranched particles with semiconducting polymers can thus fully contribute to photovoltaic conversion, when the amount of the nanoparticles is high enough. Reaching a concentration threshold necessary for a complete percolation network of organic and inorganic domains, $V_{oc}$ immediately rises to its maximum value and remains constant with increased concentration of nanoparticles. An $I_{sc}$ of 7 mA/cm$^2$, a $V_{oc}$ of 0.60 V, and an FF of 0.55 have been observed using hyperbranched CdSe nanoparticles in a P3HT matrix. The power conversion efficiency was 2.2%, respectively (Gur et al. 2007).

A critical factor for hybrid materials used in the BHJ concept is related to charge transport. The organic surfactants surrounding the nanoparticle, which prevent nanoparticle agglomeration and promote the dispersion in polymer matrices, are insulating and may negatively influence the electrical transport between the nanoparticles. Consequently, in solar cells consisting of CdSe QDs with compact surfactants—for example, pyridine instead of TOPO—in P3HT, an increase in the external yield has been reported (Huyhn et al. 2003).

Much development is still needed to prepare intimate nanocomposites of conjugated polymers and semiconductor nanocrystals in hybrid solar cells. Although organic surfactants can facilitate the dispersion of nanocrystals in polymers, their presence severely reduces the device efficiency by impeding the transfer of charges between the nanocrystal and the polymer, as well as the transport of electrons between adjacent nanocrystals.

While surfactants can be stripped from the nanocrystals during film processing to afford a direct contact between the nanocrystals and the polymer, it is difficult to control the morphology and dispersion of nanocrystals within the polymer when using this process. To address this challenge, a polythiophene derivative containing amino end-functional groups that can effectively disperse CdSe nanocrystals to afford intimate nanocomposites with favorable morphology has been synthesized. The amino group acts hereby as the surfactant for inorganic nanorods. Solar cells show efficiencies up to 1.4% using this material combination (Liu et al. 2004).

Even though most studies are concentrated on utilizing CdSe nanoparticles in photovoltaics, CdTe nanocrystals present another alternative for hybrid solar cells. The bulk absorption edge of CdTe is at 820 nm, where most of the photon flux of the sun's radiation is centered. In comparison to bulk CdSe having an absorption onset at 720 nm, the bathochrome shift of the absorbance edge of 100 nm can improve the overall absorption of sunlight by ca. 20%. However, the studies concerning CdTe nanorods in semiconducting polymers resulted in poor solar cell performances because of the morphological problems in hybrid layers up to now (Kumar and Nann 2004).

The only example of a CdTe-based solar cell showing high performance has been demonstrated in 2005 for purely inorganic material combinations processed from the solution: BHJ by CdSe/CdTe blends. CdTe nanoparticles have been used as electron donors. Thin layers of CdTe/CdSe blends have been annealed at 200°C in air, resulting in an increase of photoconductivity. *I–V* characteristics for the best device using CdSe/CdTe nanoparticles fabricated to date, which employs a Ca top contact capped with Al, have a power conversion efficiency of 2.9%, with an $I_{sc}$ of 13.2 mA/cm$^2$, a $V_{oc}$ of 0.45 V, and an FF of 0.49. As expected, the spectrum reflects a strong redshift in the onset of photocurrent to the bulk absorption edge (Gur et al. 2005).

$CuInS_2$ (CIS) is one of the classical solar cell materials for the fabrication of thin film solar cells due to its high absorption coefficient, suitable low bandgap, and radiation stability. New developments based on CISe include thin film fabrication using electrochemical methods and spin casting from the solution. Quantized CIS nanoparticles capped with triphenylphosphite (TPP) as a surfactant could be formed into thin layers via spin casting from dispersions in acetonitrile (Czekelius et al. 1999). The absorption spectrum of films, prepared via spin coating from 0.06 M acetonitrile solutions of CIS dispersions, exhibits a high shoulder at 700 nm and a weak absorbance at longer wavelengths. The IPCE onset using a device structure of ITO/PEDOT:PSS/CIS/LiF/Al was comparable with the onsets of the corresponding absorbance spectra, indicating obvious photovoltaic activity of quantized CIS nanoparticles. Thin films of CIS have been prepared from blends using PEDOT:PSS as the polymer matrix. The strong phase separation between the nanoparticles and PEDOT:PSS resulted in the fabrication of PEDOT:PSS/CIS bilayers. IPCE extended up to 900 nm has been reported by Arici et al. (2003b) by using ITO/PEDOT:PSS/CIS/LiF/Al structures. The device performance was rather poor. For a large number of series, a $V_{oc}$ of about 250–450 mV and an $I_{sc}$ of about 0.05–0.2 mA/cm$^2$ have been reported, depending on the history of the sample. The FF was about 0.25.

In a double-layer configuration CIS/PCBM, an increase in the solar cell performance has been reported: an $I_{sc}$ of 0.26 mA/cm$^2$, a $V_{oc}$ of 710–790 mV, and an FF of 0.44. The power conversion efficiency was 0.08%, respectively (Arici et al. 2003). Next, a thin ruthenium dye interlayer has been placed between $CuInS_2$ and PCBM layers to increase light absorbance within the cell. For these devices, an $I_{sc}$ of 0.6 mA/cm$^2$, a $V_{oc}$ of 500 mV, and an FF of 0.5 have been reported (Guenes et al. 2006a).

An inherent advantage of using PbS and PbSe is the increased absorption capability in the NIR region of sunlight. These particles have gained also considerable interest since reports on efficient multiple-exciton generation upon excitation with photons exceeding 2.5 $E_g$ have been published (Schaller et al. 2004). Hereby, the absorption of one photon results in the formation of more than one electron–hole pair. Solar cell configurations using PbS in P3HT blends as well as their bilayer configurations P3HT on the top of PbS as active layers display very low efficiencies (–0.02%). The absorption range was extended up to 1145 nm, but still the cell parameters were not satisfying. An $I_{sc}$ of 0.13 mA/cm$^2$, a $V_{oc}$ of 400 mV, and an FF of 0.38 have been observed for these cells (Fritz et al. 2008).

With a band edge approximately at 1500 nm, bulk PbS cover almost all NIR radiation of sunlight. Experimental results demonstrate that the NIR irradiation could be successfully utilized for light harvesting. On the other hand, the energy mismatch of the HOMO-LUMO levels of the nanocrystals and the used semiconducting polymers leads to low solar cell performance. Since conjugated polymers typically have better hole mobility than electron mobility, photoconductivity in polymer/nanocrystal mixtures favors energy alignments that allows transfer of the photogenerated hole within the nanoparticle to the semiconducting polymer. This requires that the highest occupied orbital, HOMO, of the polymer (Scharber et al. 2006) (–5.1 eV for P3HT) lies closer to vacuum than the valence band of the nanoparticle. But the low ionization potential of PbS (–4.95 eV for bulk) limits the number of available polymers and leads to low solar cell performances when mixed with typical semiconducting polymers for photovoltaic applications.

With a bandgap of 0.66 eV, Ge is another candidate to harvest the infrared portion of the solar spectrum. An external quantum efficiency of 1.4% was measured for pristine P3HT, while 4% was measured by P3HT blends with Ge nanorods (Duan et al. 2000) at 550 nm. But still, optimization of the devices is required to obtain an overall increase in cell performances of these BHJ devices.

A simple procedure for preparing hybrid BHJ devices is to create a continuous interpenetrating network of a high-bandgap material, such as $TiO_2$ or ZnO, inside a thin conjugated polymer film only for an efficient formation of photoinduced charges.

ZnO nanoparticles have been introduced as environmentally friendly n-type materials with high electron mobilities in BHJ solar cells (Beek et al. 2005b). The simultaneous observation of radical cations in poly[2-methoxy-5-(3,7′-dimethyloctyloxy)-1,4-vinylene] (MDMO-PPV) and conduction band electrons in ZnO provided direct spectral evidence of a photoinduced electron transfer between these components. The effectiveness of this material combination is demonstrated by photovoltaic cells based on MDMO-PPV/ZnO nanoparticle BHJs in a classical BHJ solar cell design. The best performance is found for a blend with 30 vol% ZnO and a thickness of ~100 nm. These devices provide a photovoltaic effect with an estimated efficiency of 1.6%. The IPCE spectrum closely resembles the absorption spectrum of the blend and reaches a value of 40% at the absorption maximum of MDMO-PPV. A short current of 3.3 mA/cm$^2$, a $V_{oc}$ of 800 mV, and an FF close to 0.6 have been estimated by the integration of this spectral response with the solar spectrum under AM 1.5 conditions. Replacing nanoparticles with nanorods does not improve the performance, most likely because ZnO nanorods are more difficult to disperse in organic solvents. One drawback of these devices is that small amounts of UV light (<420 nm) result in a rapid degradation of the photovoltaic effect. The formation of n-doped ZnO nanoparticles by direct bandgap excitation and trapping of the holes is responsible for this effect.

In a similar way, titania nanoparticles are also used for BHJ solar cells (Van Hal et al. 2003). The best performance is found for a blend with 20 vol% $TiO_2$ and a thickness of ~100 nm. The IPCE spectrum closely resembles the absorption spectrum of the blend and reaches a value of 11% at the absorptions maximum of MDMO-PPV. A short current of 0.87 mA/cm$^2$, a $V_{oc}$ of 520 mV, and an FF of 0.42 have been estimated by the integration of this spectral response with the solar spectrum under AM 1.5 conditions. For a similar control device made using MDMO-PPV without any nanoparticles, the short-circuit current was negligible (<0.02 mA/cm$^2$).

## 8.3 Conclusions

We presented a review of the state of the art in organic/inorganic hybrid solar cells using nanocrystalline inorganic materials in two different functions: as anodes and as inorganic dyes in DSSCs as well as in BHJs.

Comparing the improvement in photocurrent using stretched nanostructures over the spherical nanoparticles demonstrates the measurable advantage of the nanotube structure not only because of the increased interface for charge separation, but also because of the optimized electron transport properties of the electrodes in DSSCs.

In view of the developments in the domain of hybrid materials in the last years, it can be speculated that a large variety of new hybrid combinations will be studied to achieve optimum morphologies and interfacial structures to improve hybrid solar cells. The ability to control precisely the surface modification and shape of the nanoparticles may lead to realize well-ordered nanostructures for efficient charge transport together with increased absorbance properties in hybrid solar cells. However, most of the published work to date is dedicated to the II-VI semiconductors, in particular, to cadmium chalcogenides, which have an extremely low acceptability for technical applications due to their toxicity. In the long term, it will be necessary to develop robust synthesis methods for alternative semiconductor materials.

## References


Adachi M., Murata Y., Takao J. et al. 2004. Highly efficient dye-sensitized solar cells with a titania thin-film electrode composed of a network structure of single-crystal-like $TiO_2$ nanowires made by the "oriented attachment" mechanism. *JACS* 126:45, 14943–14949.

Arici E., Meissner D., Schaffler F. et al. 2003a. Core/shell nanomaterials in photovoltaics. *Int. J. Photoenergy* 5:4, 199–208.

Arici E., Sariciftci N. S., Meissner D. 2003b. Hybrid solar cells based on nanoparticles of $CuInS_2$ in organic matrices. *Adv. Funct. Mater.* 13:3, 165–171.

Bawendi M. G., Steigerwald M. L., Brus L. E. 1990a. The quantum mechanics of larger semiconductor clusters (Quantum dots). *Annu. Rev Phys. Chem.* 41:477–496.

Bawendi M.G., Wilson W.L., Rothberg L. et al. 1990b. Electronic structure and photoexcited carrier dynamics in nanometer size CdSe clusters. *Phys. Rev. Lett.* 65:13, 1623–1626.

Beek W. J. E., Sloof L.H., Wienk M. M. et al. 2005a. Hybrid solar cells using a zinc oxide precursor and a conjugated polymer. *Adv. Funct. Mater.* 15:10, 1703–1707.

Beek W. J. E., Wienk M. M., Kemerink M. et al. 2005b. Hybrid zinc oxide conjugated polymer bulk heterojunction solar cells. *J. Phys. Chem. B* 109:9505–9516.

Borchert H., Shevchenko E.V., Robert A. 2005. Determination of nanocrystal sizes: A comparison of TEM, SAXS, and XRD studies of highly monodisperse COPt3 particles. *Langmuir* 21:5, 1931–1936.

Chapin D.M., Fuller C.S., Pearson, G.L. 1954. A new silicon p–n junction photocell for converting solar radiation into electrical power. *J. Appl. Phys.* 25:5, 676–677.

Chiba Y., Islam A., Watanabe Y. et al. 2006. Dye-sensitized solar cells with conversion efficiency of 11.1%. *Jpn. J. Appl. Phys. Part 2 Lett. Expr. Lett.* 45:24–28, 638–640.

Colvin V. L., Schlamp M. C., Alivisatos A. P. 1994. Light-emitting diodes made from cadmium selenide nanocrystals and a semiconducting polymer. *Nature* 370:354–357.

Cui D., Xu J., Zhu T. et al. 2006. Harvest of near infrared light in PbSe nanocrystal-polymer hybrid photovoltaic cells. *Appl. Phys. Lett.* 88:18, 183111–183113.

Czekelius C., Hilgendorff M., Spanhel L. et al. 1999. A simple colloidal route to nanocrystalline ZnO/$CuInS_2$ bilayers. *Adv. Mater.* 11:8, 643–646.

Dabbusi B. O., Bawendi M. G., Onitsuka O. et al. 1995. Electroluminescence from CdSe Quantum dot polymer composites. *Appl. Phys. Lett.* 66:1316–1318.

Dennler G., Sariciftci N. S., Brabec C. J. 2006. Conjugated polymer-based organic solar cells. ed. G. Hadziioannou and G.G. Malliaras. *Semiconducting Polymers: Chemistry, Physics and Engineering*, Wiley-VCH, Weinheim.

Duan X. F., Lieber C. M. 2000. General synthesis of compound semiconductor nanowires. *Adv. Mat.* 12:4, 298–302.

Dyakonov V., Frankevich E. 1998. On the role played by polaron pairs in photophysical processes in semiconducting polymers. *Chem. Phys.* 222:1/2, 203–217.

Flegler S. L., Heckman J. W., Klomparens K. L. 1995. *Scanning and Transmission Electron Microscopy: An Introduction*. New York, Oxford University Press.

Fritz K. P., Guenes S., Luther J. et al. 2008. IV-VI nanocrystal-polymer solar cells. *J. Photochem. Photobiol. A, Chem.* 195:1, 39–46.

Fujishima A., Zhang X. T. 2005. Solid-state dye-sensitized solar cells. *Proc. Jap. Acad. B* 81:2, 34–42.

Gadisa A., Svensson M., Andersson M. et al. 2004. Correlation between oxidation potential and open-circuit voltage of composite solar cells based on blends of polythiophenes/fullerene derivative. *Appl. Phys. Lett.* 84:9, 1609–1614.

Gadisa A., Mammo W., Andersson M. et al. 2007. A new donor-acceptor-donor polyfluorene copolymer with balanced electron and hole mobility. *Adv. Funct. Mater.* 17:18, 3836–3842.

Gebeyehu D., Brabec C., Sariciftci N. S. et al. 2002. Solid-state organic/inorganic hybrid solar cells based on conjugated polymers and dye-sensitized $TiO_2$ electrodes. *Thin Solid Films* 403–404, 271–274.

Green M. A., 2008. Third generation photovoltaics: Advanced solar energy conversion, Springer Series in Photonics, V 5.
Guenes S., Arici E., Neugebauer H. et al. 2006a. Nanoporous $CuInS_2$ electrodes for hybrid solar cells, ed. A. Gombert. *Proc. SPIE* 6197, 619714.
Guenes S., Neugebauer H., Sariciftci N.S. et al. 2006b. Hybrid solar cells using HgTe nanocrystals and nanoporous $TiO_2$ electrodes. *Adv. Funct. Mater.* 16:8, 1095–1099.
Gur I., Fromer N. A., Geier L. et al. 2005. Air-stable all-inorganic nanocrystal solar cells processed from solution. *Science* 310, 462–465.
Gur I., Fromer N. A., Chen C. et al. 2007. Hybrid solar cells with prescribed nanoscale morphologies based on hyperbranched semiconductor nanocrystals. *Nanoletters* 7:2, 409–414.
Hoppe H., Sariciftci N.S. 2004. Organic solar cells: An overview. *J. Mater. Res.* 19:7, 1924–1945.
Huynh W. U., Dittmer J. J., Alivisatos A. P. 2002. Hybrid nanorod-polymer solar cells. *Science* 295:5564, 2425–2427.
Huynh W. U., Dittmer J. J., Liby W. C., G. et al. 2003. Controlling the morphology of nanocrystal-polymer composites for solar cells. *Adv. Funct. Mater.* 13:1, 73–79.
Ito S., Zakeerudin S. M., Humphry-Baker R., Liska P. et al. 2006. High-efficiency organic-dye-sensitized solar cells controlled by nanocrystalline-$TiO_2$ electrode thickness. *Adv. Mater.* 18:9, 1202–1206.
Jiu J., Isoda S., Wang F. M. 2006. Dye-sensitized solar cells based on a single-crystalline $TiO_2$ nanorod film. *J. Phys. Chem. B* 110:5, 2087–2092.
Jun Y. W., Lee S. M., Kang, N. J. et al. 2001. Controlled synthesis of multi-armed CdS nanorod architectures using mono surfactant system. *J. Am. Chem. Soc.* 123:21, 5150–5151.
Jun Y. W., Choi J. S., Cheon J. et al. 2006. Shape control of semiconductor and metal oxide nanocrystals through nonhydrolytic colloidal routes. *Angew. Chem. Int. Ed.* 45:21, 3414–3439.
Kamat P. V. 2008. Quantum dot solar cells. Semiconductor nanocrystals as light harvesters. *J. Phys. Chem. C* 112:48, 18737–18753.
Kan S., Mokari T., Rothenberg E. et al. 2003. Synthesis and size-dependent properties of zinc-blende semiconductor quantum rods. *Nat. Mater.* 2:3, 155–158.
Kang S. H., Choi S., Kang M. S. et al. 2008. Nanorod-based dye-sensitized solar cells with improved charge collection efficiency. *Adv. Mater.* 20:1, 54–58.
Kashiwa Y., Yoshida Y., Hayase S. 2008. All-metal-electrode-type dye sensitized solar cells (transparent conductive oxide-less dye sensitized solar cell) consisting of thick and porous Ti electrode with straight pores. *Appl. Phys. Lett.* 92:3, 33308–33311.
Katari J.E.B., Colvin V.L., Alivisatos A.P. 1994. X-ray photoelectron spectroscopy of CdSe nanocrystals with applications to studies of the nanocrystal surface. *J. Phys. Chem.* 98:15, 4109–4117.
Kato T., Hayase S. 2007. Quasi-solid dye sensitized solar cell with straight ion paths—Proposal of hybrid electrolytes for ionic liquid-type electrolytes. *J. Electro-Chem. Soc.* 154:1, B112–B121.
Kongkanand A., Tvrdy K., Takechi K. et al. 2008. Quantum dot solar cells: Tuning photo response through size and shape control of CdSe-$TiO_2$ architecture. *J. Am. Chem. Soc.* 130:12, 4007–4015.
Kroeze J. E., Savenije T. J., Vermeulen M. J. W. 2003. Contactless determination of the photoconductivity action spectrum, exciton diffusion length, and charge separation efficiency in polythiophene-sensitized $TiO_2$ bilayers. *J. Phys. Chem. B* 107:31, 7696–7705.
Kroon J. M., Bakker N. J., Smith H. J. P. et al. 2007. Nanocrystalline dye-sensitized solar cells having maximum performance. *Prog. Photovoltaics* 15:1, 1–18.
Kroon R., Lenes M., Hummelen J. C. 2008. Small bandgap polymers for organic solar cells (polymer material development in the last 5 years). *Polymer Rev.* 48:3, 531–582.
Kruger J., Plass R., Grätzel M. et al. 2003. Charge transport and back reaction in solid-state dye-sensitized solar cells: A study using intensity-modulated photovoltage and photocurrent spectroscopy. *J. Phys. Chem. B* 107:31, 7536–7539.
Kumar S., Nann T. 2004. First solar cells based on CdTe nanoparticle/MEH-PPV composites. *J. Mater. Res.* 19:1990–1994.

Kuo C. Y., Tang W. C., Gau C. et al. 2008. Ordered bulk heterojunction solar cells with vertically aligned $TiO_2$ nanorods embedded in a conjugated polymer. *Appl. Phys. Lett.* 93:3, 33307–33310.

Lancelle-Beltran E., Prene P., Boscher C. et al. 2006. All-solid-state dye-sensitized nanoporous $TiO_2$ hybrid solar cells with high energy-conversion efficiency. *Adv. Mater.* 18:19, 2579.

Law M., Greene L. E., Johnson J. C. et al. 2005. Nanowire dye-sensitized solar cells. *Nat. Mater.* 4:6, 455–459.

Li B., Wang L. D., Kang B. N. et al. 2006. Review of recent progress in solid-state dye-sensitized solar cells. *Sol. Energ. Mater. Sol. Cells* 90:5, 549–573.

Lin S., Lee Y., Chang C. et al. 2007. Quantum-dot-sensitized solar cells: Assembly of CdS-quantum-dots coupling techniques of self-assembled monolayer and chemical bath deposition. *Appl. Phys. Lett.* 90:14, 143517.

Liu J., Tanaka T., Sivula K. et al. 2004. Employing end-functional polythiophene to control the morphology of nanocrystal-polymer composites in hybrid solar cells. *J. Am. Chem. Soc.* 126:21, 6550–6551.

Lloyd M. T., Anthony J. E., Malliaras G. G. 2007. Photovoltaics from soluble small molecules. *Mater. Today* 10, 34–41.

Moser J. 1887. Notiz über Verstärkung photoelektrischer Ströme durch optische Sensibilisirung. *Chem. Mon.* 8, 373.

Munters T., Martens T., Goris L. et al. 2002. A comparison between state-of-the-art 'gilch' and 'sulphinyl' synthesized MDMO-PPV/PCBM bulk hetero-junction solar cells. *Thin Solid Films* 403, 247–251.

Murphy C., Jana N.R. 2002. Controlling the aspect ratio of inorganic nanorods and nanowires. *Adv. Mater.* 14:1, 80–82.

Murray C. B., Norris D.J., Bawendi M.G. 1993. Synthesis and characterization of nearly monodisperse CdE(E=S, Se, Te) semiconductor nanocrystallites. *J. Am. Chem. Soc.* 115:19, 8706–8715.

Nazeeruddin M. K., De Angelis F., Fantacci S. et al. 2005. Combined experimental and DFT-TDDFT computational study of photo-electrochemical cell ruthenium sensitizers. *J. Am. Chem. Soc.* 127:48, 16835–16847.

Paulose M., Shankar K, Yoriya S. et al. 2006. Anodic growth of highly ordered $TiO_2$ nanotube arrays to 134 mm in length. *J. Phys. Chem. B* 110:33, 16179–16184.

Peet J., Kim J. Y., Coates N. E. et al. 2007. Efficiency enhancement in low-bandgap polymer solar cells by processing with alkane dithiols. *Nat. Mater.* 6:7, 497–500.

Pelet S., Moser J. E., Grätzel, M. 2000. Cooperative effect of adsorbed cations and iodide on the interception of back electron transfer in the dye sensitization of nanocrystalline $TiO_2$. *J. Phys. Chem. B* 104:8, 1791–1795.

Peng B., Jungmann G., Jäger C. et al. 2004. Systematic investigation of the role of compact $TiO_2$ layer in solid state dye-sensitized $TiO_2$ solar cells. *Coord. Chem. Rev.* 248:13–14, 1479–1489.

Peumans P., Uchida S., Forrest S. R. 2003. Efficient bulk heterojunction photovoltaic cells using small-molecular-weight organic thin films. *Nature* 425:6954, 158–162.

Plass R., Pelet S., Krüger J. et al. 2002. Quantum dot sensitization of organic-inorganic hybrid solar cells. *J. Phys. Chem. B* 106:31, 7578–7580.

Puntes V. F., Krishnan K. M., Alivisatos A. P. 2001. Colloidal nanocrystal shape and size control: The case of cobalt. *Science* 291:5511, 2115–2117.

Rand B.P., Burk D.P., Forrest S.R. et al. 2007. Offset energies at organic semiconductor heterojunctions and their influence on the open-circuit voltage of thin-film solar cells. *Phys. Rev. B* 75:11, 115327.

Ravirajan P., Peiro A. M., Nazeeruddin M. K. et al. 2006. Hybrid polymer/zinc oxide photovoltaic devices with vertically oriented ZnO nanorods and an amphiphilic molecular interface layer. *J. Phys. Chem. B* 110:15, 7635–7639.

Robel I., Subramanian V., Kuno M. et al. 2006. Quantum dot solar cells: Harvesting light energy with CdSe nanocrystals molecularly linked to mesoscopic $TiO_2$ films. *JACS* 128:7, 2385–2393.

Rogach A. L., Eychmüller A., Hickey S. G. 2007a. Infrared-emitting colloidal nanocrystals: Synthesis, assembly, spectroscopy, and applications. *Small* 3:4, 536–557.

Rogach A. L., Franzl T., Klar T. A. 2007b. Aqueous synthesis of thiol-capped CdTe nanocrystals: State-of-the-art. *J. Phys. Chem. C* 111:14628–14637.
Sariciftci N. S., Smilowitz L., Heeger A. J. et al. 1992. Photo-induced electron transfer from a conducting polymer to buckminsterfullerene. *Science* 258:5087, 1474–1476.
Saunders B. R., Turner M. L. 2008. Nanoparticle-polymer photovoltaic cells. *Adv. Colloid Interface Sci.* 138:1, 1–23.
Schaller R.D., Klimov V.I. 2004. High efficiency carrier multiplication in PbSe nanocrystals: Implications for solar energy conversion. *Phys. Rev. Lett.* 92:18, 18661.
Scharber M. C., Mühlbacher D., Koppe M. et al. 2006. Design rules for donors in bulk-heterojunction solar cells—Towards 10% energy-conversion efficiency. *Adv. Mater.* 18:6, 789–794.
Schmidt-Mende L., Bach U., Humphry-Baker, R. et al. 2005. Organic dye for highly efficient solid-state dye-sensitized solar cells. *Adv. Mater.* 17:7, 813–816.
Shaheen S., Brabec C. J., Sariciftci N. S. 2001. 2.5% efficient organic plastic solar cells. *Appl. Phys. Lett.* 78:6, 841–843.
Shen, Y. J. and Lee Y. L. 2008. Assembly of CdS quantum dots onto mesoscopic $TiO_2$ films for quantum dot-sensitized solar cell applications. *Nanotech.* 19:045602–045609.
Snaith H. J., Zakeeruddin S. M., Schmidt-Mende L. et al. 2005. Ion-coordinating sensitizer in solid-state hybrid solar cells. *Angew. Chem. Int. Ed.* 44:39, 6413–6417.
Spanggaard H., Krebs F. C. 2004. A brief history of the development of organic and polymeric photovoltaics. *Sol. Energ. Sol. Mater.* 83:2–3, 125–146.
Svensson M., Zhang F.L., Veenstra S.C. et al. 2003. High-performance polymer solar cells of an alternating polyfluorene copolymer and a fullerene derivative. *Adv. Mater.* 15:12, 988.
Talapin D. V., Rogach A. L., Kornowski A. 2001. Highly luminescent monodisperse CdSe and CdSe/ZnS nanocrystals synthesized in a hexadecylamine-trioctylphosphine oxide-trioctylphospine mixture. *Nanoletters* 1:4, 207–211.
Tennakone K., Hewaparakkrama K. P., Dewasurendra M. et al. 1988. Dye sensitized solid state photovoltaic cells. *Semicond. Sci. Technol.* 3:4, 382–387.
Toyoda T., Kobayashi J., Shen Q. 2008. Correlation between crystal growth and photo sensitization of nanostructured $TiO_2$ electrodes using supporting Ti substrates by self-assembled CdSe quantum dots. *Thin Solid Films* 516:9, 2426–2431.
Trupke T., Green M. A., Würfel P. 2002. Improving solar cell efficiencies by down-conversion of high-energy photons. *J. Appl. Phys.* 92:2, 1668–1674.
Van Hal P. A., Wienk M. M., Kroon J. M. et al. 2003. $TiO_2$ sensitized with an oligo(p-phenylenevinylene) carboxylic acid: A new model compound for a hybrid solar cell. *J. Mater. Chem.* 13:5, 1054–1057.
Xia J., Masaki N., Lira-Cantu, M. et al. 2008. Effect of doping on poly(3,4-ethylenedioxythiophene) as hole conductors in solid-state dye-sensitized solar cells. *J. Phys. Chem. C* 112:30, 11569–11574.
Xue J., Uchida S., Rand B.R. et al. 2005. A hybrid planar-mixed molecular heterojunction photovoltaic cell. *Adv. Mater.* 17:1, 66–69.
Yin Y., Alivisatos A. P. 2005. Colloidal nanocrystal synthesis and the organic-inorganic interface. *Nature* 437:7059, 664–670.
Yu G., Gao J., Hummelen J. C., Wudl F. 1995. Polymer photovoltaic cells: Enhanced efficiencies via a network of internal donor-acceptor heterojunctions. *Science* 270, 5244, 1789–1791.
Zayats M., Kharitonov A. B., Pogorelova S. P. et al. 2003. Probing photo-electrochemical processes in Au-CdS nanoparticle arrays by surface plasmon resonance: Application for the detection of acetylcholine esterase inhibitors. *J. Am. Chem. Soc.* 125:51, 16006–16014.
Zerza G., Cravino A., Neugebauer H. 2001. Photoinduced electron transfer in donor-acceptor double-cable polymers: Polythiophene bearing tetracyano-anthraquino-dimethane moieties. *J. Phys. Chem. A* 105:17, 4172–4176.
Zhu R., Jiang C., Liu X. Z. et al. 2008. Improved adhesion of interconnected $TiO_2$ nanofiber network on conductive substrate and its application in polymer photovoltaic devices. *Appl. Phys. Lett.* 93:1, 13102–13104.

# 9

# *Inorganic Nanoparticles and Rechargeable Batteries*

**Doron Aurbach and Ortal Haik**

**CONTENTS**

## 9.1 Introduction

In recent years, the scientific and technological communities in chemistry, physics, materials science, and even life sciences have faced and advanced the "nano" revolution, namely, increasing efforts to synthesize, explore, and find uses for nano-materials, including particulated matter with a particle size of $(1–1000) \times 10^{-9}$ m (size of molecules) and composite matter with features of that size (e.g., nano-porous materials). The nano revolution was promoted by the development of high-resolution microscopy (high-resolution transmission electron microscopy (HRTEM), high-resolution scanning electron microscopy (HRSEM), and scanning probe microscopy/atomic force microscopy (SPM–AFM), scanning tunneling microscope (STM), etc.), which enables the imaging of nano-materials and, hence, the ability to monitor size, morphology, and composition (through attached techniques such as energy dispersive x-ray spectroscopy (EDS), transmission electron microscope (STEM), and electron diffraction).

As listed below, the use of nanoparticles in devices for energy storage and conversion may have clear advantages and disadvantages:

**Advantages**

1. The nanosize means a short diffusion length for transport phenomena within the bulk materials.
2. The nanosize also means a high surface area, which facilitates interfacial charge transfer processes.
   These two features of nanoparticles mean a possible enhancement of the kinetics of processes due to their use.
3. In processes in which there are pronounced volume changes, resulting in stress and strain, the use of nanosize particles may help to accommodate too drastic morphological changes.
4. There are unique conversion reactions (see later in this chapter) that proceed only when nanoparticles are involved.
5. Nano-porous structures may provide a very high surface area for electrostatic energy storage, e.g., in super-capacitors.

**Disadvantages**

1. Nanoparticles may be more surface reactive than bigger particles and hence, parasitic side reactions, which lead to self discharge, may be more pronounced.

Therefore, the relatively high surface area of nanoparticles may have the disadvantage of super reactivity.
2. Nanoparticles are a challenge for composite electrodes. There are always intrinsic problems of interparticle electrical contact and the mechanical and electrical integrity of composite electrodes comprising nano-materials.
3. Nanoparticles usually have a low density and, hence, the use of nano-materials means giving up some specific volumetric energy density.
4. The synthesis of nano-materials may be more difficult than the synthesis of regular materials comprising particles of micrometric size.

The above list of pros and cons means that it is impossible to generalize the importance of nano-materials and their use in devices for energy storage and conversion. Hence, each case, namely, different batteries, super-capacitors, fuel cells, and solar cells, has to be discussed separately. It should be emphasized that in recent years we have seen intensive and extensive efforts of synthesize and explore nano-materials in connection with energy storage and conversion. Inorganic nanoparticles are examined as electrode materials in batteries. There is an impressive work on the elaboration of new electrolyte systems in which inorganic nanoparticles play an important role in enhanced ionic conductivity: nanoparticles are critically important as catalysts for the reduction of oxygen and the oxidation of fuels ($H_2$, $CH_3OH$, $CH_3CH_2OH$, etc.) in fuel cells. In addition, the high surface area of nanoparticles plays an important role in the efficiency of sunlight harvesting in photovoltaic solar cells.

In this chapter, we discuss separately, step by step, devices for energy storage and conversion. After providing the appropriate categorization of these systems, we discuss the relevance and importance of nano-materials for them. In Sections 9.2 through 9.4, we briefly review several devices for energy storage and conversion: batteries, super-capacitors, and photovoltaic cells. Fuel cells are treated herein as primary, flow batteries. The relevance of nano-materials to the various systems is discussed in brief therein. Then, several aspects of nano-materials and their use in batteries are discussed in depth in Section 9.5, which describes components (electrodes, electrolyte systems), and in Section 9.6 that discusses syntheses of nano-materials for batteries.

Section 9.7 discusses carbon and other nanotubes.

## 9.2 Batteries

### 9.2.1 Introduction

Batteries are a very abundant commodity used by everyone on a daily basis. It is impossible to imagine modern life without batteries, the so-useful energy storage devices. While batteries are usually simple for use, transport, and even manufacturing, they are very complicated electrochemical devices in which three bulk positive and negative electrodes, an in-between electrolyte phase, and two electrode–electrolyte interfaces have to work simultaneously. The chemistry of many battery systems may be very complicated, including, in addition to the main electrochemical reactions (that are the heart of their energy storage mechanisms), side reactions, and passivation phenomena. The science behind batteries involves, in addition to electrochemistry, materials and surface chemistry, physics, and transport phenomena: diffusion processes in both solid and liquid phases, and multiphase charge transfer by ions and electrodes through bulks and interfaces. Having superior battery chemistry is never sufficient for practical commercial success. The engineering side is critical for success, as is the proper choice of materials (in terms of purity and morphology), current collectors, separators, cases, and safety features. Therefore, it is impossible to develop battery systems based on trial-and-error work. The engineering of battery systems has to be based on solid scientific work in which the major chemistry, the transport phenomena, the side reactions, and the thermal behavior are fully understood. Maintaining a high level of scientific research in the field of batteries is extremely important for thinking ahead about all the possible failure mechanisms and safety problems. This is critically important for products such as batteries that are mass produced and widely distributed. A recall of batteries from the market due to unexpected safety or operational problems may be fatal for battery manufacturers. In order to discuss the possible use of nano-materials in batteries and related devices, it is important to introduce Ragone plots: power density vs. energy storage of devices for energy storage and conversion. Figure 9.1 shows a simplified version of these plots and Figure 9.2 shows specific curves related to several battery systems. Note the location of different types of batteries, fuel cells, and super-capacitors on the charts. As can be seen, there is always a compromise between power and energy densities, because for high power, high surface area is important while for high energy the compactness (low surface area) of the active mass is important. In general, we would like to use materials judiciously and to push these devices as much as possible toward the upper right side of the chart (high power and energy densities). The use of nano-materials may be critically important for achieving high power density.

There are many ways in which batteries and related devices can be categorized. These include rechargeability, the type of electrolyte system used, temperature range, and mode of operation (e.g., flowing vs. stationary systems). Fuel cells are in fact primary (non-rechargeable) batteries that exploit electrochemical combustion reactions. We differentiate among batteries that are redox systems, electrical double layer (EDL) capacitors based on electrostatic interactions, and solar cells that are batteries in which one of the electrodes (the negative side, the anode) converts radiation to charge transfer. Major battery systems are listed in Table 9.1.

A representative scheme of secondary batteries is illustrated in Figure 9.3. For general information on a wide variety of battery systems, see references (Peled and Yamin 1979;

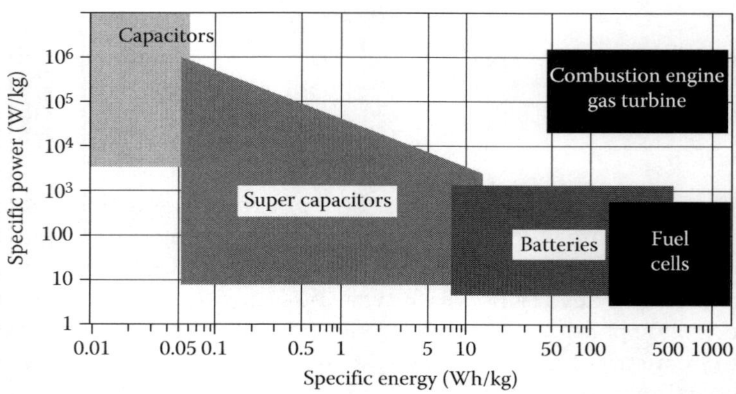


**FIGURE 9.1**
A simplified Ragone plot of the energy storage and conversion devices.

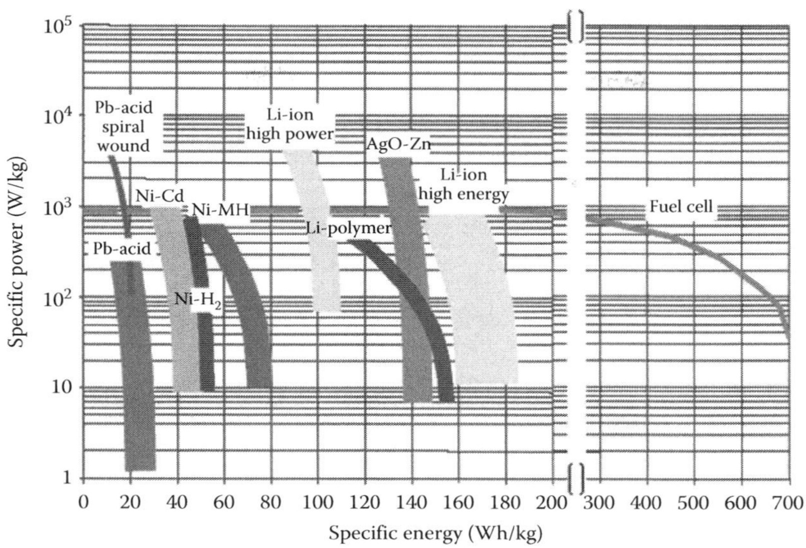


**FIGURE 9.2**
Ragone plot of various battery systems with a general comparison to fuel cells.

Linden 1994; Aurbach and Weissman 1999; Besenhard 1999; Gaberscek et al. 1999; Crompton 2000; Dell and Rand 2001; Singhal and Kendall 2003; Pistoia and Broussely 2007).

Aqueous batteries were the first to appear on the market. These systems include lead–acid, Ni–Cd, Ni–MH, and Zn-based batteries. Their electrolyte systems are based on water, which means that the voltage windows are limited by the electrochemistry of water and relatively low internal resistance due to the high ionic conductivity of aqueous solutions. Side reductions in these batteries relate to the electrolysis of water, i.e., oxygen and hydrogen evolution. The use of aqueous electrolyte solutions also means very limited passivation phenomena due to the high stability of a wide variety of metal oxides and hydroxides in water, especially in highly acidic or basic solutions. The sensitivity of aqueous systems to atmospheric contaminants is relatively low. We can mention the sensitivity of alkaline solutions to $CO_2$ (which reacts with $MO_x$ or $M(OH)_x$ to form insoluble carbonates) as a major issue. The nonaqueous systems include all types of lithium and lithium ion batteries, both primary and secondary, with liquid, polymeric, gel, and solid electrolytes. Further developments based on other reactive metals such as rechargeable magnesium and sodium batteries are also nonaqueous systems. The use of nonaqueous electrolyte systems is mandatory for high energy density systems in which highly reactive electrode materials have to be used. Reactive electrodes (Li, Li–C, Li–M, Mg, Na, low redox potential, $LiMO_x$) are stable in the relevant nonaqueous systems, due to passivation phenomena. Critical elements in these nonaqueous batteries systems are surface films that are formed on the electrodes due to spontaneous reactions with solution species, through which the active ions have to migrate. Nonaqueous batteries are highly sensitive to atmospheric contaminants (passivation phenomena). The limited ionic conductivity of many nonaqueous systems makes them critical factors that affect the internal resistance of the relevant batteries. Nonaqueous systems are usually the more complicated ones. Their characterization requires an understanding of electrode–solution interactions, surface chemistry, transport phenomena of Li-ion migration through surface films, and ion diffusion into lattices and a wide variety of possible side reactions and corrosion.

**TABLE 9.1**

Chemistry and Properties of Selected Battery Systems (Peled and Yamin 1979; Linden 1994; Aurbach and Weissman 1999; Besenhard 1999; Gaberscek and Stane 1999; Crompton 2000; Dell and Rand 2001; Singhal and Kendall 2003; Pistoia and Broussely 2007)

| Name | Chemical Reaction | Voltage (V) | Practical Energy Density (W h/kg) | Theoretical Energy Density (W h/kg) | Temperature Range $^0C$ | Cycle Life |
|---|---|---|---|---|---|---|
| *Primary batteries* | | | | | | |
| Leclanche | $Zn + 2MnO_2 \rightarrow ZnO + Mn_2O_3$ | 1.6 | 85 | 360 | −5 to 45 | — |
| Alkaline $MnO_2$ | $Zn + 2MnO_2 \rightarrow ZnO + Mn_2O_3$ | 1.5 | 125 | 335 | −20 to 55 | — |
| Magnesium | $Mg + 2MnO_2 + H_2O \rightarrow Mn_2O_3 + Mg(OH)_2$ | 2.8 | 100 | 760 | −20 to 60 | — |
| Mercury | $Zn + HgO \rightarrow ZnO + Hg$ | 1.34 | 100 | 255 | 0 to 55 | — |
| Silver oxide | $Zn + Ag_2O + H_2O \rightarrow Zn(OH)_2 + 2Ag$ | 1.6 | 120 | 290 | 0 to 55 | — |
| Zn/air | $Zn + 1/2O_2 \rightarrow ZnO$ | 1.65 | 340 | 1080 | 0 to 50 | — |
| $Li/SO_2$ | $2Li + 2SO_2 \rightarrow Li_2S_2O_4$ | 3.1 | 260 | 1175 | −55 to 70 | — |
| $Li/SOCl_2$ | $2Li + 2SOCl_2 \rightarrow 4LiCl + S + SO_2$ | 3.6 | 600 | 1500 | −55 to 70 | |
| $Li/MnO_2$ | $Li + MnO_2 \rightarrow Li^+ + MnO_2^-$ | 3.5 | 230 | 1000 | −20 to 55 | — |
| *Secondary batteries* | | | | | | |
| Lead–acid | $Pb + PbO_2 + 2H_2SO_4 \rightarrow 2PbSO_4 + 2H_2O$ | 2.1 | 35 | 250 | −40 to 60 | 200–800 |
| Nickel–cadmium | $Cd + 2NiOOH + 2H_2O \rightarrow 2Ni(OH)_2 + Cd(OH)_2$ | 1.35 | 30 | 245 | −40 to 50 | 200–700 |
| Nickel–metal hydride | $MH + NiOOH \rightarrow M + Ni(OH)_2$ | 1.35 | 50 | 280 | −20 to 50 | 300–600 |
| Sodium/sulfur (high temperature) | $2Na + 3S \rightarrow Na_2S_3$ | 2.1 | 170 | 790 | 310 to 350 | 600 |
| Li/ion | $x6C$ (graphite) $+ LiCoO_2 \rightarrow xLiC_6 + Li_{1-x}CoO_2$ | 4 | 90 | 500 | −20 to 55 | 500–1000 |

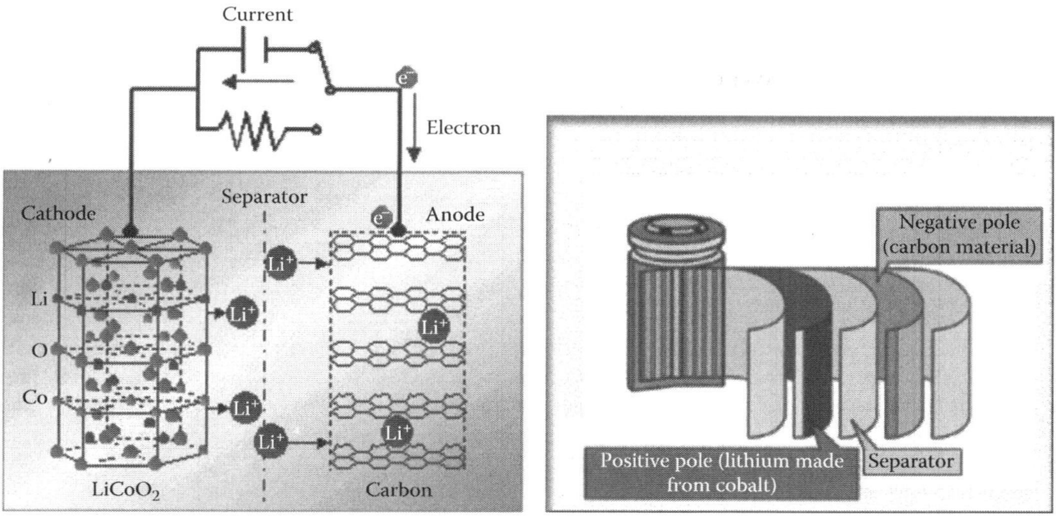


**FIGURE 9.3**
**(See color insert following page 302.)** A scheme of typical secondary battery: for example, Li-ion battery based on $Li_xC_6/Li_xCoO_2$ electrochemistry.

Most of the commonly used battery systems (both primary and secondary) are stationary. These include all types of Li and Li-ion batteries (and related active metal systems), lead–acid, Ni–Cd, Ni–MH, and alkaline batteries. The term "stationary" means that the active mass is finite, contained in the battery, and does not move during battery action. There are several types of flowing batteries:

1. Closed-cycle systems, i.e., batteries in which the active mass is finite and contained in the system. However, the active materials are flowing, continuously discharging at the working electrodes' assembly, but are charged outside the electrodes' assembly in a separate unit. Typical systems are Zn-$Br_2$ batteries (Aurbach et al. 1999) or redox batteries based on vanadium compounds ($V^{2+}/V^{3+}$ as the anodic reaction; $V^{4+}/V^{5+}$ as the cathodic reaction) (Aurbach et al. 2002). In these type of batteries, the use of nano-materials may not be important.
2. Open-cycle systems, i.e., batteries in which the active mass is not contained in the main device. All the air batteries and fuel cells belong to this category.

A critical component in batteries is the electrolyte system whose major role is to be the medium that closes the electrochemical circuit by ionic transport. In the aqueous systems, lead–acid and Ni–Cd, the solvent–water also participates in the overall electrochemical reactions of the cell. The electrolyte medium also has to act as a separator that prevents electrical short circuit between the electrodes. Batteries can be classified according to the type of electrolyte system used (i.e., normal ones operating with liquid electrolyte solutions and solid-state batteries).

The last mode of classification of batteries dealt with in this section relate to their temperature of operation. All the batteries comprising liquid electrolyte solutions, including the aqueous batteries (e.g., alkaline, lead–acid, Ni–Cd, Ni–MH) and most of the Li-ion and Li batteries, work at ambient temperatures. The implications of this fact for transport mechanisms in the electrolytic phase and across interfaces are safety, and ease of production and testing. High-temperature systems include thermal (reserve) primary batteries, sodium–sulfur rechargeable systems, Li batteries comprising solvent-free polymeric electrolytes, and high-temperature fuel cells.

### 9.2.2 Primary Systems

*An important general comment:* Most of the primary battery systems described below are relatively "old" and in use. The use of nano-materials for their electrodes may increase pronouncedly their performance in terms of rate capability. Moreover, since primary batteries are designed for a single discharging process only, many problems related to the use of nanoparticles, such as enhanced side reactions, maybe irrelevant to such systems. Nevertheless, since these systems have already found suitable applications and are commercial, we do not see intensive R&D efforts for their improvement by introducing nano-materials. The main efforts to incorporate nano-materials in batteries and related devices are devoted to rechargeable systems. Several important primary battery systems and related devices (fuel cells) are described below.

#### *9.2.2.1 Zinc Anode Batteries*

"Zinc anode batteries," a family of several members, utilizing various cathodes and electrolytes. The most important ones, commercially, are

1. "Alkaline": manganese oxide cathode. 1.5 V.

   Cathode reaction: $MnO_2 + e^- \rightarrow MnO(OH)$; electrolyte: aqueous KOH (alkaline solution).

2. "Leclanche": manganese oxide cathode. 1.5 V.

   Cathode reaction: $MnO_2 + e^- \rightarrow MnO(OH)$; electrolyte: aqueous $NH_4Cl$.

3. "Zinc–mercury": mercuric oxide cathode. 1.343 V.

   Cathode reaction: $HgO + 2e^- \rightarrow Hg$; electrolyte solution: aqueous KCl.

4. "Zinc–silver (primary)": silver oxide cathode. 1.6 V.

   Cathode reaction: $Ag_2O + H_2O + 2e^- \rightarrow Ag + 2\ OH^-$; electrolyte: aqueous KOH.

All the above cells harness the basic reaction scheme $Zn \rightarrow Zn^{2+} + 2e^-$ for the anode process.

#### *9.2.2.2 Lithium–Liquid Cathode*

"Lithium–liquid cathode," a family of several members, utilizing various liquid cathodic materials, either dissolved in the electrolyte, or constituting the electrolyte's solvent. The most important ones, commercially, are

1. "Lithium–thionyl chloride": thionyl chloride cathode. 3.67 V.

   Cell reaction: $4Li + SOCl_2 \rightarrow 4LiCl + SO_2 + S$; electrolyte: thionyl chloride as solvent, with $LiAlCl_4$.

2. "Lithium–$SO_2$": sulfur dioxide cathode. 2.9 V.

   Cathode reaction: $2Li + 2SO_2 \rightarrow 2Li_2S_2O_4$; electrolyte: acetonitrile/propylene carbonate mixture as solvent, with LiBr or $LiAsF_6$.

3. "Lithium–$SO_2Cl_2$": sulfuryl chloride cathode. 3.0 V. This system is less important at the moment.

All the above systems harness the basic reaction scheme $Li \rightarrow Li^+ + e^-$ for the anode process.

The family of lithium metal–liquid cathode batteries all exists in a metastable state. These cells are based on the passivation of the lithium metal by surface films comprising mostly LiCl. These surface films are electronically insulating, but Li-ion conducting and, hence, block the active metal from reacting with the liquid cathode materials (Peled and Yamin 1979). These systems were studied intensively and the relevant mechanisms of reaction, passivation, possible side reactions, aging, etc., were well understood. The surface films on lithium in Li–$SOCl_2$ batteries become thicker with the discharge of the battery, and thereby it was possible to develop instrumentation that measures the residual capacity based on impedance measurements (Gaberscek and Stane 1999).

#### 9.2.2.3 Lithium Batteries with Solid Cathodes

"Lithium batteries with solid cathodes," a family of several members, utilizing various solid cathodic materials as composite electrodes, as the cell depolarizer. The most important ones, commercially, are

1. "Lithium–manganese oxide": $MnO_2$ cathode. 3.0–3.6 V.

   Cathode reaction: $MnO_2 + Li^+ + e^- \rightarrow LiMnO_2$ (intercalation); electrolyte: mixtures of alkyl carbonates with ethers, e.g., propylene carbonate/dimethoxyethane, with $LiClO_4$.

2. "Lithium–copper oxide": CuO cathode. 2.4 V.

   Cathode reaction: $CuO + Li^+ + e^- \rightarrow LiCuO$ (intercalation); electrolyte: dioxolane with $LiClO_4$.

3. "Lithium–carbon monofluoride": $C_xF$ cathode. 3.2 V.

   Cell reaction: $CF_x + xLi^+ \rightarrow xLiF + C$ ($x = 0.5$–$1.0$); electrolyte: mixtures of alkyl carbonates with ethers, e.g., propylene carbonate/dimethoxyethane, with $LiBF_4$.

4. "Lithium–iodine": poly(2-vinylpyridine)-iodine complex cathode. 2.8 V.

   Cathode reaction: $P2VPnI_2 + 2Li^+ + 2e^- \rightarrow P2VP(n-1)I_2 + LiI$; electrolyte: LiI.

In these batteries, the anode is lithium metal: $Li \rightleftharpoons Li^+ + e^-$, the cathodic reaction is the reduction of the cathode material that is electrically compensated for by lithium intercalation, e.g., $MnO_2 + e^- + Li^+ \rightarrow LiMnO_2$ (except for lithium–iodine). The electrolyte solutions in all of these batteries are nonaqueous polar aprotic organic solvents with lithium salts in which the anion is big enough to allow the very good dissolution of the Li salt and ion separation in the solution (Aurbach and Weissman 1999). A high performance in terms of conductivity is obtained by using mixtures of solvents in which one of them is highly polar (e.g., cyclic esters (γ-butyrolactone) or alkyl carbonate (propylene carbonate) and, thereby, viscous, and the other one is a less polar, but low viscosity solvent, such as ether. There is special wisdom in the preparation of nonaqueous electrolyte solutions for primary Li batteries that has to take into account the voltage, rates needed, and the window temperature required.

#### 9.2.2.4 Reserve Batteries

"Reserve batteries," a family of various designs and chemistries, that, for the purpose of prolonging dramatically the shelf life, one or more of the cell components are separated until the battery is activated. The most common ones are

1. "Magnesium–cuprous chloride or silver oxide": Both are water or seawater activated.

   Anode reaction: $Mg_{(s)} \rightarrow + Mg_{(aq)}^{++} + 2e^-$

   Cathode reaction: $AgCl_{(s)} + e^- \rightarrow Ag_{(s)} + Cl_{(aq)}^-$

   Electrolyte: seawater.

2. "Zinc silver chloride": water or seawater activated

   Anode reaction: $Zn_{(s)} \rightarrow + Zn_{(aq)}^{++} + 2e^-$

   Cathode reaction: $AgCl_{(s)} + e^- \rightarrow Ag_{(s)} + Cl_{(aq)}^-$

   Electrolyte: seawater.

#### 9.2.2.5 Thermal Batteries

"Thermal batteries" are also considered as reserve batteries. They work at high temperatures using molten salts and electrolyte systems, and are designed to supply high power for short periods (mostly for military uses). Typical examples are Ca/LiCl–KCl/$WO_3$, Ca/LiCl–KCl/$CaCrO_4$, Mg/LiCl–KCl/$V_2O_5$, and Li/LiCl–KCl/$FeS_2$ batteries. At ambient temperature, the electrolytes are solid. The operation starts by igniting a thermal reaction, which melts the electrolyte, thus forming the electrochemical contact between the electrodes.

#### 9.2.2.6 Air Batteries

"Air batteries," a family of cells, utilizing various anodes, in conjugation with catalytic, semipermeable air cathodes, which use air–oxygen that diffuses as the cathode material. Hence, the cathode reaction is the catalytic reduction of oxygen. There are a variety of approaches as to how to obtain efficient oxygen reduction. Oxygen reduction has been the subject of extensive studies since the beginning of modern electrochemistry. In general, the air–electrode is a porous conductive matrix, e.g., carbon, to which catalysts are bound. These catalysts include Pt and Pt alloys (most efficient) and transition metal or transition-metal oxides or more complex compounds. The electrode has to be engineered to allow a free flow of air and good contact between the air, the electrolyte solutions, and the catalyst

(Pistoia and Broussely 2007) (three component systems). It should be emphasized that since the catalysis for oxygen reduction is heterogeneous, it is important to elaborate air electrodes with as high active surface area as possible. Therefore, the use of nano-materials (e.g., different types of metallic clusters) is highly important. The most important air batteries are Zn–air batteries, which are practical (Deiss et al. 2002). Al–air batteries are still under development (Yang and Knickle 2002). There are also recent attempts to develop even rechargeable Li–air batteries (D′ ebart et al. 2007).

### 9.2.2.7 *Fuel Cells*

"Fuel cells" are in fact air batteries in which the anode reaction is the oxidation of fuel, e.g., $H_2$ and $CH_3OH$ (which are the most important) (Andu′jar and Segura 2009; Kirubakaran et al. 2009). The main difference between fuel cells and air batteries is the fact that in the latter ones, the anode material is stationary and fixed in the battery (usually an active metal), while in fuel cells the anodic material is continuously fed to the anode side. Hence, fuel cells are primary batteries in which both the cathodic (air) and the anodic (fuel) materials are fed to the cell. The field of fuel cells is very broad, and it comprises many varieties of chemistries, electrolytes, separators, and catalytic electrodes. It is beyond the scope of this chapter to deal in depth with fuel cell technologies. In the center of fuel cells, there is the membrane–electrodes assembly (MEA) through which the fuel, air, and the combustion products flow. Here the electro-catalysis of both fuel oxidation (e.g., $H_2$, methanol, ethanol, ethylene glycol) and oxygen reduction is highly important and, hence, the use of nano-materials may be very helpful. The membranes in fuel cells, which are supposed to avoid crossover of fuel from the anode to the cathode (which neutralizes the effective electrochemical reaction that generates power), but yet have to allow the flow of ions such as protons and hydroxides, may also benefit the use of ceramic nano-materials embedded in polymeric matrices (for the promotion of high selectivity).

In general, review and discussion on fuel cell systems is far beyond the scope of this chapter. We mention here some aspects of low-medium (below 200°C) temperatures fuel cells. For both hydrogen and direct fuel cells that utilize acidic electrolyte systems, the best catalyst for fuel oxidation is platinum and its alloys (mostly platinum–ruthenium). The term direct fuel cells relate to systems that use fuels other than hydrogen, at low-medium temperatures. The most important one is methanol but there are recent efforts to use ethanol and ethylene glycol (Ren et al. 2000; Travitsky et al. 2009). For these latter systems, the amount of catalyst needed is at least an order of magnitude higher compared to hydrogen-based fuel cells. The use of nano-clusters of Pt or Pt alloys is highly important for high utility of the catalyst and for fast kinetics. However, the nanosize causes a severe problem of dissolution of the catalyst in the acidic media (Antolini et al. 2006; Basu and Choudhury 2007). The fuel cell (FC) community struggles with this problem, which is definitely one of the obstacles for wide commercialization of low-temperature fuel cells. The use of high-temperature fuel cells (e.g., solid oxide fuel cells, SOFC) alleviate this problem since at the high temperatures the Pt-based catalysts are not needed (Stambouli and Traversa 2002).

## 9.2.3 Secondary Battery Systems: How They Can Benefit from the Use of Nano-Materials

There are four major rechargeable battery technologies that are fully commercial and successful. These systems are briefly described below, with an emphasis on possible incorporation of nano-materials in them.

#### *9.2.3.1 Lead–Acid Batteries*

"Lead–acid" batteries: anode: Pb, cathode, $PbO_2$. 2.2 V. Uses sulfuric acid solution in water as the electrolyte. Electrode reactions:

$$\text{Anode: } Pb + H_2SO_4 \leftrightarrows PbSO_4 + 2H^+ + 2e^-$$

$$\text{Cathode: } PbO_2 + 2H^+ + H_2SO_4 + 2e^- \leftrightarrows PbSO_4 + 2H_2O$$

Overall reaction is

$$Pb + PbO_2 + 2H_2SO_4 \leftrightarrows 2PbSO_4 + 2H_2O$$

This technology is very old, used throughout the world, and has well-developed markets (e.g., traction of automobiles). Hence, traditional production utilizes "old," conventional formulae that do not contain nanoparticles. However, upgrading of this technology to load leveling (peak shaving) application such as storage of wind and solar energy has to bring it to prolonged cycle ability (from hundreds discharge–charge cycles possible today to thousands of possible cycles). For that purpose, novel materials have to be introduced. For instance, the use of carbon nanotubes (CNTs) as a critical component in the electrodes' active mass can improve the electrical and mechanical integrity of the electrodes and hence increase cycle ability. There is indeed work in this direction (Endo et al. 2001).

#### *9.2.3.2 Ni–Cd Batteries*

"Ni–Cd" batteries: anode: Cd, cathode, NiO(OH). 1.2 V. Uses KOH solution in water as the electrolyte. Electrode reactions:

$$\text{Anode: } Cd + 2OH^- \leftrightarrows Cd(OH)_2 + 2e^-$$

$$\text{Cathode: } 2NiOOH + 2H_2O + 2e^- \leftrightarrows 2Ni(OH)_2 + 2OH^-$$

The cathodic reaction is complicated and can be described as the insertion of a proton into protonated Ni oxide ($Ni^{3+} \rightarrow Ni^{2+}$ in the H–Ni–$O_2$ lattice). The solution is aqueous KOH, and the proton, in fact, comes from the decomposition of water:

$$H_2O \leftrightarrows H^+ + OH^-$$

This technology is well established and is used mostly for power tools. It is being replaced by Li-ion battery technology for all its applications and hence there is no room to discuss herein further development of these batteries.

#### *9.2.3.3 Ni–Metal Hydride Batteries*

"Ni–metal hydride" batteries (Ni–MH): anode: hydrogen sponge (e.g., $Ni_5La$), cathode, NiO(OH). 1.2 V. Uses KOH solution in water as the electrolyte. Electrode reactions:

Anode: $MH + 2OH^- \leftrightarrows M + H^+ + 2e^-$ (M is usually an alloy of $XY_5$, which can adsorb hydrogen and store it in a hydride state.)

$$\text{Cathode: } 2\text{NiOOH} + 2\text{H}_2\text{O} + 2\text{e}^- \leftrightarrows 2\text{Ni(OH)}_2 + 2\text{OH}^-$$

Note that in the above three systems, the solvent, $H_2O$, in fact participates in the electrochemical reaction. In addition, the electrolysis of water forming hydrogen and oxygen upon charging is possible.

This technology is still used in power tools, mobile electronic equipment, and hybrid electric vehicle (e.g., Prius/Hunda). It can benefit from the use of nanoparticles in both anode and cathode sides, in order to obtain very high rates.

However, due to the rare abundance and high price of the lanthanides used for the anode materials and the inferior energy density compared to Li-ion technology, it is expected that sooner or later these batteries will also be replaced by Li-ion batteries in most (if not all) of the applications. Hence, we do not discuss herein further development related to Ni–MH batteries.

#### 9.2.3.4 Li (Metal)—$MX_2$ Batteries

There have been several attempts to develop rechargeable Li batteries with metal as the negative electrode ($Li \leftrightarrows Li^+ + e^-$), transition metal oxides or sulfides as the positive electrodes, and liquid electrolyte solutions. Major systems suggested were

1. Li/alkyl carbonates, $LiAsF_6/MoS_2$ (MOLi Energy, Canada) (Baughman et al. 2002)
2. Li/THF, 2M-THF, Me furane, $LiAsF_6/TiS_2$ (JPL, United States) (Dan et al. 1995)
3. Li/dioxolane, tributylamine, $LiAsF_6/Li_{0.3}MnO_2$ (Tadiran, Israel) (Aurbach and Schechter 2004)

All these three systems suffered from limitations due to the high reactivity of Li metal toward the electrolyte solutions. The stability of Li electrodes in all the possible polar aprotic electrolyte solutions depends on their passivation by surface films (formed spontaneously due to reactions between Li metal and solution components). Hence, Li dissolution–deposition processes occur via Li-ion transport through surface films. This situation leads to nonuniform current distribution and dendritic Li deposition, which forms highly reactive Li deposits (Aurbach et al. 1998a). Repeated Li deposition–dissolution processes in batteries involve continuous reactions between Li and solution components. The electrolyte solution is simply consumed during prolonged charge–discharge cycling, especially at high rates. This considerably limits the cycleability of these systems, especially at high rates. This limitation prevented the commercialization of Tadiran's $Li–Li_{0.3}MnO_2$ batteries (Aurbach 2002). Attempts are now underway to develop rechargeable Li metal batteries using polymeric electrolytes (Appetecchi et al. 2005).

As discussed later in Section 9.5 (battery components), the use of nanoparticles as a component in composite electrolytes can be highly important for polymer batteries. There are two relevant polymeric systems:

1. Solvent-free, true polymeric electrolyte systems: polyethers such as polyethylene oxide (PEO) can dissolve Li salts and form electrolytic matrices. Reasonable conductivity with such systems is only obtained at elevated temperatures (practically >80°C). Fortunately, Li deposition at elevated temperatures has a much smoother morphology compared to that at ambient temperatures. Li metal electrodes can be cycled at very high efficiency many hundreds of times in systems

such as PEO–LiN($SO_2CF_3$). Relevant cathode materials for such systems are $VO_x$ compounds, e.g., $V_6O_{13}$ and $V_2O_5$, whose redox behavior (<4 V vs. Li/$Li^+$) falls well within the electrochemical window of poly-ethers (up to 4 V vs. Li/$Li^+$) (Appetecchi et al. 2005).

2. Gel electrolytes: These systems generally comprise Li salt polar aprotic solvent (usually alkyl carbonates) solutions, which are included in a polymeric matrix that is in fact used as a filler or a spaced. Typical polymers used are polyacrylonitrile (PAN) and polyvinylidene-diflouride-hexa-fluoro propene (PVdF-HFP). The use of such combinations moderates the reactivity of Li metal with the active compounds (solvent molecules, salt anions), leads to a smoother morphology of Li deposition–dissolution processes, and enables very flexible geometry for the batteries. We should also mention R&D of rechargeable Li metal batteries that use solid electrolytes such as ceramic materials or $Li_3PO_4$ (Takahara et al. 2004). Such battery systems are relevant for unique applications, in small sizes.

#### 9.2.3.5 Li-Ion Batteries

This technology has now become a leader in the rechargeable battery market. These batteries are based on two intercalation processes, as illustrated in Figure 9.3. The pristine electrodes are lithiated transition metal oxide cathodes ($LiCoO_2$ is the most commonly used cathode material), and carbonaceous anodes, usually graphitic material. The commonly used electrolyte solutions are a mixture of alkyl carbonates and $LiPF_6$ as the electrolyte (Linden 1994; Wakihara and Yamamoto 1998; Crompton 2000; Dell 2001; Singhal and Kendall 2003). The first charging process leads to the de-intercalation of Li ions from $Li_xCoO_2$ and their intercalation into graphite to form a $LiC_6$ compound. Hence, the battery processes are

$$\text{Anode: } LiC_6 \leftrightarrows C_6\,(\text{graphite}) + Li^+ + e^-$$

$$\text{Cathode: } Li_xCoO_2^+ + (1-x)Li^+ + (1-x)e^- \leftrightarrows LiCoO_2$$

The electrolyte solution is not involved in the electrochemical reactions, and thus its role is that of a carrier of Li ions, and moves repeatedly between the electrodes upon cycling (hence the name "rocking chair batteries") (Scrosati 1995). It should be noted that these systems are, in fact, very complicated. Both electrode materials ($C_6$, $Li_xCoO_2$) are reactive with all relevant electrolyte solutions, and their stability depends on passivation processes. Surface films are formed through which Li ions have to migrate during the course of the electrochemical processes. There are also very complicated solid-state diffusion processes of Li ions in these materials (Aurbach 2002).

There are complications related to both high and low temperatures, safety issues, rate capability, self-discharge, and cycle life. The development and commercialization of Li-ion batteries can be marked as one of the most important successes of modern electrochemistry. Highly intensive R&D efforts are taking place world wide by research groups, both in industry and academia, to develop new electrode materials, new salts, and new solvents (e.g., ionic liquids) in order to push these systems to very high energy and power densities and, hence, to very demanding applications such as electric vehicles and spacecraft. Thus, this area is vital and provides challenges for the materials science and electrochemical science communities. *It should be noted* that there are intensive efforts to synthesize and incorporate nano-materials in Li-ion batteries, in order to enhance their rate capability for electric vehicle (EV) applications. This is in fact one of the most important frontiers for the use of nano-technology for energy storage and conversion, as is discussed in details in Section 9.5.

#### *9.2.3.6 Other Systems*

Attempts are underway to develop rechargeable sodium–sulfur batteries (high temperature systems) in which the electrolyte system is ceramic, e.g., $\beta Al_{2O3}$ doped with $Na^0$, which conducts $Na^+$ ions at high temperatures (Oshima et al. 2004). In recent years, there have also been attempts to develop rechargeable magnesium batteries (Aurbach et al. 2002). For instance, Mg metal electrodes behave fully reversibly in solutions comprising ether solvents such as THF or glymes (e.g., $CH_3$–$(OCH_2CH_2)_4$–$OCH_3$) and complexes of the $[R_2Mg]_x \cdot [AlCl_{3-n}Rn]_y$ type, R = alkyl or aryl groups. These solutions have an electrochemical window of more than 2 V. It was also demonstrated that Chevrel phases, e.g., $Mo_6S_8$, intercalate $Mg^{2+}$ ions reversibly:

$$Mo_6S_8 + 4e^- + 2Mg^{2+} \leftrightarrows Mg_2Mo_6S_8$$

R&D of practical, rechargeable batteries is in progress. Rechargeable magnesium batteries are also an important challenge for utilizing nanotechnology. Anodes, cathodes, and electrolyte solutions for them should not develop side reactions, passivation phenomena, etc. in which nanoparticles may have a very negative role due to their high surface reactivity (as is the case in Li-ion batteries). Recent work on rechargeable Mg battery systems (Aurbach et al. 2007; Mitelman et al. 2007) demonstrates the benefit of using nanoparticles of $Mo_{6S8}$ and $Mo_6S_{8-x}Se_x$ (Chevrel phases) as superb cathodes for rechargeable Mg batteries (see details on synthesis and characterizations therein (Aurbach et al. 2007)).

## 9.3 Electrical Double Layer Capacitors

In order to provide a comprehensive discussion on battery systems, it is important to mention another type of devices for energy storage and conversion, which are very similar to batteries in terms of development, fabrication, materials, and electrochemical science, namely, EDL (super) capacitors (Conway 1999): Figure 9.4 presents schematically a typical super-capacitor. These devices are based on electrodes that store energy by electrostatic means, in liquid or gel electrolytes. In this respect, EDL capacitors usually belong to ambient temperature systems based on liquid electrolyte solutions. EDL capacitors may use aqueous or nonaqueous electrolyte solutions. The advantage of the latter systems is their possible wide electrochemical window. The energy density of EDL capacitors is a function of $E^2$ (E is the electrode's potential). However, the ionic conductivity of nonaqueous systems may be lower by two orders of magnitude compared to aqueous solutions. Thereby, EDL capacitors based on polar aprotic systems are much slower (i.e., lower power density) compared to aqueous systems (Aurbach et al. 1998a,b). We can divide super-capacitors into two categories: pure EDL systems, in which the storage and release of charges is purely electrostatic. Such systems are expected to have the major advantage of an extended cycle life (no detrimental side reactions that destroy the electrodes upon prolonged cycling).

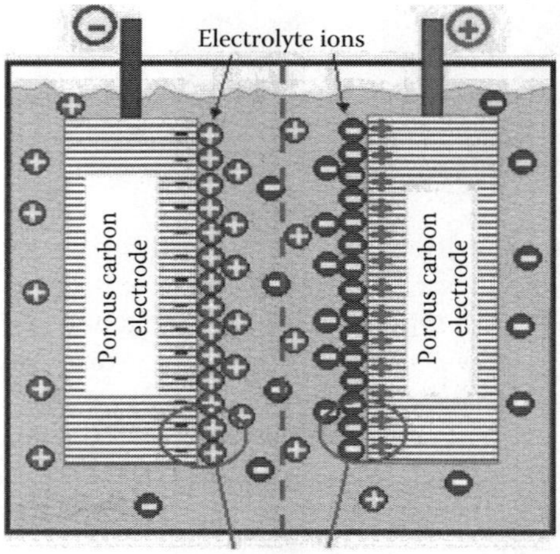


**FIGURE 9.4**
A typical scheme of a supercapacitor.

In such systems, the key factor is a high electrode surface area in order to obtain reasonable capacities. The second type involves pseudo-capacitive electrodes in which the charge storage and conversion is not purely electrostatic, and involves some degree of charge transfer between the electrode(s) and the solution. Typical advantages are super-capacitors comprising metal oxide electrodes, e.g., RuO (Zheng et al. 1995), $MnO_2$ (Subramanian et al. 2006), and other transition metal oxides, which exchange protons with solution phase in their charge–discharge processes. The performance of pseudo-capacitors can benefit a lot by the use of nanoparticles, since their high specific surface area facilitates the kinetics of the redox processes on which their energy storage is based.

Highly important are "real" EDL capacitors, which comprise activated carbon electrodes. Carbon has a number of allotropic forms including diamond, graphite, fullerenes, glassy CNT (single-wall and multi-wall forms, SWCNT and MWCNT respectively), and amorphous carbon. The latter two forms may be highly important for electrostatic energy storage. The high aspect ratio of nanotubes leads to their very high specific surface area suitable for adsorption (specially in spaghetti or brush-type arrangements). Amorphous carbons can be easily formed by carbonization of a large variety of organic precursors (compounds, polymers, residue of distillation of oil). Then, highly porous, activated carbons can be obtained by a mild oxidation. For instance, the following reaction $2C + CO_2 \rightarrow 2CO$ can produce activated carbons with a specific surface area >2000 $m^2/g$. Here, the internal nanostructure may be critically important, since it determines the type of adsorption interactions between the ions and pore walls. Figure 9.5 demonstrates schematically approaches for the preparation of activated carbons with unique properties. It is possible to start with meso-porous carbons and to obtain nano-metric fractal structure that enables effective, highly fast ions adsorption at high specific capacity (top-down approach). It is also possible to obtain activated carbons with molecular sieving properties, starting with meso-porous materials which undergo carbon vapor deposition that partially closes their pores' openings (bottom-up approach). Figure 9.6 also shows typical responses of these carbons in their voltammetric behavior, electro-adsorption kinetics (solution phase), and gas adsorption/desorption thermodynamics and kinetics.

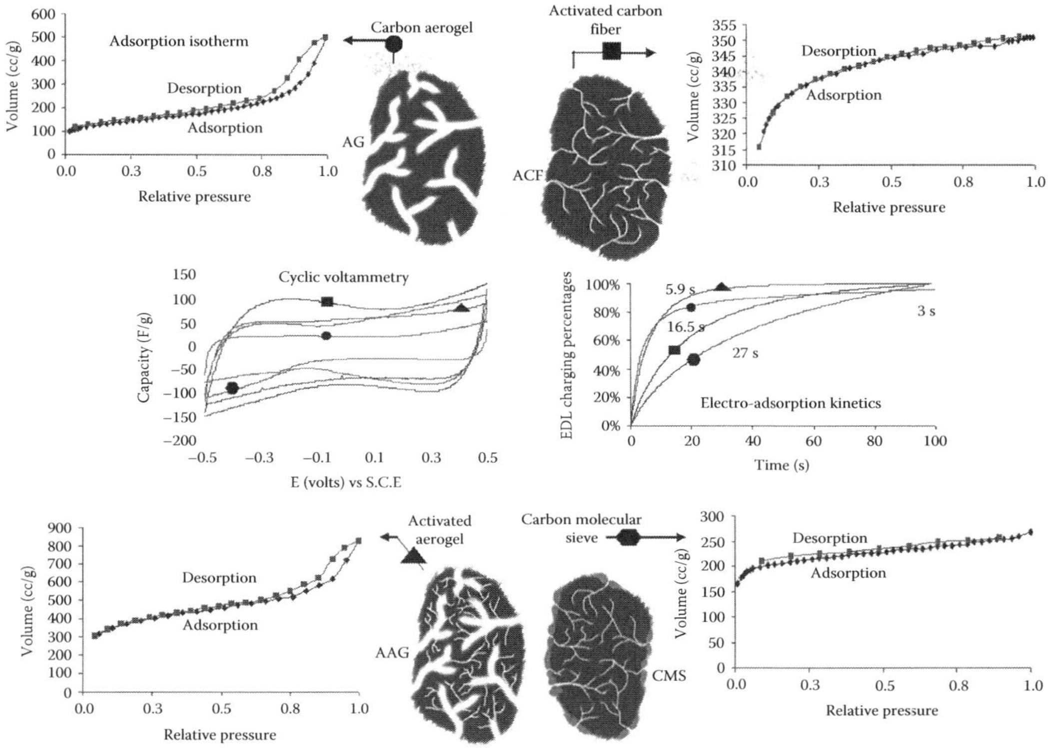


**FIGURE 9.5**
Presentation of four types of activated carbon that can be used for electrostatic energy storage and conversion, EDL capacitor devices. *Upper right*: High surface area activated carbon with micro-porous structure. *Down right*: Molecular sieving activated carbon obtained by applying carbon CVD onto the high surface area activated carbon, which partially closes their openings. *Upper left*: Meso-porous activated carbon (fast adsorption kinetics). *Down left*: High surface area, activated carbon with fractal porous structure, obtained by a further activation of the meso-porous carbon. (High surface area and fast kinetics.)

Figure 9.5 shows three types of responses: adsorption/desorption isotherms of nitrogen at 77 K, voltammograms translated to specific capacity vs. potential and electro-adsorption kinetics ($Cl^-$ ions) in aqueous NaCl solutions.

Hence, super-capacitors, EDL capacitors, provide the following challenges for nano-materials:

1. Preparation of amorphous carbon micro-particles with nano-porous structure, in which electro-adsorption should be optimized in terms of maximal surface area and ions-pores interactions → optimal capacity; fast adsorption kinetics via the appropriate fractal pores' structure. The synthetic challenges here include the right choice of carbonaceous precursor and the appropriate carbonization process. Then, it is critical to choose the appropriate activated processes and surface treatments such as deposition of thin carbon layers by CVD processes and formation or elimination of surface groups (Avraham et al. 2008; Polak et al. 2008).
2. Preparation of electrodes based on CNTs to which a separate section is dedicated later in this chapter.

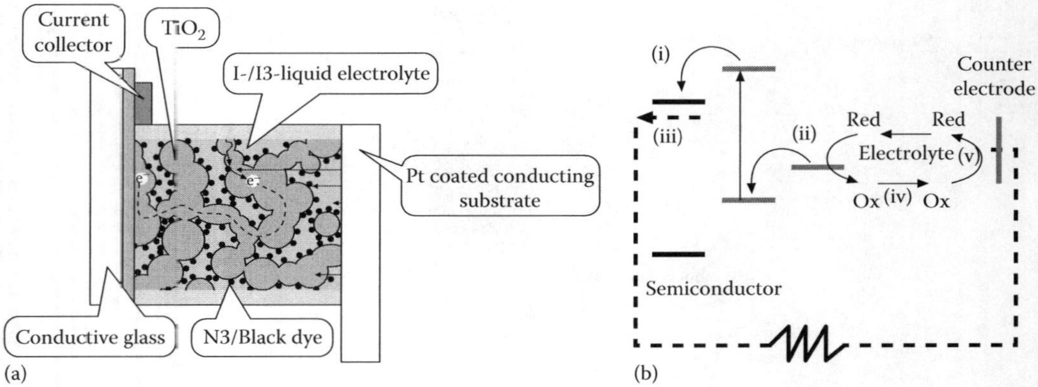


**FIGURE 9.6**
A schematic cross section of a dye-sensitized solar cell consisting of a nano-porous electrode, sensitizing dye, hole conducting mediator and a counter electrode (a) and a basic energy diagram of the dye-sensitized solar cell operation (b).

## 9.4 Photovoltaic Cells

We limit the scope of this section to photovoltaic cells that are, in fact, batteries, where one of their poles (a source of electrons, the anode) is a light-sensitive matrix that releases electrons due to interactions with light of the relevant (specific) range of wavelengths. The other components are an electrolyte solution (or solid) that contains a redox couple, and the electrochemical, cathodic pole, usually an inert metal, on which the counter reaction takes place. Highly important and promising are the dye-sensitized photovoltaic cells, which typical illustration is provided in Figure 9.6 (Grinis et al. 2008a; Dor et al. 2009).

Figure 9.6a presents their schematic structure and Figure 9.6b shows their relevant energy diagram. For this technology, the use of nanoparticles is critical, in order to obtain high surface area, high reactivity, and, hence, as high as possible efficiency of sunlight harvesting. The anode comprises sintered $TiO_2$ nanoparticles covered by a thin layer or by nano-clusters of organic dye (Grinis et al. 2008a,b). The latter absorbs light and becomes electronically excited. The excited dye transfers its excess energy to the semiconducting $TiO_2$, leading to the release of the electrons (the cell's anodic process). The excited state of the dye has to match the band gap of the major, semiconducting anode material. In turn, the relevant potentials of the redox couple in the electrolyte system (solution or polymeric) have to match the potential of the holes formed by the electrons' withdrawal. The holes thus formed in the anode matix oxidize the reductant in the solution. The circuit is closed by the reduction of the oxidant at the positive pole (produced by oxidation of the reductant).

The basic system contains a wide band gap semiconductor electrode, dye that is attached to the semiconductor, redox electrolyte and a counter electrode (platinum). Upon illumination of the sensitized solar cell, an electron is injected from the dye into the semiconductor film (process (i) in Figure 9.6b). Following the injection, a hole is transferred to the redox couple in the electrolyte, thus regenerating the dye (Figure 9.6b, process (ii)). The injected electron must cross the semiconductor layer and reach the conductive substrate (Figure 9.6b, process (iii)) while the oxidized ions diffuse toward the counter electrode (Figure 9.6b, process (iv)) where they are reduced to their original state by the electron traveling through the external circuit (Figure 9.6b, process (v)). Consequently, there is no net change in the system (Rydh and Sanden 2005).

Some major topics related to these systems are listed below:

1. A key factor is, of course, the efficiency of light trapping, namely, how much of the radiation energy can be translated to electrical energy.
2. It is important to choose the right materials with the band gaps correlated to the main radiation of the sun (visible light).
3. It is important to use high surface area materials, nanoparticles, yet still maintaining the appropriate interparticle electrical contact. The electrode matrix has to be transparent to light.
4. The choice of dyes, coating parameters, stability, and specific light absorption coefficients is critical.
5. A major problem in photovoltaic cells is the recombination between holes and electrons in the photoactive material. It is important to achieve maximal separation between electrons transmitted due to excitation by irradiation to the load circuit and holes that should interact only with the solution species.
6. In general, the optical properties of the electrodes, determined by the choice of materials and good engineering, are the most important parameters dealt with in connection with photovoltaic cells.

In summary, dye-sensitized photovoltaic cells are highly promising devices for a cost-effective harvesting of solar energy. A key factor that determines how much material should be needed for achieving a certain power is the collection efficiency: how many effective electrons that can produce electrical work are released vs. photons of sun light per area.

It should be emphasized that the use of nano-materials is critical for the possibility to develop efficient solar cells. See further descriptions and discussions in references (Andrade et al. 2009; Choi et al. 2009).

## 9.5 Battery Components: A Challenge for Inorganic Nano-Materials

### 9.5.1 Introduction

In this section, we describe components of batteries in connection to nano-materials. The main emphasis is on Li-ion batteries because this is in fact one of the most important topics related to energy storage and conversion, because of their wide applications, high energy density, and the challenge of power sources for electric vehicles, which only advanced Li-ion batteries may be able to meet. Anode and cathode materials, electrolyte systems, and even separators for Li-ion batteries are all challenges for nano-materials and nanotechnology.

### 9.5.2 Electrode Materials

#### *9.5.2.1 General Features*

Battery electrodes may be divided into four categories:

1. Metallic materials (usually electronegative, anode material).
2. Composite structures, which includes the active mass, conducting additives, binder, and current collector (the most commonly used in battery systems).
3. Catalytic materials, which include supporting porous material and dispersed catalyst particles, embedded in the supporting matrix (usually carbonaceous materials).
4. Monolithic materials, thin film of the active mass sputtered/deposited on a current collector. Such electrodes are used in microbatteries, where the amount of active mass being used is small (usually solid state systems).

Major parameters of interest are theoretical capacity (mA h/g), potential, and rate capability. The latter parameter can be expressed in capacity vs. current density (A/g) or vs. C-rates: $xC$ or $C/x$ ($xC$ means $x$ times the full capacity discharged during 1 h, and $C/x$ means discharging the full capacity during $x$ h). For rechargeable systems, stability, cycle life, and capacity-fading (e.g., % capacity loss per cycle) are also important parameters. The shape of the voltage profile during galvanostatic discharge (or charging) is also important. For instance, when the electrode processes are metal dissolution or intercalation via phase transition, the potential profile is flat (plateau), while when solid solutions or concentration gradients are formed (in the electrode and/or in solution), the potential profile is sloping. The most commonly used electrodes in batteries have a composite structure (No. 2 above), as illustrated in Figure 9.7. The active mass is a powder that has to be spread on a metallic/inert current collector, bound with a polymeric binder that keeps the system integral. The uniform electrical contact with composite electrodes is maintained by electrically conductive additives, usually carbon particles. In most cases, the percentage of active mass binder and conductive additives in composite electrodes is 85%–80%, 5%–10%, 5%–10%, respectively (by weight). Relevant technical aspects and questions related to composite electrodes are listed below.

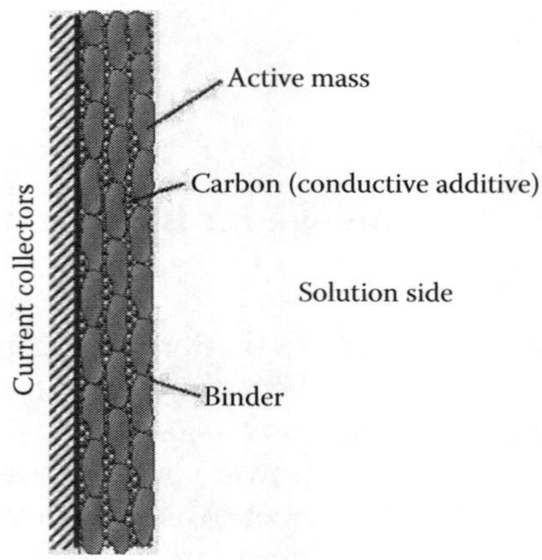


**FIGURE 9.7**
A scheme of a composite electrode.

1. The particle size and surface area are very important. There are two main processes that can be rate determining: interfacial charge transfer and bulk transport (e.g., solid-state diffusion in intercalation processes). Hence, the smaller the particles, the higher is their surface area and the smaller is their diffusion length. Therefore, the kinetics of their electrochemical reaction should be higher. However, too small particles means problems with interparticle–electrical contact, and a high surface area means a better chance for side reactions. Thereby, the particle size of the active mass in composite electrodes has to be optimized.
2. The electrode's thickness is important and has to be optimized to the relevant needs. A thick active mass means a higher capacity per unit area, but also sluggish kinetics due to a limited contact between the entire active mass and solution. For high rates, thin electrodes should be better, at the expense of lower practical capacity.
3. The porosity of composite electrodes is important. High porosity enables good contact between the active mass and solution at the expense of good interparticle contact. Hence, the average practical density of pressure to be applied during their fabrication in order to compress the active mass is a matter of optimization.
4. The electrodes active mass in many battery systems, especially when the solutions are nonaqueous (Li, Li-ion batteries), may react spontaneously with solution species, thus forming surface films that cover the particles. Hence, in many battery electrodes, the general charge transfer process includes the necessary step of ion migration through surface films (as is the case for most of the electrodes for Li and Li-ion batteries).
5. The contact between the active mass and solution is critical. Thus, the properties of both electrodes and solutions have to be optimized to obtain the sufficient wetting of the active mass. In the case of gel, polymeric, and other solid-state systems, the electrodes should also contain the electrolyte system, built in, as part of the composite structure.
6. The use of nanoparticles and/or nanostructured active mass in electrodes adds unique energetic factors due to surface tension and further electrostatic storage capability (in addition to the storage capability due to the redox activity of the active mass). These aspects were very well addressed and explored in the monumental work of Maier et al. in this area (Guo et al. 2007; Hu et al. 2007, 2008; Maier 2007).

Figure 9.8 presents a map of relevant, commonly used (state of the art) electrode materials that are used in Li-ion batteries. The coordinates in the figure are voltage and capacity.

Figure 9.9 shows the road map for development and introduction of new and novel electrodes' materials for advanced Li-ion batteries. The arrows in the figure emphasize where nano-materials are important.

The challenge for advanced rechargeable Li-ion batteries is to increase their energy and power density, in order to promote their use for EV applications. The energy density of Li-ion batteries can be increased (compared to currently used battery system) by development of high-capacity electrodes (anodes, cathodes) and high-voltage cathodes. Replacing graphite anodes by conversion reaction or silicon electrodes (see next section) can increase the capacity of the negative electrodes by two to threefold. In the positive side, there are $LiMn_{1.5}Ni_{0.5}O_4$ spinel (Park et al. 2007) and $LiCoPO_4$ olivine (Lee 2007) which redox activity approaches 5 V vs. $Li/Li^+$ and new $Li_2MnO_3 \cdot Li[MnNi]O_2$ (layered) composite materials that exhibit very high capacities (>200 mA h/g) (Sun et al. 2009). For

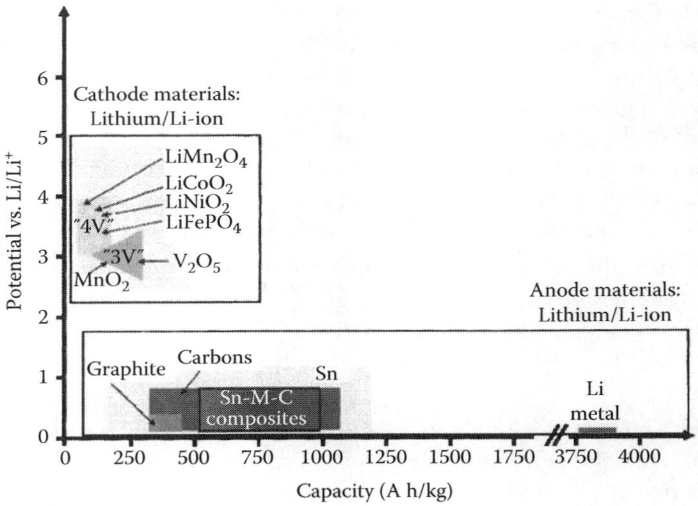


**FIGURE 9.8**
A schematic map of relevant anode and cathode materials used to date in Li-ion batteries. The ordinate presents voltage and the abscissa presents capacity.

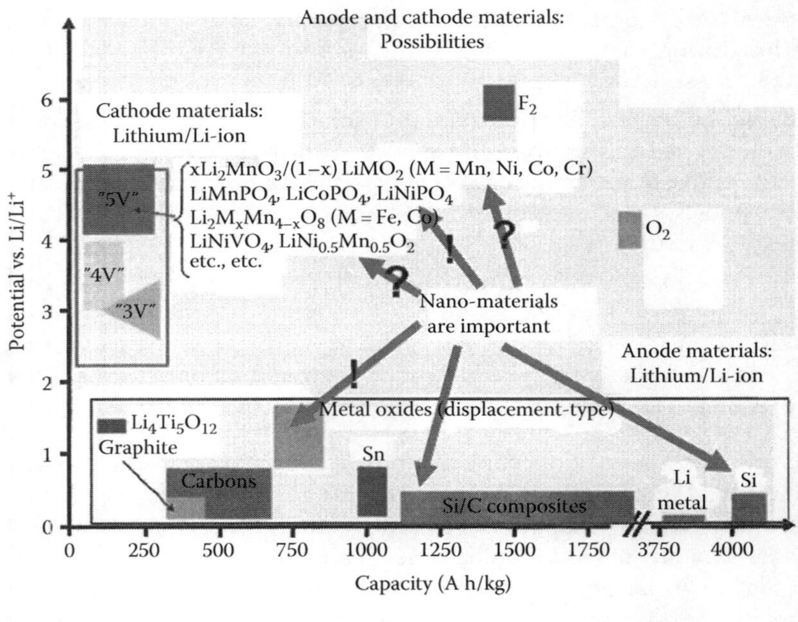


**FIGURE 9.9**
**(See color insert following page 302.)** A schematic road map of electrode's materials for advanced rechargeable Li batteries. The arrows address materials that use as nanoparticles is (marked by !) or maybe important.

obtaining high power density, the morphology (small particle size, high specific surface area) of the active mass and the electrodes' engineering may be highly important. The relevance of the use of nanoparticles for the various types of electrodes materials is discussed in detail in the following sections. In general, for olivine-type cathodes, the use of nano-materials is critical. For transition metal oxides cathodes, the use of nanoparticles may be very problematic. When dealing with the anode side, the intrinsic electrode-solution reactivity for electrodes with redox activity below 1.5 V vs. Li which leads to a rich surface chemistry and passivation phenomena is a key factor. Nevertheless, for Li–Si and MO (conversion reaction precursors, M = Co, Cu, Fe) electrodes, the use of nano-materials is crucial.

### 9.5.2.2 Negative Electrodes

The negative electrodes in batteries are the source of electrons in these systems. These electrodes are also called anodes, because in the battery discharge process, they are involved in oxidation processes. However, it should be noted that in rechargeable systems, the negative (anode) electrodes undergo reduction (e.g., cathodic) processes during the battery charging process. Hence, the main property of the negative electrodes has to be their electronegativity, namely, their ability to undergo oxidation processes and release electrons. The negative electrodes for batteries are classified as follows:

1. Active metals in aqueous batteries: Cd (Ni–Cd), Pb (lead–acid), Zn (zinc–air alkaline Zn–$MnO_2$), Al (Al–air), and Mg (reserve batteries). The anodic reactions here are the active metal dissolution: $M \rightarrow M^{Z+} + 2e^-$ (Peled and Yamin 1979; Linden 1994; Aurbach and Weissman 1999; Besenhard 1999; Gaberscek and Stane 1999; Crompton 2000; Dell and Rand 2001; Singhal and Kendall 2003; Pistoia and Broussely 2007).
2. Active metals in nonaqueous batteries: Li in primary and secondary Li batteries. For the latter systems, the relevant electrolyte systems seem to be gel, polymeric, or ceramic (Novak et al. 1997). We can also mention magnesium in rechargeable magnesium batteries (still in the R&D stage). The anodic reaction here is also the active metal dissolution (Aurbach et al. 2003).
3. Metallic compounds for Ni–MH batteries. These include the $AB_5$ type (A = La, B = Ni as major components can also include Ce, Nd, Pr, Gd, Y, and Cr) and the $AB_2$ type (A = Ti, Zn; B = Ni as major components, which can also include V, Al, and Cr). These compounds can store hydrogen in a hydride form. Hence, the electrode's reaction is: $MH + OH^- \leftrightarrows M + H_2O + e^-$ $E^o = 0.83$ (SHE) M = $AB_5$, $AB_2$. These electrodes have to be porous, comprising sintered particles, in order to facilitate the above heterogeneous reaction (Hu et al. 2004).
4. Carbonaceous electrodes for Li-ion batteries. These are composite electrodes comprising micronic-size carbon particles with a polymeric binder (90:10% by weight) (Yazami 2001). The most commonly used active mass is graphite, which inserts lithium according to the following: $Li^+ + C_6 + e^- \leftrightarrows LiC_6$ 372 mA h/g. There are a variety of types of graphite particles: synthetic flakes, natural flakes, chopped graphite fibers, and round-shaped particles such as mesocarbon microbeads (MCMB).
5. Other anodes for Li batteries: Efforts are underway to develop Li alloys and intermetallic compounds as substitutes for the highly reactive Li metal electrodes. These include Li–Sn, Li–Si, Li–Si/SnM1M2 (M1M2 are metals such as Cu, Ni, B). Li can form alloys with Sn and Si up to a stoichiometry of $Li_{4.4}Sn$ or $Li_{4.4}Si$, i.e., capacities of 900 and 4000 mA h/g, respectively (Aifantis et al. 2005). The latter systems may suffer from instability due to the huge volume increase during the lithiation of Sn or Si. There are innovative approaches as to how to stabilize

these systems (beyond the scope of this chapter), and the use of intermetallic compounds, namely, Li–Sn/Si–M1M2, etc., as one of these approaches (Tillard et al. 2005). There are also ideas of using low-potential, lithiated transition metal oxides as negative electrodes for Li-ion batteries, thus saving the irreversibility because of reactions with solution species. A good example is $Li_4Ti_5O_{12}$ spinel (1.5 V vs. $Li/Li^+$, 150 mA h/g), which shows prolonged cycleability at high rates (Mukai et al. 2005).

It does not make sense to use carbonaceous abode materials as nanoparticles because of the pronounced reactivity of *all* types of lithiated carbons with all relevant nonaqueous Li salt solutions, which are relevant for Li-ion batteries. $Li_4Ti_5O_{12}$ can be used as nanoparticles because its redox potential, around 1.5 V vs. Li, is high enough, above the potentials at which the standard electrolyte solutions for Li batteries are reduced. Nano-$Li_4Ti_5O_{12}$ is one of the fastest Li insertion materials with which the kinetic advantages of electroactive nano-materials are well expressed. There are also reports on promising anode materials for Li-ion batteries, based on nano-$TiO_2$ (Jung et al. 2009; Wilkening et al. 2009). We also discuss two classes of anodes for which the use of nanoparticles is crucial.

The first one which is dealt in here are the conversion reactions that were demonstrated as potential anodic reactions for Li-ion batteries about a decade ago by Tarascon et al. (Poizot et al. 2000). Certain transition metal oxides such as CoO, CuO, NiO, $Fe_2O_3$, and $Co_3O_4$ may undergo the following reversible reaction in Li salt solutions at low potentials: $MO + 2Li^+ + 2e^- <=> Li_2O + M^0$ (Villevieille et al. 2007; Chen et al. 2009). The condition for a reversible behavior of these reactions is the use of nanopowder of MO. Figure 9.10 illustrates the main differences between conversion and intercalation reactions in which Li ions are involved.

Conversion reactions of the type presented in Figure 9.10 were demonstrated not only with transition metal oxides but also with metal fluorides (Amatucci and Pereira 2007) and magnesium hydride (Oumellal et al. 2009). Most of them exhibit a reversible capacity between 400 and 700 mA h/g, twice higher that that of the commonly used Li–graphite anodes. Their potential profile is sloping between 2–1.5 V and 0 V vs. $Li/Li^+$. These tractions involve complicated interactions with solution species, especially at the low-voltage domain. They suffer from pronounced hysteresis: there may be a difference of more than 0.5 V between the charge and discharge potentials. There are increasing numbers of reports in the literature about these kinds of reactions due to the scientific interest in them. However, the opinion of the authors of this chapter is that these reactions are not really practical, because all types of relevant solution species are thermodynamically very unstable with the nano powders at the low potentials in which they interact with Li ions and undergo conversion reactions. Hence, it is hard to expect from electrodes based on these reactions the necessary stability in real battery systems, especially at elevated temperatures.

Another possible application for nano-materials in Li-ion batteries relates to the so-called intermetallic anodes. One of the alternatives for the problematic Li metal anode are Li alloys. Li can alloy reversibly with such Al, Mg, Sn, and Si at high capacities (e.g., around 900 and 4000 mA h/g for $Li_{4.4}Sn$ and $Li_{4.4}Si$, respectively, compared to 372 mA h/g for Li–graphite, $LiC_6$) (Anani and Huggins 1992). However, these alloying processes are accompanied by very pronounced volume changes. For instance, full alloying of Sn and Si with lithium leads to 300% volume increase. Such volume changes lead to stresses and strains that crack and disintegrate the active mass upon repeated lithiation–delithiation cycling. Moreover, these volume changes interfere very badly with the anodes' passivation, which

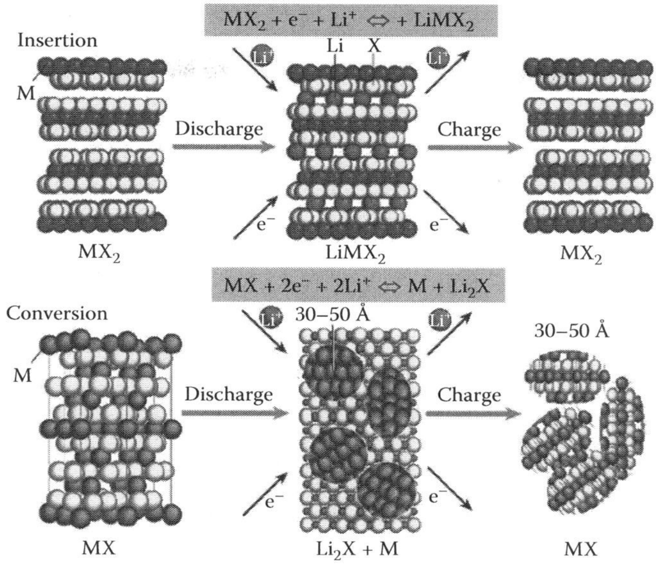


**FIGURE 9.10**
**(See color insert following page 302.)** A schematic illustration of insertion (top) and conversion (bottom) reactions in which Li ions are involved. (Reprinted from Armand, M. and Tarascon, J.-M., *Nature*, 451, 652, 2008. With permission.)

is mandatory for their operation in rechargeable Li batteries. At the low potentials of these Li alloying processes, all the relevant polar aprotic, electrolyte solutions used in Li batteries, are reduced on the electrodes and hence they are unstable. The apparent stability of most of aprotic Li salt solutions with Li, Li–C, or Li–M (any metal) anodes is because the reduction of most of aprotic Li salt solutions forms as products insoluble Li salts and oligomeric species that precipitate on the electrodes as passivating surface films. These surface films when reaching a certain thickness block electrons transfer and, hence, avoid continuous reduction of the solutions but allow Li ions transport through them. Thus, if anode materials are not stable and cannot develop steady passivation, they cannot be used in rechargeable batteries. The most important approach to improve the reversibility of Li alloying with elements such as tin and silicon (the most important candidates as alternative high capacity anode materials to graphite, for Li-ion batteries) in aprotic Li salt solutions is the use of nanoparticles. Upon lithiation of nano-powders of tin or silicon, the stresses and strains related to the volume changes are better relieved (Winter and Besenhard 1999). The pronounced surface reactivity of nanoparticles of Li–Sn or Li–Si alloys can be handled by the use of appropriate binders and additives in solutions which enhance formation of stable passivating surface films (Li et al. 2007a; Hochgatterer et al. 2008).

In addition to the use of nanoparticles, the reversibility of the Li–Sn or Li–Si alloying processes can be improved by the use of multicomponent systems. For instance, the new commercial, Nexelion advanced Li-ion batteries from Sony, contains anodes that comprise Sn, carbon, and Co composites (Wolfenstine et al. 2006). The main anode process is of course lithiation of tin, while the latter two elements act as stresses and strain relievers which keep the active mass well integrated upon cycling, by "absorbing" the volume changes due to Li–Sn alloying. There are many reports on Li–Sn–C and Li–Si–C composites as improved anode materials. Especially interesting are the development of Si–C composites with core-shell structure (Kim and Cho 2008). There are also many reports on ternary and quaternary Li–Sn(Si)–M1(M2) composites as potentially important anode materials. With all of these composite materials, there is an advantage to the use of nanoparticles (Vaughey et al. 2001).

### *9.5.2.3 Positive Electrodes*

In general, the positive electrodes have to be electrochemically active at relatively high redox potentials. The cathode materials for Li-ion batteries are lithiated transition metal oxides and sulfides ($Li_xMO_y$, $Li_xMS_y$), and $LiMPO_4$ olivine compounds (M = Fe, Mn, Co). These materials are the source of lithium in Li-ion battery systems.

A first process in these systems is always charging, in which the cathode material is delithiated and the lithium ions are intercalated in the negative electrode (the mass balance has to take into account the irreversible capacity of the negative electrodes, part of which is involved in the establishment of their passivation by surface films formation). Most of the lithiated transition metal oxides relevant as cathode materials for Li-ion batteries are reactive with alkyl carbonates/$LiPF_6$ solutions and develop surface films. Hence, their electrochemical behavior and stability are largely influenced by their surface chemistry (Aurbach et al. 1998a,b; Aurbach 1999). Thereby, most of $Li_xMO_y$ cathode materials cannot be used as nanoparticles because they become covered by too thick surface layers that impede Li ions transport. It may be possible to use nano $Li_xMO_y$ cathode materials when they are covered by carbon layers (Odani et al. 2006) or by other protecting surface layers that can act as a buffer zone that protect the active mass from detrimental interactions with solution species (e.g., acidic moieties) (Cho et al. 2000; Gnanaraj et al. 2003; Park et al. 2008).

About a decade ago, $LiFePO_4$ olivine was introduced as a promising cathode material for rechargeable Li batteries (Padhi et al. 1997). Since then, many hundreds of publications appeared in the literature on this material and it even became commercial. The electron and Li-ion transport properties of this material are very poor. However, it was discovered that by the use of nanoparticles and coating with very thin conducting layers (like carbon), it is possible to overcome the poor kinetics of Li insertion, deinsertion into/from this material, and to make it a very fast cathode material. A recent, most promising achievement is the performance of the $LiFePO_4$ cathode material described in Figure 9.11 (Kang and Ceder 2009).

The challenge with these olivine materials is to develop practical $LiMnPO_4$ and $LiCoPO_4$ as practical cathode materials because their redox potentials are 4.1 and 4.8 V vs. Li, respectively, a gain of 0.6 and 1.3 V, respectively, compared to $LiFePO_4$, yet at the same theoretical capacity (close to 170 mA h/g).

Figure 9.12 presents some recent data related to $LiMnPO_4$ that demonstrate the promising potential of this family of compounds to serve as superb cathode materials in advanced Li-ion batteries (Martha et al. 2009). This figure shows high-resolution transmission electron microscopy (TEM) images of the active mass, which comprises carbon-coated nanoparticles; the gain in potential of $LiMnPO_4$ compared to $LiFePO_4$; and some voltage profiles measured upon discharge of composite $LiMnPO_4$ cathodes at different rates (Martha et al. 2009). It should be emphasized that further modification of this cathode material can considerably improve its rate capability.

Upper right chart: comparison of the voltage profiles of $LiMnPO_4$ and $LiFePO_4$. Lower right chart: voltage profiles measured during galvanostatic (constant current) discharge processes at different rates. 5C mean discharging the electrode's capacity within 12 min.

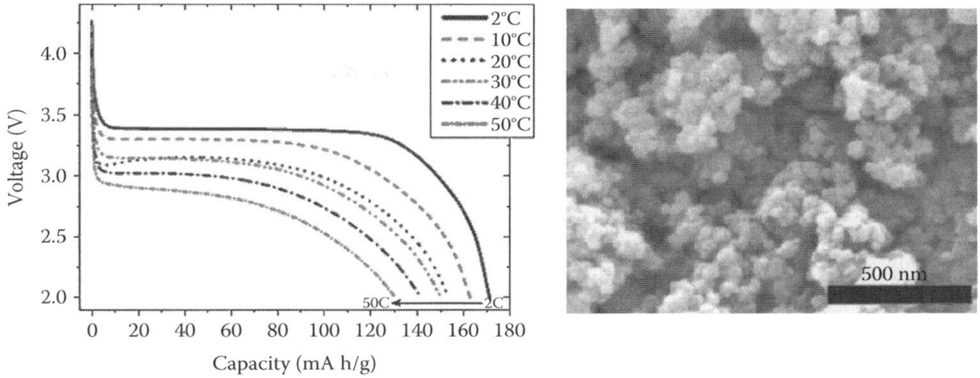


**FIGURE 9.11**
Presentation of an ultrafast $LiFePO_4$ cathode. *Left*: Voltage profiles obtained upon discharging this material at different rates. Note that a fate of 50°C means discharging most of the capacity of the cathode material within around 1.2 min. *Right*: A scanning electron microscopy (SEM) image of the $LiFePO_4$ nanoparticles. (Reprinted from Kang, B. and Ceder, G., *Nature*, 458, 190, 2009. With permission.)

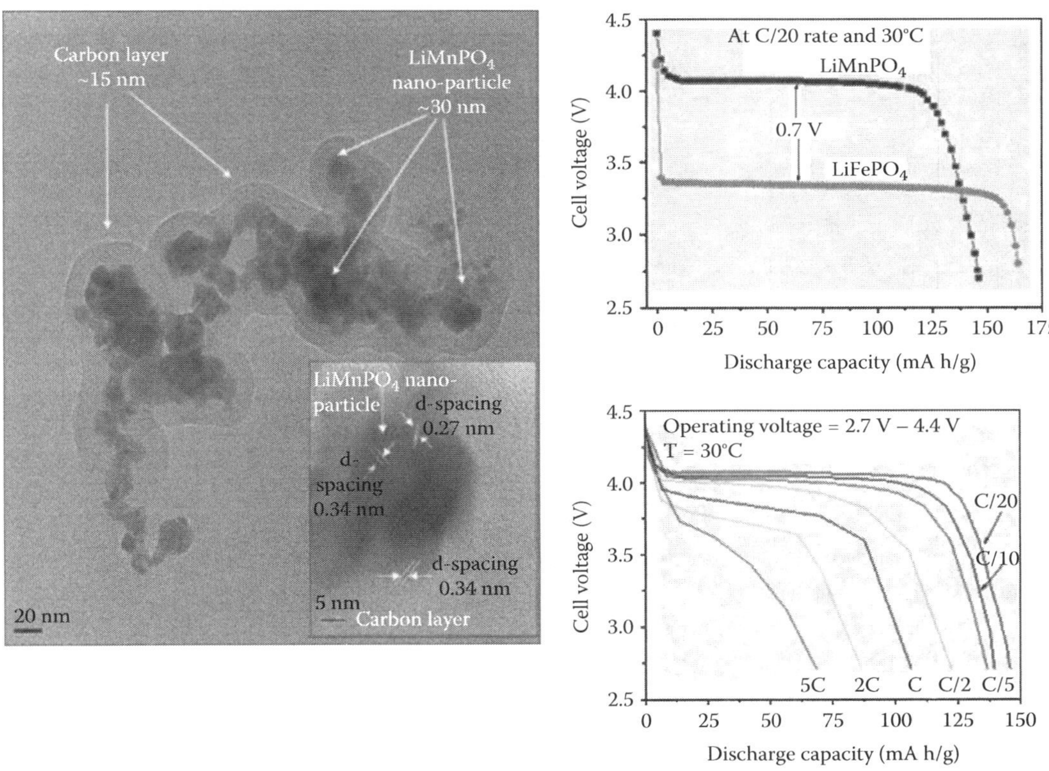


**FIGURE 9.12**
Presentation of some data related to $LiMnPO_4$ electrodes. *Left*: High resolution images of the active mass-carbon coated nanoparticles of $LiMnPO_4$. *Upper right*: Comparison of the voltage profiles of $LiMnPO_4$ and $LiFePO_4$. *Lower right*: Voltage profiles measured during galvanostatic (constant current) discharge processes at different rates. 5°C mean discharging the electrode's capacity within 12 min.

These cathodes based on the $LiMPO_4$ olivine compounds mark the most successful application of nano-materials in rechargeable batteries. The nanosize means short enough length for the solid-state diffusion of Li ions that may be a rate-determining step for Li insertion processes into inorganic hosts. It is the opinion of the authors of this chapter that it is possible to use $LiMPO_4$ compounds as nano-powder in composite cathodes for Li-ion batteries (in contrast to the case of $Li_xMO_y$ compounds), because the oxygen atoms of these compounds that are bound to $P^{5+}$ cations are not too nucleophilic or basic and, hence, there surface reactions, even when the active mass have high specific surface area, are very moderate and do not form isolating surface films that interfere badly with the inter-particle electrical contact and lead to high impedance (as is the case for electrodes comprising nanoparticles of lithiated transition metal oxides (Talyossef et al. 2007)).

#### 9.5.2.4 Electrolyte Systems

There are five major electrolyte systems relevant to batteries and related devices:

1. "Liquid solutions" based on well-dissolved electrolytes in polar solvents, with good ion separation. These liquid systems include aqueous solutions in which water is the solvent, or nonaqueous systems in which the solvents are usually polar aprotic (as dealt with later). Aqueous solutions are used in several important systems, including Zn–$MnO_2$, Zn–air, Ni–Cd, Ni–MH, and lead–acid. Except for the last system, in which the electrolyte is $H_2SO_4$, for all the other batteries in the list the solutions are alkaline, using KOH as the electrolyte. Nonaqueous electrolyte solutions are relevant mostly to Li and Li-ion batteries and related developments, e.g., rechargeable magnesium batteries. The most important families of solvents are ethers, esters, and alkyl carbonates. There may be some use of nitriles and sulfones as well (miscellaneous).
2. "Liquid systems" based on ionic liquids (ILs) (Armand et al. 2009), i.e., molten salts. Here the solvents are ionic media, thereby providing the electrolytic function. ILs are now being intensively studied in connection with Li and Mg batteries because they are stable, nonflammable, and may provide very wide electrochemical windows (and thus can be suitable for high-voltage batteries).
3. "Gel electrolytes": In these systems, the active electrolyte systems consist of solvents and salts, contained in an inert polymeric matrix. Such systems can be treated as liquid electrolyte solutions, since their ionic conductivity is similar in the order of $10^{-3}$ s $cm^{-1}$, and the interfacial properties of the electrodes are determined by their surface reactions with the solvents and the salt anions (Stephan 2006).
4. "Polymeric electrolytes": Here the solvent system consists of polymers, derivatives of polyethers. Polyethers can dissolve Li salts because of the strong interaction of Li ions with the oxygen atoms that enable the separation of charges. The room-temperature ionic conductivity of a polyether/Li salt system is lower by 2.5 orders of magnitude, compared to that of liquid solutions ($10^{-6}$–$10^{-5}$ vs. $10^{-3}$–$10^{-2}$ s $cm^{-1}$). Thus, polymeric, solvent-free electrolytes are expected to work at elevated temperatures >60°C. The reactivity of these systems toward lithium is much lower than that of liquid systems. However, they are not inert toward lithium because Li metal attacks ether linkages. Impressive innovative efforts are underway to synthesize derivatives of polyethers that facilitate charge separation and transport. Critical efforts in this field relate to increasing the low-temperature conductivity

and the transference number of Li ions (which should lower the detrimental concentration gradients) (Stephan and Nahm 2006).

5. "Ceramic electrolytes": Solids such as β-alumina, $Li_3PO_4$, and boron-based glass can transport active metal ions (e.g., $Li^+$, $Na^+$). Intensive work is underway on solid conductors for proton and oxygen ions for high-temperature fuel cells. Most of the ceramic electrolytes are planned to work at high temperatures, hundreds of degrees centigrade. However, the fabrication of microbatteries in which the electrolyte systems are thin films of solids such as $Li_3PO_4$ enable operation at ambient temperatures (Duclot and Souquet 2001; Zhang et al. 2005).

We can classify the following major differences between liquid (items 1 through 3 above) and solid (items 4 and 5 above) electrolyte systems:

1. The temperature range of interest is quite different. For liquid electrolyte solutions, the highest relevant temperatures are <80°C, and the lowest temperatures of interest are >–40°C. The range for polymeric electrolytes (gels and solvent-free systems) may be room temperature up to 60°C. The temperature range of interest for ceramic electrolytes may be from ambient temperature up to 1000°C.
2. The ion transport mechanisms are pronouncedly different when comparing liquid and solid electrolytes. Hence, while the same electrochemical methods can be applied for characterization, their interpretation is completely different for solid and liquid electrolyte systems.
3. The interfacial charge transfer is also different in terms of contact points. While the liquid–electrode interfaces are continuous and do not contain voids, the contact between electrodes and solid electrolyte matrices may not be continuous. Hence, not the entire electrode surface in contact with the solid electrolyte is really active.
4. Phenomena related to the electrodes' surface chemistry, such as corrosion, passivation, and surface film formation, are much less pronounced with solid electrolytes than with liquid electrolyte solutions. This is due to the much lower expected reactivity of solid electrolytes toward all electrode materials, as compared to that of liquid systems. Hence, the electrochemical response from electrode–liquid or electrode–solid interfaces is quite different and relates to different types of charge transfer processes.
5. The engineering aspects are, of course, much different. When liquid solutions are used, a solid separator is needed as a spacer between the electrodes (and which is usually a porous polymeric film soaked with the electrolyte solution). In solid state systems, the solid electrolyte can also serve as the inter-electrode spacer.

We can distinguish among four classes of solid electrolytes for batteries:

1. Gel electrolyte–solid polymeric matrices that are soaked with liquid electrolyte solutions: the solvent dissolves the electrolyte (Osaka et al. 1997).
2. Solvent-free polymeric electrolyte: the polymeric chains can dissolve Li salts. This is relevant mostly to poly-ethers and their derivatives (Sun and Kerr 2006).
3. Composites comprising polymer ceramic materials and electrolyte systems: in these systems, the ceramic materials that are most preferred are nano powders, e.g., $SiO_2$, $Al_2O_3$, dispersed within the polymeric matrices and provide additional conducting mechanisms of ions via migration on their high surface area. The

use of such composite systems is relevant to both gel- and solvent-free polymeric matrices (Wachtler et al. 2004).

4. Ceramic materials and solid-state electrolytes (Shahi et al. 1983).

Composite systems comprising polymers, salt, and ceramic powders (item 3 above) are very promising electrolytes because it is possible to achieve high ionic conductivity despite the solid structure. For these composite matrices, nanoparticles of ceramic materials such as silica and alumina are critical components. The surface oxygen atoms of the particles are involved in a main conducting mechanism of Li ions in the matrix and the nanosize of the particles ensures the necessary high surface area for this conducting mechanism (Peled et al. 1995; Tominaga et al. 2005; Bhattacharyya et al. 2006).

#### *9.5.2.5 Separators and Membranes*

The electrodes in batteries have to be separated in order to prevent short circuit, i.e., mechanical separation, and also to prevent the exchange of ions/compounds that interfere badly with the desired electrochemical reactions of the electrodes, i.e., chemical separation. The role of separators/membranes in batteries is critical, and it affects internal resistance, rates, stability, cycle life, temperature range, and safety features. In most of the battery systems, the active mass is contained, stable in solid electrodes, and the same ions in solution react with both electrodes. This is the case for all alkaline stationary batteries, lead acid systems, and nonaqueous Li, Li-ion, and Mg batteries. In such cases, what are needed between the electrodes are separators that maintain the mechanical stability of the systems. In Ni–Cd, Ni–MH, and L–A systems, the aqueous solutions are involved in the electrochemical reactions. Thus, they have to be thick enough to contain the appropriate amount of solution. In the case of the batteries in which the solution serves only as an ion conductor, the separator should be as thin as possible, porous, but yet strong enough to maintain the mechanical and electrical separation between two rough composite electrodes. For instance, porous polypropylene or polyethylene films (a few tens of microns thick) are used for Li-ion batteries. Although the main components in separators for batteries are of course polymeric matrices, there is a great advantage for the use of composite separators that contain ceramic nanoparticles. In Li metal batteries, a main problem is the dendrite formation during charging (Li deposition processes). Separators containing ceramic nanoparticles may be useful for preventing dendrite growth and penetration through the separator. The design of composite separators is a special art. There is a compromise between the degree of porosity and mechanical strength. In addition, the wettability of the pores is important. Ceramic particles embedded in the porous polymeric matrices of separators may facilitate their wetting by all kind of solutions for batteries application (both aqueous and nonaqueous) (Arora and Zhang 2004).

## 9.6 On the Synthesis of Nano-Materials for Rechargeable Li-Ion Batteries

There are many thousands of reports in the literature on the synthesis of nano-materials for batteries. Hence, it is impossible to cover this matter in a single chapter (or even in a single book). We describe below a few selected syntheses modes that produce nano-materials that are relevant to Li batteries.

### 9.6.1 Self-Combustion Reactions

These are simple methods for synthesizing nanosized lithiated mixed metal oxides. This method involves exothermic reaction between metal nitrates in the right stoichiometry and some inorganic or organic reducing compounds (fuel) and have been developed in recent years (Verneker 1986; Dhas and Patil 1993; Patil et al. 2002; Gopukumar et al. 2004; Hwang et al. 2006).

The main advantages are the good mixing between the atoms and the formation of nano size clusters, whose further calcination produces nanoparticles of mixed metal oxides. The heat needed for the synthesis of these metal oxides is provided by the exothermic reaction between the fuel and the oxidizers, namely, the precursors that are the metal nitrates.

For example, the self-combustion reaction (SCR) of the layered $LiNi_xMn_yCo_zO_2$ compounds, using sucrose as the fuel can be written as follows (Haik et al. 2009):

$$LiNO_3 + Ni(NO_3)_2 + Mn(NO_3)_2 + Co(NO_3)_2 + C_{12}H_{22}O_{11} \rightarrow LiNi_xMn_yCo_zO_2 + CO_2\uparrow + H_2O\uparrow + NO\uparrow$$

SCRs can be carried out in glass or ceramic vessels, in which the starting solutions are initially heated to 150°C. At this temperature, ignition of the SCR takes place leading to spontaneous exothermal reactions that involve flame and gas evolution. The final particles size (from nano to micro) can be well controlled by two parameters: further heating (up to 1000°C) in air and heating duration. It should be noted that as the calcination temperatures are higher, the particles thus formed are bigger and more ordered, possessing smoother facets. However, calcinations at too high temperatures (e.g. >950°C) may lead to oxygen evolution and, hence, to the formation of oxygen-deficient $LiNi_xMn_yCo_zO_{2-w}$ products. Fortunately, it is possible to reoxidize such species by further annealing in oxygen-containing atmosphere at lower temperatures (e.g., <700°C).

### 9.6.2 Sonochemical Reactions

Applying sonic radiation to solutions leads to the reasonably known acoustic cavitation phenomena (Dhas and Gedanken 1997; Jeevanandam et al. 2000; Kovacheva et al. 2002): bubbles of gas dissolved in the irradiated solution undergo a very fast adiabatic compression that leads to the formation of hot spots that can reach thousands of degrees. Such hot spots can direct many types of local reactions, depending on the solution composition. Due to the local nature of cavitation phenomena, when the reactions ignited by the very high (local) temperatures produce solid products, they are usually formed as nanoparticles. It is possible to produce nanoparticles of transition metal oxides that can be used as hosts for Li insertion by sonication of solutions containing organometallic compounds comprising the desirable transition metal and oxygen-containing organic ligands (e.g., alkoxides, acetates) (Odani et al. 2003). The product thus obtained (nanosized powder of metal oxide) can be crystalline or amorphous phase depending on the reaction conditions. It is also possible to obtained by sonochemistry metallic or intermetallic nanoparticles (Bhattacharyya and Gedanken 2008) and even nanoparticles of silicon (Dhas et al. 1998).

### 9.6.3 Thermal Reactions

It possible to produce useful nanosize materials that can be used for rechargeable batteries by relatively simple thermal reactions in which the precursors are heated to elevated temperatures at constant volume, under autogenic pressure (which is developed due to the high temperatures and the follow up reactions). For instance, heating organometallic compounds at constant volume under autogenic pressure forms nanoparticles of carbon-coated transition metal or transition metal oxide with core-shell structures (Pol et al. 2005).

It was possible to obtain nanoparticles of carbon-coated $V_2O_3$ by such thermal reactions of $VO(OR)_3$ that were active as Li insertion host. Further heating of this material in air produces nano-carbon-coated $V_2O_5$ that was demonstrated as a very good cathode material for Li batteries (Koltypin et al. 2007). Another interesting example relates to rechargeable Mg batteries. Nanoparticles of $CuMo_6S_{8-x}Se_x$ Chevrel phases could be produced by a simple thermal reaction at constant volume of a mixture of the relevant elements (Aurbach et al. 2007). Leaching of the Cu by a mild oxidation produced nanoparticles of $Mo_6S_{8-x}Se_x$ that were demonstrated as very promising, fast cathode materials for secondary magnesium batteries.

Dealing with simple thermal reactions that produce nanoparticles, it is important to mention the option of producing nano-materials for Li batteries (including transition metal oxides) by heating solutions of precursors with microwave radiation (Irzh and Gedanken 2009).

### 9.6.4 Mechanochemistry

In general, it is possible to reduce particle size by milling. It is possible to produce nanomaterials at relatively low temperatures, applying to relevant precursors mechanical energy via milling, which leads to chemical changes (Aymard et al. 1999).

Ball milling is the most commonly used, simple mode for mechanochemical processes. It is possible to obtain by milling mixtures of the relevant metallic elements, nanoparticles of intermetallic compounds that may be useful as high-capacity anode materials for Li batteries. Critical elements in these mixtures are tin or silicon that alloy reversibly at high capacity with lithium and, hence, serve as the main active elements in these intermetallic anode materials (Kosova et al. 1999). Lithiated transition metal oxides can also be produced by mechanochemical synthesis via ball milling of transition metal oxides and Li salt (e.g., $Li_2CO_3$) (Soiron et al. 2001). Nanoparticles of $LiMPO_4$ and carbon-coated $LiMPO_4$ olivine cathode materials can also be produced by ball milling of precursors such as phosphates, metal oxides, and Li salts (Song et al. 2007).

### 9.6.5 Sol-Gel Approaches

These are relatively cheap and low-temperature syntheses modes that allow fine control of the products' chemical composition and particle size. The sol-gel process in general is based on the transition of the system from a liquid solution into a gelatinous network. It allows homogeneous mixing at the atomic or molecular level (Brinker and Scherer 1990; Hench and West 1990). The starting material is usually a chemical solution that acts as the precursor for an integrated network of discrete particles. Typical precursors are metal salts such as acetates, chlorides, and nitrates (Sakka 2005; Jan et al. 2007), which undergo various forms of hydrolysis and condensation reactions, that may form metal oxide networks containing both liquid and solid phases. In further steps, products at the right stoichiometry are precipitated and the final morphology is achieved by calcination at high temperatures (which determines the particle size).

For example, nanoparticles of spinel $LiMn_2O_4$ are produced via sol-gel process (He et al. 2006).

Another interesting example for a sol-gel synthesis is the preparation of nano $LiCo_{1/3}Ni_{1/3}Mn_{1/3}O_2$ from $LiOH{\cdot}H_2O$ and a triple hydroxide of cobalt, nickel, and manganese (Yabuuchi and Ohzuku 2005).

### 9.6.6 The Polyol Approach

The basic idea of polyol process is dissolving of metal salts in a chelating polyalcohol (polyol) at elevated temperatures followed by controlled co-precipitation by hydrolysis at higher temperature. In most cases, diethylene glycol (DEG) is used as the chelating polyol. The polyol medium itself acts not only as a solvent in the process but also as a stabilizer, limiting particle growth and prohibiting agglomeration (Fievet et al. 1989; Wang et al. 2009). Crystal samples maybe obtained directly from the suspension without any further necessary calcination. In contrast to other methods, this route simplifies the producing procedure and diminishes the necessary cost.

In general, this synthesis approach yields well-crystallized nanoparticles. Materials prepared by the polyol method have a uniform size distribution and regular morphology. The particle size can be controlled by adjusting the experimental conditions. The usefulness of this method was nicely demonstrated by the synthesis of nano $Li_4Ti_5O_{12}$ (Kim et al. 2005) and nano $LiMnPO_4$ (Martha et al. 2009).

### 9.6.7 Carbo-Thermal Approaches

The carbo-thermal reduction (CTR) process is used to reduce metal oxides (and other compounds) to the pure metal state and relies on the application of the two carbon oxidation reactions:

$$C + O_2 \rightarrow CO_2 \tag{9.1}$$

$$2\,C + O_2 \rightarrow 2\,CO \tag{9.2}$$

This methodology allows selective and controlled reduction of appropriate metal precursors together with simultaneous incorporation of lithium (Barker et al. 2003). This is an energy-efficient method, economical, and convenient process to produce a wide range of electroactive compounds (Yang et al. 2005). As an example, we can mention the carbo-thermal synthesis of $LiFePO_4$ from $Fe_2O_3$ $LiH_2PO_4$ and carbon (the carbon reduces the $Fe^{3+}$ ions):

$$LiH_2PO_4 + 0.50\ Fe_2O_3 + 0.50\ C \rightarrow LiFePO_4 + 0.50\ CO + H_2O$$

### 9.6.8 Hydrothermal/Solvothermal Approaches

These approaches can be considered as low-temperature methods that lead to well-crystalline materials with particles in the nanometric size. In these methods, the reactants are dissolved in water or other suitable solvent and then are heated above the boiling point of the solvent (in an autoclave at high pressure) for the desired period of time (Tabuchi et al. 1999; Walton 2002; Chen et al. 2007). Upon heating, acid–base and/or hydrolysis (solvolysis) reactions occur (depending on the precursors) and the necessary mixing of ions is achieved. Mixed metal hydroxides turn to the mixed metal oxides by losing water molecules. Nanoparticles of lithiated transition metal oxides can be prepared in this way. $LiMPO_4$ olivine compounds can also be produced by hydrothermal synthesis (Kobayashi et al. 2000).

Another interesting example for a solvothermal synthesis is the preparation of nano $SnO_2$ from $SnCl_2$ in ethanol and a mild base (hexamethylene tetramine) (Fan et al. 2006).

## 9.7 Carbon (and Other) Nanotubes

### 9.7.1 Introduction

Among the nano-materials relevant to energy storage and conversion, nanotubes can be considered as promising materials for batteries, EDL (super) capacitors, and photovoltaic cells. In general, several elements and classes of compounds that form layered structures can also be crystallized by the appropriate stimuli to fullerenes and to nanotubes. Among them are (Iijima 1991), $TiS_2$ (Ivanovskayaa and Seifert 2004), $MoS_2$ (Chen et al. 2001), $WS_2$ (Tal et al. 2001),$VO_x$ (Hu et al. 2009), $TiO_2$ (Mor et al. 2006), and more. These materials may form both single or multiwall nanotube structures. They may have unique properties in terms of strength, electrical conductivity, and ability to interact with ions via intercalation or electro-adsorption. Their very high aspect ratio means the availability of high specific surface area for electrochemical interactions. Apparently, CNT seems to be the most interesting materials in this respect, and thereby a separate section is dedicated herein to this class of materials.

The credit for the discovery of CNTs is not obvious. The "official" birth date of CNTs is attributed to Sumio Iijima from NEC who, in 1991, published a paper in *Nature* (Iijima 1991) in which he reported the discovery of a "new type of finite carbon structure consisting of needle-like tubes" which comprises "coaxial tubes of graphitic sheets, ranging in number from 2 up to about 50."[4] This paper has been cited over 8000 times to date as almost every paper regarding nanotubes cites it.

Since then, progress on CNT technology and its applications have evolved at a very high pace. CNTs are one of the many allotropic forms of carbon (Wikipedia, Allotropes of carbon), which include graphite, diamond, fullerene, etc. A single-wall CNT (SWCNT) can be visualized as a rolled sheet of grapheme (sheet of carbon atoms arranged in hexagonal rings) that may be capped at the ends. A multiwall CNT (MWCNT) can be visualized as concentric rolled sheets of grapheme (see Figure 9.13). It is important to notice that, although this is a good way to describe the structure and the atomic arrangement of nanotubes, it is inconsistent with the growth mechanisms that lead to their formation.

SWCNTs have a small diameter (from 0.4 to 4 nm) and exhibit the particular property that they can be metallic (semimetals) or semiconducting, depending on their chirality. On

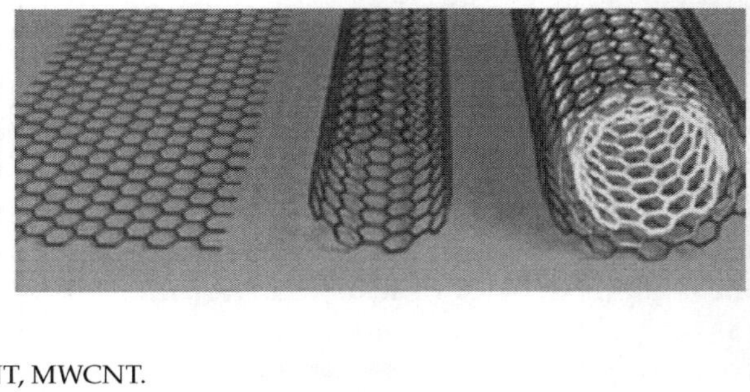

**FIGURE 9.13**
Grapheme sheet, SWCNT, MWCNT.

average, meaning without chirality control, the main synthesis modes for these materials produce 1/3 metallic and 2/3 semiconducting SWCNTs. To define chirality, one can visualize the sheet of grapheme and imagine that this sheet can be rolled in different ways.

### 9.7.2 On the Synthesis and Characterization of CNT

Prior to the nucleation and growth of CNTs, catalyst dots have to be formed on the substrate (usually a ferromagnetic metal or metallic compound). Then organic gas precursor (e.g., methane, ethane, propane) is pyrolized by heating (e.g., high temperatures in oven, laser ablation, plasma, electrical arc) and the resulting carbonaceous fragments recombine to CNTs around the catalyst dots in a similar manner as other chemical vapor deposition (CVD) processes are carried out. To utilize CNTs in industrial applications, two main approaches have been considered: "grow-in-place" and "grow-then-place" (Figure 9.14).

"Grow-in-place": this technique usually consists of preparing the sample with a catalyst present in the locations where the nanotubes is synthesized. For instance, a thin catalyst film can be deposited using e-beam evaporation or sputtering; alternatively, nanoparticles can be deposited on a substrate. Synthesis is usually performed using a CVD technique. This is the preferred (or only) approach when it is important to grow CNTs in specific locations, such as for most electronic device applications.

"Grow-then-place": this technique consists of preparing nanotubes and successively transferring them to a substrate. Arc discharge and laser ablation are the main techniques used to synthesize freestanding nanotubes. The nanotubes may be subsequently selected (e.g., SWCNTs from other CNTs) and purified prior to use. To transfer them to another substrate, CNTs are usually functionalized in a way that they will attach to prepatterned areas of the substrate, which will attract functionalized CNTs. These synthesis techniques may have a high yield of floating CNTs with high purity, which makes it suitable for producing CNTs, which can be mixed with other materials (e.g., polymers) for composite applications. Growth mechanisms for CNT on surfaces containing appropriate catalysts, by CVD processes are now well understood (Hofmann et al. 2005; Puretzky et al. 2005). Hence, it is possible to select substrates, catalysts, and process parameters in order to design different and versatile CNTs (single, bundles, arrays, carpets, etc.).

The preparation of CNT carpets (highly desirable morphology for energy storage devices) is illustrated schematically in Figure 9.15. The substrate is prepared with catalyst

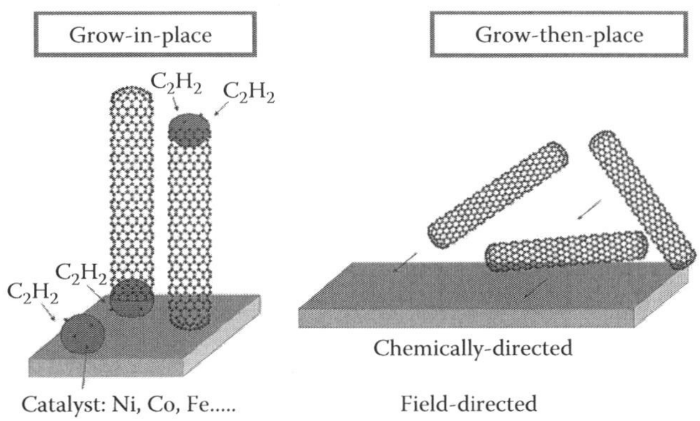


**FIGURE 9.14**
Pictorials comparing the "grow-in-place" and "grow-then-place" techniques.

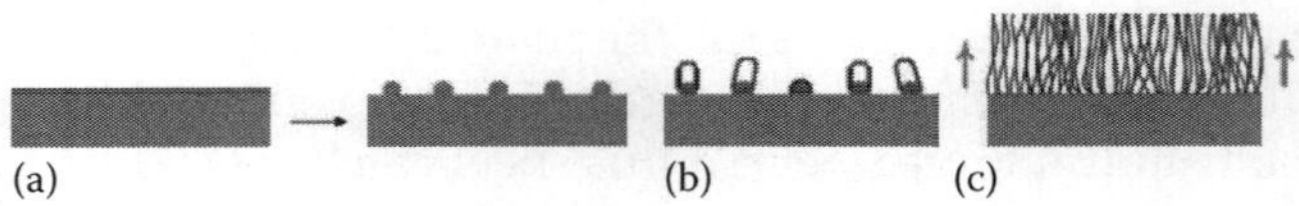


**FIGURE 9.15**
A schematic presentation of the growth of CNT carpets: the substrate is prepared with catalyst dots, around which the nanotubes grow simultaneously.

dots at the necessary surface density, and then CNTs are grown up simultaneously from all the dots on the substrate.

Control of the CNTs growth depends on a multiplicity of factors amongst which we can mention: catalyst film thickness (or nanoparticle diameter), underlayer material, gas composition (and compounds produced by the thermal gas decomposition), growth duration, and growth temperature. Among the many characterization techniques available for nanomaterials, it appears that high-resolution electron microscopy (HRSEM and HRTEM) are the most useful because they are capable in providing clear images at atomic resolution. In addition, atomic force microscopy and Raman spectroscopy are helpful as well (the latter one is effective for estimating their degree of crystallinity by the ratio between the G and D lines).

### 9.7.3 Properties and Applications of Carbon Nanotubes

Because of their exceptional and anisotropic properties (Dresselhaus et al. 2001), as illustrated in Table 9.2 (Hoenlein et al. 2003), CNTs have been widely investigated for electrical, thermal, and mechanical applications (Baughman et al. 2002), such as for microelectronic interconnects (Nihei et al. 2005), heat sinks (Tong et al. 2007), and structural composites (Garcia et al. 2008).

In addition to the properties in Table 9.2, we can mention

- High aspect ratio—the closest to ideal 1D structure
- 100 times stronger than steel
- Chemically inert
- Good electron field emitters
- Very high melting temperature

**TABLE 9.2**

Important Electrical and Mechanical Characteristics of CNTs

| | |
|---|---|
| Electrical conductivity | Metallic or semiconducting |
| Electrical transport | Ballistic, no scattering |
| Energy gap (semiconducting) | $E_g$ (eV) ~1/*d* (nm) |
| Maximum current density | ~$10^{10}$ A/cm$^2$ |
| Maximum strain | 0.11% at 1 V |
| Thermal conductivity | 6000 W/km |
| Diameter | 1–100 nm |
| Length | Up to millimeters |
| Gravimetric surface | >1500 m$^2$/g |
| *E*-modulus | 1000 GPa |

The application of these unique materials in devices for energy storage and conversion is in its first stage. However, it is possible to draw several promising directions for the use of CNT in batteries and EDL capacitors.

In general, CNT can be used as both active electrodes materials and as supporting components in composite electrodes.

MWCNTs can serve as Li insertion anode material in Li-ion batteries (Endo et al. 2004; Sheem et al. 2006). The high electrical conductivity, the very high aspect ratio, and the consequent high specific surface area make them suitable electrode's material for super (EDL) capacitors (Jung et al. 2004; Wen et al. 2006). The same properties plus their impressive mechanical strength make them very desirable components in composite electrodes of all kinds of batteries, which can enhance remarkably the mechanical and electrical integrity of electrodes for rechargeable batteries (Odani et al. 2003; Endo et al. 2008). The possibility of modifying CNT by grafting, thus attaching to them functional groups and polymeric species (Tasis et al. 2006; Piran et al. 2009), make them even more attractive to use in batteries. In recent years, we see attempts to develop new organic cathode materials for rechargeable Li batteries that may replace the inorganic host materials which are currently in use. These include polymers with S–S bonds (Deng et al. 2006; Li et al. 2007a,b) and compounds with multi C=O double bonds (Chen et al. 2008) that can be reversibly reduced and interact with Li ions at high enough potentials. Most of the organic cathode materials presented to date suffer from severe kinetic limitations, due to poor electron transfer to compounds which are electrical insulators. Functionalizing CNT, which can conducts electrons very well, by oxygenated or sulfur containing groups, may create superb, high-capacity and fast, organic, "grin" cathode materials for rechargeable Li batteries. These possible approaches can be considered as a major challenge for nano-materials in connection with energy storage and conversion.

## References

Aifantis, K.E., Dempsey, J.P., and Hackney, S.A. 2005. Cracking in Si-based anodes for Li-ion batteries. *Review of Advanced Materials Science*. 10: 403–408.

Amatucci, G.G. and Pereira, N. 2007. Fluoride based electrode materials for advanced energy storage devices. *Journal of Fluorine Chemistry*. 128: 243–262.

Anani, A. and Huggins, R.A. 1992. Multinary alloy electrodes for solid state batteries II. *Journal of Power Sources*. 38, 3: 363–372.

Andrade, L., Zakeeruddin, S.M., Nazeeruddin, M.K. et al. 2009. Influence of sodium cations of N3 dye on the photovoltaic performance and stability of dye-sensitized solar cells. *ChemPhysChem*. 10: 1117–1124.

Andu'jar, J.M. and Segura, F. 2009. Fuel cells: History and updating. *Renewable and Sustainable Energy Reviews*. 13: 2309–2322.

Antolini, E., Salgado, J.R.C., and Gonzalez, E.R. 2006. The stability of Pt–M(M=first row transition metal) alloy catalysts and its effect on the activity in low temperature fuel cells: A literature review and tests on a Pt-Co catalyst. *Journal of Power Sources*. 160: 957–968.

Appetecchi, G.B., Shin, J.H., Alessandrini, F. et al. 2005. 0.6 Ah $Li/V_2O_5$ battery prototypes based on solvent-free PEO-$LiN(SO_2CF_2CF_3)_2$ polymer electrolytes. *Journal of Power Sources*. 143: 236–242.

Armand, M., Endres, F., Macfarlane, D.R. et al. 2009. Ionic-liquid materials for the electrochemical challenges of the future. *Nature Materials*. 8: 621–629.

Arora, P. and Zhang, Z. 2004. Battery separators. *Chemical Reviews*. 104: 4419–4462.

Aurbach, D. 1999. The electrochemical behavior of active metal electrodes in nonaqueous solutions. In: *Nonaqueous Electrochemistry*, ed. D. Aurbach, pp. 289–411. New York: Marcel Dekker.

Aurbach, D. 2002. The role of surface films on electrodes in Li-ion batteries. In: *Advances in Lithium Ion Batteries*, ed. W.V. Schalkwijk and B. Scrosati, pp. 7–79. New York: Kluwer Academic/Plenum Publishers.

Aurbach, D. and Schechter, A. 2004. Advanced liquid electrolyte solutions. In: *Science and Technology of Advanced Lithium Batteries*, ed. G. Pistoia and G. Nazri, pp. 530–568. New York, London, Boston: Kluwer Academic Publishers.

Aurbach, D. and Weissman, I. 1999. Nonaqueous electrochemistry: An overview. In: *Nonaqueous Electrochemistry*, ed. D. Aurbach, pp. 1–53. New York: Marcel Dekker.

Aurbach, D., Levi, M.D., Levi, E. et al. 1998a. Common electroanalytical behavior of Li intercalation processes into graphite and transition metal oxides. *Electrochemical Soc*iety. 145, 9: 3024–3034.

Aurbach, D., Weissman, I., Yamin, H. et al. 1998b. The correlation between charge/discharge rates and morphology, surface chemistry, and performance of Li electrodes and the connection to cycle life of practical batteries. *Electrochemistry Society*. 145, 5: 1421–1426.

Aurbach, D., Markovsky, B., Levi, M.D. et al. 1999. New insight into the interactions between electrode materials and electrolyte solutions for advanced nonaqueous batteries. *Journal of Power Sources*. 81–82: 95–111.

Aurbach, D., Lu, Z., Schechter, A. et al. 2000. Prototype systems for rechargeable magnesium batteries. *Nature*. 407: 724–727.

Aurbach, D., Zinigrad, E., Teller, H. et al. 2002. Attempts to improve the behavior of Li electrodes in rechargeable lithium batteries. *Electrochemistry Society*. 149, 10: A1267–A1277.

Aurbach, D., Weissman, I., Gofer, Y. et al. 2003. Nonaqueous magnesium electrochemistry and its application in secondary batteries. *The Chemical Record 3*. 3, 1: 61–73.

Aurbach, D., Suresh, G.S., Levi, E. et al. 2007. Progress in rechargeable magnesium battery technology. *Advanced Materials*. 19, 23: 4260–4267.

Avraham, E., Soffer, A., Bouhadana, Y. et al. 2008. Developing ion electroadsorption stereoselectivity, by pore size adjustment with chemical vapor deposition onto active carbon fiber electrodes. *Journal of Physical Chemistry C*. 112, 19: 7385–7389.

Aymard, L., Lenain, C., Courvoisier, L. et al. 1999. Effect of carbon additives on the electrochemical properties of $AB_5$ graphite composite electrodes prepared by mechanical milling. *Electrochemistry Society*. 146, 6: 2015–2023.

Barker, J., Saidi, M.Y., and Swoyer, J.L. 2003. Lithium iron (II) phospho-olivines prepared by a novel carbothermal reduction method. *Electrochemical and Solid-State*. 6, 3: A53–A55.

Basu, S. and Choudhury, S.R. 2007. Phosphoric acid fuel cell technology. In: *Recent Trends in Fuel Cell and Technology*, ed. S. Basu, pp. 191–217. New Delhi, India: Anamaya.

Baughman, R.H., Zakhidov, A.A., and de Heer, W.A. 2002. Carbon nanotubes the route toward applications. *Science*. 297: 787–792.

Besenhard, J.O. 1999. *Handbook of Battery Materials*. New York: Wiley-VCH.

Bhattacharyya, S. and Gedanken, A. 2008. A template-free, sonochemical route to porous ZnO nanodisks. *Microporous and Mesoporous Materials*. 110: 553–559.

Bhattacharyya, A.J., Maier, J., Bock, R. et al. 2006. New class of soft matter electrolytes obtained via heterogeneous doping. *Solid State Ionics*. 177: 2565–2568.

Brinker, C.J. and Scherer, G.W. 1990. Sol-gel science: The physics and chemistry by a solvothermal method. *Material Science*. 41: 5696–5698.

Chen, J., Kuriyama, N., Yuan, H. et al. 2001. Electrochemical hydrogen storage in $MoS_2$ nanotubes. *American Chemical Society*. 123: 11813–11814.

Chen, J., Wang, S., and Whittingham, M.S. 2007. Hydrothermal synthesis of cathode materials. *Journal of Power Sources*. 174: 442–448.

Chen, H., Armand, M., Demailly, G. et al. 2008. From biomass to a renewable $LixC_6O_6$ organic electrode for sustainable Li ion batteries. *Chemistry & Sustainability*. 1: 348–355.

Chen, C., Ding, N., Wang, L. et al. 2009. Some new facts on electrochemical reaction mechanism for transition metal oxide electrodes. *Journal of Power Sources*. 189: 552–556.

Cho, J., Kim, Y.J., and Park, B. 2000. Novel $LiCoO_2$ cathode material with $Al_2O_3$ coating for a Li ion cell. *Chemical Materials*. 12: 3788–3791.

Choi, H., Kang, S.O., Ko, J. et al. 2009. An efficient dye-sensitized solar cell with an organic sensitizer encapsulated in a cyclodextrin cavity. *Angewandte Chemie*. 48: 5938–5941.

Conway, B.E. 1999. *Electrochemical Supercapacitors: Scientific Fundamentals and Technological Applications*. New York: Kluwer Academic/Plenum Publishers.

Crompton, T.R. 2000. *Battery Reference Book*, 3rd edn. Oxford: Newnes. *Current Opinion in Solid State and Materials Science*. 6: 507–512.

D'ebart, A., Bao, J., Armstrong, G., and Bruce, P.G. 2007. An $O_2$ cathode for rechargeable lithium batteries. *Journal of Power Sources* 174: 1177–1182.

Dan, P., Mengeritsky, E., Geronov, Y. et al. 1995. Performances and safety behaviour of rechargeable AA-Size Li/$Li_xMnO_2$., *Journal of Power Sources*. 54, 1: 143–145.

Deiss, E., Holzer, F., and Haas, O. 2002. Modeling of an electrically rechargeable alkaline Zn-/air battery. *Electrochimica Acta*. 47: 3995–4010.

Dell, R.M. and Rand, D.A.J. 2001. *Understanding Batteries*. Cambridge, U.K.: The Royal Society of Chemistry.

Deng, S.R., Kong, L.B., Hu, G.Q. et al. 2006. Benzene based polyorganodisulfide cathode materials for secondary lithium batteries. *Electrochimica Acta*. 51: 2589–2593.

Dhas, N.A. and Gedanken, A. 1997. Sonochemical synthesis of molybdenum oxide and molybdenum carbide silica nanocomposition. *Chemical Material*. 9: 3144–3154.

Dhas, N.A. and Patil, K.C.J. 1993. Combustion synthesis and properties of fine-particle rare-earth-metal zirconates, $L_{n2}Z_{r2O}7$. *Material Chemistry*. 3: 1289–1294.

Dhas, N.A., Raj, C.P., and Gedanken, A. 1998. Preparation of luminescent silicon nanoparticles. *Chemical Material*. 10: 3278–3281.

Dor, S., Grinis, L., Ruhle, S. et al. 2009. Electrochemistry in mesoporous electrodes. *Journal of Physical Chemistry* C.113, 5: 2022–2027.

Dresselhaus, M.S., Dresselhaus, G., and Avouris, P. 2001. *Carbon Nanotubes: Synthesis, Structure, Properties, and Applications*. Berlin, Germany: Springer.

Duclot, M. and Souquet, J.L. 2001. Glassy materials for lithium batteries. *Journal of Power Sources*. 97–98: 610–615.

Endo, M., Kim, Y.A., Hayashi, T. et al. 2001. Vapor-grown carbon fibers (VGCFs) Basic properties and their battery applications. *Carbon*. 39: 1287–1297.

Endo, M., Hayashi, T., Kim, Y.A. et al. 2004. Applications of carbon nanotubes in the twenty first century. *Philosophical Transactions of the Royal Society of London*. 362: 2223–2238.

Endo, M., Hayashi, T., Kim, Y.A. et al. 2008. Development and application of carbon nanotubes. *Association of Asia Pacific Physical Societies Bulletin*. 18, 1: 3–11.

Fan, C., Song, X., Yin, Z. et al. 2006. Preparation of $SnO_2$ hollow nanospheres by a solvothermal method. *Material Science*. 41: 5696–5698.

Fievet, F., Lagier, J.P., Blin, B. et al. 1989. Homogeneous and heterogeneous nucleations in the polyol process for the preparation of micron and submicron size metal particles. *Solid State Ionics* 32–33, 1: 198–205.

Gaberscek, M. and Stane, P. 1999. Time evolution of the impedance response of a passive film. *Electrochemical So*ciety. 146, 3: 933–940.

Garcia, E.J., Wardle, B.L., and Hart, A.J. 2008. Joining prepreg composite interfaces with aligned carbon nanotubes. *Composites Part A-Applied Science and Manufacturing*. 39: 1065–1070.

Gnanaraj, J.S, Pol, V.G., Gedanken, A. et al. 2003. Improving the high temperature performance of LiMn2O4 spinel electrodes by coating the active mass with MgO via a sonochemical method. *Electrochemistry Communications*. 5, 11: 940–945

Gopukumar, S.K., Chung, Y., and Kim, K.B. 2004. Novel synthesis of layered $LiNi_{1/2}Mn_{1/2}O_2$ as cathodes material for lithium rechargeable cells. *Electrochimica Acta*. 49: 803–810.

Grinis, L., Dor, S., Ofir, A. et al. 2008a. Electrophoretic deposition and compression of titania nanoparticle films for dye-sensitized solar cells. *Journal of Photochemistry and Photobiology A: Chemistry*. 198: 52–59.

Grinis, L., Ofir, A., Dor, S. et al. 2008b. Collector–shell mesoporous electrodes for dye sensitized solar cells. *Israel Journal of Chemistry*. 48: 269–275.

Guo, Y.G., Hu, Y.S., Sigle, W. et al. 2007. Superior electrode performance of nanostructured mesoporous $TiO_2$ (anatase) through efficient hierarchical mixed conducting. *Advanced Materials*. 19: 2087–2091.

Haik, O., Martha, S.K., Sclar, H. et al. 2009. Characterization of self combustion reaction for the production of nanomaterials used as advanced cathode in Li-ion batteries. *Thermochimica Acta*. 493: 96–104.

He, X., Wang, L., Pu, W. et al. 2006. Synthesis of spinel $LiMn_2O_4$ for Li-ion batteries via sol-gel process. *Electrochemical Science*. 1: 12–16.

Hench, L.L. and West, J.K. 1990. The sol-gel process. *Chemical Review*. 90, 1: 33–72.

Hochgatterer, N.S., Schweiger, M.R., Koller, et al. 2008. Silicon/graphite composite electrodes for high capacity anodes. *Electrochemical and Solid-State Letters*. 11, 5: A76–A80.

Hoenlein, W., Kreupl, F., Duesberg, G.S. et al. 2003. Carbon nanotubes for microelectronics. *Materials Science & Engineering C—Biomimetic and Supramolecular Systems*. 23: 663–669.

Hofmann, S., Csanyi, G., Ferrari, A.C., et al. 2005. Surface diffusion. *Physical Review Letters*. 95, 3.

Hu, W.K., Gao, X.P., Geng, M.M. et al. 2004. A low-cobalt $AB_5$-type hydrogen storage alloy stabilized by an inert second phase for use in low-cost, high-performance NiMH cells. *Electrochemical and Solid State Letter*. 7, 11: A439–A41.

Hu, Y.S., Guo, G., Dominko, R. et al. 2007. Improved electrode performance of porous $LiFePO_4$ using $RuO_2$ as an oxide nanoscale interconnect. *Advanced Materials*. 19: 1963–1966.

Hu, Y.S., Cakan, R.D., Titirici, M.M. et al. 2008. Superior storage performance of a Si–$SiO_x$/C nanocomposite as anode material for lithium ion batteries. *Angewandte Chemie*. 47: 1645–1649.

Hu, Y.S., Liu, X., Muller, J.O. et al. 2009. Synthesis and electrode performance of nanostructured $V_2O_5$ by using a carbon tube in tube as a nanoreactor and an efficient mixed conducting network. *Angewandte Chemistry*. 48: 210–214.

Hwang, C.C., Huang, T.H., Tsai, J.S. et al. 2006. Combustion synthesis of nanocrystalline ceria ($CeO_2$) powders by a dry route. *Material Science and Engineering B*. 132: 229–238.

Iijima, S. 1991. Helical microtubules of graphitic carbon. *Nature*. 354: 56–58.

Irzh, A. and Gedanken, A. 2009. A microwave assisted process for coating polymer and glass surfaces with semiconducting ZnO submicron particles. *Journal of Applied Polymer Science*. 113: 1773–1780.

Ivanovskayaa, V.V. and Seifert, G. 2004. Tubular structures of titanium disulfide $TiS_2$. *Solid State Communications*. 130: 175–180.

Jan, H.Y. Shu, G.D, Tie, L.G. et al. 2007. Synthesis and characterization of Li $(Ni_{1/3}Co_{1/3}Mn_{1/3})_{0.96}Si_{0.03}O_{1.96}F_{0.04}$ as a cathode material for lithium ion battery. *Materials Chemistry Physics*. 106: 354–359.

Jeevanandam, P., Koltypin, Y., Gedanken, A. et al. 2000. Synthesis of α-cobalt (II) hydroxide using ultrasound radiation. *Material Chemistry*. 10: 511–514.

Jung, M., Kim, H.G., Lee, J.K. et al. 2004. EDLC characteristics of CNTs grown on nanoporous templates. *Electrochimica Acta*. 50: 857–862.

Jung, H.G., Oh, S.W., Ce, J. et al. 2009. Mesoporous $TiO_2$ nano networks. *Electrochemistry Communications*. 11: 756–759.

Kang, B. and Ceder, G. 2009. Battery materials for ultrafast charging and discharging. *Nature*. 458: 190–193.

Kim, H. and Cho, J. 2008. Superior lithium electroactive mesoporous Si-carbon core-shell nanowires for lithium battery anode material. *Nano Letters*. 8, 11: 3688–3691.

Kim, D.H., Ahn, Y.S., and Kim, J. 2005. Polyol mediated synthesis of $Li_4Ti_5O_{12}$ nanoparticle and its electrochemical properties. *Electrochemistry Communications*. 7: 1340–1344.

Kirubakaran, A., Jain, S., and Nema, R.K. 2009. A review on fuel cell technologies and power electronic interface. *Renewable and Sustainable Energy Reviews*. 13: 2430–2440.

Kobayashi, H., Sizemore, H., Tabuchi, M. et al. 2000. Electrochemical properties of hydrothermally obtained $LiCo_{1-x}Fe_xO_2$ as a positive electrode material for rechargeable lithium batteries. *Electrochemical Society*. 147, 3: 960–969.

Koltypin, M., Gedanken, A., Pol, V. et al. 2007. The study of carbon-coated $V_2O_5$ nanoparticles as a potential cathodic material for Li rechargeable batteries. *Electrochemical Society*. 154, 7: A605–A613.

Kosova, N.V., Asanov, I.P., Devyatkina, E.T. et al. 1999. State of manganese atoms during the mechanochemical synthesis of $LiMn_2O_4$. *Solid State Chemistry*. 146: 184–188.

Kovacheva, D., Gadjov, H., Petrov, K. et al. 2002. Synthesizing nanocrystalline LiMn2O4 by a combustion route. *Material Chemistry*. 12: 1184–1188.

Lee, H., Kim, M., and Cho, G. 2007. Olivine $LiCoPO_4$ phase grown $LiCoO_2$ cathode material for high density Li batteries. *Electrochemistry Communications*. 9, 1: 149–154.

Li, J., Lewis, R.B., and Dahn, J.R. 2007a. Sodium carboxymethyl cellulose: A potential binder for Si negative electrodes for Li ion batteries. *Electrochemical and Solid-State Letters*. 10, 2: A17–A20.

Li, Y., Zhan, H., Kong, L. et al. 2007b. Electrochemical properties of PABTH as cathode materials for rechargeable lithium battery. *Electrochemical Communications*. 9: 1217–1221.

Linden, D. 1994. *Handbook of Batteries*, 2nd edn. New York: McGraw-Hill.

Maier, J. 2007. Size effects on mass transport and storage in lithium batteries. *Journal of Power Sources*. 174: 569–574.

Martha, S.K., Markovsky, B., Grinblat, J. et al. 2009. $LiMnPO_4$ as an advanced cathode material for rechargeable lithium batteries. *Electrochemical Society*. 156, 7: A541–A552.

Mitelman, A., Levi M.D., Lancry, E. et al. 2007. New cathode materials for rechargeable Mg batteries. *Chemical Communications*. 41: 4212–4214.

Mor, G.K., Varghese, O.K., Paulose, M. et al. 2006. A review on highly ordered vertically oriented $TiO_2$ nanotube arrays. *Solar Energy Materials & Solar Cells*. 90: 2011–2075.

Mukai, K., Ariyoshi, K., and Ohzuku, T. 2005. Comparative study of $Li[CrTi]O_4$, $Li[Li_{1/3}Ti_{5/3}]O_4$ and $Li_{1/2}Fe_{1/2}[Li_{1/2}Fe_{1/2}Ti]O_4$ in non-aqueous lithium cells. *Journal of Power Sources*. 146, 1–2: 213–216.

Nihei, M., Kawabata, A., Kondo, D. et al. 2005. Electrical properties of carbon nanotube bundles for future via interconnects. *Japanese Journal of Applied Physics*. 44, 4A: 1626–1628.

Novak, P., Muller, K., and Haas, O. 1997. Electrochemically active polymers for rechargeable batteries. *Chemical Reviews*. 97, 1: 207–282.

Odani A., Nimberger, A., Markovsky, B. et al. 2003. Development and testing of nanomaterials for rechargeable lithium batteries. *Journal of Power Sources*. 119–121: 517–521.

Odani, A., Pol, V.G., Gedanken, A. et al. 2006. Testing carbon coated $VO_x$ prepared via reaction under autogenic pressure at elevated temperature as Li-insertion materials. *Advanced Materials*. 18, 11: 1431–1436.

Osaka, T., Momma, T., Ito, H. et al. 1997. Performances of lithium/gel electrolyte/polypyrrole secondary batteries. *Journal of Power Sources*. 68: 392–396.

Oshima, T., Kajita, M., and Okuno, A. 2004. Development of sodium-sulfur batteries. *International Journal of Applied Ceramic Technology*. 1, 3: 269–276.

Oumellal, Y., Rougier, A., Tarascon, J.M., et al. 2009. 2LiH + M (M = Mg, Ti). *Journal of Power Sources*. 192: 698–702.

Padhi, A.K., Nanjundaswamy, K.S., Goodenough, J.B. et al. 1997. Effect of structure on the $Fe^{3+}/Fe^{2+}$ redox couple in iron phosphates. *Electrochemistry Society*. 144, 5: 1609–1613.

Park, S.H., Kang, S.H., Johnson, C.S. et al. 2007. Lithium-manganese-nickel-oxide electrode with integrated layered spinel structures for lithium batteries. *Electrochemistry Communications*. 9, 2: 262–268.

Park, B.C., Kim, H.B., Myung, S.T. et al. 2008. Improvement of structural and electrochemical properties of $AlF_3$-COATED $Li[Ni_{1/3}Co_{1/3}Mn_{1/3}]O_2$ cathode materials on high voltage region. *Journal of Power Sources*. 178: 826–831.

Patil, K.C., Aruna, S.T., and Mimani, T. 2002. Combustion synthesis: An update. *Current Opinion in Solid State and Materials Science*. 6: 507–512.

Peled, E. and Yamin, H. 1979. Solid electrolyte interphase (SEI) electrodes Part I. *Israel Journal of Chemistry*. 18: 131–135.

Peled, E., Golodnitsky, D., Ardel, G. et al. 1995. The SEI model-application to lithium polymer electrolyte batteries. *Electrochimica Acta*. 40, 13–14: 2197–2204.

Piran, M., Kotlyar, V., Medina, D.D. et al. 2009. End selective functionalization of carbon nanotubes. *Material Chemistry*. 19: 631–638.

Pistoia, G. and Broussely, M. 2007. *Industrial Applications of Batteries: From Cars to Aerospace and Energy Storage*. Amsterdam, the Netherlands: Elsevier.

Poizot, P., Laruelle, S., Grugeon, S. et al. 2000. Nano-sized transition-metal oxides as negative-electrode materials for lithium-ion batteries. *Nature*. 407: 496–499.

Pol, V.G., Pol, S.V., and Gedanken, A. 2005. Novel synthesis of high area silicon carbide by RAPET of organosilanes. *Chemical Material*. 17: 1797–1802.

Polak, E., Levy, N., Eliad, L. et al. 2008. Review on engineering and characterization of activated carbon electrodes for electrochemical double layer capacitors and separation processes. *Israel Journal of Chemistry*. 48: 287–305.

Puretzky, A.A., Geohegan, D.B., Jesse, S. et al. 2005. In situ measurements and modeling of carbon nanotube array growth kinetics during chemical vapor deposition. *Applied Physics A: Materials Science & Processing*. 81: 223–240.

Ren, X., Zelenay, P., and Thomas, S. 2000. Recent advances in direct methanol fuel cells. *Journal of Power Sources*. 86: 111–116.

Rydh, C.J. and Sanden, B.A. 2005. Energy analysis of batteries in photovoltaic systems Part I, II. *Energy Conversion and Management*. 46: 1957–2000.

Sakka, S. 2005. *Sol-Gel Science and Technology*. U.K.: Kluwer Academic Publishers.

Scrosati, B. 1995. Challenge of portable power. *Nature*. 373: 557–558.

Shahi, K., Wagner, J.B., and Owens, B.B. 1983. Solid electrolyte for lithium cells. In: *Lithium Batteries*, ed. J.P. Gabano, pp. 407–445. London, U.K.: Academic Press.

Sheem, K., Lee, Y.H., Lim, H.S. et al. 2006. High density positive electrodes containing carbon nanotubes for use in Li ion cells. *Journal of Power Sources*. 158: 1425–1430.

Singhal, S.C. and Kendall, K. 2003. *High-Temperature Solid Oxide Fuel Cells: Fundamentals, Design and Applications*. Oxford, U.K.: Elsevier.

Soiron, S., Rougier, A., Aymard, L. et al. 2001. Mechanochemical synthesis of Li-Mn-O spinels. *Journal of Power Sources*. 97–98: 402–405.

Song, M.S., Kang, Y.M., Kim, J.H. et al. 2007. Simple and fast synthesis of $LiFePO_4$-C composite for lithium rechargeable batteries by ball-milling and microwave heating. *Journal of Power Sources*. 166: 260–265.

Stambouli, A.B. and Traversa, E. 2002. Solid oxide fuel cells (SOFCs). *Renewable and Sustainable Energy Reviews*. 6: 433–455.

Stephan, A.M. 2006. Review on gel polymer electrolytes for lithium batteries. *European Polymer Journal*. 42: 21–42.

Stephan, A.M. and Nahm, K.S. 2006. Review on composite polymer electrolytes for lithium batteries. *Polymer*. 47: 5952–5964.

Subramanian, V., Zhu, H., and Wei, W. 2006. Synthesis and electrochemical characterization of amorphous manganese oxide and single walled carbon nanotube composites as supercapacitor electrode materials. *Electrochemistry Communications*. 8: 827–832.

Sun, X.G. and Kerr, J.B. 2006. Synthesis and characterization of network single ion conductors based on comb-branched polyepoxide ethers and lithium bis(allylmalonato)borate. *Macromolecules*. 39, 1: 362–372.

Sun, Y.K., Myung, S.T., Park, B.C. et al. 2009. High-energy cathodes material for long-life and safe lithium batteries. *Nature Materials*. 8: 320–324.

Tabuchi, M., Ado, K., Kobayashi, H. et al. 1999. Preparation of $LiCoO_2$ and $LiCo_{1-x}Fe_xO_2$ using hydrothermal reactions. *Material Chemistry*. 9: 199–204.

Takahara, H., Tabuchi, M., Takeuchi, T. et al. 2004. Application of lithium metal electrodes to all-solid-state lithium secondary batteries using $Li_3PO_4$-$Li_2S$-$SiS_2$ glass. *Electrochemical Society*. 151, 9: A1309–A1313.

Tal, O., Remskar, M., Tenne, R. et al. 2001. The effect of substrate topography on the local electronic structure of $WS_2$ nanotubes. *Chemical Physics Letters*. 344: 434–440.

Talyossef, Y., Markovsky, B., Lavi, R. et al. 2007. Comparing the behavior of nano- and microsized particles of $LiMn_{1.5}Ni_{0.5}O_4$ spinel as cathode materials for Li-ion batteries. *Electrochemical Society*. 154, 7: A682–A691.

Tasis, D., Tagmatarchis, N. Bianco, A. et al. 2006. Chemistry of carbon nanotubes. *Chemistry Review*. 106: 1105–1136.

Tillard, M., Belin, C., Spina, L. et al. 2005. Phase stabilities, electronic and electrochemical properties of compounds in the Li-Al-Si system. *Solid State Sciences*. 7, 9: 1125–1134.

Tominaga, Y., Asai, S., Sumita, M. et al. 2005. A novel composite polymer electrolyte. *Journal of Power Sources*. 146: 402–406.

Tong, T., Zhao, Y., Delzeit, L. et al. 2007. Dense vertically aligned multiwalled carbon nanotube arrays as thermal interface materials. *IEEE Transactions on Components and Packaging Technologies*. 30, 1: 92–100.

Travitsky, N., Burstein, L., Rosenberg, Y. et al. 2009. Effect of methanol, ethylene glycol and their oxidation by products on the activity of Pt-based oxygen–reduction catalysts. *Journal of Power Sources*. 194: 161–167.

Vaughey, J.T., Johnson, C.S., Kropf, A.J. et al. 2001. Structural and mechanistic features of intermetallic materials for lithium batteries. *Journal of Power Sources*. 97–98: 194–197.

Verneker, V.P.R. 1986. Energy Citations Database (ECD), US patent 4751070, OSTI.

Villevieille,C., Bousquet, C.M.I., Ducourant, B. et al. 2007. $NiSb_2$ as negative electrode for Li-ion batteries. *Journal of Power Sources*. 172: 388–394.

Wachtler, M., Ostrovskii, D., Jacobsson, P. et al. 2004. A study on PVDF-based $SiO_2$–containing composite gel-type polymer electrolytes for lithium batteries. *Electrochimica Acta*. 50, 2–3: 357–361.

Wakihara, M. and Yamamoto, O. 1998. *Li-Ion Batteries: Fundamentals and Performance*. New York: Wiley-VCH.

Walton, R.I. 2002. Subcritical solvothermal synthesis of condensed inorganic materials. *Chemistry Society Review*. 31: 230–238.

Wang, D., Buqa, H., Crouzet, M. et al. 2009. High performance, nano structured $LiMnPO_4$ synthesized via a polyol method. *Journal of Power Sources*. 189: 624–628.

Wen, S., Jung, M. Joo, O.S. et al. 2006. EDLC characteristics with high specific capacitance of the CNT electrodes grown on nanoporous alumina templates. *Current Applied Physics*. 6: 1012–1015.

Wikipedia, Allotropes of carbon. http://en.wikipedia.org/wiki/Allotropes_of_carbon

Wilkening, M., Lyness, C., Armstrong, A.R. et al. 2009. Diffusion in confined dimensions: Li+ transport in mixed conducting $TiO_2$-B nanowires. *Journal of Physical Chemistry C Letters*. 113, 12: 4741–4744.

Winter, M. and Besenhard, J.O. 1999. Electrochemical lithiation of tin and tin based intermetallics and composites. *Electrochimica Acta*. 45, 1–2: 31–50.

Wolfenstine, J., Allen, J.L., Read, J. et al. 2006. Chemistry and structure of Sony's Nexelion Li-ion electrode materials. *Storming Media*. 20.

Yabuuchi, N. and Ohzuku, T. 2005. Electrochemical behaviors of $LiNi_{1/3}Co_{1/3}Mn_{1/3}O_2$ in lithium batteries at elevated temperatures. *Journal of Power Sources*. 146: 636–639.

Yang, S. and Knickle, H. 2002. Design and analysis of aluminum/air battery system for electric vehicles. *Journal of Power Sources*. 112: 162–173.

Yang, M.R., Ke, W.H., and Wu, S.H. 2005. Preparation of LiFePO4 powders by co-precipitation. *Journal of Power Sources*. 146: 539–543.

Yazami, R. 2001. From room to Como. *Journal of Power Source*. 97–98: 33–38.

Zhang, S.Q., Xie, S., and Chen, C.H. 2005. Fabrication and electrical properties of $Li_3PO_4$ based composite electrolyte films. *Materials Science and Engineering B*. 121: 160–165.

Zheng, J.P., Cygan, P.J., and Jow, T.R. 1995. Hydrous ruthenium oxide as an electrode material for electrochemical capacitors. *Electrochemical Society*. 142, 8: 2699–2703.

# 10

# *Quantum Dots Designed for Biomedical Applications*

**Andrea Ragusa, Antonella Zacheo, Alessandra Aloisi, and Teresa Pellegrino**

**CONTENTS**

## 10.1 Introduction

Colloidal semiconductor nanocrystals, or quantum dots (QDs), are inorganic semiconductor nanocrystals with sizes of a few nanometers (Alivisatos 1996; Schmidt 2006). The unique optical properties of these nanoparticles (NPs) brought researchers to exploit them in many biomedical fields, from labeling and imaging to detection and sensoring, as well as gene and drug delivery. The huge interest in QDs is due to their peculiar optical properties, their relatively cheap cost of fabrication, and the easy functionalization of their surfaces for bioconjugation. There is a great interest in trying to develop better synthetic strategies able to yield new nanocrystals with a precise control over shape and composition, from which the optical properties depend. Colloidal synthesis is able to give the best results from this point of view, also with great control over monodispersity (Ozin and Arsenault 2005). Still, nanocrystals synthesized in organic solvents have to be transferred into aqueous solvents before being exploited for biological applications. In fact, biomedicine is a field where semiconductor nanocrystals, and QDs in particular, are being employed the most, substituting more traditional organic dyes (Doshi and Mitragotri 2009; Kim and Dobson 2009; Peteiro-Cattelle et al. 2009).

In this chapter, we first give a general overview of the physicochemical properties of QDs and the procedures used to synthesize and functionalize them for exploitation in a biological environment. We then review the state of the art of the QD-related literature examining the many biological fields where they have found application and where they are bringing important innovations, from biosensoring to labeling and imaging, both *in vitro* and *in vivo*, before exploring new areas where they are being exploited, such as photodynamic therapy (PDT) and multimodal imaging techniques.

## 10.2 Physical Properties, Preparation, and Functionalization of Quantum Dots

### 10.2.1 Optical Properties

The main characteristic of semiconductor QDs is that their physical properties and their optical properties are different from those of the corresponding bulk material (Klimov 2003). This derives from the very tiny dimensions of QD nanocrystals, from a few nanometers up to a few tens of nanometers, which then obeys the laws of quantum physics and not "classical" physics as is the case for bulk semiconductor crystals. In a QD, the electronic energy levels are not as many as in the bulk semiconductor, and the difference between them can be relatively large. The energy gap between the valence band and the conduction band depends on the size, i.e., the number of atoms of the QD (Alivisatos 1996; Schmidt 2006). The different distribution of the energy levels can be clearly noticed in the absorption spectra of QDs (Figure 10.1). The typical absorption spectrum of a QD presents a broad absorption, covering hundreds of nanometers, which allows exciting many QDs with just a single source. Many peaks can also be also observed in a QD absorption spectrum, corresponding to the various allowed electronic transitions (which create electron–hole pairs, also known as "excitons").

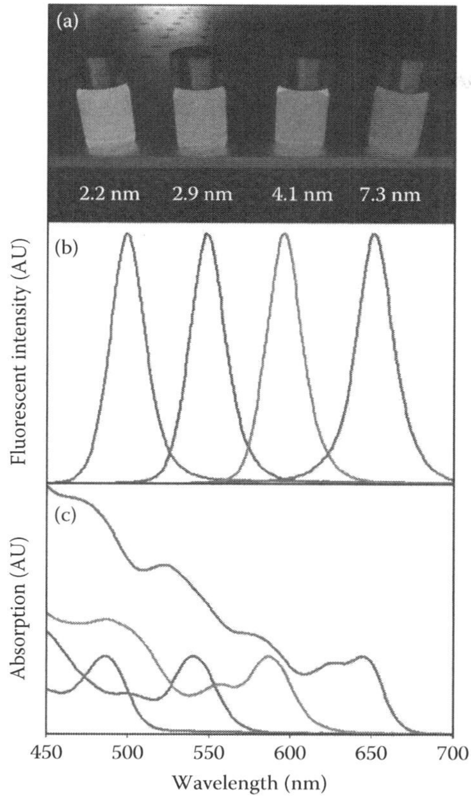


**FIGURE 10.1**
**(See color insert following page 302.)** (a) Size-dependent optical properties of cadmium selenide QDs with diameters ranging from 2.2 to 7.3 nm and corresponding emission (b) and absorption (c) spectra. (Adapted from Smith, A.M. et al., *Adv. Drug Deliv. Rev.*, 60, 1226, 2008. With permission.)

However, the smaller the dot, the larger the energy gap between the conduction and the valence band will be, thus yielding a bigger blue-shifting of the first exciton peak (corresponding to a higher energy) in the absorption spectrum. When excited, because of the radiative recombination of the electron–hole pairs, the QD emits with a narrow and symmetric peak slightly redshifted compared to the lowest energy absorption one. Since the position (i.e., the wavelength) of the emission peak is also dependent on the size of the QD, it is possible to obtain a variety of colors with one type of QD just by modifying its diameter. Furthermore, if we take into account the different types of chemical composition that can be used to fabricate QDs, it is possible to obtain any type of color ranging from the ultraviolet to the infrared region. CdSe are the most common type of QDs in biological applications because, by tuning their size, almost any region of the visible spectrum can be covered (Murray et al. 1993; Rogach et al. 1998). Moreover, addition of an outer shell of inorganic nanocrystals with a higher band gap, usually ZnS or CdS, improves their optical properties yielding the so-called core@shell QDs (Hines and Guyotsionnest 1996; Dabbousi et al. 1997; Peng et al. 1997). This shell improves the robustness against photo-oxidation and enhances the photoluminescence (PL) quantum yield (QY) of the core. Also, addition of a shell of ZnS has been proven to reduce toxicity, probably by decreasing the leakage of Cd ions from the surface of the QD into the solution.

Compared to traditional organic dyes, QDs present many advantages. As already mentioned, one of them is the possibility of exciting many QDs, each with a different color, with a single excitation source. This is possible because of the broad absorption spectra of the QDs compared to those of the organic dyes. On the other hand, the emission spectra are very narrow and symmetric, with the absence of the red tail observed in the typical spectra of organic dyes. Nowadays, high-temperature colloidal syntheses allow for the generation of nearly monodisperse semiconductor nanocrystals with size distribution below 5%, which results in emission spectra with peak widths as narrow as 25–30 nm. This allows for the simultaneous use of different-color QDs excited with a single source and whose peaks can be spectrally resolved in a quite limited range of wavelengths, with tremendous impact on many biological applications. Furthermore, the PL lifetime of QDs can be up to 100 ns, while traditional organic dyes have PL lifetimes of few nanoseconds, allowing the time-gated collection of photons, which increases enormously the signal-to-noise ratio. On the other hand, QDs present a phenomenon, also presented by many organic fluorophores, called "blinking" which is usually undesired, especially for tracking studies (Neuhauser et al. 2000; Schuster et al. 2005). This phenomenon refers to the intermittency in light emission; it is related to the "Auger recombination" and it is directly proportional to the power of the excitation source, with "off times" ranging from hundreds of microseconds to hundreds of seconds (Efros et al. 1995; Neuhauser et al. 2000; Zegrya and Samosvat 2007). Still, progress in chemical synthesis and surface functionalization allowed to reduce this phenomenon considerably (Hohng and Ha 2004; Fomenko and Nesbitt 2008).

An interesting property of fluorophores is that the energy employed to excite a fluorophore can be transferred to another fluorophore in close proximity if their corresponding emission and absorption spectra overlap. This process is called "Förster (of fluorescence) resonance energy transfer" (FRET) and occurs through dipole–dipole interactions, with all the limitations and conditions of this type of interaction (Lakowicz 2006). When this process occurs, the first fluorophore, the donor, transfers a certain amount of energy to a second fluorophore, the acceptor, thus quenching itself. On the other hand, the acceptor has now sufficient energy to relax emitting a photon. QDs perfectly adapt to FRET applications, in particular, as FRET donors although some examples of QDs as FRET acceptors have been also reported, and they have already been widely employed in many biological applications, especially in biosensoring, as we see in the following sections.

### 10.2.2 Synthesis of Quantum Dots

For the chemical preparation of QDs, procedures that aim at the synthesis of nanoparticles with an accurate control over size, size distribution, and crystallinity of the nanoparticles are highly desired, as all of those parameters define the optical properties of QDs. In the last decade, significant advances have been made on the colloidal synthesis of nanocrystals in high boiling point organic solvents (Ozin and Arsenault 2005; Schmidt 2006). This procedure usually relies on the decomposition at high temperatures of organometallic precursors in presence of an appropriate mixture of surfactant molecules. These amphiphilic surfactant molecules in a nonpolar medium act as coordinating agents for both the atomic species and for growing the nanocrystals, and allow to control their reactivity, such that high-quality samples can be prepared. By these methods QDs that comprise a combination of elements from II and VI groups (such as CdSe, CdS, and ZnO) as well as less commonly from III and V groups (InAs, InP, InSb) can be prepared. This approach can be also exploited for the synthesis of different geometries of QD-based heterostructures in which the QDs nanocrystals are grown within the same nano-object together with other inorganic nanoparticles (for instance, magnetic–fluorescent nanostructures of iron oxide QDs can be prepared by this method). In addition, with this procedure reasonable amounts of product are produced and the approach can be scaled up. This method delivers QDs coated by organic surfactant molecules and they are thus soluble in organic solvent (such as toluene, hexane, and chloroform). Hence, further procedures to transfer the as synthesized QDs from the nonpolar environment to aqueous solutions are required.

### 10.2.3 Water-Transferring Procedures

Chemical engineering of the QDs surface is crucial in order to obtain NPs, which preserves at the best the optical and physical properties of the nanoparticles and, at the same time, allow obtaining NPs colloidally stable in physiological conditions. As documented by the relevant number of papers published in this field over the last 12 years, since the first experiment has demonstrated the use of QDs in biology, several groups have been working on the development of procedures for transferring the nanoparticles from the organic phase, in which the nanoparticles are synthesized, to the aqueous phase, typical of biological systems. The surface of the QD is crucial not only for the preservation of its optical properties, which are influenced by the type of ligands and for the manipulation in biological media, but also for rendering biocompatible the nanoparticles. Later in this chapter, the correlation between the surface of the nanoparticles and toxicity will be taken for consideration. In this section, we focus on the procedures developed for the transfer of QDs into aqueous solution. It is well accepted that two major techniques are now available for the water solubilization of nanoparticles: (1) the ligand-exchange procedure and (2) the encapsulation of nanoparticles within a protecting shell (Tomczak et al. 2009). The ligand exchange procedure is based on the replacement of the original surfactant molecules at the nanoparticles surface with hydrophilic ligands. The ligands are chosen such that they carry moieties that have strong affinity for the atoms at the QD surface. Molecules bearing thiols (Nabiev et al. 2007), phosphonic acids (Milliron et al. 2003), and pyridine (Skaff and Emrick 2003) groups have been proven to be good ligands that bind tightly to the QD surface and can thus easily replace the surfactant molecules. At the same time, the ligands possess molecular portions able to stabilize the nanoparticles in water (Figure 10.2). The ligand stabilization can be conferred by mainly choosing molecular portions able (1) to introduce charges at the nanoparticle surface (Aldana et al. 2001; Goldman et al. 2005a,b; Nabiev et al. 2007) or (2) to act as polymer brushes. Examples of electrostatic stabilization are given by the use of ligands bearing functional groups, for example, amines (e.g., mercaptoalkyl amine) or carboxylic acids (e.g. mercaptoalkyl carboxylic acid), which, depending on the pH of the media, can be, respectively, in a protonated/unprotonated form thus repelling each other and disfavoring QDs aggregation. On the other hand, the exchange with ligands bearing polyethylene glycol (PEG) moieties are examples of polymer brush molecules that can stabilize the nanoparticles by exploiting the steric stabilization or the hydrogen bonding formation (the ether groups in the case of PEG molecules) (Goldman et al. 2005a,b; Susumu et al. 2007; Chen et al. 2008). It is important to underline that the PEG-functionalized nanoparticles have shown reduced nonspecific binding to biological components, and PEG molecules are also exploitable as spacers for further nanocrystal functionalization.

However, the ligand exchange procedure has some limitations. One of the main critical aspects is the limited stability in water. Small ligands linked at the surface via chemisorption

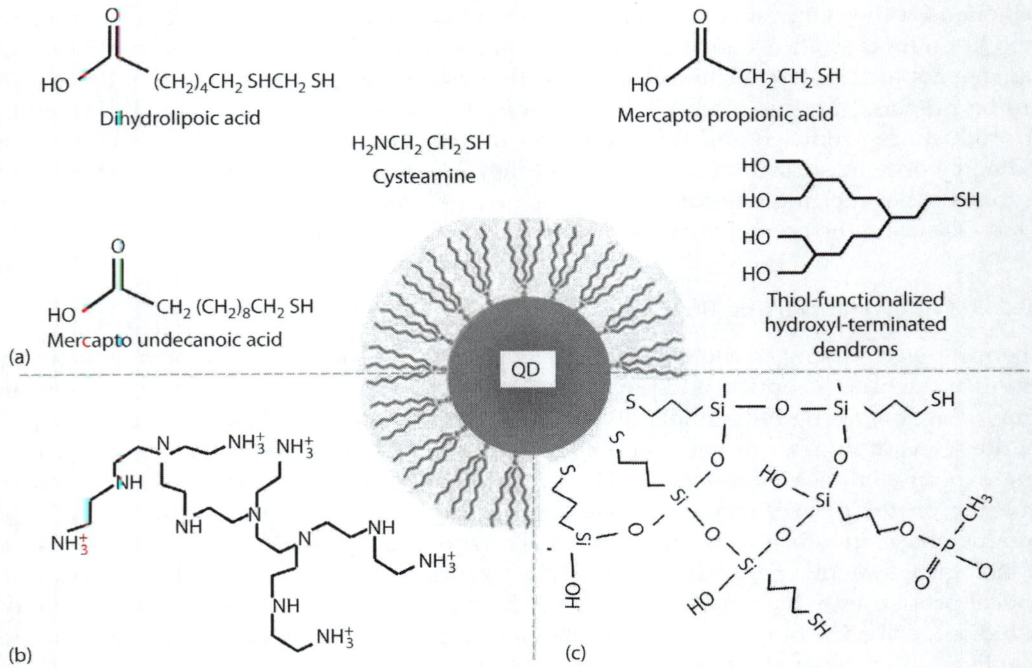


**FIGURE 10.2**
Examples of hydrophilic molecules used in the ligand exchange procedure for transferring QDs from the organic to the aqueous phase: (a) mercapto-based ligands, (b) polydentate ligands based on poly(amido amine), (c) silane-based ligands.

of the mercapto groups are easily oxidized to disulfides and tend to come off from the QD surface. This affects also the photochemical stability because, when the ligands are removed, the QD surfaces are exposed to water and oxygen species, thus quenching the PL and facilitating the precipitation of the sample on a long-term scale. To overcome these drawbacks, polydentate ligands have been also tried to stabilize the ligands at the QD surface by the presence of several bridges between the ligand and the nanoparticle (Xiong et al. 2002; Chen et al. 2004; Pinaud et al. 2004; Nann 2005). To cite some examples, poly(dimethylamino ethyl metacrylate) molecules, amine-containing polymeric ligands, have been used to replace trioctylphosphine oxide (TOPO) surfactant at the nanoparticles surface, and the resulting nanoparticles were stable both in water and toluene (Chen et al. 2004; Cai et al. 2006). Other amine-based linkers which proved to be good multidentate ligands are amine-containing dendrimers, which are second generation of poly(amido amine) (G2-PAMAM) dendrimeric ligands (Xiong et al. 2002). In this view also, hyperbranched polyethylenimine (PEI) have been proven to be good polydentate stabilizer for transferring QD nanoparticles into water (Nann 2005). Also, organic hydroxyl terminated dendrons functionalized with thiols have been used as good ligands in ligand exchange procedure, since they were shown to render the QDs water soluble and to efficiently protect the nanoparticles surface against oxidation compared to thiol-based ligands. The advantage of such shell is to be very thin and closely packed, and thus difficult to be removed (Rosenthal et al. 2002). In order to make the ligand exchange shell strongly bound to the QD surface, a reasonable alternative to the multidentate ligands is to cross-link, by chemical reactions, the ligand molecules exchanged at the nanoparticles surface. This is the case of dendrons carrying terminal vinylic functionalities, which are cross-linked via ring-closing metathesis reactions (Guo et al. 2003). However, in this example of cross-linking reaction, the QDs were soluble only in organic surfactant due to the characteristics of the shell molecules. An alternative to water-soluble dendrons cross-linked shell was obtained by cross-linking OH-terminated dendrons with amine-terminated dendrimers (Guo et al. 2003). Another alternative example of ligand exchange followed by cross-linking is given by the growth around each individual QD of a shell of silica (Correa-Duarte et al. 1998; Alejandro-Arellano et al. 2000; Parak et al. 2002b; Holzinger et al. 2006). Priming the surface first with a derivate molecule of the silane, such as mercaptopropyltrimethoxysilane, allows the displacement of the original TOPO surfactant molecules. Basification and heating allow for the hydrolysis of the trimethoxysilane and consequently the cross-linking through the formation of siloxane bonds (Si–O–Si). This procedure, although has been extended to different types of QDs and has provided QDs much more stable in water than the original ligand exchange procedure, is laborious and time consuming.

An approach for the water solubilization of nanoparticles that relies on a different principle is the encapsulation of surfactant-coated QDs within an amphiphilic polymeric shell (Figure 10.3) (Dubertret et al. 2002; Gao et al. 2004; Feng et al. 2005; Geissbuehler et al. 2005; Jin et al. 2005; Tortiglione et al. 2007). With respect to the ligand exchange procedure, in this case, the surfactant molecules are not replaced at the QD surface; they are instead wrapped by the polymer units. The affinity between the surfactant molecules and the hydrophobic portions of the encapsulating molecules is exploited in order to maintain tightly the enwrapping polymer molecules at the nanoparticles surface, while the polar head of the polymer are used for giving functionalities that make the nanoparticles charged. Different polymer molecules have been exploited for such transfer. In some cases, block copolymer carrying a more hydrophobic polymeric portion and a distinct polymer portion more hydrophilic have been used. This is the case, for instance, of the block copolymer used by Durbrertet et al., which is based on a mixture of *n*-poly(ethylene glycol)

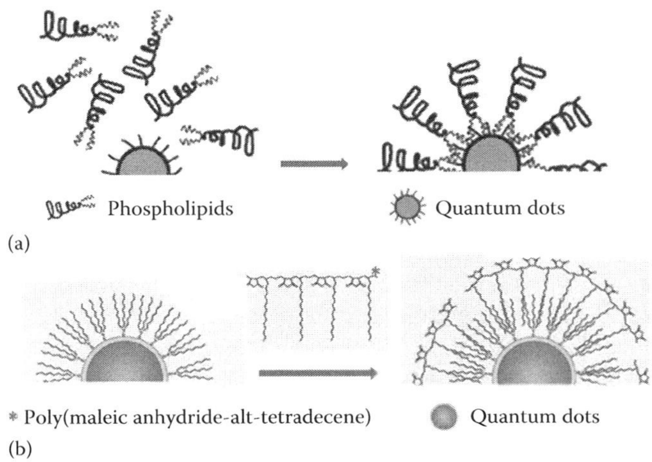


**FIGURE 10.3**
Schematic picture of the water-transferring procedures based on the polymer encapsulation. The amphiphilic molecules, either (a) the phospholipids (Adapted from Dubertret, B. et al., *Science*, 298, 1759, 2002. With permission.) or (b) the amphiphilic polymer, such as the poly(maleic anhydride-alt-tetradecene) in the scheme, are used to enwrap the hydrophobic coated QDs within the polymeric shell bearing some hydrophilic units which stabilize the QDs in water.

phosphatidylethanolamine and phosphatylcholine (Dubertret et al. 2002). This is also the case of the tri-block copolymer used by another group based on two hydrophobic sections, the polybutylacrylate and the polymethylacrylate, an additional hydrocarbon chain, and a hydrophylic section based on polymethylacrylic acid (Gao et al. 2004).

In this case, the strong hydrophobic interaction given by the side chains of the block copolymer yielded a spontaneous assembly of the polymer around each QD and resulted in a strong fluorescence and good pH and salt stability of the resulting nanoparticles. As an alternative choice, the polymer molecules could be made of monomer units consisting of a polar head group, which represents the hydrophilic portion of the polymer and alkyl chain units, which instead are the hydrophobic portion of the polymer monomer unit. Several alkyl poly(maleic anhydride)-based polymers (poly(maleic *alt*-tetradecene), poly(maleic *alt*-octadecene), and poly(styrene-co-maleic anhydride) (Tortiglione et al. 2007; Di Corato et al. 2008; Lees et al. 2009) have been proven to be useful for such transferring procedure and easy to extend to QDs of different compositions.

The alkyl chains of the polymer intercalate among the surfactant molecules of the nanoparticles and the further cross-linking of the anhydride groups through diamine or triamine facilitate the opening of the anhydride and the formation of a compact shell around each nanoparticle (Tortiglione et al. 2007). Sometimes the cross-linking step has been demonstrated to be not necessary (Cai et al. 2007; Di Corato et al. 2008; Lees et al. 2009). The choice of the molecular weight of the polymer together with the choice of the lateral alkyl chain allow to tune from few to some tens of nanometers (2–10 nm) the coating thickness at the QD surface. In addition, most of these poly(maleic anhydride)-derivated polymers are commercially available, and it is thus an easily applicable procedure.

As a general consideration, it is important to underline that different procedures can introduce different molecules with different functional groups (carboxy, amines, thiols, etc.) available at the QD surface, which could allow further conjugation with almost any biomolecule. Given the number of papers appeared in the last decade in the field of water transferring and solubilization, it is straightforward the importance of achieving control over the surface of the nanoparticles in order to better manipulate and further process the nanoparticles.

## 10.3 Quantum Dots as Biosensors

The optical properties of QDs combined with their ability to be functionalized with a variety of biomolecules make them ideal nanosensors for bioanalytical purposes. Since the PL of the QDs is highly dependent to their surface states, any chemical or physical modification occurring at their surface modify the efficiency of the radiative recombination leading to its enhancement or quenching (Murphy 2002). Following this principle, the changes induced by the specific interaction between a multitude of ions and molecules and the QDs surface, or the ligands bound to the QDs surface, have been widely exploited to develop ultrasensitive nanosensors able to detect analytes even at picomolar concentrations.

### 10.3.1 pH Nanosensors

Chromogenic molecules are able to change their ability to absorb electromagnetic radiation in response to chemical stimulations, and their conjugation to the surface of QDs allowed the development of pH-sensitive QDs. Tomasulo et al. linked to CdSe/ZnS QDs a [1,3]oxazine ring which, in a basic aqueous environment, opens up to generate a phenolate chromophore which quenches the QD, reducing its QY by 85% (Tomasulo et al. 2006). On the other hand, in acidic conditions, the phenol is protonated and there is no electron transfer with the QD, additionally increasing its QY by 33%. As a matter of fact, the authors claimed the QDs could probe the pH of aqueous solutions by adjusting their luminescence intensity to pH changes in the 3–11 range.

A different approach was exploited by Snee et al. who linked a pH-sensitive squaraine dye to the polymeric backbone of CdSe/ZnS QDs (Snee et al. 2006). Owing to the pH dependence of the dye absorption spectrum, the spectral overlap between the dye absorption and the QD emission increases in a linear manner as the pH is lowered, thus modulating the FRET efficiency of the system and allowing to determine the pH.

Thioglycolic acid–capped CdTe QDs also showed a linear decrease of the PL intensity in aqueous solutions when the pH was lowered from 6 to 4 (Susha et al. 2006). At a pH of 3.3, the QDs were completely quenched while no differences were observed in the 6–12 range. The same system also showed to be very sensitive to various biologically important cations, such as calcium(II), manganese(II), iron(III), silver (I), and mercury (II). Liu et al. also observed a monotonically linear decrease of the PL intensity of CdSe/ZnSe/ZnS core/shell/shell QDs capped with mercaptoacetic acid (MAA) with the lowering of the pH (Liu et al. 2007). The same effect could be detected both in living SKOV-3 human ovarian cancer cells, where the fluorescence intensity of internalized MAA-QDs was enhanced by 10-fold when the pH changed from 4 to 10.

Yu et al. used mercaptopropionic acid (MPA)-capped CdTe/ZnS QDs to determine the acidity in aqueous solutions (Yu et al. 2007). The fluorescence intensity of the QDs decreased linearly as the pH decreased in the range of 8.0–5.0, and this phenomenon was exploited to follow the kinetics of the enzymatic hydrolysis of glycidyl butyrate catalyzed by *porcine pancreatic* lipase. Similarly, Wang et al. used MAA-capped CdTe QDs to determine tiopronin in solution by measuring the pH change reaching, under optimal conditions, a limit of detection (LOD) of 0.15 μg/mL (Y. Q. Wang et al. 2008). Also, Huang et al. prepared mercaptosuccinic acid (MSA)–capped CdSe/ZnS QDs and exploited the linear increase of their PL intensity in the pH range of 8–11.5 to detect the amount of urea (Huang et al. 2007). By monitoring its urease-catalyzed hydrolysis, which releases hydroxide anions, they were able to determine the urea concentration in a range of 0.01–100 mM.

### 10.3.2 Ions Detection

Chen et al. first studied the influence of many physiologically important metal cations on CdS QDs capped with different ligands (Chen and Rosenzweig 2002). The best selectivities were observed with L-cysteine-QDs, which showed a PL enhancement when chelating $Zn^{2+}$, and with thioglycerol-QDs, which were quenched by $Cu^{2+}$. Later, Lin et al. developed a CdSe/ZnS QD functionalized with bovine serum albumin (BSA), which was able to selectively detect $Cu^{2+}$ ions with a detection limit of 10 nM (Lin et al. 2007). Apart from the just described copper sensors, Konishi et al. reported a CdS QD functionalized with a cluster molecule whose fluorescence increased upon complexation of $Cu^{+}$ ions, supposedly because of the formation of a network structure with S–Cu–S bridges (Konishi and Hiratani 2006). On the other hand, Singh et al. recently reported the synthesis of CdSe/ZnS QDs functionalized with a Schiff base, which allowed the selective and simultaneous detection of $Cu^{+}$ and $Fe^{3+}$ in semi-aqueous solution (Singh et al. 2008).

The fact that the ligand plays a fundamental role in ions recognition was recently evidenced by Zhang et al. (2009). Three thiolated ligands, MAA, L-cysteine, and reduced glutathione, were capped to CdTe QDs and the influence of their complexation ability on the PL intensity was studied systematically. The three types of QDs were all quenched by $Cu^{2+}$ and $Hg^{2+}$ ions, probably because of the formation of CuTe and HgTe particles on the surface of the nanocrystals, while were insensitive to many other physiologically important cations.

Normally, ion complexation induces a fluorescence quenching of the QD through inner-filter effects, nonradiative recombination pathways, and electron transfer processes (Wang et al. 2007). However, ions can also cause passivation of trap states or defects on the surface of the QD, thus inducing a PL enhancement, as observed by Moore et al. on CdS QDs (Moore and Patel 2001).

Determination of mercury ions has been also largely investigated because of its toxicity on humans (Clarkson 1997). Cai et al. was able to determine $Hg^{2+}$ in aqueous solution with L-cysteine-functionalized CdS QDs with a detection limit of 2.4 nM (Cai et al. 2006). Instead, Shang et al. prepared triethanolamine-capped CdSe QDs responsive only to the simultaneous presence in solution of $Hg^{2+}$ and $I^-$ (Shang et al. 2009).

Li et al. exploited an innovative supramolecular system to detect mercury ions with a detection limit in the nanomolar range (Li et al. 2007). Sulfur calixarenes were conjugated to CdSe/ZnS QDs whose fluorescence was quenched upon complexation of $Hg^{2+}$. Ruedas-Rama et al. also exploited a supramolecular system to recognize zinc ions (Ruedas-Rama and Hall 2008). Azamacrocycles-functionalized CdSe/ZnS QDs recovered the fluorescence upon complexation of $Zn^{2+}$ ions, probably by disrupting the interactions between the lone pair electrons of the nitrogens and the holes on the QD surface.

Another elegant approach exploited a FRET system between thioglycolic acid–CdTe QDs and butyl-rhodamine B (Li et al. 2008). The changes in the PL spectrum of the QD upon binding $Hg^{2+}$ induced a change in the PL spectrum of the organic dye, which was used to determine the metal cation up to nM concentrations.

### 10.3.3 Detection of Organic Molecules

Recently, the detection of organic compounds, such as explosives and pesticides, by using QDs is attracting much attention due to the need for fast, highly specific, and reliable tests.

Goldman et al. described the use of CdSe/ZnS QD-IgG antibody conjugates in a fluoroimmunoassay for the detection of 2,4,6-trinitrotoluene (TNT) (Goldman et al. 2002). The lowest detection limit based on a plate-based competition assay was reported to be 0.01 μg/mL, while a detection limit of 10 ng/mL total TNT was reported for a flow displacement assays. The same group later derivatised the same QDs with dihydrolipoic acid (DHLA) and recombinant anti-TNT antibodies engineered with a poly-histidine tag, although a higher concentration could be detected with this system (41 ng/mL) (Goldman et al. 2005a,b). Goldman et al. also exploited FRET to generate a TNT nanosensor (Goldman et al. 2005a,b). Anti-TNT-specific antibody fragments were conjugated to the QDs and a dye-labeled TNT analogue prebound in the antibody binding site quenched the QD PL. When TNT was added to the solution, it displaced the dyed analogue, eliminating FRET and allowing the recovery of the fluorescence in a concentration-dependent way.

More recently, Wilson et al. reported the use of multiplexed assay for the detection of explosives (Wilson et al. 2007). Three different types of explosives, TNT, pentaerythritol tetranitrate (PETN), and 1,3,5-trinitro-1,3,5-triazacyclohexane (RDX), could be simultaneously detected in a competitive immunoassay by magnetic microbeads encoded with

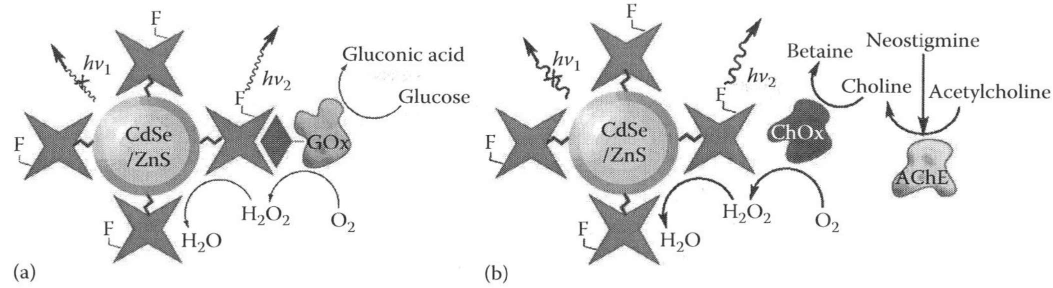


**FIGURE 10.4**
(a) Ratiometric analysis of glucose by biotinylated GOx associated with the fluorophore-functionalized avidin bound to the CdSe/ZnS QDs. (b) Ratiometric analysis of the activity of AChE by the fluorophore-modified avidin-capped CdSe/ZnS QDs, and its inhibition by neostigmine. (Adapted from Gill, R. et al., *Angew. Chem. Int. Ed.*, 47, 1676, 2008. With permission.)

particular spectral codes by loading them with red, green, and yellow QDs through a layer-by-layer technique.

Accurate detection of pesticides is also needed since they are still extensively used in agriculture despite the health and environmental problems they can cause when accumulated. Ji et al. electrostatically conjugated an organophosphorous (OP) hydrolase to MAA-CdSe/ZnS QDs and they were able to detect trace amounts of paraoxon in solution by measuring the quenching of the QDs upon its binding with a detection limit of 10nM (Ji et al. 2005). Vinayaka et al. were able to detect 2,4-dichlorophenoxyacetic acid (2,4-D) by conjugating it to the enzyme alkaline phosphatase (ALP)–functionalized CdTe QDs and quantitatively analyzing it by competitive fluoroimmunoassay (Vinayaka et al. 2009).

Gill et al. exploited the sensitivity of CdSe/ZnS QDs to $H_2O_2$ to monitor the activities of oxidases and to detect their substrates (Gill et al. 2008). As proof of principle, they were able to analyze glucose in the presence of glucose oxidase (Figure 10.4a). In a more complex system, they employed the fluorescent nanosensors to monitor the inhibition of acetylcholine esterase (AChE) (Figure 10.4b). The hydrolysis of acetylcholine by AChE generated choline, which was subsequently oxidized to betaine by choline oxidase, thus generating $H_2O_2$ and quenching the QDs. However, in the presence of the inhibitor neostigmine, the biocatalytic activity was interrupted and the QDs fluorescence was not quenched.

The quenching of QDs when exposed to tetraalkylammonium and alkyl sulfate salts, due to nonradiative recombination with deep-traps, less mobile holes, and stabilization of electron on the QD surface, was also exploited by Hamity et al. and, more recently, by Diao et al. to detect various cationic surfactants with good selectivity and sensitivity (Hamity et al. 1998; Diao et al. 2007). On the other hand, Qu et al. functionalized CdTe QDs with cyclodextrins to recognize polycyclic aromatic hydrocarbons in aqueous solutions through a supramolecular approach, with detection limits of 580 and 85nM for phenanthrene and acenaphtene, respectively (Qu and Li 2009).

### 10.3.4 Detection of Biomolecules

#### *10.3.4.1 Nucleic Acid Detection*

The specificity of the hybridization process between two complementary nucleic acid sequences is at the basis of many DNA-sensing approaches. From a nanotechnological point of view, the approaches usually involve the conjugation of a single-strand DNA fragment (ssDNA) to a QD, which is able to recognize a fluorescent target probe in solution, resulting in the quenching of QD fluorescence. However, many other variants have been developed such as fluorescence recovery upon hybridization, fluorescence in situ hybridization (FISH), and bioluminescence resonance energy transfer (BRET).

A basic example of a simple nanosensor was reported in 2005 by Zhang et al. who used CdSe/ZnS QDs linked to DNA sequences which were able to detect a dye-labeled reporter strand with high sensitivity and selectivity through FRET quenching (Zhang et al. 2005). Later, a similar assembly was exploited but using two QDs with different emission wavelengths in a donor-acceptor configuration in order to overcome the limitations of the organic dyes (Zhang and Johnson 2006). In another similar approach, ssDNA was linked to the QD while the complementary strand was linked to Au nanoparticles, leading to fluorescence quenching when hybridization occurs (Dyadyusha et al. 2005; Zhao et al. 2007).

An alternative approach was investigated by Gill et al., who hybridized nucleic acid-functionalized QDs and the dye-labeled complementary DNA sequence, thus leading to FRET, and monitored the fluorescence recovery upon exposure to DNase I (Gill et al. 2005). Similarly, Patolsky et al. were able to follow the dynamics of telomerization occurring between the DNA primer on the QD and the dye-labeled complementary strand by FRET enhancement as they were brought in close proximity (Patolsky et al. 2003). The same scheme was also exploited to follow DNA replication.

Peng et al. prepared thioglycolic acid (TGA) capped CdTe QDs and used electrostatic interactions to add a cationic polymer, which in turn acted as a bridge to link the dye-labeled ssDNA (Peng et al. 2007). Hybridization was recognized by the various FRET efficiencies, which were dependent on the different strengths of the electrostatic interaction between single-stranded and double-stranded DNA and the polymer. Electrostatic interactions have been also exploited by Yuan et al. who adsorbed mitoxantrone (MXT, an anticancer drug) on the QD surface quenching their fluorescence (Yuan et al. 2009). In the presence of DNA that could bind MTX, the QD PL could be recovered, allowing the detection of the nucleic acid with good sensitivity.

Nucleic acid–capped CdSe/ZnS QDs were also successfully exploited by Pathak et al. as probes for a modified version of FISH, allowing the detection of chromosome abnormalities or mutations with high sensitivity (Pathak et al. 2001). Similarly, Gerion et al. used QD-DNA conjugates as efficient probes for single nucleotide polymorphism and for multiallele detection in a microarray format (Gerion et al. 2003).

#### *10.3.4.2 Proteins and Enzymes*

The first example of protein conjugation to QDs was reported in 2000 by Mattoussi et al. who electrostatically bound a chimeric fusion protein based on the maltose binding protein (MBP) of *Escherichia coli* (Mattoussi et al. 2000). Since then, much progress has been made and several QD-based antibody conjugates have been prepared and applied in fluoroimmunoassay. Goldman et al. extended that concept to multiple detection by linking four different-color QDs to antitoxin antibodies, thus allowing the simultaneous detection of the corresponding toxins (Goldman et al. 2004). Lao et al. developed a simple method for the direct conjugation of IgG to CdSe/ZnS by using a genetically engineered fusion protein with protein L (a cell-wall component of *Peptostreptococcus magnus*), thus generating a sensitive immunofluorescent probe for the detection of a representative tumor antigen (Lao et al. 2006). Huang et al. developed a FRET-based probe for the quantitative determination of micrococcal nuclease (MNase) by conjugating dye-labeled ssDNA to CdSe/ZnS QDs through biotin-avidin linkage (Huang et al. 2008). Upon digestion of the ssDNA by the MNase in solution, the QD fluorescence was restored indirectly allowing the determination in culture medium of *Staphylococcus aureus*. Very recently, Bae et al. designed nickel–nitriloacetic acid (Ni–NTA)-functionalized CdTe/CdS QD clusters for localizing and isolating histidine-tagged fusion proteins (Bae et al. 2009). This Ni–NTA–QD cluster demonstrated to be very efficient, especially for targeting the 6x histidine region of tagged proteins, due to its high affinity, site specificity, and reversibility.

Many peptide-capped QDs have been also employed to monitor enzymatic activity, which can be related to important biological processes and diseases. For example, proteases activities and their modulation (e.g., collagenes, thrombin, and chymotrypsin) were detected upon specific cleavage of the dye- or AuNP-labeled peptide conjugated to the QDs, thus restoring their fluorescence (Chang et al. 2005; Medintz et al. 2006a). Recently, Boeneman et al. conjugated a modified fluorescent mCherry protein to QDs and this system was able to detect the presence of caspase enzymes, which cleaved the target peptide sequence thus reducing the FRET efficiency (Boeneman et al. 2009).

An alternative approach was used by Yildiz et al. who electrostatically bound to the surface of CdSe/ZnS fluorescent nanocrystals a biotin-functionalized bypiridinium molecule acting as QD quencher via positron emission tomography (PET) (Yildiz et al. 2006). Upon addition of streptavidin, the quencher was removed and the QD fluorescence restored.

Other elegant approaches are based on BRET, which is a naturally occurring phenomenon in which a light-emitting protein (the donor) transfers energy in a nonradiative (dipole–dipole) way to a suitable fluorescent protein (Ward and Cormier 1978; Wilson and Hastings 1998). In this way, there is no need for an external source of light for exciting the donor, as instead occurs with FRET.

So et al. conjugated a mutant *Renilla* luciferase with eight mutations (RLuc8) to CdSe/ZnS QDs and, since the corresponding emission and absorption spectra of the portions perfectly overlapped, QDs were efficiently excited in the absence of external light when RLuc8 bound to its substrate coelenterazine (So et al. 2006). Later, the same group successfully applied a similar system to create a highly sensitive nanosensor that could detect the activity of matrix metalloproteinases (MMPs) (Yao et al. 2007). On the other hand, Huang et al. created a modified BRET sensor by employing a chemiluminescent donor instead of a bioluminescent one (Huang et al. 2006). The chemiluminescent oxidation of luminol by hydrogen peroxide, catalyzed by horseradish peroxide, excited the QD acceptor generating a simple and sensitive immunoassay.

#### *10.3.4.3 Other Biomolecules*

Great interest has attracted the development of nanosensors for the detection of amino acids. Wang et al. linked *p*-sulfonatocalix(*n*)arene to CdSe QDs allowing the recognition in physiological buffer of methionine and phenylalanine (X. Wang et al. 2008). Instead, Han et al. conjugated cyclodextrins to the surface of CdSe/ZnS QD allowing the enantioselective recognition of tyrosine and methionine (Han and Li 2008). Furthermore, within a certain concentration range, one enantiomer of the chiral amino acid enhanced the QD fluorescence while the other had no effect. Recently, Huang et al. also noticed a fluorescence enhancement of MAA-capped QDs upon the selective binding of L-cysteine, with a detection limit in the nM range (Huang et al. 2009).

QD probes have been also exploited for monitoring carbohydrates. Medintz et al. engineered an MBP with a oligo-histidine tag and electrostatically bound it to negatively charged QD (Medintz et al. 2003). Maltose could be easily detected by observing the FRET changes upon its binding. Later, similar alternative approaches were also investigated but immobilized on glass substrates (Sapsford et al. 2004; Medintz et al. 2006b). More recently, Cao et al. developed a CdTe QD-based glucose nanosensor (Cao et al. 2008). Glucose oxidase (GOx) was covalently bound to the surface of the QD to catalyze the glucose oxidation thus generating $H_2O_2$ which quenched the QD fluorescence.

Apart from amino acids and carbohydrates, other biologically relevant molecules have been detected. Jin et al. developed a supramolecular system covering CdSe/ZnS QDs with *p*-sulfonatocalix[4]arene receptors hosting acetylcholine, which induced FRET quenching (Jin et al. 2005). Upon recognition by the neurotransmitter acetylcholine, the QD fluorescence was recovered. He et al. used Mn-doped ZnS QDs to detect enoxacin, a quinolone antibiotic, which was able to quench the QD phosphorescence (He et al. 2008). Liu et al. developed a QD-based nanosensor able to simultaneously detect multiple analytes by using two QDs with different emission peaks (Liu et al. 2007). Conjugation with AuNPs through selected aptamers quenched the QDs fluorescence, which was restored when the target analyte disassembled the assembly. Recently, Zhang et al. prepared an aptameric nanosensor for cocaine based on dual FRET among QD, fluorophore, and quencher (Zhang and Johnson 2009). The aptamer was sandwiched between a fluorophore-labeled oligonucleotide and a quencher-labeled oligonucleotide, and the whole system was bound to the QD. Upon cocaine binding, the quencher-labeled oligonucleotide was released, thus activating the fluorescence of the fluorophore.

## 10.4 Quantum Dots for Cellular Imaging: The *In Vitro* Studies

### 10.4.1 Nonspecific and Specific Targeting of Cells

Since the first demonstrations of QD cell labeling by Alivisatos (Bruchez et al. 1998) and Nie (Chan and Nie 1998), QDs have been extensively used as fluorescent cells markers for high-resolution imaging, for the study of intracellular processes at the single-molecule level, and for high-speed applications such as flow cytometry. Furthermore, narrow and symmetric fluorescence emission spectra allow QDs to be exploited in simultaneous multicolor labeling of different structures in living cells.

Due to the unique photophysical properties of QDs, already elucidated in Section 10.2.1, different research groups have achieved considerable success in using them for *in vitro* bioassays (Smith and Nie 2004), for labeling fixed cells (Wu et al. 2003) and tissue specimens (Ferrara et al. 2006; Fountaine et al. 2006), and for imaging proteins on living cells (Rosenthal et al. 2002; Dahan et al. 2003; Lidke et al. 2004; Young and Rozengurt 2006; Roullier et al. 2009). QDs have been shown to be able to measure the action of individual molecular motors in the cytoplasm (Courty et al. 2006), to monitor antigen uptake by dendritic cells (DCs) (Cambi et al. 2007) and the membrane fusion of synaptic vesicles in neurons (Zhang et al. 2007; Zhang and Johnson 2009). Several methods have been exploited to efficiently delivery QDs to cells; however, they can be organized into four main groups: passive nonspecific uptake (Parak et al. 2002a; Hanaki et al. 2003); receptor-mediated internalization (Jaiswal et al. 2003; Derfus et al. 2004; Lidke et al. 2004); chemical transfection (Chen et al. 2004; Mattheakis et al. 2004); and mechanical delivery (Dubertret et al. 2002) (reviewed by Medintz et al. (2008) and Smith et al. (2008)). In this section, we skip some topics already extensively discussed in the literature, for instance, the feasibility of using QDs for antigen detection in fixed cellular monolayers (Bruchez et al. 1998), and we focus more specifically on intracellular staining aspects of living cells, such as nonspecific targeting for stem cells and specific organelles and protein tagging in eukaryotic cells. For example, we review the use of bioconjugated QDs for understanding mechanisms of action in virus, bacteria, and model organisms, such as yeast, zebrafish, and hydra. Moreover, we discuss on the availability of these imaging agents also as RNA interference and gene delivery tools for gene therapy.

### 10.4.2 Quantum Dots as Nonspecific Targeting Probes for Stem Cells Imaging

Stem cells offer an attractive new branch of therapy to treat numerous diseases. Thus, it is essential to develop systems to monitor cells survival, proliferation, as well as differentiation.

Results of several studies address QD effects on proliferating mesenchymal stem cells (MSCs), cell tracking, and engraftment in *in vitro* cocultures. Recently, QDs were shown to have potential advantages as fluorescent probes for stem cells labeling over traditional organic fluorophores (Voura et al. 2004; Gao et al. 2005). Murasawa et al. used two different protein-conjugated QDs ($\lambda_{max}$ 565 and 655 nm) to study progenitor cell fusion in long-term cocultures (Murasawa et al. 2005). Given the high proliferative nature of human MSCs (hMSCs) and their phenotypic changes during differentiation, Moioli et al. determined the efficiency of long-term labeling on hMSCs during proliferation and differentiation with bioconjugated QDs (Moioli et al. 2006). By means of CdSe/ZnS QDs functionalized with arginine-glycine-aspartic acid (RGD) peptides on their surface, the authors demonstrated that QDs are capable of labeling hMSCs during population doubling for the tested 22 days. Similarly, Seleverstov et al. explored the suitability of QDs for stem cell labeling by testing two QDs with identical chemical components but which differed in size by almost a factor of 2 (Giepmans et al. 2005; Seleverstov et al. 2006). In co-labeling experiments, they verified the different distribution of these two types of particles: QD525 fluorescence disappeared rapidly (after 2–7 days of culture) even in the presence of QD605 fluorescent signal (visible after 52 days) in the same hMSC cell. The authors also showed that the size-dependent uptake of QD is mediated by autophagy. Similarly, Hsieh et al. used CdSe/ZnS QDs for labeling MSCs maintained in differentiation medium supplemented with transforming growth factor β (Hsieh et al. 2006). Perinuclearly distributed QDs were visible for at least 2 weeks in cultured cells, without affecting chondrogenic differentiation, even though specific condrocyte protein expression was inhibited. Stem cells long-term tracking by means of QDs was also confirmed by identifying exogenous hMSCs in histological sections. Rosen et al. evidenced that MSC in culture retained QDs for more than 6 weeks, and 8 weeks after QD-loaded MSCs injection into infarcted myocardium, QDs fluorescence was still observable in tissue sections (Rosen et al. 2007). Furthermore, some of the labeled cells showed an endothelial phenotype. Similar findings to those of Seleverstov et al. and Rosen et al. were reported by Muller-Borer et al. who found that QDs tend to form large intracellular aggregates in the MSCs and that labeled MSCs coupled functionally with cardiomyocytes in coculture, indicating that QDs hold a promise as cell-labeling agents for tracking studies on the fate of MSCs in culture (Muller-Borer et al. 2007). In fact, they verified that QDs are inherited by daughter cells for at least 6 generations (~15 days). F. Laco et al. highlighted that only multiple QD events in cells are representative markers for locating QD-labeled MSCs (Laco et al. 2009). They suggest that QDs are not exocytosed from the endosomal vesicles in live cells and phagocytosis of dead QDs-labeled MSCs by other cocultured cells was not found in mixed culture studies. However, during cell splitting or mechanical disturbance, QD-positive dead cells can break up and free QDs will be re-endocytosed by the nearby cells, which were shown as single QD events. Therefore, in this work, only multiple QD events in cells have been considered as representative markers for locating QD-labeled MSCs. With this, the authors showed that the QD-labeled MSCs visibly proved the participation of MSCs within different epidermal layers on day 14. In a recent study, Wang et al. showed integration and differentiation of QD-loaded lineage negative bone marrow cells ($Lin^-$ BMCs) 4 weeks after transplantation into laser-induced retinal trauma, evidencing the ability to differentiate into retinal pigment epithelium (RPE), endothelial cells, pericytes, and photoreceptors (Wang et al. 2010). Indeed, it is becoming more and more clear that the visualization of stem cells trafficking is a critical point in following tissue regeneration from transplanted cells, and these studies, and others not mentioned herein, demonstrate the aptitude of QDs for high-quality tracking.

### 10.4.3 Quantum Dots as Specific Markers for Organelles (and Protein) Targeting in Eukaryotic Cells

Several QD delivery methods have been explored and several procedures are currently available to encapsulate and solubilize semiconductor QDs for biological applications (see Section 10.2.3 for details) (Michalet et al. 2005). If the designed destination of the cargo is a precise organelle, sequestration of the particles into vesicles, such as endosomes and lysosomes, is a critical factor for the successful delivery of single QDs into the cytoplasm of living cells, thus avoiding aggregation (Delehanty et al. 2006). In a recent work, Duan et al. have shown the use of surface-coating chemistry based on multivalent and endosome-disrupting (endosomolytic) surface coatings, such as PEG-grafted PEI, to deliver QD probes across the plasma membrane and to facilitate their release from subcellular organelles (Smith et al. 2008). Due to the cationic charges and the "proton sponge effect" (Neuhauser et al. 2000; Pack et al. 2005) associated with multivalent amine groups, these QDs were able to pass through the cell membrane and upset endosomal organelles in living cells. In particular, while only PEG coating led to QD entrapment in vesicles, the PEI-g-PEG coated QDs were able to escape from endosomes and free to move into the cytoplasm (reviewed by Smith et al. (2008)). Another approach was used by Kim et al. (2008), who designed a bioresponsive delivery system that underwent endolysosomal to cytosolic translocation via pH-dependent reversal of nanocomposite (poly(D,L-lactide-*co*-glycolide) (PLGA)) surface charge polarity incorporating antibody-coated QDs within biodegradable polymeric nanospheres (Figure 10.5).

In contrast, the endosome cargo confinement was exploited as a novel approach to analyze structural assembly, stability, and dynamics of axonal microtubules, which is of great interest for understanding neuronal functions and pathologies. In fact, Mudrakola and colleagues used nerve growth factor-activated receptor tyrosine kinase (NGF-TrKA) NGF-QD sequestration in endosomes to resolve more than six microtubules in an axon of 1 mm in diameter by real-time tracking of endosomic vesicles containing QDs (Mudrakola et al. 2009). They positioned the centers of moving endosomes labeled with NGF conjugated QDs ($\lambda_{max}$ 605 nm) with high accuracy at each time point. Time-lapse positions of a moving endosome reveal the unlabeled microtubule track along which the endosome travels by exploiting the fact that the centre of the point spread function (PSF) of a single emitter can be determined up to a few nanometers, a precision significantly greater than the diffraction limit of 200–300 nm (Yildiz et al. 2003; Kural et al. 2005; Moerner 2006). They measured that the vast majority (>80%) of the endosomes contain a single QD–NGF complex, as identified by QD photoblinking (Gao et al. 2004) taking advantage of the axonal transport process to separate single fluorophores. In another work, single-molecule

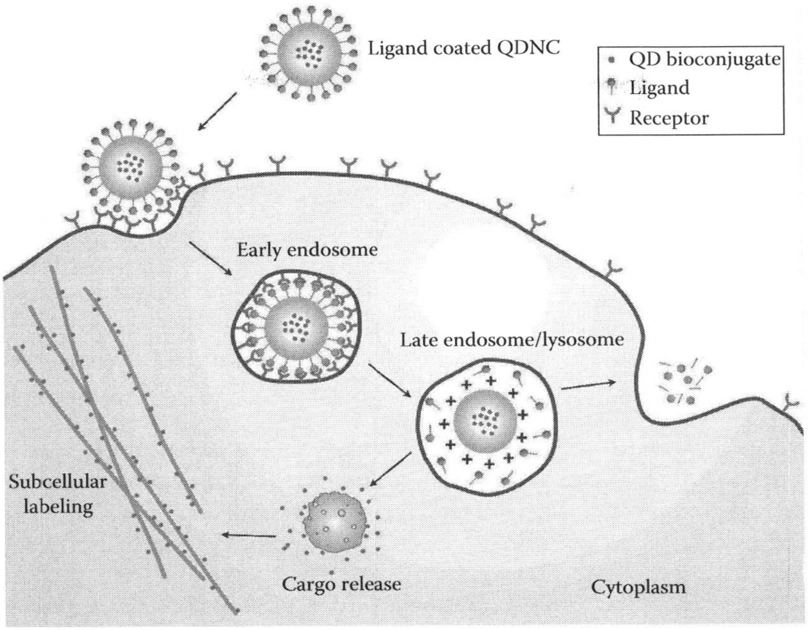


**FIGURE 10.5**
Mechanism of cytosolic delivery and subcellular targeting of QDs nanocomposite (QDNCs). Schematic representation depicting QDNC escape from the endolysosomal compartment upon cellular internalization with cytosolic release of the encapsulated cargo. Antibody-conjugated QDs can be delivered in this manner to allow the labeling of subcellular organelles or other molecular targets. (Adapted from Kim, B.Y. et al., *Nano Lett.*, 8, 3887, 2008. With permission.)

tracking using QDs has provided direct evidence of the clustering of acetylcholine receptors (AChRs) in muscle cells in response to synaptogenic stimuli by two distinct cellular processes: the Brownian motion of the receptors in the membrane and their trapping, and immobilization at the synaptic specialization (Geng et al. 2009).

This and other similar investigations, such as glycine receptors tracking (Dahan et al. 2003) and ion channels tracking (Haggie et al. 2006; O'Connell et al. 2006) highlight that, when specific proteins are coupled to their surface, QDs become powerful imaging agents for highly specific recognition and tracking of plasma membrane antigens and excellent probes for molecular localization in whole living cells. Previously, Lidke et al. attached red-light emitting CdSe/ZnS QDs to epidermal growth factor (EGF), a protein with a specific affinity for the erbB/HER membrane receptor (Lidke et al. 2004). Adding these conjugates to cultured human cancer cells, receptor-bound QDs could be branded at a single-molecule level. Specific QDs bio-conjugation has been also demonstrated to be an attractive method for targeting cancers cells. For instance, folic acid (FA) is widely used for the selective delivery of anticancer agents to cells over-expressing folate receptors (FR), which are present on the cellular wall of many types of human cancer cells, such as ovarian, breast, and prostate cancer cell (Sudimack and Lee 2000; Hilgenbrink and Low 2005). In a recent paper, Pan et al. proved that QDs encapsulated in FA-decorated nanoparticles of poly(lactide)-alpha-tocopheryl polyethylene glycol succinate (vitamin E TPGS) and vitamin E TPGS-carboxyl (PLA-TPGS/TPGS-COOH) copolymer mix, are viable tools for the targeted imaging of cancer cells also improving specificity and sensitivity as well as reducing the cytotoxic side effects of bare QDs to normal cells (Fischer et al. 2006). They investigated the *in vitro* cellular uptake of such nanoparticles and observed a much higher internalization of the FA-decorated QDs-loaded polymeric NPs by MCF-7 breast cancer cells, over-expressing high levels of FA receptors, compared to NIH 3T3 fibroblast cells, which express lower levels of FA receptors. In an analogue study, Manzoor et al. (2009) examined the potential use of FA–ZnS QDs for the targeted cancer imaging in FR+ KB oral cancer cells compared with four different negative controls. They tested a new class of heavy-metal-free QD bio-probes based on single-phase "doped ZnS," and they carried out cytotoxicity studies using bare and FA-conjugated ZnS (FA–ZnS) QDs. They found that the nature of the interaction was considerably altered when the FA–QDs were incubated with FR+ KB cells and observed a very specific aggregation of larger concentration of QDs on the cell membrane, unlike the case of control experiments. Indeed, this type of cancer cells demarcation offers very useful applications in fluorescent histopathology of tissue samples.

Aiming at appreciating the QDs cellular uptake mechanism to a broader level and at achieving a disclosure of the mechanisms of nanoparticle interaction with specific intracellular structures, size- and charge-selective nuclear delivery of QDs in human cells was analyzed due to its outstanding potential for bioimaging and therapeutics. Lovric et al. reported that very small QDs (2.2nm) coated with cysteamine translocated to the nuclei of murine microglial cells following cellular uptake through passive endocytosis (Lovric et al. 2005). In contrast, larger QDs (5.5nm) and small QDs bound to albumin only remained in the cytosol. Live human macrophages were shown to be able to rapidly uptake and accumulate QDs in distinct cellular compartment depending on QDs size and charge. Nabiev et al. studied a size-dependent QD segregation trend in human macrophages and found that small QDs may mark histones in cell nuclei by a multistep process involving endocytosis, active cytoplasmic transport, and entering the nucleus via nuclear pore complexes (Nabiev et al. 2007). In addition, they observed that treating the cells with the anti-microtubule agent nocodazole precludes QDs cytoplasmic transport, whereas the nuclear import inhibitor thapsigargin blocks QD import into the nucleus. A year later, Conroy et al. discussed the unmodified CdTe QDs particular tropism to the histone proteins, which resulted in a dramatic shift of the absorption band, and decrease in the PL intensity of the QDs (Nabiev et al. 2007). The possible reason for the QDs lifetime reduction observed in the nucleus and nucleoli could be caused by aggregation of the QDs, mediated by the binding of the negatively charged QDs to the core histones, which are approximately positively charged (Hansen et al. 1998).

Certainly, QD–histone interactions could provide the basis for QD nuclear localization downstream of intracellular transport mechanisms, and it is clear that unfunctionalized QDs exploit the cell's active transport machineries for the delivery to specific intranuclear destinations (Nabiev et al. 2007). In two different experimental works, Ruan and Biju used two small peptides for nuclear targeting and, while HIV TAT peptide-conjugated QDs failed, the attachment of insect neuropeptide allostatin to the QDs showed an efficient transfection of 3T3 and A431 cells (Ruan et al. 2007; Biju et al. 2009).

Yum et al. specifically targeted the QD delivery to the nucleus of living HeLa cells by means of a nanoscale mechanochemical method (Yum et al. 2009). They used a membrane-penetrating Au-coated nanoneedle to deliver QDs to the cytoplasm and to the nucleus of living cells. Since the nucleus has a reducing environment (Arrigo 1999; Schafer and Buettner 2001), a delivery strategy based on the reductive cleavage of disulfide bonds is also applicable. Similarly, more invasive microinjection of peptide-conjugated PEG-QD showed the ability to direct QD to specific sites, such as mitochondria (28mer mitochondrial localization sequences (MLS)) or nucleus (23mer nuclear localization sequence (NLS)) from SV40 T antigen (Derfus et al. 2004).

QDs appear to be useful tools not only to target specific subcellular structures, but also to conduct studies on cellular transport. One example is reported in the work carried out by Nelson et al. (2009). Introducing QD-labeled Myosin Va into mammalian COS-7 cells by pinocytosis, they monitored the intracellular motion paths associated with single myoVa motors using total internal reflection fluorescence (TIRF) microscopy, with imaging restricted to the subplasmalemmal actin cortex. By this system, the authors were able to observe the individual and sequential steps of a single myoVa heavy meromyosin motor as it processively carried its QD cargo through the actin cytoskeleton, similar to the QD-labeled myosin V heads processivity visualized by Warshaw et al. (2005). In another recent work on the vesicular secretion of neurotransmitters, the authors reported a new approach for studying the process of membrane fusion and retrieval. By loading individual synaptic vesicles with single QDs and monitoring the pH-dependent PL alteration, the authors were able to distinguish kiss-and-run (K&R) from full-collapse fusion and to track single vesicles without affecting the vesicle cycle (Zhang et al. 2009). Indeed, using QDs, Zhang et al. developed a method that enabled them to make clear the uncertainty about K&R at small nerve terminals of the central nervous system (Aravanis et al. 2003; Fernandez-Alfonso and Ryan 2004; Harata et al. 2006; Balaji and Ryan 2007), showing once more that these fluorescent agents offer the prospect to get sharp optical signals at single-event resolution (Figure 10.6).

### 10.4.4 Quantum Dots for siRNA and Gene Therapy

Gene therapy is a method by which proper DNA sequences are inserted into target cells as corrective genetic material. On the other hand, at a posttranscriptional level, the delivery of short RNA sequences, so-called interference-RNA (RNAi), which inhibits gene expression (gene silencing) primarily by targeting messenger-RNA sequences (mRNAs), is exploited in short interfering RNA (siRNA) therapy. RNAi was first observed in the nematode worm *Caenorhabditis elegans* (Fire et al. 1998). RNAi has soon become a promising tool for sequence-specific gene silencing when Tushl et al. showed that RNAi in mammalian cells was mediated by 21-22-nucleotides RNA sequences (Elbashir et al. 2001). In recent years, this kind of therapeutic modality reaches particular relevance because it has the prospective to modulate "non-druggable" targets (Troy et al. 2004; Uprichard 2005; Dykxhoorn et al. 2006). A gene encoding the antisense RNA can be introduced into the cell organisms by using different vectors, including plasmid vector and lipofectamine.

In order to elucidate more deeply the siRNA process, organic dyes have been used to tag siRNA to the delivery vehicles (Hoshino et al. 2004; Troy et al. 2004; Rieger et al. 2005). However, the photobleaching of the dye fluorophores has limited the long-term tracking of RNAi. In addition, as it has been underlined in numerous reports, one critical issue in siRNA therapy is the transfection efficiency, which is too low (Itaka et al. 2004; Muratovska and Eccles 2004). These needs have pushed research to test new materials which could be used as cargo to improve RNAi delivery, but also to switch to inorganic fluorophores, such as QDs, for imaging the entire RNAi delivery process.

In 2005, Chen et al. have co-delivered green QDs and siRNA for silencing the lamin a/c gene into murine fibroblasts by using standard transfection systems, such as cationic liposomes (Chen et al. 2005). Flow cytometry analysis, based on the uptaken intracellular QDs, showed that gene silencing of co-transfected cells correlated directly with intracellular fluorescence level, allowing selection of a uniformly silenced cell cluster by fluorescence-activated cell sorting. Given the optical properties of QDs, they were particular

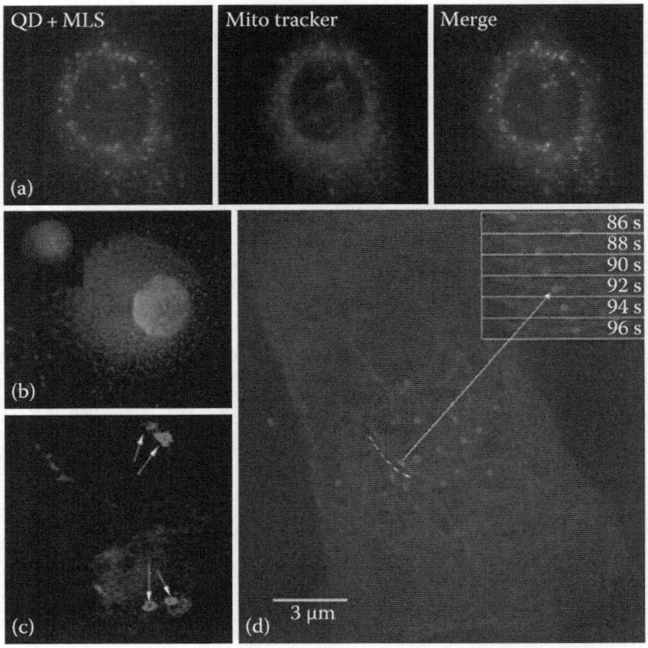


**FIGURE 10.6**
**(See color insert following page 302.)** Fluorescence imaging of distinct cellular compartments labeled with differently colored QDs. (a) Fluorescence micrograph of 3T3 fibroblast after 24 h injection of MLS-QDs; co-localization with Mitotracker Red confirmed mitochondrial labeling. (Adapted from Derfus, A.M. et al., *Adv. Mater.*, 16, 961, 2004. With permission.) (b) Mixture of red and green CdTe QDs injected into the cytoplasm of macrophages separates into two distinctive cytosolic and nuclear compartments within 60 min of observation. In the inset, same cell 15 min post-microinjection. (c) Confocal optical section of macrophages showing nucleolar localization of green-emitting QDs (evidenced by the arrows). (Adapted from Nabiev, I. et al., *Nano Lett.*, 7, 3452, 2007. With permission.) (d) Real-time imaging and tracking of siRNA-QD nanocomplexes in living breast cancer cells by using spinning disk confocal microscopy The QD-siRNA complexes were found to undergo active and directional motion, and their trajectory and velocity were similar to active vesicle transport medicated by molecular motors. The red dots in the image are QD clusters and not of single QDs. In the inset a time-lapsed image series showing a single nanocomplex moving along a microtubule. (Adapted with permission from Yezhelyev, M.V. et al., *J. Am. Chem. Soc.*, 130, 9006, 2008. With permission.)

appealing for multiplexed monitoring and sorting cells that were transfected at the same time with different siRNA/QD pairs. Limiting factors for this method were represented by endosomal escape, dissociation of siRNA and QDs from the carrier (unpacking), and coupling ability of the delivered RNAi with the multiprotein RNA-induced silencing complex (RISC) (Figure 10.7).

An improved approach with respect to the one just cited was reported by Qi and Gao, who exploited the electrostatic interaction between the siRNA double-strand sequence and the surface of QDs coated by an amphipol polymer (poly(maleic anhydride-*alt*-1-decene) modified with dimethylamino propylamine (PMAL)) to associate RNAi and QDs (Qi and Gao 2008). The authors have unexpectedly observed that, once the nanostructures were endocytosed, both the tertiary amines and carboxylic groups on the QD surface played important roles in endosome RNAi escape leading to an increased siRNA-mediated knockdown. Furthermore, they suggested that the association of the siRNA onto these QDs surfaces provides a mechanism for siRNA protection from the enzyme degradation.

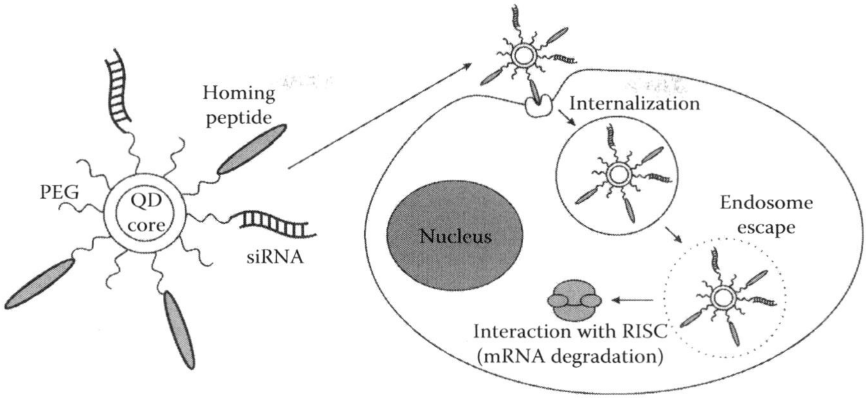


**FIGURE 10.7**
Design of a multifunctional nanoparticle for siRNA delivery. Because of their photostable fluorescence and multivalency, QDs are suitable vehicles for ferrying siRNA into live cells *in vitro* and *in vivo*. Conjugation of homing peptides (along with the siRNA cargo) to the QD surface allows targeted internalization in tumor cells. Once internalized, these particles must escape the endolysomal pathway and reach the cytoplasm to interact with the RNA-induced silencing complex (RISC), which leads to degradation of mRNA homologous to the siRNA sequence. (Adapted from Derfus, A.M. et al., *Adv. Mater.*, 16, 961, 2004. With permission.)

In these studies, they could follow the endosomal uptake and subsequently the endosomal escape by exploiting the QDs fluorescence (through a FRET process between the QDs and a siRNA bearing a dye at one end). The silencing activity of the delivered siRNA was also observed by the suppression of the Her-2 expression level of about 36% under serum-free conditions. This activity was compared to that of standard siRNA delivery systems, such as lipofectamine, and it was by far more efficient.

In another study, Derfus et al. investigated the delivery efficiency of siRNA–QD conjugates in which the siRNA molecules were covalently linked through labile disulfide bond to the QD surface (Derfus et al. 2004). As a control, the siRNA was also linked to the QDs surface through a linker, which did not have a thiol–thiol bond and thus was not cleavable. In a proof-of-concept experiment on the knockdown of enhanced green fluorescent protein (EGFP) gene on EGFP-transfected HeLa cells, the disulfide-bearing siRNA–QD conjugate proved to have greater silencing efficiency. Most probably, the release of siRNA from the QD surface was required for the incorporation of siRNA into RISC (the hindrance of siRNA–QD if covalently linked did not favor the siRNA–QD/RISC interaction).

More recently, Klein et al. used 2-vinylpyridine-functionalized silicon QDs as carrier to achieve high gene-transfection efficiency for ABCB1 siRNA, which they delivered to the cytosol of Caco-2 cells (Klein et al. 2009). Release and incorporation of siRNA into the RISC were tracked by detecting a 50% reduced ABCB1 mRNA level and there upon the transient down-regulation of the Pgp translation of successfully transfected Caco-2 cells. In another study, Ishihama and colleagues illustrated the use of QD-mRNAs as fluorescent tag to observe, with elevated spatial resolution over long observation time, the mRNA dynamics in cell, which had not been achieved yet by conventional labeling with fluorescent dyes or fluorescent proteins (Ishihama and Funatsu 2009). By using QDs, the authors succeeded in observing the movement of individual mRNAs for more than 60 s, with a temporal resolution of 30 ms. These results provided direct evidence of channel mRNA diffusion into interchromatin regions. Those multifunctional, compact, and higher traceable QD-based nanocarriers are now expected to yield important information on gene silencing and at the same time to improve delivery efficiency for supplementing a deficient gene directly to the nucleus. Indeed, although some successful results have been obtained in specific gene delivery by using viral vectors and liposomes as cargo, most of these methods have limited efficiency. Few more examples in these directions of QDs for gene delivery have been proposed by Srinivasan et al. who first showed that QDs could be covalently conjugated to plasmid DNA for transfection studies (Srinivasan et al. 2006). Likewise, Hoshino et al. reported that QDs conjugated with nuclear localizing signal peptides (NLSP) successfully introduced gene fragments with promoter elements, which induced the expression of the EGFP gene in mammalian cells (Hoshino et al. 2008).

### 10.4.5 Quantum Dots for Labeling Virus, Bacteria, and Model Organisms (Yeast, Zebrafish, and *Hydra*)

Cambi et al. exploited ligand-conjugated, virus-sized, highly photostable QDs aiming at monitoring in living cells antigen binding, entry, and trafficking modulated by DCs (Cambi et al. 2007). DC-SIGN, a receptor responsible for the binding and uptake of several pathogens among which HIV-1, forms nanoclusters at the cell membrane that assist virus capture (Geijtenbeek et al. 2000). The authors preloaded streptavidin-PEG-QDs with biotinylated ligands of DC-SIGN, such as LewisX, the HIV-1 envelope protein gp120, and the anti-DC-SIGN mAb AZN-D1. After incubation of CHO cells expressing human DC-SIGN with ligand-conjugated QDs, the authors monitored the internalization of virus-sized QDs (~40 nm diameter after conjugation) at different timepoints (Figure 10.8) and concluded that the cellular endocytotic machinery is the rate-limiting process for the internalization mediated by DC-SIGN.

Encapsulating QDs in viral capsids is a smart alternative to virus labeling. This idea has been first proposed by QD encapsidation in two plant viruses (Dixit et al. 2006; Loo et al. 2007). Recently, Li et al. designed QD-containing virus-like particles (VLPs) of simian virus 40 (SV40) by using the *in vitro* self-assembly system of the mammalian virus, yielding a type of new inorganic–organic hybrid particles, simian virus 40 like particles QDs (SVLP-QDs). When incubated with living cells, SVLP-QDs are shown to enter the cells by caveolar endocytosis, travel along the microtubules, and accumulate in the endoplasmic reticulum mimicking the early infection steps of SV40 (Li et al. 2009).

A work of Edgar et al. on high-sensitivity bacterial detection exploited the *in vivo* biotinylation of engineered host-specific bacteriophage and conjugated the phage to streptavidin-coated QDs to rapidly detect different types of bacteria (Edgar et al. 2006). The

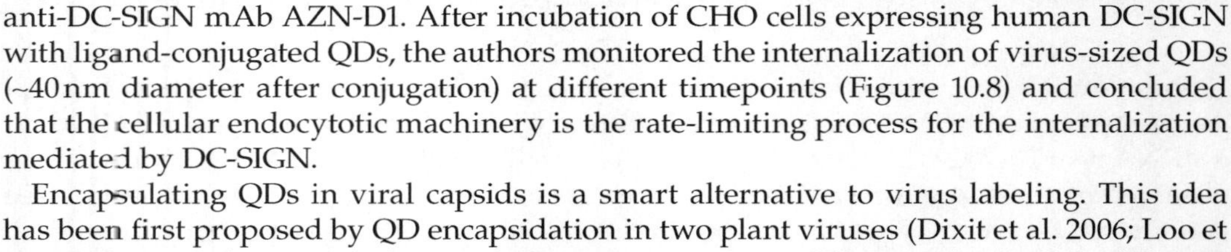


**FIGURE 10.8**
**(See color insert following page 302.)** Quantitative analysis of real-time binding and internalization of ligand-coated QDs. Time series of CHO-DC-SIGN cells incubated with 2 nM gp120-QD655. Co-transfection of erbB1-EGFP labeled the cell membrane and allows delineation from cytoplasm. (Adapted from Cambi, A. et al., *Nano Lett.*, 7, 970, 2007. With permission.)

authors considered that since phage could not be seen previously by light microscopy and QD-labeled phages are infective, their method opens up new avenues to address phage biology-related questions on topics such as initial binding, phage localization, distribution, and more.

Kloepfer et al. investigated on the specific QD labeling of bacteria and, since prokaryotes do not endocytose like mammalian cells, the researchers investigated possible mechanisms through which the nanocrystals could pass through bacterial cell walls and membranes (Kloepfer et al. 2005). They found that adenine- and adenosine mono phosphate-conjugated QDs are able to label bacteria only if the particles are <5 nm in diameter and that labeling is dependent upon purine-processing system. Furthermore, since QDs are both fluorescent and electron dense, fluorescence and electron micrographs could be directly evaluated for high-resolution imaging of internal mechanisms in a single bacteria.

Coulon et al. demonstrated that galactose- and mannose-functionalized QDs associate, respectively, with *Kluyveromyces bulgaricus* and *Saccharomyces cerevisiae* yeast strains due to specific saccharide/lectin recognition (Coulon et al. 2010). On the other hand, glucose-functionalized CdTe QDs, which were not recognized by cell lectins, preferentially localized in the bud scars of *S. cerevisiae*. Friedrich et al. (2009) exploited the selective plan illumination microscopy (SPIM) technique to achieve single QD imaging of both larvae of living fruit flies (*Drosophila melanogaster*) and also living embryos of zebrafish (*Danio rerio*) with penetrative depths going beyond 300 μm. They were also able to follow individual QDs in developing zebrafish at those depths. As QDs cannot pass through gap junctions between blastomeres, the closure of cytoplasmic bridges can account for the observed transition of the QDs to a more confined motility, which is also in agreement with previous studies (Rieger et al. 2005). As zebrafish is a recognized model system for investigations on physiological development and disease mechanisms at *in vivo* cellular and subcellular levels (Beis and Stainier 2006); success in embryo labeling is an important goal for developmental biology research. The coelenterate *Hydra Vulgaris* represents another used model organism with the particularity of having regenerating capabilities. Tortiglione et al. suggest that temporal dynamics of remodeling and the plasticity phenomena involved can be best appreciated by means of QDs as they show very photostable fluorescence (Tortiglione et al. 2007).

## 10.5 Toxicity of Quantum Dots

Toxicity of QDs is an important parameter to consider, especially in the biomedical field. Several concerns about QDs toxicity have been raised in view of the exploitation of QDs as diagnostic agents for medical and imaging devices and as therapeutic agents in gene or drug delivery in human patients. So far, no QDs have been approved for therapeutic and diagnostic purposes and no regulation is still available about them. However, there is already a significant amount of academic knowledge describing the toxicity of QDs at a cellular level reviewed in many manuscripts (Hardman 2006; Lewinski et al. 2008; Pelley et al. 2009; Rzigalinski and Strobl 2009). Although the comparison between those studies is complicated because of the diversity of the QDs employed (different QD composition, coatings, and functionalization), the cell lineages used, and the differences in conditions of exposure (e.g. time, environmental conditions, etc.), some conclusions can be already made. All these aspects are analyzed here.

QDs are made of heavy metals, such as cadmium, selenium, lead, and tellurium, and the toxicity to humans of these elements is already well known (Vinceti et al. 2001; Bertin and Averbeck 2006; Ferrara et al. 2006). Exposure to cadmium, for instance, has been correlated to prostate and lung cancer development (Bertin and Averbeck 2006). Known the intrinsic toxicity of the elements that constitute the QDs at a bulk level, it is now crucial to clarify if the nanoscale manipulation of these materials minimizes or enhances the toxicity of the obtained nanoparticles. Several works have shown that the leakage of Cd ions from the nanoparticles surface is responsible for cellular toxicity of QDs (Waisberg et al. 2003; Derfus et al. 2004; Kirchner et al. 2005; Lewinski et al. 2008). For instance, Derfus et al. demonstrated that oxidative and photolytic environmental conditions, such as exposure to air or UV light, can indeed degrade the ligand shell at the QD surface and promote the leakage of Cd ions (Derfus et al. 2004). However, the Cd release is linked to the type of coating present at the nanoparticle surface. Coating of CdSe nanoparticles with a shell of another semiconductor material, such as ZnS, besides increasing the PL, also strongly reduces the QD toxicity. It is also known that the environmental stability of the nanocrystal is also dependent from the capping ligand (Hoshino et al. 2004). Aldana et al. observed photochemical instability in thiol-coated CdSe QDs, although this stability could be varied by changing the packing at the QDs' surface of the nanoparticles (Aldana et al. 2001). Also, Lovric et al., by comparing the toxicity of MPA- and cysteamine-capped QDs to the same uncoated CdTe nanoparticles, showed that the former were more toxic than the latter probably because of the intrinsic toxicity of the capping ligands (Lovric et al. 2005). The toxicity of the capping agents, in this case a mercaptoundecanoic acid, was also confirmed in a similar study by Hoshino et al. on a murine cell lineage, the T-cell lymphoma (EL-4) (Hoshino et al. 2004). The authors found that mercaptoundecanoic acid alone (without QDs) was highly toxic at a concentration of 100 μg/mL after exposure of 12 h. On the other hand, cysteamine at the same concentration and for the same exposure time resulted to be less toxic.

In a comparative study, Kirchner et al. demonstrated the effects of different types of coating on the same batch of QDs, confirming the results obtained by Aldana (Kirchner et al. 2005). Silica-coated QDs were found to be less toxic then polymer-coated ones due to better physiological stability in cell-culture medium (Kirchner et al. 2005). The glass silica shell grown around the QDs seemed to reduce the toxicity also in the study reported by Chen and Gerion, who attributed the lack of observable genotoxicity of QDs to the silica coating, which prevents the interaction of Cd, Se, and Zn with proteins and the DNA in the nucleus (Chen et al. 2004).

The toxicity of QDs has been also related to their size. Lovric et al. observed that 5.2 nm cationic CdTe QDs accumulate only into the cytoplasm of murine microglial cell lines (Lovric et al. 2005). On the other hand, smaller CdTe QDs of about 2.2 nm were found in the nucleus within the same timescale. However, nuclear pores also allowed the passage of macromolecules of 9 nm in diameter and therefore the size of the QDs cannot be the only explanation for the migration of the smaller QDs through the nuclear membrane. The different bioactivity of the BSA on the two QDs surfaces, which in the case of the big CdTe QDs was higher, might be one of the possible explanations for such different cellular localization. Finally, at equal administered concentration, the smaller QDs localized inside the nucleus were found to be more toxic than the bigger ones. The origin of this toxicity was not established but some hypotheses were advanced, such as Cd leakage from the surface, inactivation of some cellular functions due to interaction of the QDs with specific cellular compartments, and production of free radicals. It has been proven that QDs can be both electron donors and acceptors; they can thus take part in redox reactions. They are able to generate highly reactive radical species, which can produce reactive oxygen species (ROS) that are considered really toxic for cells. It is important to point out that the intrinsic QDs toxicity toward cells might be exploited for the PDT, as discussed in detail in Section 10.7 (Samia et al. 2003; Lovric et al. 2005; Bakalova et al. 2008).

Together with the toxicity of the QDs on cellular models, also the *in vivo* biological fate of QDs is of paramount importance in order to elucidate pathway, biodistribution, clearance, metabolism, and degradation of the QDs and is discussed in the following section.

However, the quest for the synthesis of nanoparticles that are nontoxic is still an active field. So far, most high-temperature colloidal syntheses of QDs have used toxic and expensive chemicals, while recent research efforts have made cheaper these type of syntheses, less dangerous, and "greener." QDs based on the elements of the III and V groups, such as InP is one possibility, although much effort still needs to be done in order to improve the photo-physical properties of the resulting QDs to obtain the performances similar to those of the more common type II–VI QDs. Also, promising candidates for the generation of cadmium-free QDs are ZnO (Demir et al. 2006), ZnSe (Chen et al. 2004), and ZnSe/ZnS nanocrystals (Edgar et al. 2006). However, these materials emit in the UV–blue range of the spectrum but emission in the visible range would be possible by doping them with transition metal ions (Norris et al. 2001; Pradhan et al. 2005; Pradhan and Peng 2007; Diagaradjane et al. 2008).

## 10.6 *In Vivo* Cellular Imaging and Tracking with Quantum Dots

QDs have been widely exploited as fluorescent probes for *in vivo* imaging of normal and tumor tissues. They can target either actively a desired tissue with antibodies or passively via the enhanced permeability and retention (EPR) effect. They can bind specifically to a particular molecular target and document its activity during whole-body observation. Several studies on animal models based on near-infrared (NIR)-emitting QDs, which coincides with a range of optical transparency for living tissue, have been also reported.

Researchers have begun to investigate what happens to QDs when administered *in vivo*, what may affect their biological properties, and if they preferentially accumulate in certain cells or tissues. In this section, all those aspects are discussed in detail.

### 10.6.1 *In Vivo* Applications of Quantum Dots for Specific Targeting Cells and Tissues

One of the greatest potentials of QDs in *in vivo* applications is the selective targeting of specific cells and/or tissues for imaging and/or therapeutic purposes (Michalet et al. 2005; Pathak et al. 2007). Currently, one of the most studied areas concerns tumor cells tracking and detection (Akerman et al. 2002; Gao et al. 2004). Two different approaches can be exploited for *in vivo* tumor targeting. In the passive targeting mechanism, macromolecules and also nanometer-size particles accumulate preferentially at the tumor site through EPR effects, which can arise from angiogenic tumor products, such as vascular endothelial growth factor (VEGF). At the same time, the lack of an effective lymphatic drainage system leads to subsequent macromolecule or nanoparticle accumulation (Dolmans et al. 2003; Duncan 2003). Few examples of QDs accumulation through passive targeting have been reported (Bertolini et al. 2008).

In comparison to passive targeting, active targeting exploits the use of biomolecule-driven specific binding to a target cell type or tissue.

Imaging of tumor vasculature in living animals is a primary requisite for the early detection and treatment of cancer. For this purpose, Åkerman et al. used QDs functionalized with three peptides for active targeting: CGFECVRQCPERC peptide (denoted as GFE) for a specific receptor (a membrane dipeptidase) on the endothelial cells in lung blood vessels; KDEPQRRSARLSAKPAPPKPEPKPKKAPAKK (F3) that preferentially binds to blood vessels and tumor cells in various tumors; and CGNKRTRGC (LyP-1) to recognize lymphatic vessels and tumor cells in certain tumors, for the active targeting of a specific receptor expressed on the endothelial cells in lung blood vessels (Akerman et al. 2002).

After intravenous injection in living mice, each of the peptide-coated QDs efficiently targeted the appropriate vascular sites with an extraordinary specificity and causing only moderate toxicity. Brain, heart, kidney, or skin do not contain detectable QDs, indicating their regional specificity and the feasibility of functionally targeting distinct components of the tumor (e.g., blood vessels vs. lymphatic vessels). In addition to the target tissues, QDs accumulated both in the liver and spleen, indicate an uptake by the reticuloendothelial system (RES). However, adding PEG to the surface of QDs, an accumulation in the liver and spleen was reduced by about 95% and the PEG coating did not alter noticeably the QD accumulation in the tumor tissue.

Cai and coworkers developed QDs functionalized with RGD tripeptide for targeting integrin $\alpha v\beta 3$-positive tumor vasculature imaging in a murine *Xenograft* model (Cai et al. 2006). Integrin $\alpha v\beta 3$, which binds to RGD-containing components of the interstitial matrix, plays a key role in tumor angiogenesis and metastasis (Xiong et al. 2002). This protein is significantly up-regulated in invasive tumor cells of certain cancer types (glioblastoma, melanoma, breast, ovarian, and prostate cancer, and in almost all tumor vasculatures) but not in quiescent endothelium and normal tissues (Hood and Cheresh 2002; Jin and Varner 2004). The authors of this study observed that after injection through tail vein in athymic nude mice bearing U87MG tumor, the uptake of RGD-QDs significantly increased in the tumor at 20 min postinjection, and after 6 h it reached its maximum intensity. Confocal microscopy images showed the presence of QD705-RGD in the tumor tissue, while *ex vivo* analysis confirmed that the fluorescence signal colocalized with the CD31 staining, indicating that the fluorescence signal came mainly from the tumor vasculature.

In another study, streptavidin-conjugated QDs were used for specific tumor targeting *in vivo* (Chen et al. 2004). Chen et al. adopted a mouse mammary gland tumor model to show the target efficiency of a bifunctional RGD-4C/R5C2 phage that displays an integrin-binding motif (RGD) and a streptavidin-binding motif (R5C2 peptide). Mouse-bearing tumors received phage-streptavidin-QD complexes intravenously *in vivo*. Immunohistochemistry showed strong staining of blood vessels revealing localization in the tumor after 5–8 min of circulation.

Gao et al. reported a new class of multifunctional QD probes for simultaneous cancer targeting and imaging in animal models (Gao et al. 2004). QDs were functionalized with prostate-specific membrane antigen (PSMA) monoclonal antibody and multiple PEG molecules to improve biocompatibility and circulation time. The QD-antibody bioconjugates were injected intravenously in mice and reached the tumor site in a much faster and efficient way compared to nonconjugated QDs.

In an alternative approach, CdSe QDs entrapped in a lipidic shell (referred to as lipodots, with about 100 nm of total diameter) were labeled with FA for targeting mouse tumor cells expressing the FR (Schroeder et al. 2007). Selective uptake of the lipodots was observed *in vivo* after intraperitoneal injection in mice bearing ascitic J6456-FR lymphoma cells.

QDs were also delivered by peptide-induced transport, e.g., using the protein transduction domain of HIV-1 Tat peptide (Tat-PTD) (Santra et al. 2005). This strategy could represent a way to overcome the blood-brain barrier (BBB) and to rapidly transport QDs in the brain tissue. After infusion of a QD suspension by cervical carotid artery, a craniotomy was performed on Sprague-Dawley rats, the brain was removed and immediately analyzed, confirming the presence of fluorescence. TAT peptide-mediated QD delivery allowed rapid and high dose of QDs accumulation in the brain tissue in rats. Histological analysis on paraffin sections revealed that TAT-conjugated QDs reached the nucleus of brain cells.

A more recent report documented the ability to image tumor-related receptors *in vivo* using QDs conjugated to EGF, of the receptor EGFR, a glycoprotein over-expressed on the surface of many tumors that controls proliferation and invasiveness of tumor cells (Diagaradjane et al. 2008). In cancer therapy, it is crucial to have tools for discriminating EGFR on tumor tissues from surrounding normal tissues, where it is expressed at a lower level. To this end, Krishnan et al. utilized NIR-emitting QDs coupled to EGF as a noninvasive optical imaging of EGFR expression in human colorectal cancer xenografts in mice (Diagaradjane et al. 2008). *In vivo* optical imaging of EGFR-expressing tumors has been done at different time points after injection. The small size of the peptide-QD conjugates increased tumor penetration with a quantifiable accumulation of the nanoprobes at an early time point (4h) and a more rapid clearance via the kidneys. Compared with QDs lacking the EGFR binding domain, EGF-QDs presented a more stable fluorescence in tumor tissue between 1 and 6h after injection with normalization at 24h, a good affinity to EGFR, and favorable pharmacokinetic properties.

The use of QDs conjugated to EGF has been previously described by Lidke et al. for signal transduction studies (Lidke et al. 2004). Biotinylated EGF was bound to commercial QDs functionalized with streptavidin. When administered to living cells, EGF-QDs bound to the cell surface and readily activated erbB1, the EGF receptor, as revealed by the rapid internalization of the complex into endosomes. Thus, the small nanoparticles did not alter the biological function of the attached growth factor.

In order to understand the molecular mechanism of proteins *in vivo*, Tada et al. reported the first example of real-time single particle tracking in living animals (Tada et al. 2007). The authors employed HER-2 over-expressing human breast cancer cell line KPL-4 that was transplanted subcutaneously to the dorsal skin of female BALB/c mice. Several weeks after tumor inoculation, QDs loaded with the monoclonal anti-HER2 antibody were injected into the tail vein of mice bearing a tumor volume of 100–200 $mm^3$. By tumor imaging in living mice with a high-resolution imaging system, single QD complexes with a Brownian motion were clearly observed 6h after injection on the membrane of KPL-4 cells and intracellularly localized 24h postinjection. Three-dimensional images of the tumor allowed the observation on the tumor vessels of single QD complexes. A random movement, in speed and orientation, of the QD complexes from tumor vessels to the interstitial space was also observed.

### 10.6.2 Biodistribution

QDs injected in animal models can be systemically distributed and accumulated in specific organs or tissues. The possibility to control their biodistribution and to target specific sites is of remarkable importance in biomedical applications. In cancer therapy, lymph node (LN) metastasis is one of the critical prognostic signs of cancer. The use of QD as fluorescent probes for LN mapping is a promising application for staining certain types of cancers, thus helping clinicians to locate and dissect samples for biopsy.

Two basic methodologies have been used in the literature to examine the biological fate of QDs after *in vivo* administration in model animals, either by tracking the fluorescent signal (Gopee et al. 2007; Hama et al. 2007; Kobayashi et al. 2007) or by using radiolabeled QDs (Kennel et al. 2008; Schipper et al. 2009). Kennel et al. used a quantitative method to follow the *in vivo* biodistribution of radiolabeled Cd $^{125m}$Te QDs coupled to monoclonal antibody (MAb), a targeting agent that binds lung vasculature lumen in a murine model (Kennel et al. 2008). As directly compared with iodinate antibody attached to CdTe NP, $^{125m}$Te QDs could be tracked by following the radioactivity for months. Standard biodistribution studies showed a uniform distribution and accumulation of $^{125m}$Te QDs coupled with MAb specifically in lung. The distribution of the $^{125m}$Te NP was also documented by single photon emission computed tomography (SPECT) imaging and tissue autoradiography.

One advantage of radiolabeling is that it allows for the derivation of quantitative data. On the other hand, the tracking of radioactivity does not allow for distinguishing between QDs remaining active and those that become inactive, including those degraded into their component molecules. In contrast, only intact QD particles fluoresce.

Fluorescence imaging of LN in animal models using QDs was first demonstrated by Frangioni and coworkers (Frangioni 2003). Kim et al. also prepared NIR-emitting type II CdTe/CdSe QDs, with a hydrodynamic diameter of 15–20 nm, to selectively map sentinel LN (Kim et al. 2004). When injected intradermally in a mouse model, these QDs entered the lymphatics and migrated within minutes to the axillary. NIR fluorescence signal was confirmed histologically to originate from the sentinel lymph node (SLN). To prove that NIR QDs could be also used in bigger animals, the nanocrystals were injected intradermally in the thigh of pig. Localization of the SLN required only 3–4 min and the NIR QDs allowed its identification even when it was part of a large LN cluster. Furthermore, it was possible to identify SLNs approximately 1 cm below the skin surface by means of reflectance imaging.

Successively, Hama et al. demonstrated simultaneous imaging with QDs of two different lymphatic drainages *in vivo* and their trafficking to an LN (Hama et al. 2007). They simultaneously injected QDs having distinct emissions ($\lambda_{max}$ = 705 and 800 nm) into the mammary gland and the skin of the middle phalange in the upper extremity in mice. After injection, two-color NIR lymphagiography successfully visualized the lymphatic drainage territory. The axillary LN received the two different lymphatic flows, simultaneously from the mammary pad and the upper extremity, as indicated by mixed contributions from both QDs. LNs, as well as the lymphatic vessels, were clearly visualized through the skin. Similarly, for the lymphatic imaging *in vivo*, Kobayashi et al. applied multicolor imaging by using five different colors of QDs with distinct emissions (QD565, QD605, QD655, QD705, and QD800) injected intracutaneously into five different sites in the upper body of athymic mice (Kobayashi et al. 2007). Within 5 min after injection, a spectral imaging was carried out using a fluorescence lymphangiography system. All major drainage trunks of the lymphatics in the upper body could be depicted individually and simultaneously with different colors through the skin in the *in vivo* image (Figure 10.9).

Remarkably, similar results in the level of migration to the LN were obtained by Gopee and coworkers who used PEG-coated CdSe QDs (with a diameter of 37 nm) to investigate the biodistribution of these nanocrystals after intradermally (i.d.) injection in healthy mice (Gopee et al. 2007). The QDs were localized in LN soon after the injection. Mice were sacrificed at 24 h after i.d. injection and the presence of QDs in tissues was confirmed by

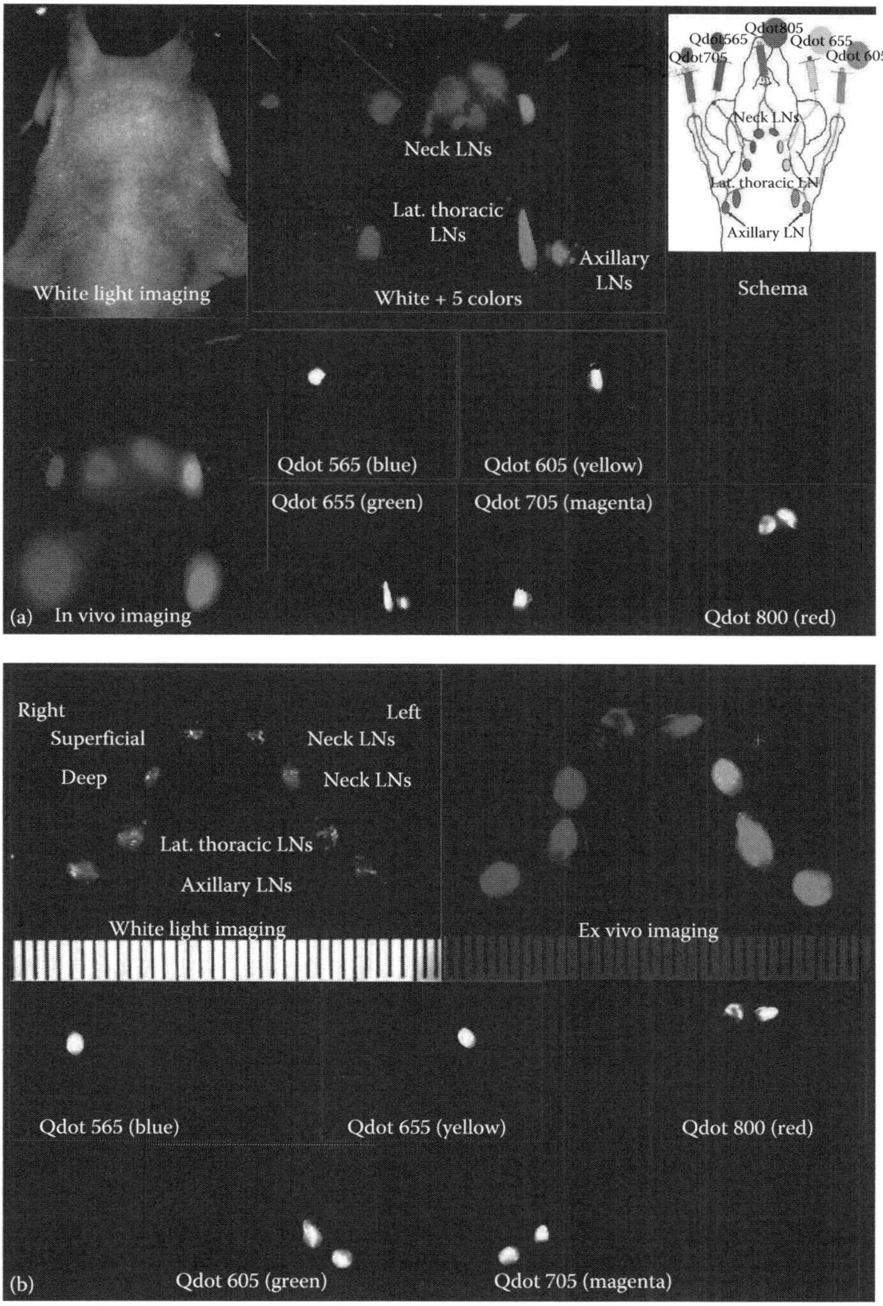


**FIGURE 10.9**
**(See color insert following page 302.)** *In vivo* imaging of lymphatic drainage by using five QDs emitting at different colors: (a) Intrasurgical spectral fluorescence imaging of a mouse injected intracutaneously with five QDs (565 nm, blue; 605 nm, green; 655 nm, yellow; 705 nm, magenta; 800 nm, red) into five different sites in the upper body of athymic mice. Five primary draining LNs were simultaneously visualized *in vivo*. (b) *Ex vivo* spectral fluorescence imaging of the eight draining LNs after surgical resection. (Adapted from Kobayashi, H. et al., *Nano Lett*., 7, 1711, 2007. With permission.)

fluorescence microscopy. Fluorescent QDs were clearly visible at the site of injection and distributed to regional draining LNs and to other organs presumably via the lymphatic system. In parallel, the Cd concentration in different tissues was measured by inductively coupled plasma mass spectroscopy (ICP-MS) and correlated to the QD concentration. Significant increase in cadmium level was detected in several tissues (LNs, kidney, heart, liver, and spleen) at 12 and 24 h postinjection. These data demonstrate that regional LN, liver, kidney, and spleen can be used as sentinel organs for monitoring transdermal penetration of nanoparticles into animals.

This work is in agreement with the study reported by Fischer et al. where the liver is the primary organ of nanoparticle deposition, although the QDs used were coated with mercaptoundecanoic acid and BSA (Fischer et al. 2006). These observations support the use of the liver as a sentinel organ in nanoparticle studies.

In a work of Smith et al., the authors utilized a variety of NIR and visible QDs for noninvasive imaging of large vasculature and capillary structure of the chick chorioallantoic membrane (CAM), a popular assay for studying blood vessels development (Smith et al. 2007). The QDs had robust biocompatibility and after intravital injection in 10 days-old chick embryo, no effect on development was observed during all the time course experiment. QDs were kept in circulation for days and clearly imaged CAM vasculature. Compared to FITC-dextran, a dye commonly used for CAM vessel imaging, QDs were brighter and showed comparable illumination across vessels using a much lower concentration and longer residence time to endothelial cell uptake.

NIR-emitting QDs were also used for noninvasive *in vivo* tracking of injected DCs to LNs through lymphatic vessels, without sacrificing the animal model (Noh et al. 2008). The DCs, antigen-presenting cells, are uniquely capable of initiating a primary immune response activating T-cells into secondary lymphoid tissues. DCs, labeled using QD800 Q-tracker cell labeling kit, were injected subcutaneously in the hind-leg footpad of C57/BL6 syngeneic mice. The injection site was preconditioned for 24 h with TNF-$\alpha$ to enhance the DC migration rate. After 2 days, NIR fluorescence images of mouse LNs were acquired. Despite control mice injected with unlabeled DCs, a strong NIR fluorescence signal was observed into popliteal and inguinal LN areas resulted from the migration of injected labeled DCs. The NIR signal intensity started to decrease after 72 h.

It has also been shown that QDs can map SLNs in model tumors *in vivo* (Ballou et al. 2007). QDs were coated with differently charged terminal groups, methoxy-PEG (neutral surface), carboxy-PEG (negatively charged surface), and amino-PEG (positively charged PEG surface) to examine their ability to access sentinel nodes after injection into a tumor. To demonstrate that transfer of QDs from tumors occurred through the lymphatics, LNs were labeled with QDs of specific wavelength followed by injection in the tumor of QDs emitting at different wavelengths. Migration from the tumor to the surrounding LNs rapidly occurred and no significant differences were noted among the differently charged QDs. The QD path could be dynamically followed through the skin and *post mortem* examination confirmed either the presence or the absence of the tumor in the LNs.

Soltesz and coworkers developed a completely portable NIR fluorescence imaging system that allowed for real-time intraoperative SLN mapping (Soltesz et al. 2005). They investigated the feasibility of QDs for mapping pulmonary lymphatic drainage and guiding excision of the SLN after intraparenchymal injection of NIR-QDs in a pig model. Lymphatic flow was visualized in real time using the mapping system. QDs remained localized to the subcapsular and intermediate sinuses of the SLN and migration of QDs beyond the identified SLN was not observed after 4 h. Similarly, NIR-emitting QDs were tested for identifying SLNs of the esophagus (Parungo et al. 2005). Direct injection of QDs into the esophageal wall of pigs and QD migration to a single SLN was visible in real time. A single SLN was identified within 5 min in 100% of pigs and SLNs did not diminish in fluorescence nor did additional SLNs appear after 4 h.

In a recent paper, Robe and colleagues proposed carboxyl-coated QDs to monitor both localization and clearance in mice (Robe et al. 2008). QDs with diameters ranging between 15 and 20 nm showed to be optimal probes for allowing the QDs to travel through the lymphatic vessels and to be trapped by SLN (Parungo et al. 2005).

While vascular endothelial targets may be reached *in vivo*, it appears less likely that extravascular targets, such as tumor cells, can be reached, since longer circulation time and extravasation of the imaging agent is required. Zimmer et al. produced DHLA-PEG-coated NIR-emitting InAs/ZnSe QDs with a small size (hydrodynamic diameter less than 10 nm) (Zimmer et al. 2006). Once injected s.c. in a murine model, QDs were not trapped completely in the SLN but migrated through the lymphatic system, imaging sequentially five LNs and the lymphatic channels. The decreased size of the QDs also allowed their extravasation from blood vessels into the interstitial fluid in rat. No extravasation was observed in QDs without the DHLA coating, probably because of an increased size caused by nonspecific serum protein binding.

Papaggiannaros et al. utilized long-circulating NIR QD-PEGs incorporated into micelles (QD-Ms) for both tumor detection and for biodistribution studies (Papagiannaros et al. 2009). 4T1 murine breast cancer cells were injected subcutaneously in mice to produce tumor. After 2 weeks, QD-Ms or commercially QD-PEGs were injected into the tumor-bearing mice via the tail vein. Comparing the two treatments, the authors observed significant differences in their biodistribution patterns. Within 1 h after injection, QD-Ms accumulated rapidly in the organs and tumor resulting in high intensity of the signal while after the whole 6 h observation period, a limited fluorescence was observed. The QD-PEGs produced the maximum signal-to-noise ratio only after 3 h, with a diffuse fluorescence over the entire body of the animal and poorly contrasted images due to their longer time circulation in blood vessels (in tissues around tumor). These results clearly demonstrated that functionalized QD-Ms were deposited faster in organs and tumors and were also cleared faster than QD-PEGs.

The research group of Gambhir investigated the influence of particle size, PEGylation, and peptide coating of radiolabeled QDs on quantitative biodistribution, RES uptake, and excretion of nanoparticles (Schipper et al. 2009). After tail-vein injection, a dominant uptake was observed in liver and spleen for the majority of QDs, even if they showed differences in biodistribution: polymer- or peptide-coated QD uptake was faster than PEGylated counterparts. Also, surface coating with peptides instead of polymer influenced circulation time: a decreased affinity of the peptide coating to opsonizing proteins prolonged the serum half-life. The authors could observe that small QDs presented a different biodistribution pattern when compared with bigger ones, with the former filtered through the glomeruli and excreted renally. These findings are consistent with those reported by Choi et al. (2007) confirming that these QDs have a hydrodynamic diameter small enough to extravasate from vessels thus reaching tumor or interstitial target and can be easily cleared from the organism.

An efficient regional LN detection and whole-body biodistribution of NIR-QD encapsulated lipid has been reported by Pic et al. (2009). After subcutaneous injection in mice, right axillary LN (RALN) and right thoracic LN (RTLN) were visualized as early as 5 min up to 10 days after injection by NIR fluorescence images. However, a stronger fluorescence signal was observed in RALN, probably due to the fact that LTLNs are deeper in the tissue, while ALNs are under the skin. After 10 days postinjection, no sign of toxicity was observed in living mice. Similar results were observed by ICP-MS by determining the cadmium concentration in major organs, blood, and excretions, thus monitoring accumulation kinetics. The rapid increase in the blood cadmium concentration suggested that QDs could enter blood vessels immediately after injection. After 4 h, the decreased concentration in blood was due to progressive capture by various organs. These results are in agreement with Kobayashi's and Hama's study, who showed lymphatic communication between ALN and LTLN after injection of QD and NIR dyes, respectively (Hama et al. 2007; Kobayashi et al. 2007).

Together, these findings provide guidelines for the use of QDs for noninvasive fluorescence detection of superficial LNs in a clinical setting.

### 10.6.3 Clearance

In view of potential clinical applications of QDs, QD clearance, i.e., their elimination from the body has to be taken carefully into consideration. Hydrodynamic size and surface coating of the nanoparticles QDs are under evaluation in several studies as it has been suggested that these features affect their *in vivo* behavior, their excretion, endocytosis, and accumulation in tissue (Choi et al. 2007).

An improved understanding on the metabolism of QDs in living organism comes from Frangioni and coworkers, who synthesized a series of QDs at different hydrodynamic diameters and surface charges and explored the effect of those parameters on their body elimination (Choi et al. 2007; Liu et al. 2007). Fluorescent CdSe/ZnCdS QDs having a hydrodynamic diameter less than 6 nm were prepared choosing as coating cysteine ligands (Cys). When injected intravenously in rats, 4 h postinjection, QD-Cys accumulated mainly in the bladder indicating that the QDs were near or below the size threshold for renal clearance. Injection of protein-bound QD-Cys showed no renal clearance after the same time postinjection, suggesting that protein-QD complexes were stable over the timescale study and also that renal clearance was largely dependent on the size of the QD (Liu et al. 2007).

Moreover, depending on the surface charge of QDs, it is expected that the adsorption of serum protein will be more or less enhanced thus influencing the hydrodynamic diameter and *in vivo* localization.

Moreover, in the same study the surface charge of QDs was observed to affect the adsorption of serum protein thus influencing the hydrodynamic diameter and *in vivo* localization of QDs.

Frangioni et al. successfully coated the QDs (QD515–574) surface with anionic (DHLA), cationic (cysteamine), zwitterionic (Cys), or neutral (DHLA-PEG) molecules and incubated them with serum (Choi et al. 2007). While cationic or anionic QDs showed an increase in hydrodynamic diameter to almost 15 nm, QD-Cys prevented protein adsorption and showed the smallest hydrodynamic diameter (less than 5.5 nm). A hydrodynamic diameter less than 5.5 nm of the organic-coated QDs resulted in efficient renal filtration and urinary excretion from the body 4 h after i.v. injection, while QDs with an hydrodynamic diameter larger than 15 nm were delayed and trapped in the RES. Recently, the same group used ultrasmall NIR-QDs conjugated to different PEG molecules having different molecular weight to control the QDs distribution and organ localization *in vivo* (Figure 10.10) (Choi et al. 2009). The authors tuned the QDs size in the range from 4.5 to 16 nm by changing the number of PEG repeating units ($CH_2$–$CH_2$–O–) from 0 to 22. These QDs were stable in rat serum and urine for over 4 h at 37°C. The series of PEG-conjugated QDs were i.v. injected in rat and the biodistribution and clearance evaluated 4 h after injection by NIR fluorescence

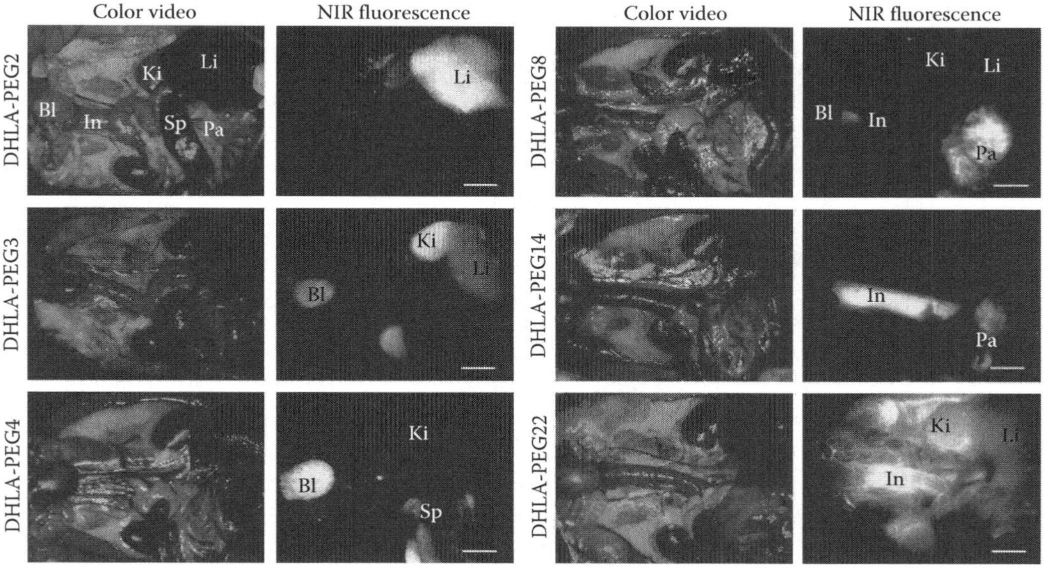


**FIGURE 10.10**
**(See color insert following page 302.)** *In vivo* NIR fluorescence imaging of NIR PEG-conjugated QDs. QDs coated with DHLA-PEG at different molecular weight of the PEG (DHLA-PEG2, DHLA-PEG3, DHLA-PEG4, DHLA-PEG8, DHLA-PEG14, and DHLA-PEG22) were administered to the rat and imaged 4 h postinjection. Color video is shown in the left panel and NIR fluorescence in the right panel. (Abbreviations used are: Ki, kidneys; Bl, bladder; Li, liver; Pa, pancreas; Sp, spleen; and In, intestine.) NIR fluorescence images have identical exposure times and normalizations. Scale bar = 500 μm. (Adapted from Choi, H.S. et al., *Nano Lett.*, 9, 2354, 2009. With permission.)

imaging. The authors found a significant effect of the hydrodynamic diameters on their behavior *in vivo*. Ultrashort PEG chain resulted in rapid uptake in liver, while relatively long PEG chains increased the blood half-life for long period of time; intermediate length chain exhibited specific tissue/organ distribution and clearance.

So far, few nanometer-sized objects have being actively applied for clinical purposes. However, those studies on the one hand indicated to reduce the QDs diameter, in order to allow complete QDs elimination from the body after diagnosis or therapy, and, on the other hand, to prepare new formulations of QDs made of completely nontoxic and biodegradable elements, which could be metabolized by the body and which at the state of the art is still challenging.

### 10.6.4 Kinetic

Quantitative kinetic studies regard the determination of dose, fate (i.e., clearance from plasma and the sequestration within organs), and circulation lifetime of exogenous substances administered to a living organism. The physicochemical properties of QDs, including chemical composition, size, and surface chemistry, govern their bio-kinetics. The biological behavior of QDs is strongly affected by the complex biological environment they encounter once they enter the body. In order to design QD-based nanoparticles and/or nanostructures with improved features for *in vivo* biomedical applications, besides considering biodistribution and clearance full characterization of the QDs behavior in living animals is crucial to determine the kinetic of QDs *in vivo* and the long-term stability, toxicity, and tissue deposition of QDs.

Ballou et al. conducted the first study on QD *in vivo* kinetics by using whole-animal fluorescence imaging (Ballou et al. 2004). Amphiphilic poly(acrylic acid) polymer-coated QDs (amp-QDs) embedded within an additional shell of methoxy- or carboxy-terminal PEG molecules were independently injected into the tail vein of nude mice. In contrast to small organic dyes, which were eliminated from circulation few minutes after injection, even in this study, PEG molecules were proven to increase the QDs circulation time in the blood up to 4 months. Even after 4 months, the QDs were still fluorescent, thus demonstrating the outstanding stability of these probes while the unique structural properties of these QDs allowed long circulation time. Additionally, the PEG-coated QDs had the right size, indeed they were small enough and sufficiently hydrophilic to delay opsonization and to escape the RES uptake, while on the other hand, they were large enough to avoid renal filtration. These results provided a qualitative assessment of surface chemistry–dependent kinetics of the QDs.

An important contribution to the use of QD probes to track metastatic tumor cells in a living animal model came from Voura et al. (2004). Using lipofectamine associated to DHLA-capped QDs, the fluorescent probes were efficiently delivered to B16 melanoma cells. QD-labeled tumor cells preserved the ability to form tumor, as observed at 40 days after injection in a mouse model. In this study, multiplexing imaging of five different populations of cells *in vivo* was also proven by using multiphoton and emission scanning microscopy for deep tissue-imaging. Metastatic cancer cells present abnormal membrane fluidity. This feature allows the tumor cells to detach from the parental tumor, migrate through the bloodstream, and invade organs and tissues. A contribution in understanding the mechanism of cancer progression by employing QDs was just reported by Gonda et al., who developed single-molecule imaging method to detail protein dynamics, membrane fluidity, and morphology in metastatic cancer cells *in vivo,* with a high spatial precision (Gonda et al. 2010). They have used anti-PAR1 antibody-conjugated QDs (anti-PAR1-QDs), specific for a tumor cell membrane protein (Protease-activated receptor 1 (PAR1)) that plays a crucial role in metastatic process. Transformed PAR1-KPL breast cancer cells were transplanted subcutaneously in female SCID mice, while anti-PAR1-QDs were injected into the tail vein 5 to 10 weeks after transplantation. Labeled anti-PAR1-QD cancer cells were observed in different locations: far from the blood vessel in tumors, in the bloodstream within tumors, near the vessel and adherent to the inner vascular surface in normal tissue near tumors, thus demonstrating the process of cancer metastasis.

In stem cell therapy, the possibility to monitor cell survival and location after transplantation is important in determining their efficacy. QDs provide an excellent tool for imaging stem cell therapy. These nanoparticles have been efficiently used for long-term tracking of the fate of delivered stem cells once implanted *in vivo* (Rosen et al. 2007). Exogenous hMSCs were labeled by passive loading of QDs and could be easily identified in histological sections to determine their location up to 8 weeks after *in vivo* delivery to canine heart. Intracellular QDs did not interfere with cellular function or proliferation, because QD-hMSCs were shown to have similar proliferative and differentiation capacities to that of control hMSC (without QDs). To document the spatial distribution of the stem cells after injection in the heart, the authors imaged the QD fluorescence sequentially each 10 μm transverse section and reconstructed digitally the locations of the stem cells in a 3D image into a rat heart.

Lin and coworkers report the successful labeling of embryonic stem cells (ES cells) with different QDs (having maximum emission wavelength respectively at 525, 565, 605, 705, and 800 nm) and imaging the labeled cells *in vivo*, when subcutaneously injected into various locations on the back of athimyc nude mice (Lin et al. 2007). The QDs did not affect the viability, proliferation, and differentiation capacity of ES cells. The authors have also shown that QD800 offered greater fluorescent intensity with respect to the other QDs tested. ES cells labeled with QD525, 565, 605, and 705 were imaged up to 2 days after implantation. A likely cause for such behavior could be the loss of signal due to the rapid division of ES cells (doubling time of 12–15 h) or serum instability of the QDs, as reported by Cai et al. By contrast, QD 800 signal was detected in the animal up to 14 days postinjection. These results provide another example of long-term imaging of QD-based tool exploitable in stem cell therapy, for noninvasive tracking QD-labeled stem cells within deep tissue.

Sen et al. used QDs as imaging tool for both tracking DCs (antigen-presenting cells of the immune-system) inside the LN and triggering T-cell activation *in vivo* (Sen et al. 2008). QDs were conjugated with biotinylated ovalbumin ($QD^{ova}$) and subcutaneously injected in eYFP-CD11c reporter mice. In these transgenic mice, the expression of yellow fluorescent protein (YFP) is driven by a CD11c promoter that regulates the expression of CD11c, the characteristic DC marker. The system allowed evaluation of QD together with YFP fluorescence in DC. QDs were selectively endocytosed by skin-resident $eYFP^{+}$DCs and were exploited as fluorescent markers for DCs that had migrated, via lymphatics, from the site of immunization to draining LNs. Thus, QD-labeled DCs were imaged depth inside LNs 12–16 h after s.c. injection.

The pharmacokinetics of QDs have been also studied by Fischer and coworkers (2006). In order to determine how surface chemistry affects plasma clearance and organ uptake, the authors have used two QDs having different chemical functionalization: QDs coated with mercaptoundecanoic acid (QD-LM) and QDs coated by BSA (QD-BSA). After the injection into the jugular vein of the rat, to determine how fast the QDs left the circulation and entered organs, the decay of nanoparticles, in terms of Cd determination, in plasma was examined by ICP-AES. The plasma clearance of the QD-LM was significantly lower than that for QD-BSA. A big quantitative difference was also appreciated in tissue distribution, as measured by ICP-AES and by fluorescence imaging. The uptake of QD-BSA into the liver was significantly higher (99%) than that of QD-LM (40%), 90 min after injection. Also, in bone marrow, the proportion of QD-BSA was higher, while lung and kidney shower a higher accumulation of QD-LMs. A small amount of both QDs appear in spleen and LNs. Moreover, QDs were not detected in urine and feces for up to 10 days after injection, suggesting that they are not degraded or metabolized in blood and tissue within the experimental time. These results differ from those of Ballou et al., who also demonstrated *in vivo* stability of QDs fluorescence after several months. However, in contrast, they observed by imaging and monitoring QDs organs distribution fluorescent signals arising from the intestine, concluding that the QDs were excreted in feces. Comparing the two reports, the main difference between the QDs used in the two studies is the surface coating of QDs employed, thus confirming the influence of the QDs surface chemistry on the kinetics and accumulation.

### 10.6.5 Bioluminescence Resonance Energy Transfer

In principle, properly functionalized QDs will only mark specific targeting molecules, although the utility of now available QDs for *in vivo* imaging is limited as indeed, in order to fluoresce, they require excitation from external illumination sources. Furthermore, such external excitation source produces a strong auto-fluorescence background from ubiquitous endogenous chromophores, such as collagens, porphyrins, and flavins (Troy et al. 2004).

BRET is a naturally occurring phenomenon in which a light-emitting protein (the donor) transfers energy in a non-radiative (dipole–dipole) way to a suitable fluorescent protein (the acceptor) in close proximity (Ward and Cormier 1978; Wilson and Hastings 1998). By substituting the acceptor protein with a QD, a bioluminescent system which will fluoresce without the need for an external illuminating source can be obtained, with potential applications for *in vivo* imaging.

So et al. exploited the BRET process for *in vivo* imaging in mice based on a QD bioluminescent probe (So et al. 2006). To construct bioluminescent QD conjugates, an eight-mutation variant of *R. reniformis* luciferase, designated Luc8, was used as BRET donor, emitting blue light upon addition of its substrate coelenterazine. The polymer-coated CdSe/ZnS QD655 was covalently coupled to Luc8 and the bioluminescence energy of Luc8-catalyzed oxidation of coelenterazine was transferred to the QD, resulting in QDs emission. A solution of QD655-Luc8 was injected subcutaneously in a nude mouse, which was imaged sequentially after tail-vein injection of coelenterazine. The images collected showed a strong signal, indicating that BRET between the Luc8 and the QD could occur in animals at superficial depths.

Evaluation of BRET emission in deeper tissues was also carried out by injecting a solution of QD655-Luc8 intramuscularly at a depth of 3 mm. Bioluminescent QD conjugates emitting at different wavelengths were prepared (QD705-Luc8; QD800-Luc8; QD605-Luc8). BRET occurred in each conjugate and it was even possible to selectively distinguish *in vivo* the bioluminescence emission of each conjugate from the others, whether it was present alone or in a mixture. Finally, in order to assess if QD655-Luc8-labeled cells could be detected once injected in animals, polycationic peptides which improve the cell uptake efficiency, were conjugated to QD655-Luc8 and the resulting functionalized QDs were administered to cells. The bioluminescence emission was also observed on labeled cells *in vitro* and it was still detected once the QD655-Luc8-polycation complex-labeled cells were injected via tail vein in a nude mouse.

One of the limitations of this first generation of QD-BRET was their moderate stability in blood and serum, which led the authors to improve the stability of QD-BRET probes for prolonged animal imaging purposes (Xing et al. 2008). They indeed adopted an alternative strategy for the encapsulation of QD655-Luc8 within a polyacrylamide polymer shell, which rendered the QD probe more stable. The QD conjugates were injected subcutaneously into 4–6 week-old nude mice and 2 min after coelenterazine was injected i.v. through the tail vein. Bioluminescence and fluorescence images were captured at 6, 12, 24, 48, 72, and 196 h postinjection. Results showed that encapsulated QDs remained stable and bioluminescent for longer time inside living body (up to 196 h), while unmodified control conjugates became barely detectable at 48 h. The QD fluorescence inside living animal was also improved by polymeric encapsulation, still detectable after 196 h postinjection, by contrast, signal from the unmodified conjugates disappear by 72 h.

The works above discussed introduce a few examples of modified QD-based conjugates which have overcome some limitations of standard fluorescent imaging. However, it is also worth highlighting the drawbacks of this technique. Although the bioluminescent of QDs described by Rao and colleagues solved the problem of tissue auto-fluorescence, this technique requires systemic administration of a foreign protein (luciferase) and a foreign enzyme substrate (coelenterazine), both of which have immunogenic potential and can alter the biological system under study. Furthermore, light generation is completely dependent on the biodistribution of the enzyme substrate, and thus on the relative perfusion of the particular organ, tissue, or tumor to be imaged.

## 10.7 Photodynamic Therapy

PDT is a biomedical treatment modality that exploits interactions relating light, photosensitizers (PSs) (light-activated drugs), and oxygen (Dougherty et al. 1998).

At the most fundamental level, the photodynamic process depends on the PS molecule itself (Y. Q. Wang et al. 2008). Briefly, following the absorption of light, the PS is transformed from its ground state into an electronically excited state (triplet state) via intersystem crossing (Henderson and Dougherty 1992). As it returns to the ground state, it releases energy that, if transferred to tissue oxygen, produces cytotoxic species, such as singlet oxygen ($^1O_2$), hydroxyl radical ($\cdot OH$), and super oxide anion ($^-O_2$) through a series of energy and electron transfer reactions between the PS and dissolved oxygen ($^3O_2$) (Figure 10.11) (Ochsner 1997; Oleinick and Evans 1998; Dolmans et al. 2003).

PSs are compounds able of absorbing light with a specific wavelength and transforming it into functional energy. Moreover, PSs have the ability to preferentially accumulate in diseased tissues and to generate cytotoxic agents to induce the required biological outcome (Sharman et al. 1999). Bakalova and coworkers highlighted eight different characteristics of an ideal PS: (1) composition uniformity; (2) ease in synthesizing and open availability of the starting materials; (3) non-toxicity in absence of light; (4) target specificity and localization; (5) high triplet state yield, triplet state energy larger than 94 kJ/mol = 0.97 eV (energy of singlet oxygen) and efficient energy transfer for the formation of singlet oxygen; (6) rapid clearance; (7) minimal self aggregation in body (which decreases triplet state yield); and (8) guaranty of photostability (Bakalova et al. 2008).

The specific subcellular targets damaged by PDT depend on the photosensitizer's distribution and localization within the cell, which varies among sensitizers and cell lines (Y. Q. Wang et al. 2008). Since only cells in close proximity to the PS are affected (Schulke et al. 2003), the surrounding healthy cells should be preserved. Moreover, the PS is not cytotoxic until illuminated, allowing clearance of unbound reagent from the system without perturbation.

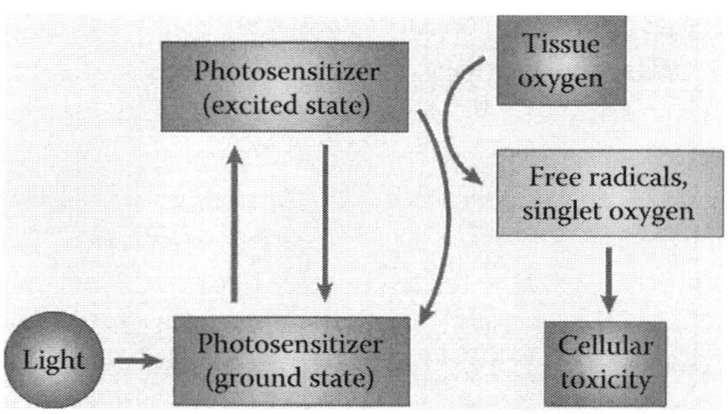


**FIGURE 10.11**
Scheme describing mechanism of action of PDT. PDT requires three elements: light, a photosensitizer, and oxygen. When the photosensitizer (PS) is exposed to specific wavelengths of light, it becomes activated from a ground to an excited state. As it returns to the ground state, it releases energy, which is transferred to oxygen to generate ROS, such as singlet oxygen and free radicals. These ROS mediate cellular toxicity. (Adapted from Dolmans, D. et al., *Nat. Rev. Cancer*, 3, 380, 2003. With permission.)

### 10.7.1 Quantum Dots as Photosensitizers for Cancer Therapy

In 2003, Samia et al. proposed the idea of semiconductor QDs as PSs for PDT (Samia et al. 2003). Several papers have explored their potential as PSs on their own, because QDs alone already meet the first five criteria previous reported (Bakalova et al. 2008). In parallel, studies on the ability of QDs to transfer energy to organic dyes have attracted much attention (Goldman et al. 2004, 2005b). Inspecting the two-step energy-transfer system from CdSe QDs to attached PS, Samia and colleagues discovered that QDs alone can generate $^1O_2$, without a mediating PS molecule, in toluene, most likely because of the intercalation of dissolved oxygen within the TOPO layer at the QD surface (Samia et al. 2003). They suggested that, since the lowest excited state of CdSe QDs is a triplet state, the generation of $^1O_2$ from $^3O_2$ is dependent on the triplet energy transfer (TET).

In particular, broad absorption band and large two-photon absorption cross section of QDs are advantageous for photoactivation using various visible and NIR light sources. In fact, absorbance in the NIR range is desirable for medical applications, as these wavelengths penetrate more deeply into tissues than visible light (Soltesz et al. 2005).

Targeted delivery of *ad hoc* functionalized QDs in cancer cells have become possible in recent times, although compared with 43% region of interest (ROI) generation efficiency of conventional PS drugs (Dabbousi et al. 1997), QDs efficiency was lower (5%) (Samia et al. 2003). To enhance singlet oxygen generation efficiency, several efforts have been made to covalently conjugate PSs to CdSe/ZnS via organic bridges, and their properties were widely investigated (Samia et al. 2003; Hsieh et al. 2006; Shi et al. 2006). Advantages over conventional PS drugs are various, for example, indirect photoactivation of PS drugs by using photostable QDs offers long-lasting imaging and PDT without photobleaching. Moreover, the surface of the QDs allows enough space for conjugating multiple PS and cancer markers for efficient and targeted malignant cells imaging and PDT (see reference (Biju et al. 2009) for a review). To improve their intracellular radical generating ability, escape of QDs into cytosol must be stimulated by UV (Silver and Ou 2005) or blue light (Juzenas et al. 2008).

For *in vivo* condition, cells could be loaded with QDs together with dipeptides, such as ala-ala, ser-tyr, and tyr-ala, which were able to induce lysosomal rupture due to osmotic imbalance (Bird and Lloyd 1995) (reviewed by Juzenas et al. (2008)). Several groups already studied the cytotoxicity of QDs upon exposure to UV radiation (Bakalova et al. 2008). In a study, CdTe QD and aluminum tetrasulfophthalocyanine (QD-AlSPc) nanocomposites were shown to generate singlet oxygen with QY around 15%, high compared to the 1% of bare QDs in $D_2O$ but still low compared to that of AlSPc alone (36%) (Ma et al. 2008). The determined FRET efficiency of QD-AlSPc nanocomplexes was around 58% (Idowu et al. 2008). An enhancement in efficiency of the energy transfer was observed when QDs were stabilized by carboxylate–thiol ligands. A mixture of MPA-capped QDs with AlTSPc resulted in long triplet lifetime, high energy transfer efficiency, and high triplet yield of the latter, showing to be most suitable as a potential candidate for PDT of cancer studies.

The results obtained from Clark et al. suggest that catechols may serve as useful sensitizers for QD PDT (Clarke et al. 2006). Different peptide-coated QD-PS conjugates were prepared by Tsay et al. using novel strategies (Tsay et al. 2007). Rose bengal and chlorine e6 PSs were covalently attached to phytochelatin related peptides. These conjugates were next used to overcoat green- and red-emitting CdSe/CdS/ZnS QDs. With these complexes, it was possible to achieve ROI generation via indirect excitation through FRET from the nanocrystals to PSs, or by direct excitation of the PSs.

Organically modified silica NPs were recently reported to be used for two-photon PDT energy-transferring (Kim et al. 2004). The authors concluded that their approach of two-photon induced intra-particle FRET, based on the use of two-photon fluorescent aggregates as donors and a photosensitizing drug as acceptor, offers a simple and proper methodology for developing formulations of drug-carrier nanoassemblies applicable in two-photon activated PDT.

A more recent study emphasizes the potential of CdTe QDs as energy donors to activate *meso*-tetra-4-sulfonato-phenyl porphine dihydrochloride or applications in two-photon excitation PDT. By means of nanosecond time-resolved technique, the authors investigated the role of the excitonic and trapping states of QDs, respectively, in the energy transfer process. In this way, they tried to find a way of improving the energy transfer efficiency and, consequently, to increase the singlet oxygen QY for PDT application.

### 10.7.2 Quantum Dots in Photodynamic Therapy: An Alternative to Antibiotic Therapy

The use of PSs to kill bacteria has been investigated since the 1950s (Oginsky et al. 1959). Due to upcoming chemoresistance, PDT has recently attracted a renewed interest as an effective alternative to antibiotic therapy in the treatment of local infections (Demidova and Hamblin 2004; Jori et al. 2006).

Targeting of the infecting microorganisms can be attained by chemical modification of the PS and/or by choosing proper light energy doses that do not damage host cells (Demidova and Hamblin 2004; Embleton et al. 2005). As an example, Narband et al. recently proved that addition of CdSe/ZnS PEG QDs with an emission maximum of 627 nm, close to the absorption maximum of Toluidine Blue O (TBO) at 630 nm, can significantly enhance the efficacy of TBO-mediated lethal photosensitization of two significant human pathogens *Staphylococcus aureus* and *Streptococcus pyogenes* in a concentration-dependent manner, with the greatest kills at the higher TBO to QD ratios (Narband et al. 2008).

## 10.8 From Quantum Dots Toward Multifunctional Quantum Dots–Based Materials for Multimodal Imaging

Research in nanoscience is now focused on the development and miniaturization of structures made of different functional entities, each of them able to carry out specific tasks. In the view to design multifunctional nanostructures based on fluorescent QDs that might serve as new medical diagnostic tools, it is crucial to identify nanomaterials to associate to the QDs that have peculiar features.

While the optical imaging provides detailed information at subcellular level, the QD biodistribution in living animals cannot be quantified accurately from fluorescence measurement. In contrast, scanning techniques like molecular resonance imaging (MRI) and PET, respectively, based on magnetic contrast agents and positron emitter molecules can be applied to follow the distribution of molecules *in vivo*. These techniques offer potential applications in many diseases for monitoring tissue implants, studying the dynamics of tumor metastasis, or biochemical mediators. Therefore, the combination of QDs with magnetic materials or positron emitters would be definitely more advantageous.

Perhaps, the most common association so far has been the combination of fluorescent QDs with magnetic nanoparticles, based mainly on superparamagnetic iron oxide, which are T2 MRI contrast agents.

While several works have reported on the fabrication of fluorescent-magnetic nanostructures, only few have shown the main advantage of having both nanoparticles within the same nanostructures in *in vivo* animal models (Tortiglione et al. 2007; Corr et al. 2008; Corsi et al. 2009; Thakur et al. 2009). With respect to the preparation of QDs-iron oxide nanocomposites three main approaches have been followed: (1) the binding of the two domains, the magnetic and fluorescent through spacer molecules (Corr et al. 2006; You et al. 2007); (2) the encapsulation of both domains within a common matrix, such as silica (Hoshino et al. 2004; Veiseh et al. 2005; Salgueirino-Maceira et al. 2006; Thakur et al. 2009) or polymer materials (Chu et al. 2006; Edgar et al. 2006; Sathe et al. 2006; Nabiev et al. 2007); (3) the direct preparation through colloidal synthesis of heterostructures composed of two distinct spherical domains one fluorescent and one magnetic attached though a small interface (Groc et al. 2004; Veiseh et al. 2005; Du et al. 2006). Depending on the preparation route followed quenching problems due to the proximity of the QDs to the magnetic nanoparticles could be encountered, however to the state of the art, several studies have shown the preservation of still good optical and magnetic performances, even when the systems have been applied on cell cultures (Mandal et al. 2005; Zebli et al. 2005; Sathe et al. 2006; Sun et al. 2010).

QD-iron oxide nanostructures have been applied for the simultaneous detection and separation of tumor cells (Wang et al. 2004; Chu et al. 2006) and apoptotic cells (Lin et al. 2007).

Wang et al. have demonstrated the sorting of MCF-7 breast cancer cells by using iron oxide–CdSe/ZnS nanocomposites functionalized with the antibody anticycline E for the sorting of breast tumor cells MCF-7 (Wang et al. 2004). After applying the nanocomposites of 30–50 nm in size to a suspension of MCF-7 cells for an incubation time of 15 min, the authors were able to separate the tumor cells from the suspension by applying a permanent magnet. A similar response to the magnet was observed by another group that used fluorescent-magnetic nanocomposites based on iron oxide–CdTe polystyrene nanospheres with an average size of 300 nm (Chu et al. 2006). In that case, the nanocomposites functionalized with ligand for EGF receptors had been used for targeting MDA-MB-435S cell populations.

The work reported by Xie et al. represents a proof of principle for the quantitative separation and detection of apoptotic cells with respect to live cells (Lin et al. 2007). In this study, avidin-functionalized nanocomposites based on iron oxide and QDs embedded in copolymer nanospheres have been employed. Cell apoptosis was first induced by means of UV irradiation. Apoptotic cells express on their surface phosphatidylserine, which can be recognized by annexin V group molecules. The cell suspensions, after UV irradiation, were first treated with annexin-biotin and then with avidin-functionalized nanocomposites (avidin recognizes biotin and therefore can label the apoptotic cells). By applying a magnet, the authors of the study were able to sort out apoptotic cells. Those kinds of magnetic-fluorescent nanostructures might represent a valuable tool for the investigation of diseases that trigger apoptosis.

Within the study of fluorescent-magnetic nanostructures, QDs have also been associated to T1 MRI contrast agents. Bakalova and colleagues developed silica-shell QD that incorporates within the silica matrix the gadolinium, thus combining fluorescent and paramagnetic properties that could be detected *in vivo* in a single manner, simultaneously using optical imaging and MRI (Bakalova et al. 2008). The incorporation of gadolinium did not degrade the fluorescence QY as well as their paramagnetic characteristics of the resulting nanocomposites. The QD-based nanocomposites injected intravenously in rat did not affect the physiological parameters, such as blood pressure, heart rate, and microcirculation. The probe did not change the diameter and structure of the vasculature in the brain, visualized by two-photon microscope, which makes QDs an appropriate tracer in the visualization of neurons, astrocytes, and blood vessels.

As an alternative study in which fluorescent QDs have been associated to positron emitters, Cai et al. in order to obtain an accurate quantification of tumor-targeting efficacy together with a complete understanding of pharmacokinetics have developed a dual-function probe combining NIR-QD with a PET isotope (Cai et al. 2007). In this paper, RGD peptide-labeled CdTe QDs were conjugated with 1,4,7,10-tetraazacyclodocecane-N,N′,N″,N‴-tetraacetic acid (DOTA), a chelating agent for $^{64}Cu$, the positron emitter species. This new probe allows both PET and QD fluorescence imaging and overcomes the tissue penetration limitation of optical imaging. The bone marrow, LN, liver and U87MG tumor were clearly visualized 5 h after injection by *in vivo* PET imaging, obtaining an accurate quantification of the probe distribution in mice. Also, the results from *ex vivo* small-animal PET and near infrared fluorescence imaging, performed on harvested tissues at the same time point, were quite similar. The *ex vivo* histological data indicate minimal extravasation of this dual-function probe, which in turn leads to relatively low tumor uptake.

Finally, it is worth citing a recently reported bimodal imaging system proposed as a probe for the optical and ultrasound detection (Schipper et al. 2009). The nanocomposites used in this study derived by the deposition of CdTe QDs via the electrostatic layer-by-layer technique onto the outer surface of ST68 (Span60 and Tween 80) microbubbles, a well-studied ultrasound agent.

The QD-functionalized microbubbles were characterized *in vitro* for what concern their optical and ultrasound contrast by using confocal microscopy and ultrasonography techniques, showing the preservation of both features.

On an *in vivo* animal model, after the intravenously injection of the QD-modified microbubbles through ear vein of the mouse, ultrasound contrast enhancement of the kidneys was clear observed, indicating that the QD-modified microbubbles were stable and small enough and could pass through pulmonary capillary, to achieve systemic enhancement. Interesting as previously observed, under the diagnostic ultrasound conditions, microbubbles could be destroyed and release their content (which could be a drug or a gene) (Hernot and Klibanov 2008).

This feature was also observed in the case of QD-modified microbubbles, as indeed enhancement of the fluorescent signal could be observed for a solution of QD-modified microbubbles exposed to ultrasound irradiation. Indeed, the authors proposed their system as a controlled delivery system for QD to a site where ultrasound irradiation is applied.

It is significant to underline that the works here cited are only some examples of bimodal imaging nanocomposites based on QDs′ multifunctional medical platforms. Those works just highlight the great potential of QDs and surely more studies are still under development.

## 10.9 Perspectives

The advent of semiconductor QDs has brought a tremendous impact in many fields. Their unique characteristics and their peculiar optical properties have induced the employment of these semiconductor nanocrystals in an ever increasing variety of applications. In the last years, the use of QDs for biological applications already showed how these new compounds can bring a considerable improvement to traditional strategies by enhancing old performances and also adding new features. Their physicochemical characteristics perfectly adapt to the requirements of biological applications. New synthetic methods and ever-increasing functionalization strategies are allowing the successful exploitation of QDs in a biological environment, from detection and biosensing to labeling and imaging. Their unique optical properties overcome those of traditional organic dyes in almost any of these fields. Broad absorption spectra and narrow emission spectra with a robust signal strength, resistance to photobleaching, the possibility of multiplexing are only some of the reasons why these new fluorophores are quickly substituting organic dyes in almost every biological application. Every year in the literature, there is an ever increasing number of publications related to QDs, and always new research groups make contact with nanotechnology and the potentiality of these fluorescent nanocrystals. Many challenges, such as technical and toxicity problems, still remain to be faced and overcome. Reproducibility of synthetic procedures, which always yield small differences in the physicochemical characteristics of the nanocrystals, the best procedure to transfer the QDs into aqueous solutions, conjugation of targeting and therapeutic ligands, such as antibodies, nucleic acids, peptides, and drugs, the influence of the size and steric bulk of the nanocrystals on the biological activity of the conjugated compounds, and the biodistribution and the fate of the bio-nanosystems, all these are only some questions researchers are trying to address and answer. However, the impact that QDs has had on biology and medicine is already enormous and surely it will be still growing in the next future.

## References


Akerman, M. E., Chan, W. C. W., Laakkonen, P., Bhatia, S. N., and Ruoslahti, E. 2002. Nanocrystal targeting in vivo. *Proceedings of the National Academy of Sciences of the United States of America* 99: 12617–12621.

Aldana, J., Wang, Y. A., and Peng, X. G. 2001. Photochemical instability of CdSe nanocrystals coated by hydrophilic thiols. *Journal of the American Chemical Society* 123: 8844–8850.

Alejandro-Arellano, M., Ung, T., Blanco, A., Mulvaney, P., and Liz-Marzan, L. M. 2000. Silica-coated metals and semiconductors. Stabilization and nanostructuring. *Pure and Applied Chemistry* 72: 257–267.

Alivisatos, A. P. 1996. Semiconductor clusters, nanocrystals, and quantum dots. *Science* 271: 933–937.

Aravanis, A. M., Pyle, J. L., and Tsien, R. W. 2003. Single synaptic vesicles fusing transiently and successively without loss of identity. *Nature* 423: 643–647.

Arrigo, A. P. 1999. Gene expression and the thiol redox state. *Free Radical Biology and Medicine* 27: 936–944.

Bae, P. K., Kim, K. N., Lee, S. J. et al. 2009. The modification of quantum dot probes used for the targeted imaging of his-tagged fusion proteins. *Biomaterials* 30: 836–842.

Bakalova, R., Zhelev, Z., Aoki, I. et al. 2008. Multimodal silica-shelled quantum dots: Direct intracellular delivery, photosensitization, toxic, and microcirculation effects. *Bioconjugate Chemistry* 19: 1135–1142.

Balaji, J. and Ryan, T. A. 2007. Single-vesicle imaging reveals that synaptic vesicle exocytosis and endocytosis are coupled by a single stochastic mode. *Proceedings of the National Academy of Sciences of the United States of America* 104: 20576–20581.

Ballou, B., Lagerholm, B. C., Ernst, L. A., Bruchez, M. P., and Waggoner, A. S. 2004. Noninvasive imaging of quantum dots in mice. *Bioconjugate Chemistry* 15: 79–86.

Ballou, B., Ernst, L. A., Andreko, S. et al. 2007. Sentinel lymph node imaging using quantum dots in mouse tumor models. *Bioconjugate Chemistry* 18: 389–396.

Beis, D. and Stainier, D. Y. R. 2006. In vivo cell biology: Following the zebrafish trend. *Trends in Cell Biology* 16: 105–112.

Bertin, G. and Averbeck, D. 2006. Cadmium: Cellular effects, modifications of biomolecules, modulation of DNA repair and genotoxic consequences (a review). *Biochimie* 88: 1549–1559.

Bertolini, G., Paleari, L., Catassi, A. et al. 2008. In vivo cancer imaging with semiconductor quantum dots. *Current Pharmaceutical Analysis* 4: 197–205.

Biju, V., Mundayoor, S., Omkumar, R. V., Anas, A., and Ishikawa, M. 2009. Bioconjugated quantum dots for cancer research: Present status, prospects and remaining issues. *Biotechnology Advances* 28: 199–213.

Bird, S. J. and Lloyd, J. B. 1995. Mechanism of lysosome rupture by dipeptides. *Cell Biochemistry and Function* 13: 79–83.

Boeneman, K., Mei, B. C., Dennis, A. M. et al. 2009. Sensing caspase 3 activity with quantum dot-fluorescent protein assemblies. *Journal of the American Chemical Society* 131: 3828–3829.

Bruchez, M., Moronne, M., Gin, P., Weiss, S., and Alivisatos, A. P. 1998. Semiconductor nanocrystals as fluorescent biological labels. *Science* 281: 2013–2016.

Cai, W. B., Shin, D. W., Chen, K. et al. 2006. Peptide-labeled near-infrared quantum dots for imaging tumor vasculature in living subjects. *Nano Letters* 6: 669–676.

Cai, W. B., Chen, K., Li, Z. B., Gambhir, S. S., and Chen, X. Y. 2007. Dual-function probe for PET and near-infrared fluorescence imaging of tumor vasculature. *Journal of Nuclear Medicine* 48: 1862–1870.

Cambi, A., Lidke, D. S., Arndt-Jovin, D. J., Figdor, C. G., and Jovin, T. M. 2007. Ligand-conjugated quantum dots monitor antigen uptake and processing by dendritic cells. *Nano Letters* 7: 970–977.

Cao, L., Ye, J., Tong, L., and Tang, B. 2008. A new route to the considerable enhancement of glucose oxidase (GOx) activity: The simple assembly of a complex from CdTe quantum dots and GOx and its glucose sensing. *Chemistry—A European Journal* 14: 9633–9640.

Chan, W. C. W. and Nie, S. M. 1998. Quantum dot bioconjugates for ultrasensitive nonisotopic detection. *Science* 281: 2016–2018.

Chang, E., Miller, J. S., Sun, J. et al. 2005. Protease-activated quantum dot probes. *Biochemical and Biophysical Research Communications* 334: 1317–1321.

Chen, Y. F. and Rosenzweig, Z. 2002. Luminescent CdS quantum dots as selective ion probes. *Analytical Chemistry* 74: 5132–5138.

Chen, L. M., Zurita, A. J., Ardelt, P. U. et al. 2004. Design and validation of a bifunctional ligand display system for receptor targeting. *Chemistry & Biology* 11: 1081–1091.

Chen, A. A., Derfus, A. M., Khetani, S. R., and Bhatia, S. N. 2005. Quantum dots to monitor RNAi delivery and improve gene silencing. *Nucleic Acids Research* 33: e190.

Chen, Z., Chen, H., Meng, H. et al. 2008. Bio-distribution and metabolic paths of silica coated CdSeS quantum dots. *Toxicology and Applied Pharmacology* 230: 364–371.

Choi, H. S., Liu, W., Misra, P. et al. 2007. Renal clearance of quantum dots. *Nature Biotechnology* 25: 1165–1170.

Choi, H. S., Ipe, B. I., Misra, P. et al. 2009. Tissue- and organ-selective biodistribution of NIR fluorescent quantum dots. *Nano Letters* 9: 2354–2359.

Chu, M. Q., Song, X., Cheng, D., Liu, S. P., and Zhu, J. 2006. Preparation of quantum dot-coated magnetic polystyrene nanospheres for cancer cell labelling and separation. *Nanotechnology* 17: 3268–3273.

Clarke, S. J., Hollmann, C. A., Zhang, Z. J. et al. 2006. Photophysics of dopamine-modified quantum-dots and effects on biological systems. *Nature Materials* 5: 409–417.

Clarkson, T. W. 1997. The toxicology of mercury. *Critical Reviews in Clinical Laboratory Sciences* 34: 369–403.

Corr, S. A., O' Byrne, A., Gun'ko, Y. K. et al. 2006. Magnetic-fluorescent nanocomposites for biomedical multitasking. *Chemical Communications*: 4474–4476.

Corr, S. A., Rakovich, Y. P., and Gun'ko, Y. K. 2008. Multifunctional magnetic-fluorescent nanocomposites for biomedical applications. *Nanoscale Research Letters* 3: 87–104.

Correa-Duarte, M. A., Giersig, M., and Liz-Marzan, L. M. 1998. Stabilization of CdS semiconductor nanoparticles against photodegradation by a silica coating procedure. *Chemical Physics Letters* 286: 497–501.

Corsi, F., De Palma, C., Colombo, M. et al. 2009. Towards ideal magnetofluorescent nanoparticles for bimodal detection of breast-cancer cells. *Small* 5: 2555–2564.

Coulon, J., Thouvenin, I., Aldeek, F., Balan, L., and Schneider, R. 2010. Glycosylated quantum dots for the selective labelling of *Kluyveromyces bulgaricus* and *Saccharomyces cerevisiae* yeast strains. *Journal of Fluorescence* 20: 591–597.

Courty, S., Luccardini, C., Bellaiche, Y., Cappello, G., and Dahan, M. 2006. Tracking individual kinesin motors in living cells using single quantum-dot imaging. *Nano Letters* 6: 1491–1495.

Dabbousi, B. O., RodriguezViejo, J., Mikulec, F. V. et al. 1997. (CdSe)ZnS core-shell quantum dots: Synthesis and characterization of a size series of highly luminescent nanocrystallites. *Journal of Physical Chemistry B* 101: 9463–9475.

Dahan, M., Levi, S., Luccardini, C. et al. 2003. Diffusion dynamics of glycine receptors revealed by single-quantum dot tracking. *Science* 302: 442–445.

Delehanty, J. B., Medintz, I. L., Pons, T. et al. 2006. Self-assembled quantum dot-peptide bioconjugates for selective intracellular delivery. *Bioconjugate Chemistry* 17: 920–927.

Demidova, T. N. and Hamblin, M. R. 2004. Photodynamic therapy targeted to pathogens. *International Journal of Immunopathology and Pharmacology* 17: 245–254.

Demir, M. M., Munoz-Espi, R., Lieberwirth, I., and Wegner, G. 2006. Precipitation of monodisperse ZnO nanocrystals via acid-catalyzed esterification of zinc acetate. *Journal of Materials Chemistry* 16: 2940–2947.

Derfus, A. M., Chan, W. C. W., and Bhatia, S. N. 2004. Intracellular delivery of quantum dots for live cell labeling and organelle tracking. *Advanced Materials* 16: 961–966.

Di Corato, R., Quarta, A., Piacenza, P. et al. 2008. Water solubilization of hydrophobic nanocrystals by means of poly(maleic anhydride-*alt*-1-octadecene). *Journal of Materials Chemistry* 18: 1991–1996.

Diagaradjane, P., Orenstein-Cardona, J. M., Colon-Casasnovas, N. E. et al. 2008. Imaging epidermal growth factor receptor expression in vivo: Pharmacokinetic and biodistribution characterization of a bioconjugated quantum dot nanoprobe. *Clinical Cancer Research* 14: 731–741.

Diao, X. L., Xia, Y. S., Zhang, T. L., Li, Y., and Zhu, C. Q. 2007. Fluorescence-detecting cationic surfactants using luminescent CdTe quantum dots as probes. *Analytical and Bioanalytical Chemistry* 388: 1191–1197.

Dixit, S. K., Goicochea, N. L., Daniel, M. C. et al. 2006. Quantum dot encapsulation in viral capsids. *Nano Letters* 6: 1993–1999.

Dolmans, D., Fukumura, D., and Jain, R. K. 2003. Photodynamic therapy for cancer. *Nature Reviews Cancer* 3: 380–387.

Doshi, N. and Mitragotri, S. 2009. Designer biomaterials for nanomedicine. *Advanced Functional Materials* 19: 3843–3854.

Dougherty, T. J., Gomer, C. J., Henderson, B. W. et al. 1998. Photodynamic therapy. *Journal of the National Cancer Institute* 90: 889–905.

Du, G. H., Liu, Z. L., Lu, Q. H. et al. 2006. Fe3O4/CdSe/ZnS magnetic fluorescent bifunctional nanocomposites. *Nanotechnology* 17: 2850–2854.

Dubertret, B., Skourides, P., Norris, D. J. et al. 2002. In vivo imaging of quantum dots encapsulated in phospholipid micelles. *Science* 298: 1759–1762.

Duncan, R. 2003. The dawning era of polymer therapeutics. *Nature Reviews Drug Discovery* 2: 347–360.

Dyadyusha, L., Yin, H., Jaiswal, S. et al. 2005. Quenching of CdSe quantum dot emission, a new approach for biosensing. *Chemical Communications*: 3201–3203.

Dykxhoorn, D. M., Palliser, D., and Lieberman, J. 2006. The silent treatment: siRNAs as small molecule drugs. *Gene Therapy* 13: 541–552.

Edgar, R., McKinstry, M., Hwang, J. et al. 2006. High-sensitivity bacterial detection using biotin-tagged phage and quantum-dot nanocomplexes. *Proceedings of the National Academy of Sciences of the United States of America* 103: 4841–4845.

Efros, A. L., Kharchenko, V. A., and Rosen, M. 1995. Breaking the phonon bottleneck in nanometer quantum dots – Role of Auger-like processes. *Solid State Communications* 93: 281–284.

Elbashir, S. M., Lendeckel, W., and Tuschl, T. 2001. RNA interference is mediated by 21-and 22-nucleotide RNAs. *Genes & Development* 15: 188–200.

Embleton, M. L., Nair, S. P., Heywood, W. et al. 2005. Development of a novel targeting system for lethal photo sensitization of antibiotic-resistant strains of Staphylococcus aureus. *Antimicrobial Agents and Chemotherapy* 49: 3690–3696.

Feng, J., Ding, S. Y., Tucker, M. P. et al. 2005. Cyclodextrin driven hydrophobic/hydrophilic transformation of semiconductor nanoparticles. *Applied Physics Letters* 86: 033108.

Fernandez-Alfonso, T. and Ryan, T. A. 2004. The kinetics of synaptic vesicle pool depletion at CNS synaptic terminals. *Neuron* 41: 943–953.

Ferrara, D. E., Weiss, D., Carnell, P. H. et al. 2006. Quantitative 3D fluorescence technique for the analysis of en face preparations of arterial walls using quantum dot nanocrystals and two-photon excitation laser scanning microscopy. *American Journal of Physiology—Regulatory, Integrative and Comparative Physiology* 290: R114–R123.

Fire, A., Xu, S. Q., Montgomery, M. K. et al. 1998. Potent and specific genetic interference by double-stranded RNA in *Caenorhabditis elegans. Nature* 391: 806–811.

Fischer, H. C., Liu, L. C., Pang, K. S., and Chan, W. C. W. 2006. Pharmacokinetics of nanoscale quantum dots: In vivo distribution, sequestration, and clearance in the rat. *Advanced Functional Materials* 16: 1299–1305.

Fomenko, V. and Nesbitt, D. J. 2008. Solution control of radiative and nonradiative lifetimes: A novel contribution to quantum dot blinking suppression. *Nano Letters* 8: 287–293.

Fountaine, T. J., Wincovitch, S. M., Geho, D. H., Garfield, S. H., and Pittaluga, S. 2006. Multispectral imaging of clinically relevant cellular targets in tonsil and lymphoid tissue using semiconductor quantum dots. *Modern Pathology* 19: 1181–1191.

Frangioni, J. V. 2003. In vivo near-infrared fluorescence imaging. *Current Opinion in Chemical Biology* 7: 626–634.

Friedrich, M., Nozadze, R., Gan, Q. et al. 2009. Detection of single quantum dots in model organisms with sheet illumination microscopy. *Biochemical and Biophysical Research Communications* 390: 722–727.

Gao, X. H., Cui, Y. Y., Levenson, R. M., Chung, L. W. K., and Nie, S. M. 2004. In vivo cancer targeting and imaging with semiconductor quantum dots. *Nature Biotechnology* 22: 969–976.

Gao, X., Yang, L., Petros, J. A. et al. 2005. In vivo molecular and cellular imaging with quantum dots. *Current Opinion in Biotechnology* 16: 63–72.

Geijtenbeek, T. B., Kwon, D. S., Torensma, R. et al. 2000. DC-SIGN, a dendritic cell-specific HIV-1-binding protein that enhances trans-infection of T cells. *Cell* 100: 587–597.

Geissbuehler, I., Hovius, R., Martinez, K. L. et al. 2005. Lipid-coated nanocrystals as multifunctionalized luminescent scaffolds for supramolecular biological assemblies. *Angewandte Chemie-International Edition* 44: 1388–1392.

Geng, L., Zhang, H. L., and Peng, H. B. 2009. The formation of acetylcholine receptor clusters visualized with quantum dots. *BMC Neuroscience* 10: 80.

Gerion, D., Chen, F., Kannan, B. et al. 2003. Room-temperature single-nucleotide polymorphism and multiallele DNA detection using fluorescent nanocrystals and microarrays. *Analytical Chemistry* 75: 4766–4772.

Giepmans, B. N., Deerinck, T. J., Smarr, B. L., Jones, Y. Z., and Ellisman, M. H. 2005. Correlated light and electron microscopic imaging of multiple endogenous proteins using quantum dots. *Nature Methods* 2: 743–749.

Gill, R., Willner, I., Shweky, I., and Banin, U. 2005. Fluorescence resonance energy transfer in CdSe/ZnS-DNA conjugates: Probing hybridization and DNA cleavage. *Journal of Physical Chemistry B* 109: 23715–23719.

Gill, R., Bahshi, L., Freeman, R., and Willner, I. 2008. Optical detection of glucose and acetylcholine esterase inhibitors by H2O2-sensitive CdSe/ZnS quantum dots. *Angewandte Chemie—International Edition* 47: 1676–1679.

Goldman, E. R., Anderson, G. P., Tran, P. T. et al. 2002. Conjugation of luminescent quantum dots with antibodies using an engineered adaptor protein to provide new reagents for fluoroimmunoassays. *Analytical Chemistry* 74: 841–847.

Goldman, E. R., Clapp, A. R., Anderson, G. P. et al. 2004. Multiplexed toxin analysis using four colors of quantum dot fluororeagents. *Analytical Chemistry* 76: 684–688.

Goldman, E. R., Medintz, I. L., Hayhurst, A. et al. 2005a. Self-assembled luminescent CdSe-ZnS quantum dot bioconjugates prepared using engineered poly-histidine terminated proteins. *Analytica Chimica Acta* 534: 63–67.

Goldman, E. R., Medintz, I. L., Whitley, J. L. et al. 2005b. A hybrid quantum dot-antibody fragment fluorescence resonance energy transfer-based TNT sensor. *Journal of the American Chemical Society* 127: 6744–6751.

Gonda, K., Watanabe, T. M., Ohuchi, N., and Higuchi, H. 2010. In vivo nano-imaging of membrane dynamics in metastatic tumor cells using quantum dots. *The Journal of Biological Chemistry* 285: 2750–2757.

Gopee, N. V., Roberts, D. W., Webb, P. et al. 2007. Migration of intradermally injected quantum dots to sentinel organs in mice. *Toxicological Sciences* 98: 249–257.

Groc, L., Heine, M., Cognet, L. et al. 2004. Differential activity-dependent regulation of the lateral mobilities of AMPA and NMDA receptors. *Nature Neuroscience* 7: 695–696.

Guo, W. H., Li, J. J., Wang, Y. A., and Peng, X. G. 2003. Luminescent CdSe/CdS core/shell nanocrystals in dendron boxes: Superior chemical, photochemical and thermal stability. *Journal of the American Chemical Society* 125: 3901–3909.

Haggie, P. M., Kim, J. K., Lukacs, G. L., and Verkman, A. S. 2006. Tracking of quantum dot-labeled CFTR shows near immobilization by C-terminal PDZ interactions. *Molecular Biology of the Cell* 17: 4937–4945.

Hama, Y., Koyama, Y., Urano, Y., Choyke, P. L., and Kobayashi, H. 2007. Simultaneous two-color spectral fluorescence lymphangiography with near infrared quantum dots to map two lymphatic flows from the breast and the upper extremity. *Breast Cancer Research and Treatment* 103: 23–28.

Hamity, M., Lema, R. H., and Suchetti, C. A. 1998. The effect of tetraalkylammonium and alkyl sulfate salts on the fluorescence bands of quantum-sized CdS. *Journal of Photochemistry and Photobiology A—Chemistry* 115: 163–168.

Han, C. and Li, H. 2008. Chiral recognition of amino acids based on cyclodextrin-capped quantum dots. *Small* 4: 1344–1350.

Hanaki, K., Momo, A., Oku, T. et al. 2003. Semiconductor quantum dot/albumin complex is a long-life and highly photostable endosome marker. *Biochemical and Biophysical Research Communications* 302: 496–501.

Hansen, J. C., Tse, C., and Wolffe, A. P. 1998. Structure and function of the core histone N-termini: More than meets the eye. *Biochemistry* 37: 17637–17641.

Harata, N. C., Choi, S., Pyle, J. L., Aravanis, A. M., and Tsien, R. W. 2006. Frequency-dependent kinetics and prevalence of kiss-and-run and reuse at hippocampal synapses studied with novel quenching methods. *Neuron* 49: 243–256.

Hardman, R. 2006. A toxicologic review of quantum dots: Toxicity depends on physicochemical and environmental factors. *Environmental Health Perspectives* 114: 165–172.

He, Y., Wang, H.-F., and Yan, X.-P. 2008. Exploring Mn-doped ZnS quantum dots for the room-temperature phosphorescence detection of enoxacin in biological fluids. *Analytical Chemistry* 80: 3832–3837.

Henderson, B. W. and Dougherty, T. J. 1992. How does photodynamic therapy work. *Photochemistry and Photobiology* 55: 145–157.

Hernot, S. and Klibanov, A. L. 2008. Microbubbles in ultrasound-triggered drug and gene delivery. *Advanced Drug Delivery Reviews* 60: 1153–1166.

(a)

(b) (c)

**FIGURE 2.1**
Pictures of mural paintings in (a) Mayapan (Mexico) treated with acrylic/vinyl-acetate copolymer films to protect the surfaces. The pictures show the dramatic effects due to the aging of the polymer film that generates (b) whitening and (c) shrinkage of the paint layer.

(a) (b)

**FIGURE 2.3**
The Maya ruins of Calakmul: (a) a view of structure I (*Acropolis Chik Naab*), (b) a view of the mural paintings discovered inside the building.

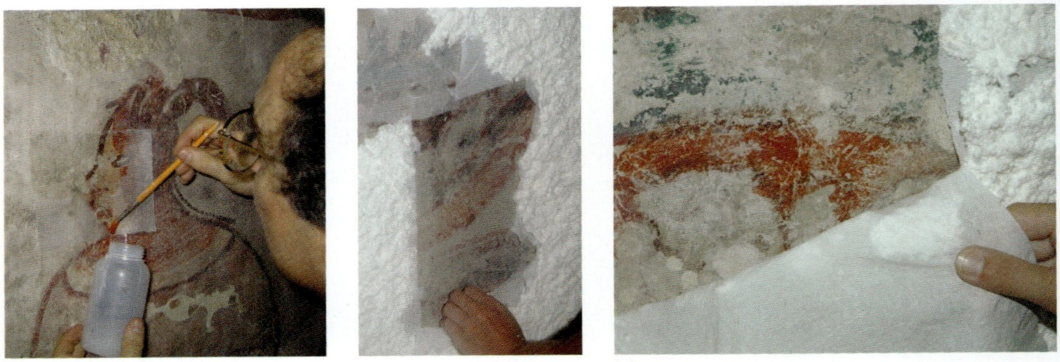


**FIGURE 2.4**
Different restoration steps of paintings consolidation by means of $Ca(OH)_2$ nanoparticles. (a) Application of particles dispersion and (b–c) removal of wet cellulose compresses used to maintain surfaces humid, in order to favor a slow calcium carbonate formation.

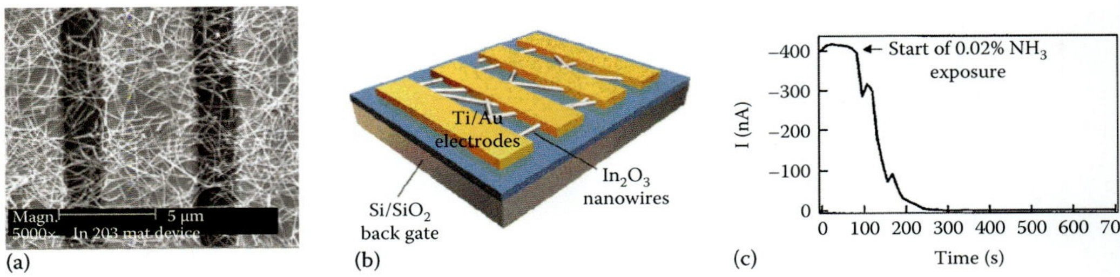


**FIGURE 4.18**
(a) SEM image of the $In_2O_3$ nanowires device having electrode structure, (b) multi-nanowire gas sensor device, and (c) response of $In_2O_3$ nanowire sensor to 0.02% $NH_3$. (Reprinted with permission from Zhang, D. et al., *Nano Lett.*, 4, 1919, 2004. Copyright 2004, American Chemical Society; Li, C. et al., *Ann. N. Y. Acad. Sci.*, 1006, 104, 2003. Copyright 2004, American Chemical Society.)

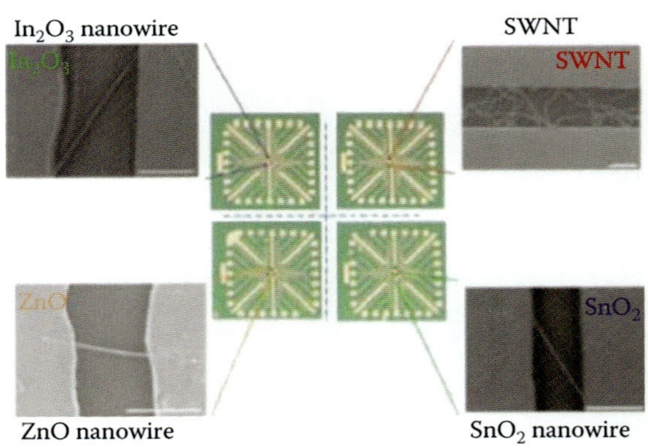


**FIGURE 4.19**
Hybrid chemical sensor array chip composed of four individual chemical sensors. (Reprinted with permission from Chen, P.-C. et al., *IEEE Trans. Nanotechnol.*, 7, 668, 2008. Copyright 2008, American Institute of Physics.)

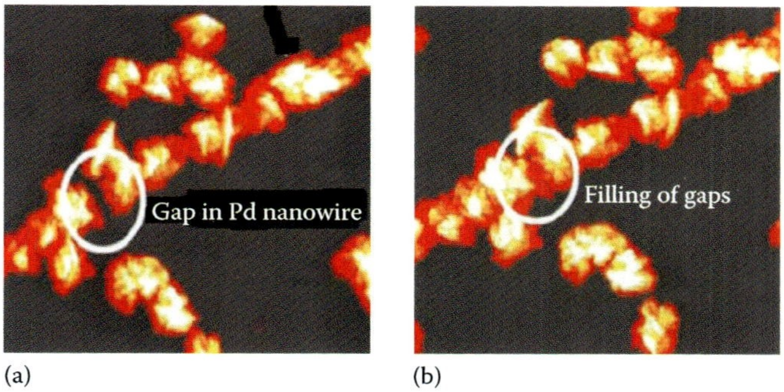


**FIGURE 4.29**
In situ AFM of Pd mesowire in (a) air and (b) hydrogen. (From Walter, E.C., *Anal. Chem.* 74, 1546, 2002. With permission.)

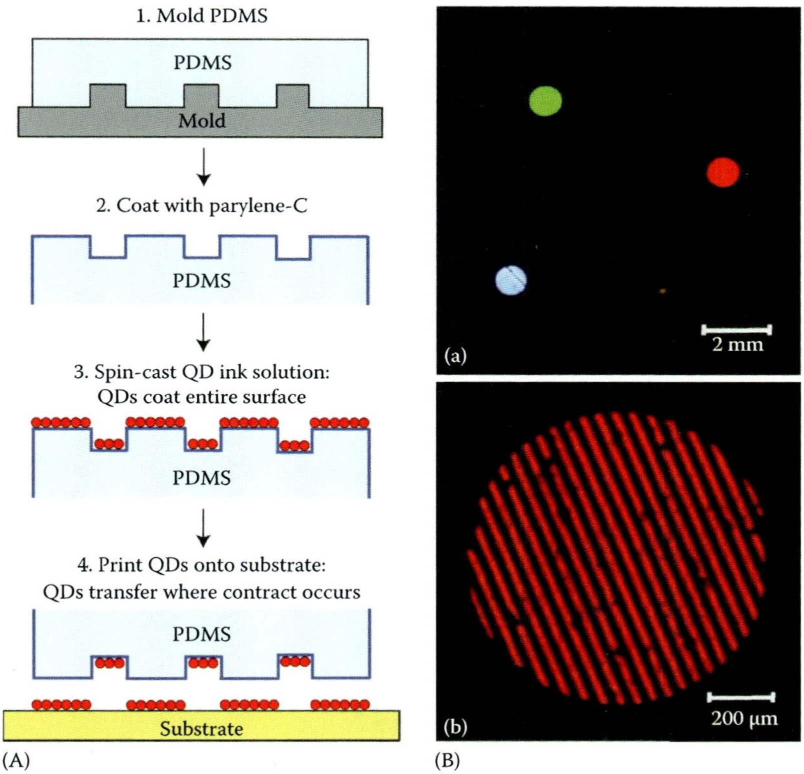


**FIGURE 5.6**
(A) The schematic of the contact printing process; (B) (a) electroluminescent red and green QD pixels, fabricated on the same substrate. A blue pixel is a result of TPD emission in the area without QDs; (b) QD-LED pixel, patterned with 25 μm wide stamp features. (Reprinted with permission from Kim, L.A. et al., *Nano Lett.*, 8, 4513, 2008. Copyright [2008], American Chemical Society.)

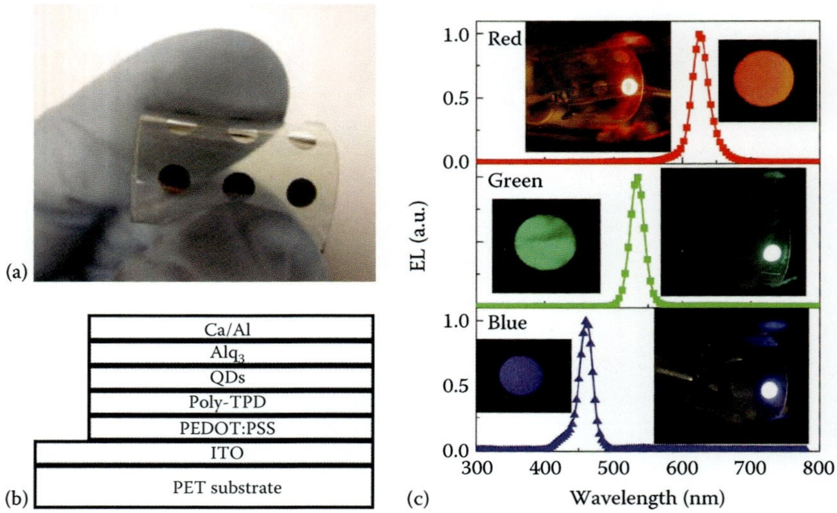


**FIGURE 5.9**
(a) Photograph, (b) device configuration, and (c) images of the relaxed and bent flexible RGB-emitting QD-LEDs, operated at maximum brightness according to EL spectra. (Reprinted with permission from Tan, Z. et al., *J. Appl. Phys.*, 105, 034312, 2009. Copyright [2009], American Institute of Physics.)

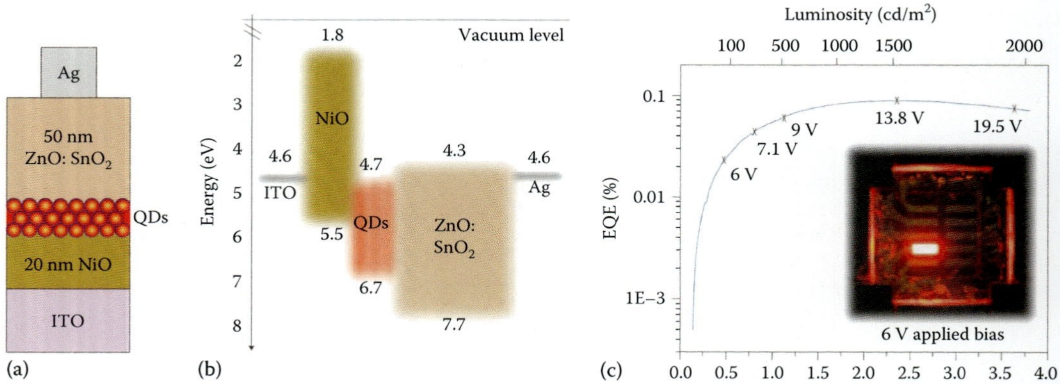


**FIGURE 5.11**
(a) Schematic of the device structure, (b) band diagram determined from UV photoemission spectroscopy and optical absorption measurements, and (c) QD-LED EQE as a function of the current density J. The inset shows a photograph of the bright and uniform pixel at 6 V applied bias. (Reprinted with permission from Caruge, J.M. et al., *Nat. Photon.*, 2, 247, 2008. Copyright [2008], Macmillan Publishers Ltd.)

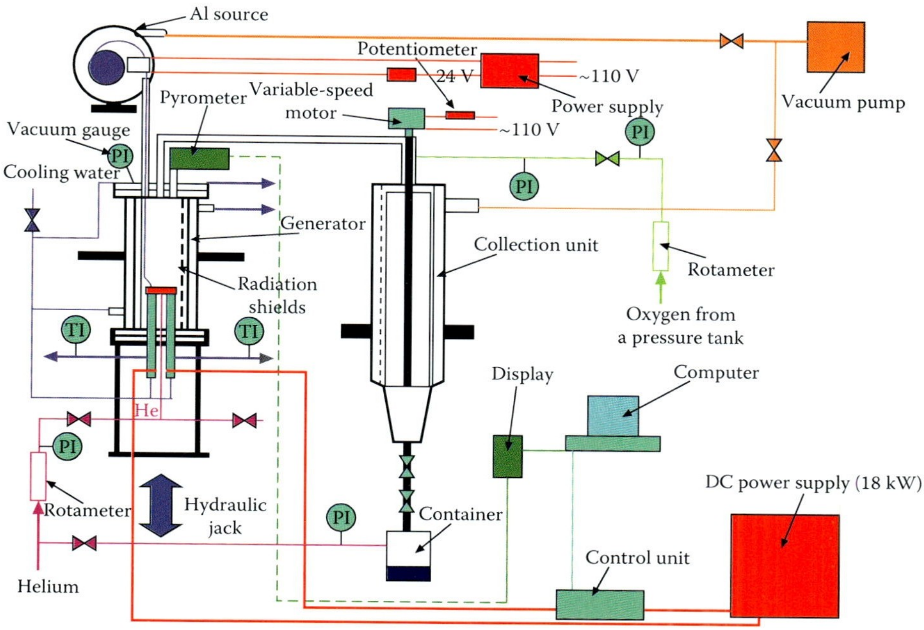


**FIGURE 6.3**
Schematic of the nanosized aluminum generator and collection systems.

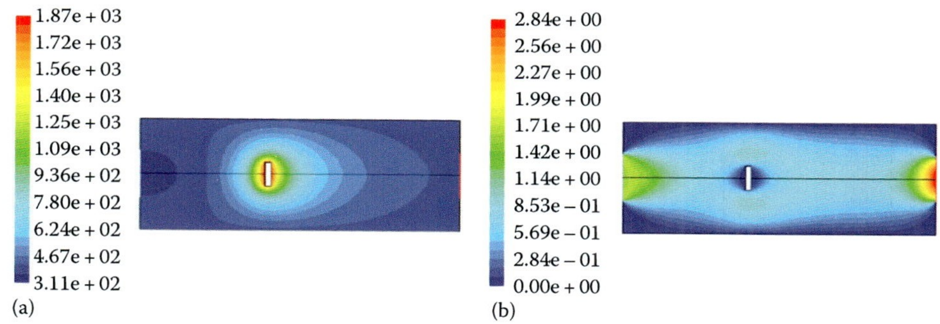


**FIGURE 6.5**
(a) Temperature (in K) and (b) velocity (in m/s) profiles in the generator filled with helium. (From Puszynski, J.A., Formation, passivation, and reactivity of energetic nanopowders, in Kuo, K.K. and Rivera, J.D. (eds.) *Advancements in Energetic Materials and Chemical Propulsion*, Begell House Inc., New York, 3–21, 2007. With permission.)

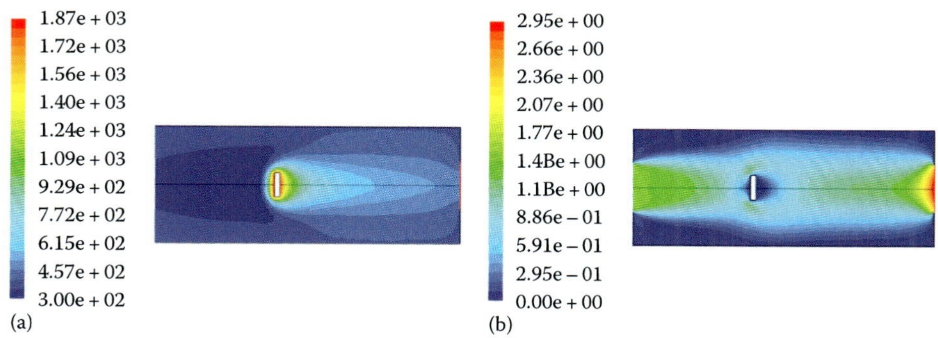


**FIGURE 6.6**
(a) Temperature (in K) and (b) velocity (in m/s) profiles in the generator filled with argon. (From Puszynski, J.A., Formation, passivation, and reactivity of energetic nanopowders, in Kuo, K.K. and Rivera, J.D. (eds.) *Advancements in Energetic Materials and Chemical Propulsion*, Begell House Inc., New York, 3–21, 2007. With permission.)

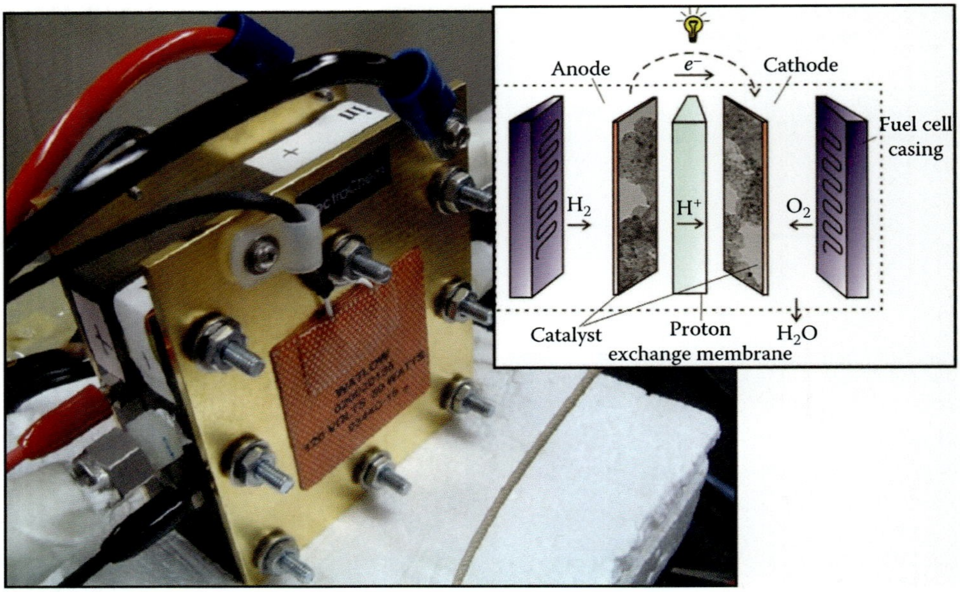


**FIGURE 7.1**
The picture of a PEMFC. Inset: a simplified and dissected scheme of the basic components of PEMFC.

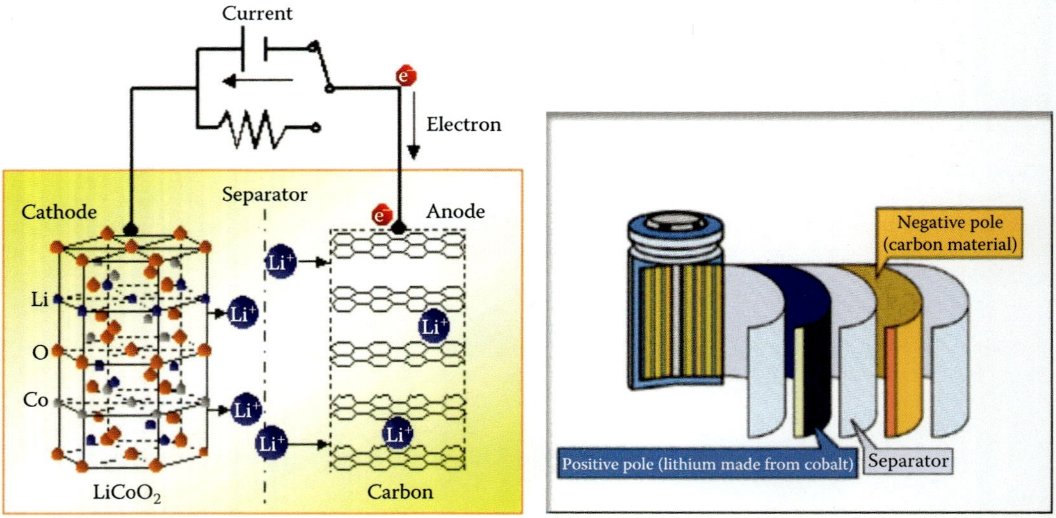


**FIGURE 9.3**
A scheme of typical secondary battery: for example, Li-ion battery based on $Li_xC_6/Li_xCoO_2$ electrochemistry.

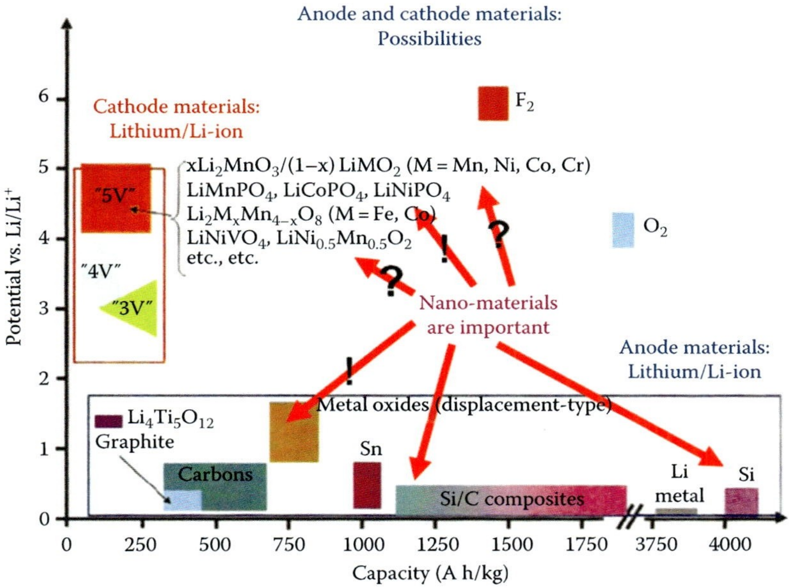


**FIGURE 9.9**
A schematic road map of electrode's materials for advanced rechargeable Li batteries. The arrows address materials that use as nanoparticles is (marked by !) or maybe important.

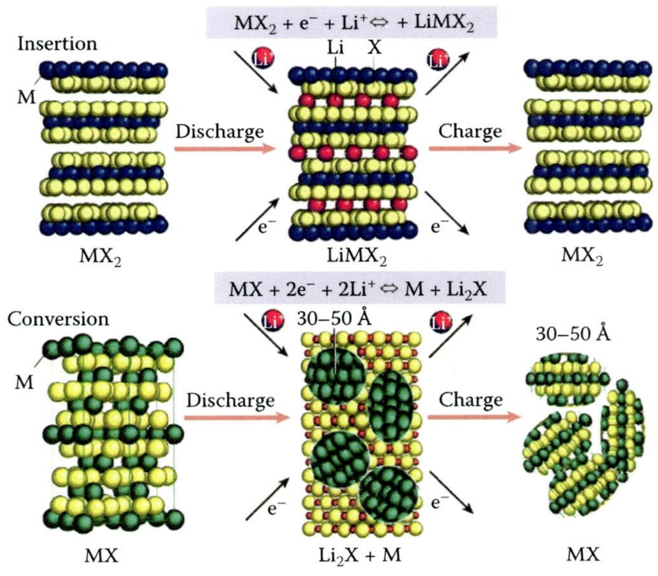


**FIGURE 9.10**
A schematic illustration of insertion (top) and conversion (bottom) reactions in which Li ions are involved. (Reprinted from Armand, M. and Tarascon, J.-M., *Nature*, 451, 652, 2008. With permission.)

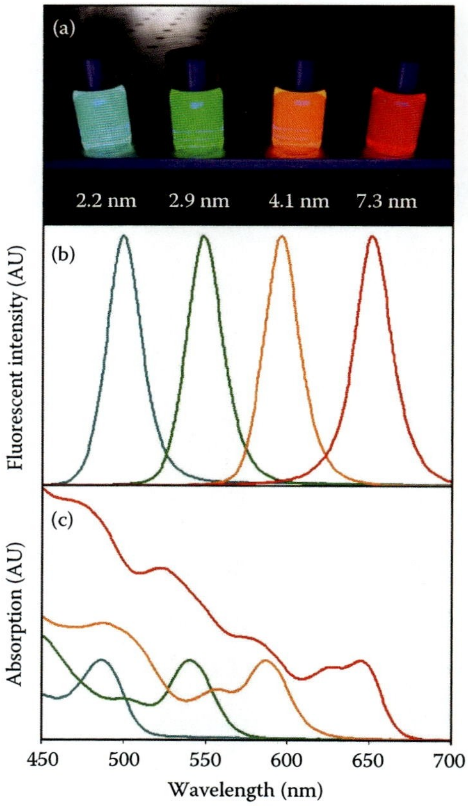


**FIGURE 10.1**
(a) Size-dependent optical properties of cadmium selenide QDs with diameters ranging from 2.2 to 7.3 nm and corresponding emission (b) and absorption (c) spectra. (Adapted from Smith, A.M. et al., *Adv. Drug Deliv. Rev.*, 60, 1226, 2008. With permission.)

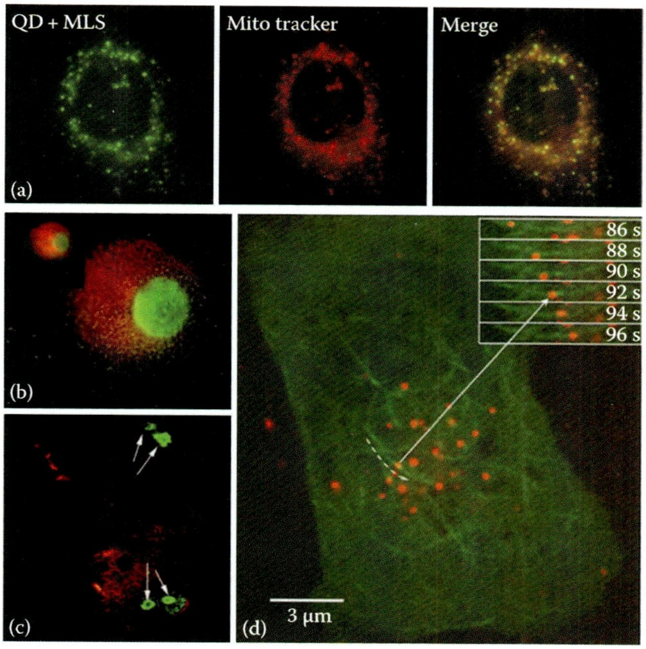


**FIGURE 10.6**
Fluorescence imaging of distinct cellular compartments labeled with differently colored QDs. (a) Fluorescence micrograph of 3T3 fibroblast after 24 h injection of MLS-QDs; co-localization with Mitotracker Red confirmed mitochondrial labeling. (Adapted from Derfus, A.M. et al., *Adv. Mater.*, 16, 961, 2004. With permission.) (b) Mixture of red and green CdTe QDs injected into the cytoplasm of macrophages separates into two distinctive cytosolic and nuclear compartments within 60 min of observation. In the inset, same cell 15 min post-micro-injection. (c) Confocal optical section of macrophages showing nucleolar localization of green-emitting QDs (evidenced by the arrows). (Adapted from Nabiev, I. et al., *Nano Lett.*, 7, 3452, 2007. With permission.) (d) Real-time imaging and tracking of siRNA-QD nanocomplexes in living breast cancer cells by using spinning disk confocal microscopy The QD-siRNA complexes were found to undergo active and directional motion, and their trajectory and velocity were similar to active vesicle transport medicated by molecular motors. The red dots in the image are QD clusters and not of single QDs. In the inset a time-lapsed image series showing a single nanocomplex moving along a microtubule. (Adapted with permission from Yezhelyev, M.V. et al., *J. Am. Chem. Soc.*, 130, 9006, 2008. With permission.)

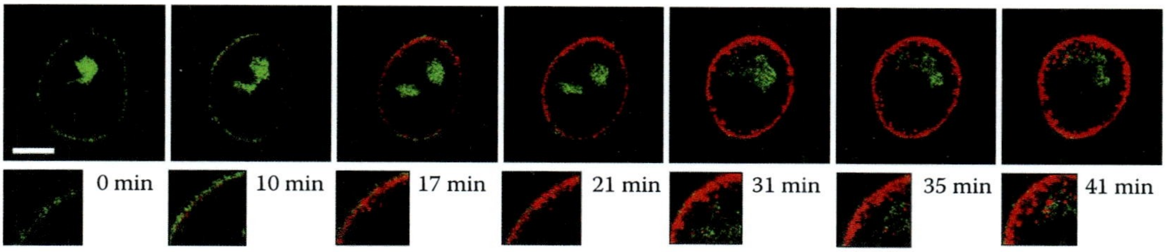


**FIGURE 10.8**
Quantitative analysis of real-time binding and internalization of ligand-coated QDs. Time series of CHO-DC-SIGN cells incubated with 2 nM gp120-QD655. Co-transfection of erbB1-EGFP labeled the cell membrane and allows delineation from cytoplasm. (Adapted from Cambi, A. et al., *Nano Lett.*, 7, 970, 2007. With permission.)

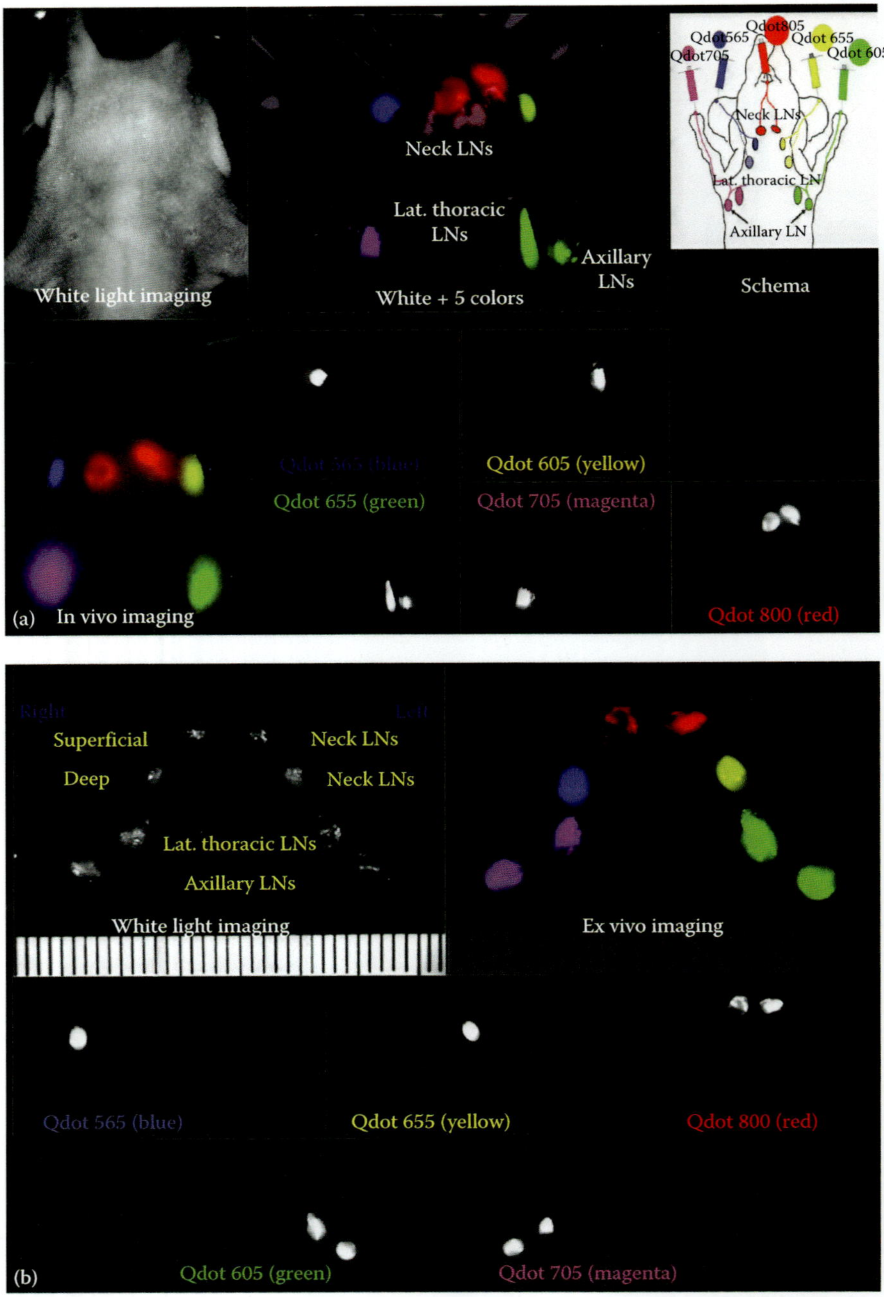


**FIGURE 10.9**
*In vivo* imaging of lymphatic drainage by using five QDs emitting at different colors: (a) Intrasurgical spectral fluorescence imaging of a mouse injected intracutaneously with five QDs (565 nm, blue; 605 nm, green; 655 nm, yellow; 705 nm, magenta; 800 nm, red) into five different sites in the upper body of athymic mice. Five primary draining LNs were simultaneously visualized *in vivo*. (b) *Ex vivo* spectral fluorescence imaging of the eight draining LNs after surgical resection. (Adapted from Kobayashi, H. et al., *Nano Lett.*, 7, 1711, 2007. With permission.)

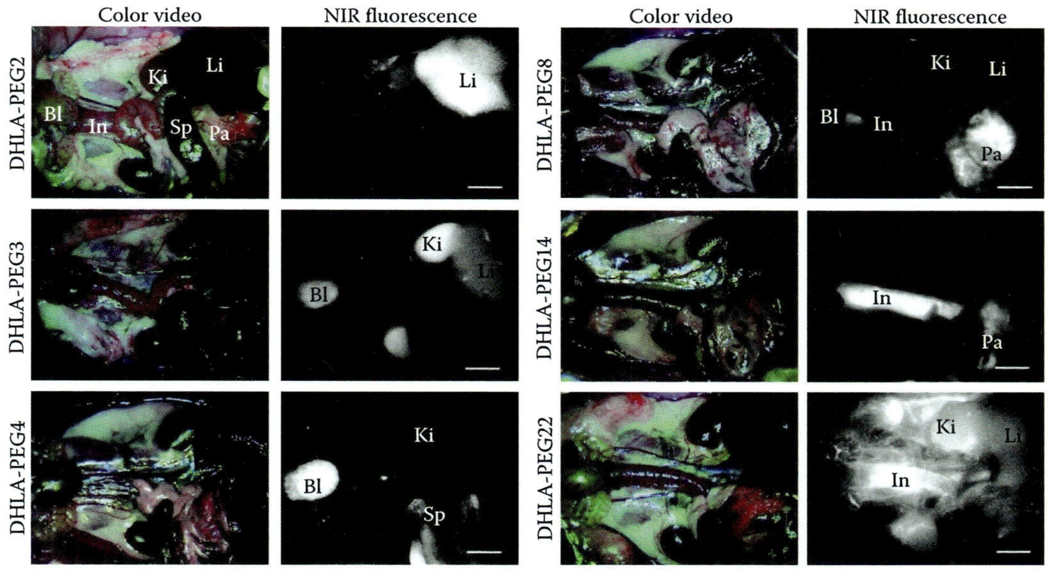


**FIGURE 10.10**
*In vivo* NIR fluorescence imaging of NIR PEG-conjugated QDs. QDs coated with DHLA-PEG at different molecular weight of the PEG (DHLA-PEG2, DHLA-PEG3, DHLA-PEG4, DHLA-PEG8, DHLA-PEG14, and DHLA-PEG22) were administered to the rat and imaged 4h postinjection. Color video is shown in the left panel and NIR fluorescence in the right panel. (Abbreviations used are: Ki, kidneys; Bl, bladder; Li, liver; Pa, pancreas; Sp, spleen; and In, intestine.) NIR fluorescence images have identical exposure times and normalizations. Scale bar = 500 μm. (Adapted from Choi, H.S. et al., *Nano Lett.*, 9, 2354, 2009. With permission.)

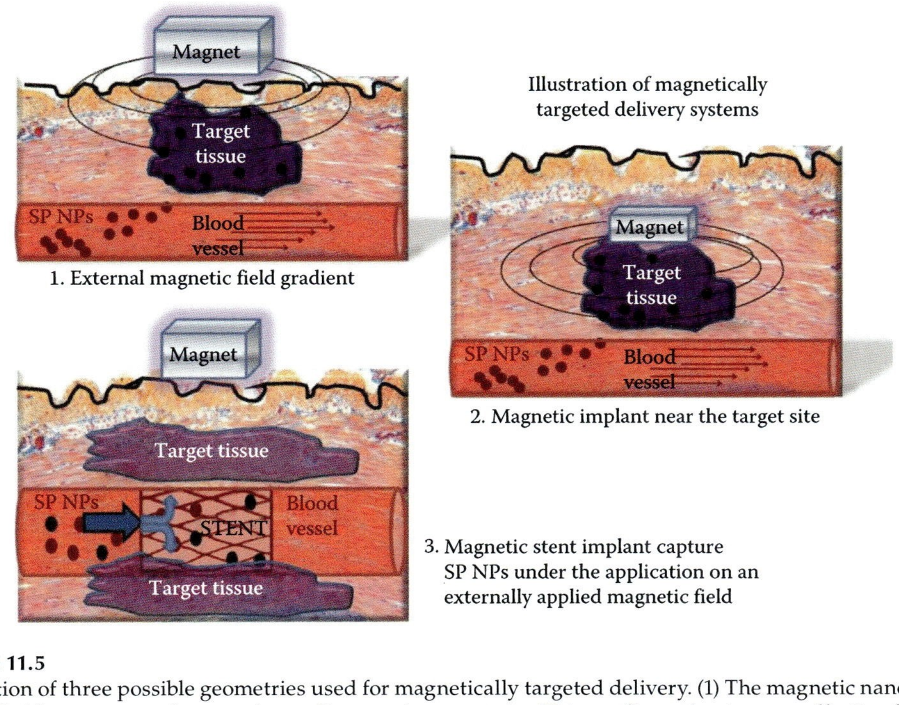


**FIGURE 11.5**
Illustration of three possible geometries used for magnetically targeted delivery. (1) The magnetic nanocarriers are guided by an external magnetic gradient on the target site. This configuration is more effective for target sites that are near the body surface. (2) A magnetic implant near the target tissue attracts the magnetic particles passively carried by the blood flow. The magnetic field generated by the implant is not sufficient to magnetically drive nanoparticles from the injection site to the target tissue if they are not close. (3) Magnetic implant stents attract magnetic drug carriers under the external guidance of a magnetic field. The particles are entrapped in the network stent and the release of drug is controlled.

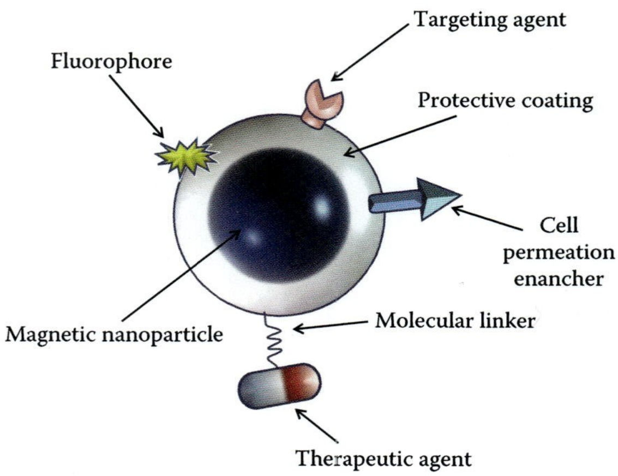


**FIGURE 11.6**
Hypothetical representation of a multifunctional hybrid magnetic nanoparticle for biomedical applications.

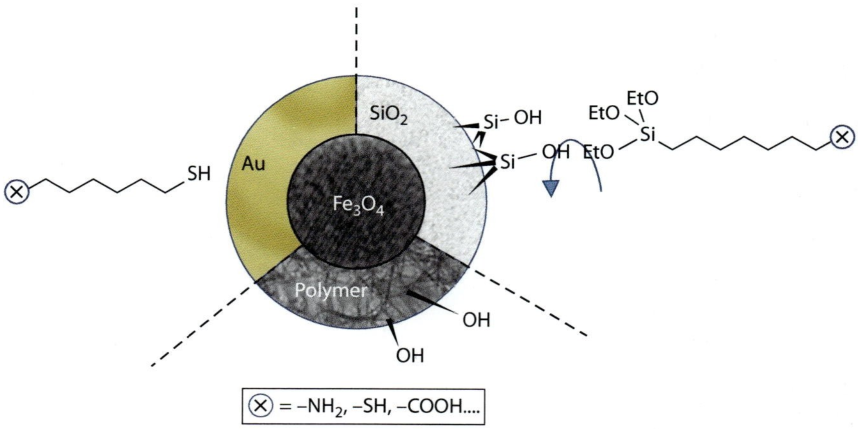


**FIGURE 11.10**
Illustrative scheme of possible interactions of molecular linkers with materials used for nanoparticle coatings.

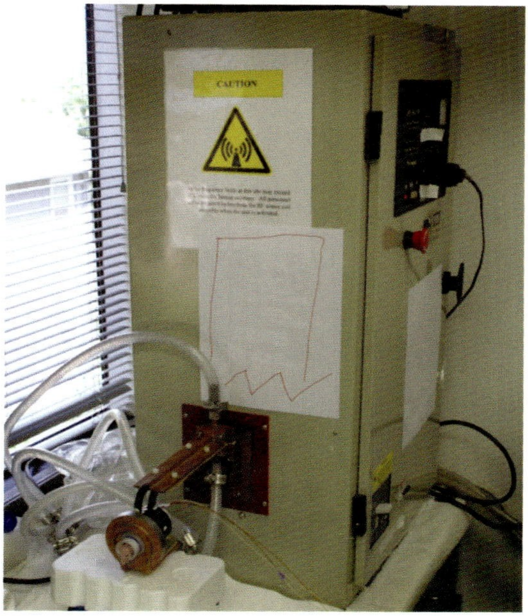

**FIGURE 12.2**
AMF generator for mouse and cell studies by Triton BioSystems, Inc. (Chelmsford, MA). The system consists of a water-cooled induction coil, the mouse version of which is shown here; a capacitance network; and the power supply, both of which are in the steel closet.

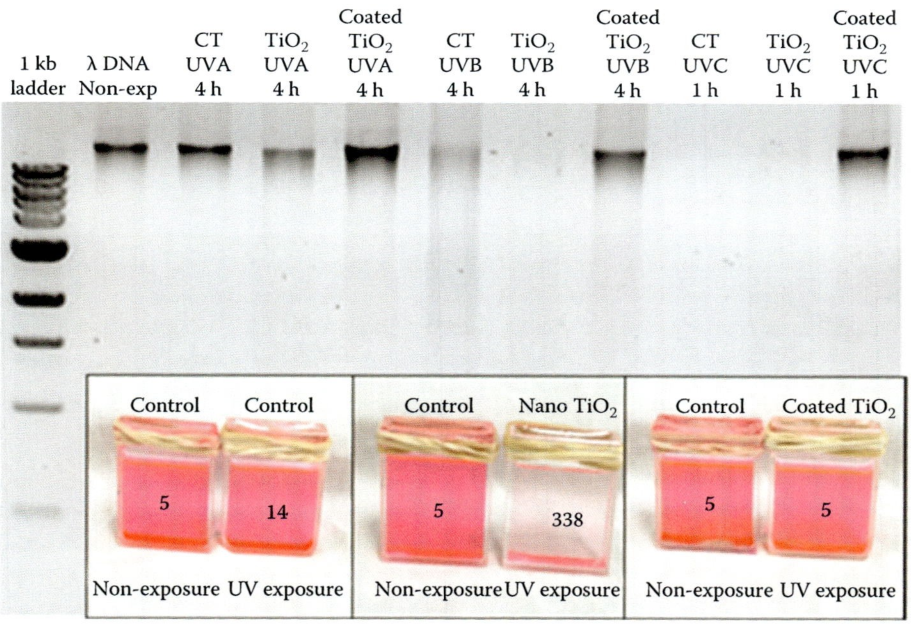


**FIGURE 13.3**
Lambda-DNA gel electrophoresis for the control (no UV exposure), after exposure to UVA or UVB or UVC with $TiO_2$ or with coated $TiO_2$. The results show that coated $TiO_2$ #2 prevents λ-DNA from catalysis under 4h of UVA, 4h of UVB, and 1h of UVC illumination. (a–c) Show solution with Red 28, which is an assay for photodegradation. The absorbance after 27h is written on each cuvette (absorbance is increasing as photodegradation is increasing). (a) Spectrophotometer assay exhibits slight photodegradation of the Red 28 ($1.2 \times 10^{-4}$ M) with UV illumination for the control. (b) Spectrophotometer assay exhibits drastic photodegradation in presence of $TiO_2$ nanoparticles after UV illumination. (c) Same conditions but now with the coated $TiO_2$ nanoparticles show no changes in photodegradation after UV illumination.

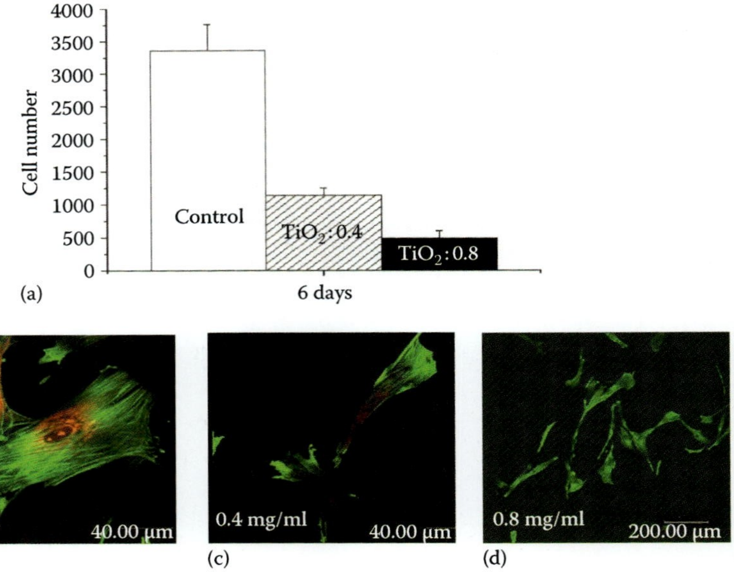


**FIGURE 13.5**
(a) Cell number as a function of $TiO_2$ concentration. (b) Confocal image of human dermal fibroblasts incubated for 6 days. (c) Confocal image of human dermal fibroblasts incubated with 0.4 mg/mL of rutile $TiO_2$. (d) Confocal image of human dermal fibroblasts incubated with 0.8 mg/mL of rutile $TiO_2$.

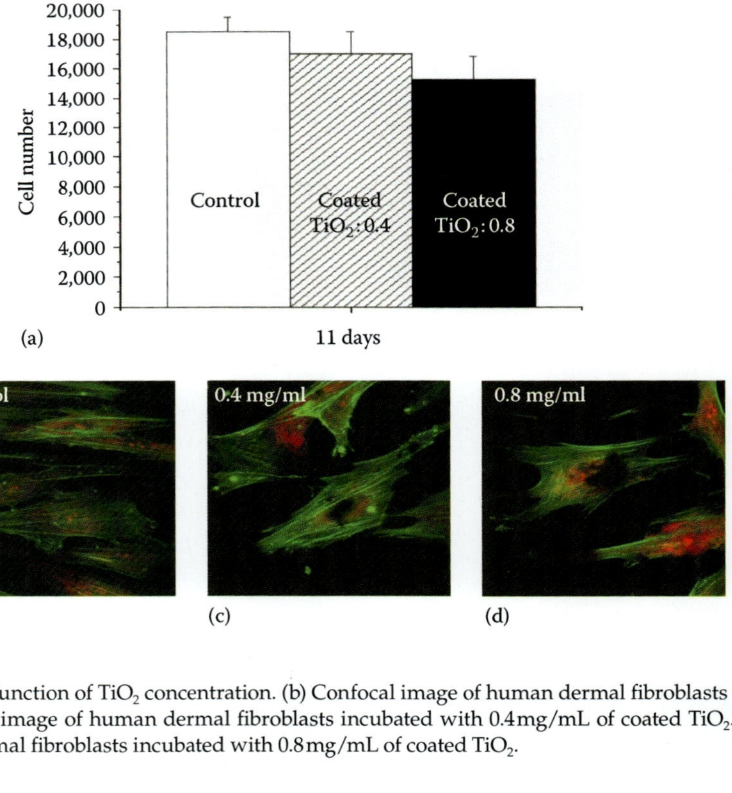


**FIGURE 13.6**
(a) Cell number as a function of $TiO_2$ concentration. (b) Confocal image of human dermal fibroblasts incubated for 11 days. (c) Confocal image of human dermal fibroblasts incubated with 0.4 mg/mL of coated $TiO_2$. (d) Confocal image of human dermal fibroblasts incubated with 0.8 mg/mL of coated $TiO_2$.

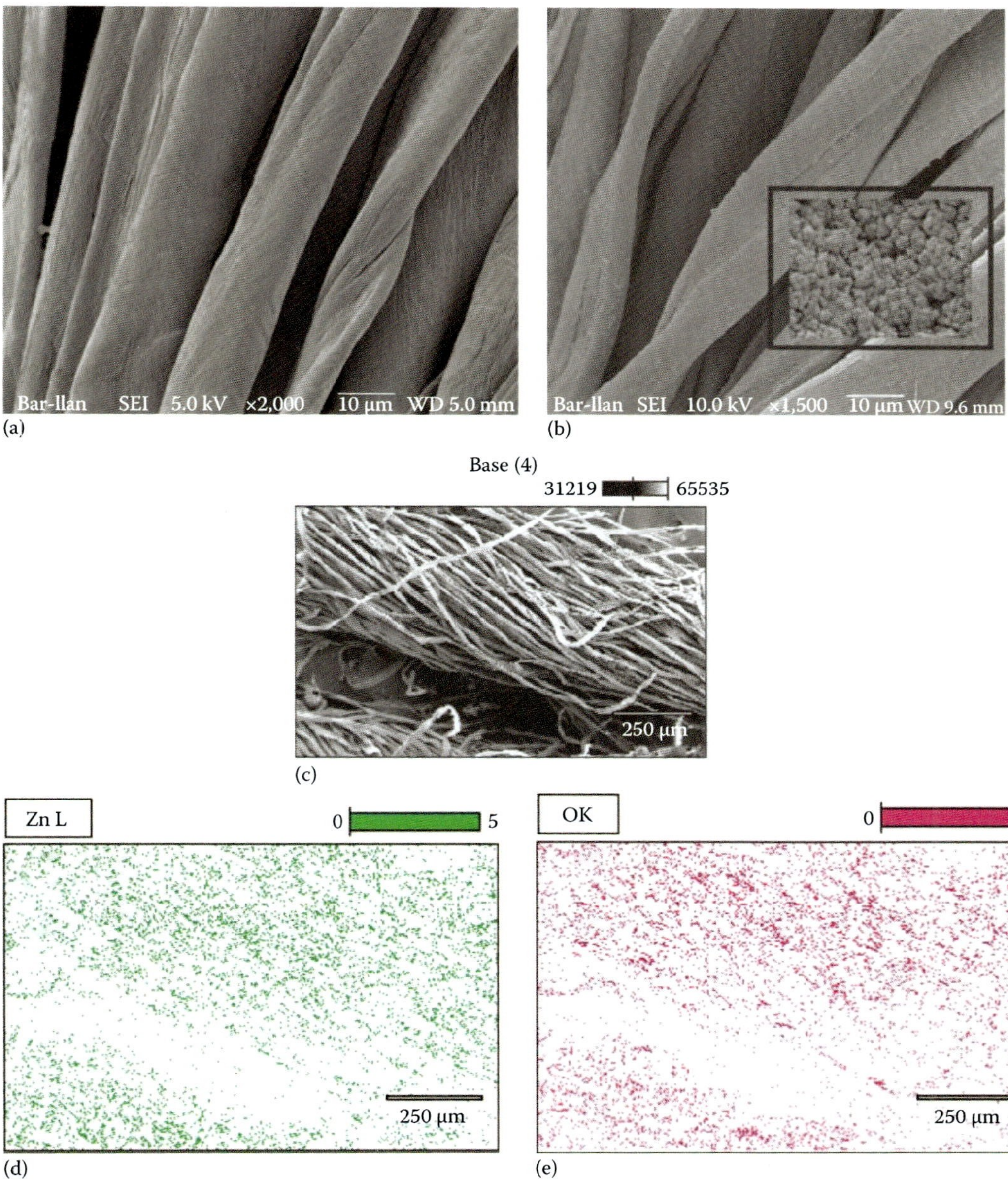


**FIGURE 14.3**
HR SEM images of (a) pristine bandage fibers (magnification ×2,000), (b) bandage coated with ZnO NPs (magnification ×1,500; the inset shows a magnified image (×50,000) of the ZnO NPs on the fabric, (c) selected image for x-ray dot mapping, (d) x-ray dot mapping for zinc, and (e) x-ray dot mapping for oxygen.

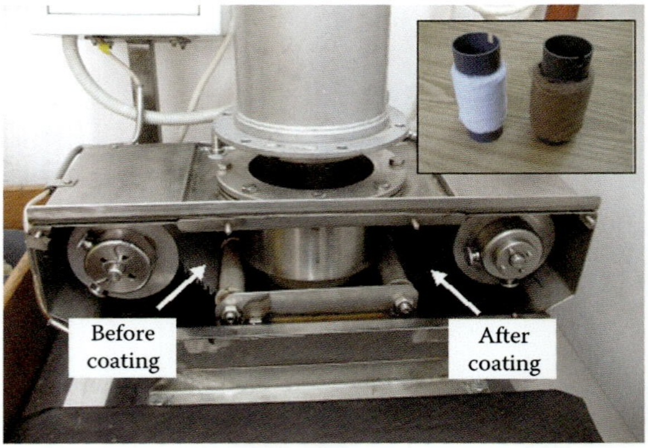


**FIGURE 14.10**
Photo image of the pilot installation. In the inset of the figure, the CuO-coated and uncoated spools are shown.

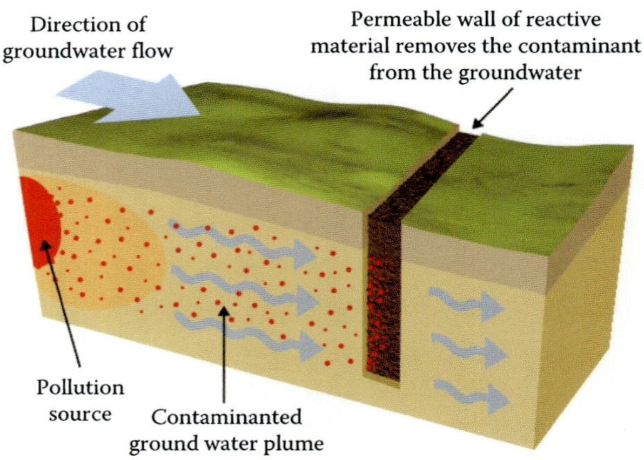


**FIGURE 15.2**
Simplified model of a permeable reactive barrier (PRB) system.

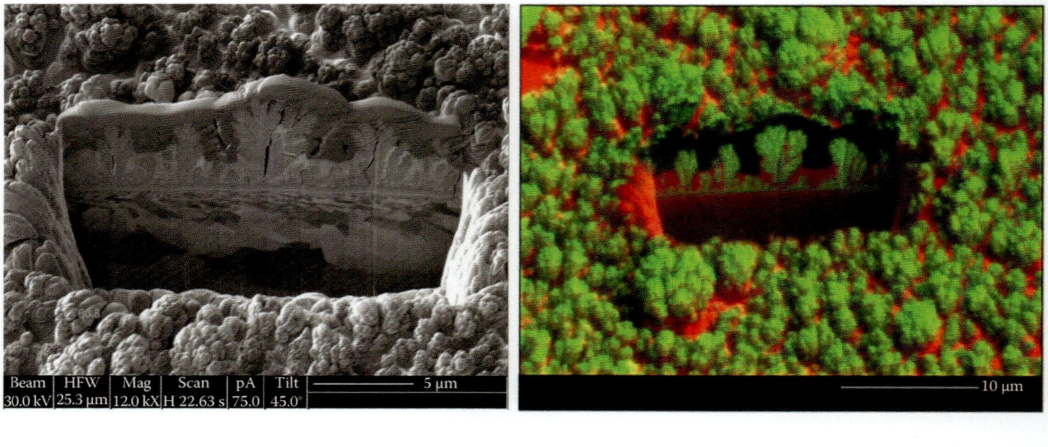


**FIGURE 15.7**
Secondary electron (left) and ion map (right) taken from a section cut through an iron surface. The iron was reacted for 24 h in water containing 100 ppm uranium and displayed significant precipitation of uranium oxide ($UO_2$). The ion map depicts U in green and Fe in red, with the black area representing platinum deposited as a protective coating during sectioning.

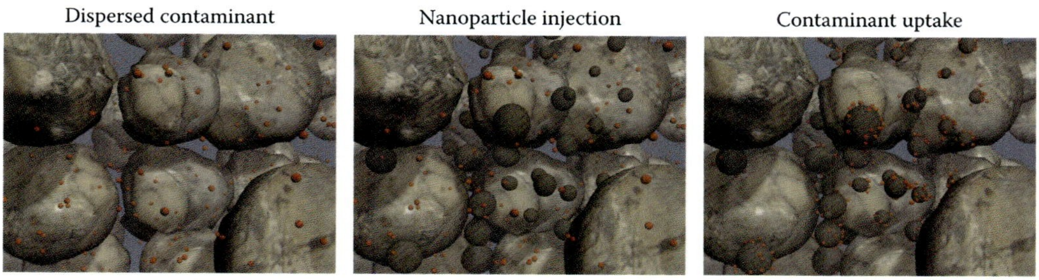


**FIGURE 15.10**
Conceptual use of nanoparticles to cleanup polluted sediments. Particles are injected into the contaminated area where they rapidly sequester contaminant species present in the groundwater either immobilizing or destroying them.

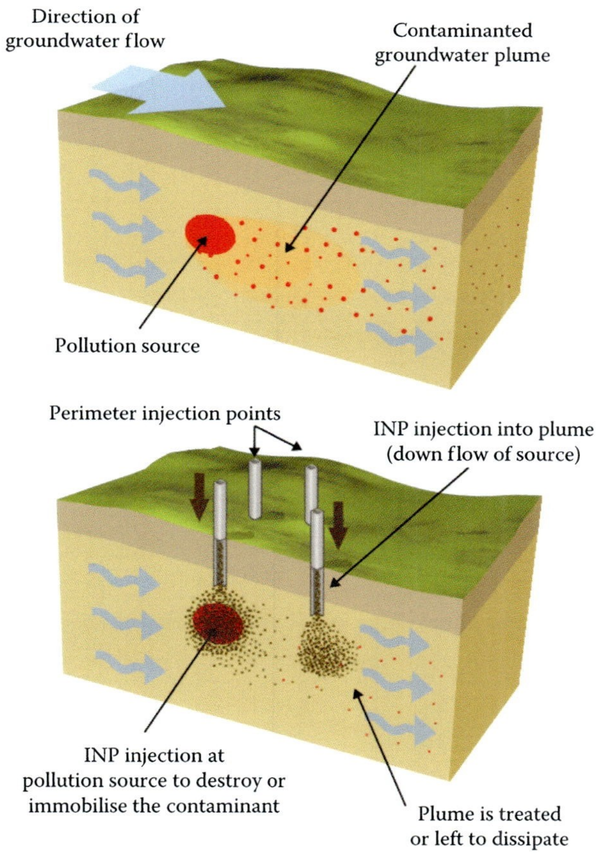


**FIGURE 15.11**
Conceptual use of INP injection technology for treating subsurface pollution sources. The location and number of particle injection wells is site specific and tailored to achieve the best possible remediation.

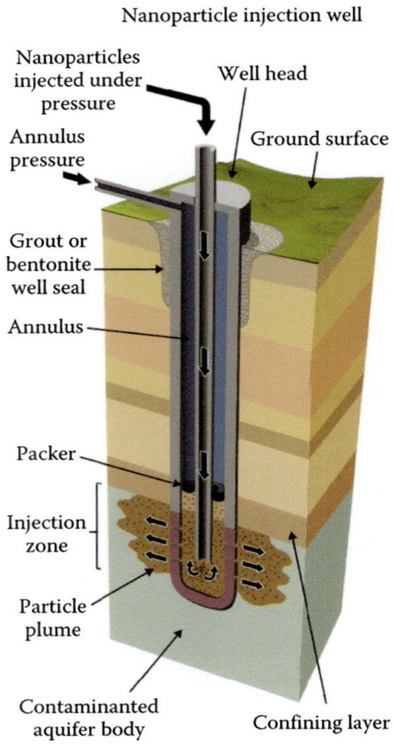


**FIGURE 15.12**
Diagram showing the typical layout of a nanoparticle injection well. The technology is very similar to that used for injection of $CO_2$ into subterranean storage reservoirs.

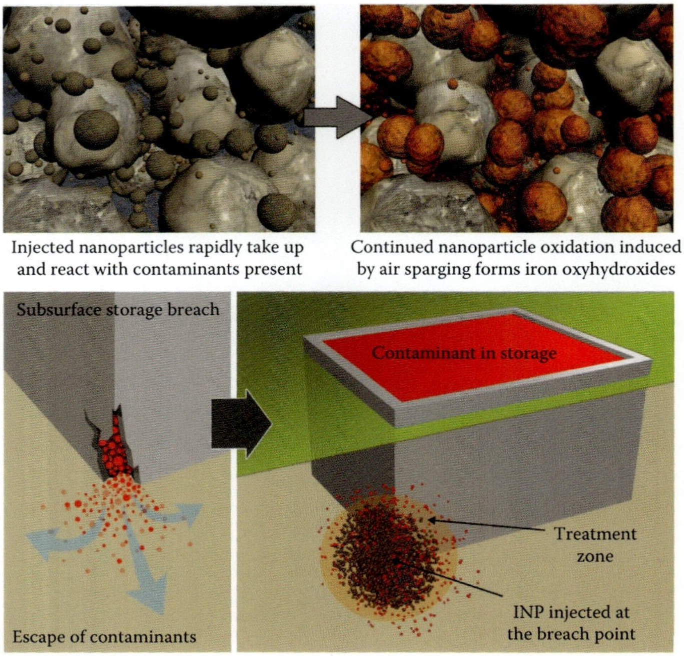


**FIGURE 15.13**
Application of nanoparticle injection technology for remediation of subsurface zones contaminated by uranium and other pollutant heavy metals.

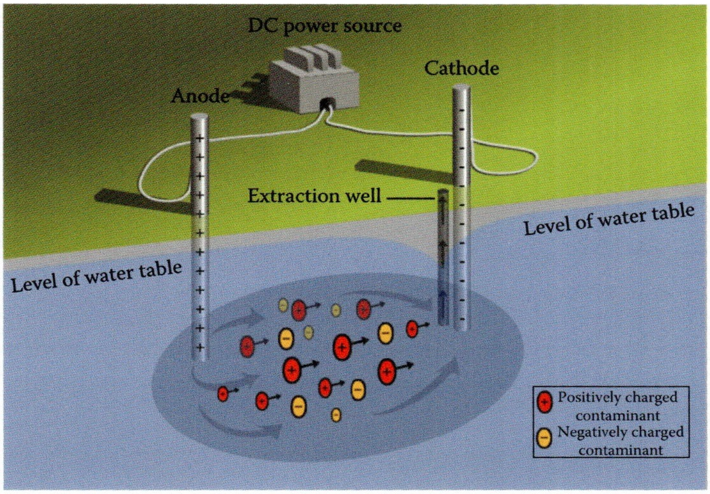


**FIGURE 15.14**
Environmental remediation of contaminants in the subsurface using electrokinetic processes to drive contaminant migration.

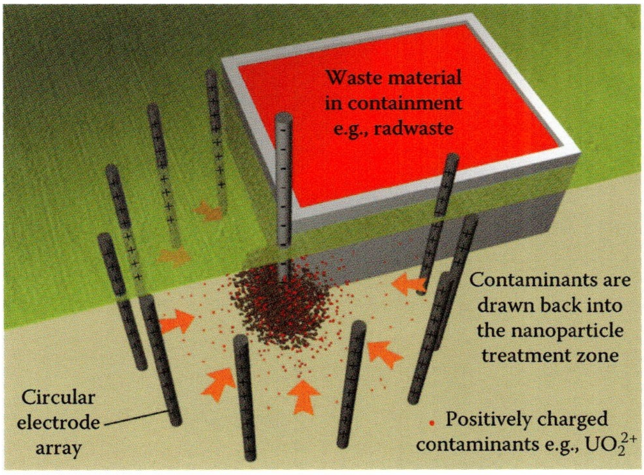


**FIGURE 15.15**
Application of combined nanoparticle and electrokinetic techniques for enhanced remediation of subsurface zones contaminated by uranium and other pollutant heavy metals.

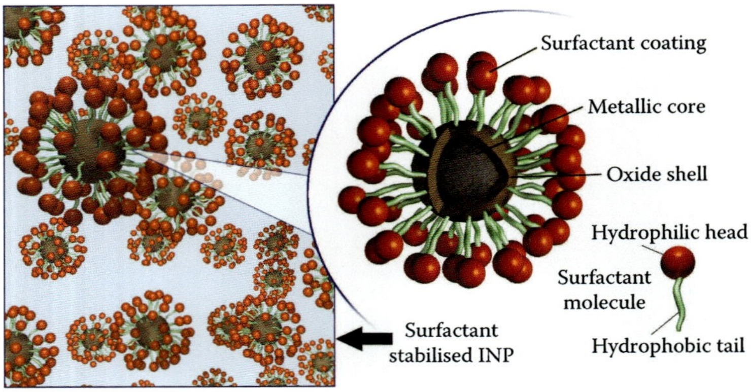


**FIGURE 15.21**
An idealized representation of surfactant stabilized INPs in aqueous solution, showing (inset) the association of surfactants on INP surfaces to form hemimicelles. Constituent components are also labeled.

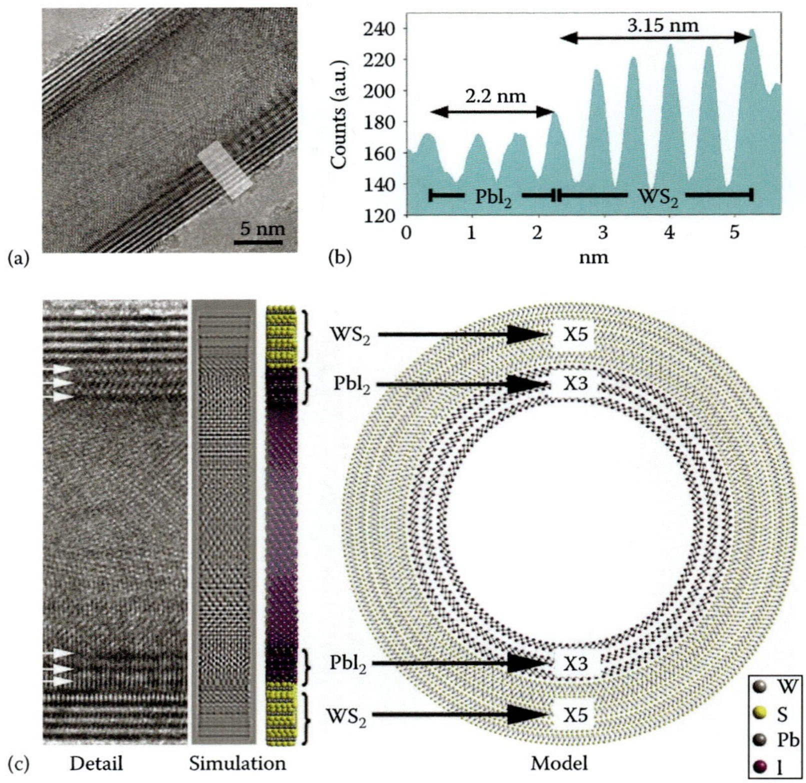


**FIGURE 16.8**
(a) HRTEM micrograph showing a core-shell $PbI_2@WS_2$ composite nanotube. (b) Line profile obtained from the indicated region in (a) showing two types of nanotube layers; five "outer" $WS_2$ layers with sharper contrast and an average spacing of 0.63 nm and three "inner" layers with more complex contrast and an average spacing of 0.70 nm, corresponding to three concentric $PbI_2$ nanotubes. (c) Detail from (a) showing the complex contrast of the inner $PbI_2$ layers (arrowed) relative to the outer $WS_2$ layers. To the right of the detail is a simulation and a cutaway space filling model (left) and cross-sectional structure model (right) with both $WS_2$ (aba stacking) and $PbI_2$ layers (abc stacking) indicated. (Reprinted from Kreizman, R. et al., *Angew. Chem. Int. Ed.*, 48, 1230, 2009. With permission. Copyright Wiley-VCH Verlag GmbH & Co. KGaA.)

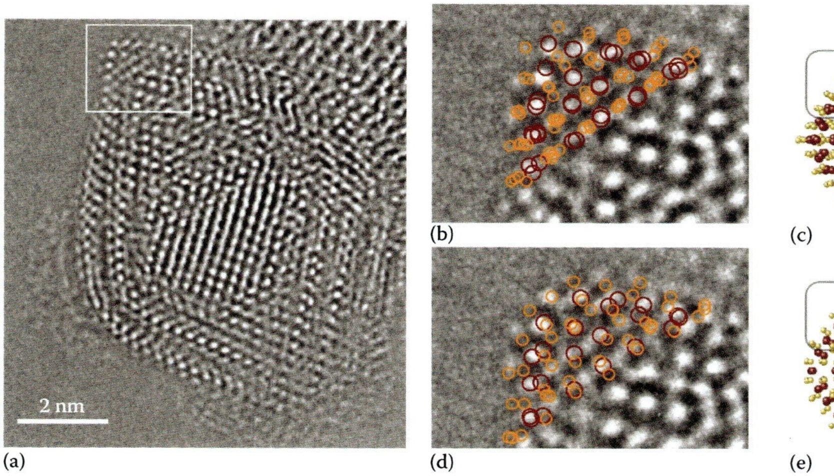


**FIGURE 16.9**
(a) Atomic resolution image of a $MoS_2$ nanooctahedron taken in an image-side aberration-corrected FEI Titan 80–300 under NCSI conditions. (b, d) A magnified part of the tip of the octahedron revealing the sulfur atoms at the surfaces with (c, e) a superposition of models for two of the 15 hypothetical structures proposed by Parilla et al. (2004). Mo atoms are displayed in red, S atoms in yellow. (b) One of the most stable models (no. 5 of Bar-Sadan et al. 2006a) coincides, while (d) the less stable structure fails to match. (Reprinted from Bar Sadan, M. et al., *Proc. Natl. Acad. Sci.*, 105, 15643, 2008a. With permission. Copyright the National Academy of Sciences.)

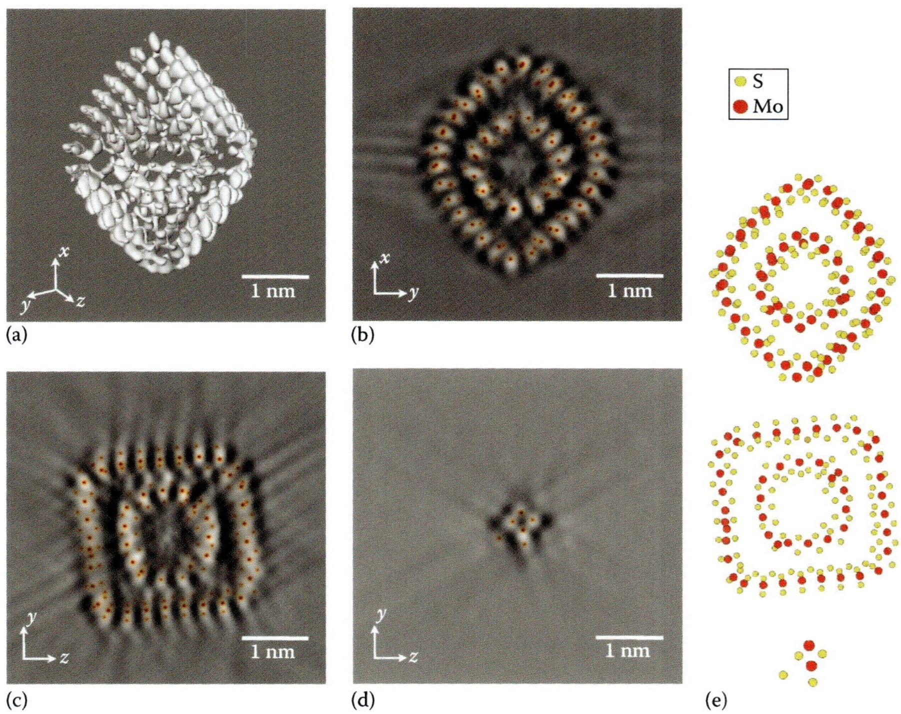


**FIGURE 16.10**
Single-axis tomographic reconstruction of a $MoS_2$ nanooctahedron from a tilt series of simulated images. A tilt range of 120° and an angular increment of 1° were chosen. By convention, $z$ refers to the axis of transmission and $x$ is parallel to the tilt axis in an orthogonal base system ($x$,$y$,$z$). (a) Isosurface visualization of the reconstructed tomogram of the closed-cage structure model. (b–d) Slices taken in different planes and at different positions. Corresponding slices of the projected electrostatic potential of the model structure are superimposed in a red temperature scale. Strong dots represent Mo atoms; the weaker dots correspond to S atoms. (b) Tomogram slice in the $xy$-plane through the center of the octahedron. (c) Central tomogram slice in the $yz$ plane. (d) Tomogram slice in the $yz$ plane through the apex of the structure showing the square defect in the seaming of the triangular faces. (e) The corresponding structure model slices. (Reprinted from Bar Sadan, M. et al., *Nano Lett.*, 8, 891, 2008b. With permission. Copyright the American Chemical Society.)

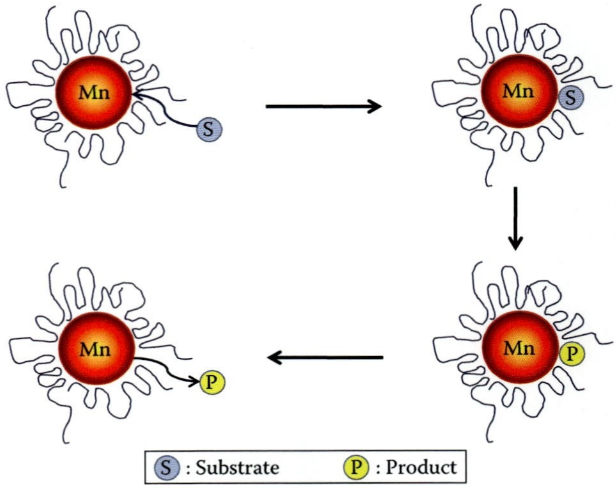


**FIGURE 17.6**
Catalysis process on the surface of polymer-protected metal nanoparticles.

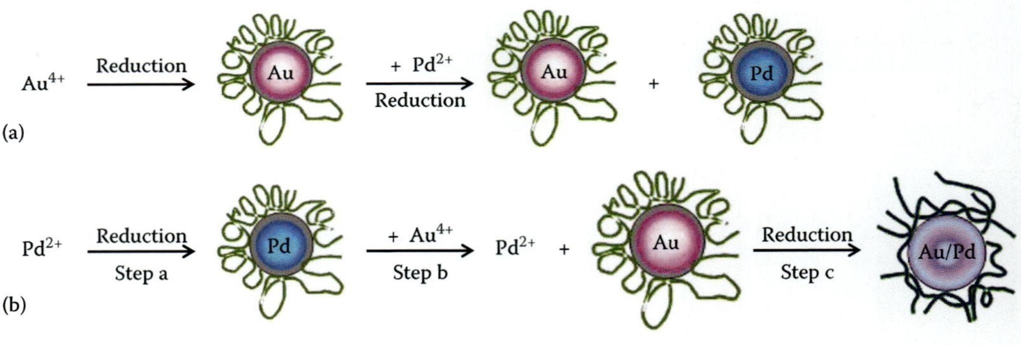


**FIGURE 17.10**
Formation of (a) mixture of distinct Au and Pd monometallic NPs and (b) Au-Pd bimetallic nanoparticles by the successive reduction of Pd and Au ions with ethanol in the presence of PVP.

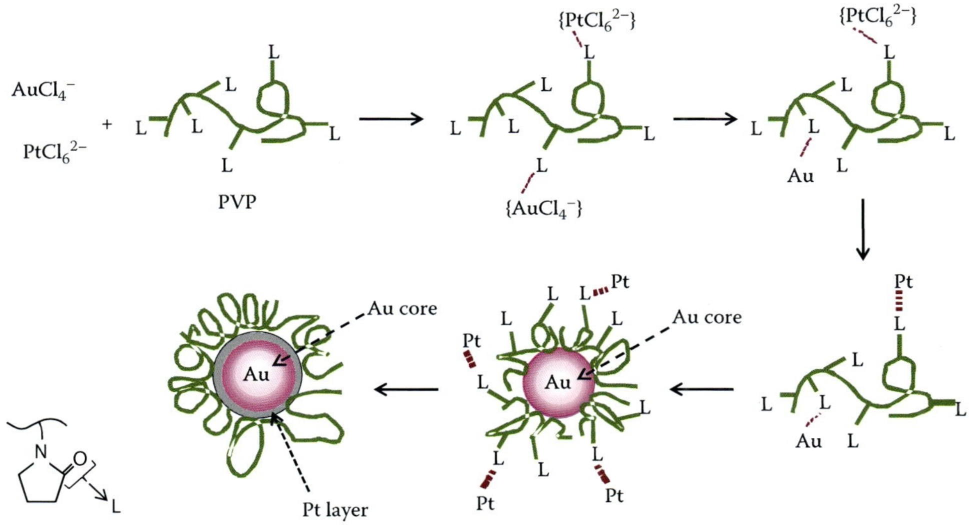


**FIGURE 17.11**
Proposed formation process of PVP-protected AuPt bimetallic nanoparticle. (Reprinted from Yonezawa, T. and Toshima, N., *J. Mol. Catal.*, 83, 167, 1993. With permission from Rightslink.)

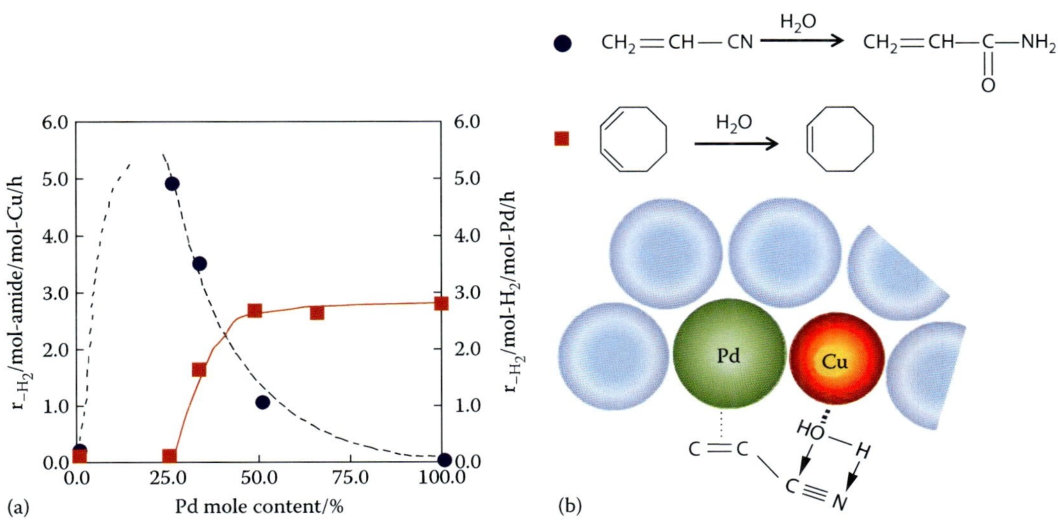


**FIGURE 17.19**
(a) Dependence of catalytic activities (hydrolysis of nitrile and hydrogenation of diene) upon the composition of CuPd bimetallic nanoparticle catalysts. (Reprinted from Toshima, N. and Wang, Y., *Langmuir*, 10, 4574, 1994b. With permission from Rightslink.) (b) Inspirative explanation of an ensemble effect of adjacent Pd and Cu atom in CuPd bimetallic nanoparticle catalyst for hydrolysis of acrylonitrile. (Reprinted from Toshima, N. and Wang, Y., *Adv. Mater.*, 6, 245, 1994a. With permission from Wiley-VCH.)

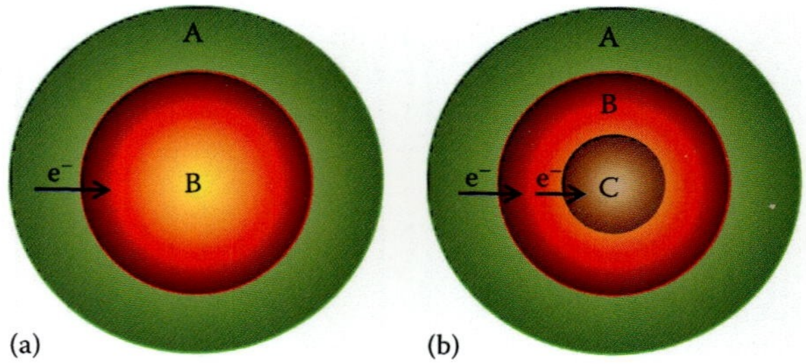


**FIGURE 17.22**
Schematic illustration for cross sections of charge transfers between different metals A, B, and C in core/shell-structured (a) bimetallic and (b) trimetallic nanoparticles.

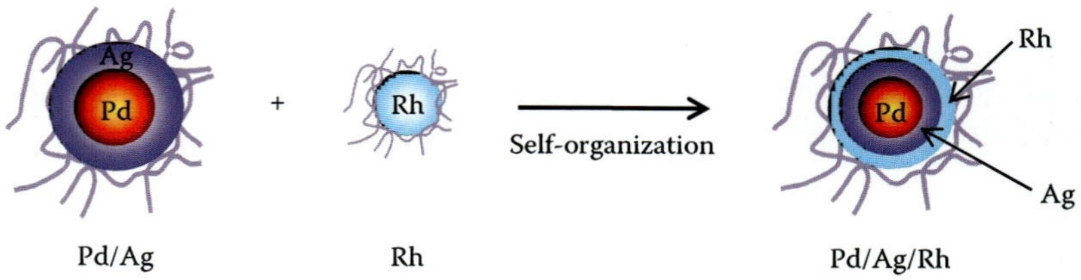


**FIGURE 17.23**
Schematic illustration of spontaneous formation of Pd/Ag/Rh triple core/shell-structured trimetallic nanoparticles by self-organization at room temperature. (Reprinted from Matsushita, T. et al., *Bull. Chem. Soc. Jpn.*, 80, 1217, 2007. With permission from The Chemical Society of Japan.)

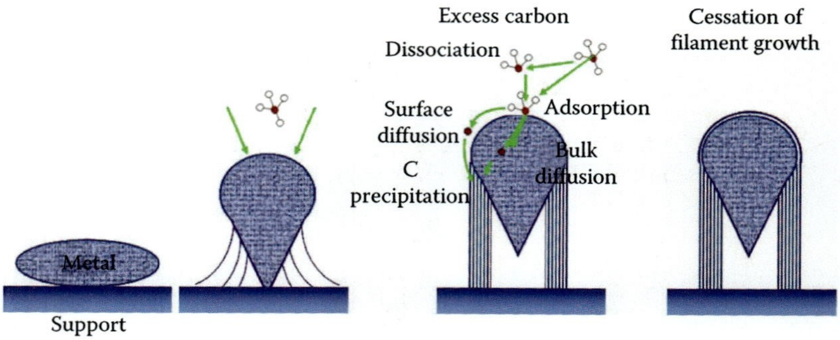


**FIGURE 18.13**
Schematic representation of the tip mechanism of the filamentous carbon growth.

Hilgenbrink, A. R. and Low, P. S. 2005. Folate receptor-mediated drug targeting: From therapeutics to diagnostics. *Journal of Pharmaceutical Sciences* 94: 2135–2146.

Hines, M. A. and Guyotsionnest, P. 1996. Synthesis and characterization of strongly luminescing ZnS-capped CdSe nanoparticles. *Journal of Physical Chemistry B* 100: 468–471.

Hohng, S. and Ha, T. 2004. Near-complete suppression of quantum dot blinking in ambient conditions. *Journal of the American Chemical Society* 126: 1324–1325.

Holzinger, D., Liz-Marzan, L. M., and Kickelbick, G. 2006. Synthesis of core-shell nanoparticles using organic surface-functionalized gold and Au@SiO2 nanoparticies as multifunctional initiators in atom transfer radical polymerization. *Journal of Nanoscience and Nanotechnology* 6: 445–452.

Hood, J. D. and Cheresh, D. A. 2002. Role of integrins in cell invasion and migration. *Nature Reviews Cancer* 2: 91–100.

Hoshino, A., Fujioka, K., Oku, T. et al. 2004. Quantum dots targeted to the assigned organelle in living cells. *Microbiology and Immunology* 48: 985–994.

Hoshino, A., Manabe, N., Fujioka, K. et al. 2008. GFP expression by intracellular gene delivery of GFP-coding fragments using nanocrystal quantum dots. *Nanotechnology* 19: 495102.

Hsieh, S. C., Wang, F. F., Hung, S. C., Chen, Y., and Wang, Y. J. 2006. The internalized CdSe/ZnS quantum dots impair the chondrogenesis of bone marrow mesenchymal stem cells. *Journal of Biomedical Materials Research Part B: Applied Biomaterials* 79B: 95–101.

Huang, X., Li, L., Qian, H., Dong, C., and Ren, J. 2006. A resonance energy transfer between chemiluminescent donors and luminescent quantum-dots as acceptors (CRET). *Angewandte Chemie—International Edition* 45: 5140–5143.

Huang, C. P., Li, Y. K., and Chen, T. M. 2007. A highly sensitive system for urea detection by using CdSe/ZnS core-shell quantum dots. *Biosensors & Bioelectronics* 22: 1835–1838.

Huang, S., Xiao, Q., He, Z. K. et al. 2008. A high sensitive and specific QDs FRET bioprobe for MNase. *Chemical Communications*: 5990–5992.

Huang, S., Xiao, Q., Li, R. et al. 2009. A simple and sensitive method for L-cysteine detection based on the fluorescence intensity increment of quantum dots. *Analytica Chimica Acta* 645: 73–78.

Idowu, M., Chen, J. Y., and Nyokong, T. 2008. Photoinduced energy transfer between water-soluble CdTe quantum dots and aluminium tetrasulfonated phthalocyanine. *New Journal of Chemistry* 32: 290–296.

Ishihama, Y. and Funatsu, T. 2009. Single molecule tracking of quantum dot-labeled mRNAs in a cell nucleus. *Biochemical and Biophysical Research Communications* 381: 33–38.

Itaka, K., Kanayama, N., Nishiyama, N. et al. 2004. Supramolecular nanocarrier of siRNA from PEG-based block catiomer carrying diamine side chain with distinctive pK(a) directed to enhance intracellular gene silencing. *Journal of the American Chemical Society* 126: 13612–13613.

Jaiswal, J. K., Mattoussi, H., Mauro, J. M., and Simon, S. M. 2003. Long-term multiple color imaging of live cells using quantum dot bioconjugates. *Nature Biotechnology* 21: 47–51.

Ji, X. J., Zheng, J. Y., Xu, J. M. et al. 2005. (CdSe)ZnS quantum dots and organophosphorus hydrolase bioconjugate as biosensors for detection of paraoxon. *Journal of Physical Chemistry B* 109: 3793–3799.

Jin, H. and Varner, J. 2004. Integrins: Roles in cancer development and as treatment targets. *British Journal of Cancer* 90: 561–565.

Jin, T., Fujii, F., Sakata, H., Tamura, M., and Kinjo, M. 2005. Amphiphilic p-sulfonatocalix[4]arene-coated CdSe/ZnS quantum dots for the optical detection of the neurotransmitter acetylcholine. *Chemical Communications*: 4300–4302.

Jori, G., Fabris, C., Soncin, M. et al. 2006. Photodynamic therapy in the treatment of microbial infections: Basic principles and perspective applications. *Lasers in Surgery and Medicine* 38: 468–481.

Juzenas, P., Generalov, R., Juzeniene, A., and Moan, J. 2008. Generation of nitrogen oxide and oxygen radicals by quantum dots. *Journal of Biomedical Nanotechnology* 4: 450–456.

Kennel, S. J., Woodward, J. D., Rondinone, A. J. et al. 2008. The fate of MAb-targeted (CdTe)-Te-125m/ZnS nanoparticles in vivo. *Nuclear Medicine And Biology* 35: 501–514.

Kim, D. K. and Dobson, J. 2009. Nanomedicine for targeted drug delivery. *Journal of Materials Chemistry* 19: 6294–6307.

Kim, S., Lim, Y. T., Soltesz, E. G. et al. 2004. Near-infrared fluorescent type II quantum dots for sentinel lymph node mapping. *Nature Biotechnology* 22: 93–97.

Kim, B. Y., Jiang, W., Oreopoulos, J. et al. 2008. Biodegradable quantum dot nanocomposites enable live cell labeling and imaging of cytoplasmic targets. *Nano Letters* 8: 3887–3892.

Kirchner, C., Liedl, T., Kudera, S. et al. 2005. Cytotoxicity of colloidal CdSe and CdSe/ZnS nanoparticles. *Nano Letters* 5: 331–338.

Klein, S., Zolk, O., Fromm, M. F. et al. 2009. Functionalized silicon quantum dots tailored for targeted siRNA delivery. *Biochemical and Biophysical Research Communications* 387: 164–168.

Klimov, V. 2003. *Semiconductor and Metal Nanocrystals: Synthesis and Electronic and Optical Properties*. Boca Raton, Florida: CRC Press.

Kloepfer, J. A., Mielke, R. E., and Nadeau, J. L. 2005. Uptake of CdSe and CdSe/ZnS quantum dots into bacteria via purine-dependent mechanisms. *Applied and Environmental Microbiology* 71: 2548–2557.

Kobayashi, H., Hama, Y., Koyama, Y. et al. 2007. Simultaneous multicolor imaging of five different lymphatic basins using quantum dots. *Nano Letters* 7: 1711–1716.

Konishi, K. and Hiratani, T. 2006. Turn-on and selective luminescence sensing of copper ions by a water-soluble $Cd_{10}S_{16}$ molecular cluster. *Angewandte Chemie—International Edition* 45: 5191–5194.

Kural, C., Kim, H., Syed, S. et al. 2005. Kinesin and dynein move a peroxisome in vivo: A tug-of-war or coordinated movement? *Science* 308: 1469–1472.

Laco, F., Kun, M., Weber, H. J., Ramakrishna, S., and Chan, C. K. 2009. The dose effect of human bone marrow-derived mesenchymal stem cells on epidermal development in organotypic co-culture. *Journal of Dermatological Science* 55: 150–160.

Lakowicz, J. R. 2006. *Principles of Fluorescence Spectroscopy*. New York: Springer.

Lao, U. L., Mulchandani, A., and Chen, W. 2006. Simple conjugation and purification of quantum dot-antibody complexes using a thermally responsive elastin-protein L scaffold as immunofluorescent agents. *Journal of the American Chemical Society* 128: 14756–14757.

Lees, E. E., Nguyen, T. L., Clayton, A. H. A., Muir, B. W., and Mulvaney, P. 2009. The preparation of colloidally stable, water-soluble, biocompatible, semiconductor nanocrystals with a small hydrodynamic diameter. *Acs Nano* 3: 2049–2049.

Lewinski, N., Colvin, V., and Drezek, R. 2008. Cytotoxicity of nanoparticles. *Small* 4: 26–49.

Li, H., Zhang, Y., Wang, X., Xiong, D., and Bai, Y. 2007. Calixarene capped quantum dots as luminescent probes for $Hg^{2+}$ ions. *Materials Letters* 61: 1474–1477.

Li, J., Mei, F., Li, W.-Y., He, X.-W., and Zhang, Y.-K. 2008. Study on the fluorescence resonance energy transfer between CdTe QDs and butyl-rhodamine B in the presence of CTMAB and its application on the detection of Hg(II). *Spectrochimica Acta Part A: Molecular and Biomolecular Spectroscopy* 70: 811–817.

Li, F., Zhang, Z. P., Peng, J. et al. 2009. Imaging viral behavior in mammalian cells with self-assembled capsid-quantum-dot hybrid particles. *Small* 5: 718–726.

Lidke, D. S., Nagy, P., Heintzmann, R. et al. 2004. Quantum dot ligands provide new insights into erbB/HER receptor-mediated signal transduction. *Nature Biotechnology* 22: 198–203.

Lin, S., Xie, X. Y., Patel, M. R. et al. 2007. Quantum dot imaging for embryonic stem cells. *BMC Biotechnology* 7: 67.

Liu, J., Lee, J. H., and Lu, Y. 2007a. Quantum dot encoding of aptamer-linked nanostructures for one-pot simultaneous detection of multiple analytes. *Analytical Chemistry* 79: 4120–4125.

Liu, W. H., Choi, H. S., Zimmer, J. P. et al. 2007b. Compact cysteine-coated CdSe(ZnCdS) quantum dots for in vivo applications. *Journal of the American Chemical Society* 129: 14530–14531.

Liu, Y. S., Sun, Y. H., Vernier, P. T. et al. 2007c. pH-sensitive photoluminescence of CdSe/ZnSe/ZnS quantum dots in human ovarian cancer cells. *Journal of Physical Chemistry C* 111: 2872–2878.

Loo, L., Guenther, R. H., Lommel, S. A., and Franzen, S. 2007. Encapsidation of nanoparticies by Red Clover Necrotic Mosaic Virus. *Journal of the American Chemical Society* 129: 11111–11117.

Lovric, J., Bazzi, H. S., Cuie, Y. et al. 2005. Differences in subcellular distribution and toxicity of green and red emitting CdTe quantum dots. *Journal of Molecular Medicine* 83: 377–385.
Ma, J., Chen, J. Y., Idowu, M., and Nyokong, T. 2008. Generation of singlet oxygen via the composites of water-soluble thiol-capped CdTe quantum dots-sulfonated aluminum phthalocyanines. *Journal of Physical Chemistry B* 112: 4465–4469.
Mandal, S. K., Lequeux, N., Rotenberg, B. et al. 2005. Encapsulation of magnetic and fluorescent nanoparticles in emulsion droplets. *Langmuir* 21: 4175–4179.
Manzoor, K., Johny, S., Thomas, D. et al. 2009. Bio-conjugated luminescent quantum dots of doped ZnS: A cyto-friendly system for targeted cancer imaging. *Nanotechnology* 20: 65102.
Mattheakis, L. C., Dias, J. M., Choi, Y. J. et al. 2004. Optical coding of mammalian cells using semiconductor quantum dots. *Analytical Biochemistry* 327: 200–208.
Mattoussi, H., Mauro, J. M., Goldman, E. R. et al. 2000. Self-assembly of CdSe-ZnS quantum dot bioconjugates using an engineered recombinant protein. *Journal of the American Chemical Society* 122: 12142–12150.
Medintz, I. L., Clapp, A. R., Mattoussi, H. et al. 2003. Self-assembled nanoscale biosensors based on quantum dot FRET donors. *Nature Materials* 2: 630–638.
Medintz, I. L., Clapp, A. R., Brunel, F. M. et al. 2006a. Proteolytic activity monitored by fluorescence resonance energy transfer through quantum-dot-peptide conjugates. *Nature Materials* 5: 581–589.
Medintz, I. L., Sapsford, K. E., Clapp, A. R. et al. 2006b. Designer variable repeat length polypeptides as scaffolds for surface immobilization of quantum dots. *Journal of Physical Chemistry B* 110: 10683–10690.
Medintz, I. L., Mattoussi, H., and Clapp, A. R. 2008. Potential clinical applications of quantum dots. *International Journal of Nanomedicine* 3: 151–167.
Michalet, X., Pinaud, F. F., Bentolila, L. A. et al. 2005. Quantum dots for live cells, in vivo imaging, and diagnostics. *Science* 307: 538–544.
Milliron, D. J., Alivisatos, A. P., Pitois, C., Edder, C., and Frechet, J. M. J. 2003. Electroactive surfactant designed to mediate electron transfer between CdSe nanocrystals and organic semiconductors. *Advanced Materials* 15: 58–61.
Moerner, W. E. 2006. Single-molecule mountains yield nanoscale cell images.' *Nature Methods* 3: 781–782.
Moioli, E. K., Shah, B., Clark, P. A., Stroscio, M., and Mao, J. J. 2006. Quantum dot labeling of stem cells during proliferation and differentiation. *MCB* 3: 153–155.
Moore, D. E. and Patel, K. 2001. Q-CdS photoluminescence activation on $Zn^{2+}$ and $Cd^{2+}$ salt introduction. *Langmuir* 17: 2541–2544.
Mudrakola, H. V., Zhang, K., and Cui, B. 2009. Optically resolving individual microtubules in live axons. *Structure* 17: 1433–1441.
Muller-Borer, B. J., Collins, M. C., Gunst, P. R., Cascio, W. E., and Kypson, A. P. 2007. Quantum dot labeling of mesenchymal stem cells. *Journal of Nanobiotechnology* 5: 9.
Murasawa, S., Kawamoto, A., Horii, M., Nakamori, S., and Asahara, T. 2005. Niche-dependent translineage commitment of endothelial progenitor cells, not cell fusion in general, into myocardial lineage cells. *Arteriosclerosis Thrombosis and Vascular Biology* 25: 1388–1394.
Muratovska, A. and Eccles, M. R. 2004. Conjugate for efficient delivery of short interfering RNA (siRNA) into mammalian cells. *FEBS Letters* 566: 63–68.
Murphy, C. J. 2002. Optical sensing with quantum dots. *Analytical Chemistry* 74: 520A–526A.
Murray, C. B., Norris, D. J., and Bawendi, M. G. 1993. Synthesis and characterization of nearly monodisperse CdE (E = sulfur, selenium, tellurium) semiconductor nanocrystallites. *Journal of the American Chemical Society* 115: 8706–8715.
Nabiev, I., Mitchell, S., Davies, A. et al. 2007. Nonfunctionalized nanocrystals can exploit a cell's active transport machinery delivering them to specific nuclear and cytoplasmic compartments. *Nano Letters* 7: 3452–3461.
Nann, T. 2005. Phase-transfer of CdSe@ZnS quantum dots using amphiphilic hyperbranched polyethylenimine. *Chemical Communications*: 1735–1736.

Narband, N., Mubarak, M., Ready, D. et al. 2008. Quantum dots as enhancers of the efficacy of bacterial lethal photosensitization. *Nanotechnology* 19: 445102.

Nelson, S. R., Ali, M. Y., Trybus, K. M., and Warshaw, D. M. 2009. Random walk of processive, quantum dot-labeled Myosin Va molecules within the actin cortex of COS-7 cells. *Biophysical Journal* 97: 509–518.

Neuhauser, R. G., Shimizu, K. T., Woo, W. K., Empedocles, S. A., and Bawendi, M. G. 2000. Correlation between fluorescence intermittency and spectral diffusion in single semiconductor quantum dots. *Physical Review Letters* 85: 3301–3304.

Noh, Y. W., Lim, Y. T., and Chung, B. H. 2008. Noninvasive imaging of dendritic cell migration into lymph nodes using near-infrared fluorescent semiconductor nanocrystals. *FASEB Journal* 22: 3908–3918.

Norris, D. J., Yao, N., Charnock, F. T., and Kennedy, T. A. 2001. High-quality manganese-doped ZnSe nanocrystals. *Nano Letters* 1: 3–7.

O'Connell, K. M., Rolig, A. S., Whitesell, J. D., and Tamkun, M. M. 2006. Kv2.1 potassium channels are retained within dynamic cell surface microdomains that are defined by a perimeter fence. *Journal of Neuroscience* 26: 9609–9618.

Ochsner, M. 1997. Photophysical and photobiological processes in the photodynamic therapy of tumours. *Journal of Photochemistry and Photobiology B—Biology* 39: 1–18.

Oginsky, E. L., Green, G. S., Griffith, D. G., and Fowlks, W. L. 1959. Lethal photosensitization of bacteria with 8-methoxypsoralen to long wave length ultraviolet radiation. *Journal of Bacteriology* 78: 821–833.

Oleinick, N. L. and Evans, H. H. 1998. The photobiology of photodynamic therapy: Cellular targets and mechanisms. *Radiation Research* 150: S146–S156.

Ozin, G. A. and Arsenault, A. C. 2005. *Nanochemistry: A Chemical Approach to Nanomaterials*. London, U.K.: Royal Society of Chemistry.

Pack, D. W., Hoffman, A. S., Pun, S., and Stayton, P. S. 2005. Design and development of polymers for gene delivery. *Nature Reviews Drug Discovery* 4: 581–593.

Papagiannaros, A., Levchenko, T., Hartner, W., Mongayt, D., and Torchilin, V. 2009. Quantum dots encapsulated in phospholipid micelles for imaging and quantification of tumors in the near-infrared region. *Nanomedicine: Nanotechnology, Biology and Medicine* 5: 216–224.

Parak, W. J., Boudreau, R., Le Gros, M. et al. 2002a. Cell motility and metastatic potential studies based on quantum dot imaging of phagokinetic tracks. *Advanced Materials* 14: 882–885.

Parak, W. J., Gerion, D., Zanchet, D. et al. 2002b. Conjugation of DNA to silanized colloidal semiconductor nanocrystalline quantum dots. *Chemistry of Materials* 14: 2113–2119.

Parungo, C. P., Ohnishi, S., Kim, S. W. et al. 2005. Intraoperative identification of esophageal sentinel lymph nodes with near-infrared fluorescence imaging. *Journal of Thoracic and Cardiovascular Surgery* 129: 844–850.

Pathak, S., Choi, S.-K., Arnheim, N., and Thompson, M. E. 2001. Hydroxylated quantum dots as luminescent probes for in situ hybridization. *Journal of the American Chemical Society* 123: 4103–4104.

Pathak, S., Davidson, M. C., and Silva, G. A. 2007. Characterization of the functional binding properties of antibody conjugated quantum dots. *Nano Letters* 7: 1839–1845.

Patolsky, F., Gill, R., Weizmann, Y. et al. 2003. Lighting-up the dynamics of telomerization and DNA replication by CdSe-ZnS quantum dots. *Journal of the American Chemical Society* 125: 13918–13919.

Pelley, J. L., Daar, A. S., and Saner, M. A. 2009. State of academic knowledge on toxicity and biological fate of quantum dots. *Toxicological Sciences* 112: 276–296.

Peng, X. G., Schlamp, M. C., Kadavanich, A. V., and Alivisatos, A. P. 1997. Epitaxial growth of highly luminescent CdSe/CdS core/shell nanocrystals with photostability and electronic accessibility. *Journal of the American Chemical Society* 119: 7019–7029.

Peng, H., Zhang, L., Kjällman, T. H. M., Soeller, C., and Travas-Sejdic, J. 2007. DNA hybridization detection with blue luminescent quantum dots and dye-labeled single-stranded DNA. *Journal of the American Chemical Society* 129: 3048–3049.

Peteiro-Cattelle, J., Rodriguez-Pedreira, M., Zhang, F. et al. 2009. One example on how colloidal nano- and microparticles could contribute to medicine. *Nanomedicine* 4: 967–979.

Pic, E., Pons, T., Bezdetnaya, L. et al. 2009. Fluorescence imaging and whole-body biodistribution of near-infrared-emitting quantum dots after subcutaneous injection for regional lymph node mapping in mice. *Molecular Imaging and Biology*: in press.

Pinaud, F., King, D., Moore, H. P., and Weiss, S. 2004. Bioactivation and cell targeting of semiconductor CdSe/ZnS nanocrystals with phytochelatin-related peptides. *Journal of the American Chemical Society* 126: 6115–6123.

Pradhan, N. and Peng, X. G. 2007. Efficient and color-tunable Mn-doped ZnSe nanocrystal emitters: Control of optical performance via greener synthetic chemistry. *Journal of the American Chemical Society* 129: 3339–3347.

Pradhan, N., Goorskey, D., Thessing, J., and Peng, X. G. 2005. An alternative of CdSe nanocrystal emitters: Pure and tunable impurity emissions in ZnSe nanocrystals. *Journal of the American Chemical Society* 127: 17586–17587.

Qi, L. and Gao, X. 2008. Quantum dot-amphipol nanocomplex for intracellular delivery and real-time imaging of siRNA. *ACS Nano* 2: 1403–1410.

Qu, F. G. and Li, H. B. 2009. Selective molecular recognition of polycyclic aromatic hydrocarbons using CdTe quantum dots with cyclodextrin as supramolecular nano-sensitizers in water. *Sensors and Actuators B—Chemical* 135: 499–505.

Rieger, S., Kulkarni, R. P., Darcy, D., Fraser, S. E., and Koster, R. W. 2005. Quantum dots are powerful multipurpose vital labeling agents in zebrafish embryos. *Developmental Dynamics* 234: 670–681.

Robe, A., Pic, E., Lassalle, H. P. et al. 2008. Quantum dots in axillary lymph node mapping: Biodistribution study in healthy mice. *BMC Cancer* 8: 111.

Rogach, A. L., Eychmuller, A., Kornowski, A., and Weller, H. 1998. Thiol-stabilized CdSe and CdTe nanocrystals in the size quantization regime: Synthesis, optical and structural properties. *Macromolecular Symposia* 136: 87–89.

Rosen, A. B., Kelly, D. J., Schuldt, A. J. T. et al. 2007. Finding fluorescent needles in the cardiac haystack: Tracking human mesenchymal stem cells labeled with quantum dots for quantitative in vivo three-dimensional fluorescence analysis. *Stem Cells* 25: 2128–2138.

Rosenthal, S. J., Tomlinson, A., Adkins, E. M. et al. 2002. Targeting cell surface receptors with ligand-conjugated nanocrystals. *Journal of the American Chemical Society* 124: 4586–4594.

Roullier, V., Clarke, S., You, C. et al. 2009. High-affinity labeling and tracking of individual histidine-tagged proteins in live cells using Ni2+ tris-nitrilotriacetic acid quantum dot conjugates. *Nano Letters* 9: 1228–1234.

Ruan, G., Agrawal, A., Marcus, A. I., and Nie, S. 2007. Imaging and tracking of tat peptide-conjugated quantum dots in living cells: New insights into nanoparticle uptake, intracellular transport, and vesicle shedding. *Journal of the American Chemical Society* 129: 14759–14766.

Ruedas-Rama, M. J. and Hall, E. A. H. 2008. Azamacrocycle activated quantum dot for zinc ion detection. *Analytical Chemistry* 80: 8260–8268.

Rzigalinski, B. A. and Strobl, J. S. 2009. Cadmium-containing nanoparticles: Perspectives on pharmacology and toxicology of quantum dots. *Toxicology and Applied Pharmacology* 238: 280–288.

Salgueirino-Maceira, V., Correa-Duarte, M. A., Spasova, M., Liz-Marzan, L. M., and Farle, M. 2006. Composite silica spheres with magnetic and luminescent functionalities. *Advanced Functional Materials* 16: 509–514.

Samia, A. C. S., Chen, X. B., and Burda, C. 2003. Semiconductor quantum dots for photodynamic therapy. *Journal of the American Chemical Society* 125: 15736–15737.

Santra, S., Yang, H., Stanley, J. T. et al. 2005. Rapid and effective labeling of brain tissue using TAT-conjugated CdS: Mn/ZnS quantum dots. *Chemical Communications*: 3144–3146.

Sapsford, K. E., Medintz, I. L., Golden, J. P. et al. 2004. Surface-immobilized self-assembled protein-based quantum dot nanoassemblies. *Langmuir* 20: 7720–7728.

Sathe, T. R., Agrawal, A., and Nie, S. M. 2006. Mesoporous silica beads embedded with semiconductor quantum dots and iron oxide nanocrystals: Dual-function microcarriers for optical encoding and magnetic separation. *Analytical Chemistry* 78: 5627–5632.

Schafer, F. Q. and Buettner, G. R. 2001. Redox environment of the cell as viewed through the redox state of the glutathione disulfide/glutathione couple. *Free Radical Biology and Medicine* 30: 1191–1212.

Schipper, M. L., Iyer, G., Koh, A. L. et al. 2009. Particle size, surface coating, and PEGylation influence the biodistribution of quantum dots in living mice. *Small* 5: 126–134.

Schmidt, G. 2006. *Nanoparticles: From Theory to Applications*. Weinheim, Germany: Wiley.

Schroeder, J. E., Shwek, I., Shmeeda, H., Banin, U., and Gabizon, A. 2007. Folate-mediated tumor cell uptake of quantum dots entrapped in lipid nanoparticles. *Journal of Controlled Release* 124: 28–34.

Schulke, N., Varlamova, O. A., Donovan, G. P. et al. 2003. The homodimer of prostate-specific membrane antigen is a functional target for cancer therapy. *Proceedings of the National Academy of Sciences of the United States of America* 100: 12590–12595.

Schuster, J., Cichos, F., and von Borczyskowski, C. 2005. *Optics and Spectroscopy* 98: 712–717.

Seleverstov, O., Zabirnyk, O., Zscharnack, M. et al. 2006. Quantum dots for human mesenchymal stem cells labeling. A size-dependent autophagy activation. *Nano Letters* 6: 2826–2832.

Sen, D., Deerinck, T. J., Ellisman, M. H., Parker, I., and Cahalan, M. D. 2008. Quantum dots for tracking dendritic cells and priming an immune response in vitro and in vivo. *Plos One* 3: e3290.

Shang, Z. B., Wang, Y., and Jin, W. J. 2009. Triethanolamine-capped CdSe quantum dots as fluorescent sensors for reciprocal recognition of mercury(II) and iodide in aqueous solution. *Talanta* 78: 364–369.

Sharman, W. M., Allen, C. M., and van Lier, J. E. 1999. Photodynamic therapeutics: Basic principles and clinical applications. *Drug Discovery Today* 4: 507–517.

Shi, L., Hernandez, B., and Selke, M. 2006. Singlet oxygen generation from water-soluble quantum dot-organic dye nanocomposites. *Journal of the American Chemical Society* 128: 6278–6279.

Silver, J. and Ou, W. 2005. Photoactivation of quantum dot fluorescence following endocytosis. *Nano Letters* 5: 1445–1449.

Singh, N., Mulrooney, R. C., Kaur, N., and Callan, J. F. 2008. A nanoparticle based chromogenic chemosensor for the simultaneous detection of multiple analytes. *Chemical Communications*: 4900–4902.

Skaff, H. and Emrick, T. 2003. The use of 4-substituted pyridines to afford amphiphilic, pegylated cadmium selenide nanoparticles. *Chemical Communications*: 52–53.

Smith, A. M. and Nie, S. 2004. Chemical analysis and cellular imaging with quantum dots. *Analyst* 129: 672–677.

Smith, J. D., Fisher, G. W., Waggoner, A. S., and Campbell, P. G. 2007. The use of quantum dots for analysis of chick CAM vasculature. *Microvascular Research* 73: 75–83.

Smith, A. M., Duan, H., Mohs, A. M., and Nie, S. 2008. Bioconjugated quantum dots for in vivo molecular and cellular imaging. *Advanced Drug Delivery Reviews* 60: 1226–1240.

Snee, P. T., Somers, R. C., Nair, G. et al. 2006. A ratiometric CdSe/ZnS nanocrystal pH sensor. *Journal of the American Chemical Society* 128: 13320–13321.

So, M. K., Xu, C. J., Loening, A. M., Gambhir, S. S., and Rao, J. H. 2006. Self-illuminating quantum dot conjugates for in vivo imaging. *Nature Biotechnology* 24: 339–343.

Soltesz, E. G., Kim, S., Laurence, R. G. et al. 2005. Intraoperative sentinel lymph node mapping of the lung using near-infrared fluorescent quantum dots. *Annals of Thoracic Surgery* 79: 269–277.

Srinivasan, C., Lee, J., Papadimitrakopoulos, F. et al. 2006. Labeling and intracellular tracking of functionally active plasmid DNA with semiconductor quantum dots. *Molecular Therapy* 14: 192–201.

Sudimack, J. and Lee, R. J. 2000. Targeted drug delivery via the folate receptor. *Advanced Drug Delivery Reviews* 41: 147–162.

Sun, P., Zhang, H. Y., Liu, C. et al. 2010. Preparation and characterization of Fe3O4/CdTe magnetic/fluorescent nanocomposites and their applications in immuno-labeling and fluorescent imaging of cancer cells. *Langmuir* 26: 1278–1284.

Susha, A. S., Javier, A. M., Parak, W. J., and Rogach, A. L. 2006. Luminescent CdTe nanocrystals as ion probes and pH sensors in aqueous solutions. *Colloids and Surfaces A—Physicochemical and Engineering Aspects* 281: 40–43.

Susumu, K., Uyeda, H. T., Medintz, I. L. et al. 2007. Enhancing the stability and biological functionalities of quantum dots via compact multifunctional ligands. *Journal of the American Chemical Society* 129: 13987–13996.

Tada, H., Higuchi, H., Wanatabe, T. M., and Ohuchi, N. 2007. In vivo real-time tracking of single quantum dots conjugated with monoclonal anti-HER2 antibody in tumors of mice. *Cancer Research* 67: 1138–1144.

Thakur, D., Deng, S., Baldet, T., and Winter, J. O. 2009. pH sensitive CdS-iron oxide fluorescent-magnetic nanocomposites. *Nanotechnology* 20: 485601.

Tomasulo, M., Yildiz, I., Kaanumalle, S. L., and Raymo, F. M. 2006. pH-sensitive ligand for luminescent quantum dots. *Langmuir* 22: 10284–10290.

Tomczak, N., Janczewski, D., Han, M. Y., and Vancso, G. J. 2009. Designer polymer-quantum dot architectures. *Progress in Polymer Science* 34: 393–430.

Tortiglione, C., Quarta, A., Tino, A. et al. 2007. Synthesis and biological assay of GSH functionalized fluorescent quantum dots for staining Hydra vulgaris. *Bioconjugate Chemistry* 18: 829–835.

Troy, T., Jekic-McMullen, D., Sambucetti, L., and Rice, B. 2004. Quantitative comparison of the sensitivity of detection of fluorescent and bioluminescent reporters in animal models. *Molecular Imaging* 3: 9–23.

Tsay, J. M., Trzoss, M., Shi, L. X. et al. 2007. Singlet oxygen production by peptide-coated quantum dot-photosensitizer conjugates. *Journal of the American Chemical Society* 129: 6865–6871.

Uprichard, S. L. 2005. The therapeutic potential of RNA interference. *FEBS Letters* 579: 5996–6007.

Veiseh, O., Sun, C., Gunn, J. et al. 2005. Optical and MRI multifunctional nanoprobe for targeting gliomas. *Nano Letters* 5: 1003–1008.

Vinayaka, A. C., Basheer, S., and Thakur, M. S. 2009. Bioconjugation of CdTe quantum dot for the detection of 2,4-dichlorophenoxyacetic acid by competitive fluoroimmunoassay based biosensor. *Biosensors & Bioelectronics* 24: 1615–1620.

Vinceti, M., Rovesti, S., Bergomi, M. et al. 2001. Risk of birth defects in a population exposed to environmental lead pollution. *Science of the Total Environment* 278: 23–30.

Voura, E. B., Jaiswal, J. K., Mattoussi, H., and Simon, S. M. 2004. Tracking metastatic tumor cell extravasation with quantum dot nanocrystals and fluorescence emission-scanning microscopy. *Nature Medicine* 10: 993–998.

Waisberg, M., Joseph, P., Hale, B., and Beyersmann, D. 2003. Molecular and cellular mechanisms of cadmium carcinogenesis. *Toxicology* 192: 95–117.

Wang, D. S., He, J. B., Rosenzweig, N., and Rosenzweig, Z. 2004. Superparamagnetic Fe2O3 Beads-CdSe/ZnS quantum dots core-shell nanocomposite particles for cell separation. *Nano Letters* 4: 409–413.

Wang, X., Ruedas-Rama, M. J., and Hall, E. A. H. 2007. The emerging use of quantum dots in analysis. *Analytical Letters* 40: 1497–1520.

Wang, X., Wu, J., Li, F., and Li, H. 2008a. Synthesis of water-soluble CdSe quantum dots by ligand exchange with *p*-sulfonatocalix(*n*)arene ($n=4$, 6) as fluorescent probes for amino acids. *Nanotechnology* 19: 205501.

Wang, Y. Q., Ye, C., Zhu, Z. H., and Hu, Y. Z. 2008b. Cadmium telluride quantum dots as pH-sensitive probes for tiopronin determination. *Analytica Chimica Acta* 610: 50–56.

Wang, H. C., Brown, J., Alayon, H., and Stuck, B. E. 2010. Transplantation of quantum dot-labelled bone marrow-derived stem cells into the vitreous of mice with laser-induced retinal injury: Survival, integration and differentiation. *Vision Research* 50: 665–673.

Ward, W. W. and Cormier, M. J. 1978. Energy transfer via protein-protein interaction in *Renilla* biolumiscence. *Photochemistry and Photobiology* 27: 389–396.

Warshaw, D. M., Kennedy, G. G., Work, S. S. et al. 2005. Differential Labeling of myosin V heads with quantum dots allows direct visualization of hand-over-hand processivity. *Biophysical Journal* 88: L30-L32.

Wilson, T. and Hastings, J. W. 1998. Bioluminescence. *Annual Review of Cell and Developmental Biology* 14: 197–230.

Wilson, R., Spiller, D. G., Prior, I. A., Bhatt, R., and Hutchinson, A. 2007. Magnetic microspheres encoded with photoluminescent quantum dots for multiplexed detection. *Journal of Materials Chemistry* 17: 4400–4406.

Wu, X. Y., Liu, H. J., Liu, J. Q. et al. 2003. Immunofluorescent labeling of cancer marker Her2 and other cellular targets with semiconductor quantum dots (vol 21, pg 41, 2003). *Nature Biotechnology* 21: 452–452.

Xing, Y., So, M. K., Koh, A. L., Sinclair, R., and Rao, J. H. 2008. Improved QD-BRET conjugates for detection and imaging. *Biochemical and Biophysical Research Communications* 372: 388–394.

Xiong, J. P., Stehle, T., Zhang, R. G. et al. 2002. Crystal structure of the extracellular segment of integrin alpha V beta 3 in complex with an Arg-Gly-Asp ligand. *Science* 296: 151–155.

Yao, H., Zhang, Y., Xiao, F., Xia, Z., and Rao, J. 2007. Quantum dot/bioluminescence resonance energy transfer based highly sensitive detection of proteases. *Angewandte Chemie—International Edition* 46: 4346–4349.

Yezhelyev, M. V., Qi, L. F., O'Regan, R. M., Nie, S., and Gao, X. H. 2008. Proton-sponge coated quantum dots for siRNA delivery and intracellular imaging. *Journal of the American Chemical Society* 130: 9006–9012.

Yildiz, A., Forkey, J. N., McKinney, S. A. et al. 2003. Myosin V walks hand-over-hand: Single fluorophore imaging with 1.5-nm localization. *Science* 300: 2061–2065.

Yildiz, I., Tomasulo, M., and Raymo, F. M. 2006. A mechanism to signal receptor-substrate interactions with luminescent quantum dots. *Proceedings of the National Academy of Sciences of the United States of America* 103: 11457–11460.

You, X. G., He, R., Gao, F. et al. 2007. Hydrophilic high-luminescent magnetic nanocomposites. *Nanotechnology* 18(3): 035701.

Young, S. H. and Rozengurt, E. 2006. Qdot nanocrystal conjugates conjugated to bombesin or ANG II label the cognate G protein-coupled receptor in living cells. *American Journal of Physiology—Cell Physiology* 290: C728–C732.

Yu, D. H., Wang, Z., Liu, Y. et al. 2007. Quantum dot-based pH probe for quick study of enzyme reaction kinetics. *Enzyme and Microbial Technology* 41: 127–132.

Yuan, J., Guo, W., Yang, X., and Wang, E. 2009. Anticancer drug–DNA interactions measured using a photoinduced electron-transfer mechanism based on luminescent quantum dots. *Analytical Chemistry* 81: 362–368.

Yum, K., Na, S., Xiang, Y., Wang, N., and Yu, M. F. 2009. Mechanochemical delivery and dynamic tracking of fluorescent quantum dots in the cytoplasm and nucleus of living cells. *Nano Letters* 9: 2193–2198.

Zebli, B., Susha, A. S., Sukhorukov, G. B., Rogach, A. L., and Parak, W. J. 2005. Magnetic targeting and cellular uptake of polymer microcapsules simultaneously functionalized with magnetic and luminescent nanocrystals. *Langmuir* 21: 4262–4265.

Zegrya, G. G. and Samosvat, D. M. 2007. *Journal of Experimental and Theoretical Physics* 104: 951–965.

Zhang, C. Y. and Johnson, L. W. 2006. Homogenous rapid detection of nucleic acids using two-color quantum dots. *Analyst* 131: 484–488.

Zhang, C.-Y. and Johnson, L. W. 2009. *Analytical Chemistry* 81: 3051–3055.

Zhang, C. Y., Yeh, H. C., Kuroki, M. T., and Wang, T. H. 2005. Single-quantum-dot-based DNA nanosensor. *Nature Materials* 4: 826–831.

Zhang, Q., Cao, Y. Q., and Tsien, R. W. 2007. Quantum dots provide an optical signal specific to full collapse fusion of synaptic vesicles. *Proceedings of the National Academy of Sciences of the United States of America* 104: 17843–17848.

Zhang, Q., Li, Y., and Tsien, R. W. 2009a. The dynamic control of kiss-and-run and vesicular reuse probed with single nanoparticles. *Science* 323: 1448–1453.

Zhang, Y.-H., Zhang, H.-S., Ma, M., Guo, X.-F., and Wang, H. 2009b. The influence of ligands on the preparation and optical properties of water-soluble CdTe quantum dots. *Applied Surface Science* 255: 4747–4753.

Zhao, D., Jimei, Z., Quanxi, D. et al. 2007. Adaption of Au nanoparticles and CdTe quantum dots in DNA detection. *Chinese Journal of Chemical Engineering* 15: 791–794.

Zimmer, J. P., Kim, S. W., Ohnishi, S. et al. 2006. Size series of small indium arsenide-zinc selenide core-shell nanocrystals and their application to in vivo imaging. *Journal of the American Chemical Society* 128: 2526–2527.

# 11

## *Magnetic Nanoparticles for Drug Delivery*

**Claudia Altavilla**

**CONTENTS**

## 11.1 Introduction

Nowadays, standard therapeutic methods for the treatment of different diseases use the systemic distribution of molecular drugs. The most common pharmaceuticals are generally fast-acting chemical compounds that are either dispersed orally (pills or liquid) or injected. Moreover, a lot of protocols require the use of highly toxic chemicals, as in cancer therapy.

Therefore, controlled release as well as the targeting of delivery systems must evolve in parallel to drug research. The main problems caused by systemic bio-distribution are due to the scarce specificity of drugs that may never reach the target organ and it being difficult to achieve the required level of the drug in the body. For this reason, pharmacokinetics imposes the administration of high drug doses in order to obtain the desired therapeutic effect. Large doses can result in serious side effects, damaging normal cells and organs as well as diseased cells (Durán et al. 2008).

Over the past few decades, research in modern pharmaceutical technology supported by the exponential growth of biotechnology and nanotechnology has revolutionized the approach to drug delivery (Robinson and Lee 1987). In particular, a controlled-release drug delivery system should be able to satisfy the following requisites: (1) maintenance of optimum active drug molecules concentration in the blood; (2) reproducible and predictable release rate for long time; (3) improvement of activity duration for short half-life drugs; (4) elimination of side effects, frequent dosing, and wastage of drug; and (5) optimized therapy and patient compliance.

In order to achieve these benefits, the design of a controlled-release system is a multifactor and multidisciplinary process, which includes consideration of the chemical and physical properties of the drug, the route of administration, the nature of the delivery vehicle, the drug release mechanism, and biocompatibility (Martín del Valle et al. 2009).

Biological polymers, liposomes, hydrogels, viruses, and other controlled-release carriers have been widely investigated. These vehicles release therapeutic agents under the influence of ultrasound, pH, temperature, or chemical interactions. However, in many cases, the drug delivery vehicles do not have a mechanism for localization, where it is possible to deliver high concentrations of drugs with minimally invasive techniques (Pillai and Panchagnula 2001, Hatefi and Amsden 2002, Park 2002, Georgens et al. 2005, Lin and Metters 2006, Satish et al. 2006, Crampton and Simanek 2007, Samad et al. 2007, Udit et al. 2010).

Thanks to the interdependence of these factors, the possibilities of designing a specific controlled-release drug delivery system are enormous.

An interesting strategy, with immense potentiality, that can be used to remotely control the delivery of a drug or gene, is the use of magnetic nanoparticles manipulated by an external magnetic field.

In this chapter, the design and use of magnetic nanoparticles as delivery vehicles will be reviewed and discussed.

## 11.2 Historical Background

The concept of using a small magnet, guided by an external magnetic field through blood vessels to gain access to parts of the body that are otherwise inaccessible except via major surgery, dates back to the 1960s. Yodh et al. proposed the use of a movable electromagnet to propel and guide a small Pt-Co magnet through the intravascular system. They also suggested the possibility of a magnet vehicle that either incorporated or towed minute diagnostic or therapeutic devices such as a flexible catheter, in order to take fluids to or from the target site (Yodh et al. 1968). Similar studies carried out by Asken in the same period on the therapy of intracranial aneurism used a magnetically controlled intravascular catheter (Asken 1968).

At the end of the 1970s, the pioneering work on the use of magnetic targeting to deliver drugs to tumors that inspired many subsequent studies was carried out.

In 1978, Widder et al. proposed the synthesis and characterization of a novel parenteral drug carrier capable of area-specific localization by magnets. The carrier consisted of human serum albumin microspheres (AMS), with an average diameter of 1 μm, in which magnetite ($Fe_3O_4$) particles and a prototype drug (doxorubicin) (DXR) were entrapped. The in vitro drug release rate was tested after the stabilization of the microsphere matrix with formaldehyde, 2,3-butanedione, and heat. The released doxorubicin was then tested in order to verify that it was chemically identical to the starting material (Widder et al. 1978a,b,c). Only a few years later, AMS-DXR was tested in vivo on Yoshida sarcoma tumors in rats by using an extracorporeal magnet to guide the system. Of the 12 animals treated with a single dose of AMS-DXR, nine exhibited total remission of the tumor, while, marked tumor regression was observed in the remaining three rats, and no deaths or metastases occurred in the experimental group.

In contrast, significant increases in tumor size with widespread metastases occurred in all the control groups (placebo or non magnetic AMS-DXR), with most of the rats dying.

These experiments clearly indicated that the targeting of oncolytic agents to solid neoplasms by magnetic microspheres may be a means of increasing the efficacy and decreasing the toxicity of antitumor agents (Widder et al. 1981).

The promising results lead the researcher to entrap other drugs in magnetic-AMS and test the systems in vitro and in vivo on guinea pigs, for efficacy and toxicity (Widder et al. 1978b,c, 1979, Ovadia et al. 1982, Gupta and Hung 1990, Ghassabian et al. 1996). For example, Janhua et al. tested the toxicity of ADM-MAM on mice and guinea pigs, demonstrating that there was no macroscopically and microscopically direct cytotoxic damage of the compound to the organs or cells of the animals. Moreover, the LD50 value of the compound was higher than that of the single adriamycin used, indicating that the compound was less toxic and quite safe in its therapeutic dosage (Jianhua et al. 2000).

The success of this hybrid system (organic–inorganic) was due to the physical and chemical properties of both counterparts (Morimoto et al. 1980, 1981). In particular, albumin (a natural polymer) has been used in the preparation of microspheres (Kramer 1974), (spherical particles ranging in size from 1 to 1000 μm), due to its biodegradability (Bernard et al. 1980) as well as other desirable characteristics such as nontoxicity (Schafer et al. 1994, Müller et al. 1996) and biocompatibility as an ideal drug carrier. The association with magnetic particles increased the interest in AMD studies because by restricting the drug to the desired site, a much lower drug concentration can be administered when compared to systemic therapy. Significantly lower dosages can minimize or eliminate the toxic side effects associated with high-dose systemic chemotherapy or other drugs. More than 100 therapeutic and diagnostic agents have been incorporated into albumin microspheres including drugs from various therapeutic categories such as nifedipine (Chuo et al. 1996a,b), mitoxantrone (Luftensteiner et al. 1999), dexamethasone (Ghassabian et al. 1996), salbutamol sulfate (Karunakar and Singh 1994), which have been prepared and characterized as albumin microsphere delivery systems. Studies are still being carried out (Chen et al. 2008).

In addition to AMS, since the 1980s, several authors have developed this strategy to deliver different drugs using magnetic microcapsules and microspheres. Ethyl cellulose microcapsules containing mitomycin (an anticancer drug), and magnetic zinc ferrite particles (mean size = 1.6 μm) were successfully used for the treatment of VX2 tumors in the hind limbs and bladder of a rabbit. Pharmacokinetics confirmed that the magnetic control markedly enhanced drug absorption into the surrounding tumor tissues for a prolonged period compared to non magnetic systems (Kato et al. 1983). In 1994, Häfeli et al. prepared biodegradable poly(lactic acid) microspheres that incorporated magnetite and the β-emitter $^{90}Y$ for targeted radiotherapy and successfully applied them to subcutaneous tumors (Häfeli et al. 1995, 1997).

Magnetic oil-in-water emulsions were also evaluated as a carrier system for site-specific and sustained delivery of lipophilic antitumor agents (Akimoto et al. 1985). Many drugs were stable in oil and degradable in water. The microemulsion containing magnetic particles could be directed to the specific site by the external magnetic guidance or localized to the predetermined site (lungs) by application of an electromagnet to the target site (Akimoto and Morimoto 1983),and the release of the drugs was longer than the free drug. However, many of these approaches were micrometric.

The success of this technology was strictly dependent on the availability of powerful magnets capable of producing high magnetic field gradients at the target sites. Most of the inhomogeneous fields available, were only strong enough to manipulate particles against the diffusion and blood stream velocities found in living systems over a distance of only a few centimeters from the sharp edge of a magnet pole (Seyei et al. 1978). It was difficult to build up sufficient field strengths that focused on a small area and were able to counteract the linear blood-flow rates in the tissue (>10 cm/s in arteries and >0.05 cm/s in capillaries). Therefore, in order to effectively retain the magnetic drug carrier, magnetic forces must be high enough to reach that goal. The size and magnetic properties of the magnetic counterpart, overlooked for many years, for the same reasons must therefore be carefully optimized to draw the particles through the endothelial wall of the capillary bed as well as prolong the circulation time in the human organism (Lübbe et al. 2001).

The first data on magnetic drug targeting in human patients were reported by Lube et al. in 1996 (Lübbe et al. 1996b), after a previous study on animals (Lübbe et al. 1996a). A colloidal dispersion of multidomain iron oxide ($Fe_3O_4$) was used, as a ferrofluid, with a size range of 50–150 nm, made using a wet chemical method. The particles were surrounded with anhydroglucose polymers in order to promote stabilization under various physiological conditions as well as the chemisorption of drug. The oncolytic, used in the Phase I Study in 14 patients with advanced (but near to the body surface) solid tumors, was 4′-epidoxorubicin (Lübbe et al. 1996b). Since the binding, between drug and magnetic particles coating, was reversible, desorption of the drug that had been bound to the surface occurred according to the physiological environment (pH, osmolality, and temperature).

The magnets consisted of rare earths, the majority being neodymium, arranged according to the individually shaped tumor of the patient. The magnetic field strengths were between 0.5 and 0.8 T, with the distance between the tumor surface and the magnet being less than 0.5 cm (Lübbe et al. 1996a,b, 2001). The results were promising, but many obstacles had to be overcome.

## 11.3 Physical Principle of Magnetic Targeting

Some of the relevant basic concepts of magnetism will be reviewed and discussed, in order to understand the potential applications available in biomedicine as a result of the particular properties of magnetic nanoparticles. Moreover, how a magnetic field gradient can exert a force at distance to move the particles and what strategies are used to control the addressing in the specific site will be also described.

### 11.3.1 Classification of Material Magnetism

If a magnetic material is placed in a magnetic field, **H**, the magnetic induction, **B**, due to the overall response of all the individual atomic moments is equal to

$$\mathbf{B} = \mu_0 (\mathbf{H} + \mathbf{M})$$

where

$\mu_0$ is the permeability of the free space
$\mathbf{M} = \mathbf{m}/V$ is the magnetic moment per unit of volume of the material

Conventionally, to classify the response of a material in presence of a magnetic field, the magnetic susceptibility $\chi$ is used:

$$\mathbf{M} = \chi \mathbf{H}$$

Materials that have a very weak and negative susceptibility to magnetic fields are classified as *diamagnetic* (DM) ($-10^{-6} \leq \chi \leq -10^{-3}$). They are slightly repelled by a magnetic field and the material does not retain the magnetic properties when the external field is removed. This behavior is due to the realignment of the electron orbits under the influence of an external magnetic field. Most elements in the periodic table, including copper, silver, and gold, are DM. In the blood vessels, the response of proteins is DM.

Materials that have a small and positive susceptibility to magnetic fields are classified as *paramagnetic* ($10^{-6} \leq \chi \leq 10^{-1}$). They are slightly attracted by a magnetic field and the material does not retain the magnetic properties when the external field is removed. Paramagnetic properties are due to the presence of some unpaired electrons, as well as from the realignment of the electron orbits caused by the external magnetic field. Many organometallic coordination compounds of transition metals are paramagnetic. In the blood, an example of a paramagnet is hemoglobin (Pauling and Coryell 1936).

*Ferro*, *ferri*, and *antiferromagnetic* materials have unpaired electrons, therefore their atoms have a net magnetic moment with different arrangements (see Figure 11.1). Their magnetic properties are due to the presence of magnetic domains. In particular, in a ferromagnet, the magnetic moments of the atoms are aligned in parallel so that the magnetic force within the domain is strong. Iron, nickel, and cobalt are examples of ferromagnetic materials (FM). In an antiferromagnet, the magnetic moments of the atoms are aligned in anti-parallel ($\mathbf{M} = 0$). Generally, an antiferromagnetic order may exist at sufficiently low temperatures and vanishes above the Néel temperature (Néel 1948). Above this temperature, the thermal energy is sufficient to remove the magnetic order, and the material is paramagnetic. In a ferrimagnet, the magnetic moments are aligned in anti-parallel, but

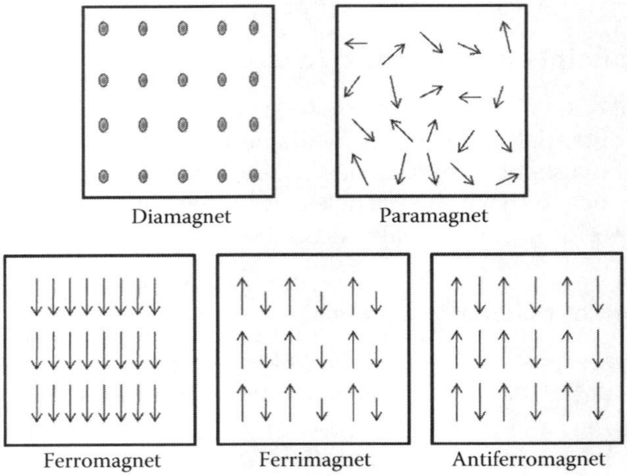


**FIGURE 11.1**
Scheme of the possible magnetic arrangements of materials.

have an unequal strength, producing a strong overall magnetization. Iron ferrites such as magnetite are typically ferrimagnet due to the inverse spinel structure of crystalline lattice (Greenwood and Earnshaw 1984).

### 11.3.2 Multidomain, Single-Domain, and Superparamagnetic Particles

Ferro and ferrimagnets exhibit a strong attraction to magnetic fields and are able to retain their magnetic properties after the external field has been removed. When a ferromagnetic material is in an un-magnetized state, the domains are randomly organized and the net magnetic field for the part as a whole is zero. When a magnetizing force is applied, the domains become aligned in order to produce a strong magnetic field within the part. The susceptibility in ordered materials depends on the temperature as well as the **H** applied. The curve **M–H** is characteristic of magnetic materials. In ferro- and ferrimagnetic materials, the open **M–H** curve is called a hysteresis loop, which is an irreversible magnetization process that is related to the progressive alignment of magnetic domains in respect to **H**. It depends on the magnetic anisotropy of the crystalline lattice as well as the impurities contained (Figure 11.2).

The shape of these loops is also determined by particle size. As the particle size decreases, the number of magnetic domains per particle decreases down to the limit where it is energetically unfavorable for a domain wall to exist (Wohlfarth 1983). Below a critical diameter, the magnetic particles have a single-domain nature. As the particle size further decreases below the single-domain value, the magnetic moment of the particles will be gradually affected by thermal fluctuation and they will behave "paramagnetically with giant moments". This phenomenon is known as superparamagnetism and has zero coercivity (i.e., the intensity of the applied magnetic field required to reduce the magnetization to zero after the magnetization of the sample has been driven to saturation), with it occurring above the blocking temperature at which thermal energy is sufficient for the moment to relax during the time of the measurement (O'Connor et al. 2001). Particles with relaxation times greater than 100 s or with diameters larger than the critical values are called blocked. The blocking temperature ($T_b$) of a material is given by

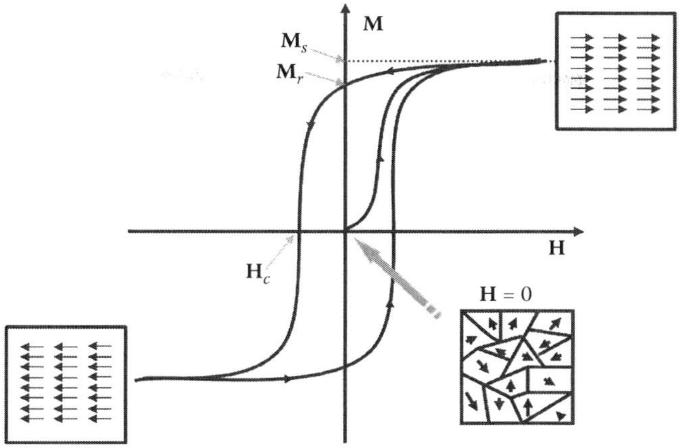


**FIGURE 11.2**
Hysteresis cycle of a multidomain magnetic material. **H** is the magnetic field amplitude and **M** is the magnetization of the material. $\mathbf{M}_s$ is the saturation magnetization and $\mathbf{M}_r$ the remanent magnetization. Squares symbolize multidomain material with magnetization of each domain.

$$T_b = \frac{KV}{25k_B}$$

where
*K* is the nanoparticle magnetic anisotropy
*V* its volume

*KV* can be thought of as the energy barrier ΔE, associated with the magnetization moving from its initial "easy axis" direction, through a "hard axis", ending at another easy axis. $k_B$ is the Boltzmann constant.

In the presence of an applied magnetic field, the spin orientation and subsequent magnetic saturation is achieved with lower field strengths than with the analogous bulk materials. The magnetic moment of each particle is ~100 times larger than for transition metal ions and saturation magnetization is reached at applied magnetic fields as low as 1 kOe. When the field is decreased, demagnetization is dependent on coherent rotation of the spins, which results in large coercive forces (Leslie-Pelecky and Rieke 1996).

The evolution of coercivity, as a function of particle size is illustrated in Figure 11.3.

Above the blocking temperature, the nanoparticles are superparamagnetic, the magnetic moment is free to fluctuate in response to thermal energy, and the result is the anhysteretic (Pankurst 2003), but still sigmoidal, **M**–**H** curve shown in Figure 11.3.

### 11.3.3 Superparamagnetism: An Essential Requisite for Biological Applications

The anhysteretic behavior of superparamagnetic nanoparticles is highly appealing for a wide range of biomedical applications. In fact, in order to avoid the aggregation of the particles (both in the step previous to the injection of the particles and after the drug release), it is required that: (1) the particles are protected against irreversible aggregation by a protective coating and (2) the remanent magnetization $\mathbf{M}_r$ of the particles is null or negligible at room temperature (Lu et al. 2007, Durán et al. 2008). When, in fact, the external magnetic field is removed, the magnetization of superparamagnetic nanoparticles disappears; thus

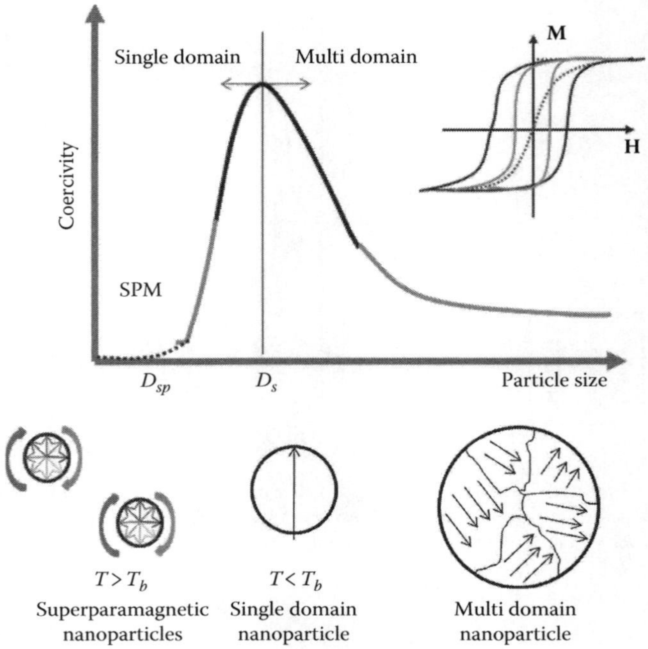


**FIGURE 11.3**
Schematic representation of coercivity as a function of the size of a magnetic particle. Above a critical particle size, $D_S$, the particles are multidomain. The coercivity increases as the particle size decreases. Below, $D_S$, the particles are single domain. When the average particle size decreases further below $D_{SP}$ (i.e., function of blocking temperature $T_b$), the particles become superparamagnetic with unstable magnetic moments and vanishing coercivity. In this condition, the curve **M–H** is a sigmoid but anhysteretic.

the agglomeration and possible embolization of the capillaries is avoided (Arruedo et al. 2007) (Figure 11.4).

### 11.3.4 Heating with Magnetic Nanoparticles: The Principle of Hyperthermia for Cancer Therapy and of Drug Release under the Influence of Thermal Energy

**M–H** hysteretic or anhysteretic curves behavior can be used to produce heat. For multidomain ferro- and ferrimagnetic materials, heating is due to hysteresis loss. In these materials, the "domain wall displacement" is not reversible, i.e., the magnetization curves for increasing and decreasing magnetic fields do not coincide. The energy to overcome the barrier to domain walls motion is delivered by an AC magnetic field and produces heat.

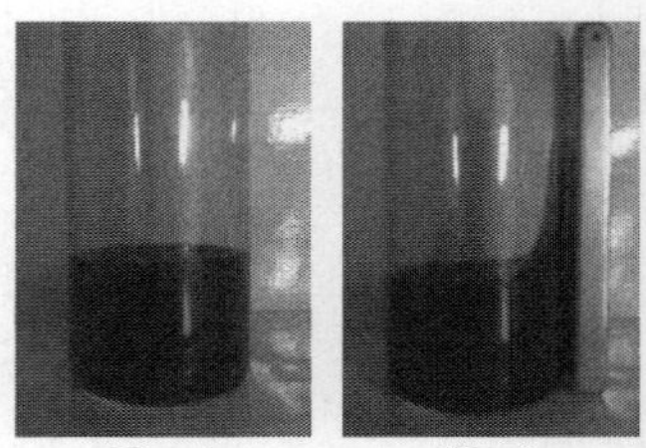


**FIGURE 11.4**
Superparamagnetic response of a ferrofluid of monodispersed cobalt ferrite nanoparticles with an average size of 6 nm. (Courtesy of Altavilla, C.) Magnetization disappears when the magnetic field is removed.

In superparamagnetic nanoparticles (when $T < T_b$), it is possible to observe two relaxation phenomena that can be used to produce heat: the Néel relaxation, which involves the flipping of the magnetic moment, and the Brown relaxation, which involves the rotation of the particle as a whole.

In particular, an AC magnetic field supplies energy and assists magnetic moments in rotating to overcoming the energy barrier $\Delta E = KV$ (in the simplest case as an uniaxial form) where K is the anisotropic constant and V the nanoparticles volume. This energy is dissipated when the particle moment relaxes to its equilibrium orientation characterized by the Néel relaxation time, $\tau_N$:

$$\tau_N = \tau_0 e^{\frac{\Delta E}{kT}}$$

where
- $T$ is the temperature
- $k$ is the Boltzman constant
- $\tau_0$ ~$10^{-9}$ s (Mornet et al. 2004)

Rotational Brownian motion (observed also in multidomain nanoparticles) within a carrier liquid (blood) is due to the torsion exerted by the external AC magnetic field on the magnetic moment that produces the rotation of particle as a whole and determines the friction with the surrounding liquid. The Brown relaxation time, $t_B$, is correlated with the viscosity of the liquid ($\eta$) and the hydrodynamic volume of the particles ($V$) through the equation

$$t_B = \frac{3\eta V}{kT}$$

The frequency ($\nu_B$) for maximum heating via Brown rotation is given by the equation (Fannin et al. 1987, Mornet et al. 2004)

$$\nu_B = \frac{1}{2\pi t_B}$$

For hyperthermia treatment, particles with a size around monodomain–multidomain transition, i.e., particles below a diameter of 50 nm, have been found to produce the maximum specific absorption rate (SAR) (Roca et al. 2009).

The same principle can be used for drug release after addressing in the specific site the magnetic vehicle (Liu et al. 2007). Recently, Hu et al. proposed the controlled rupture of magnetic sensitive polyelectrolyte microcapsules for drug delivery. The system was prepared using $Fe_3O_4$/poly (allylamine) to construct the shell. The presence of magnetic particles was used to produce heat under the influence of a high frequency magnetic field, thanks to Brown and Néel relaxations, and triggered the release of drugs from the microcapsule (Hu et al. 2008).

### 11.3.5 Addressing of Magnetic Particles under the Influence of Magnetic Field Gradient

Magnetic targeting is based on the attraction of magnetic particles to a magnetic field source. This is one of the most attractive methods for localizing drugs in the body, because magnetic forces act at relatively long ranges and magnetic fields do not affect most biological tissues.

In the presence of a magnetic field gradient, a translational force will be exerted on the particle/drug complex that is trapped in the field and addressed to the targeted site.

It is important to recognize that a uniform field gives rise to a torque but not to a translational motion. This magnetic force is governed by the equation

$$F = \Delta\chi V \frac{1}{\mu_0} \mathbf{B}(\nabla \mathbf{B})$$

where

**B** is the magnetic field strength

$\nabla$**B** is the field gradient

$\Delta\chi$ is the difference between the magnetic susceptibility of the particles with respect to the medium (biological fluids have a very small $\chi$)

*V* is the volume of the particles (Pankhurst et al. 2003, Dobson et al. 2006b)

Methods and devices proposed to deliver drugs encapsulated within magnetic carriers to specific locations in the body have relied on a single source of magnetic fields, both to magnetize the carriers as well as pull them by magnetic force to specific locations in the body. These single magnetic field sources are usually applied externally on the surface body near the tumor site or near the specific site through the use of an internal implant. On the one hand, sources applied externally to the body are excellent for magnetizing the carriers, but they provide only weak magnetic field gradients to attract them. On the other, an internal magnetic implant provides strong magnetic field gradients to attract the carriers, but its fields decay too quickly to magnetize the bulk of the injected carriers, especially if the injection site is far from the target site.

In 2005, Yellen et al. proposed magnetic implants placed directly in the cardiovascular system to attract injected magnetic carriers (Yellen et al. 2005, Forbes et al. 2008). Theoretical simulations and experimental results supported the assumption that using magnetic implants in combination with externally applied magnetic fields will optimize the delivery of magnetic drugs to selected sites within a subject (Yellen et al. 2005).

In 2007, an innovative method of manipulating magnetic carriers was proposed by Cha et al. The magnetic device used pulsed-field solenoid coils with high-Tc superconductor inserts in the form of cylindrical disks strategically located outside the body. Preliminary experimental results demonstrated that the proposed method can (1) move magnetic particles, ranging in size from a few millimeters to 10 μm, with strong enough forces over a substantial distance; (2) hold the particles at a designated position as long as needed; and (3) reverse the processes and retrieve the particles (Cha et al. 2007).

A review of the state of the art of the field of targeted drug delivery with internal magnets to concentrate magnetic nanoparticles near tumor locations and the different approaches to this task performed in vitro and in vivo, was recently published (Fernández-Pacheco et al. 2009).

A scheme of possible geometries for the magnetic drug delivery is reported in Figure 11.5.

## 11.4 Design and Synthesis of Magnetic Nanoparticles for Biomedical Applications

The advantages of using nanotechnology for biomedical purposes essentially come from the versatility of the different synthesis methods that now allow for the precise engineering of the critical features of the wide variety of nanoparticles (NPs).

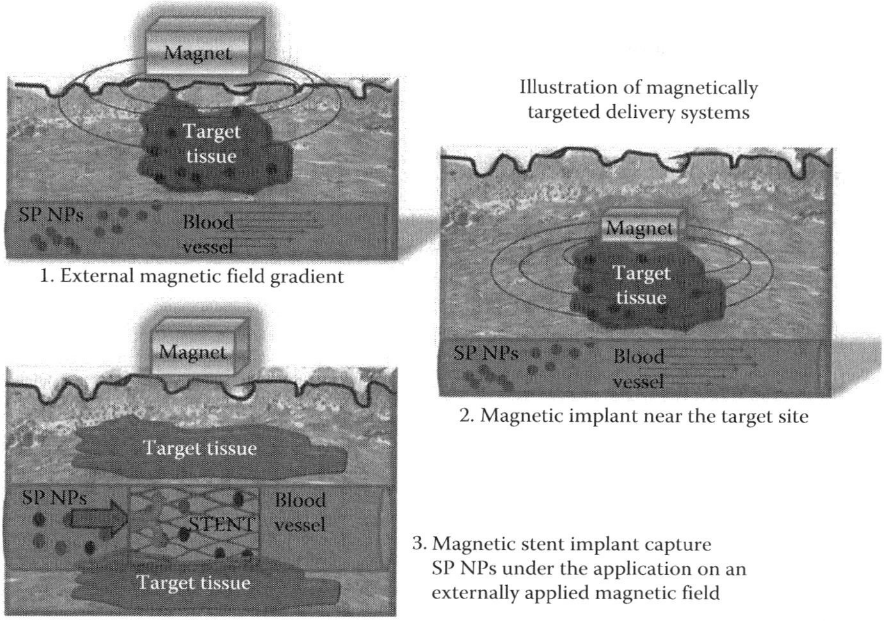


**FIGURE 11.5**
**(See color insert following page 302.)** Illustration of three possible geometries used for magnetically targeted delivery. (1) The magnetic nanocarriers are guided by an external magnetic gradient on the target site. This configuration is more effective for target sites that are near the body surface. (2) A magnetic implant near the target tissue attracts the magnetic particles passively carried by the blood flow. The magnetic field generated by the implant is not sufficient to magnetically drive nanoparticles from the injection site to the target tissue if they are not close. (3) Magnetic implant stents attract magnetic drug carriers under the external guidance of a magnetic field. The particles are entrapped in the network stent and the release of drug is controlled.

Numerous forms of magnetic nanoparticles (MNPs) have been proposed and evaluated for biomedical applications in order to exploit the nanoscale magnetic phenomena, such as enhanced magnetic moments and superparamagnetism. It is well known that composition, size, morphology, and surface chemistry, not only improve magnetic properties, but also influence the bio-application of magnetic nanoparticles in vivo (Corot et al. 2006, Sun et al. 2008).

In its simplest form, a biomedical MNP platform needs to satisfy several essential features:

1. Nanoscale dimensions, to pass through the narrowest blood vessel but also to penetrate through the cell membrane when necessary (Berry 2005).
2. A magnetic or superparamagnetic (SP) core (SP NPs avoid aggregation due to magnetic attraction) to be manipulated by the magnetic field as well as be driven to the target site (Arruedo et al. 2007) and/or produce heat for hyperthermia treatment or for the controlled release of drugs.
3. Biocompatible surface coating to provide stabilization under physiological conditions.
4. Suitable surface chemistry for the integration of functional ligands to perform multiple functions simultaneously (Frullano and Meade 2007), such as drug delivery and real-time monitoring (Liong et al. 2008, Riehemann et al. 2009), as well as

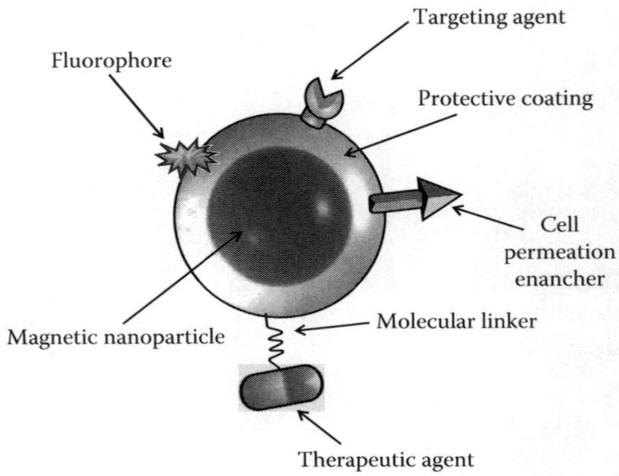


**FIGURE 11.6**
**(See color insert following page 302.)** Hypothetical representation of a multifunctional hybrid magnetic nanoparticle for biomedical applications.

combined therapeutic approaches, i.e., hyperthermia and chemotherapy (Ciofani et al. 2009).

5. Finally, but no less important, magnetic carriers should be biodegradable once their function is completed, and the decay products should be rapidly excreted and nontoxic.

A schematic representation of a multifunctional hybrid magnetic nanoparticle for biomedical applications is reported in Figure 11.6.

### 11.4.1 Synthetic Strategies for Production of Magnetic Core Materials

There are many magnetic materials available with a wide range of magnetic properties, but many of them are highly toxic and cannot be used without efficient protective coating for in vivo applications. Iron oxide–based materials are relatively safe, and commercial superparamagnetic iron oxide (SPIO) such as FERIDEX (Cantillon-Murphy et al. 2009) and Resovis® (Reimer et al. 2005) are currently being used as magnetic resonance imaging (MRI) contrast agents.

The magnetic nanoparticles reported as potential candidates for biomedical applications are magnetite ($Fe_3O_4$), maghemite ($\gamma$-$Fe_2O_3$), ferrite of general formula $MFe_2O_4$ (M=Co, Ni, Zn), iron, and iron based alloys such as iron-platinum (FePt).

A brief review of the most important chemical methods for the syntheses of magnetic nanoparticles will be described in this section.

### 11.4.2 Coprecipitation

Coprecipitation is an easy and economical way to synthesize iron oxides (Massart 1981), nanoparticles ($Fe_3O_4$, $\gamma$-$Fe_2O_3$), or ferrites of transition metals (Rajendran et al. 2001, Li et al. 2002), from aqueous $M^{2+}/Fe^{3+}$ salt solutions ($M^{2+}=Fe^{2+}$, $Co^{2+}$, $Ni^{2+}$, $Zn^{2+}$, $Mn^{2+}$...) added to a base ($NH_4OH$, NaOH, $Na_2CO_3$...) under inert atmosphere at either room temperature or elevated temperatures. The size distribution, shape, and composition of magnetic nanoparticles depends on the pH value (Kim et al. 2001, Jolivet et al. 2002), temperature (Kim et al. 2003), $M^{2+}/Fe^{3+}$ ratio (Lu et al. 2007), ionic strength (Hanh et al. 2003), as well as the presence of oxidant species (Chinnasamy et al. 2003).

Maghemite nanoparticles can be obtained by controlled oxidation of the magnetite nanoparticles (Jolivet et al. 2002). The use of autoclave systems to increase reaction temperature and pressure were also reported for the coprecipitation synthesis of magnetite sub micro-nanoparticles (Fan et al. 2001, Khollam et al. 2002).

Particles prepared by coprecipitation are generally polydispersed, and this condition is non-ideal for many applications. In fact, it is well known that a short burst of nucleation and subsequence-controlled growth is crucial to produce monodispersed particles. Moreover, the blocking temperature ($T_b$) depends on the particle size, with a large distribution size resulting in a wide range of $T_b$. The blocking temperature can be estimated from zero field cool–field cool (ZFC–FC) curves (Hansen and Mørup 1999). The curves describe the temperature dependence of the magnetic susceptibility from room temperature down to 2 K.

The ZFC magnetization curve is typically obtained by cooling, in zero field, from room temperature, where all the particles show superparamagnetic, to a low temperature (2–3 K) and measuring the magnetization at stepwise-increasing temperatures in a small applied field. At each temperature, measurements are taken after time, *t*. If a random assembly of nanoparticles is cooled down in zero field, at the equilibrium, the magnetic moment will be frozen in all the directions. Then, if a small external field is applied and the temperature is raised, the magnetic moment will tend to align along the external field, thus causing magnetization to increase. On the other hand, upon increasing the temperature, relaxation becomes progressively more efficient, so that above a certain temperature magnetization will decrease.

The FC magnetization curve is typically obtained by measuring at stepwise-decreasing temperatures in the same small applied field after time *t* at each temperature. In this case, if the random assembly is cooled down with an applied external field, the magnetic moment will tend to be frozen parallel to the applied field. If the temperature is then increased, relaxation will cause the magnetization to decrease, until finally, when all the particles will be in the superparamagnetic state, the FC curve will collapse into the ZFC one.

The temperature of the maximum in the ZFC curve ($T_{max}$) indicates the temperature at which the superparamagnetic relaxation sets in, i.e., can be considered as the blocking temperature.

The temperature at which the ZFC and FC curves start to separate ($T_{sep}$) corresponds to the blocking of the largest particles.

Figure 11.7 reports the ZFC–FC curves of a sample of magnetite nanoparticles obtained at room temperature by coprecipitation reaction of $Fe^{2+}/Fe^{3+}$ in an aqueous solution of NaOH under an inert atmosphere and vigorous stirring.

The absence of a maximum in the ZFC curve and the increasing trend of magnetization also at 300 K, clearly indicate that there is a large dispersion of dimensions in the sample and that at room temperature, there are micrometric particles that are blocked.

Significant developments in preparing magnetite nanoparticles with controlled dimensions have been obtained by adding stabilizing agents during the coprecipitation synthesis. Disodium tartrate has been used as a stabilizing agent in the synthesis of $CoFe_2O_4$ and added to the mixture of Co and Fe nitrites before the addition of NaOH. By varying the amount of organic ligands, nanoparticles in a large size range were obtained. The mean diameter varied from 3 to 10 nm (Neveu et al. 2002). Ultrafine magnetic nanoparticles with an average diameter of 4–7 nm were prepared by precipitation at 80°C of an aqueous solution of ferrous and ferric ions in a polyvinyl alcohol (PVA) using NaOH as a base (pH 13.8), (Lee et al. 1996). An interesting approach to preparing very small magnetic nanoparticles

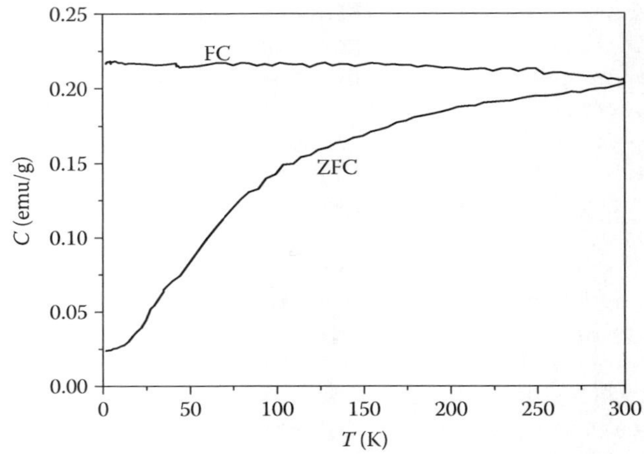


**FIGURE 11.7**
ZFC–FC of magnetite nanoparticles obtained by coprecipitation reaction. The absence of $T_{max}$ in the ZFC curve and the increasing trend also at 300 K clearly suggest a poly dispersion of the sample as well as the presence of large blocked particles at room temperature. (Courtesy of Altavilla, C.)

was proposed by Bonacchi et al. that obtained nanoparticles of 2–3 nm by adding NaOH to a ferrous chloride solution in the presence of γ-cyclodextrin (Bonacchi et al. 2004).

### 11.4.3 Microemulsion Water-in-Oil: The Reverse Micelles

Microemulsions are thermodynamically stable systems composed of two immiscible liquids (usually, water and oil) and a surfactant. Droplets of water-in-oil (W/O) (reverse micelles) or oil-in-water (O/W) (micelles) are stabilized by surfactants when small amounts of water or oil are used, respectively.

These nanodroplets can be used as nanoreactors to carry out chemical reactions. In particular, a reverse micelle is a self-organized aggregation of surfactant molecules formed in an apolar solvent and has the ability to solubilize a relatively large amount of water in the polar core to form a nanometer-sized waterpool.

It was initially assumed that these nanodroplets could be used as templates to control the final size of the particles. However, research carried out over the last few years has shown that besides the droplet size, several other parameters play an important role in the final size distribution (López-Quintela et al. 2004).

The surfactants generally used to produce nanoparticles with the reverse micelles synthesis are anionic (i.e., Sodium bis(2-ethylhexyl) sulfosuccinate(AOT)), cationic (i.e., cetyltrimethylammonium bromide (CTAB)) zwitterionic [Dipalmitoyl-phosphatidylcholine (lecithin)] and non-ionic [Polyoxyethylene(4) lauryl ether (Brij 30)] species (Simmons et al. 2002, Gupta and Gupta 2005).

The formation of reverse micelles in solution is a function of the concentration of surfactant. There is a relatively small range of concentrations separating the limit below which virtually no micelles are detected and the limit above which virtually all additional surfactant molecules form micelles. This concentration is called critical micelle concentration (CMC) (McNaught and Wilkinson 1997).

The dimension of reverse micelles is a function of the water amount. The size of the droplets can be controlled very precisely, by merely changing the ratio R = [water or oil]/[surfactant] in the nanometer range (Kinugasa et al. 2001). Another important factor is the packaging parameter (P) or shape factor that determines the shape of micelle in the solution:

$$P = \frac{V}{al}$$

where

*V* is the chain volume (volume of hydrocarbon tail)

*l* is the critical chain length (the longest effective length that the chain can be extended to in the fluid)

*a* is the optimal head-group area of hydrophilic head (Israelachvili 1991)

In order to obtain spherical micelles, $V/al < 1/3$ while for reverse micelles, $V/al > 1$.

After the mixing of the two microemulsions containing respectively, metal ions ($Fe^{2+}/Fe^{3+}$) and reactive anionic counterparts ($OH^-$, $S^{2-}$...), continuous collision, coalescence, and exchange of water content produce the desired precipitate that can be extracted by filtration or centrifugation after the addition of ethanol or acetone (Gupta and Gupta 2005). For example, cobalt ferrite nanoparticles with different sizes have been synthesized from the microemulsion of a metallic salt and $Na_2CO_3$ as precipitating agent, using AOT as the surfactant and isooctane as the oil phase (Ahn et al. 2003) (Figure 11.8).

Interesting studies have been carried out over the last 20 years by Pileni, which have lead to a fundamental understanding of the kinetics and mechanisms in colloidal solutions as well as the controlled synthesis of colloidal nanocrystals with different sizes and shape using a reverse micelles approach. Some of the most important articles are reported in references section (Pileni 1993, 1997, 2001, 2003, 2007).

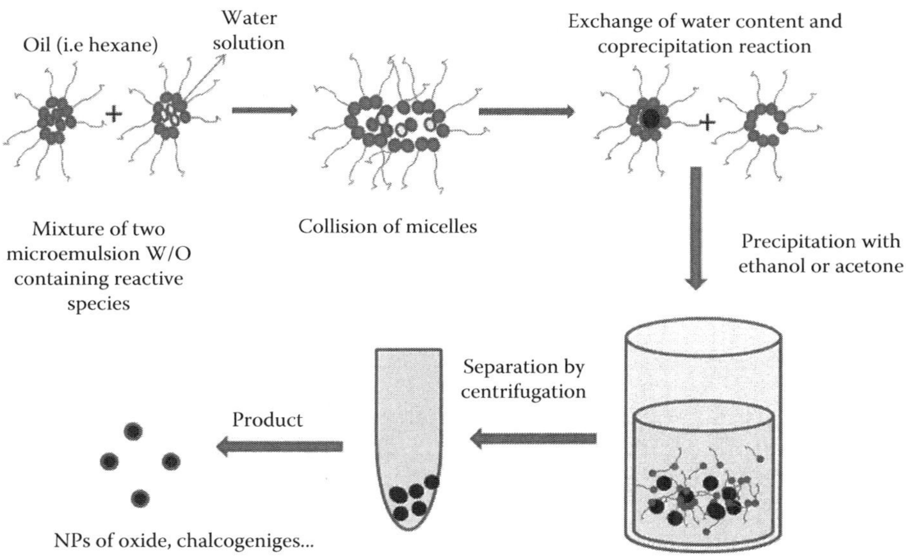


**FIGURE 11.8**
Schematic representation of reverse micelles synthesis of inorganic nanoparticles.

Although many types of magnetic nanoparticles have been successful synthesized in a controlled manner using the microemulsion method, the particle size and shape usually vary over a relatively wide range. Moreover, the yield of nanoparticles is lower than that obtained by other synthetic strategies such as coprecipitation or thermal decomposition. Large amounts of solvent are required in order to obtain appreciable quantities of the product. These issues make it difficult to scale up the process (Lu et al. 2007).

### 11.4.4 Thermal Decomposition

Smaller sized monodispersed magnetic nanocrystals can be synthesized through the thermal decomposition of transition metal complexes in a high boiling point organic solvent, in the presence of surfactant molecules that have the function of stabilizing dispersion as well as controlling the growth of nanoparticles. Reaction temperature, reaction time, and aging time are crucial parameters for the size of the nanoparticles.

Monodispersed nanometric ferrite $MFe_2O_4$ (M=Fe, Co, Mn) can be obtained through the thermal decomposition of metal acetylacenotates [M(acac)n] in the presence of different surfactant molecules (fatty acids, long chain amines, diols) (Sun et al. 2004). Hu et al. recently demonstrated the role of oleylamine as both a stabilizer and reductive agent in the synthesis of $Fe_3O_4$ (Xu et al. 2009b).

A TEM image of $Fe_3O_4$ magnetite nanoparticles obtained with the Sun method (Altavilla et al. 2005) is reported in Figure 11.9. In the insert, the ZFC–FC of the sample is characteristic of monodipsersed nanoparticles as suggested by the presence of a maximum in the ZFC curve corresponding to the blocking temperature ($T$=54K). The synthetic strategy produced high crystalline particles as is clearly shown in the HR-TEM image insert of the same figure. If the metal in the precursor is zerovalent such as in iron carbonyl, the thermal decomposition gives metal nanoparticles that can be successively oxidized (Hyeon et al. 2001, Lu et al. 2003). Iron nanocubes were synthesized by thermal decomposition in mesitylene of $Fe[N(SiMe_3)_2]_2$ with $H_2$ as a reductive agent, in the presence of oleic acid and hexadecylamine (Dumestre et al. 2004). Similar reactions can be used to obtain nanoparticles of iron alloy such as FePt. Synthesis of monodispersed FePt nanoparticles, from 3 to 10nm with a standard deviation of less than 5%, by reduction of platinum acetylacetonate and decomposition of iron pentacarbonyl in the presence of oleic acid and oleylamine stabilizers has also been reported (Sun et al. 2000). The FePt particle composition is readily controlled, and the size is of a tunable diameter.

Using the general approach of thermal decomposition, some authors have reported the synthesis of monodispersed oxide nanoparticles by pyrolysis of metal–fatty acid complexes in different solvents (Jana et al. 2004). Monodispersed cobalt ferrite nanoparticles coated by undecanoic acid were produced by decomposition in octyl ether of the undecanoic complexes of iron and cobalt. The presence of a terminal vinyl group in a fatty acid chain was used to anchor nanoparticles onto a silicon substrate (Altavilla et al. 2007).

The thermal decomposition of the metal–fatty acid complexes at reduced pressure and high temperature (about 300°C) was carried out without any solvents to produce monodispersed $Fe_3O_4$. The size and shape of the nanoparticles depends on the amount of Na-oleate. After the reduction process, α-Fe nanoparticles were obtained (Cha et al. 2006). Recently, the effects of annealing time and vacuum pressure on the shape and size of $Fe_3O_4$ nanocrystals obtained by thermal decomposition have been reported (Cha et al. 2008).

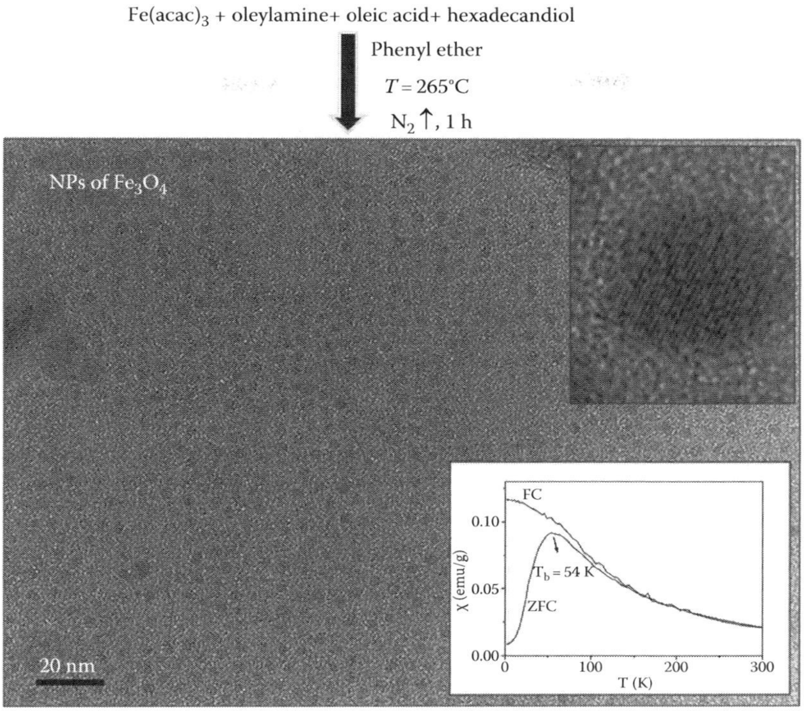


**FIGURE 11.9**
TEM image of monodispersed magnetite nanoparticles obtained by thermal decomposition in phenyl ether of $Fe(acac)_3$ in the presence of a mixture of surfactant molecules. In the insert, an HR-TEM image of a single nanocrystal is reported (Courtesy of Altavilla, C.) The ZFC–FC of the sample is in the bottom insert.

### 11.4.5 Hydrothermal Synthesis

This method exploits the solubility of almost all inorganic substances in water at elevated temperatures and pressures, as well as the subsequent crystallization of the dissolved material from the fluid. Water at elevated temperatures plays an essential role in the transformation of precursor materials, due to vapor pressure being much higher and the structure of water being different at elevated temperatures from that at room temperature. The properties of the reactants, including their solubility and reactivity, also change at high temperatures. Thanks to these peculiarities, this method produces a wide range of high-quality nanocrystals, which is not possible at low temperatures. During the synthesis, the parameters such as water pressure, temperature, and reaction time can be tuned to maintain a high simultaneous nucleation rate as well as good size distribution (Burda et al. 2005). A general strategy based on a general phase transfer and separation mechanism occurring at the interfaces of the liquid, solid, and solution phases present during the synthesis was reported by Wang et al. in order to produce metal, semiconductor, and metal oxide nanoparticles (Wang et al. 2005a,b). Magnetite particles with an average size of 39 nm and good monodispersity have been synthesized by coprecipitation at 70°C from ferrous $Fe^{2+}$ and ferric $Fe^{3+}$ ions by a $(N(CH_3)_4OH)$ solution, followed by hydrothermal treatment at 250°C by Daou et al. (2006).

With the same principle, but in different solvents, magnetite nanoparticles were prepared through a solvothermal reduction approach in the presence of ethylene glycol, oleic acid, and trioctylphosphine oxide or hexadecylamine. The size of the magnetite nanoparticles was modulated through tailoring surfactants in the system (Hou et al. 2003).

An ultrafast hydrothermal synthesis of highly crystalline and water-soluble magnetite nanoparticles (with an average diameter of 6.5 nm) was recently proposed. The reaction was conducted using an aqueous $FeCl_3+KOH$ solution in the presence of L-ascorbic acid as modifier and reductant, by a flow-through hydrothermal method with a microreactor at 673 K and 3 MPa during a residence time of $2 \times 10^{-3}$ s (Sue et al. 2009).

All the general methods described have both advantages and disadvantages. Coprecipitation is the simplest method to produce a large quantity of polydispersed particles. Thermal decomposition can control the size and crystallinity, as well as produce monodispersed nanoparticles directly coated by surfactant molecules (fatty acid or long chain amine). Microemulsion can control shape and size, but a great number of parameters need to be controlled. Hydrothermal synthesis is a relatively unexplored technique for the synthesis of magnetic nanoparticles.

## 11.5 Protection and Stabilization and Functionalization of Magnetic Nanoparticles

One of the most important requirements for almost all the applications of magnetic nanoparticles is stability. Thanks to the high surface-to-volume ratio, the surface energy of nanoparticles is significantly higher compared to that of the bulk. Surface energy promotes the coalescence between closer grains and, especially for metals and metal alloys, the oxidation phenomena are considerable. The first consequence of aggregation is the loss of monodispersion, while oxidation produces a modification in the nature of the material. The global result is an alteration of the magnetic properties that are not only a function of the size, but also of the chemical composition (i.e. a metal has different magnetic properties with respect to its oxide/s). Moreover, the physiological environment could also promote the leaching of potentially toxic components during in vivo applications of magnetic nanoparticles. Many strategies have been developed to cover naked particles with a shell that improves the chemical stability and biocompatibility as well as protect the magnetic core from oxygen or erosion by acid or base. The materials generally used for the coating can be divided into three classes: (1) polymers and surfactants, (2) inorganic compounds such as silica or carbon, and (3) precious metals. Another possibility to prevent agglomeration and oxidation is to embed or disperse nanoparticles in a dense matrix made by the same kinds of aforementioned materials.

Moreover, without surface modification, biomolecules may not bind to the magnetic core. If the interaction between the active molecules and the magnetic vehicle is weak, the result is an instant release of the drug before it reaches the target site. For this reason, the choice of a opportune coating is the key to controlling the release mechanism of the chemicals. In fact, the materials proposed to surround MNPs have a particular affinity to the functional group that allow them to interact directly with either the biomolecules or molecular linkers that anchor the active system. Polymer coatings are generally rich in hydroxylic groups (–OH), (dextran, PVA…), or amino group ($-NH_2$) (chitosan). Silica coatings can be easily functionalized with alkoxy-silane molecules containing various functional groups (–COOH, $-NH_2$…). Analogous possibilities are offered by precious metal

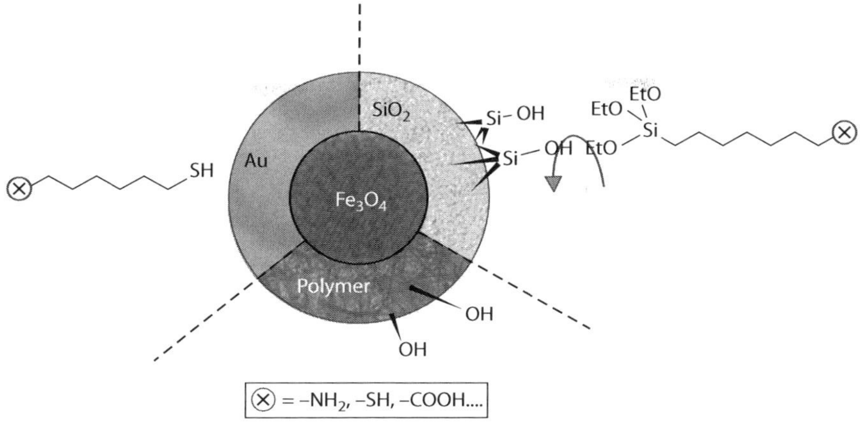


**FIGURE 11.10**
**(See color insert following page 302.)** Illustrative scheme of possible interactions of molecular linkers with materials used for nanoparticle coatings.

coatings that are suitable substrates for the self assembly of long chain thiols. Also in this case, the end functional group can be used to link biomolecules (Figure 11.10).

### 11.5.1 Polymer and Surfactant Coatings

Polymers and surfactants are commonly used to protect nanoparticles from oxidation or agglomeration, during or after synthesis, forming a stable colloidal dispersion. They can be physisorbed or chemically bonded to the magnetic nanoparticles surface, forming single or double layers, which create a steric repulsion in order to stabilize the ferrofluid suspension.

The synthesis of nanoparticles by thermal decomposition (described above) directly produces nanoparticles covered by a monolayer of surfactant molecules (long chain amines, carboxylic acids, diols…) that form stable suspensions in organic solvents for a very long time, and protect the magnetic core from oxidation. Moreover, this kind of coating can be easily modified by monolayer exchange reactions with other molecules containing different functional groups (Altavilla et al. 2005, Bogani et al. 2010). In the case of naked nanoparticles, produced via coprecipitation or microemulsion, natural polymers such us carbohydrates (dextran, chitosan) or synthetic polymers (polyvinyl alcohol (PVA), polyethyleneglycol (PEG)….) have been widely used to improve the biocompatibility and stability as well as promote the interaction and/or release mechanism of biomolecules.

For example, a general strategy for obtaining polysaccharide-coated iron oxide particles is coprecipitation in alkaline Fe(II) and Fe(III) salt solutions in the presence of a colloid stabilizing agents such as dextran (a polysaccharide of D-glucose monomers) or its derivatives (Berry et al. 2003, Lemarchand et al. 2004). Chitosan, a natural linear polysaccharide molecule that contains amino groups, as a protective coating of magnetite nanoparticles, is another interesting alternative (Makha et al. 2006).

The synthesis conditions such as pH and temperature could influence the structure of the polymer coating as well as the magnetic properties of the final system (Hong et al. 2009). The thickness of the polymer coating can have a strong influence on the final diameter of the core–shell particles as well as their persistence time in the blood (Weissleder et al. 1995).

One of the disadvantages of using natural polymers is the lack of mechanical strength as well as the water solubility that can be prevented by cross-linking. Synthetic polymers such as polyethylene glycol (PEG), polyvinyl alcohol (PVA), poly-L-lactic acid (PLA) are interesting alternatives to natural polymer coatings (Zhang et al. 2002, McBain et al. 2008).

### 11.5.2 Silica Coating

There are many advantages to using silica as a protective coating for magnetic nanoparticles. Silica is in fact, an amorphous material with high stability under aqueous conditions (low pH) and high mechanical strength. Moreover, silica is hydrophilic and successive functionalizations can be easily introduced thanks to the presence of silanol groups (–Si–OH) on the surface. One of the most used strategies to coat nanoparticles with a silica shell is the well-known Stöber process, which includes the hydrolysis and polycondensation of tetraethoxysilane (TEOS) under alkaline conditions in ethanol (Stöber et al. 1968). The thickness of the coating can be tuned by varying the concentration of ammonia and TEOS in $H_2O$. A systematic investigation of the reaction parameters, including the type of alcohol, the volume ratio of alcohol to water, the amount of catalyst, as well as the amount of precursor in the formation of silica-coated magnetite particles via sol–gel approach was carried out by Deng et al. (2005). An interesting version of this synthesis is the introduction, during or after the condensation of TEOS, of aminopropyl-triethoxy-silane molecules that, thanks to the silane group, are able to condense on the silica matrix and also give a functional group ($-NH_2$) that easily interacts with the biomolecules.

Alternative strategies to the synthesis of the silica coating on magnetic nanoparticles have also been explored, such as the arc-discharge (Fernández-Pacheco et al. 2006) and microemulsion.

An overview of the recent progress on the silica coating of nanoparticles was recently published (Guerrero-Martínez et al. 2010) but the synthesis of uniform silica shells with controlled thickness on the nanometer scale still remains challenging.

The disadvantages of using silica coatings are also due to the instability of the system under basic conditions, in addition to the presence of pores in the amorphous layer, which could allow oxygen and other species to be diffused and reach the magnetic core.

### 11.5.3 Gold Coating

The use of gold as a protective coating to avoid oxidation is justified by the low reactivity of the precious metal. In addition, gold, due to the surface chemistry and a lack of toxicity, has been proven to be an excellent candidate for conjugation with numerous biomolecules. In fact, it is easily functionalized with thiol (-SH) organic molecules that form in the air and at room temperature form compact and ordered self-assembled monolayers (SAM) on its surface. SAMs are able to modify the surface properties of a materials but also to anchor other systems such as molecules, proteins, DNA, and nanoparticles, thanks to the presence of the exposed end group ($-NH_2$, –COOH, –SH, –Cl, $-CH_3$…). The chemistry of a self-assembled monolayer on gold surfaces has been widely explored and many examples of the linking of biomolecules on these systems have been reported (Luderer and Walschus 2005). The core–shell nanoparticles of magnetite completely covered by gold can be obtained by reverse micelles as constrained reactors for both particle synthesis and gold (or silver) coatings (Mikhaylova et al. 2004, Mandal et al. 2005). Such reverse micelle methods are able to form gold-coated particles but are of low yield and are fairly difficult to reproduce. Another approach is the coprecipitation synthesis of magnetite nanoparticles followed by the reduction of chloroauric acid to form a gold coating (Lo et al. 2007). These aqueous methods are simple, quick, and produce particles that are dispersible in water. The thermal decomposition of iron (III) oleate to magnetite nanoparticles followed by coating via the reduction of gold acetate in the presence of capping agents has also been tested. $Fe_3O_4$ nanoparticles of selected sizes were used as seeding materials for the reduction of gold precursors to produce gold-coated $Fe_3O_4$ nanoparticles (Wang et al. 2005a,b).

A separate class of gold-magnetite composites has also been reported involving the attachment of discrete gold nanoparticles onto magnetite. These composites may be useful in applications such as protein separation, optical imaging or catalysis, where a full coating is not required (Bao et al. 2007). Saturation of the magnetite surface with gold seeds can be used to facilitate the subsequent overlaying of gold, which forms a protective layer that is resistant to chemical attacks and increases particle stability against the aggregation of particles (Goon et al. 2009). In fact, even though gold seems the ideal coating for magnetic nanoparticles, there are many difficulties that need to be overcome. The direct coating of iron oxide with gold is problematic because the dissimilar nature of the two crystalline surface make the coating weak. $TiO_2$ was suggested as a bridging material for magnetite nanoparticles coated by gold (Oliva et al. 2006).

## 11.6 Applications

Some of the most relevant approaches to the synthesis of hybrid magnetic systems for biological applications are reported below.

### 11.6.1 Superparamagnetic and Fluorescent Multisystems for Targeting, Imaging, and Drug Delivery

The combination of magnetic and fluorescent entities is of extraordinary interest because it may give a new "two-in-one" multifunctional nanomaterial with a broad range of potential biomedical applications. The potentialities are enormous: in vitro and in vivo for bioimaging applications such as MRI and fluorescence microscopy and bimodal anticancer therapy, encompassing photodynamic and hyperthermic capabilities. Another exciting application of magnetic-fluorescent nanocomposites is in cell tracking and magnetic separation, which could be easily controlled and monitored using fluorescent microscopy (Coor 2008).

Some recent explicative and significant studies of these systems are briefly reported.

Li et al. proposed magnetic and fluorescent chitosan nanoparticles obtained by microemulsion W/O. Water soluble superparamagnetic $Fe_3O_4$ NPs, fluorescent CdTe quantum dots (QDs), and cefradine, used as a model drug, were simultaneously incorporated in chitosan nanoparticles, with the size, morphology, surface properties, and drug release being tailored. The composite cross-linking nanoparticles were promoted with glutaraldehyde. Superparamagnetic and fluorescent properties can be used, respectively, to address and detect the system. Moreover, control of drug release is possible because the vehicle showed pH-sensitive drug release for a very long time (Li et al. 2007).

An alternative system was proposed by Zhou et al., who synthesized $Fe_3O_4$@poly(caprolactone)-carbazole ($Fe_3O_4$@PCL-CAA). In particular, magnetite nanoparticles, obtained by coprecipitation methods, were surrounded by PCL-CAA *via* surface-initiated ring-opening polymerization. Combined with the advantages of the superparamagnetic core, the biodegradability of the polymer shell and the fluorescence of the carbazol group, this multifunctional hybrid system was proposed as a novel potential carrier for drug delivery. The relate rate of the drug loaded in the shell was also tested in vitro using progesterone as a drug model (Zhou et al. 2009b).

"Inorganic versions" of analogous systems were also proposed.

Silica is in fact a good physical barrier to prevent the direct interaction of the inorganic core with dye molecules. It is widely known that one of the particular difficulties in preparing magnetic-fluorescent nanocomposites is the risk of quenching the fluorophore on the surface of the particle induced by the iron oxide core. The problem can be partially solved by providing the magnetic nanoparticle with a stable shell prior to the introduction of the fluorophore (Corr et al. 2008).

For example, superparamagnetic magnetite nanoparticles, obtained by the thermal decomposition of the iron oleate complex, were encapsulated inside mesostructurated silica spheres that were labeled with fluorescent dye molecules (fluorescein isothiocyanate) and coated with a hydrophilic group to prevent aggregation using 3-(trihydroxysilyl)-propyl-methylphosphonate by Liong et al. The mesoporous silica sphere was modified further with folic acid as targeting ligands because α-folate receptors seems to be regulated in various human cancers. Finally, the pores were loaded with an oncolytic drug (camptothecin or paclitaxel).The system was tested in vitro on human cancer cell lines, PANC-1 and BxPC3, and on foreskin fibroblasts, with the results confirming that the efficacy of the targeted drug delivery using folate modifier nanoparticles is higher than using NPs alone. Thanks to the dual–imaging capability, the system has been detected both by MRI and fluorescent microscopy (Liong et al. 2008).

### 11.6.2 An Interesting Case of Targeting Photodynamic Therapy System

Sun et al. have reported the first in vivo magnetic drug delivery system with chitosan nanoparticles for targeting photodynamic therapy (PDT) monitored by MRI. Briefly, in PDT, drug action is controlled by a light source (laser) transferred to a fiber. The photosensitization in situ of a nontoxic sensitizer produces cytotoxic reactive oxygen species (ROS), causing the tumors cells to die with minimal damage to the surrounding tissue. The system proposed by Sun et al. contains an iron oxide core surrounded by a chitosan shell (MTCNPs) functionalized with the photosensitizer 2,7,12,18-tetramethyl-3,8-di-(1-propoxyethyl)-13,17-bis-(3-hydroxypropyl) porphyrin called PHPP.

The PHPP-MTCNPs (quasi-spherical with an average size of $20 \pm 5$ nm), were used in MRI-monitored targeting PDT with excellent targeting and imaging. The high photodynamic efficacy tested in vitro and in vivo on SW480 carcinoma cells was demonstrated. The level of nanoparticles in the skin and liver was also significantly lower than in the tumor tissue, confirming that hepatotoxicity and photosensitivity can be minimized in conventional PDT protocols (Sun et al. 2009), thanks to the strategic targeting system.

### 11.6.3 Drug Release under the Influence of a Magnetic Field or pH

Gong et al. proposed the synthesis of core–shell microspheres for smart drug delivery using an MFF (microfluidic flow-focusing) approach. The chitosan shells were embedded with magnetite nanoparticles and an aspirin solution, used as a model drug, was encapsulated inside the microspheres. The drug release was controlled by varying the frequency and magnitude of an applied AC magnetic field that produced the deformation of the chitosan shell, acting on the superparamagnetic nanoparticles (Gong et al. 2009).

The controlled drug release of pH-sensitive hybrid magnetite nanoparticles coated with poly((2-dimethylamino)ethyl methacrylate) [PDMAEMA] was proposed by Zhou et al. The method consists of two steps: synthesis of N-bromo isobutyric acid–functionalized $Fe_3O_4$ NPs and the subsequent polymerization of the monomer of PDMAEMA in the presence of DL-ethyl 2-bromobutyrate as a co-initiator. Drug release experiments were carried out using phenolphthalein as a model drug. Under the conditions of pH 3.0 and pH 7.0, the drug release was respectively 26.4% and 56.4%. The results indicated that the drug release rate could be effectively controlled by altering the pH values of the environment (Zhou et al. 2009a).

An interesting approach to provoke the death of cancer cells is the use of FePt nanoparticles as an Fe reservoir for controlled Fe release at a low pH. The release of iron ions within cancer cells catalyzes $H_2O_2$ decomposition into ROS, causing the rapid deterioration of the cellular membrane. The system was recently proposed by Xu et al., who synthetized (via a thermal decomposition route) monodispersed, chemically disordered (face-centered cubic) fcc-FePt nanoparticles (9 nm). They demonstrated that the system releases $Fe^{2+}$ ions under an acid pH (4.8) and remains chemically stable under neutral conditions (pH=7.4). For in vitro experiments, nanoparticles were coated with a phospholipid shell in order to improve their stability in aqueous solutions. The release of $Fe^{2+}$ and the consequent increase in the concentration of ROS species inside the cell culture was detected by fluorescence microscopy. In order to maximize the therapeutic efficacy, fcc-FePt NPs can be, in principle, functionalized with any kind of cancer-targeting agent. In this study, luteinizing hormone-releasing hormone (LHRH) peptide via phospholipid interaction was proposed as a targeting agent to successfully bind fcc-FePt NP preferentially to the human ovarian cancer cell line (A2780) (Xu et al. 2009a,b).

Gao et al. developed FePt@$CoS_2$ yolk shell nanoparticles as potential controlled-release nanodevices. The authors suggested a mechanism to the DNA damage that seems to be due to Pt(II) species as a toxic agent. After cellular uptake through the endocythosis pathway, under the acid environment inside the secondary lysosomes, the FePt core is oxidized and destroyed (probably by $O_2$ due to the presence of oxydase inside the cell) and releases Pt(II). Thanks to the permeability of $CoS_2$ shells, the Pt(II) species diffuses into cytoplasm and enters the nucleus and mitochondria. The damage on the DNA and the apoptosis of HeLa cells (in vitro) produced by FePt@$CoS_2$ yolk shell nanoparticles is similar to the effect produced by cisplatin, a well-known cancer drug.

FePt@$Fe_2O_3$ yolk shell nanoparticles were also tested and offer many advantages: first, the biocompatibility of iron oxide, as well as the possibility of functionalizing the shell with a targeting agent that would reduce the side effects; second, MR relaxation enhancement effects of $Fe_2O_3$ may provide a useful and direct monitoring facility to evaluate the treatment efficacy during the therapy (Gao et al. 2009).

### 11.6.4 Magnetic Targeting of Nucleic Acids

Another interesting aspect of magnetic delivery is nucleic acid delivery (Dobson et al. 2006a, Pan et al. 2008). Recent developments in this intriguing field have been achieved by Namiki et al. They synthesized magnetite nanoparticles coated with oleic acid, surrounded by a cationic lipid shell, which are named LipoMag. The system was functionalized with small interfering RNA (siRNA) designed to silence the expression of the epidermal growth factor in the blood vessels of the tumor. When LipoMag optimized with siRNA was injected into mice with gastric tumors and delivered on the target site under the influence of a magnet, the nucleic acid was able to block the growth of tumor (Namiki et al. 2009, Plank 2009).

Many other applications of magnetic nanoparticles have been proposed and successfully used in vitro and/or in vivo over the last few years such as cell labeling and magnetic separation, hyperthermia, and MRI contrast enhancement (Jing et al. 2006, Shi et al. 2010). Recent progress of hyperthermia is dealt with in Chapter 12 by Jeorg Lehman and Brita Lehman. A general overview of the other applications can be found in the review of the articles by Lu, Berry, and Riehamanne (Berry et al. 2003, Lu et al. 2007, Riehemann et al. 2009).

## 11.7 Conclusions

Research has continued to be carried out since the initial studies that used functionalized magnetic nanoparticles for drug delivery, and though progress in clinical trials has been slow since the first studies in 1996 (Lubbe 1996), the potentiality for this technique remain great.

The advantages due to the shift from micro to nano are enormous. This fascinating field of research requires a multidisciplinary approach to cover the different aspects of this many-sided subject.

Many efforts have been made in the design and optimization of synthetic strategies in order to obtain magnetic nanoparticles that can be used in MRI, magnetic fluid hyperthermia, cell sorting and targeting, sensing, as well as bio-separation.

Thanks to new chemical methodologies that are being continuously updated, scientists are now able to control the size, shape, and composition of nanoparticles as well as their magnetic properties.

Multifunctional systems, that include a magnetic core (for drive guidance), protective coating (to improve biocompatibility), drug molecules, fluorophore (for imaging), targeting agents (to improve the specificity of drug delivery), etc., all concentrated in only a few dozen nanometers, are the new frontiers of nanobiotechnology. A further important aspect is the understanding of physics and engineering required to improve the technology in relation to the power and control of magnetic fields of new medical devices. In fact, magnetic field strength falls off rapidly with distance, and inner sites of the body become more difficult to hit. Finally, but no less important, preclinical and clinical trials are required to test the advanced nanomedical products on animals as well as human patients. All these potential aspects justify the exponential growth in the number of publications on nanoparticles for drug delivery applications over the last 10 years.

## References

Ahn, Y., Choi, E.J., and Kim, E.H. 2003. Superparamagnetic relaxation in cobalt ferrite nanoparticles synthesized from hydroxide carbonate precursors. *Rev. Adv. Mater. Sci.* 5:477–480.

Akimoto, M. and Morimoto, Y. 1983. Use of magnetic emulsion as a novel drug carrier for chemotherapeutic agents. *Biomaterials* 4:49–51.

Akimoto, M., Sugibayashi, K., and Morimoto, Y. 1985. Application of magnetic emulsions for sustained release and targeting of drugs in cancer chemotherapy. *J. Control. Release* 1:205–215.

Altavilla, C., Ciliberto, E., and Gatteschi, D. et al. 2005. A new route to fabricate monolayers of magnetite nanoparticles on silicon. *Adv. Mater.* 8:1084–1087.
Altavilla, C., Ciliberto, E., and Aiello, A. et al. 2007. A shortcut to organize self-assembled monolayers of cobalt ferrite nanoparticles on silicon. *Chem. Mater.*19:5980–5985.
Arruedo, M., Fernàndez-Phacheco, R., and Ibarra, M.R. et al. 2007. Magnetic nanoparticles for drug delivery. *Nanotoday* 2:22–32.
Asken, J.F. 1968. Magnetically controlled intravascular catheter. *Surgery* 64:339–345.
Bao, J., Chen, W., and Liu, T. 2007. Bifunctional Au-$Fe_3O_4$ nanoparticles for protein separation. *J. Am. Chem. Soc.* 129:8698–8699.
Bernard, N.G., Shaw, S.M., and Kessler, W.V. et al. 1980. Distribution and degradation of I-125 albumin microspheres and technetium 99m sulphur colloid. *J. Pharm. Sci.* 15:30–34.
Berry, C.C. 2005. Possible exploitation of magnetic nanoparticle–cell interaction for biomedical applications. *J. Mater. Chem.* 15:543–547.
Berry, C.C., Wells, S., and Charles, S. et al. 2003. Dextran and albumin derivatised iron oxide nanoparticles: Influence on fibroblasts in vitro. *Biomaterials* 24:4551–4557.
Bogani, L., Maurand, R., and Marty, L. et al. 2010. Effect of sequential grafting of magnetic nanoparticles onto metallic and semiconducting carbon-nanotube devices: Towards self-assembled multi-dots. *J. Mater. Chem.* 20:2099–2107.
Bonacchi, D., Caneschi, A., and Gatteschi, D. et al. 2004. Synthesis and characterisation of metal oxides nanoparticles entrapped in cyclodextrin. *J. Phys. Chem. Sol.* 65:719–722.
Burda, C., Chen, X., and Radha, N. et al. 2005. Chemistry and properties of nanocrystals of different shapes. *Chem. Rev.* 105:1025–1102.
Cantillon-Murphy, P., Wald, L.L., and Zahn, M. 2009. Measuring SPIO and Gd contrast agent magnetization using 3 T MRI. *NMR Biomed.* 22:891–897.
Cha, H.G., Kim, Y.H., and Kim, C.W. et al. 2006. Synthesis of α-Fe nanoparticles by solventless thermal decomposition. *J. Nanosci. Nanotech.* 6:3412–3416.
Cha, Y., Chen, L., and Askewa, T. et al. 2007. Manipulation of magnetic carriers for drug delivery using pulsed current high Tc superconductors. *J. Magn. Magn. Mater.* 311:312–317.
Cha, H. G., Lee, D.K., and Kim, Y.H. 2008. Solventless nanoparticles synthesis under low pressure. *Inorg. Chem.* 47:21–127.
Chen, C.Y., Long, Q, Li, X.H., and Xu, J. 2008. Microwave-assisted preparation of magnetic albumin microspheres. *J. Bioact. Compat. Polym.* 23:490–500.
Chinnasamy, C.N., Senoue, M., and Jeyadevan, B. et al. 2003. Synthesis of size-controlled cobalt ferrite particles with high coercivity and squareness ratio. *J. Colloid Interface Sci.* 263:80–83.
Chuo, W.H., Tsai, T.R., Hsu, S.H., and Cham, T.M. 1996a. Preparation and in-vitro evaluation of nifedipin-loaded albumin microspheres cross-linked by different glutaraldehyde concentrations. *Int. J. Pharm.* 144:241–245.
Chuo, W.H., Tsai, T.R., Hsu, S.H., and Cham, T.M. 1996b. Development of nifedipin-loaded albumin microspheres using a statistical factorial design. *Int. J. Pharm.* 134:247–251.
Ciofani, G., Riggio, C., and Raffa, V. et al. 2009. A bi-modal approach against cancer: Magnetic alginate nanoparticles for combined chemotherapy and hyperthermia. *Med. Hypotheses* 73:80–82.
Corot, C. Robert, P., and Idée, J.-M. et al. 2006. Recent advances in iron oxide nanocrystal technology for medical imaging. *Adv. Drug Del. Rev.* 58:1471–1504.
Corr, S.A., Rakovich, Y.P., and Gun'ko, Y.K. 2008. Multifunctional magnetic-fluorescent nanocomposites for biomedical applications. *Nanoscale Res. Lett.* 3:87–104.
Crampton, H.L. and Simanek, E.E. 2007. Dendrimers as drug delivery vehicles: Non-covalent interactions of bioactive compounds with dendrimers. *Polym. Int.* 56:489–496.
Daou, T.J., Pourroy, G., and Bégin-Colin, S. et al. 2006. Hydrothermal synthesis of monodisperse magnetite nanoparticles. *Chem. Mater.* 18:4399–4404.
Deng, Y.-H., Wang, C.-C., and Hu, J.-H. et al. 2005. Investigation of formation of silica-coated magnetite nanoparticles via sol–gel approach. *Colloids and Surfaces A: Physicochem. Eng. Aspects* 262:87–93.

Dobson, J. 2006a. Gene therapy progress and prospects: Magnetic nanoparticle-based gene delivery. *Gene Ther*. 13:283–287.
Dobson, J. 2006b. Magnetic nanoparticles for drug delivery. *Drug Dev. Res*. 67:55–60.
Dumestre, F., Chaudret, B., and Amiens, C. 2004. Superlattices of iron nanocubes synthesized from $Fe[N(SiMe_3)_2]_2$. *Science* 303:281–283.
Durán, J.D.G., Arias, J.L., and Gallardo, D. 2008. Magnetic colloids as drug delivery vehicles. *J. Pharm. Sci*. 97:2948–2983.
Fan, R., Chen, X.H., and Gui, Z. et al. 2001. A new simple hydrothermal preparation of nanocrystalline magnetite $Fe_3O_4$. *Mater. Res. Bull*. 36:497–502.
Fannin, P.C., Scaife, B.K.P., and Charles, S.W. 1987. A study of the complex susceptibility of ferrofluids and rotational Brownian motion. *J. Magn. Magn. Mater*. 65:279–281.
Fernández-Pacheco, R., Arruebo, M., and Marquina, C. et al. 2006. Highly magnetic silica-coated iron nanoparticles prepared by the arc-discharge method. *Nanotechnology*. 17:1188–1192.
Fernández-Pacheco, R., Valdivia, J. G., and Ibarra, M. R. 2009. Magnetic nanoparticles for local drug delivery using magnetic implants in *Micro and Nano Technologies in Bioanalysis, Methods in Molecular Biology*, Vol. 544, chap. 35, James Weifu Lee and Robert S. Foote (Eds.), Humana Press, a part of Springer Science+Business Media, LLC.
Forbes, Z.G., Yellen, B.B., and Halverson, D.S. 2008. Validation of high gradient magnetic field based drug delivery to magnetizable implants under flow. *IEEE Trans. Biomed. Eng*. 55:643–649.
Frullano, L. and Meade, T.J. 2007. Multimodal MRI contrast agents. *J. Biol. Inorg. Chem*. 12:939–949.
Gao, J., Gu, H., and Xu, B. 2009. Multifunctional magnetic nanoparticles: Design, synthesis, and biomedical applications. *Acc. Chem. Res*. 42:1097–1107.
Georgens, C., Weyermann, J., and Zimmer, A. 2005. Recombinant virus like particles as drug delivery system. *Curr. Pharm. Biotech*. 6:49–55.
Ghassabian, S., Ehtezazi, T., and Forutan S.M. et al. 1996. Dexamethasone-loaded magnetic albumin microspheres: Preparation and in vitro release. *Int. J. Pharma*. 130:49–55.
Gong, X., Peng, S., and Wen, W. 2009. Design and fabrication of magnetically functionalized Core/shell microspheres for smart drug delivery. *Adv. Funct. Mater*. 19:292–297.
Goon, I.Y., Lai, L.M.H., and Lim, M. et al. 2009. Fabrication and dispersion of gold-shell-protected magnetite nanoparticles: Systematic control using polyethyleneimine. *Chem. Mater*. 21:673–681.
Greenwood, N.N. and Earnshaw, A. 1984. *Chemistry of the Element*, New York, Pergamon press Ltd.
Guerrero-Martínez, A., Pérez-Juste, J., and Liz-Marzán, L.M. 2010. Recent progress on silica coating of nanoparticles and related nanomaterials. *Adv. Mater*. Early View (Articles online in advance of print) Published Online: January 4, 2010.
Gupta, A.K. and Gupta, M. 2005. Synthesis and surface engineering of iron oxide nanoparticles for biomedical applications. *Biomaterials* 26:3995–4021.
Gupta, P.K. and Hung, C.-T. 1990. Targeted delivery of low dose doxorubicin hydrochloride administered via magnetic albumin microspheres in rats. *J. Microencapsul*. 7:85–94.
Häfeli, U.O., Sweeney, S.M., and Beresford, B.A. et al. 1995. Effective targeting of magnetic radioactive $^{90}$Y-microspheres to tumor cells by an externally applied magnetic field. Preliminary in vitro and in vivo results. *Nucl. Med. Bio*. 22:147–155.
Häfeli, U.O., Pauer, G.J., and Roberts, W.K. et al. 1997. Magnetically targeted microspheres for intracavitary and intraspinal $^{90}$Y radiotherapy. In: *Scientific and Clinical Applications of Magnetic Carriers,* U. Häfeli, W. Schutt, J. Teller, and M. Zborowski (Eds.), pp. 501–516, New York, Plenum Press.
Hanh, N., Quy, O.K., and Thuy, N.P. et al. 2003. Synthesis of cobalt ferrite nanocrystallites by the forced hydrolysis method and investigation of their magnetic properties. *Physica B*. 327:382–384.
Hansen, M.F. and Mørup, S. 1999. Estimation of blocking temperatures from ZFC/FC curves. *J. Magn. Magn. Mater*. 203:214–216.
Hatefi, A. and Amsden, B. 2002. Biodegradable injectable in situ forming drug delivery systems. *J. Control. Release* 80:9–28.
Hong, R.Y., Li, J.H., and Qu, J.M. et al. 2009. Preparation and characterization of magnetite/dextran nanocomposite used as a precursor of magnetic fluid. *Chem. Eng. J*. 150:572–580.

Hou, Y., Yua, J., and Gao, S. 2003. Solvothermal reduction synthesis and characterization of superparamagnetic magnetite nanoparticles. *J. Mater. Chem.* 13:1983–1987.

Hu, S.H., Tsai, C.H., and Liao, C.-F. et al. 2008. Controlled rupture of magnetic polyelectrolyte microcapsules for drug delivery. *Langmuir*. 24:11811–11818.

Hyeon, T., Lee, S.S., and Park, J. et al. 2001. Synthesis of highly crystalline and monodisperse maghemite nanocrystallites without a size-selection process. *J. Am. Chem. Soc.* 123:12798–12801.

Israelachvili, J.N. 1991. *Intermolecular & Surface Forces,* 2nd edn., London, U.K., Academic Press.

Jana, N.R., Chen, Y., and Peng, X. 2004. Size- and shape-controlled magnetic (Cr, Mn, Fe, Co, Ni) oxide nanocrystals via a simple and general approach. *Chem. Mater.* 16:3931–3935.

Jianhua, M., Daoda, C., Yuan, T., and Kaixiong, T. 2000. Toxicity of magnetic albumin microspheres bearing adriamycin. *J. Huazhong Univ. Sci. Technolog. Med. Sci.* 20:261–262.

Jing, Y., Moore, L.R., and Williams, P.S. 2006. Blood progenitor cell separation from clinical leukapheresis product by magnetic nanoparticle binding and magnetophoresis. *Biotechnol. Bioeng.* 96:1139–1154.

Jolivet, J.-P., Tronc, É., and Chanéac, C. 2002. Synthesis of iron oxide-based magnetic nanomaterials and composites. *C. R. Chimie* 5:659–664.

Karunakar, S. and Singh, J. 1994. Preparation, characterization and in-vitro release kinetics of salbutamol sulfate loaded albumin microspheres. *Drug Dev. Ind. Pharm.* 20:1377–1399.

Kato, T., Nemoto, R., and Mori, H. et al. 1983. Magnetic microcapsules for targeted Delivery of anticancer drugs. *Appl. Biochem. Biotech.* 10:199–211.

Khollam, Y.B., Dhage, S.R., and Potdar, H.S. 2002. Microwave hydrothermal preparation of submicron-sized spherical magnetite ($Fe_3O_4$) powders. *Mater. Lett.* 56:571–577.

Kim, D.K., Zhang, Y., and Voit, W. et al. 2001. Superparamagnetic iron oxide nanoparticles for biomedical applications. *Scripta Mater.* 44:1713–1717.

Kim, Y.I., Kim, D., and Lee, C.S. 2003. Synthesis and characterization of $CoFe_2O_4$ magnetic nanoparticles prepared by temperature-controlled co-precipitation method. *Phys. B.* 337:42–51.

Kinugasa, T., Kondo A., and Nishimura, S. et al. 2002. Estimation for size of reverse micelles formed by AOT and SDEHP based on viscosity measurement. *Colloids and Surfaces A: Physicochem. Eng. Aspects* 204:193–199.

Kramer, P.A. 1974. Albumin microspheres as vehicles for achieving specificity in drug delivery. *J. Pharm. Sci.* 63:1646–1647.

Lee, J., Isobe, T., and Senna, M. 1996. Preparation of ultrafine $Fe_3O_4$ particles by precipitation in the presence of PVA at high pH. *J. Colloid Interface Sci.* 177:490–494.

Lemarchand, C., Grefa, R., and Couvreur, P. 2004. Polysaccharide-decorated nanoparticles. *Eur. J. Pharm. Biopharm.* 58:327–341.

Leslie-Pelecky, D.L. and Rieke, R.D. 1996. Magnetic properties of nanostructured materials. *Chem. Mater.*8:1770–1783.

Li, J., Dai, D., and Zhao, B. 2002. Properties of ferrofluid nanoparticles prepared by coprecipitation and acid treatment. *J. Nan. Res.* 4:261–264.

Li, L., Chen, d., and Zhang, Y. et al. 2007. Magnetic and fluorescent multifunctional chitosan nanoparticles as a smart drug delivery system. *Nanothecnology* 18:405102–405108.

Lin, C.C. and Metters, A.T. 2006. Hydrogels in controlled release formulations: Network design and mathematical modelling. *Adv. Drug Del. Rev.* 58:1379–1408.

Liong, M., Lu, J., and Kovochich, M. et al. 2008. Multifunctional inorganic nanoparticles for imaging, targeting, and drug delivery. *ACS Nano* 2:889–896.

Liu, T.-Y., Hu, S.H., and Hu, S.H. et al. 2007. Preparation and characterization of thermal-sensitive ferrofluids for drug delivery application. *J. Magn. Magn. Mater.* 310:2850–2852.

Lo, C.K., Xiao, D., and Choi, M.M.F. 2007. Homocysteine-protected gold-coated magnetic nanoparticles: Synthesis and characterisation. *J. Mater. Chem.* 17:2418–2427.

López-Quintela, M.A., Rivas, J., and Blanco, M.C. et al. 2003. Synthesis of nanoparticles in microemulsions. In: *Nanoscale Materials,* Ed., Liz-Marzán L.M. and Kamat P.V., pp. 135–155, New York, Springer.

Lu, J., Fan, J., and Xu, R. et al. 2003. Synthesis of alkyl sulfonate/alcohol-protected $\gamma$-$Fe_2O_3$ nanocrystals with narrow size distributions. *J. Colloid Interface Sci.* 258:427–431.

Lu, A.-H., Salabas, E.L., and Schüth. 2007. Magnetic nanoparticles: Synthesis, protection, functionalization and application. *Angew. Chem. Int. Ed.* 46:1222–1244.

Lübbe, A.S., Bergemann, C., and Huhnt, W. et al. 1996a. Preclinical experiences with magnetic drug targeting: Tolerance and efficacy. *Cancer Res.* 56:4694–4701.

Lübbe, A.S., Bergemann, C., and Riess, H. et al. 1996b. Clinical experiences with magnetic drug targeting: A phase I study with 4″-epidoxorubicin in 14 patients with advanced solid tumors. *Cancer Res.* 56:4686–4693.

Lübbe, A.S., Alexiou, M.D., and Bergemann, C. 2001. Clinical applications of magnetic drug targeting. *J. Surg. Res.* 95:200–206.

Luderer, F. and Walschus, U.2005. Immobilization of oligonucleotides for biochemical sensing by self-assembled monolayers: Thiol-organic bonding on gold and silanization on silica surfaces. In: *Immobilisation of DNA on Chips* I, pp. 37–56, Ed., Wittmann. C., Berlin Heidelberg, Germany, Springer-Verlag.

Luftensteiner, C.P., Schwedenwien, I., and Paul, B. et al. 1999. Evaluation of mitoxantrone-loaded albumin microspheres following intra-peritoneal administration to rats. *J. Control. Release* 57:35–44.

Mandal, M., Kundu, S., and Ghosh, S.K. et al. 2005. Magnetite nanoparticles with tunable gold or silver shell. *J. Colloid Interface Sci.* 286:187–194.

Martín del Valle, E.M., Galán, M.A., and Carbonell, R.G. 2009. Drug delivery technologies: The way forward in the new decade. *Ind. Eng. Chem. Res.* 48:2475–2486.

Massart, R. 1981. Preparation of aqueous magnetic liquids in alkaline and acidic media. *IEEE Trans. Magn.* 17:1247–1248.

McBain, S.C., Yiu, H.H.P., and Dobson, J. 2008. Magnetic nanoparticles for gene and drug delivery. *Int. J. Nanomed.* 3:169–180.

McNaught, A.D. and Wilkinson, A. 1997. *IUPAC. Compendium of Chemical Terminology,* 2nd edn. (the "Gold Book"), Oxford, U.K., Blackwell Scientific Publications.

Mikhaylova, M., Kim, D.K., and Bobrysheva, N. 2004. Superparamagnetism of magnetite nanoparticles: Dependence on surface modification. *Langmuir.* 20:2472–2477.

Morimoto, Y., Sugibayashi, K., Okumura, M., and Kato, Y. 1980. Biomedical applications of magnetic fluids. Magnetic guidance of ferro-colloid-entrapped albumin microsphere for site specific drug delivery in vivo. *J. Pharmacobiodyn.* 3:264–267.

Morimoto, Y., Okumura, M., Sugibayashi, K., and Kato, Y. 1981. Biomedical applications of magnetic fluids II. 1) preparation and magnetic guidance of magnetic albumin microsphere for site specific drug delivery in vivo. *J. Pharmacobiodyn.* 4:624–631.

Mornet, S., Vasseur, S., and Grasset, F. et al. 2004. Magnetic nanoparticle design for medical diagnosis and therapy. *J. Mater. Chem.* 14:2161–2175.

Müller, B.G., Leuenberger, H., and Kissel, T. 1996. Albumin nanospheres as carriers for passive drug targeting: An optimized manufacturing technique. *Pharm. Res.* 13:32–37.

Namiki, Y., Namiki, T., and Yoshida, H. et al. 2009. A novel magnetic crystal–lipid nanostructure for magnetically guided in vivo gene delivery. *Nat. Nanothecnol.* 4:598–606.

Néel, L. 1948. Propriétées magnétiques des ferrites: Férrimagnétisme et antiferromagnétisme. *Ann. Phys.* 3:137–198.

Neveu, S, Bee, A., and Robineau, M. et al. 2002. Size-selective chemical synthesis of tartrate stabilized cobalt ferrite ionic magnetic fluid. *J. Colloid Interface Sci.* 255:293–298.

O'Connor, C.J., Tang, J., and Zhang, J.H. 2001. Nanostructurated magnetic materials magnetism in *Nanosized Magnetic Materials* vol. 3: Molecules to materials (Magnetism: Molecules to materials) by Joel S. Miller and Marc Drillon, pp. 1–31, Weinheim, Germany, Wiley-VCH.

Oliva, B.L., Pradhan, A., and Caruntu, D. et al. 2006. Formation of gold-coated magnetic nanoparticles using $TiO_2$ as a bridging material. *J. Mater. Res.* 21:1312–1316.

Ovadia, H, Carbone, A.M, and Paterson, P.Y. 1982. Albumin magnetic microspheres: A novel carrier for myelin basic protein. *J. Immunol. Methods* 53:109–122.

Pan, X., Guan, J., and Yoo, J.-W. et al. 2008. Cationic lipid-coated magnetic nanoparticles associated with transferrin for gene delivery. *Inter. J. Pharm.* 358:263–270.
Pankhurst, Q.A., Connolly, J., and Jones, S.K. et al. 2003. Application of magnetic nanoparticles in biomedicine, *J. Phys. D: Appl. Phys.* 36:R167–R181.
Park, J.W. 2002. Liposome-based drug delivery in breast cancer treatment. *Breast Cancer Res.* 4:95–99.
Pauling, L. and Coryell, C.D. 1936. The magnetic properties and structure of hemoglobin, oxyhemoglobin and carbonmonoxyhemoglobin. *Proc. Natl. Acad. Sci.* 22:210–216.
Pileni, M.P. 1993. Reverse micelles as microreactors *J. Phys. Chem.* 97:6961–6973.
Pileni, M.P. 1997. Nanosized particles made in colloidal assemblies. *Langmuir*.13:3266–3276.
Pileni, M.P. 2001. Ferrite magnetic fluids: A new fabrication method and magnetic properties of nanocrystals differing by their size and composition. *Adv. Funct. Mater.* 11:323–336.
Pileni, M.P., 2003. Role of soft colloidal templates in the control of size and shape of inorganic nanocrystals. *Nat. Mater.* 2:145–150.
Pileni, M.P. 2007. Control of the size and shape of inorganic nanocrystals at various scales from nano to macrodomains. *J. Phys. Chem. C.* 111:9019–9038.
Pillai, O. and Panchagnula, R. 2001. Polymers in drug delivery. *Curr. Opin. Chem. Biol.* 5:447–451.
Plank, C. 2009. Nanomedicine: Silence the target. *Nat. Nanotechnol.* 4:544–545.
Rajendran, M., Pullar, R.C., and Bhattacharya A.K. 2001. Magnetic properties of nanocrystalline $CoFe_2O_4$ powders, prepared at room temperature: Variation with crystallite size. *J. Magn. Magn. Mater.* 232:71–83.
Reimer, P., Schuierer G., and Balzer, T. et al. 2005. Application of a superparamagnetic iron oxide (Resovis®) for MR imaging of human cerebral blood. *Magn. Reson. Med.* 34:694–697.
Riehemann, K., Schneider, S.W., and Luger, T.A. et al. 2009. Nanomedicine-challenge and perspectives. *Angew. Chem. Int. Ed.*48:872–897.
Robinson, J. and V.H.L. Lee. 1987. *Controlled Drug Delivery: Fundamentals and Applications*, Drugs and the Pharmaceutical Sciences, Second Edition, New York: Marcel Dekker, Inc.
Roca, A.G., Costo, R., and Rebolledo, A.F. et al. 2009. Progress in the preparation of magnetic. nanoparticles for application in biomedicine. *J. Phys. D: Appl. Phys.* 42:224002–224013.
Samad, A., Sultana, Y, and Aqil, M. 2007. Liposomal drug delivery systems: An update review. *Curr. Drug Deliv.* 4:297–305.
Satish, C.S., Satish, K.P., and Shivakumar, H.G. 2006. Hydrogels as controlled drug delivery systems: Synthesis, crosslinking, water and drug transport mechanism. *Indian J. Pharm. Sci.* 68:133–140.
Schafer, V, Briesen, H, Rubsamen-Waigman, H, Steffan, A.M, Royer, C, and Kreuter, J. 1994. Phagocytosis and degradation of human serum albumin microspheres and nanoparticles in human macrophages. *J. Microencapsul.* 11:261–269.
Seyei, A., Widder, K., and Czerlinski, G. 1978. Magnetic guidance of drug carrying microspheres. *J. Appl. Phys.* 49:3578–3583.
Shi, B., Wang, Y., and Ren, J. 2010. Superparamagnetic aminopropyl-functionalized silica core-shell microspheres as magnetically separable carriers for immobilization of penicillin G acylase. *J. Mol. Catal. B: Enzym.* 63:50–56.
Simmons, B.A., Li, Si., and John, V.T. et al. 2002. Morphology of CdS nanocrystals synthesized in a mixed surfactant system. *Nano. Lett.* 2:263–268.
Stöber, W., Fink, A., and Bohn, E. 1968. Controlled growth of monodisperse silica spheres in the micron size range. *J. Colloid Interface Sci.* 26:62–69.
Sue, K., Hattori, H., and Sato, T. 2009. Super-rapid hydrothermal synthesis of highly crystalline and water-soluble magnetite nanoparticles using a microreactor. *Chem. Lett.* 38:792–793.
Sun, S., Murray, C. B., and Weller, D. 2000. Monodisperse FePt nanoparticles and ferromagnetic FePt nanocrystal superlattices. *Science* 287:1989–1992.
Sun, S., Zeng, H., and Robinson, D.B. 2004. Monodisperse $MFe_2O_4$ (M=Fe,Co,Mn) nanoparticles. *J. Am. Chem. Soc.* 126:273–279.
Sun, C., Lee, J.S.H., and Zhang, M. 2008. Magnetic nanoparticles in MR imaging and drug delivery. *Adv. Drug Del. Rev.* 60:1252–1265.

Sun, Y., Chen, Z.-l., and Yang, X-x. et al. 2009. Magnetic chitosan nanoparticles as drug delivery system for targeting photodynamic therapy. *Nanotechnology* 20:135102–135110.

Udit, A.K., Hackenberger, C.P.R., and O'Reilly, M.K. 2010. Chemically tailored multivalent virus platforms: From drug delivery to catalysis. *ChemBioChem* 10.1002/cbic.201000001 Early View (Articles online in advance of print) Published Online: February 1, 2010.

Wang, L., Wang, L., and Luo, J. et al. 2005a. Monodispersed core?shell $Fe_3O_4$@Au nanoparticles. *J. Phys. Chem. B*. 109:21593–21601.

Wang, X., Zhuang, J., and Peng, Q. et al. 2005b. A general strategy for nanocrystal synthesis. *Nature* 437:121–124.

Weissleder, R., Bogdanov, A., and Neuwelt, E.A. et al. 1995. Long-circulating iron oxides for MR imaging. *Adv. Drug Del. Rev*. 16:321–334.

Widder, K., Flouret G., and Senyei, A. 1978a. Magnetic microspheres: Synthesis of a novel parenteral drug carrier. *J. Pharm. Sci*. 68:79–82.

Widder, K.J., Senyei, A.E., and Scarpelli, D.G. 1978b. Magnetic microspheres: A model system for site specific drug delivery in vivo. *Proc. Soc. Exp. Biol. Med*. 758. 141.

Widder, K.J., Senyei, A.E., Reich, S.D., and Ranney. D.F. 1978c. Magnetically responsive microspheres as a carrier for site-specific delivery of Adriamycin. *Proc. Am. Assoc. Cancer Res*. 19:17.

Widder. K.J., Senyei, A.E., and Ranney, D.F. 1979. Magnetically responsive release of adriamycin by magnetic albumin microspheres and other carriers for the biophysical targeting of antitumor agents. In: *Advances in Pharmacology and Chemotherapy* Vol. 16, ed. S. Garottini, A. Goldin. F. Hawking. I.J. Kopin, and R.J. Schnitzer, pp. 213–271, New York, Academic Press, Inc.

Widder, K.J., Morris, R.M., and Poore, G. et al. 1981.Tumor remission in Yoshida sarcoma-bearing rats by selective targeting of magnetic albumin microspheres containing doxorubicin. *Proc. Natl. Acad. Sci. USA* 78(1):579–581

Wohlfarth, E.P. 1983. Magnetic properties of single domain ferromagnetic particles. *J. Magn. Magn. Mater*. 39:39–44.

Xu, C., Yuang, Z., and Kohler, N. et al. 2009a. FePt nanoparticles as an Fe reservoir for controller Fe release and tumor inhibition. *J. Am. Chem. Soc*.131:15346–15351.

Xu, Z., Shen, C., and Hou, Y. et al. 2009b. Oleylamine as both reducing agent and stabilizer in a facile synthesis of magnetite nanoparticles. *Chem. Mater*. 21:1778–1780.

Yellen, B.B., Forbesb, Z.G., and Halversona, D.S. et al. 2005.Targeted drug delivery to magnetic implants for therapeutic applications. *J. Magn. Magn. Mater*. 293:647–654.

Yodh, S.B., Pierce, N.T., Weggel, R.J., and Montgomery, D.B. 1968. A new magnet system for 'Intravascular Navigation'. *Med. Biol. Eng*. 6:143–147.

Zhang, Y., Kohler, N., and Zhang, M. 2002. Surface modification of superparamagnetic magnetite nanoparticles and their intracellular uptake. *Biomaterials* 23:1553–1561.

Zhou, L., Yuan, J., and Yuan, W. et al. 2009a. Synthesis, characterization, and controllable drug release of pH-sensitive hybrid magnetic nanoparticles. *J. Magn. Magn. Mater*. 321:2799–2804.

Zhou, L., Yuan, J., and Yuan, W. et al. 2009b. Synthesis, characterization, of multifunctional hybrid magnetite nanoparticles with biodegradability, superparamagnetism, and fluorescence. *Mater. Lett*. 63:1567–1570.

# 12

# *Nanoparticle Thermotherapy: A New Approach in Cancer Therapy*

**Joerg Lehmann and Brita Lehmann**

**CONTENTS**



## 12.1 General Introduction to Nanoparticle Thermotherapy

Based on earlier understanding that heat can be used to kill cancer cells, nanoparticle thermotherapy (NPTT) provides a new approach to deliver lethal amounts of heat to cancer cells while keeping surrounding tissues at lower temperatures. NPTT is based on exciting of magnetic nanoparticles, which are placed in or near cancer cells by means of an externally applied alternating magnetic field (AMF). NPTT overcomes problems of earlier thermotherapy, also referred to as hyperthermia (tissue temperature > 40°C–41°C), which was often spatially rather unspecific in its heat delivery and, therefore, limited in the amount of heat deliverable to the cancer by the effects of heat on the surrounding tissues. Thermotherapy is a physical therapy, and is typically combined with chemo- and radiation therapy.

Depending on the specific type of NPTT, two mechanisms to make the heat delivery specific are employed: selective placement of the nanoparticles and selective application of the AMF to excite them for heat delivery. The principle is illustrated in Figure 12.1, which shows the schematic process of NPTT for antibody-based NPTT of breast cancer. Bioprobes containing nanoparticles are injected into the bloodstream and localize at the cancer site through binding of the specific antibody on the bioprobe to the antigen at the tumor cell (Figure 12.1a). (The figure is drawn much out of scale; the size of the bioprobes is in the range of a few nanometers.) Once this process has been completed, an external AMF is applied and selectively heats these cancer cells by exciting the nanoparticles attached to the bioprobes (Figure 12.1b).

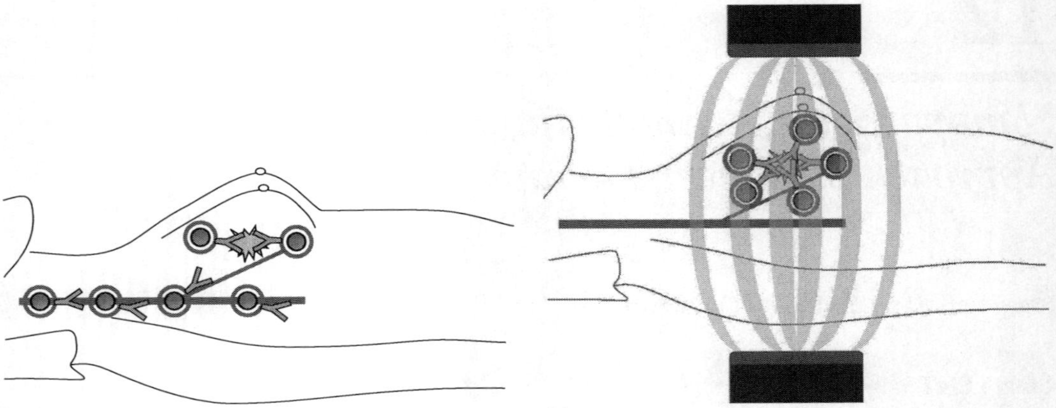

(a) (b)

**FIGURE 12.1**
Principle of nanoparticle thermotherapy (NPTT). (a) Step 1: Bioprobes containing nanoparticles are injected into the bloodstream and localize at the cancer site through binding of the specific antibody on the bioprobe to the antigen at the tumor cell. (b) Step 2: An external alternating magnetic field (AMF) is applied and selectively heats the cancer cells by exciting the nanoparticles attached to the bioprobes. Illustrations are far off scale. The size of the bioprobes is about 20 nm.

Other delivery mechanisms have been developed and are detailed below. Depending on the size of the nanoparticles used, and the thereby determined heating mechanism, rather large magnetic field strengths are needed to reach sufficient temperatures.

## 12.2 Historical Background

Using heat to kill cancer cells has had a long history (Campbell 2007, Gazeau et al. 2008, Hilger et al. 2005, Overgaard 1985, Streffer and van Beuningen 1987, Hildebrandt et al. 2002, Moroz et al. 2002b, Corry and Amour 2005). Temperatures between 42°C and 46°C lead to the inactivation of normal cellular processes in a dose-dependent manner. Cell kill occurs here, particularly in cells that are resistant to radiation—cells in the S phase of the cell cycle and hypoxic cells (Jordan et al. 2006). Such effect made the use of heat attractive as a treatment regimen given in addition to radiation therapy (Corry and Amour 2005). Temperatures above 46°C cause extensive necrosis, and are, therefore, termed thermoablation.

Thermotherapy methods differ in energy sources used for generating heat in tumor tissue, e.g. tubes with hot water, ultrasonic sound, radiofrequency-/microwave-hyperthermia (electromagnetic waves radiated by antennas), and magnetically excited thermoseeds (Wust et al. 2006, Van der Zee 2002).

Methods to heat cancerous tissue in humans include inductive heating and submersion of limbs into water. This form of cancer therapy is generally referred to as hyperthermia. However, with a few notable exceptions (Moroz et al. 2002a,b), the hyperthermia as a treatment for cancer has not found widespread use as a stand-alone therapy or as a combination therapy with radiation or chemotherapy, due to several technical problems.

The major challenge for hyperthermia therapy is to selectively treat especially deep-seated tumors with a more or less homogeneous heat distribution while sparing the surrounding tissue, which is also sensitive to heat.

A second challenge has been to measure the temperature in the target tissue. Without accurate measurement of the temperature, it is difficult to quantify the intervention and to perform meaningful treatment planning. A method of 3D thermal analysis based on computed tomography (CT) of prostates was developed by Johannsen et al. (2007b). They correlated their analysis with invasive and intraluminal temperature measurements. They concluded that a noninvasive thermometry method does not yet fully replace direct temperature measurements and monitoring, as the differences in modeled versus actual temperature (of several degrees Celsius at times) are still significant. Progress in the field of MRI-based thermometry has been made (Wust et al. 2006). But magnetic resonance imaging (MRI) is not suitable for thermotherapy with magnetic nanoparticles because imaging is disturbed by signal loss in the regions of interest (Gneveckow et al. 2004).

Using the AMF component of electromagnetic fields in the radiofrequency spectrum to localize and concentrate ablative heat for cancer treatment can be done by either directly heating the tissue or activating a susceptor material (Ivkov et al. 2005). Using nanoparticles as such, susceptor material is the basis of NPTT. Nanoparticles allow selective heat delivery to cancer cells. The therapy involves two main steps, both of which offer means to make the therapy highly selective. The first step encompasses in bringing nanosized magnetic particles to the cancer cells. In the second step the nanoparticles are excited using an AMF, which causes them to create very localized heating of the cancer cells.

NPTT has, at this point, been successfully used on cell lines, in mouse studies, and in initial human studies (DeNardo et al. 2005, Johannsen et al. 2007a,b, Thiesen and Jordan 2008).

## 12.3 Material and Properties of Nanoparticles

Nanoparticles used for NPTT are magnetic particles. Usually, the particles are in the submicroscopic range of 1–100 nm (Praetorius and Mandal 2007).

Their size and shape determine their response to the AMF, that is, how much of the magnetic energy is transformed into local heating. Several studies have shown that the specific absorption rate of magnetic nanoparticles depends on the diameter of the particle core (Ma et al. 2004). The heat produced by magnetic nanoparticles exposed to AMF can be attributed to magnetic hysteresis losses and Brownian relaxation losses.

In multidomain ferro- or ferrimagnetic materials, heating under exposure to an AMF is caused by hysteresis losses (Andrä and Nowak 1998). Single domain particles of magnetite can also generate heat by relaxation loss. The boundary size to allocate particles between single domain and multidomain is 10 nm of diameter (Motoyama et al. 2008). According to Jordan et al. (2006) hysteresis losses play a role in larger particles, while relaxation losses play a role in particles below 20 nm.

Biological factors also need to be considered when selecting the optimal particle size. On one hand, the smaller the particles, the easier they can be maneuvered and placed near, or potentially even inside cancer cells. However, on the other hand, the larger the particles the more heat production can be expected for the same AMF strength. With the technical challenges in delivering sufficient AMF strengths in clinical settings, finding the right trade off is crucial. Another factor is the nonspecific heating of the tissue surrounding the cancer cells, which should be kept low. This can be done by focusing the field to the target area and by using optimal field strength—nanoparticles combinations.

Most particles used clinically consist of magnetic iron oxides (Jordan et al. 2006, Natarajan et al. 2008a,b,c), because they are less toxic and because of a profound knowledge of their metabolism pathways already exists. There are two materials that are clinically relevant—magnetite ($Fe_3O_4$) and maghemite (gamma-$Fe_2O_3$). The crystal structure of both oxides is based on a cubic dense packing of oxide atoms, but they differ in the distribution of Fe ions in the crystal lattice. The most common ferrite is magnetite with its inverse spinel structure (Jordan et al. 2006).

Since pure iron oxide particles tend to stick together, nanoparticles for treatment are generally coated, e.g., with dextran, starch, or polyethylene glycol. This coat, in addition to preventing cluster formation, also provides a means to attach carrier and connector molecules to the nanoparticle.

The influence of collective behavior on the magnetic and heating properties of iron oxide nanoparticles has been investigated (Dennis et al. 2008).

The control of particle size and shape on the nanoscale level is also still a synthetic challenge. Some work suggests that deviations form regular shape increases the heat production.

While initially nanoparticles already used in other areas of medicine (i.e., for contrast enhancement for MRI) have been utilized for NPTT, research is now ongoing in multidisciplinary teams to determine the theoretically optimal particle size and shapes in addition to the development of manufacturing technology for such particles (Grüttner et al. 2007, Natarajan et al. 2008a).

## 12.4 Applications

### 12.4.1 Step 1: Positioning of Nanoparticles Near Cancer Cells

Since the concept of the therapy is to create heat by exciting nanoparticles, the first goal is to position the nanoparticles close to the target, i.e., the cancer cells.

A mechanical approach to this is to inject a solution with nanoparticles directly into the tumor using a syringe and an appropriate needle (Salloum et al. 2008, van Landeghem et al. 2008). Injection can be also done under stereotactic guidance or during surgery. While striking in its simplicity, there are some limitations to direct needle injection of nanoparticles fluid. First, it is difficult to assure even distribution of the nanoparticles throughout the cancer tissue. In cancer therapy, it is crucial to eliminate any and all cancer cells to defeat the disease. If only a small portion of the cells is not treated sufficiently, the chances are high that the cancer will grow back. Injecting the treatment mediator with a needle will, therefore, have limited success. The distribution problems have been shown in postmortem pathological studies, which found injected particles restricted in distribution to the sites of instillation (van Landeghem et al. 2008).

A related problem is the retention of the nanoparticles in the position near the cancer cells. Since the treatment will take an extended amount of time, probably several minutes to half an hour, and is sometimes given in multiple fractions spread out over several days, it is important that the nanoparticles do not migrate away. Injecting the nanoparticles into the cancerous tissue does not by itself provide mechanisms to prevent their migration, although stable localizations over several days have been reported (Maier-Hauff et al.

2007). An approach taken to keep magnetic nanoparticles in the target area is to use a magnetic field to hold the particles in the area of interest.

Another method researched is ferromagnetic embolization hyperthermia. A feeding artery is used to carry ferromagnetic particles into certain body regions as a pathway to a tumor. It has been proven to be successful in a number of preclinical studies with liver cancers. Hepatic cancers receive their blood supply from the arterial system while the healthy liver parenchyma receives it from the portal venous system. Substances infused into the artery, therefore, target the liver cancer cells while sparing the healthy tissue (Archer and Gray 1989, 1990). This embolization technique has also been studied in the renal artery with selective heating of the embolized kidney (Mitsumori et al. 1994). In another study, arterial infusion of iron oxide particles suspended in lipidol into hepatic carcinomas of the liver showed that an iron concentration of 2–3 mg per gram tumor was necessary to produce a heating rate up to 11.5 times greater than those in adjacent normal hepatic tissue (Moroz et al. 2001). The same researcher demonstrated that for a given iron concentration, larger tumors heat at a greater rate than smaller tumors after arterial embolization. This was explained by the better heat conduction and poorer tissue cooling in the necrotic regions of large carcinoma. In a later study (Moroz et al. 2003), the clearance of ferromagnetic particles from hepatic tissue was studied. Arterial embolization of pig livers was investigated with gamma-$Fe_2O_3$ particles (150 nm) suspended in lipidol and polymer microspheres (32 μm) containing ferromagnetic particles suspended in 1% Tween-water. Both particles were well phagozytozed in the hepatic tissue. Twenty-eight days after embolization, no significant reduction in iron concentration of the liver in either treatment group could be found. The suspension of 32 μm spheres was safe and well tolerated, while the 150 nm particle suspension in lipidol proved to be too vaso-occlusive for use in the liver.

Since the heat production with nanoparticles is very localized, positioning the nanoparticles as close as possible to the cancer cell (or even inside it) increases the cancer cell killing effect of the NPTT treatment. A method to deliver the nanoparticles directly to the cancer cells is to attach them to cancer-cell specific bioprobes. These bioprobes contain an antibody, which is specific for the cancer to be treated. Given systemically via injection into the bloodstream, bioprobes will localize at the cancer cells where the antibody in the bioprobe will bind to the antigen at the surface of the cancer cell.

Since the antibody in the bioprobe is specific to the antigen in the cancer cell membrane (surface protein), bioprobes will generally accumulate on the cancer cell membranes. While antibodies have been used for many years in delivering drugs and also radioactive isotopes to cancer calls, the process is not perfect. Some bioprobes will be located elsewhere in the body, in particular in the liver. Since the AMF application will also be localized, as described in the next section, bioprobes outside the target area will not heat and, therefore, not cause lethal damage.

A crucial component of the bioprobes approach for bringing nanoparticles close to the cancer cells is the chemistry of the binding of the nanoparticle to the antibody. Several groups have done significant work in this area. For example, 111In-chimeric L6 monoclonal antibody-linked iron oxide nanoparticles (DeNardo et al. 2005, Natarajan et al. 2008b), magnetite cationic liposomes (MCLs) (Matsuoka et al. 2004, Motoyama et al. 2008), or luteinizing hormone releasing hormone (LHRH), which has high affinity to breast cancer, can be used for tumor-specific targeting (Jin et al. 2008). One group tested the internalization and biocompatibility of iron oxide nanoparticles coated with differently charged carbohydrates in the human cervical carcinoma cell line (HeLa) in which cationic magnetic nanoparticles showed promising properties for possible in vivo biomedical applications such as cell tracking by MRI and cancer treatment by hyperthermia (effective access into cells, localization in endosomes, easy detection by optical microscopy, lack of cytotoxicity) (Villanueva et al. 2009).

### 12.4.2 Step 2: AMF to Excite Nanoparticles and Create Heat Locally

Once the nanoparticles have been positioned near the cancer cells, an AMF is applied to excite the nanoparticles. Nanoparticle heating is attributed to magnetic hysteresis losses and Brownian relaxation losses, the extent of which are dependent on particle size and domains, as discussed above. Heating nanoparticles will lead to very localized heat production near each of the nanoparticles. Depending on the proximity of a given nanoparticle to a cancer cell and the temperature increase resulting from the heat, the cell will be destroyed by thermoablation or, for lower temperatures, by inactivation of normal cellular processes in a dose-dependent manner.

The heat production around the nanoparticles is different from the heat production through eddy currents, which occurs nonspecifically in all tissue that is exposed to the AMF. The tissue is also susceptible to AMFs since it is a conductive material. The mechanism that dominates this type of heating results from the production of electric eddy currents producing heat that scales with the square of the frequency and the amplitude of the AMF as well as with the square of the radius of the area exposed to the field (Ivkov et al. 2005). Heat production through eddy currents needs to be monitored and limits the field strength that can be applied.

Descriptions of heating mechanisms and their dependence on the particle size can be found in the literature (Jordan et al. 1999, Johannsen et al. 2007a,b).

Depending on the size of the nanoparticles used, and the thereby determined heating mechanism, rather large magnetic field strengths are needed to reach sufficient temperatures.

Figure 12.2 shows a device by Triton BioSystems, Inc. (Chelmsford, MA) that has been used for the application of AMFs to mice and to cell cultures.

The system consists of three main components: (1) a water-cooled induction coil, or inductor; (2) a capacitance network that, when combined with the inductor, forms a resonant circuit; and (3) the power supply (DeNardo et al. 2005, Ivkov et al. 2005, Quang et al. 2007, Lehmann et al. 2008).

The shown induction coil is designed to deliver high magnetic field strength to the hind limbs of the animal, where the tumors were implanted for the studies, and as little as possible to the remainder of its body. Maximum field strength is 103 kA $m^{-1}$ at 153 kHz. Nonmetallic fiber-based temperature probes are used to monitor the temperature at different parts of the animal's body.

A pulse-timer circuit (Giltron, Inc., Medfield, MA) enables pulsed delivery of the AMF. The ratio of AMF on-time versus total time is referred to as duty cycle. A typical treatment time with the system using 20 nm iron oxide particles is 20 min with a duty cycle between 50% and 100% (DeNardo et al. 2005, Quang et al. 2007).

Animal studies with female BALB/c athymic nude mice investigated the feasibility of delivery of AMFs of up to 103 kA $m^{-1}$ for varying duty cycles. They found that no adverse effects were observed for AMF amplitudes of 55 kA $m^{-1}$ even at continuous power application (100% duty) for up to 20 min. High-amplitude AMF (up to 103 kA $m^{-1}$) was well tolerated, provided the duty was adjusted to dissipate heat (Ivkov et al. 2005).

The system can also be utilized for the application of AMF to cell cultures (Lehmann et al. 2008). Here, a different induction coil (4.5 cm internal diameter, 15 cm long) is used,

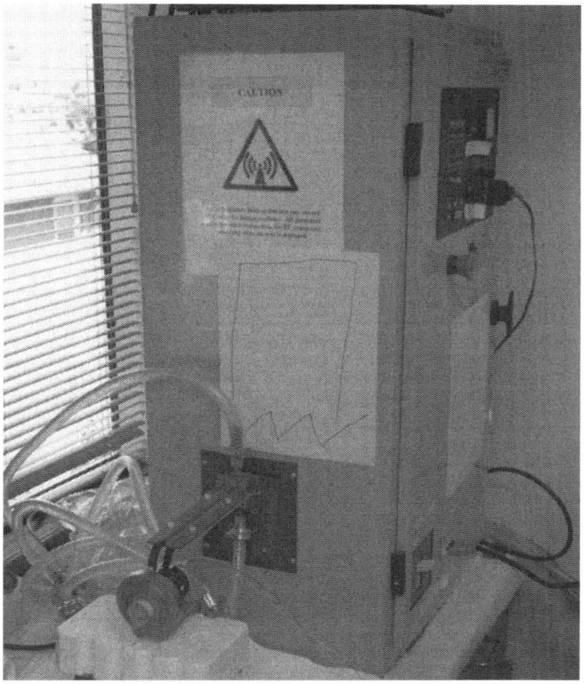


**FIGURE 12.2**
**(See color insert following page 302.)** AMF generator for mouse and cell studies by Triton BioSystems, Inc. (Chelmsford, MA). The system consists of a water-cooled induction coil, the mouse version of which is shown here; a capacitance network; and the power supply, both of which are in the steel closet.

specifically constructed to apply a homogenous field (28 kA $m^{-1}$ at 136 kHz) to an assembly of 16 wells arranged in 2 layers. To illustrate the might of the magnetic field, a small iron file inserted into the field of the coil will immediately glow red.

Salloum et al. (2008) reported on rat studies with the radiofrequency generator Hotshot 2 by Ameritherm Inc., Rochester, NY. The system uses 2-turn water-cooled coil of 20 cm diameter and 7 cm height. The animal was placed on a platform in the center of the coil. The limb to be treated was extended from the body toward the middle of the coil, where the magnetic field is at maximum. Field strengths of up to 3 kA $m^{-1}$ were reported at an operating frequency of 184 kHz.

The first system for human use has been built and successfully used in Berlin, Germany. The MFH® 300 F by MagForce Nanotechnologies AG, Berlin, Germany (Gneveckow et al. 2004, Johannsen et al. 2007a,b) creates a 100 kHz magnetic field with variable field strength of 0–18 kA $m^{-1}$. The system has been applied in clinical studies for several body sites, which are summarized in Jordan et al. (2006). Field strengths of 10–14 kA $m^{-1}$ are technically achievable in the patient and have been tolerated well.

The system features a cylindrical treatment area of 20 cm diameter and an aperture height up to 300 mm. The magnetic field strength is controlled during the treatment to optimize the specific absorption rate (SAR). The team uses nanoparticles in the form of injected magnetic fluid and reports that the achievable energy absorption rates of the magnetic fluid distributed in the tissue are sufficient for either hyperthermia or thermoablation (Gneveckow et al. 2004). Treatment planning is based on imaging prior to the treatment. Therapeutic concentrations of the magnetic fluid are visible in CT images. Lower concentrations (down to 0.01 g/L Fe) can be visualized using MRI. The authors report that the relationship between magnetic field strength and the iron normalized SAR (SARFe) is only slightly dependent on the concentration of the magnetic fluid and can be used for planning the target SAR.

## 12.5 Quantification of Nanoparticle Thermotherapy—Dosimetry

In order to develop a new form of treatment and to safely apply it to patients, quantification of the treatment is important. It is needed for consistency and to best determine "how much" of the therapy to apply to which condition and disease site. It is also crucial when comparing delivery mechanisms and systems. Given the nature of NPTT, temperature measurements seem to be the logical choice.

As reported above, recently, progress has been made in this area of hyperthermia that had traditionally been weak. A method of 3D thermal analysis for human treatment based on CT of the target organ was developed by Johannsen et al. (2007a,b). It correlates the imaging data with invasive and intraluminal temperature measurements to find the temperature distribution.

For very localized heat delivery with NPTT using bioprobes attached to the cancer cell, direct temperature measurements at the site of the cancer cell have not been found feasible at this point. An indirect approach to quantify the heat has been developed by DeNardo et al. (2007). The total heat dose (THD) is a measure for the amount of heat deposited per mass unit of tissue (commonly, per gram of tumor). THD refers to the heat created by a given AMF amplitude over a given time, working on a known amount of bioprobes deposited in the tumor. It can be obtained by using the measured heat response of the bioprobes in vitro and the concentration of bioprobes in vivo or in the cell culture. THD is expressed in units of Joules per gram of tissue (J/g) and provides a valuable tool for the comparison of different heat treatment regimens.

The above-described imaging-based planning reported by Gneveckow et al. (2004) is a very good practical example.

Another indirect measure, which has been used for nontargeted, directly injected nanoparticles, uses the local blood perfusion rate and the amount of nanofluid delivered to the target region to determine the temperature distribution in tissue. The effects of these factors on the heating pattern and temperature elevations in the muscle tissue of rat hind limbs induced by intramuscular injections of magnetic nanoparticles have been evaluated during in vivo experiments (Salloum et al. 2008). The temperature measurements together with the measured blood perfusion rate, ambient air temperature, and limb geometry, were used as inputs into an inverse heat transfer analysis for the evaluation of the SAR by the nanoparticles.

## 12.6 Conclusion and Prospective

NPTT has revived the treatment of cancer with heat. The concept has been proven successful in the laboratory and has been applied in first clinical trials.

Groups from several parts of the world work in this truly interdisciplinary field, involving chemists, biologists, physicians, and physicists. New findings are being published constantly (Dennis et al. 2009, Latorre and Rinaldi 2009, Le Renard et al. 2009, Li et al. 2009, Tseng et al. 2009, Vorotnikova et al. 2006, Wang et al. 2009).

There is evidence that the technology is capable of providing a serious hit to cancer, possibly even complete remission (Dennis et al. 2009); however, to this point, only mice have been cured.

Many aspects of NPTT remain to be developed. These include the selective delivery of the magnetic nanoparticles to the cancer cells as well as the heating mechanisms. Even the name of the treatment keeps changing with new publications.

Clinical studies will need to be performed to find the best regimes for NPTT as either a monotherapy or a combined therapy with radiation therapy or chemotherapy.

## References


Andrä, W. and Nowak, H. 1998. *Magnetism in Medicine: A Handbook*. New York: Wiley-VCH.

Archer, S.G. and Gray, B.N. 1989. Vascularization of small liver metastases. *Br. J. Surg*. 76: 545–548.

Archer, S. and Gray, B. 1990. Intraperitoneal 5-fluorouracil infusion for treatment of both peritoneal and liver micrometastases. *Surgery*. 108: 502–507.

Campbell, R.B. 2007. Battling tumors with magnetic nanotherapeutics and hyperthermia: Turning up the heat. *Nanomedicine*. 2: 649–652.

Corry, P.M. and Amour, E.P. 2005. The heat shock response: Role in radiation biology and cancer therapy. *Int. J. Hypertherm*. 21: 769–778.

DeNardo, S.J., DeNardo, G.L., Miers, L.A. et al. 2005. Development of tumor targeting bioprobes ((111)In-chimeric L6 monoclonal antibody nanoparticles) for alternating magnetic field cancer therapy. *Clin. Cancer Res*. 11: 7087–7092.

DeNardo, S.J., DeNardo, G.L., Natarajan, A. et al. 2007. Thermal dosimetry predictive of efficacy of 111In-ChL6 nanoparticle AMF-induced thermoablative therapy for human breast cancer in mice. *J. Nucl. Med*. 48: 437–444.

Dennis, C.L., Jackson, A.J., Borchers, J.A. et al. 2008. The influence of collective behavior on the magnetic and heating properties of iron oxide nanoparticles. *J. Appl. Phys*. 103: A319–A321.

Dennis, C.L., Jackson, A.J., Borchers, J.A. et al. 2009. Nearly complete regression of tumors via collective behavior of magnetic nanoparticles in hyperthermia. *Nanotechnology*. 20: 395103–395110.

Gazeau, F., Lévy, M., and Wilhelm, C. 2008. Optimizing magnetic nanoparticle design for nanothermotherapy. *Nanomedicine*. 3: 831–844.

Gneveckow, U., Jordan, A., Scholz, R. et al. 2004. Description and characterization of the novel hyperthermia- and thermoablation-system MFH 300F for clinical magnetic fluid hyperthermia. *Med. Phys*. 31: 1444–1451.

Grüttner, C., Müller, K., Teller, J. et al. 2007. Synthesis and antibody conjugation of magnetic nanoparticles with improved specific power absorption rates for alternating magnetic field cancer therapy. *J. Magn. Magn. Mater*. 311: 181–186.

Hildebrandt, B., Wust, P., Ahlers, O. et al. 2002. The cellular and molecular basis of hyperthermia. *Crit. Rev. Oncol. Hematol*. 43: 33.

Hilger, I., Hergt, R., and Kaiser, W.A. 2005. Use of magnetic nanoparticle heating in the treatment of breast cancer. *IEE Proc. Nanobiotechnol*. 152: 33–39.

Ivkov, R., DeNardo, S.J., Daum, W. et al. 2005. Application of high amplitude alternating magnetic fields for heat induction of nanoparticles localized in cancer. *Clin. Cancer Res*. 11: 7093s.

Jin, H., Hong, B., Kakar, S.S., and Kang, K.A. 2008. Tumor-specific nano-entities for optical detection and hyperthermic treatment of breast cancer. *Adv. Exp. Med. Biol*. 614: 275–284.

Johannsen, M., Gneveckow, U., Taymoorian, K. et al. 2007a. Thermal therapy of prostate cancer using magnetic nanoparticles. *Actas. Urol. Esp*. 31: 660–667.

Johannsen, M., Gneveckow, U., Thiesen, B. et al. 2007b. Thermotherapy of prostate cancer using magnetic nanoparticles: Feasibility, imaging, and three-dimensional temperature distribution. *Eur. Urol*. 52(6): 1661–1662.

Jordan, A., Scholz, R., Wust, P. et al. 1999. Magnetic fluid hyperthermia (MFH): Cancer treatment with AC magnetic field induced excitation of biocompatible superparamagnetic nanoparticles. *J. Magn. Magn. Mater*. 201: 413–419.

Jordan, A., Maier-Hauff, K., Wust, K. et al. 2006. Nanoparticles for thermotherapy. In: *Nanomaterials for Cancer Therapy*, ed. Challa S.S.R. Kumar. Weinheim, Germany: Wiley-VCH.

Latorre, M. and Rinaldi, C. 2009. Applications of magnetic nanoparticles in medicine: Magnetic fluid hyperthermia. *P. R. Health Sci. J*. 28: 227–238.

Le Renard, P.E., Buchegger, F., Petri-Fink, A. et al. 2009.Local moderate magnetically induced hyperthermia using an implant formed in situ in a mouse tumor model. *Int. J. Hypertherm*. 25(3): 229–239.

Lehmann, J., Natarajan, A., DeNardo, G.L. et. al. 2008. Nanoparticle thermotherapy and external beam radiation therapy for human prostate cancer cells. *Cancer Biother. Radio*. 23: 265–271.

Li, F.R.,Yan, W.H., Guo, Y.H. et al. 2009. Preparation of carboplatin-Fe@C-loaded chitosan nanoparticles and study on hyperthermia combined with pharmacotherapy for liver cancer. *Int. J. Hypertherm*. 25: 383–391.

Ma, M., Wu, Y., Zhou, H. et al. 2004. Size dependence of specific power absorption of $Fe_3O_4$ particles in AC magnetic field, *J. Magn. Magn. Mater*. 268: 33–39.

Maier-Hauff, K., Rothe, R., Scholz, R. et al. 2007. Intracranial thermotherapy using magnetic nanoparticles combined with external beam radiotherapy: Results of a feasibility study on patients with glioblastoma multiforme. *J. Neurooncol*. 81: 53–60.

Matsuoka, F., Shinkai, M., Honda, H. et al. 2004. Hyperthermia using magnetite cationic liposomes for hamster osteosarcoma. *Biomagn. Res. Technol*. 2: 3.

Mitsumori, M., Hiraoka, M., Shibata, T. et al. 1994. Development of intra-arterial hyperthermia using a dextran-magnetite complex. *Int J. Hyperthermia*. 10: 785–793.

Moroz, P., Jones, S.K., Winter, J. et al. 2001. Targeting liver tumors with hyperthermia: Ferromagnetic embolization in a rabbit tumor model. *J. Surg. Oncol*. 78: 22–29; discussion 30–21.

Moroz, P., Jones, S.K., and Gray, B.N. 2002a. The effect of tumor size on ferromagnetic embolization hyperthermia in a rabbit tumor model. *Int. J. Hypertherm*. 18: 129–140.

Moroz, P., Jones, S.K., and Gray, B.N. 2002b. Magnetically mediated hyperthermia: Current status and future directions. *Int. J. Hypertherm*. 18: 267.

Moroz, P., Jones, S.K., Metcalc, C. et al. 2003. Hepatic clearance of arterially infused ferromagnetic particles. *Int. J. Hypertherm*. 19: 23–34.

Motoyama, J., Yamashita, N., Morino, T. et al. 2008. Hyperthermic treatment of DMBA-induced rat mammary cancer using magnetic nanoparticles. *Biomagn. Res. Technol*. 25(6): 2.

Natarajan, A., Gruettner, C., Ivkov, R. et al. 2008a. NanoFerrite particle based radioimmunonanoparticles and in vivo pharmacokinetics. *Bioconjugate Chem*. 19: 1211–1218.

Natarajan, A., Xiong, C.-Y., Gruettner, C. et al. 2008b. Development of multivalent radioimmunonanoparticles for cancer imaging and therapy. *Cancer Biother. Radio*. 23: 82–91.

Natarajan, A., Xiong, C.-Y., Gruettner, C. et al. 2008c. Development of 111-In-DOTA-di-scFv-NP (Bioprobes) for cancer therapy. *J. Nucl. Med*. 48: 71 P.

Overgaard, J. 1985. History and heritage: An introduction. In: *Hyperthermia Oncology*, ed. J. Overgaard. London, U.K.: Taylor & Francis.

Praetorius, N.P. and Mandal, T.K. 2007. Engineered nanoparticles in cancer therapy. *Recent Pat. Drug Deliv. Formul*. 1: 37–51.

Quang, T., Lehmann, J., DeNardo, G.L. et al. 2007. Novel immunotargeted nanoparticle-AMF thermotherapy in mice with human breast cancer xenografts. *UC Davis Cancer Center Symposium*, Sacramento, CA.

Salloum, M., Ma, R., and Zhu, L. 2008. An in-vivo experimental study of temperature elevations in animal tissue during magnetic nanoparticle hyperthermia. *Int. J. Hypertherm.* 24: 589–601.
Streffer, C. and van Beuningen, D. 1987. The biological basis for tumor therapy by hyperthermia and radiation. In: *Hyperthermia and the Therapy of Malignant Tumors*, ed. J. Streffer, 24–79. Berlin, Germany: Springer.
Thiesen, B. and Jordan, A. 2008. Clinical applications of magnetic nanoparticles for hyperthermia. *Int. J. Hypertherm.* 24: 467–474.
Tseng, H.Y., Lee, G.B., Lee, C.Y. et al. 2009. Localised heating of tumours utilising injectable magnetic nanoparticles for hyperthermia cancer therapy. *IET Nanobiotechnol.* 3: 46–54.
Van der Zee, J. 2002. Heating the patient: A promising approach? *Ann. Oncol.* 13: 1173–1184.
van Landeghem, F.K., Maier-Hauff, K., Jordan, A. et al. 2008. Post-mortem studies in glioblastoma patients treated with thermotherapy using magnetic nanoparticles. *Biomaterials.* 30: 52–57.
Villanueva, A., Cañete, M., Roca, A.G. et al. 2009. The influence of surface functionalization on the enhanced internalization of magnetic nanoparticles in cancer cells. *Nanotechnology.* 20: 115103.
Vorotnikova, E., Ivkov, R., Foreman, A. et al. 2006. The magnitude and time-dependence of the apoptotic response of normal and malignant cells subjected to ionizing radiation versus hyperthermia. *Int. J. Radiat. Biol.* 82: 549.
Wang, Z.-Y., Song, J., and Zhang, D.S. 2009. Nanosized $As_2O_3/Fe_2O_3$ complexes combined with magnetic fluid hyperthermia selectively target liver cancer cells. *World J. Gastroenterol.* 15: 2995–3002.
Wust, P., Gneveckow, U., Johannsen, M. et al. 2006. Magnetic nanoparticles for interstitial thermotherapy—feasibility, tolerance and achieved temperatures. *Int. J. Hypertherm.* 22: 673–685.

# 13

# *Inorganic Particles against Reactive Oxygen Species for Sun Protective Products*

**Wilson A. Lee and Miriam Raifailovich**

**CONTENTS**



## 13.1 Introduction

Using sunscreen to protect our skin from UV assault is an essential regimen of our daily life. Without sunscreen protection, ultraviolet (UV) radiations can induce dimerization of thymine bases and sometimes breakage of the sugar-phosphate backbone of DNA.

UV light between 280 and 400 nm is known to cause most photodamage to the skin. UV light can be classified into three levels: UVA (320–400 nm) is the longest wavelength component and, consequently, can penetrate into the dermis where melanoma is situated (Hidaka et al. 1997); UVB (280–320 nm) is mostly stopped within the epidermis and causes the inflammation known as "sunburn cell formation"; and UVC (200–280 nm), also called the germicidal rays, is the most energetic and, therefore, can cause the highest amount of damage. Fortunately, it is mostly filtered by the stratospheric ozone layer, and, therefore, is not a factor near earth's level. However, with continual usage of chlorofluoro hydrocarbons (CFC), the depletion of this layer will eventually allow this ray to reach our skin. Nevertheless, continual exposure to the sun will eventually lead to photocarcinogenesis, which involves the suppression of the immune system, as well as photoaging of the skin, and subsequently leads to the development of basal cell carcinoma (Matsumara and Ananthaswamy 2004).

However, finding an ideal sunscreen in the market that can resist photodegradation and avoid possible penetration into the skin is not an easy task. The active ingredients used in sunscreen products are either made of organic, inorganic, or even both materials, which function to mitigate the amount of UV illumination reaching the skin by absorbing or scattering the radiation.

The organic molecules are mostly sophisticated aromatic compounds, which are functionalized to delocalize electrons and absorb radiation in the wavelength range of 280–400 nm. Unfortunately, the UV radiation can also facilitate the decomposition of these molecules, and the subunits are easily absorbed through the skin where they can potentially cause allergic reactions (Perugine et al. 2002). Subsequently, inorganic particulates, such as $TiO_2$ and zinc oxide, were introduced in sunscreen formulations in order to reflect the UV radiation and reduce the amount of the organic molecules required in order to achieve the desired sun protection factors (SPFs) for enhanced protection.

Nevertheless, using these inorganic materials to replace or complement organic materials is not a perfect solution. According to our finding (Lee et al. 2007) and Wu et al. (Wu et al. 1999, 2000), (Horikoshi et al. 2001), (Chen et al. 2004) showed that photodegradation of squarylium cyanline dyes was accelerated when $TiO_2$ nanoparticles were dispersed in the aqueous medium, prior to illumination.

It is well known that when $TiO_2$ is illuminated with UV light, the energy is greater than its band gap, which promotes electrons from the valence to the conduction band. These electrons then migrate quickly to the particle's surface and react with oxygen to form superoxide and hydroxyl radicals. Further evidence from our surface electrophoresis on λ-DNA and Dunford et al. (1997) showed that when plasmid DNA was exposed to simulated sunlight—UVA and UVB rays—in the presence of $TiO_2$ particles, the hydroxyl radicals were instrumental in accelerating the breakage of the chains and reducing supercoiled DNA to linear fragments.

Additional concern with inorganic materials is the penetration of nanosize particles. Using nanosize (<100 nm) inorganic materials, like $TiO_2$ and zinc oxide, are mostly preferred by cosmetic companies because its small sizes can achieve the most desired coverage without showing opacity. However, the health risk implication from the nanoparticles may potentially pose an issue in many different topical products. On the other hand, using sunscreen containing organic UV components instead of using $TiO_2$ or zinc oxide in the formulation would most likely run the risk of getting 20% more UVA exposure, as per environmental working group (EWG) et al. (http://www.ewg.org). Although studies done by the EWG has not shown any unequivocal evidence that either nanosized zinc oxide or $TiO_2$ would penetrate through pig skin, healthy human skin, or the skin of patients with skin disorders (NanoDerm 2007), however, we have conducted an in vitro penetration experiment on the human dermal fibroblasts (Zhi et al. 2009) and shown that the $TiO_2$ nanoparticles could penetrate into the cell and impair cell function by decreasing cell area, cell proliferation, mobility, and ability to contract collagen, eventually leading the cell to rupture.

The overall effect from the UV decomposition of the organic materials, free radical generation after UV exposure of $TiO_2$, and penetration risk factor from the nanoparticles significantly attenuates the protection attributes provided by the product. Even when $TiO_2$ is >100 nm, which is considered safe in respect to skin penetration by most industries, it can still drive various chemical reactions when exposed to UV illumination due to its strong oxidizing and reducing ability. It has been suggested by Cai et al. (1992) that $TiO_2$ particles produce hydroxyl radical and hydrogen peroxide when exposed to UV light. As a result, consumers rely on the belief that sunscreen containing $TiO_2$ is supposed to protect them from UV damages, but, on the contrary, the free radical generated from $TiO_2$ after UV illumination actually oxidizes their skin. Dondi et al. (2006) has further shown the loss of UV protection when sunscreen containing both $TiO_2$ and organic sunscreens, like octyl methoxycinnamate (OMC) or avobenzone, Parsol 1789, after exposure to UV light. As a result, fragmentation of OMC, or avobenzone could potentially react with DNA pyrimidines, if sunscreen agents indeed penetrate the skin (Pflucker et al. 1999).

As a result, new organic materials were introduced. These molecules were designed with a polymer backbone (Parsol SLX, Mexoryl) or were encapsulated in glass (Octinoxate Pearls). This new generation of UV filters has provided with least amount of penetration into the skin than conventional organic materials, but still did not provide with a full warrant (Shaath 2007). Working concurrently to resolve the photocatalytic reactive issue produced by $TiO_2$ or zinc oxide, many emerging industries have tried to reduce or eliminate free radicals generation by coating the surface of $TiO_2$ with different film formers (Shafi et al. 2001) or using sol-gel method to coat Silica particles onto $TiO_2$ (Okada 2008). However, those methods do not perform any quenching activity around the $TiO_2$ or zinc oxide.

We therefore proposed that the photocatalytic activity and penetration into the cells could be nearly prevented simply by blocking the emission of the surface electron and attached anionic polymer chemically onto the surface of the $TiO_2$.

Here, we demonstrate that this could be accomplished by chemical grafting of antioxidant molecules with anionic polymer directly onto the $TiO_2$ particle surface, using sonochemistry. This would minimize free radical formation and create repulsion away from the cells, while still providing protection against UV irradiation. It is known that fully stretched polymer brushes do not attach to surfaces of the cells easily, since the value of the grafting density is nearly the size of the intermolecular spacing. As a result, the particles are more entropically hindered from interdigitating and creating a repulsion effect. Furthermore, we show that grafting an additional hydrophobic polymer coating, can stabilize the antioxidant without increasing the local pH of the solution, thereby, allowing these particles to be further tested in tissue culture.

## 13.2 Materials and Methods

Ultrafine rutile $TiO_2$ (U.S. Cosmetics) nanoparticles were used in the coating process. The average size of the particles was measured by dispersing them in water and depositing a small drop of solution onto a carbon coated TEM grid. The images were captured with a Phillips Transmission Electronic Microscope and are shown in Figure 13.1a and b.

Imaging tool was used for calculating the size distribution and the results are plotted as a histogram in the inset. We can easily distinguish individual particles from the figure even though they are densely aggregated. The average particle size distribution in Figure 13.1a before coating is approximately $30.2 \pm 6.9$ nm and the average particle size distribution in Figure 13.1b after coating is approximately $30.5 \pm 8.8$ nm. Antioxidant is derived from grape seed extracts (oligomeric proanthocyanidins) (Rosch 2004) and anionic polymer-poly (methyl vinyl ether/maleic acid) were solubilized in an equal ratio and then dispersed in a 0.05% ethanol solution with a lightening mixer at 25°C. After the solution became uniform, a new mixture was prepared composed of 30% weight percent of the antioxidant/anionic polymer solution, 22% weight percent deionized water, 43% weight percent $TiO_2$, and 5% triethoxysilylethyl polydimethylsiloxyethayl dimethicone from Shin-Etu Chemical Co., Ltd. The entire slurry was then sonicated for 30 min with amplitude intensity set at 50 at room temperature using an ultrasonic probe (Sonicor Instrument Co.). This last sequence was performed to ensure antioxidant and anionic molecules are chemically bonded onto the surface of the particles to avoid them from dissociating, dissolving, or affecting pH of the solution. In order to remove any excess polymers from the coated particles, the resultant solution, was then centrifuged at 9000 rpm for 15 min. This washing step was repeated three times before drying at 110°C under vacuum for 16–20 h.

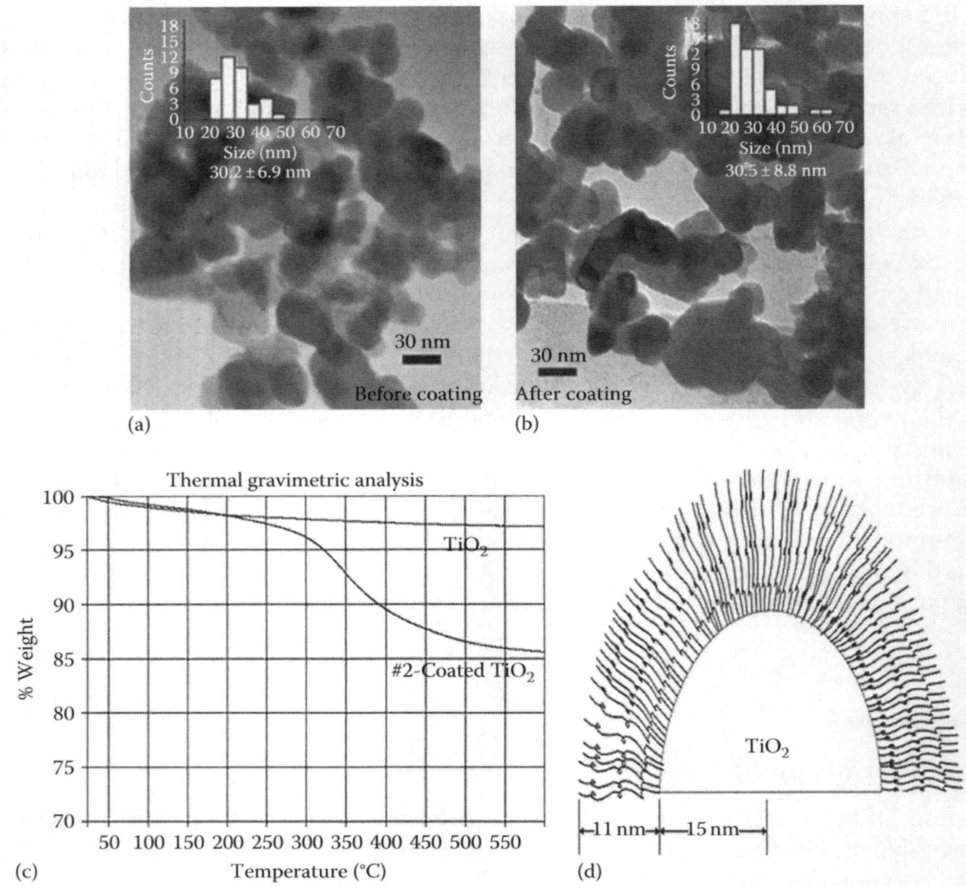


**FIGURE 13.1**
(a, b) show there is no significant difference in size and size distribution after sonication. The average nano $TiO_2$ that we used is $30.2 \pm 6.9$ nm and after surface modified is $30.5 \pm 8.8$ nm. We have derived approximately 12% of polymer coated onto the $TiO_2$ from the TGA analysis in (c). We can then calculate the thickness of the coating to be approximately 11 nm in (d).

Figure 13.1c shows that in order to determine the amount of the polymer coating on the particles, we used thermal gravimetric analysis on both the coated and uncoated particles, where the heating rate was set at 10°C/min. We found that the $TiO_2$ particles are stable for all temperatures studied, as expected. The coating begins to decompose at $T > 300$°C. Complete decomposition occurred at $T = 500$°C, where we found that the total mass fraction of coating was approximately 12%. The density of coated and uncoated $TiO_2$ were measured with a pycnometer and found the value to be 0.986 g/cm$^3$ and 4.23 g/cm$^3$, respectively.

The thickness of the polymer coating could be calculated with the following derived equation: $\alpha = \beta (r_1 + r_s)^3 / r_p^3$, where $\alpha = \rho_1/\rho_2$, is the ratio of the densities of the uncoated $TiO_2$ particles, $\rho_1$, and the coated particles, $\rho_2$ and $\beta = M_1/M_2$, where $\beta$ is the ratio of mass between the uncoated $TiO_2$, $M_1$, and coated $TiO_2$, $M_2$, particles, respectively. Using the densities in Table 13.1, we find $\alpha = 4.32$.

The mass ratio of the functionalized particles can be obtained from the TGA measurements, while the mean mass of the bare particles can be estimated from the mean particle

**TABLE 13.1**

Molecular Weight, pH, and Density of Each Individual Component

| | MW (g/mol) | pH | Density (ρ) ($g/cm^3$) |
|---|---|---|---|
| Hydrophobic polymer | 4000 | — | 0.96 |
| Antioxidant (OPC) | 582.59 | 2.1 | 1.021 |
| Anioinc polymer | 155.1696 | 2.8 | 1.017 |
| $TiO_2$ | 221.39 | — | 4.23 |
| Functional $TiO_2$ | — | — | 0.98 |

radii, $r_1 = 15$ nm and the density to obtain, $\beta = 0.88$. Substituting into the previous relation, we find the radius of the shell $r_s = 11$ nm, see Figure 13.1d. Then calculating the mass of the shell, $m = 6.8\,E - 17$ g, dividing by the mass per chain and the mean area of the bare $TiO_2$ particle core, we obtain a grafting density of $\alpha \approx 0.5$ chains per $nm^2$. This value is nearly the size of the intermolecular spacing; hence, the chains in the coating are fully stretched.

Figure 13.2 shows the proposed mechanism for binding the antioxidant, anionic polymer, and the dimethicone derivative polymers to the $TiO_2$ nanoparticles. Fourier transform infrared (FTIR) analysis was performed on the solution after each step in the

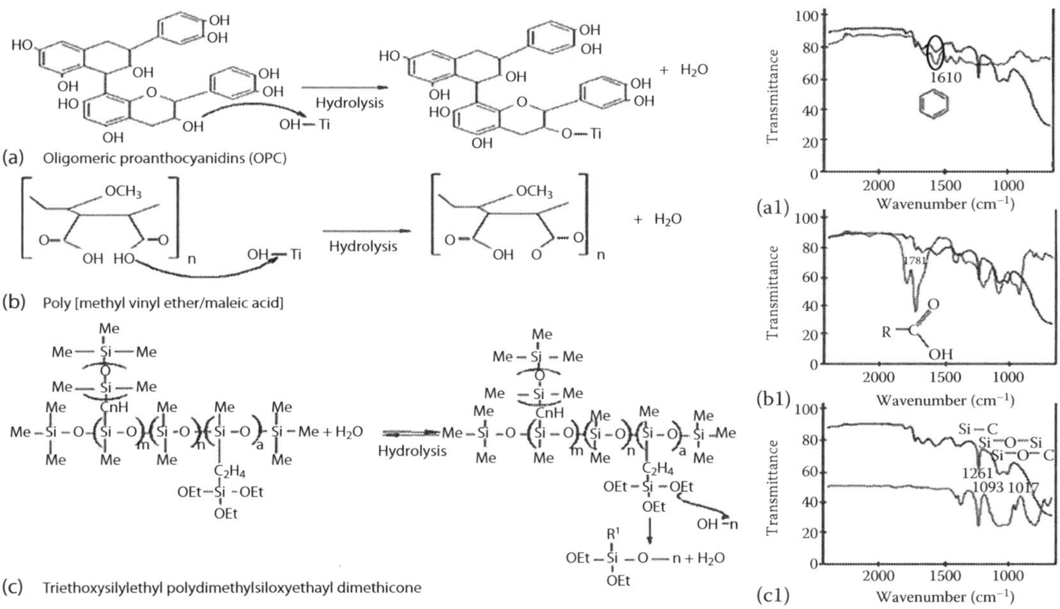


**FIGURE 13.2**
(a) Chemical reaction pathway of the oligomeric proanthocyanidins (OPC) with $TiO_2$ nanoparticles. (b) Chemical reaction pathway of the poly[methyl vinyl ether/maleic acid] with $TiO_2$ nanoparticles. (c) Chemical reaction pathway of the triethoxysilylethyl polydimethylsiloxyethayl dimethicone with $TiO_2$ nanoparticles. (a1) The raw material (light gray curve) vs. coated $TiO_2$ (dark gray curve). The raw material corresponds to pure oligomeric proanthocyanidins (OPC), which is a molecule with six benzene rings. The circle highlights the key component corresponding to the benzene ring absorption peak at $1610\,cm^{-1}$. (b1) the presence of the anionic polymer is confirmed by the presence of the carboxylic acid absorption peak at $1781\,cm^{-1}$. (c1) The two spectra show the presence of silicone derivative peaks at 1261, 1093, and $1017\,cm^{-1}$. This confirms that the silicone derivative has reacted and attached to $TiO_2$.

reaction. The dark gray trace in the corresponding FTIR spectra corresponds to the spectra associated with the nanoparticles after thorough washing of excess solution, while the lower light gray spectra is obtained from the pure antioxidant or polymer molecules. In each case, we see that the appropriate step was successful and the molecules remained bound to the nanoparticles surface even after centrifugation.

In order to verify whether these particles still maintain their activities, we also tested whether the sonication process affects their ability to screen against UV radiation. Equal concentrations of coated and uncoated particles were dispersed in a oil phase emulsion, similar to the base of suntan lotion and spread on a slide at a concentration of 2.0 mg/cm$^2$ (http://www.fda.gov). The material was allowed to dry for 15 min and the Sun Protection Factor (SPF) was calculated from the absorption curve, as measured using a UV spectrophotometer, in the wavelength region of UVB (280–320 nm). Result of SPF = 22 was obtained for both types of particles, showing that no degradation of UV screening occurred upon sonication.

A supplementary experiment was performed to determine whether the coating was effective against reducing the electron emission from the particles upon exposure to UV illumination. Sample of disodium 2′,4′,5′,7′-tetrabromo-4,5,6,7-tetrachlorofluorescein also known as Red Dye 28 was dissolved in water. A quantity of 0.15 g of either coated or uncoated $TiO_2$ particles was added to cuvette containing dye solution and the samples were exposed to 5.42 μw/cm$^2$ in the UV range between 280 and 400 nm.

The cuvettes are shown as an inset to Figure 13.3a through c, after exposure for 27 h. From Figure 13.3b we see that in the cuvette with $TiO_2$ particles most of the dye is removed after irradiation, while the color in the cuvette (Figure 13.3c) containing the coated $TiO_2$ particles, is identical to the color of the unexposed control sample. Since the quenching of the dye florescence is known to result from the electron emission from the $TiO_2$ particles surface, these results indicate that even though the coating does not effect the SPF value, or the UV absorption, it is very effective in preventing the emission of the electrons in the solution surrounding the particles and producing free radicals.

A more dramatic magnitude of the free radical formation has been shown by Serpone et al. (2006) to result in damage to DNA. A solution containing λ-phage DNA (48,502 bp) in a concentration of 50 μg/mL in 1 X TE buffers, and added 2 mg/mL of either nano $TiO_2$ (rutile) or surface modified nano $TiO_2$ was prepared. Samples were placed 3 cm below UVA, UVB, and UVC light sources. The exposure times ranged from 1 to 4 h for different wavelengths. The gel electrophoresis was prepared with 0.8% (w/v) Agarose in 1 × TAE buffer and 5 V/cm of electric field was applied for 30 min. The results are shown in Figure 13.3. The control run on the far left column containing 1 kb ladder, which shows normal separation of digested DNA fragments. The adjacent column shows λ-DNA, which was not exposed. All the intensity remains in the input because λ-DNA is too large to elute through the gel. Exposure of the λ-DNA to UVA for 4 h does not seem to affect the intensity of the signal. A significant reduction is observed after exposure for 4 h in the presence of $TiO_2$ uncoated particles, followed by a diffuse tail. This indicates that the DNA was broken forming short fragments, which were eluted rapidly in the channel. No effect is observed in the input containing the coated $TiO_2$ particles. Exposure to UVB radiation for 4 h produced significant breakage in the column containing DNA and uncoated $TiO_2$ particles; no DNA remains. Similarly, exposure to UVC for 1 h dramatically destroys DNA with and without the presence of $TiO_2$ particles. On the other hand, it is surprising that the intensity of signal from the DNA remains nearly unchanged in the column where the coated $TiO_2$ particles were added prior to irradiation.

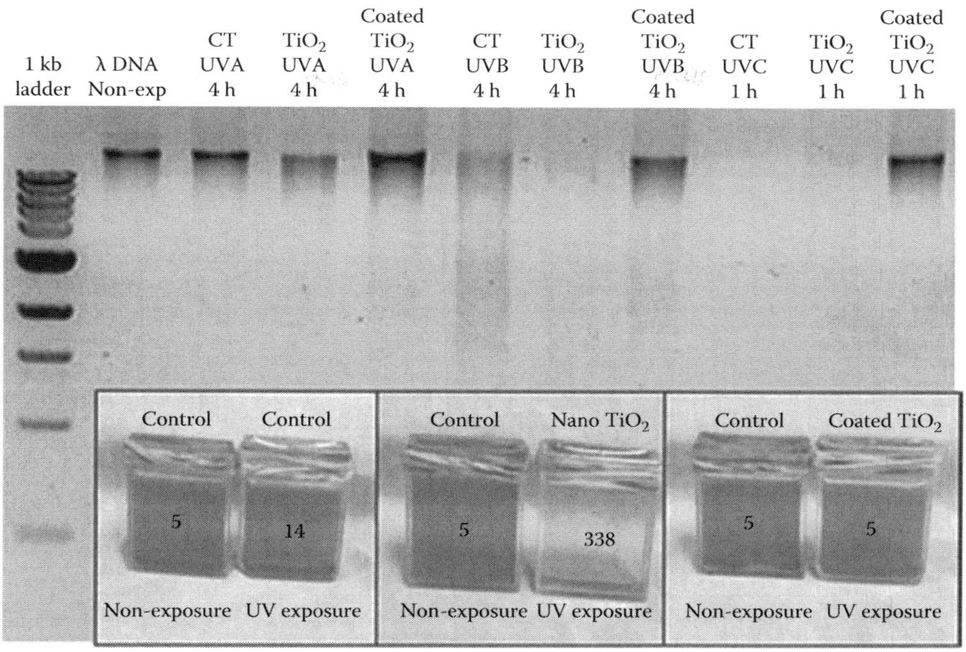


**FIGURE 13.3**
**(See color insert following page 302.)** Lambda-DNA gel electrophoresis for the control (no UV exposure), after exposure to UVA or UVB or UVC with $TiO_2$ or with coated $TiO_2$. The results show that coated $TiO_2$ #2 prevents λ-DNA from catalysis under 4h of UVA, 4h of UVB, and 1h of UVC illumination. (a–c) Show solution with Red 28, which is an assay for photodegradation. The absorbance after 27h is written on each cuvette (absorbance is increasing as photodegradation is increasing). (a) Spectrophotometer assay exhibits slight photodegradation of the Red 28 ($1.2 \times 10^{-4}$ M) with UV illumination for the control. (b) Spectrophotometer assay exhibits drastic photodegradation in presence of $TiO_2$ nanoparticles after UV illumination. (c) Same conditions but now with the coated $TiO_2$ nanoparticles show no changes in photodegradation after UV illumination.

In order to further confirm that the absence of intensity in the UVC column is the cause of chain scission, we also confirmed these results by performing surface electrophoresis. This is a new technique that has been described previously where a droplet of DNA is deposited upon a silicon wafer and the migration time of individual chains is measured at a fixed distance from the injection point. Since this technique does not use a sieving medium, it has the advantage that it could detect simultaneously DNA chains that vary by more than six orders of magnitude in the number of base pairs. The results of the measurements are similar to the samples illuminated with UVC, as shown in Figure 13.4 a through d.

A single peak is eluted in the control sample, which is not exposed to UV radiation. The peak position is similar to that reported in the literature, corresponding to λ-DNA (Pernodet et al. 2000). Exposure with and without uncoated $TiO_2$ nanoparticles results in a complex spectrum with multiple peaks eluting faster than the central λ peak, which correspond to short broken fragments. On the other hand, a large single peak, arriving at the same time as the one in the unexposed control sample, is observed for the DNA where the coated $TiO_2$ particles were added prior to illumination, confirming the electrophoresis results, which do not show breakage of the λ-DNA. In addition, the fact that the mobility

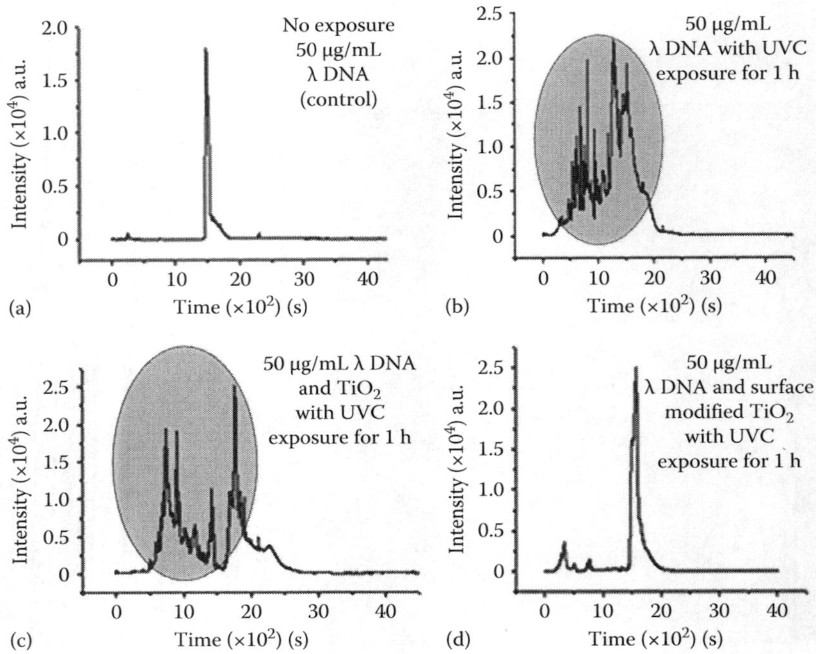


**FIGURE 13.4**
(a) Surface electrophoresis of λ-DNA for the control. (b) Surface DNA electrophoresis of λ-DNA after exposure to UVC for 1 h. Results exhibit DNA breakage with many fragments. (c) Surface DNA electrophoresis of λ-DNA with $TiO_2$ nanoparticles after exposure to UVC for 1 h. In this case also, we can observe DNA breakage with many fragments. (d) Surface DNA electrophoresis of λ-DNA with coated $TiO_2$ nanoparticles after 1 h of UVC exposure. Coated $TiO_2$ nanoparticles protected λ-DNA from breaking.

of the chains is unaltered also indicates that no detectable hydrolysis of the chains occurs with the coated particles. The mobility of the DNA chains on the surface is not only a function of the chain length, but also the interaction of the chains with the substrate and the chain rigidity. Hence, if the DNA became hydrolyzed, as a result of the irradiation, even in the absence of chain scission, the surface mobility and interactions would have been altered.

To further demonstrate the impact of $TiO_2$ nanoparticles on human cells, we incubated human dermal fibroblasts with the rutile particles at two different concentrations, 0.4 mg/mL and 0.8 mg/mL. The cell actin is stained (green), and the nuclei (red) are shown in Figure 13.5b through d.

The control cells, in Figure 13.5b, are healthy and well-spread on the surface, while the cells incubated with 0.4 mg of $TiO_2$ in Figure 13.5c and 0.8 mg of $TiO_2$ in Figure 13.5d are stretched and detached from the surface. In Figure 13.6a, the graph has shown roughly 60% cell reduction at 0.4% mg/mL and 85% cell reduction at 0.8 mg/mL.

In order to demonstrate the effectiveness of preventing penetration in the human dermal fibroblast, we have repeated the same experiment as above with our coating technology (Lee 2007).

In Figure 13.6c and d, the confocal images are shown indistinguishable from the control in Figure 13.6b. In Figure 13.6a, cell numbers after 11 days of incubation are still comparable to the control.

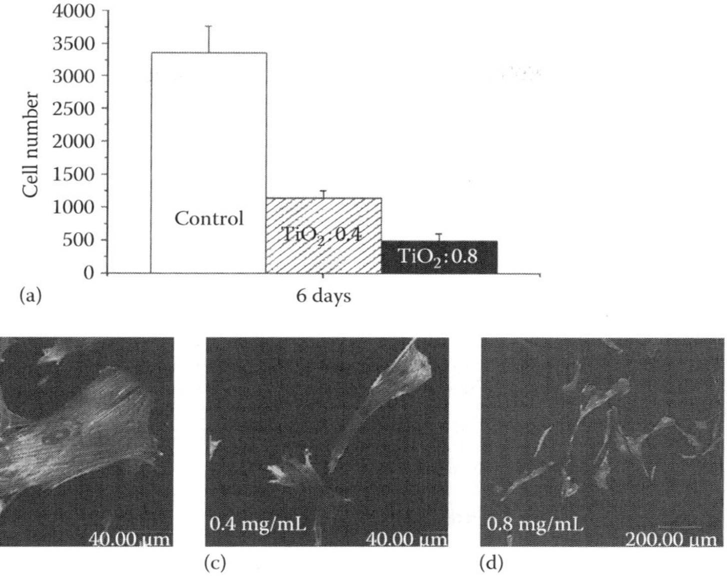


**FIGURE 13.5**
**(See color insert following page 302.)** (a) Cell number as a function of $TiO_2$ concentration. (b) Confocal image of human dermal fibroblasts incubated for 6 days. (c) Confocal image of human dermal fibroblasts incubated with 0.4 mg/mL of rutile $TiO_2$. (d) Confocal image of human dermal fibroblasts incubated with 0.8 mg/mL of rutile $TiO_2$.

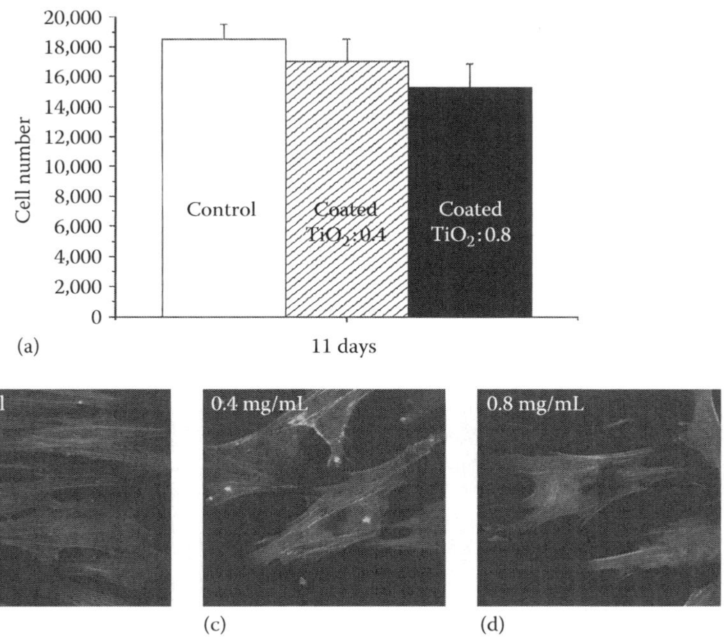


**FIGURE 13.6**
**(See color insert following page 302.)** (a) Cell number as a function of $TiO_2$ concentration. (b) Confocal image of human dermal fibroblasts incubated for 11 days. (c) Confocal image of human dermal fibroblasts incubated with 0.4 mg/mL of coated $TiO_2$. (d) Confocal image of human dermal fibroblasts incubated with 0.8 mg/mL of coated $TiO_2$.

## 13.3 Conclusion

We have developed a new coating which is grafted onto $TiO_2$ particles using sonochemistry. The coating consists of a densely grafted polymer, an anionic polymer, and a free radical scavenger. Addition of the coated particles prevents scission and possible even hydrolysis of DNA after exposure to UVA, UVB, and even UVC radiation.

## References


Cai, R., Kubota, Y., Shuin, T., Sakai, H., Hashimoto, K., and Fujishima, A. 1992. Induction of cytotoxicity by photoexcited $TiO_2$ particles. *Cancer Res.* 52: 2346–2348.

Chen, C., Zhao, W., Lei, P., Zhao, J., and Serpone, N. 2004. Photosensitized degradation of dyes in polyoxometalate solutions versus $TiO_2$ dispersions under visible-light irradiation: Mechanistic implications. *Chem. Eur. J.* 10: 1956–1965.

Dondi, D., Albini, A., and Serpone, N. 2006. Interactions between different solar UVB/UVA filters contained in commercial suncreams and consequent loss of UV protection. *Photochem. Photobiol. Sci.* 5: 835–843.

Dunford, R., Salinaro, A., Cai, L., Serpone, N., Horikoshi, S., and Hidaka, H. 1997. Chemical oxidation and DNA damage catalysed by inorganic sunscreen ingredients. *FEBS Lett.* 418: 87–90.

Hidaka, H., Horikoshi, S., Serpone, N., and Knowland, J. 1997. In vitro photochemical damage to DNA, RNA and their bases by an inorganic sunscreen agent on exposure to UVA and UVB radiation. *J. Photochem. Photobiol. A* 111: 205–213.

Horikoshi, S., Watanabe, N., Mukae, M., Hidaka, H., and Serpone, N. 2001. Mechanistic examination of the titania photocatalyzed oxidation of ethanolamines. *New J. Chem.* 25: 999–1005. http://www.ewg.org/cosmetics/report/sunscreen09/investigation/Nanotechnology-Sunscreens http://www.fda.gov/ohrms/dockets/dailys/00/Sep00/090600/c000573_10_

Lee, W., Pernodet, N., Li, B., Lin, C., Hatchwell, E., and Rafailovich, M. 2007. Multicomponent polymer coating to block photocatalytic activity of $TiO_2$ nanoparticles. *Chem. Comm.* 45: 4815–4817.

Matsumara, Y. and Ananthaswamy, H. 2004. Toxic effects of ultraviolet radiation on the skin. *Toxicol. Appl. Pharmacol.* 195: 298–308.

NanoDerm. Quality of skin as a barrier to ultra-fine particles QLK4-CT-2002-02678. Final Report 2007. Available at: http://www.uni-leipzig.de/~nanoderm/Downloads/Nanoderm_Final_Report.pdf

Okada, H., Ida, J., Yoshikawa, T., Matsuyama, T., and Yamamoto, H. 2008. Use of the sol-gel method for titania coating and the effect of support silica particle size. *Adv. Powder Tech.* 19: 39–48.

Pernodet, N., Samuilov, V., Shin, K., Sokolov, J., Rafailovich, M. H., and Gersappe, D. 2000. DNA electrophoresis on a flat surface. *Phys. Rev. Lett.* 85: 11794–2275.

Perugine, P., Simeoni, S., Scalia, S., Genta, I., Modena, T., and Conti, B. 2002. Effect of nanoparticle encapsulation on the photostability of the sunscreen agent 2-ethylhexyl-pmethoxycinnamate. *Int. J. Pharmac.* 246: 37–45.

Pflucker, F., Hohenberg, H., Holzle, E., Will, T., Pfeiffer, S., and Wepf, R. 1999. The outermost stratum corneum layer is an effective barrier against dermal uptake of topically applied micronized titanium dioxide. *Int. J. Cosmet. Sci.* 21: 399–411.

Rosch, D., Mugge, C., Fogliano, V., and Kroh, L. 2004. Antioxidant oligomeric proanthocyanidins from sea buckthorn (*Hippophae rhamnoides*) Pomace. *J. Agric. Food Chem.* 52: 6712–6718.

Serpone, N., Salinaro, A.,Horikoshi, S., and Hidaka, H. 2006. Beneficial effects of photo-inactive titanium dioxide specimens on plasmid DNA, human cells and yeast cells exposed to UVA/UVB simulated sunlight. *J. Photochem. Photobiol. A.* 179: 200–212.

Shaath, N. 2007. *The Encyclopedia of Ultraviolet Filters*. Carol Stream, IL: Allured Publishing Corp.
Shafi, K., Ulman, A., Yan, X., Yang, N., Himmelhau M., and Grunze, M. 2001. Sonochemical preparation of silane-coated titania particles. *Langmuir* 17: 1726–1730.
Wu, T. X., Lin, T., Zhao, J. C., Hidaka, H., and Serpone, N. 1999. $TiO_2$-assisted photodegradation of dyes. 9. Photooxidation of a squarylium cyanine dye in aqueous dispersions under visible light irradiation. *Environ. Sci. Technol*. 33: 1379–1387.
Wu, T., Liu, G., Zhao, J., Hidaka H., and Serpone, N. 2000. Mechanistic study of the $TiO_2$-assisted photodegradation of squarylium cyanine dye in methanolic suspensions exposed to visible light. *New J. Chem*. 24: 93–98.
Zhi, P., Lee, W., Slutsky, L., Clark, R., Pernodet, N., and Rafailovich, M. 2009. Adverse effects of titanium dioxide nanoparticles on human dermal fibroblasts and how to protect cells. *Small* 5: 511–520.

# 14

# *Innovative Inorganic Nanoparticles with Antibacterial Properties Attached to Textiles by Sonochemistry*

**Nina Perkas, Aharon Gedanken, Eva Wehrschuetz-Sigl, Georg M. Guebitz, Ilana Perelshtein, and Guy Applerot**

CONTENTS



## 14.1 Introduction—The Sonochemical Method

This chapter is devoted to the research that has been done using the sonochemical method for the deposition of metal oxide NPs on textiles. Sonochemistry is the scientific area where chemical reactions occur under ultrasound irradiation. Liquids irradiated with ultrasound produce bubbles. The reaction is dependent on the development of an acoustic bubble in the solution. Ultrasonic waves with the frequency range of 20 kHz–1 MHz are responsible for the process of acoustic cavitation, which means the formation, growth, and explosive collapse of the bubbles. There are a number of theories that explain how 20 kHz ultrasonic radiation can break chemical bonds (Suslick et al., 1986; Doktycz and Suslick, 1990; Mason, 1990). The first question that arises is how such a bubble can be formed, considering the fact that the forces required to separate water molecules to a distance of two van der Waals radii would require a power of $10^5$ W/cm. On the other hand, it is well known that in a sonication bath with a power of 0.3 W/cm, water is readily converted into hydrogen peroxide. Different explanations have been offered, and they are all based on the existence of unseen particles or gas bubbles that decrease the intermolecular forces, enabling the creation of the bubble. The experimental evidence for the importance of unseen particles in sonochemistry is that when the solution undergoes ultrafiltration before the application of ultrasonic power, there is no chemical reaction and chemical bonds are not ruptured. The second stage is the growth of the bubble, which occurs through the diffusion of solvent and/or solute vapors into the volume of the bubble. The third stage is the collapse of the bubble, which occurs when the bubble size reaches its maximum value. From here on, we will adopt the hot spot mechanism, one of the theories that explain why, upon the collapse of a bubble, chemical bonds are broken. This theory claims that very high temperatures (5,000–25,000 K) (Suslick et al., 1986) are obtained upon the collapse of the bubble. Since this collapse occurs in less than a nanosecond, very high cooling rates in excess of $10^{11}$ K/s are obtained. These extreme conditions develop when the bubble's collapse causes the chemical reactions to occur. The high cooling rate prevents the crystallization of the products. This is the reason why amorphous NPs are formed when volatile precursors are used and the gas-phase reaction is predominant. However, from this explanation, the reason for the formation of nanostructured material is not clear. Our explanation for the creation of nanoproducts is that the fast kinetics does not permit the growth of the nuclei, and in each collapsing bubble, a few nucleation centers are formed whose growth is limited by the short collapse. If the precursor is a nonvolatile compound, the reaction occurs in a liquid phase in a 200 nm ring surrounding the collapsing bubble (Doktycz and Suslick, 1990). The products are sometimes nano-amorphous particles and in other cases nanocrystalline, depending on the temperature in the ring region where the reaction takes place. In fact, when the sonochemical reactions were used for the synthesis of inorganic products, nanomaterials were obtained.

In previous reviews by our group we described the development of the sonochemical technique for the fabrication of various kinds of nanomaterials (Gedanken, 2004) and for the doping of NPs into ceramic and polymer bodies (Gedanken, 2007). Other review articles on similar topics have also been published (Suslick and Price, 1999; Vajnhandl and Le Marechal, 2005; Mason, 2007).

In this chapter, the unique properties that make ultrasound radiation an excellent technique for adhering the NPs to a large variety of substrates will be described. We will concentrate on the deposition of inorganic metal nanooxides (ZnO, CuO, MgO) on textile fabrics and explain the mechanism of their antibacterial activity. The chapter will compare the deposition of NPs formed during the sonochemical process and NPs purchased from a commercial source, on textiles. The advantages of sonochemistry as a one-step, environmentally friendly method for the deposition of NPs on different kinds of textiles such as cotton, wool, nylon, polyester, etc., will be demonstrated. The chapter will scan the works performed by different authors using ultrasound irradiation.

## 14.2 Deposition of Nanosized Metal Oxides on Solid Surfaces by the Sonochemical Technique

The increasing interest in the synthesis of different kinds of nanomaterials is caused by their large specific surface area, and their new size-dependent physical and chemical properties in comparison with the bulk structures. The potential application of the NPs can be significantly extended by their deposition on the different types of substrates. The interest in coatings of different types of substrates with metal oxides lies in the possibility to combine the properties of the two (or more) materials involved in the process, namely, the substrate and the coated layer, with emphasis on the fact that one of the materials will determine the surface properties of the composite, while the other can be responsible for other (optical, catalytic, magnetic, antibacterial, etc.) properties of the system (Rao and Cheetham, 2001; Schmidt, 2001; Liz-Marzan and Kamat, 2003). The development of scalable methods that effectively bind NPs to surfaces and provide precise patterns is a key step toward the commercial exploitation of the distinctive properties of nanostructured materials (Klimov et al., 2000; Tricoli et al., 2008).

As mentioned above and discussed previously, many inorganic nanomaterials have been prepared sonochemically. The dynamics of cavity growth and collapse during sonication are strictly dependent on the local environment. Cavity collapse in a homogeneous liquid is very different from cavitation near a liquid–solid interface. Suslick and Price (1999) demonstrated that microjets and shock waves produced by acoustic cavitation are able to drive metal particles together at sufficiently high velocities to induce melting upon collision. Metal particles that were irradiated in hydrocarbon liquids with ultrasound underwent collisions at roughly half the speed of sound, and generated localized effective temperatures between 2600°C and 3400°C at the point of impact of the particles. This approach was developed further in our experiments on the deposition of NPs on different types of solid substrates. Typically, the solid substrate was introduced into the sonication cell containing the precursor's solution leading to the fabrication of NPs under ultrasonic waves. The ultrasonic irradiation passes through the sonication slurry under an inert or oxidizing atmosphere for a specified time. This synthetic route is a single-step effective procedure. The microjets formed after the collapse of the bubble throw the just-formed NPs at the surface of the substrate at such a high speed that they strongly adhere to the surface, either via physical or chemical interactions, depending on the nature of the substrate, i.e., ceramic, polymer, or textile. The excellent adherence of the NPs to the substrate is reflected, e.g., in the lack of leaching of the NPs from the substrate surfaces after many washing cycles. If instead of forming the NPs sonochemically we purchase them and use ultrasonic radiation just for throwing stones at a solid surface, a good adherence is still obtained, but the amount of the deposited material found on the surface is smaller by a factor of 3–4. The mechanism of NPs deposition on the substrate under ultrasound irradiation schematically is illustrated in Figure 14.1. From this scheme, one can suppose that the NPs formed in the precursor solution under ultrasound irradiation are thrown immediately at the solid surface by the microjets. When the suspension of the preliminary synthesized NPs is irradiated, not all the particles are pushed by shock waves and a large part of them remains in the slurry. Thus, in the last case, the coating is less effective.

When the NPs hit the solid surface, sintering of the particles and/or interparticle collision between the solid surface and inorganic NPs changes the surface morphology and reactivity, resulting finally in the coating of these particles. Using the sonochemical approach, we studied the coating of the metal oxide NPs on various kinds of surfaces. Ultrasonic irradiation of a decalin solution of iron pentacarbonyl in the presence of the alumina submicrospheres resulted in the coating of highly dispersed iron oxides on alumina (Figure 14.2) (Zhong et al., 1999). The strong interaction between adhered iron particles and an alumina substrate can hinder the transformation of $\gamma$-$Fe_2O_3$ to $\alpha$-$Fe_2O_3$, even at temperatures higher than 700°C. Conversely, the presence of $\alpha$-$Fe_2O_3$ can induce the formation of $\alpha$-$Al_2O_3$ at high temperatures.

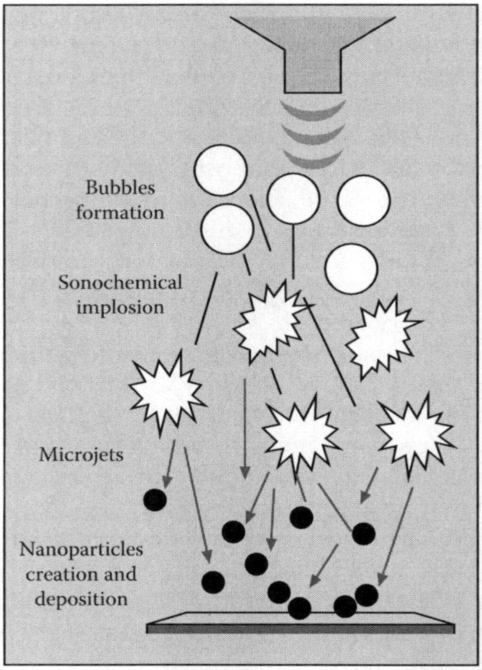


**FIGURE 14.1**
Supposed mechanism of NPs adhesion to the substrate.

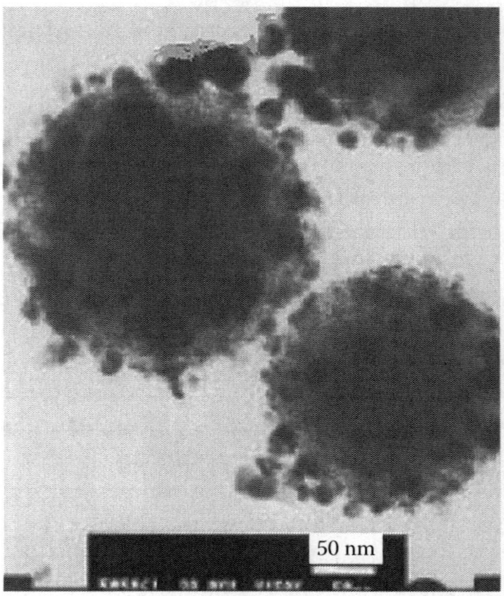


**FIGURE 14.2**
TEM micrographs of $\gamma$-$Fe_2O_3$ on alumina microspheres.

The 20–40 nm rare earth oxides, $Eu_2O_3$, $Tb_2O_3$, were deposited on a number of ceramic materials such as silica alumina, titania, and yttria-stabilized zirconia (Patra et al., 1999; Gedanken et al., 2000; Pol et al., 2002a,b). These materials demonstrated significant photoluminescence properties and may have a wide application in optical and luminescent devices and photocatalysis. A similar sonochemical approach was later used for the synthesis of another photocatalyst, $TiO_2/SiO_2$ (Guo et al., 2006). $TiO_2$ clusters in the size range of about 4 nm were deposited on silica particles by the high-intensity ultrasound radiation of a suspension containing $TiCl_4$ as a precursor and silica particles in water. It was demonstrated that the $TiO_2$/silica photocatalysts exhibited a higher reactivity than bulk $TiO_2$ (P25, Degussa) in the photo-oxidation of methyl orange. These few examples, and the many others that are not related to textiles, show that sonochemistry as a deposition method can avoid the agglomeration of the newly formed NPs. Moreover, the fear that NPs will disperse in air is also inhibited. The strong adherence to the substrate prevents their scattering to air. The mode of distribution of the NPs on the substrate is always that of nanoclustering, and not of nanolayering. Namely, we coat well-dispersed individual NPs and not a full continuous layer of the deposited material. A special feature of the sonochemical method for the coating of various substrates is the opportunity to regulate the particle size and the formation of the thickness of the coated layer. The size of the particles formed in the sonochemical reaction depends very much on the reaction conditions. When the sonoreactor is chilled to a low-ambient temperature and a low concentration of the precursor solution is used, 5–10 nm inorganic compounds could be obtained and deposited on the surface of the substrate by the sonochemical method. All the above-mentioned properties make the sonochemical method a very perspective technique that provides a homogeneous coating and strong adhesion of the nanostructure materials to the surface of the supporting substrate.

The sonochemical method was developed to deposit NPs on flat and curved surfaces of ceramics (Landau et al., 2001; Perkas et al., 2001; Pol et al., 2002a,b, 2006), polymers (Kotlyar et al., 2007, 2008), metals (Perkas et al., 2009), and paper. It is worth mentioning that polymer and glass surfaces could realize antibacterial properties by the deposition of nanosilver (Perkas et al., 2007, 2008a,b) or ZnO nanooxide (Applerot et al., 2009a,b). The structure and morphology of the ZnO NPs deposited sonochemically on the glass slides were studied as a function of the synthesis time. The adjustment of processing time allowed the attainment of ZnO films with various thicknesses. The ZnO nanocrystals were obtained with a mean diameter of 300 nm. The high temperature and the speed of the NPs thrown at the solid surface by sonochemical microjets cause their strong and stable attachment to the glass, with a unique sphere structure of the ZnO. The antibacterial activities of the ZnO-glass composites were tested against *Escherichia coli* (Gram negative) and *Staphylococcus aureus* (Gram positive) cultures. A significant bactericidal effect, even in a 0.13 wt% ZnO-coated glass, was demonstrated.

This chapter will cover the research carried out on textiles that were made antibacterial by the deposition of NPs on their surfaces and report on the ultrasound-assisted deposition of antibacterial NPs on textiles.

## 14.3 Functionalization of Textiles with Nanoparticles

Nowadays, there is a growing need for high-quality textiles with antibacterial properties for hygienic clothing, active wear, and wound healing. It is recognized that neither synthetic nor natural fibers are resistant to bacteria and pathogenic fungi (Purwar and Joshi, 2004). The control of microorganisms extends from medical institutions to ordinary households. Consumer demands caused a significant growth in the production of antibacterial textiles. An explosive growth is also expected in wound-care production. The wound-care market of the US healthcare system was in excess of $7 billion in 2007 (Zalesky, 2008). Worldwide industry reports estimate the wound-care market to exceed $11.8 billion by 2009 and project a yearly growth for all products (devices for wound closure such as sutures and staples, dressings, etc.) in excess of 7%. European markets have accounted for about half of the expenditure (Petrulyte, 2008). The high demands encourage intensive research and the development of new methods for the antimicrobial treatment of textile fabrics and fibers. Recent achievements in the field of antibacterial textiles were briefly described in a review (Gao and Cranston, 2008). According to this data, metal and metal salts are one of the major classes of antibacterial agents, as are quaternary ammonium compounds, triclosan, chitosan, chlorine-containing N-halamine compounds, etc.

The fast development of nanotechnology opens up various possibilities for the synthesis of new materials whose properties are influenced very much by their nanosized structure. Nanometal and nanometal oxides have a larger surface than the conventional powders, and can be finely spread on the surface of fibers and fabrics. These unique properties have found wide application in the textile industry, namely, in the antibacterial treatment of textiles (Qian and Hinestroza, 2004; Craighead and Leong, 2006; Wong et al., 2006). The market for textiles using nanotechnologies is predicted to climb dramatically from $13.6 billion in 2007 to $115 billion by 2012 (Coyle et al., 2007).

Nanosilver is one of the most widely used antibacterial agents in general textiles and in wound dressing (Duran et al., 2007; Gorenšek and Recelj, 2007; Wang et al., 2007). The antibacterial properties of silver have been known and used for centuries (Searle, 1919). A unique and available source of silver has long been mineral salts. A new way for the delivery of silver into the bacterial-killing medium is the formation of organic–inorganic nanocomposites combining the properties of textile substrates with antibacterial activity (Shukla et al., 2001; Yeo et al., 2003; Dubas et al., 2006). To achieve the optimum antibacterial effect of nanocomposite fibers, a high concentration of silver ions must be available in the solution. Despite the small number of silver ions released from metallic silver nanocrystals, about 30 times less than that from silver complexes (e.g., silver sulfadiazine), a stronger antimicrobial property has been observed with nanocrystals (Richard et al., 1994).

Different methods have been used for the deposition of silver NPs on fabrics. For example, a poly(ethylene terephthlate) fabric (meadox double velour) was coated with metallic silver using a patented ion beam–assisted deposition process developed by the Spire Corporation (Bedford, MA) (Klueh et al., 2000). Antimicrobial fibers were produced by the implementation of nanoscaled silver particles into a solution of cellulose and N-methylmorpholine-N-oxide (Wendler et al., 2007). Other methods were constant pressure padding (Lee et al., 2003), impregnation in the colloid silver solution (Duran et al., 2007), immersion of the fabric in the silver precursor solution in ethanol or propanol following the boiling procedure for the reduction of silver ions (Yuranova et al., 2006), the magnetron sputter technique (Scholz et al., 2005), etc. Some of the methods are based on reactions in the liquid medium and require surfactants, reducing agents, or templates for the synthesis of silver NPs, resulting in the presence of toxic impurities in the final products. This method has some disadvantages with regard to the environment. Recently, we reported on the simple and effective ultrasound-assisted deposition of silver NPs on wool fibers (~5–10 nm in size) (Hadad et al., 2007) and on different kinds of fabrics (nylon, polyester, and cotton) of about 80 nm in size (Perelshtein et al., 2008). The excellent antibacterial activity of these Ag–fabric composites against *Escherichia coli* (Gram-negative) and *Staphylococcus aureus* (Gram-positive) cultures was demonstrated.

Among the heavy metals, silver is considered as nontoxic, in spite of claims that it kills many different disease organisms. Its low toxicity to mammalian cells was demonstrated (Joeger et al., 2001). Detailed investigations showed that nanosilver is skin friendly and does not cause skin irritation (Lee et al., 2005); although some work warned of the danger of nanosilver if its particles are smaller than 55 nm (Carlson, 2006). Thus, once reaching nanoscale, these are questions that need to be imperatively answered before people rush to participate in the nanosilver boom (Chen and Schluesener, 2008). Lately, more attention is being paid to the application of inorganic metal oxides for the antibacterial finishing of textiles. This report will not cover the biocidal properties of silver-coated fabrics because (1) this subject has already been exhausted by many papers and reviews, (2) the FDA has started recently to limit the use of silver; and (3) in contrast to silver, research on metal oxides as antibacterial agents is novel. This is the reason why this chapter will concentrate on ZnO, MgO, and CuO NPs coated on various textiles.

Some metal oxides like $TiO_2$, ZnO, MgO, and CuO, are recognized by the FDA as nontoxic for the human body. Nanosized particles of $TiO_2$, ZnO, and MgO possess photocatalytic ability, UV absorption, and a photooxidizing capacity against chemical and biological species. During the last decade, research involving metal oxide NPs was intensified, focusing on the production of textiles with antibacterial, self-decontaminating and UV-blocking functions (Daoud and Xin, 2004; Shi et al., 2008; Sojka-Ledakowicz et al., 2008; Uddin, 2008).

Nanostructured metals and inorganic oxides can be incorporated into textiles by various methods, e.g., high energy $\gamma$-radiation and thermal treatment–assisted impregnation (El-Naggar et al., 2003; Zohdy et al., 2003). In this work, cotton and cotton/polyester fabrics were immersed in an antimicrobial formulation based on zinc oxide (ZnO), Impron MTP (binder), and Setamol WS (dispersing agent), and subjected to fixation by $\gamma$-radiation techniques. The effect of this treatment on the growth of bacteria (*Bacillus subtilis*) was estimated. On the basis of microbial detection, it was found that the ZnO formulation causes a net reduction in the bacterial cells amounting to 78% and 62% in the case of treated cotton and cotton/polyester fabrics. However, it was found that treatment with the ZnO formulation caused a reduction in the thermal stability of the fabrics, as indicated by thermogravimetric analysis.

One of the widely used techniques for coating textile substrates is the combination of the sol-gel synthetic procedure with the "pad-dry-cure" method (Schollmeyer, 2007; Wang et al., 2008; Xue et al., 2008). The synthesis process usually involves two main steps. For instance, the hexagonally ordered ZnO nanorod arrays might be grown on fiber substrates in the same way as zerogel ZnO (Wang et al., 2004). The growing seeds were formed by coating ZnO nanosol using dip-coating, dip-pad-curing, or spraying methods by natural solvent evaporation. In order to stabilize the precursor solution, triethenamine, with the same molar ratio as zinc acetate, was added to form a transparent homogeneous solution. The $TiO_2$ and $TiO_2/SiO_2$ nanocomposites prepared by the low temperature sol-gel synthesis were coated on cotton fabrics by a dip-pad–dry-cure process (Daoud et al., 2005; Qi et al., 2007). The sol-gel immobilization and controlled release of various bioactive liquids from modified silica coatings were investigated in (Haufe et al., 2008). The deposition of nano-ZnO onto cotton fabric was performed by padding the textiles in the colloid formulation of the zinc oxide–soluble starch nanocomposite to pass on to the material the antibacterial and UV-protection functions (Vigneshwaran et al., 2006).

Very recently, there have been some new publications on the deposition of *in situ*-formed metal oxide NPs on fabrics. A superhydrophobic ZnO nanorod array film on cotton substrate was fabricated via a wet chemical route and subsequent modification with a layer of *n*-dodecyltrimethoxysilane (Xu and Cai, 2008). ZnO NPs were grown *in situ* on a $SiO_2$-coated cotton fabric through the hydrothermal method. After water treatment at 100°C or higher, the cotton fabric was covered with approximately 24 nm diameter needle-shaped ZnO nanorods, which had an excellent UV-blocking property (Mao et al., 2006). ZnO particles were prepared by the wet chemical method using zinc nitrate and sodium hydroxide as precursors and solubilized starch as a stabilizing agent (Kathirvelu et al., 2009). These NPs were impregnated onto cotton fabrics by the "pad-dry-cure" method using an acrylic binder.

Copper is one of a relatively small group of metallic elements that are essential to human health. These elements, along with amino and fatty acids and vitamins, are required for normal metabolic processes (Toxicological profile of copper. 2004 U.S. Department of Health and Human Services). Copper is considered safe to humans, as demonstrated by the widespread and prolonged use of copper intrauterine devices (IUDs) (Bilian, 2002). However, except for the work of Gabbay and coworkers (Borkow and Gabbay, 2005; Gabbay et al., 2006), there are not many publications on the production and application of CuO-textile composites. The copper-containing fibers of cotton and polyester prepared by these authors demonstrated significant antifungal and antimicrobial properties. They inserted the preliminary synthesized copper oxide powder into the polymer fibers during the master-batch stage, and impregnated them into the cotton by a multi-phase soaking procedure, including treatment in formaldehyde.

In summary, most of the methods for the deposition of nanostructured materials on textiles are based on a multistage procedure, including the preliminary synthesis of NPs and the application of some templating agents for anchoring them to the substrates. This approach is rather complicated and can result in the release of some toxic compounds into the wastes. The sonochemical method prevents the use of toxic binders and makes the coating procedure shorter, effective, and environmentally friendly.

## 14.4 Deposition of Metal Oxide Nanoparticles on Textiles by Ultrasound Irradiation

### 14.4.1 Synthesis and Deposition of ZnO

The antibacterial activity of ZnO depends on the particle size: decreasing the particle size leads to an increase in the antibacterial activity (Yamamoto, 2001). We have developed a simple new method for the preparation of cotton bandages with antibacterial properties by immobilizing ZnO NPs on the fabric's surface via ultrasound irradiation (Perelshtein et al., 2009). The aim of this work was to obtain a homogeneous coating of small ZnO NPs on fabrics with a narrow size distribution and to reach a minimal effective ZnO concentration, which will still demonstrate antibacterial activity. This process involves the *in situ* generation of ZnO under ultrasound irradiation and its deposition on fabrics in a one-step reaction. The sonication was performed in a water–ethanol solution in the presence of a cotton bandage. Zinc acetate was used as a precursor, and the pH was adjusted to 8–9 with the addition of $NH_3 \cdot H_2O$.

Previous studies indicated that the product yield and particle size are strongly dependent on the rate of interparticle collision and on the concentration of the reagents during the sonochemical synthesis (Gedanken, 2004). That is why experimental parameters such as time and concentration of the precursor were selected as important factors for the optimization of the sonochemical reaction. The XRD demonstrated that the ZnO NPs on the coated bandage are crystalline, and the diffraction patterns matched the hexagonal phase of the wurtzite ZnO structure. No peak characteristics of any impurities were detected. The particle size estimated by the Debye–Scherrer equation is 30 nm. The morphology of the coated bandage before and after the deposition of ZnO NPs studied by high-resolution scanning electron microscopy (HR SEM) is presented in Figure 14.3. Figure 14.3a demonstrates the smooth texture of the pristine cotton bandage. After sonication, the fibers of bandage are homogeneously coated with NPs (Figure 14.3b). The inset image in Figure 14.3b was taken at a higher magnification in order to obtain the particles' size distribution. The distribution of the particles is quite narrow, and primary particles are in a very low nanometric range (~30 nm) that matches well with the XRD results. The selected-area HR SEM image studied with the elemental dot-mapping technique is shown in Figure 14.3c. The distribution of zinc and oxygen in the mapped area is presented in parts d and e of Figure 14.3, respectively. These images verify a homogeneous coating of the fibers with ZnO NPs.

We considered the coating mechanism as follows: it involves the *in situ* generation of ZnO NPs and their subsequent deposition on fabrics in a one-step reaction via ultrasound irradiation. Zinc oxide is formed during the irradiation according to the following reactions:

$$Zn^{2+}_{(aq)} + 4NH_3H_2O_{(aq)} \rightarrow [Zn(NH_3)_4]^{2+}_{(aq)} + 4H_2O \quad (14.1)$$

$$[Zn(NH_3)_4]^{2+}_{(aq)} + 2OH^-_{(aq)} + 3H_2O \rightarrow ZnO_{(s)} + 4NH_3H_2O_{(aq)} \quad (14.2)$$

Ammonia works as the catalyst of the hydrolysis process, and the formation of zinc oxide takes place through the ammonium complex $[Zn(NH_3)_4]^{2+}$. The ZnO NPs produced by this reaction are thrown at the surface of the bandages by the sonochemical microjets resulting from the collapse of the sonochemical bubble.

As already mentioned above, the sonochemical irradiation of a liquid causes two primary effects, namely, cavitation (bubble formation, growth, and collapse) and heating. When the microscopic cavitation bubbles collapse near the surface of the solid substrate, they generate powerful shock waves and microjets that cause the effective stirring/mixing of the adjusted liquid layer. The after effects of the cavitation are several hundred times greater in heterogeneous systems than in homogeneous systems (Suslick, 1989). In our case, the ultrasonic waves promote the fast migration of the newly formed zinc oxide NPs to the fabric's surface. This fact might cause a local melting of the fibers at the contact sites, which may be the reason why the particles strongly adhere to the fabric.

Here the question rises as to whether sonication doesn't damage the fabric's substrate. Thus, the tensile mechanical properties of a cotton-impregnated fabric were studied on a universal testing machine, Zwick 1445. Fourfolded fabric sample with a gauge length of 60 mm and a width of 40 mm was placed in special grips. The tensile force for the zinc oxide–coated sample was observed to be ~11% less than that of the pristine bandage (Figure 14.4). The observed changes in the mechanical behavior of the yarn are in a range that is acceptable for standard cotton fabrics. According to this result, we conclude that the sonochemical treatment of the bandage doesn't cause any significant change in the structure of the yarn.

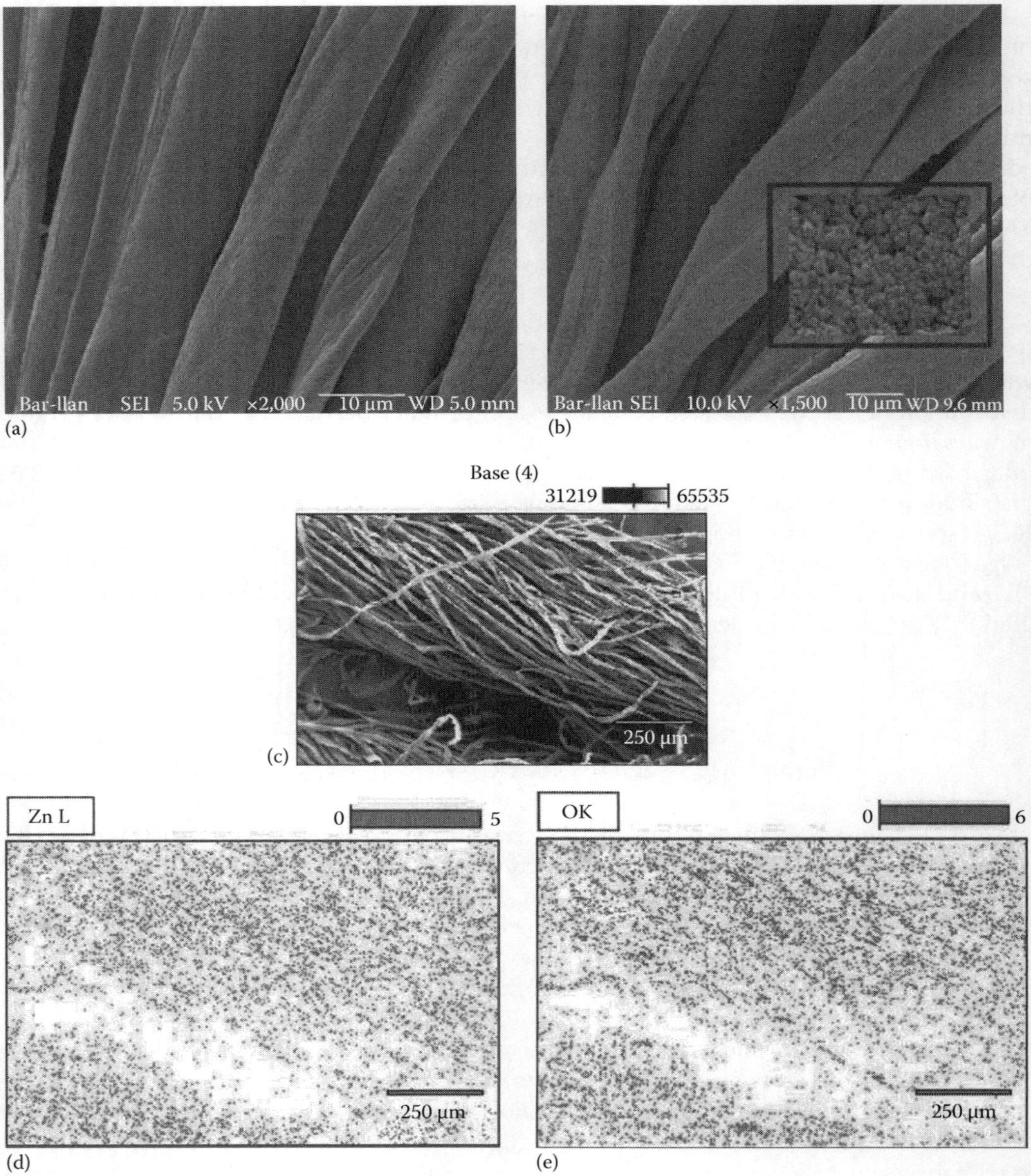


**FIGURE 14.3**
**(See color insert following page 302.)** HR SEM images of (a) pristine bandage fibers (magnification ×2,000), (b) bandage coated with ZnO NPs (magnification ×1,500; the inset shows a magnified image (×50,000) of the ZnO NPs on the fabric, (c) selected image for x-ray dot mapping, (d) x-ray dot mapping for zinc, and (e) x-ray dot mapping for oxygen.

One of the factors influencing the commercial exploitation of the antibacterial bandages is the release of NPs into the surrounding environment. In light of a recent paper (Benn and Westernoff, 2008) that found that silver NPs of 10–500 nm in diameter leach from sock fabric, we attempted to find the leached ZnO NPs in the washing solution. In the control experiments, we treated the coated bandages with an aqueous solution of 0.9 wt% NaCl

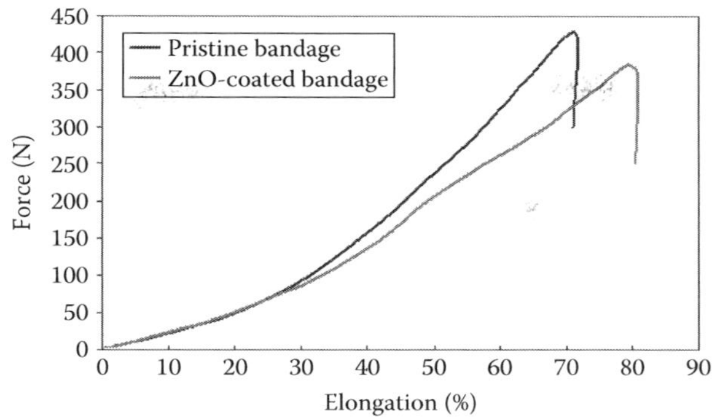


**FIGURE 14.4**
Mechanical properties of the cotton bandage before and after the deposition of ZnO NPs.

overnight at 37°C. The leaching experiment indicated that only 16% of the deposited zinc was removed under these conditions in an ionic form that is dictated by the low $K_{sp}$ of zinc oxide in water. The dynamic light scattering (DLS) and TEM studies did not reveal the presence of any NPs in the washing solution. This means that the sonochemically deposited ZnO NPs are strongly anchored to the textile substrate.

### 14.4.2 Synthesis and Deposition of CuO

We extended the sonochemical approach for the deposition of CuO NPs on the textile fabrics, and, as in the case with ZnO NPs, the formation of copper oxide takes place through the ammonium complex. Copper ions react with a solution of ammonia to form a deep blue solution containing $[Cu(NH_3)_4]^{2+}$ complex ions. This complex is hydrolyzed and crystalline CuO NPs are obtained. The CuO NPs produced by these reactions are thrown at the surface of the fabric by the above-described mechanism of the sonochemical microjets, and are deposited on the surface of substrate.

The morphology of the fibers' surface area before and after the deposition of copper oxide was studied by XRD and HR SEM methods. The XRD revealed the monoclinic structure of CuO nanocrystals. The difference between pristine and coated cotton fabric is clearly demonstrated in Figure 14.5. The insert image in Figure 14.5b at higher magnification shows that the primary particles are in a very low nanometric range (~10–20 nm).

While $Cu^{2+}$ is considered an environmentally safe ion, a much more important and serious issue is the leaching of CuO NPs. DLS and TEM studies of the washing solution after treatment of the CuO-coated fabrics in 0.9 wt% NaCl did not reveal the presence of any NPs. This means that the sonochemically deposited CuO NPs are strongly anchored to the textile substrate, probably due to a local melting of the fibers at the contact sites. Similar results were obtained for the coating of various types of textiles such as nylon, polyester, and composite types of textiles with ZnO and CuO NPs.

### 14.4.3 Deposition of MgO

MgO is well known to have a strong antibacterial activity (Huang et al., 2005; Ohira et al., 2008). Different methods have been reported on the synthesis of magnesium oxide NPs, such as the controlled speed of formation following the heating procedure

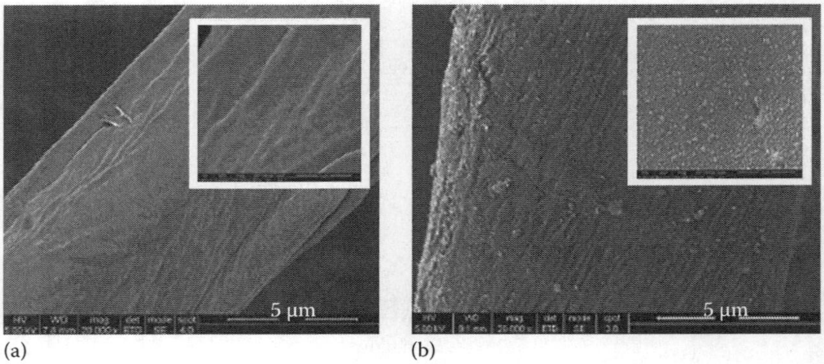


**FIGURE 14.5**
HR SEM images of cotton fibers: (a) pristine cotton (magnification ×20,000), inset image (magnification ×1,00,000); (b) cotton coated with CuO NPs (magnification ×20,000), inset image (magnification ×1,00,000).

(Huang et al., 2005), microwave-assisted synthesis (Makhluf et al., 2005), formation of MgO from aqueous droplets in a flame spray pyrolysis reactor (Seo et al., 2003), sonochemically enhanced hydrolysis followed by supercritical drying (Stengl et al., 2003), etc. However, there is nothing in the literature related to the deposition of magnesium oxide on textiles.

We have developed a method for depositing MgO NPs on fabrics by ultrasound irradiation. In this case, ultrasound is used just for "throwing stones" at the fabric, namely, we took commercial MgO nanopowder (Aldrich, <25 nm) and sonicated it in the presence of a cotton fabric. The microjets formed after the collapse of the bubble throw the NPs at high speed at the cotton yarn. This process is different from the deposition of ZnO and CuO where the NPs were *in situ* generated and deposited on fabrics in a one-step reaction.

The morphology of the fiber surface after the deposition of magnesium oxide NPs studied by HR SEM is presented in Figure 14.6. Fibers after sonication are homogeneously coated with NPs. The particle size is in the low nanometric range (~10–15 nm) and no aggregation was observed.

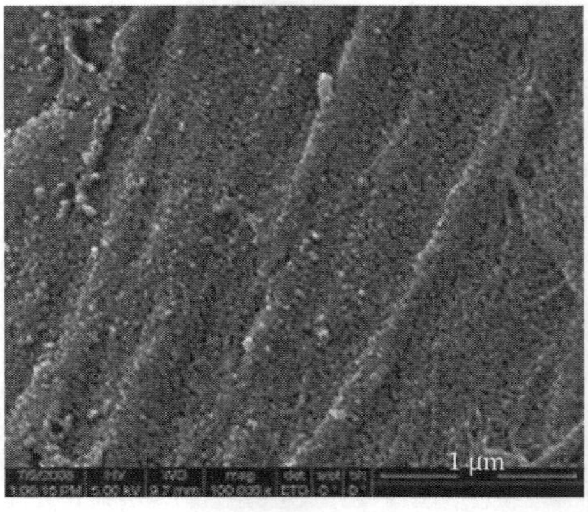


**FIGURE 14.6**
HR SEM images of 0.8 wt% MgO-coated cotton fabric.

## 14.5 Mechanism of the Antibacterial Activity of Metal Nanooxides

In this section, we will elucidate the mechanisms and factors that govern antibacterial activity in metal-oxide, nanoparticle-engineered textile systems. It has been well established that some metal oxide NPs can cause damage to bacterial membranes and, as a consequence, exhibit clear antibacterial properties. In this respect, the key advantages of metal oxides as inorganic antimicrobial agents are improved safety and stability, as compared to classic organic antimicrobial agents. Moreover, the spreading of antibiotic-resistant pathogens is still a growing concern globally. Therefore, the development of new bactericidal preparations is called for (Anagnostakos et al., 2008).

The mechanism of the antibacterial activity of these metal oxides is poorly understood, and is still controversial. Suggested mechanisms in the literature include the role of reactive oxygen species (ROS) generated on the surface of the particles, ion release, membrane dysfunction, and NP internalization (Li et al., 2008; Neal, 2008; Hu et al., 2009). However, both CuO and $TiO_2$ are scarcely soluble in water, whereas the $Zn^{2+}$ and $Mg^{2+}$ ions that might be released from oxide NPs are not considered to be harmful to biological systems (Vallee and Falchuk, 1993; Sawai, 2000). Moreover, our studies aimed at checking the leaching of NPs from the sonochemically coated textiles did not reveal any escape of the NPs to the washing solution (Perelshtein et al., 2009). These findings make the mechanism of ROS-mediated toxicity the key event in the antibacterial activity of the nanoparticle-engineered textile systems, and worthy of further detailed evaluation. ROS are probably best known in biology for their ability to cause oxidative stress. They can damage DNA, cell membranes, and cellular proteins, and may lead to cell death (Shen et al., 2009). Hydroxyl radicals (•OH) are highly reactive for the oxidation of organic substances, and are able to attack and decompose polyunsaturated phospholipids in bacteria. Other ROS, such as superoxide ions $\bullet O_2^-$ or hydroperoxyl radicals $HO_2^\bullet$ are less effective against bacteria (Imlay, 2008). The negative charge of superoxide ions prevents them from penetrating bacteria cell membranes. Therefore, these species must be in direct contact with the outer surface of bacteria for inducing the killing effect. Hydrogen peroxide ($H_2O_2$), as well as hydroperoxyl radicals $HO_2^\bullet$, can enter the cell despite their being less harmful compared to hydroxyl radicals. Furthermore, hydrogen peroxide can cause cell damage via hydroxyl radicals produced by the Fenton reaction (Wright, 2007). Furthermore, hydrogen peroxide in the presence of $\bullet O_2^-$ can generate singlet oxygen, which is very toxic (Rosen and Klebanoff, 1979):

$$\bullet O_2^- + H_2O_2 \rightarrow OH^- + \bullet OH + {}^1O_2 \qquad (14.3)$$

The mechanism underlying the generation of ROS from the metal oxide surface isn't yet well understood, but it has been hypothesized that it is derived from the highly reactive nature of defect sites (such as oxygen vacancies) on a wet metal oxide surface. In this respect, it was also suggested that the intimate contact possible between the NPs' surface and the bacteria might cause physical damage to the bacterial cell wall, and thus enhance an antibacterial effect.

### 14.5.1 Antibacterial Activity of ZnO

The excellent studies by Sawai et al. (1996a,b, 1998) clearly showed that ROS concentrations increased with the ZnO content of slurries. Following the same paradigm, Applerot et al. (2009a,b), in an innovative study using electron spin resonance (ESR) coupled with the spin

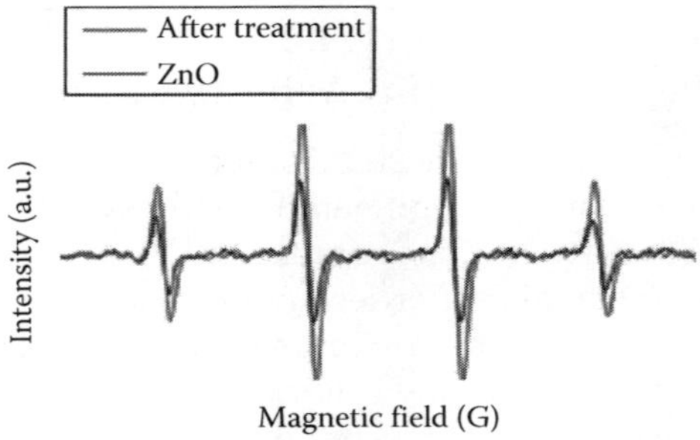


**FIGURE 14.7**
ESR spectra demonstrating changes in hydroxyl radical concentration upon antibacterial treatment of *E. coli* with a water suspension of ZnO.

trapping probe technique, monitored ROS, namely, hydroxyl radicals, and production in water suspensions of ZnO NPs. Their findings showed that hydroxyl radicals were present in water suspensions of ZnO, and their amount was closely related to the size of the ZnO particles, with smaller sizes having greater amounts of •OH on the basis of equivalent ZnO mass content. These results were correlated with the increase in the antibacterial effect of the small NPs. Thus, the small size and large specific surface area endow them with high chemical reactivity and intrinsic toxicity. Interestingly, combining Gram-negative bacterium *E. coli* and ZnO NPs suspensions immediately enhanced the generation of •OH for up to an average of 142% (Figure 14.7). It appears that sub-lethal amounts of ROS may induce excessive intracellular oxidative stress, probably by disturbing the balance between oxidant and antioxidant processes.

We determined the antibacterial activity of the sonochemically coated cotton fabrics with 0.75 wt% ZnO (Perelshtein et al., 2009) using the Gram-negative bacterium *E. coli* and the Gram-positive bacterium *S. aureus*. As shown in Table 14.1, treatment for 1 h with the coated cotton leads to the complete inhibition of *E. coli* growth. Regarding *S. aureus*, a 100% reduction in viability was reached after 3 h, while after 1 h of treatment, a reduction of 60% could be seen.

Zinc is an essential micronutrient for prokaryotic organisms. However, at super physiological levels, zinc inhibits the growth of many bacteria (Soderberg et al., 1989). According to a leaching experiment with 0.9 wt% NaCl, the concentration of $Zn^{2+}$ in solution corresponds to 36.7 μM/L (Perelshtein et al., 2009). Compared to the minimum inhibitory concentration reported in the literature of 4–8 mM/L (Lansdown et al., 2007), the amount of zinc released from fabrics in our work is lower at least by a factor of 2. Therefore, we assume that the Zn ions have a minor influence on antibacterial activity. The major components responsible for the bactericidal effect are the ZnO NPs. Although ZnO NPs were not found in the solution, they can generate some species of oxyradicals (Sengupta et al., 1979). In light of these results, the difference in the antibacterial action of ZnO-coated bandages on two strains of bacteria can be explained by the difference in their sensitivity toward oxyradicals. In this respect, it has been found recently that *S. aureus* contains a large amount of cartenoid pigment, which promotes a higher resistance to oxidative stress (Liu et al., 2005).

### 14.5.2 Antibacterial Activity of CuO

We tested the antimicrobial activity of the cotton bandages coated with CuO *via* ultrasonic irradiation against the *E. coli* and the *S. aureus*. Detailed investigations showed that after

**TABLE 14.1**

Antibacterial Activity Test of ZnO-Coated Cotton [a]SSS

| | Duration of Treatment | | | | | |
|---|---|---|---|---|---|---|
| | 1 h | | | 2 h | | |
| **Sample** | **CFU ($mL^{-1}$)** | ***N/N*₀** | **% Reduction in Viability** | **CFU ($mL^{-1}$)** | ***N/N*₀** | **% Reduction in Viability** |
| *E. coli* | | | | | | |
| Pristine fabric | $1.02 \times 10^{-7}$ | 0.98 | 0.98 | $1.34 \times 10^{-7}$ | 1.28 | −28.23 |
| No fabric | $1.17 \times 10^{-7}$ | 1.14 | −28.57 | $1.23 \times 10^{-7}$ | 1.35 | −35.16 |
| 0.75 wt% ZnO on fabric | $1.71 \times 10^{-7}$ | $1.58 \times 10^{-3}$ | 99.84 | 0 | $0.9 \times 10^{-8}$ | 100 |
| | 1 h | | | 3 h | | |
| *S. aureus* | | | | | | |
| Pristine fabric | $0.7 \times 10^{-7}$ | 0.71 | 20.46 | $0.99 \times 10^{-7}$ | 1.125 | −12.5 |
| No fabric | $0.98 \times 10^{-7}$ | 1.10 | −10.11 | $0.67 \times 10^{-7}$ | 0.75 | 24.72 |
| 0.75 wt% ZnO on fabric | $3.9 \times 10^{-6}$ | 0.34 | 66.4 | $7.6 \times 10^{-3}$ | $6.55 \times 10^{-4}$ | 99.93 |

[a] The viable bacteria were monitored by counting the number of colony-forming units (CFU); $N/N_0$: Survival fraction.

1 h, the growth of both strains was completely inhibited. One of the factors influencing the antibacterial activity of the developed coating is the release of the active phase into the surrounding medium, namely, copper ions or/and copper oxide NPs. The examination of the leaching of copper ions indicated that only a very low amount (namely, ~1.3%) of the deposited copper was removed by washing with a 0.9 wt% NaCl solution that corresponds to a concentration of $Cu^{2+}$ 0.15 ppm. The slight solubility of copper oxide can be explained by the very low $K_{sp}$ of CuO that dictates the very low concentration of $Cu^{2+}$ in the solution. In order to examine the influence of copper ions on the antibacterial effect, a control antibacterial test was performed using a supernatant with the same concentration of $Cu^{2+}$ instead of the coated cotton. After incubation for 24 h at 37°C, no reduction in *E. coli* after 2 h was observed (Figure 14.8).

This result indicates that the $Cu^{2+}$ ions have no influence on the antibacterial activity. Thus, the antibacterial effect can be attributed to the copper oxide NPs. It should be emphasized that no leaching of CuO NPs to the environment was detected in these experiments. Although CuO NPs were not found in the solution, they can generate some active species that are responsible for damaging the bacteria's cells. These active species were detected in ESR studies conducted with and without the bacteria present in the ESR tube.

### 14.5.3 Antibacterial Activity of $TiO_2$

There is some controversy in the literature regarding the effective inactivation of microorganisms by $TiO_2$, primarily due to the different experimental conditions, (UV/vis) irradiance, length of exposure, photocatalyst present in suspension or powder, range of concentration, and the different $TiO_2$ photocatalysts and microorganisms employed, although some effect is generally acknowledged (Caballero et al., 2009). However, a bactericidal effect was reported on food-pathogenic bacteria such as *Salmonella choleraesuis*, *Vibrio parahaemolyticus*, and *Listeria monocytogenes* (Kim et al., 2003), as well as *Pseudomonas*

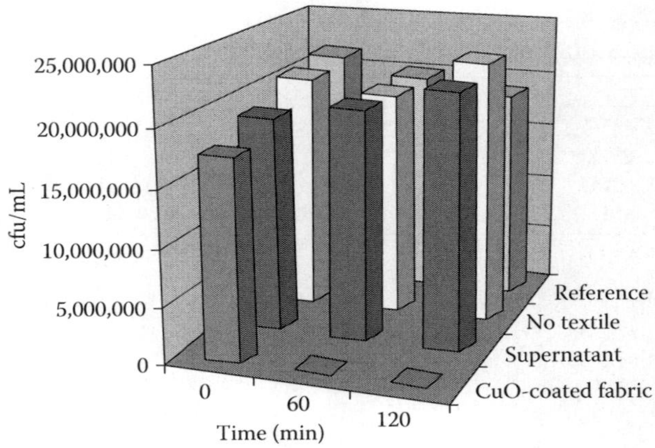


**FIGURE 14.8**
Antibacterial test of CuO-coated cotton against *E. coli*.

*aeruginosa* (Madrid et al., 2002). The formation of a well-adhering bactericidal surface of $TiO_2$ on organic cellulose fibers was reported by (Daoud et al., 2005). In presence of UV radiation and $O_2$, ROS are produced by $TiO_2$ NPs according to the following reactions (Li et al., 2008):

$$TiO_2 + hv \rightarrow TiO_2(h^+ + e^-) \quad (14.4)$$

$$e^- + O_2 \rightarrow \bullet O_2^- \quad (14.5)$$

$$\bullet O_2^- + 2H^+ + e^+ \rightarrow H_2O_2 \quad (14.6)$$

$$H_2O_2 + O_2^- \rightarrow \bullet OH + OH^- + O_2 \quad (14.7)$$

$$h^+ + H_2O \rightarrow \bullet OH + H^+ \quad (14.8)$$

At the same time, upon the irradiation of a $TiO_2$ surface with photons of wavelength ≤385 nm, an electron is promoted from the valence band to the conduction band, thus forming an electron–hole pair. The photogenerated holes and electrons react with water molecules attached to $TiO_2$ surfaces in the presence of oxygen to form hydroxyl radicals and other ROS such as superoxide ions. It is still a subject of investigation as to which of these ROS are directly involved in the damage to bacteria cells. Moreover, it remains unclear what conditions of irradiated light necessitate the activation of ROS generation.

Kangwansupamonkon et al. (2009) have reported on the antibacterial performance of apatite-coated $TiO_2$, which was fixed on cotton textiles by a dip-coating technique. Their study indicated that the photocatalytic activity of an apatite-coated $TiO_2$ suspension can help in microbial decomposition in textile applications. Its effectiveness was clearly confirmed against *S. aureus*, *E. coli* ATCC 25922, MRSA DMST 20627, and *M. luteus* strains of bacteria. The effect of an irradiation source on the antimicrobial activity of cotton fabrics coated with apatite-coated $TiO_2$ was examined. The highest antibacterial activity was found from black-light irradiation compared with visible light and dark conditions. This could be explained by the presence of more ROS on the surface of $TiO_2$ particles after irradiation with black light (at wavelengths < 385 nm), whereas the amount of ROS with visible light and dark condition would be less pronounced.

### 14.5.4 Antibacterial Activity of MgO

During the application of MgO NPs as a household paint, it was reported that it exhibited good bactericidal activity against *Bacillus subtilis var. niger* and *Staphylococcus aureus* strains of bacteria and is effective even in the absence of light irradiation (Huang et al., 2005). In contrast, nano-$TiO_2$, a common type of bactericidal material, is photoactive and requires light irradiation in order to exhibit effective bactericidal activity. Makhluf et al. (2005) investigated the nature of MgO nanocrystalline as a bacteriocide. They demonstrated a clear nanoscale effect, where the amount of eradicated bacteria increased with the decrease in the NP size. No publications were found in the literature concerning an antibacterial textile coated with MgO.

We determined the antibacterial activity of cotton fabrics sonochemically coated with 0.8 wt% MgO against *E. coli* and *S. aureus*. As shown in Figure 14.9a, treatment for 1 h with the coated cotton leads to the complete inhibition of *E. coli* growth. Regarding *S. aureus* (Figure 14.9b), a 99% reduction in viability was reached after 3 h, while after 1 h of treatment a reduction of 60% could be seen.

These results confirmed that the sonochemical method is effective for the production of antibacterial fabrics irrespective of the NP deposition, i.e., the creation of NPs in the sonochemical reaction and the simultaneous deposition or sonochemical-assisted deposition of the preliminary synthesized NPs.

By using the chemiluminescence method, Sawai et al. (2000) showed that the dominant ROS generated on MgO surfaces were superoxide anions. As mentioned above, these ROS are not very reactive themselves to bacterial cells; however, these species are in equilibrium, as shown in the following reaction:

$$\bullet O_2^- + H^+ \rightleftarrows HO_2^\bullet \tag{14.9}$$

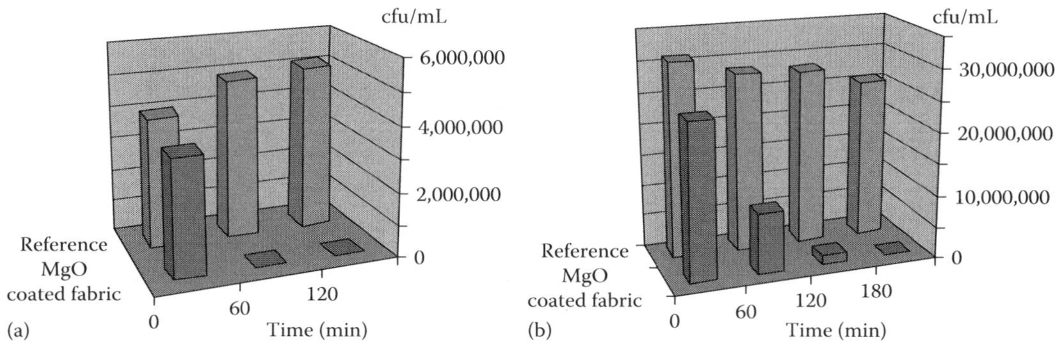


**FIGURE 14.9**
Antibacterial test of MgO-coated cotton against: (a) *E. coli* and (b) *Staphylococcus. aureus.*

When $\bullet O_2^-$ capture hydrogen ions are generated near the bacterial membrane during the respiratory process, hydroperoxyl radicals ($HO_2^\bullet$) are produced. The $HO_2^\bullet$ is much more reactive than $\bullet O_2^-$ and is able to penetrate the cell membrane. This concept will support the fact that contact between bacterial cells and MgO powders are important for MgO-induced bacterial death. Moreover, as a consequence of this contact, a local alkaline effect due to basic sites on MgO surfaces may also enhance the antibacterial activity of MgO (Ardizzone et al., 1998).

## 14.6 Pilot Installation for the Deposition of Nanoparticles on Textiles

The industrial use of many advanced coating techniques has limitations related to the combination of materials for the substrate and the coating. Another important consideration is the reproducibility of the sonochemical synthesis, which can often be traced to a failure in the specification of the exact sonication conditions, e.g., the frequency of ultrasonic irradiation, the precise power entering the reaction system, the geometry of the reaction vessel, the presence of a bubbled gas, or even the temperature of the reaction (Mason et al., 1992; Abramov, 2000). Thus, the design of sonochemical devices for up-scaled processes that provide good reproducibility is a very important task. Usually, the fabrics are produced and supplied in the form of rolls, which has dictated the choice of design. The purpose of our work was to design a pilot installation for the roll-to-roll continuous sonochemical coating of cotton bandages with NPs. The assumption of the new instrument was that the fabric band runs between two plane magnetorestrictive ultrasound transducers and is collected on a roller. The magnetorestrictive transducers were chosen because of the additional benefit of a wide range of tolerance for resonance conditions and a lower price, in comparison with the piezoelectric transducers. The speed of the moving ribbon could be regulated within limits of 0.004–0.050 m/s. The ZnO or CuO NPs were deposited on cotton and polyester fabrics according to the procedure described in Section 14.4, using the water/ethanol solution of the corresponding acetate salts as precursors with the addition of ammonia as a catalyst.

In Figure 14.10, the photo image of the pilot installation during the coating of the bandage by CuO is presented. The front panel was open to show the change in the color of bandage after coating with CuO NPs. In the inset, one can observe that with the deposition of CuO NPs, the color of the bandage changed from white to brown. When the ZnO was deposited on the bandage, the white color was retained. Using various initial concentrations of the precursor, we obtained different percentages of ZnO and CuO NPs in the cotton bandages.

During the study of the optimal conditions for the coating of CuO NPs on the cotton bandage, it was revealed that after the first 10 min of the reaction, the content of the CuO in the coated bandage decreased. The explanation for this was related to the reducing number of crystallization centers in the working solution over time. As was reported previously (Suslick and Price, 1999; Gedanken, 2004), the yield of the nanomaterials formed in solution under ultrasonic irradiation depends on the concentration of the precursor. During the sonochemical deposition, the nuclei, as they are formed, are thrown toward the solid substrate by the microjets created after the collapse of the cavitation bubbles. At the same time, the concentration of the precursor in the solution decreases. To prevent any drop in the deposition of the metal oxides on the bandages, the necessary quantity of the precursor dissolved in a small amount of water was added during the deposition procedure.

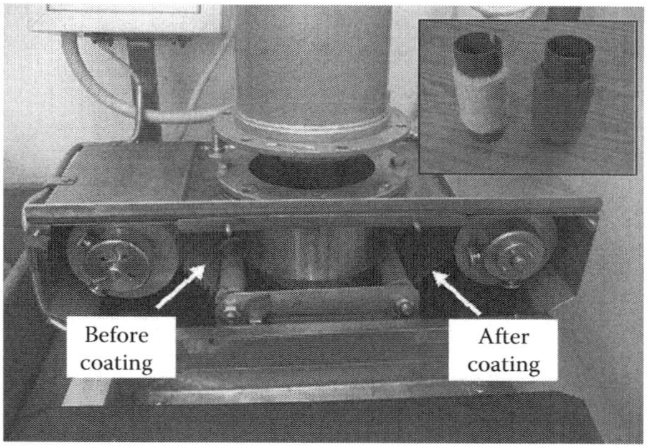


**FIGURE 14.10**
**(See color insert following page 302.)** Photo image of the pilot installation. In the inset of the figure, the CuO-coated and uncoated spools are shown.

During the experimental conditions with the speed feeding fixed as 0.004 m/s, the pilot installation produced 10 m of bandage (10 cm width) in 45 min. The original laboratory scale device produced only a single piece of fabric (10 cm × 10 cm) in 60 min. For a comparison of the efficiency of the laboratory scale and the pilot sonochemical installations, we calculated the energy consumption for coating cotton bandages with metal oxide NPs using the following formula:

$$E = P \cdot e \cdot t, \tag{14.10}$$

where
$E$ is the energy consumption (kW·h)
$P$ is the sonicator output power (kW)
$e$ is the efficiency (%)
$t$ is the time (h) necessary for coating a 10 cm$^2$ piece of fabric

For the pilot installation, the $E_p$ value was 0.042 kW·h, and for the laboratory device this value was 0.36 kW·h. Thus, the efficiency of the up-scaled device is almost one order of magnitude higher than that of the laboratory reactor. The metal oxide NPs are strongly attached to the fabric, and their concentration in the cotton bandages after 20 washing cycles was unchanged.

The working solution remaining after the reaction can be used several times with good reproducibility of the results, as long as a small amount of the precursor (10% of the starting content) is added to the reaction slurry to generate new crystallization centers. The concentration of the metal oxide on the cotton substrate can be controlled either by the initial concentration of the precursor in the solution, or by the velocity of the bandage passing between the transducers.

The XRD revealed that the coated phase has a crystalline nanophase structure of metal oxides (ZnO or CuO, respectively). HR SEM studies demonstrated that the coating performed on the pilot installation is homogeneous, and the particle distribution is quite narrow. The primary particles are in a very low nanometric range (~10–20 nm) (Figure 14.11). These results are in good agreement with the ZnO and CuO particles size observed by HR SEM on the textiles coated at laboratory scale conditions (see Figures 14.3 and 14.5).

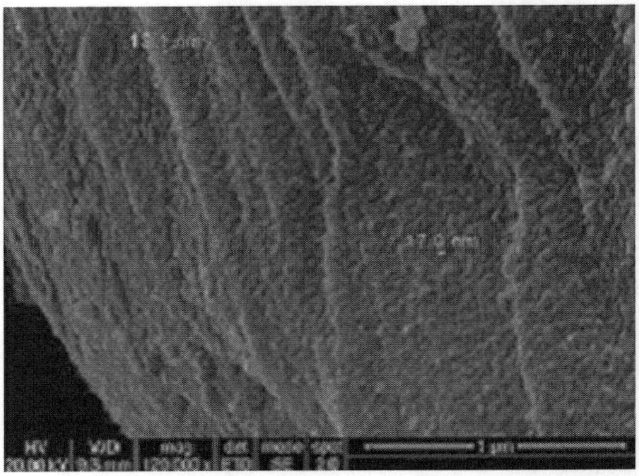

**FIGURE 14.11**
HR SEM images of the CuO/cotton bandage prepared on the pilot sonochemical installation.

The highly homogeneous deposition of CuO NPs on the polyester was also achieved in this pilot installation (Figure 14.12). The size of the individual particles is in the range of 20–30 nm. The concentrations of ZnO and CuO on the coated fabric in the cotton bandages were close to those obtained under laboratory conditions with a good reproducibility of the results.

This instrument has the potential for coating different kinds of the bandages with various types of NPs. The homogeneity of the coating can be controlled by varying either the solution concentration or the speed of the bandage through the reactor. Compared with laboratory experiments, this pilot scale system significantly increased productivity, and at the same time decreased the energy consumption by one order of magnitude.

(a) (b)

**FIGURE 14.12**
HR SEM images of the CuO/polyester fabric prepared in the pilot installation: (a) magnification ×20,000; (b) magnification ×1,00,000.

## 14.7 Conclusion

Metal oxide NPs can be uniformly deposited onto the surface of different kinds of textiles by the sonochemical method. The coating was performed by a simple, efficient, one-step procedure using environmentally friendly reagents. The physical and chemical analyses demonstrated that nanocrystals of ~20–30 nm in size are finely dispersed onto the fabric surfaces without significant damage to the structure of the yarn. The mechanism of nano-oxide formation and adhesion to the textile was discussed. It is based on the local melting of the substrate due to the high rate and temperature of NPs thrown at the solid surface by sonochemical microjets. The strong adhesion of metal nanooxides to the substrate was demonstrated in terms of the absence of the leaching of the NPs into the washing solution. The performance of fabrics coated with a low content of nanooxides (<1 wt%) as an antibacterial agent was investigated, and their excellent bactericidal effect was demonstrated. The method was up-scaled using a pilot sonochemical device suitable for coating various textile bandages with metal nanooxides.

## Acknowledgment

This research was carried out as part of the activities of the LIDWINE Consortium, Contract No. NMP2-CT-2006–026741 LIDWINE is an IP project of the 6th EC program.

## References

Abramov, O. 2000. Nonlinear acoustic effects induced by propagation of intense ultrasound through liquid. *Theor. Foundam. Chem. Eng*. 34: 324–333.

Anagnostakos, K., Hitzler, P., Pape, D. et al. 2008. Persistence of bacteria growth on antibiotic-loaded beads: Is it actually a problem? *Acta Orthop*. 79: 302–307.

Applerot, G., Lipovsky, A., Dror, R. et al. 2009a. Enhanced antibacterial activity of nanocrystalline ZnO due to increased ROS-mediated cell injury. *Adv. Funct. Mater*. 19: 842–852.

Applerot, G., Perkas, N., Amirian, G. et al. 2009b. Coating of glass with ZnO via ultrasonic irradiation and a study of its antibacterial properties. *Appl. Surf. Sci*. 256S: S3–S8.

Ardizzone, S., Bianchi, C.L., and Vercelli, B. 1998. MgO powders: Interplay between adsorbed species and localization of basic sites. *Appl. Surf. Sci*. 126: 169–175.

Benn, T. and Westernoff, P. 2008. Nanoparticle silver released into water from commercially available sock fabrics. *Environ. Sci. Technol*. 42: 4133–4139.

Bilian, X. 2002. Intrauterine devices. *Best. Pract. Res. Clin. Obstet. Gynaecol*. 16: 155–168.

Borkow, G. and Gabbay, J. 2005. Copper as a biocidal tool. *Curr. Med. Chem*. 12: 2163–2175.

Caballero, L., Whitehead, K.A., Allen, N.S. et al. 2009. Inactivation of *Escherichia coli* on immobilized $TiO_2$ using fluorescent light. *J. Photochem. Photobiol. A* 202: 92–98.

Carlson, K. 2006. In vitro toxicity assessment of silver nanoparticles in rat alveolar macrophages. MS thesis, Wright State University, Dayton, OH.

Chen, X. and Schluesener, H.J. 2008. Nanosilver: A nanoproduct in medical application. *Toxicol. Lett*. 176: 1–12.

Coyle, S., Wu, Y., Lau, K.-T., Diamond D. et al. 2007. Smart nanotextiles: A review of materials and applications. *MRS Bull.* 32: 434–442.

Craighead, H. and Leong, K. 2000. *Nanotechnology Research Direction: Biotechnology, Medicine, and Healthcare*, ed. M.C. Roco, R.S. Williams, and P. Alivisatos, pp. 153–72. The Netherlands: Kluwer Academic Publishers.

Daoud, W.A. and Xin, J.H. 2004. Low temperature sol-gel processed photocatalytic titania coating. *J. Sol-Gel Sci. Technol.* 29: 25–29.

Daoud, W.A., Xin, J.H., and Zhang, Y.H. 2005. Surface functionalization of cellulose fibers with titanium dioxide nanoparticles and their combined bactericidal activities. *Surf. Sci.* 599: 69–75.

Doktycz, S.J. and Suslick, K.S. 1990. Interparticles collisions driving by ultrasound. *Science* 247: 1067–1069.

Dubas, S.T., Kumlangdudsana, P., and Potiyaraj, P. 2006. Layer-by-layer deposition of antimicrobial silver nanoparticles on textile fibers. *Coll. Surf. A: Physicochem. Eng. Aspects* 289: 105–109.

Duran, N., Marcato, P.D., De Souza, G.I. H. et al. 2007. Antibacterial effect of silver nanoparticles produced by fungal process on textile fabrics and their effluent treatment. *J. Biomed. Nanotechnol.* 13: 203–208.

El-Naggar, A.M., Zohdy, M.H., Hassan, M.S. et al. 2003. Antimicrobial protection of cotton and cotton/polyester fabrics by radiation and thermal treatments. I. Effect of ZnO formulation on the mechanical and dyeing properties. *J. Appl. Polym. Sci.* 88: 1129–1137.

Gabbay, J., Borkow, G., Mishal, J. et al. 2006. Copper oxide impregnated textiles with potent biocidal activities. *J. Ind. J. Ind. Textiles* 35: 323–335.

Gao, Y. and Cranston, R. 2008. Recent advances in antimicrobial treatments of textiles. *Textile Res. J.* 78: 60–72.

Gedanken, A. 2004. Using sonochemistry for the fabrication of nanomaterials. *Ultrason. Sonochem.* 11: 47–55.

Gedanken, A. 2007. Doping nanoparticles into polymers and ceramics using ultrasound radiation. *Ultrason. Sonochem.* 14: 418–430.

Gedanken, A., Reisfeld, R., Sominski, E. et al. 2000. Sonochemical preparation and characterization of europium oxide doped in and coated on $ZrO_2$ and yttrium-stabilized zirconium (YSZ). *J. Phys. Chem. B.* 104: 7057–7065.

Gorenšek, M. and Recelj, P. 2007. Nanosilver functionalized cotton fabric. *Textile Res. J.* 77: 138–141.

Guo, W.-L., Yang, Z.-X., and Wang, X.-K. 2006. Sonochemical deposition and characterization of nanophasic $TiO_2$ on silica particles. *Mater. Res. Innovat.* 10: 11–20.

Hadad, L., Perkas, N., Gofer, Y. et al. 2007. Sonochemical deposition of silver nanoparticles on wool fibers. *J. Appl. Polym. Sci.* 104: 1732–1737.

Haufe, H., Muschter, K., Siegert, J. et al. 2008. Bioactive textiles by sol-gel immobilised natural active agents. *J. Sol-Gel Sci. Technol.* 45: 97–101.

Hu, X., Cook, S., Wang, P. et al. 2009. In vitro evaluation of cytotoxicity of engineered metal oxide nanoparticles. *Sci. Total Environ.* 407: 3070–3072.

Huang, L., Li, D.Q., Lin, Y.J. et al. 2005. Controllable preparation of nano-MgO and investigation of its bactericidal properties. *J. Inorg. Biochem.* 99: 986–993.

Imlay, J.A. 2008. Cellular defenses against superoxide and hydrogen peroxide. *Annu. Rev. Biochem.* 77: 755–776.

Joeger, T.K., Joeger, R., Olsson, E. et al. 2001. Bacteria as workers in the living factory: Metal-accumulating bacteria and their potential for materials science. *Trends Biotechnol.* 19: 15–20.

Kangwansupamonkon, W., Lauruengtana, V., Surassmo, S. et al. 2009. Antibacterial effect of apatite-coated titanium dioxide for textiles applications. *Nanomedicine.* 5: 240–249.

Kathirvelu, S., D'Souza, L., and Dhurai, B. 2009. A study on functional finishing of cotton fabrics using nano-particles of zinc oxide. *Mater. Sci.* (*Medziagotyra*) 15: 75–79.

Kim, B., Kim, D., Cho, D. et al. 2003. Bactericidal effect of $TiO_2$ photocatalyst on selected food-borne pathogenic bacteria. *Chemosphere* 52: 277–281.

Klimov, V.I., Mikhailovsky, A.A., Xu, S. et al. 2000. Optical gain and stimulated emission in nanocrystal quantum dots. *Science* 290: 314–317.

Klueh, U., Wagner V., Kelly, S. et al. 2000. Efficacy of silver-coated fabric to prevent bacterial colonization and subsequent device-based biofilm formation. *J. Biomed. Mater. Res.* 53: 621–631.
Kotlyar, A., Perkas, N., Amiryan, G. et al. 2007. Coating silver nanoparticles on poly(methyl methacrylate) chips and spheres via ultrasound irradiation. *J. Appl. Polym. Sci.* 104: 2868–2876.
Kotlyar, A., Perkas, N., Koltypin, Y. et al. 2008. Coating zinc oxide submicron crystals on poly(methyl methacrylate) chips and spheres via ultrasound irradiation. *Ultrason. Sonochem.* 15: 839–845.
Landau, M.V., Vradman, L., Herskowitz, M. et al. 2001. Ultrasonically controlled deposition–precipitation Co–Mo HDS catalysts deposited on wide-pore MCM material. *J. Catal.* 201: 22–36.
Lansdown, A.B.G., Mirastschijski, U., Stubbs, N. et al. 2007. Zinc in wound healing: Theoretical, experimental, and clinical aspects. *Wound Rep. Reg.* 15: 2–16.
Lee, H.J. and Jeong, S.H. 2005. Bacteriostasis and skin innoxiousness of nanosize silver colloids on textile fabrics. *Textile Res. J.* 75: 551–556.
Lee, H.J., Yeo, S.Y., and Jeong, S.H. 2003. Antibacterial effect of nanosized silver colloidal solution on textile fabrics. *J. Mater. Sci.*, 38: 2199–2204.
Li, N., Xia, T., and Nel, A.E. 2008. The role of oxidative stress in ambient particulate matter-induced lung diseases and its implications in the toxicity of engineered nanoparticles. *Free Radic. Biol. Med.* 44: 1689–1699.
Liu, G.Y., Essex, A., Buchanan, J.T. et al. 2005. *Staphylococcus aureus* golden pigment impairs neutrophil killing and promotes virulence through its antioxidant activity. *J. Exp. Med.* 202: 209–215.
Liz-Marzan, L.M. and Kamat, P.V. 2003. *Nanoscale Materials.* Norwell, MA: Kluwer Academic Publishers.
Madrid, P.A., Moorillon, G.V.N., Borunda, E.O. et al. 2002. Photoinduced bactericidal activity against *Pseudomonas aeruginosa* by $TiO_2$ based thin films. *FEMS Microbiol. Lett.* 211: 183–188.
Makhluf, S., Dror, R., Nitzan, Y. et al. 2005. Microwave-assisted synthesis of nanocrystalline MgO and its use as a bacteriocide. *Adv. Funct. Mater.* 15: 1708–1715.
Mao, Z., Shi, Q., Zhang, L. et al. 2006. The formation and UV-blocking property of needle-shaped ZnO nanorod on cotton fabric. *Thin Solid Films* 517: 2681–2686.
Mason, T.J. 1990. *Sonochemistry.* London, U.K.: Royal Society of Chemistry Press.
Mason, T.J. 2007. Development in ultrasound non-medical. *Prog. Biophys. Molec. Biol.* 93: 166–175.
Mason, T.J., Lorimer, J.P., and Bates, D.M. 1992. Quantifying sonochemistry: Casting some light on a 'black art'. *Ultrasonics* 30: 40–42.
Neal, A.L. 2008. What can be inferred from bacterium–nanoparticle interactions about the potential consequences of environmental exposure to nanoparticles? *Ecotoxicology* 17: 362–371.
Ohira, T., Kawamura, M., Iida, Y. et al. 2008. Influence of the mixing ratio on antibacterial characteristics of MgO-ZnO solid solution in two phase coexistence region. *J. Ceram. Soc. Jpn.* 116: 1234–1237.
Patra, A., Sominska, E., Ramesh, S. et al. 1999. Sonochemical preparation and characterization of $Eu_2O_3$ and $Tb_2O_3$ doped in and coated on silica and alumina nanoparticles. *J. Phys. Chem. B.* 103: 3361–3365.
Perelshtein, I., Applerot, G., Perkas, N. et al. 2008. Sonochemical coating of silver nanoparticles on textile fabrics (nylon, polyester and cotton) and their antibacterial activity. *Nanotechnology* 19: art. 245705 (6 pp).
Perelshtein, I., Applerot G., Perkas, N. et al. 2009. Antibacterial properties of an in situ generated and simultaneously deposited nanocrystalline ZnO on fabrics. *ACS Appl. Mater. Interfaces* 1: 361–366.
Perkas, N., Wang, Y., Koltypin, Y. et al. 2001. Mesoporous iron-titania catalysts for cyclohexane oxidation. *Chem. Commun.* 988–989.
Perkas, N, Amirian, G., Dubinsky, S. et al. 2007. Ultrasound - assisted coating of nylon 6,6 with silver nanoparticles, and its antibacterial activity. *J. Appl. Polym. Sci.* 104: 1423–1430.
Perkas, N., Amirian, G., Applerot, G. et al. 2008a. Depositing silver nanoparticles on/in a glass slide by the sonochemical method. *Nanotechnology* 19: art. 435604 (9 pp).

Perkas, N., Shuster, M., Amirian, G. et al. 2008b. Sonochemical immobilization of silver nanoparticles on porous polymers. *J. Polym. Sci. Part A: Polym. Chem.* 46: 1719–1729.

Perkas, N., Amirian, G., Rottman, C. et al. 2009. Sonochemical deposition of magnetite on silver nanocrystals. *Ultrason. Sonochem.* 16: 132–135.

Petrulyte, S. 2008. Advanced textile materials and biopolymers in wound management. *Dan. Med. Bull.* 55: 72–77.

Pol, V.G., Reisfeld, R., and Gedanken, A. 2002a. Sonochemical synthesis and optical properties of europium oxide nanolayer coated on titania. *Chem. Mater.* 14: 3920–3924.

Pol, V.G., Srivastava, D.N., Palchik, O. et al. 2002b. Sonochemical deposition of silver nanoparticles on silica spheres. *Langmuir* 18: 3352–3357.

Pol, V.G., Wildermuth, G., Felsche, J. et al. 2006. Sonochemical deposition of Au nanoparticles on titania and the significant decrease in the melting point of gold. *J. Nanosci. Nanotechnol.* 5: 975–979.

Purwar, R. and Joshi, R.M. 2004. Recent developments in antimicrobial finishing of textiles. *AATCC Rev.* 4: 22–25.

Qi, K.H., Chen, X. Q., Liu, Y.Y. et al. 2007. Facile preparation of anatase/$SiO_2$ spherical nanocomposites and their application in self-cleaning textiles. *J. Mater. Chem.* 17: 3504–3508.

Qian, L. and Hinestroza, J.P. 2004. Application of nanotechnology for high performance textile. *J. Textile Apparel. Technol. Manage.* 4: 1–7.

Rao, C.N.R. and Cheetham, A.K. 2001. Science and technology of nanomaterials: Current status and future prospects. *J. Mater. Chem.* 11: 2887–2894.

Richard, P., LeFloch, R., Chamoux, C. et al. 1994. *Pseudomonas aeruginosa* outbreak in a burn unit: Role of antimicrobials in the emergence of multiply resistant strains. *J. Inf. Dis.* 170: 377–383.

Rosen, H. and Klebanoff, S.J. 1979. Bactericidal activity of a superoxide anion-generating system. *J. Exp. Med.* 149: 27–39.

Sawai, J., Igarashi, H., Hashimoto, A. et al. 1996a. Effect of particle size and heating temperature of ceramic powders on antibacterial activity of their slurries. *J. Chem. Eng. Jpn.* 29: 251–256.

Sawai, J., Kawada, E., Kanou, F. et al. 1996b. Detection of active oxygen generated from ceramic powders having antibacterial activity. *J. Chem. Eng. Jpn.* 29: 627–633.

Sawai, J., Shoji, S., Igarashi, H. et al. 1998. Hydrogen peroxide as an antibacterial factor in zinc oxide powder slurry. *J. Ferment. Bioeng.* 86: 521–522.

Sawai, J., Kojima, H., Igarashi, H. et al. 2000. Antibacterial characteristics of magnesium oxide powder. *World J. Microb. Biot.* 16: 187–194.

Schmidt, H. 2001. Nanoparticles by chemical synthesis, processing to materials and innovative applications. *Appl. Organometal. Chem.* 15: 331–343.

Schollmeyer, E. 2007. Nanotechnology for functionalization of textile materials. Third international textile clothing and design conference, 2006, Dubrovnik, Croatia. *Tekstil* 56: 75–80.

Scholz, J., Nocke, G., Hollstein, F. et al. 2005. Investigations on fabrics coated with precious metals using the magnetron sputter technique with regard to their anti-microbial properties. *Surf. Coat. Technol.* 192: 252–256.

Searle, A. 1919. *The Use of Metal Colloids in Health and Disease.* Sutton, New York.

Sengupta, G., Ahluwalia, H. S., Banerjee, S. et al. 1979. Chemisorption of water-vapor on zinc oxide. *J. Colloid. Interface Sci.* 69: 217–224.

Seo, D.J., Park, S.B., Kang, Y.C. et al. 2003. Formation of ZnO, MgO and NiO nanoparticles from aqueous droplets in flame reactor. *J. Nanopart. Res.* 5: 199–210.

Shen, Q., Nie, Z., Guo, M. et al. 2009. Simple and rapid colorimetric sensing of enzymatic cleavage and oxidative damage of single-stranded DNA with unmodified gold nanoparticles as indicator. *Chem. Commun.* 929–931.

Shi, L., Zhao, Y., Zhang, X. et al. 2008. Antibacterial and anti-mildew behavior of chitosan/nano-$TiO_2$ composite emulsion. *Korean J. Chem. Eng.* 25: 1434–1438.

Shukla, S., Seal, S., Schwartz, S. et al. 2001. Synthesis and characterization of nanocrystalline silver coating of fly ash cenosphere particles by electroless process. *J. Nanosci. Nanotechnol.* 417–424.

Soderberg, T., Agren, M., Tengrup, I. et al. 1989. The effects of an occlusive zinc medicated dressing on the bacterial flora in excised wounds in the rat. *Infection* 17: 81–85.

Sojka-Ledakowicz, J., Lewartowska, J., Kudzin, M. et al. 2008. Modification of textile materials with micro- and nano-structural metal oxides. *Fibers Text. East. Eur. A* 16: 112–116.

Stengl, V., Bakardjieva, S., and Marikova, M. et al. 2003. Magnesium oxide nanoparticles prepared by ultrasound enhanced hydrolysis of Mg-alkoxides. *Mater. Lett.* 57: 3998–4003.

Suslick, K.S. 1989. The chemical effects of ultrasound. *Sci. Am.* 260: 80–86.

Suslick, K.S. and Price, G.J. 1999. Application of ultrasound to materials chemistry. *J. Annu. Rev. Mater.* 29: 295–326.

Suslick, K.S., Hammerton, D.A., and Cline R.E. 1986. Sonochemical hot spot. *J. Am. Chem. Soc.* 108: 5641–5642.

Toxicological profile of copper. 2004. U.S. department of health and human services, public health service agency for toxic substances and disease registry. *Textiles* 35: 323–335.

Tricoli, A., Graf, M., Mayer, F. et al. 2008. Micropatterning layers by flame aerosol deposition annealing. *Adv. Mater.* 20: 3005–3010.

Uddin, F. 2008. Some aspects in advancement of textile nanofinishing. *86th Textile-Institute World Conf.* 2:U291–U307. Hong Kong: Polytechnic Univercity, Institute of Textiles and Clothing Press.

Vajnhandl, S. and Le Marechal, A.M. 2005. Ultrasound in textile dyeing and the decolouration/mineralization of textile dyes. *Dyes Pigm.* 65: 89–101.

Vallee, B.L. and Falchuk, K.H. 1993. The biochemical basis of zinc physiology. *Physiol. Rev.* 73: 79–118.

Vigneshwaran, N., Kumar. S., Kathe, A.A. et al. 2006. Functional finishing of cotton fabrics using zinc oxide-soluble starch nanocomposites. *Nanotechnology* 17: 5087–5095.

Wang, R., Xin, J.H., Tao, X.M. et al. 2004. ZnO nanorods grown on cotton fabrics at low temperature. *Chem. Phys. Lett.* 398: 250–255.

Wang, H., Wang, J., Hong, J. et al. 2007. Preparation and characterization of silver nanocomposite textile. *J. Coat. Technol. Res.* 4: 101–106.

Wang, C., Li, M., Wu, M. et al. 2008. Cotton fabric properties with water-repellent finishing via sol-gel process. *Sur. Rev. Lett.* 15: 833–839.

Wendler, F., Meister, F., Montigny, R. et al. 2007. A new antimicrobial ALCERU fibre with silver nanoparticles. *Fibers Text. East. Eur.* 15: 64–65.

Wong, Y.W.H., Yuen, C.W.M., Leung, M.Y.S. et al. 2006. Selected applications of nanotechnology in textiles. *AUTEX Res. J.* 6: 1–8.

Wright, G.D. 2007. On the road to bacterial cell death. *Cell* 130: 781–783.

Xu, B. and Cai, Z.S. 2008. Fabrication of a superhydrophobic ZnO nanorod array film on cotton fabrics via a wet chemical route and hydrophobic modification. *Appl. Sur. Sci.* 254: 5899–5904.

Xue, C.H., Jia, S.T., Chen, H.Z. et al. 2008. Superhydrofobic cotton fabrics prepared by sol-gel coating of $TiO_2$ and surface hydrofobization. *Sci. Technol. Adv. Mater.* 9: art. 035001 (5 pp).

Yamamoto, O. 2001. Influence of particle size on the antibacterial activity of zinc oxide. *Int. J. Inorg. Mater.* 3: 643–646.

Yeo, S.Y., Lee, H.J., and Jeong, S.H. 2003. Preparation of nanocomposite fibers for permanent antibacterial effect. *J. Mater. Sci.* 38: 2143–2147.

Yuranova, T., Rincon, A. G., Pulgarin., C. et al. 2006. Performance and characterization of Ag-cotton and Ag/$TiO_2$ loaded textiles during the abatement of *E. coli*. *J. Photochem. Photobiol. A.* 181: 363–369.

Zalesky, T. 2008. "Active" products drive wound-care market. *GEN Biobusiness Biomarket Trends* 28(9): 1–2, May 1. http://genengnewa.com/articles/chitem

Zhong, Z.Y., Prozorov, T., Felner, I. et al. 1999. Sonochemical synthesis and characterization of iron oxide coated on submicrospherical alumina: A direct observation of interaction between iron oxide and alumina. *J. Phys. Chem. B.* 103: 947–956.

Zohdy, M.H., Kareem, H.A., El-Naggar, A.M. et al. 2003. Microbial detection, surface morphology, and thermal stability of cotton and cotton/polyester fabrics treated with antimicrobial formulations by a radiation method. *J. Appl. Polym. Sci.* 89: 2604–2610.

# 15

## *Inorganic Nanoparticles for Environmental Remediation*

**Thomas B. Scott**

**CONTENTS**



## 15.1 Overview

In the last 2000 years, industrial activity and associated technological developments have led to the world of computers, space exploration, and nanotechnology that we live in today. An unfortunate by-product has been environmental pollution, the majority of which has occurred since the onset of the industrial revolution in the 1800s, but examples can be traced back to Roman and Greek settlements. The quantity and type of emissions have changed over the past 50–60 years, with pollution from complex chemicals, dense non-aqueous liquid phases (DNAPLs), and radioactive metals adding to those released into the environment. Only in recent years has their fate, transport, and toxicology begun to be thoroughly investigated, revealing their potential impact on the environment and ultimately the human population.

Human populations rely on clean water, and it is generally accepted that pollution poses the greatest risk to human populations through its contamination. For example, in the Sindh Province of Pakistan, 36% of the population is exposed to arsenic in drinking water at levels of over 10 ppb and 16% is exposed to levels over 50 ppb. In extremis, arsenic levels of up to 800 ppb have been detected in drinking water supplies in the city of Johi in the Dadu District (Ahmad et al., 2004). Arsenic is well recognized as a persistent, bioaccumulative toxin and these recorded values are far in excess of the 10 ppb maximum permissible concentration (MAC) limit set by the World Health Organization (WHO).

## 15.2 Methods for Addressing Environmental Pollution

The goal of environmental remediation is clean water, and various technical measures are available to achieve this end. Remedial measures may act to remove or destroy the polluting contaminant (either in situ or ex situ), confine the contaminant (to prevent migration), or, in cases where pollution is extensive, provide groundwater treatment at the point of use.

Traditional methods of site remediation have typically involved soil excavation and disposal without treatment, or groundwater extraction with ex situ treatment known as pump-and-treat technology (Simon et al., 2002). These straightforward methods are well established but rather costly.

Currently, the trend in the development of remediation technologies is away from pump-and-treat processes in favor of in situ applications. For pollutant heavy metals and radionuclides, in situ methods involve immobilization of the contaminant by physically, chemically, or biologically mediated binding or precipitation within the soil or sediment. For organic contaminants such as herbicides, pesticides, and VOCs, in situ methods typically involve destruction or degradation of pollutant species to form harmless breakdown products. Overall, these treatments tend to be cheaper and more environmentally compatible than pump-and-treat processes, but, the primary driver for alternative technologies lies with the requirement for greater efficiency in terms of both the speed and the level of pollution reduction. The most notable methods for the remediation of pollutants in the near-surface geosphere are outlined in Table 15.1.

Many classical remediation methods are regarded as being too costly for extensive deployment in the developing world and well beyond economic feasibility for most rural communities. Nanotechnology seems well placed to restructure the remediation industry, offering significant reductions in the overall costs of cleanup of large-scale contaminated sites while also presenting the prospect of reduced cleanup times and near total removal of contaminants from groundwater (Masciangioli and Zhang, 2003; Whatlington, 2005, Sheppard, 2007).

**TABLE 15.1**

Most Notable Methods for Environmental Remediation of Subsurface Contaminants

| Remediation Type | Remediation Technology | Brief Description |
|---|---|---|
| Traditional remediation methods | Pump-and-treat | Contaminated material or fluids are removed and treated ex situ using processes such as washing, ion exchange, reverse osmosis, reductive precipitation, heating, and bioremediation to remove contaminants |
| | Excavation and disposal | Contaminated soil and sediment are removed and disposed without any treatment |
| Methods for enhancing pump-and-treat technology | In situ flushing | Injection or infiltration of an aqueous solution into a zone of contaminated soil/groundwater, followed by downgradient extraction and treatment of the elutriate—an improvement technology for pump-and-treat systems |
| | Hydraulic fracturing | Used to increase the removal rate of contaminants by other methods, such as pump-and-treat |
| Chemical in situ methods | Permeable reactive barriers | In situ treatment that uses different materials to chemically filter contaminants from groundwater |
| | In situ redox manipulation | Works in a similar way to a PRB but with injection of chemical reductants |
| | Nanoparticle injection | Particles are injected to the subsurface to immobilize or destroy contaminants through chemical interactions |
| Electrical in situ methods | Solidification and Stabilization | Processes designed to reduce the mobility of contaminants by reducing the contaminant solubility or the permeability of the medium |
| | In situ vitrification | An immobilization and destruction technology, where soils are heated and melted by applying an alternating electrical current and left to cool, forming an insoluble crystalline or vitrified material |
| | Electrokinetic processes | Concentration of inorganic contaminants through application of an electric DC field to a contaminated subsurface volume |
| Biological methods | Phytoremediation | The use of plants to extract metals and metalloids from contaminated soils |
| | Bioremediation | Processes through which contaminants are degraded, mobilized, or immobilized as a direct result of microbiological activity |

## 15.3 Why Nanotechnology?

So how can engineered nanomaterials deliver better, faster, and cheaper remediation of contaminated land? The answer lies primarily with the size of the materials, which contributes a number of advantageous properties. The high surface area to volume ratio and high surface energy (Zhang et al., 1998) exhibited by nanosized structures (particles, tubes, wires, etc.) means that they offer similar or slightly enhanced reactivity to conventional materials but at a fraction of the mass. By using a smaller mass of reactive material to achieve the same objective (i.e., site remediation), both raw materials and energy are conserved during manufacturing (Masciangioli and Zhang, 2003), resulting in potentially significant cost savings. Enhanced reactivity over macroscale materials of similar composition is commonly attributed to a greater density of reactive surface sites or surface sites of higher intrinsic reactivity (Klabunde et al., 1996; Mulvaney, 2001). Below approximately 10 nm, properties such as the free energy and work function have been observed to change significantly with decreasing size, following classical equations such as the Gibbs–Thompson relation between particle size and energy (Campbell and Parker, 2002). At such scales, quantum size effects that alter the electronic structure of the material can also be observed (Brus, 1986; Wang and Herron, 1991; Elliott, 1998). For metallic nanostructures, quantum effects are observed to influence physical and chemical properties when the size is less than approximately 5 nm (Klimenkow et al., 1997; Nepijko et al., 1998, 1999; Gan et al., 2001; Campbell and Parker, 2002). For materials with lower electron density, such as semiconductors and oxides, quantum effects are observed at larger nanostructure sizes (10–150 nm) (Punoose et al., 2001; Baer et al., 2003). With decreasing structural size at the nanoscale, quantum effects can cause changes in the Fermi level and band gap of a material, leading to measurable increases in reactivity (Sharma et al., 2003). In relation to aqueous contaminant interactions for remedial applications, the enhanced reactivity of different nanomaterials (relative to bulk comparators) is reported to frequently result in rapid removal or degradation of the contaminant (Cantrell et al., 1995; Wang and Zhang, 1997; Li et al., 2006).

## 15.4 Natural Nanoparticles in the Environment

Nanoparticles have existed in the environment since the formation of the Earth itself. For example, it is well recognized that nanoparticles can be produced during volcanic eruptions (Aiken, 1884), at deep sea hydrothermal vents (Luther and Rickard, 2005), and in numerous other ways (Figure 15.1). With the advent of human activity, three types of nanoparticles can be considered to exist in environmental systems: natural, incidental, and engineered. This chapter is primarily concerned with the engineered particles, but it should also be recognized that naturally occurring nanoparticles may also exhibit an effect on the transport and fate of contaminants in the environment. Naturally formed submicron-scale particles (termed colloids) are typically a mixture of crystalline silicate clays (smectite, kaolinite, chlorite, vermiculite), noncrystalline silicate clays (allophane and imogolite), iron and aluminum oxides (gibbsite, goethite, magnetite), and organic colloids (humus). For groundwater contaminants with high affinity to the solid phase (e.g., pesticides, HOCs etc.), natural nanoparticles can provide a low energy surface for binding and subsequently enhance the spread of pollution by acting as a vehicle for contaminant transport in the subsurface (Vinten et al., 1983; Sabatini and Knox, 1992; Grolimund et al., 1996; Saiers and Hornberger, 1996; Roy and Dzombak, 1998; McCarthy and McKay, 2004; Grolimund and Borkovec, 2005; Frimmel et al., 2007). Acting against colloidal transport of contaminants, other natural subsurface processes may have a remedial effect on contaminants when conditions are favorable. The "natural attenuation processes" include a variety of naturally occurring physical, chemical, or biological processes that act to reduce mass, toxicity, mobility, volume, or concentration of contaminants in soil or groundwater (CGER, 2000; Yong and Mulligan, 2003). These in situ processes include biodegradation,

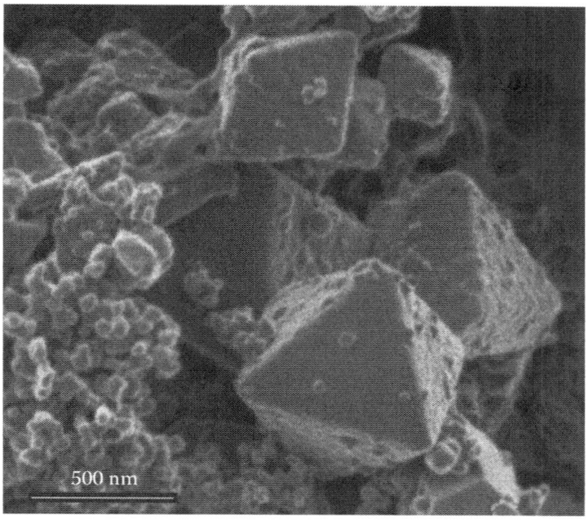


**FIGURE 15.1**
Colloidal-sized particles of naturally formed magnetite ($Fe_3O_4$).

dispersion, dilution, sorption, volatilization, and chemical or biological stabilization, transformation, or destruction of contaminants (Domenico and Schwartz, 1990), some of which would also be applicable to engineered nanoparticles when present in environmental systems.

## 15.5 Engineered Nanoparticles for Environmental Remediation

Although the array of available environmental nanotechnologies is quickly growing, for regular commercial applications, the principle nanomaterials used for remediation of contaminated land are nanoscale iron particles and their associated derivatives. According to the "Project on Emerging Nanotechnologies" (PEN, 2009), there are currently 42 site-remediation projects in the United States utilizing nanotechnology as a remedial solution, with 33 of these sites employing iron-based nanoparticles, hereafter termed INPs. Resultantly, much of the focus for the present chapter is on describing the reactivity, technological development, and applications of INPs in land remediation.

The use of iron-based reagents for the removal or degradation of contaminants or toxicants in industrial waste streams has been frequently studied over the past 40 years. For land remediation, interest in the use of metallic iron, also termed "zero-valent iron" (ZVI) began in the mid-1990s, heralded by the first *Symposium on Contaminant Remediation with Zero-Valent Metals* which took place as part of the *209th American Chemical Society National Meeting* in Anaheim, CA in April 1995.

Early work by Khudenko (1991) demonstrated that zero-valent iron could be used for the destruction of organics in industrial wastewater, while others demonstrated the removal of halogenated compounds via chemical reduction (Gillham and O'Hannessin, 1994; Matheson and Tratnyek, 1994; Gavaskar et al., 1998). Laboratory tests also indicated that iron surfaces could be used to remove a wide range of inorganic contaminants from water, including As, Cd, Cr, Cu, Hg, Fe, Mn, Mo, Ni, Pb, Se, Tc, U, V, $NO_3$, $PO_4$, and $SO_4$ (see Tratynek, 1996; Bigg and Judd, 2000; Blowes et al., 2000; Cundy et al., 2008 for reviews). More recently, metallic iron has been identified as useful for the uptake of radionuclides from aqueous solution (Allen et al., 2004; Scott et al., 2005a,b).

### 15.5.1 Decontamination of Groundwater Using Zero-Valent Iron

For contaminated sites, the initial focus was on the conceptual and demonstrative use of scrap iron or iron filings as a reactive material in Permeable Reactive Barrier (PRB) systems (Morrison and Spangler, 1993; Smyth et al., 1997). PRB technology offers an alternative cost-effective in situ approach to classical methods of site remediation by using a permeable wall of reactive material to filter contaminants from groundwater (Figure 15.2).

Barrier systems are placed in the path of a migrating plume of contaminated groundwater and physical, chemical, or biological processes occurring within the reactive wall (Scherer et al., 2000) act to remove or destroy the target contaminants and prevent their downgradient transport. The most commonly used processes are redox reactions, precipitation, adsorption, ion exchange, and biodegradation, with metallic iron most commonly used as the reactive material, with others including limestone ($CaCO_3$), lime (CaO), ferric oxides (FeOOH, $Fe_2O_3$ and $Fe_3O_4$), zeolite, activated carbon, coal, lignite, and peat (Simon et al., 2002). Sustainable waste materials such as flyash, oyster shells (nominally $CaCO_3$), and recycled concrete have also been investigated (Golab et al., 2006). The most important factors for selecting a reactive material are the type of contaminants to be remediated and the reactive longevity of the material i.e., a suitable material must be able to treat water for periods of years to decades (Waybrant et al., 1998).

Over the last decade, PRBs using metallic iron have been used successfully on a significant number of polluted sites and proven effective for treatment of acid-mine drainage, dissolved nutrients, numerous heavy metals, radionuclides, and other inorganics such as phosphorus, arsenic, and selenium (Blowes et al., 2000 and references therein). At relatively shallow depths, these passive treatment systems are simple and cost effective

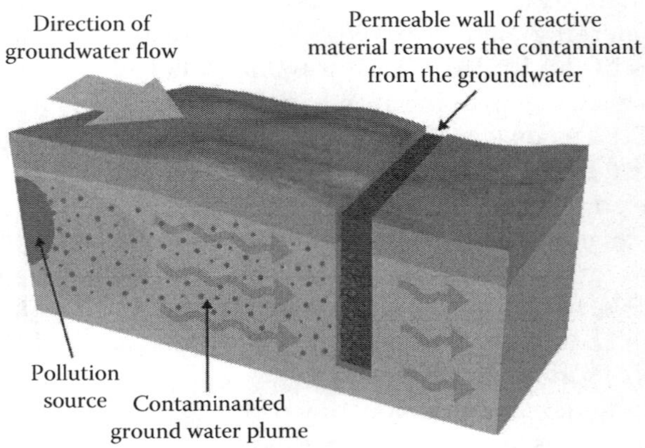


**FIGURE 15.2**
**(See color insert following page 302.)** Simplified model of a permeable reactive barrier (PRB) system.

to implement, offering remediation costs up to 50% lower than those of pump-and-treat methods (Simon et al., 2002) and causing only limited environmental disruption.

### 15.5.2 Iron Nanoparticles: Scaling Down from the Microscale to Nanoscale

The relative success of PRBs as a novel and cost-effective tool for site remediation led logically to the development and use of nanoscale iron. The high surface area to volume ratio and high surface energy (Zhang et al., 1998) of INPs means that they offer a greater reactivity than the surfaces of bulk granular iron used in PRBs and have been repeatedly demonstrated to degrade contaminants more rapidly (Wang and Zhang, 1997; Lien and Zhang, 1999; Choe et al., 2000; Ponder et al., 2000; Lien and Zhang, 2001; Schrick et al., 2002; Glazier et al., 2003; Li and Zhang, 2007) (Figure 15.3). Because a smaller mass of material is needed to have the same remedial effect, there are also significant cost savings to be made (in terms of raw materials and manufacture), with the added bonuses of improved versatility and further reduced environmental disruption.

To date, iron nanoparticles have been shown to be effective remediators of a range of contaminants including chlorinated organics (Wang and Zhang, 1997; Zhang et al., 1998, Lien and Zhang, 1999; Kim and Carraway 2000; Elliot and Zhang, 2001; Lien and Zhang, 2001; Schrick et al., 2002; Liu et al., 2005, Nurmi et al., 2005; Cheng et al., 2007) and inorganic anions (Choe et al., 2000; Alowitz and Scherer, 2002; Mondal et al., 2004; Cao et al., 2005; Kanel et al., 2005) among others (Miehr et al., 2004; Shimotori et al., 2004; Feitz et al., 2005; Nurmi 2005; Joo et al., 2005). In addition, INP have also been shown to successfully remediate solutions contaminated with a range of metals, including Lead (Ponder et al., 2001; Li and Zhang, 2007), Chromium, (Ponder et al., 2001; Alowitz and Scherer, 2002; Miehr et al., 2004; Shimotori et al., 2004), Copper (Miehr et al., 2004; Li and Zhang, 2007; Karabelli et al., 2008), Cobalt (Uzum et al., 2008), Arsenic (Kanel et al., 2005, 2007; Burghardt et al., 2007), Nickel (Li and Zhang, 2007), Zinc (Li and Zhang, 2007), Cadmium (Li and Zhang, 2007), and Silver (Li and Zhang, 2007). The application of zero-valent INP for the remediation of radionuclides remains less widely researched than that for the aforementioned heavy metals and organic contaminants with studies limited to the radioisotopes of Ba (Çelebi et al., 2007), $TcO_4$ (Ponder et al., 2001; Darab et al., 2007), and Uranium (Scott, 2005; Riba et al., 2008). The list of contaminants known to be treatable with INP is extensive and still growing. New investigations are reported on a frequent basis and highlight the great weight

**FIGURE 15.3**
Electron microscopy images of (left) aggregated microscale metallic iron particles and (right) aggregated nanoscale iron particles.

of interest in this field of technology, which may have transferrable application to other industries such as medicine or nuclear decommissioning.

### 15.5.3 Iron Nanoparticle Synthesis

INP can be synthesized by a variety of methods including the reduction of goethite (FeOOH) with heat (Nurmi et al., 2005), hydrogenation of bis(ditrimethylsilyl)amido iron complexes (Margeat et al., 2005), decomposition of iron pentacarbonyl, $Fe(CO)_5$, in organic solvents or argon (Elihn et al., 1999; Choi et al., 2001; Karlsson et al., 2005), vacuum sputtering (Kuhn et al., 2002), or chemical vapor deposition (Zaera, 1989; Elihn et al., 2001). These methods have been more comprehensively reviewed elsewhere (Huber, 2005). However, the most commonly used method, and the one utilized in the majority of environmental applications, is via the reduction of Fe(II) or Fe(III) to a metallic state using sodium borohydride ($NaBH_4$). This method was first described by Wang and Zhang (1997) and occurs via the following reaction:

$$4Fe^{3+} + 3BH_4^- + 9H_2O \rightarrow 4Fe(0)\downarrow + 3H_2BO_3^- + 12H^+ + 6H_2$$

INP produced by this method (Figure 15.4) are typically characterized by a metallic Fe core with diameter of approximately 20–80 nm (Zhang, 2003; Nurmi et al., 2005; Sun et al., 2006) surrounded by an oxide layer/shell. Although the shell predominantly consists of Fe oxides, a small percentage of oxidized B is generally present as a consequence of the manufacturing process (Nurmi et al., 2005; Sun et al., 2006; Li and Zhang, 2007). The process also results in the formation of particles that are highly polydispersed, ranging over tens to hundreds of nanometers in size and prone to agglomeration.

By forming INPs via chemical reduction in the presence of certain chemicals or compounds, it is possible to tune INP synthesis to produce nanoparticles with a tightly controlled particle size distribution. For example, work by Huang et al. (2008) used supported polyelectrolyte multilayers to demonstrate tunable synthesis of INP, yielding nanoparticles

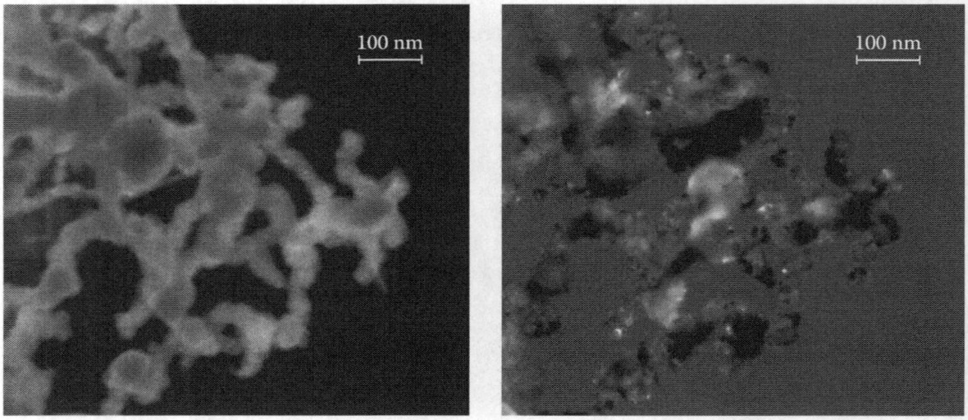


**FIGURE 15.4**
Secondary electron (left) and transmitted electron (right) images of a fractal cluster of INP showing the poorly formed but polycrystalline nature of individual particles formed by sodium borohydride reduction of an iron salt solution.

of uniform sizes by altering the multilayer assembly conditions and the number of reaction cycles. Shao et al. (2005) demonstrated tunable INP synthesis using thermal decomposition of iron pentacarbonyl mixed with oleylamine (as a surfactant), where particle size was controlled by limiting the decomposition time and varying the surfactant concentration. For the borohydride reduction method, the addition of poly acrylic acid (PPA) to the iron salt solution prior to reduction is also found to be a method for tuning particle size (Yang et al., 2007).

INP produced via all these different methods is known to have slightly different surface chemistries (Kuhn et al., 2002; Signorini et al., 2003; Nurmi et al., 2005), which can subtly alter their efficiency for contaminant interaction. If INP remediation technology gains industrial popularity then competition between emergent INP producers is likely to be based on particle reactivity and price. Consequently, much research effort is currently being invested into finding methods for mass production of INP in multikilogram or even tonne quantities, while maintaining or improving reactivity.

### 15.5.4 Environmental Reactivity of Metallic Iron Nanoparticles

While the list of contaminants that can be removed from water by INPs is extensive, the mechanism by which they work is not always the same (Miehr et al., 2004). In fact, multiple interactions such as surface complexation, sorption, chemical reduction, and (co)precipitation are all possible (Shokes and Möller, 1999; Miehr et al., 2004). The underlying basis for all of these reactions is the aqueous corrosion, or rusting, of the metallic iron.

#### *15.5.4.1 Aqueous Corrosion of Iron Nanoparticles*

Metallic iron ($Fe^0$), also referred to as zero-valent iron (ZVI), is recognized as being very chemically reactive. When added to a groundwater system (as PRB material or injected INPs), water ($H_2O$) is the primary species available for interaction. It is well accepted that in moist air or aqueous systems the corrosion of the metal will rapidly occur. Corrosion is considered to occur via either chemical or electrochemical oxidation, with the latter providing the dominant mechanism in environmental systems (see Noubactep, 2008 and references therein). Electrochemical oxidation consists of both anodic and cathodic components. The anodic process is the oxidation and dissolution of iron (Equation 15.1):

$$Fe^0 \rightarrow Fe^{2+} + 2e^- \tag{15.1}$$

The associated cathodic process can follow two different paths depending on the nature of the aqueous environment. Under aerobic conditions, typical of vadose zone soils or shallow oxygenated groundwaters, corrosion of metallic iron can occur through reaction with dissolved oxygen:

$$2Fe^0_{(s)} + 4H^+_{(aq)} + O_{2(aq)} \rightarrow 2Fe^{2+}_{(aq)} + 2H_2O_{(1)} \tag{15.2}$$

The standard electrode potential ($E^0$) for this reaction is +1.71 V, indicating that from a thermodynamics perspective the reaction is strongly favorable and likely to be spontaneous (Snoeyink and Jenkins, 1980). Implicit in the above reaction is the associated increase of solution pH as protons are consumed, a process that is commonly observed during the early stages of laboratory-scale nanoparticle experiments. Under certain conditions, where dissolved oxygen is plentiful, the ferrous iron would then undergo a further oxidation step to ferric iron, $Fe^{3+}$ (Equation 15.4), and form iron hydroxide–based precipitates (e.g., FeOOH).

Alternatively, in groundwaters that are oxygen poor or anaerobic (e.g., deep aquifers), cathodic iron corrosion will proceed through direct reduction of water:

$$Fe^0_{(s)} + 2H_2O_{(1)} \rightarrow Fe^{2+}_{(aq)} + H_2(g) + 2OH^-_{(aq)} \quad (15.3)$$

The standard electrode potential ($E^0$) for this reaction is −0.39 V (Snoeyink and Jenkins, 1980), indicating that it would not be favorable in environmental systems. Indeed in natural waters the kinetics of these and other reactions are reported to be sluggish, and, consequently, chemical equilibrium is generally not attained (Langmuir, 1997). However, it should also be noted that the anoxic reaction evolves hydrogen gas, a reductant species that may subsequently undergo desirable remedial interactions with aqueous contaminants.

Each of the above reactions (Equations 15.1 through 15.3) assumes direct contact between metallic iron and water. However, in reality, direct interaction is not always possible due to the obstructive presence of an oxide film at the surface of the metal that is produced by the aforementioned corrosion reactions. In aqueous environments, the first iron corrosion product to form is amorphous ferrous hydroxide ($Fe(OH)_2$), which then gets converted to the iron oxide magnetite ($Fe_3O_4$) (Odziemkowski et al., 1998) and forms a thin film on the metal surface. Other mixed-valent iron salts known as green rusts and ferric hydroxides (FeOOH) are also known to form, which can further cover the metal surface. The film of surface oxide then limits direct metal–$H_2O$ and metal–contaminant interactions and is recognized as the most significant factor in controlling the corrosion of metallic iron in natural waters (Noubactep, 2008).

It should also be noted that because the kinetics of the initial stages of Fe(0) oxidation are rapid, any metallic iron that is introduced to an environmental system (whether as granular or nano-iron) will already have a film of surface oxide acquired directly after synthesis, which may imbue some initial degree of corrosion resistance (Figure 15.5). Hereafter, the

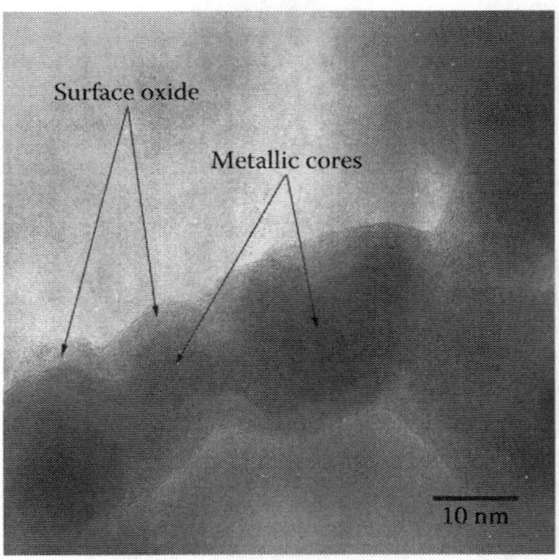


**FIGURE 15.5**
High resolution transmitted electron image of an iron nanoparticle cluster. The surface oxide surrounding the metallic core of each particle can be readily defined.

term "iron surface" is inclusive and implicit of surface oxide (predominantly magnetite) at the outermost surface of the metal.

#### *15.5.4.2 Remedial Reactions between Iron and Aqueous Contaminants*

The previous section established that, in the environment, metallic iron (with its strong tendency to donate electrons in the presence of oxygen and water) will undergo electrochemical oxidation, with the attributed surface-mediated reactions typically causing rapid consumption of dissolved oxygen and release of $Fe^{2+}$ into solution. It is also well established that based on the same electrochemical activity metallic iron in water will also readily react with a wide variety of redox-amenable contaminants. The mechanism for contaminant removal is observed to vary for different contaminants (Miehr et al., 2004) with numerous possible pathways for removal including adsorption, precipitation, coprecipitation, and different types of Fe-linked chemical reduction (Noubactep, 2008) that are generally surface mediated (Table 15.2). Specific examples for tetrachloroethene (TCE) and uranium are provided in the following section.

Aqueous adsorption of contaminants onto iron (and mineral) surfaces is well recognized (Domenico and Schwartz, 1990). The process is most strongly dependent on solution pH, temperature, contaminant concentration, and the number of available adsorption sites. Although the process is reversible in natural systems, engineered systems using reactive materials provide zones of significant chemical disequilibrium in which contaminant adsorption is dominant over desorption. Adsorption is incredibly important because without it, surface-mediated chemical reduction cannot occur. Surface reduction is *always* preceded by contaminant adsorption.

Precipitation and coprecipitation are closely related phenomena and are applicable for the removal of inorganic contaminants (e.g., heavy metals and radionuclides) from water. Direct precipitation of contaminant-based compounds is dependent on the aqueous concentration of the contaminant with respect to its solubility limit and will only occur if the solution (in this case groundwater) becomes oversaturated with the contaminant. Precipitation is not likely to occur quantitatively in groundwater systems but may occur in a reactive barrier system resulting from changing water chemistry (pH and Eh) and permeability (Morrison et al., 2006). Contaminant removal by coprecipitation involves entrapment of contaminant ions within the structure of other mineral phases as they precipitate from solution and grow e.g., $CaCO_3$ or FeOOH (Sridharan and Lee, 1972; De Carlo et al., 1981; Crawford et al., 1993). Initial trapping is considered to occur via surface adsorption

**TABLE 15.2**

Possible Reaction Pathways for Contaminant (Ox) Removal from the Aqueous Phase in a $Fe^0$-$H_2O$ System and Their Reversibility under Natural Conditions

| Mechanism | Reaction | Reversibility | Equation |
|---|---|---|---|
| Adsorption | $S_{(sorption\ site)} + Ox \Leftrightarrow S\text{-}Ox$ | Reversible | (15.4) |
| Precipitation | $Ox_{(aq)} + n_{\bar{O}H} \Leftrightarrow Ox(HO)_{n(s)}$ | Reversible | (15.5) |
| Coprecipitation | $Ox + nFe_x(OH)_y^{(3x-y)} \Leftrightarrow Ox - [Fe_x(OH)_y^{(3x-y)}]_n$ | Irreversible | (15.6) |
| $Fe^0$ reduction | $Fe^0 + Ox_{(aq)} \Rightarrow Red_{(s)} + Fe^{2+}$ | Irreversible | (15.7) |
| $Fe^{2+}_{(aq)}$ reduction | $Fe^{2+}_{(aq)} + Ox_{(aq)} \Rightarrow Red_{(s)} + Fe^{3+}_{(aq)}$ | Irreversible | (15.8) |
| $Fe^{2+}_{(s)}$ reduction | $Fe^{2+}_{(s)} + Ox_{(aq)} \Rightarrow Red_{(s)} + Fe^{3+}_{(s)}$ | Irreversible | (15.9) |
| $Fe^{2+}_{(org)}$ reduction | $Fe^{2+}_{(org)} + Ox_{(aq)} \Rightarrow Red_{(s)} + Fe^{3+}_{(org)}$ | Irreversible | (15.10) |

on nascent material. Once entrapped, the rerelease of contaminant ions into solution becomes entirely dependent on the dissolution of the host mineral phase. When using ZVI as a reactive material for remediation, this pathway is most significant in aerated waters at neutral pH, where iron supersaturation in the vicinity of the Fe(0) surfaces yields precipitation of amorphous and crystalline iron oxyhydroxides ($Fe_3O_4$, $Fe_2O_3$, FeOOH, $Fe(OH)_2$, $Fe(OH)_3$), which are well-known adsorbents of organic and inorganic compounds (Cornell and Schwertmann, 1996; Brown et al., 1999; and references therein). Iron oxide precipitation is recognized as a dynamic process (Schmuki, 2002), with some inflowing contaminants adsorbing onto both aged and nascent iron oxyhydroxides, with resultant structural entrapment (coprecipitation) occurring on the nascent phases with continued ageing.

By far, the most recognized mechanism by which ZVI removes contaminants from groundwater is via chemical reduction. Reduction is possible by either Fe(0) or Fe(II) (Equations 15.7 through 15.10) and requires the contaminant to be adsorbed on to or in close proximity (electronic range) of the iron surface. Until recently, contaminant reduction was considered to be predominantly driven by the direct interaction of Fe(0). The limiting effect provided by the omnipresent layer of surface oxide was largely ignored until a compelling volume of data had accumulated (Noubactep et al., 2005; Noubactep, 2008) to recognize that since oxide films are initially porous, reduction can occur either at the $Fe^0$ surface or within the oxide film.

On comparison of $E^0$ values for the two implicated redox couples, the Fe(II):Fe(0) couple ($E^0 = -0.44$ V) has a negative potential compared to the aqueous Fe(III):Fe(II) couple ($E^0 = +0.77$ V), indicating that thermodynamically it will be much more favorable for degradation or immobilization of redox-labile compounds (Gillham and O'Hannessin, 1994; Matheson and Tratnyek, 1994; Weber, 1996; Singh and Singh, 2003). However, surface-coordinated $Fe^{2+}$ species either adsorbed on mineral surfaces or as "structural" $Fe^{2+}$ ions within an oxide/hydroxide lattice (magnetite or green rust) has also been identified as a strong reductant for various contaminants (Stucki, 1988; Fendorf and Li, 1996; White and Peterson, 1996; Liger et al., 1999; Scott et al., 2005a,b). Work by White and Peterson (1996) demonstrated that the oxidation potential of adsorbed or structural Fe(II) ($E^0 = -0.34$ to $-65$ V) can be more powerful for contaminant reduction than the surface of Fe(0) ($E^0 = -0.44$ V). Similarly, Stumm and Sulzberger (1992) showed that at a pH of ≥7, inner-sphere complexation of $Fe^{2+}$ to metal oxides can also create a stronger reductant.

Additionally, recent results from Naka and coworkers (Naka et al., 2006) have also demonstrated that, when complexed with organic substances, aqueous Fe(II) is a significantly more powerful reductant ($+0.520 \geq E^0(\text{V}) \geq -509$) than uncomplexed aqueous Fe(II) ($E^0 = 0.77$ V).

Therefore, Fe(0) is not necessarily the most powerful reductant in an Fe-$H_2O$ system (Table 15.3) and competition for contaminant reduction and removal is likely to occur, with different mechanisms predominating as the system ages. The activity of Fe(0) as a reductant is likely to occur early in the lifecycle of the system before significant volumes of corrosion products can accumulate at the iron surface and limit the migration of contaminants through the oxide to access underlying Fe(0). Subsequently, contaminant reduction by Fe(II) may become more thermodynamically favorable within the oxide film than at the Fe(0) interface, with the oxide acting as both an ionic and electronic conductor to mediate contaminant reduction, (Noubactep, 2008).

Overall, numerous different removal pathways are possible for contaminant remediation by ZVI. The mechanism of remediation varies depending on the nature of the contaminant (Miehr et al., 2004), e.g., the oxidation of $Fe^{(0)}$ is recognized to drive the reductive transformation of chlorinated organics, such as the carcinogenic trichloroethylene

**TABLE 15.3**

Standard Electrode Potentials of all Possible Redox Couples of Iron Relevant for INP Corrosion, Ubiquitous Groundwater Oxidants ($H^+$, $O_{2(aq)}$), and Two Selected Contaminants ($C_2Cl_4$, $UO_2^{2+}$)

| Redox Couple | $E^0$ (V) (SHE) | Equation |
|---|---|---|
| $Fe^0 \Leftrightarrow Fe^{2+} + 2\,e^-$ | −0.44[a] | (15.1) |
| $Fe^{2+}_{(s)} \Leftrightarrow Fe^{3+}_{(s)} + e^-$ | −0.34/−0.65[b] | (15.4) |
| $Fe^{2+}_{(org)} \Leftrightarrow Fe^{3+}_{(org)} + e^-$ | 0.52/−0.51[c] | (15.5) |
| $2\,H^+ + 2\,e^- \Leftrightarrow H_2$ (g) | 0.00[a] | (15.6) |
| $UO_2(OH)_2 + 2\,H^+ + 2\,e^- \Leftrightarrow UO_{2(s)} + 2H_2O$ | 0.57/0.43[d] | (15.7) |
| $C_2Cl_4 + H^+ + 2\,e^- \Leftrightarrow C_2HCl_3 + Cl^-$ | 0.59 | (15.8) |
| $Fe^{2+} \Leftrightarrow Fe^{3+} + e^-$ | 0.77[a] | (15.9) |
| $O_2 + 2\,H_2O + 4\,e^- \Leftrightarrow 4\,OH^-$ | 0.81[a] | (15.10) |

[a] Gerasimov et al. (1985).
[b] White and Peterson (1996).
[c] Naka et al. (2006).
[d] Billon et al. (2005).

(TCE), to relatively innocuous hydrocarbons (Sirk et al., 1995). For waters containing heavy metals and/or radionuclides, decontamination occurs via sorption and/or reduction onto the surface of the iron (Riba et al., 2008). In a study by Li and Zhang (2007), it was demonstrated that for metal ions such as Zn(II) and Cd(II), which have standard potentials ($E^0$) very close to, or more negative than, that of iron (−0.44 V), the removal mechanism is sorption and/or surface complexation, while for metals such as Cu(II), Ag(I), and Hg(II), which have $E^0$ much more positive than iron, removal occurs predominantly via reduction. For those metals with $E^0$ only slightly more positive than iron, e.g., Ni(II) and Pb(II), both sorption and reduction have been shown to occur at the iron surface, depending on water chemistry.

#### *15.5.4.3 Iron Nanoparticles for the Remediation of Chlorinated Solvents*

Trichloroethylene (TCE) is one of the most ubiquitous pollutants of groundwater. Exposure to TCE has been linked to liver damage, birth defects, and cancer in humans. Many studies have demonstrated that INPs are very effective for the rapid degradation of TCE and other chlorinated hydrocarbons through reduction reactions. The generalized form of the Fe(0)-mediated reductive reaction of tetrachloroethene (TCE) can be represented as

$$C_2Cl_4 + H_2O + Fe^0 \rightarrow C_2HCl_3 + Fe^{2+} + OH^- + Cl^- \quad E^0 = 1.02\,V \qquad (15.11)$$

From a thermodynamics perspective, the strongly positive $E^0_{rxm}$ of the overall reaction implies a likelihood of spontaneous reactions. This also applies for many other chlorinated hydrocarbons, which have standard electron potentials ($E^0_{rxm}$) for their reduction on the order of +0.5 to +1.5 V at 25°C (Vogel et al., 1987; Matheson and Tratnyek, 1994).

During reduction by Fe(0), most of the TCE is converted to ethene and chloride by beta-elimination reaction (Equation 15.19), which proceeds with the formation of short-lived intermediates, such as acetylene ($HC_2H$). A small portion of TCE decomposes by hydrogenolysis, a sequential reduction pathway that results in the formation of longer-lived intermediates, such as *cis*-1,2-dichloroethene (*cis*-1,2-DCE) and vinyl chloride (VC) (Equations 15.20 through 15.22).

$$C_2HCl_3 + 3H^+ + 3Fe^0 \rightarrow C_2H_4 + 3Fe^{2+} + 3Cl^- \quad (15.12)$$

$$C_2HCl_3 + H_2O + Fe^0 \rightarrow C_2H_2Cl_2 + Fe^{2+} + OH^- + Cl^- \quad (15.13)$$

$$C_2H_2Cl_2 + H_2O + Fe^0 \rightarrow C_2H_3Cl + Fe^{2+} + OH^- + Cl^- \quad (15.14)$$

$$C_2H_3Cl + H_2O + Fe^0 \rightarrow C_2H_4 + Fe^{2+} + OH^- + Cl^- \quad (15.15)$$

Although Equation 15.18 implies direct reduction of TCE by Fe(0), this is not necessarily the case due to the presence of surface oxide on the metallic iron. Matheson and Tratnyek (1994) proposed three possible mechanisms for TCE reduction: (1) direct reduction by Fe(0) at the metal surface, (2) reduction by aqueous Fe(II), and (3) reduction by hydrogen with catalysis. Of these three mechanisms, direct reduction at the metal surface was reported to be prevalent and rapid (Matheson and Tratnyek, 1994), with reduction by aqueous ferrous iron occurring much more slowly (in concert with certain ligands) and reduction by dissolved hydrogen gas (produced by iron corrosion in anoxic waters) only occurring in the presence of a suitable catalytic surface.

Following the discussion provided in the previous section, work by Sivavec and Horney (1995) and Charlet et al. (1998) has postulated that surface-bound Fe(II) is involved in the reduction reactions, implying that an electronically conductive surface oxide (magnetite) might mediate the reduction reaction rather than direct reaction with Fe(0) (Equation 15.23).

$$C_2HCl_3 + 3H^+ + 6Fe^{2+} \rightarrow C_2H_4 + 6Fe^{3+} + 3Cl^- \quad (15.16)$$

A summary of published kinetic data for reduction of chlorinated solvents by iron metal can be found in Johnson et al. (1996). On comparison of the reactivity of bulk ZVI and INP for TCE reduction, the mass-normalized reaction rates are observed to be much higher for INP due to the larger surface area for reaction (Wang and Zhang, 1997; Liu et al., 2005). Example data for the reduction of TCE in natural soil samples by INP and annealed INP is shown in Figure 15.6 (Barnes et al., 2009).

#### 15.5.4.4 Iron Nanoparticles for the Remediation of Uranium

One of the most infamous sources of heavy metal pollution is uranium mining. The radionuclide has proven toxic when ingested (Hyne et al., 1992; Pavlakis et al., 1996; ATSDR, 1999) and the potential risk of chronic uranium poisoning is not to be taken lightly for populations located close to uranium mining sites, processing plants, or storage and disposal sites for nuclear waste. Effective methods for groundwater remediation are consequently important, with metallic iron and iron-based minerals well recognized as highly effective scavengers of uranium from water (Hsi and Langmuir, 1985; Lenhart and Honeyman, 1999; Scott, 2005; Scott et al., 2007). However, views are mixed regarding the actual mechanism for the removal of soluble U from water by iron. Removal has been attributed to both the adsorption of uranyl ions ($UO_2^{2+}$) onto iron corrosion products (Fiedor et al., 1998; Farrell et al., 1999; Scott et al., 2005a,b) and the reductive precipitation of soluble U(VI) into

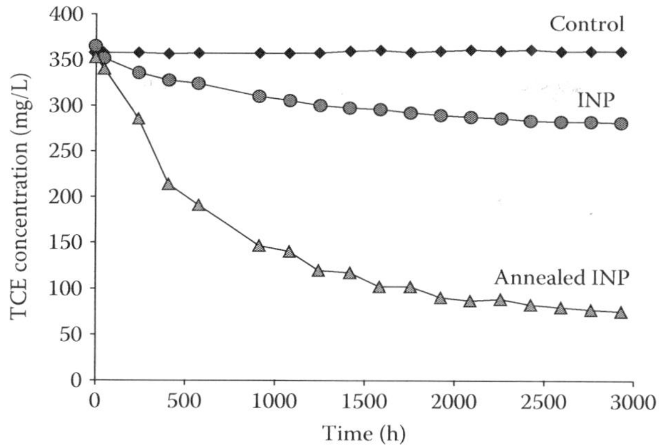


**FIGURE 15.6**
Dehalogenation of 350 mg/L TCE over a period of 2928 h by annealed and nonannealed INP. Results from a control batch experiment containing no Fe particles are also included. (From Barnes, R.J., Dechlorinating bacterial strains and nano-scale iron particles for the remediation of CAH contaminated sites. D.Phil dissertation, Department of Earth Sciences, University of Oxford, Oxford, U.K., 2009.)

insoluble U(IV) oxides, driven by coupled corrosion of Fe(0) or Fe(II) (Cantrell et al., 1995; Charlet et al., 1998; Gu et al., 1998; Liger et al., 1999; Morrison et al., 1998, 2001) (Figure 15.7).

Following Gu et al. (1998), direct uranium reduction by Fe(0) is considered to proceed via the following reaction:

$$Fe^0_{(s)} + 1.5UO_2^{2+} + 6H^+ \Leftrightarrow Fe^{2+} + 1.5U^{4+} + 3H_2O \ (E^0 = 0.17 \text{ V}) \qquad (15.17)$$

As Fe(0) is a stronger reducing agent than aqueous Fe(II), it was previously thought that contaminant reduction was driven by the oxidation of Fe(0) to Fe(II) (Powell et al., 1995).

**FIGURE 15.7**
**(See color insert following page 302.)** Secondary electron (left) and ion map (right) taken from a section cut through an iron surface. The iron was reacted for 24 h in water containing 100 ppm uranium and displayed significant precipitation of uranium oxide ($UO_2$). The ion map depicts U in green and Fe in red, with the black area representing platinum deposited as a protective coating during sectioning.

However, White and Peterson's 1996 study determined structural Fe(II) to be a stronger reducing agent than aqueous Fe(II) with similar capability to Fe(0) and implying that Fe(II)/Fe(III) oxidation may also be a significant pathway for uranium reduction. A number of separate studies have since shown the reduction of U(VI) to U(IV) by structural Fe(II) (Wersin et al., 1994; Charlet et al., 1998; Fiedor et al., 1998; Liger et al., 1999; Missana et al., 2003; O'Loughlin et al., 2003; Scott, 2005; Scott et al., 2005a,b) with the reaction taking the basic form:

$$2Fe^{2+}_{(s)} + UO_2(OH)_2 + 2H^+ \Leftrightarrow 2Fe^{3+}_{(s)} + UO_2 + 2H_2O \qquad (15.18)$$

Based on the work of Charlet et al. (1998), it is generally accepted that Fe-driven reduction of uranium in environmental systems is driven by the Fe(II)/Fe(III) couple rather than the Fe(0)/Fe(II) couple. More recently, INPs have been demonstrated as highly efficient for the removal of uranium from laboratory and groundwater solutions (Riba et al., 2008; Crane et al., 2009; Dickinson and Scott, 2010), with further work extending to the removal of other actinides and treatment of polluted water containing numerous contaminants including uranium (Figure 15.8). The latter work has used a combination of aqueous and surface analytical techniques to demonstrate rapid partial reduction of U(VI) to U(IV) on the surfaces of INP concurrent with the oxidation of Fe(0) to Fe(II) and resulting in the precipitation of a partially reduced uranium oxide ($U_3O_8$).

In open laboratory INP experiments where levels of dissolved oxygen could be naturally replenished, the initial period of significant actinide uptake and retention was followed by a period of gradual rerelease after approximately 100–160 h, ascribed to oxidative redissolution of the uranium driven by the ingress of atmospheric oxygen into the systems. Significant transformation of surface Fe(II) to Fe(III) was then observed over extended periods and coincided with a further (but gradual) decrease in aqueous uranium concentrations. This final stage of uranium removal is attributed to entrapment (coprecipitation) in iron hydroxide phases (FeOOH) formed from INP corrosion (Riba et al., 2008; Dickinson and Scott, 2010). In anoxic INP systems, the initial phase of uranium retention was observed to persist for significantly longer (>300 h) (Dickinson and Scott, 2010).These recent studies have provided a clear example of how the mechanism for contaminant uptake by INP can change with system ageing.

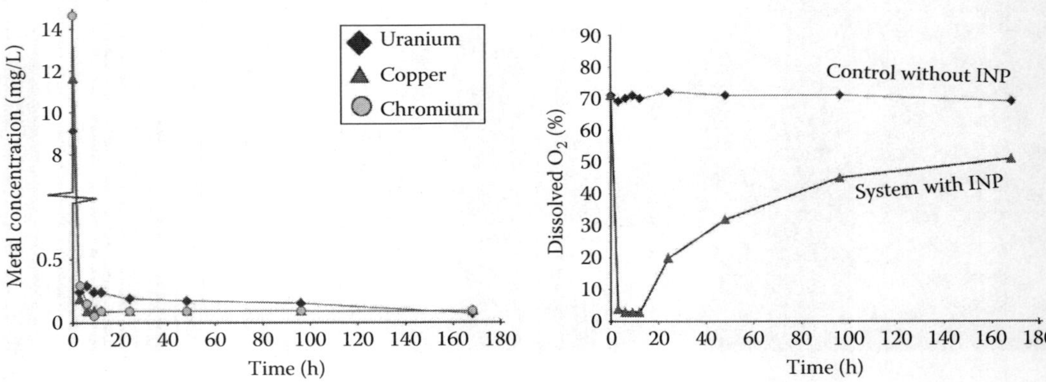


**FIGURE 15.8**
(Left) Removal and retention of uranium, copper, and chromium from solution by INPs at 0.5 mg/L concentration. (Right) Variation in dissolved oxygen levels for the same reaction system compared to a control run without INP (Scott, Popescu, Crane—prepared for publication).

### 15.5.5 Field Application of Nano-Iron Technology

The main appeal of using INP for land remediation comes from the potential for providing a better, faster, and cheaper remediation option that can be used for a wide range of environmental contaminants and that is also versatile when considering the possible options for deployment (Zhang, 2003). The choice of deployment is site specific, with two generic options, in situ or ex situ application. In situ technologies involve treatment of contaminants in place, whereas ex situ refers to treatment that is performed after removing the contaminated material (soil, sediment, or groundwater) to a more convenient location such as a surface treatment facility. Associated with ex situ treatment is also the possibility for treating polluted waters at the point of use (e.g., a village standpipe or water storage tank). Full-scale commercial use of INP for land remediation is rapidly becoming common, and competitive markets already exist for suppliers of reactive INP materials and geotechnical companies specializing in their deployment.

The following sections describe the possible in situ and ex situ applications of nanoscale iron, some of which are already in use and others which have yet to be commercially realized. In some cases, there is the possibility that developments arising from other nanomaterials may be applied in retrospect to nano-iron.

#### *15.5.5.1 Ex Situ Nanotechnology*

Although INPs are most obviously suited to in situ applications for land and groundwater remediation, they may also be utilized ex situ in engineered water treatment systems. Such systems typically conform to either batch or flow-through designs.

Batch systems, of which many different designs are possible, all involve the treatment of confined volumes of polluted water in a tank or other container to which INP can then be added. During treatment, the particle–fluid system is stirred or agitated to ensure thorough and turbulent mixing, which in turn increases the rate of cleanup by (1) promoting and enhancing particle–contaminant interactions, (2) maintaining particles in suspension, and (3) limiting particle aggregation. At the end of the treatment, the clean water must be effectively separated from the INP and associated surface-bound contaminants. This may be achieved most simply by allowing a period of settling, although other methods such as magnetic extraction or filtration may provide faster alternatives. Based on experimental data, the optimum treatment time when using INP and related derivatives is between 1 and 24 h. Due to the gradual but continual ingress of dissolved oxygen into batch systems (unless specifically controlled), the aqueous immersion of INP for periods greater than approximately 48 h is observed to result in the development of significant volumes of ferric oxyhydroxide phases and the associated partial rerelease of some heavy metal contaminants back into solution. The magnetic properties of INP are also observed to degrade significantly from 24 h onward, related to the oxidation of magnetite and resultantly impact the efficacy of magnetic removal.

Flow-through treatment systems offer the possibility for processing a continuous flow of contaminated water, which is a desirable advantage over batch methods. Treatment systems would most typically employ INPs by incorporating them within a porous filter (fixed nanoparticle supports) although other means (e.g., magnetic fields) may be used to maintain the particles within a specific part of the system while still letting water through (Penchev and Hristov, 1990; Hristov and Fachikov, 2007). Numerous filters and porous membranes incorporating nanotechnology are available for purchase with many more in development. Porous "filter" materials may be manufactured in different ways (Adebajo

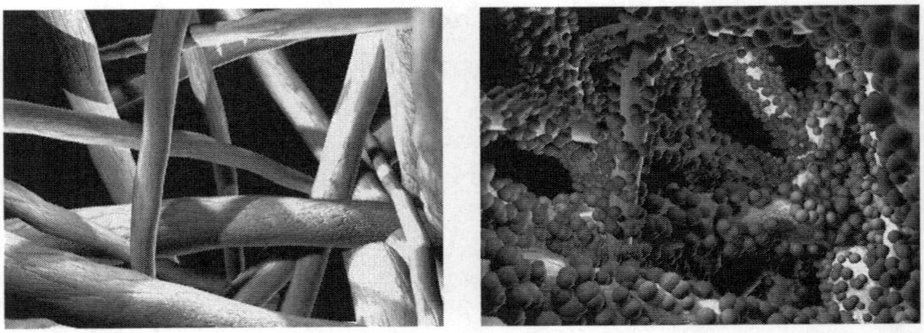

**FIGURE 15.9**
Conceptual nanofilter material involving attachment of reactive nanoparticles to a substrate mesh.

et al., 2003), but in order to incorporate nanoparticles, the porous material must either be nanostructured itself or have nanoparticles chemically bound to its internal surfaces as a reactive coating (Figure 15.9). For example, Ponder et al. (1999) investigated the use of various high surface area INP supports for the reduction of soluble metal ions [Cr(VI), Hg(II), Pb(II), Tc(VII)] and $ReO_4$ (as a surrogate for $TcO_4^-$) to insoluble forms under strongly alkaline conditions. The studies were undertaken with INP impregnated onto support materials including silica, polymeric resin, and metal oxide powders (e.g., $ZrO_2$) and produced promising results.

To this end, many existing technologies may be modified for the purpose of supporting INP in a fixed and porous structure. For example, biopolymer Ca-alginate hydrogels and microbeads have been a proven technology in the food, beverage, and pharmaceutical industries for a number of years, used for cell and compound isolation. Drawing on this extant technology, recent work by Bezbaruah et al. (2009) has demonstrated the successful entrapment of INP in calcium (Ca)-alginate beads for use as a supported remedial material. Data from remediation experiments using nitrate (as a model contaminant) revealed that the alginate gel clusters bound the INP together without causing any significant decrease in the reactivity of the supported particles compared to untreated INP.

Other research in the United States has drawn from ink and toner technology to investigate the use of carbon nano-platelets as a support for INP (Ponder et al., 2000, 2001). Similarly, Shimotori et al. (2004) developed a polyvinyl alcohol (PVA) membrane containing INP as a model barrier for contaminant containment and subsequently proved it practical through a series of tests involving carbon tetrachloride, copper, nitrobenzene, 4-nitroacetophenone, and chromate. Other INP-containing polymer membrane materials have received investigation (Ponder et al., 2000; Meyer et al., 2004; Wu et al., 2005; Shimotori et al., 2006; Wu and Ritchie 2008; Surdo et al., 2009), showing reactivity similar to that of unsupported INPs. Wu et al. (2005) successfully incorporated INP into a porous polymeric membrane 100 μm thick, with an aim of producing a support material that would preserve the chemical nature of the INP for contaminant interactions by inhibiting iron oxidation by nontarget compounds. The researchers found that while TCE adsorbed to the membrane it did not inhibit TCE dechlorination, compared to unsupported INP. In fact, at lower contaminant loadings, the authors observed TCE to preferentially concentrate upon the membrane surface: a process that could prove important for the remediation of waters containing low contaminant concentrations.

It may also be possible to adapt other existing environmental nanotechnology to incorporate INP. Materials for INP incorporation might include SAMMS fabrics (Fryxell et al.,

2004) or other porous nanoscale support structures that can be produced in different ways including templating and nanofiber electrospinning (Park and Xia, 1998; Johnson et al., 1999; Li, Wang and Xia, 2004; Li et al., 2004; Zhang et al., 2005; Thomson, 2008; Yuan et al., 2008; Xiao et al., 2009).

Related to these aforementioned ex situ technologies for INP are those that utilize nanoscale particles of photoactive compounds such as $TiO_2$ (Hoffmann et al., 1995). The potential benefits of quantum-sized (<10–15 nm) photocatalysts have long been recognized for contaminant degradation applications (Kamat and Meisel, 2003; Obare and Meyer, 2004; Abrams and Wilcoxon, 2005). However, the reliance on effective illumination (by UV light) of these compounds to work dictates that they can only be suited for ex situ use and only with a specifically engineered system. While generally unsuited for use in land remediation, there exists a strong potential for these materials to be used in atmospheric cleanup or surface water treatment.

### *15.5.5.2 In Situ Nanotechnology*

As previously outlined, the in situ degradation or immobilization of contaminants is often preferred over ex situ approaches because fewer resources, infrastructure, and services are potentially required to implement the remedial solution, translating to reduced costs. Permeable reactive barriers using granular iron (or scrap iron) are a prominent example of the successful use of in situ land remediation technology. However, a primary limitation for PRBs is that they are practicable only for relatively near-surface land contamination. By comparison, INPs can be used to treat plumes of subsurface contamination at almost any possible depth or location because they can be delivered into the subsurface by injection. Once present in the subsurface in large numbers, the INP can rapidly react with contaminant species, removing them from groundwater by either destroying or immobilizing them in solid precipitates (Figure 15.10).

This highly advantageous aspect of the in situ technology is based on the premise that INPs introduced to groundwater systems will idealistically behave as other naturally occurring colloids, being effectively transported through the subsurface by the flow of groundwater and remaining suspended for extended periods unless removed by aggregation, microorganisms, or reactive subsurface materials such as organic carbon or oxide minerals. In this way, injected INPs potentially have free reign to travel through the subsurface following groundwater flow paths and destroying or immobilizing aqueous contaminants as they migrate (Figure 15.11).

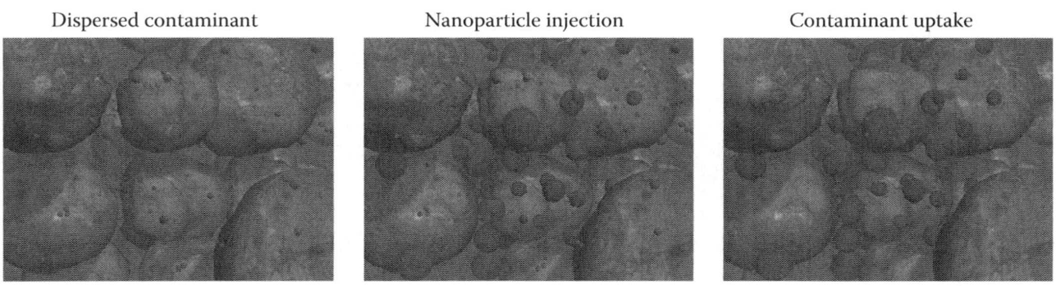


**FIGURE 15.10**
**(See color insert following page 302.)** Conceptual use of nanoparticles to cleanup polluted sediments. Particles are injected into the contaminated area where they rapidly sequester contaminant species present in the groundwater either immobilizing or destroying them.

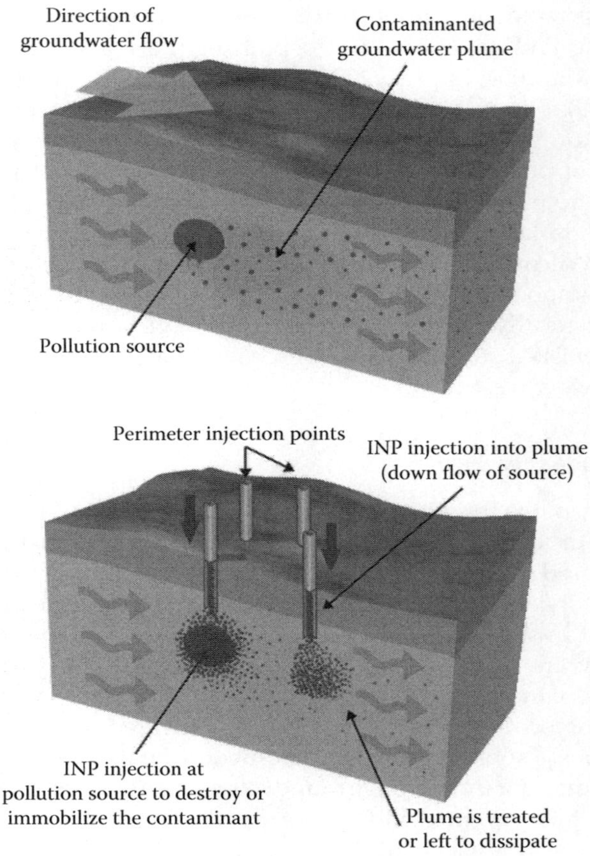


**FIGURE 15.11**
**(See color insert following page 302.)** Conceptual use of INP injection technology for treating subsurface pollution sources. The location and number of particle injection wells is site specific and tailored to achieve the best possible remediation.

Nanoscale iron particles can be injected (through drilled holes) into the subsurface as either a slurry (in suspension) or as a dry powder. Slurries are generally favored because the supporting liquid can often be chemically tailored to aid dispersion by maintaining particle suspension and providing added moisture when the treatment zone lies in under-saturated sediments. To aid the delivery of the INP, nitrogen gas or compressed air is typically used, with the turbulent and pressurized injection flow preventing INP aggregation and ensuring dispersion to significant radial distances, even at relatively low pressures (<150 psi). To prevent backflow up the injection well during delivery, a counter-pressurized system incorporating a "packer" or "blocker" is used (Figure 15.12) that ensures maximized dispersion of the INP into the sediment.

The use of compressed gas also has the secondary benefit of inducing pneumatic fracturing of the subsurface. The resultant network of fractures can provide preferential transport pathways to promote rapid and effective dispersal of the INPs throughout the contaminated zone.

Assessment of injected INPs used to treat chlorinated solvents at three Navy sites in the United States (Gavaskar et al., 2005) have indicated that the key geochemical target for INP

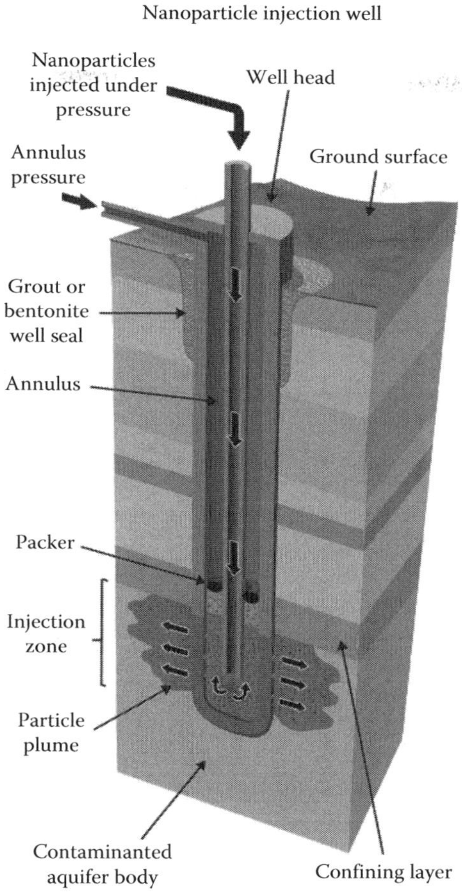


**FIGURE 15.12**
**(See color insert following page 302.)** Diagram showing the typical layout of a nanoparticle injection well. The technology is very similar to that used for injection of $CO_2$ into subterranean storage reservoirs.

remediation, which is likely applicable to other contaminants, should be the generation of strongly reducing groundwater conditions within the treatment zone (Eh < −0.4 V). Under such conditions, rapid abiotic degradation reactions were observed to account for the majority of solvent degradation, with minimal production of unwanted partially dechlorinated by-products (see previous section on TCE degradation). An Fe:soil ratio of 0.004 within the target zone was determined to be sufficient to achieve suitable Eh conditions with lower amounts of INP observed to produce only mildly reducing conditions in which solvents are degraded more slowly via hydrogenolysis and anaerobic biodegradation. Additionally, in order to limit the amount of INP oxidation by dissolved oxygen, it was suggested that the water/liquid used for particle delivery be deoxygenated prior to use. In relation to heavy metal and radionuclide remediation, these geochemical conditions (low Eh and oxygen) match those more commonly attributed to the formation of metal-bearing ores, which is essentially what INP injection sets out to achieve as a means of long-term contaminant immobilization.

An alternative possibility for in situ treatment involves injection of INP (through drilled holes) that are engineered to be relatively immobile such that a static subsurface treatment zone may be established, providing something akin to a subsurface PRB. Static treatment zones may be further enhanced, by drawing on observations made from the long-term performance assessments of iron-based PRB systems relating to flow reduction through the barriers caused by pore-space congestion from the oxidative development of ferric hydroxide (rust) and precipitation of contaminant phases (O'Hannesin and Gillham, 1998; Henderson and Demond 2007; Johnson et al., 2008).

Pumping oxygen-bearing gases into the subsurface treatment zone after INP injection (more commonly known as air sparging) may be used to deliberately induce oxidation of the INPs to form voluminous ferric hydroxides and cause pore clogging in the host rock. Extensive clogging may render the treatment zone impermeable, trapping the contaminants and causing upstream uncontaminated groundwater to flow around the pollution source rather than through it. This "treat and plug" approach may prove very useful for application at aging waste storage sites, landfills, or even nuclear disposal sites, which have developed subsurface containment breaches. Reactive nanoparticle suspensions could be injected directly at the point of breach, rapidly immobilizing contaminant metals present and degrading chlorinated hydrocarbons. Subsequently, air sparging could be used to promote the formation of aforementioned rust that would clog the pore spaces in surrounding sediments to form an impermeable seal around the breach (Figure 15.13).

A complimentary technique that may extend the in situ capability of INP technology for remediation is electrokinetics. The use of electrokinetic (EK) phenomena for environmental remediation has been actively developed since the 1990s and typically involves passing a 50–150 V direct current through a contaminated subsurface zone using a pair or an array of electrodes, typically made from carbon, graphite, or platinum. Given that the contaminated soil or sediment contains sufficient moisture (>10%), the electric field induces movement of water and migration of contaminant ions to electrodes of opposite charge. Essentially, the power supply can be regarded as an electron pump, pushing electrons from the cathode to the anode and driving the movement of contaminants (Acar and Alshawabkeh, 1993; Probstein and Hicks, 1993; Kovalic, 1995) (Figure 15.14). Within a DC electric field, the surface charge of colloidal particles suspended in groundwater will determine whether they migrate toward the anode or cathode. The isoelectric point (IEP), also known as the point of zero charge (PZC) is the critical pH value at which the net surface charge of a particle is zero. For INP, this is typically at around pH 6.8 assuming a magnetite ($Fe_3O_4$) surface or pH 8.4 assuming a hematite ($Fe_2O_3$) surface and is found to be independent of iron concentration (Koshmulski, 2001). Resultantly, in environmental systems, INPs will have a positive charge when the pH is less than 6.8 and will tend to migrate toward the cathode under an electrical potential. Because of the positive potential, INP will also be attracted to the negatively charged surfaces of clay minerals and, resultantly, INP mobility in clayey soils is expected to be lower than that in other sediments (Sun et al., 2006).

A recent bench-scale study by Pamuka et al. (2008) has demonstrated that electrokinetics might successfully be integrated with INP technology. Tests were conducted in clay soils in which NP transport time and process efficiency were considered problematic, with results clearly showing that INPs were both transported and activated by the electric field.

Further pilot-scale work is required to validate these claims but if proved successful, electrokinetic processes combined with INPs may be harnessed in order to draw unretarded contaminants into static INP treatment zones or, conversely, injected screens of

**FIGURE 15.13**
**(See color insert following page 302.)** Application of nanoparticle injection technology for remediation of subsurface zones contaminated by uranium and other pollutant heavy metals.

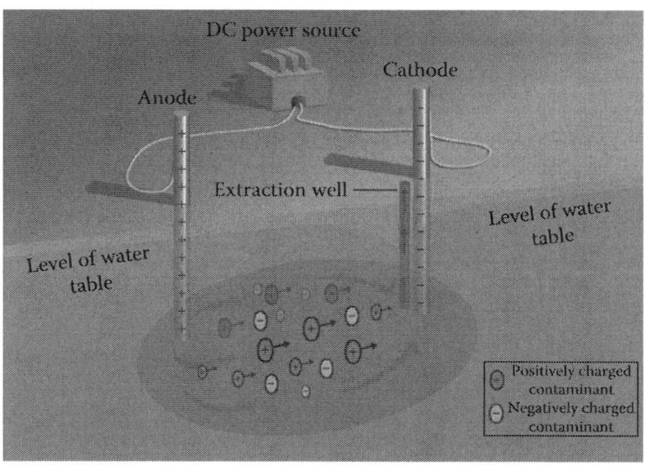


**FIGURE 15.14**
**(See color insert following page 302.)** Environmental remediation of contaminants in the subsurface using electrokinetic processes to drive contaminant migration.

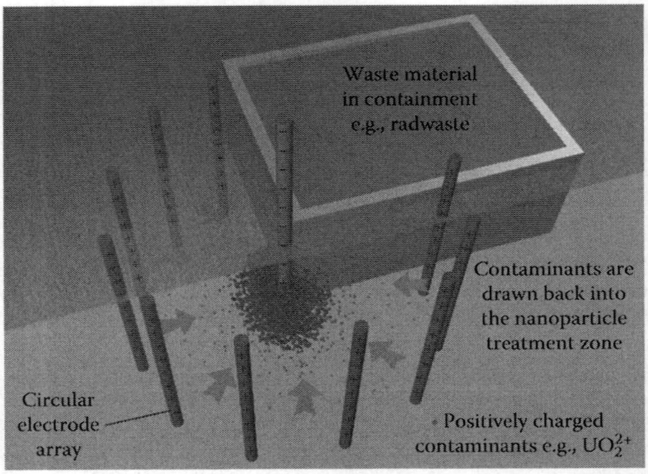


**FIGURE 15.15**
**(See color insert following page 302.)** Application of combined nanoparticle and electrokinetic techniques for enhanced remediation of subsurface zones contaminated by uranium and other pollutant heavy metals.

mobile INPs may be directed into an inaccessible contaminant plume, where contaminants can be effectively remediated. A radial electrical gradient generated by a circular array of electrodes centered about the point of nanoparticle injection would be used to do this (Figure 15.15). Following the observations made by Pamuka et al. (2008), the development of a reduced zone around the central cathode might increase the effectiveness of the nanoscale iron particles in removing contaminants.

#### *15.5.5.3 Summary of Iron Nanoparticle Remediation Technology*

Compared to other potential nanomaterials for land remediation, INP technology has shifted from preliminary laboratory testing (Wang and Zhang, 1997) to pilot-scale field trials (Elliott and Zhang, 2001) over an extremely short time period. This has only been possible because much of the environmental research needed for validation and assurance had already been carried out in relation to PRB technology (Tratnyek et al., 2003, ITRC, 2005).

It is clear from the literature that both in situ and ex situ treatments using INP can be used to effectively remove or destroy contaminants from groundwater. However, when applied as an in situ solution two problem issues remain:

1. When treating heavy metals and radionuclides, the contaminants are not physically removed from the subsurface. Although these contaminants are removed from groundwater onto INP surfaces, they still physically remain in the subsurface, leaving the potential for future remobilization. Due to a lack of field data (and trials), it is not yet clear whether remobilization might occur on human or geological timescales; only time will tell.
2. When injecting significant concentrations of nanoparticles into the subsurface, dispersion of particles will be limited by aggregation, mineral sorption, and microbiological processes that may even lead to pore clogging and significantly reduce the extent of remediation.

The following section looks more closely into how INP technology has matured to improve reactivity, mobility, and longevity.

### 15.5.6 Maturation of Nano-Iron Technology

Since the early 1990s, permeable reactive barrier (PRB) technology, using metallic iron as a reactive material, has provided an improved alternative to more classical methods of land remediation.

PRB systems provide a passive, in situ, low energy method for site remediation that is effective both in terms of cost and contaminant removal. In turn, iron nanoparticle technology looks set to offer a further improvement on PRBs by offering improved reactivity and versatility with reduced costs and remediation time.

For example, a cost comparison provided by PARS Environmental (2009) for a contaminated manufacturing site in New Jersey, United States, estimated cleanup costs using INP technology to be 11% of that for remediation by pump and treat and 20% of that for PRB cleanup. Although improvements in reactivity and cost are the primary drivers for the development of new technologies for land remediation, other factors are also worthy of consideration. Gavaskar et al. (1997) suggested six criteria for the selection of reactive media for PRB systems, each of which is equally applicable to the selection of engineered nanomaterials for environmental remediation.

- Reactivity: It is desirable to have high reaction rates for contaminant interactions such that they are rapidly removed from groundwater and removed/destroyed. The reactivity of the material can be quantitatively evaluated in laboratory experiments.
- Stability: The material is expected to remain active for long periods of time (longevity) because its replacement is not easily achieved. Particle stability during changes of pH, temperature, pressure, and antagonistic factors is also required.
- Environmental compatibility: It is important that the reactive media do not form any by-products when reacting with the contaminants and that it is not a source of contamination itself by solubilization or other mobilization mechanisms.
- Hydraulic performance: The hydraulic conductivity of the material depends on its particle size distribution and its value must be greater or equal to the value of the surrounding soil. However, an optimum particle size that would provide appropriate permeability and sufficient contact time must be determined.
- Availability and cost: The amount of reactive material required for remediation may be large and therefore it is essential to have considerable quantities in low prices.
- Safety: Handling of the material should not generate any risks for the workers' health.

While demonstrative laboratory tests have clearly indicated that INPs are effective for the rapid and significant removal of a wide range of contaminants from polluted waters, there are a number of fundamental issues concerning the use of INP in environmental systems that are not yet fully understood. The most significant issues of note are (1) the mobility of engineered nanoparticles, (2) the kinetics and products of contaminant degradation/removal, (3) the fate and persistence of INP, and finally, (4) the possible ecotoxicity of INPs and associated impact they may have when introduced to the environment.

Perhaps preempting some of these partially resolved issues, scientists have made the logical and progressive step of questioning whether modified INP or nanoparticles made from other materials can offer improved remedial performance over monometallic INP for a number of the aforementioned selection criteria. The following sections examine the current "state of play" for developmental advances in nanoparticle technology under the interrelated headings of particle chemistry, longevity, and mobility.

#### *15.5.6.1 Improved Compositional Chemistry of Nanoparticle Materials*

##### *15.5.6.1.1 Bimetallic Nanoparticles*

Over the past 5–10 years, there has been much interest in the development of physiochemical alterations, alternative metals, and bimetallic combinations, to possibly improve upon monometallic INPs. Bimetallic nanoparticles combining iron with a more noble metal (e.g., Pd, Pt, Ag, Ni, Cu) have received significant attention and have recently seen commercial deployment at a number of sites in the United States (PEN, 2009).

Much of the ongoing development of INP materials has been based upon improving reactivity and also increasing environmental longevity by limiting aqueous corrosion. Conceptually, the two properties are directly opposed and mutually exclusive. In comparison to standard iron nanoparticles (especially those produced by methods where particle formation is very rapid), the contaminant reactivity of bimetallic or thermally treated particles has generally been demonstrated to be greater. Some reports suggest a difference in reactivity by orders of magnitude (Tee et al., 2009) and can be explained by examining the catalytic properties of the different materials, which determine the overall reactivity of any given nano-iron particle.

For bimetallic nanoparticles (BNPs), the addition of a noble metal results in the formation of a galvanic system in which the noble metal behaves catalytically to transform contaminants, and the iron is sacrificially oxidized to galvanically protect the more noble metal. The noble metal is incorporated into the particles at concentrations typically below about 20 atomic percent as either alloyed or core-shell particles. Regardless of particle structure, the Fe(0) present in the particle is considered to behave anodically, supplying electrons for contaminant transformations via the noble metal that behaves cathodically and catalyzes reactions by transferring electrons while remaining chemically unchanged (Elliott and Zhang, 2001; Schrick et al., 2002). Chemical reduction of sorbed contaminants at the particle surface is considered to occur through either direct electron transfer with the noble metal or through reaction with hydrogen produced by oxidation of Fe(0). Hydrogen is likely present as a dissolved gas, some of which is adsorbed to the particle surfaces, with an undetermined fraction possibly present as active metal hydride having undergone diatomic dissociation and reaction with the exposed noble metal. (Figure 15.16) (Wang and Zhang 1997; Zhang et al., 1998; Schrick et al., 2002). For chlorinated organic contaminants such as TCE or PCP, hydrogen is observed to be the predominant driver for degradation (Cheng et al., 1997; Liang et al., 1997; Li and Klabunde, 1998; Dabro et al., 2000; Nyer and Vance, 2001; Schrick et al., 2002) by breaking C–Cl bonds and swapping itself for chlorine, which is liberated as a gas (see Equations 15.11 through 15.15). It is important to note that these hydrogen reduction reactions are considered only to occur in the presence of a suitable catalytic surface (Matheson and Tratnyek, 1994) and are therefore not applicable to monometallic INP.

This is exemplified by comparing TCE degradation rates between bimetallic and monometallic nanoparticles (Cheng and Wu, 2000). For BNPs, an advantageous consequence of the inferred catalysis is that dichloro-ethelyenes (DCE) and vinyl chloride (VC), which

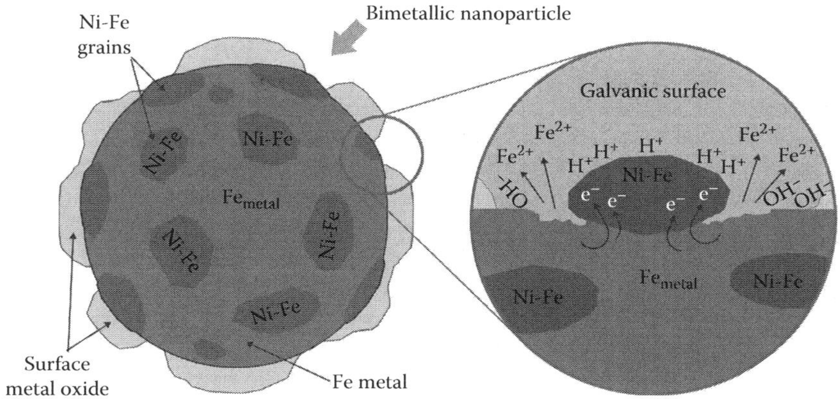


**FIGURE 15.16**
Structure and galvanic behavior of an idealized bimetallic iron-nickel nanoparticle.

are generated by TCE breakdown, are observed to be rapidly hydrogenated at the particle surfaces and do not accumulate in the reaction. By comparison, DCE and VCs are often observed to accumulate during TCE reduction by monometallic INP. This occurs because direct reduction of water by Fe(0) proceeds more slowly without a galvanic influence and, resultantly, the hydrogenolysis of contaminant species is limited (Schrick et al., 2002).

When assessing BNPs in environmental systems, there exists a play-off between contaminant reactivity and longevity, which is not applicable to monometallic INPs because they lack a galvanic system. The efficiency of the cathodic reaction is found to determine the particle corrosion rate. If the rate is high, contaminant reactivity will also be high but longevity will be poor. The most favorable case for most limited corrosion occurs when the anodic surface area is very large and the cathodic area is small. Resultantly, particle compositions close to monometallic INPs should exhibit least corrosion and greatest longevity, but only exhibit contaminant reactivity close to that of standard INP. Consequently, it has been suggested that bimetallic particles such as Ni-Fe and Pd-Fe are more suitable for short-term remediation applications, such as injection into contaminated groundwater, than in reactive barriers that are intended to remain active for years or decades (Schrick et al., 2002).

#### *15.5.6.1.2 Thermal Treatments*

Recent efforts to improve the longevity of both INP and BNP nanoparticles have examined the use of thermal treatments in vacuo to refine physiochemical particle structure and improve galvanic behavior.

It is well established (Dickinson et al., 2009; Scott et al., 2010) that INP and BNP synthesis via aqueous chemical reduction (among others) is observed to result in physiochemical imperfections, which may alter particle reactivity. Unmodified particles are typically observed to have a highly disordered crystalline structure within both the bulk metal and the surface oxide layer (Bonin et al., 2000). For BNPs, the inherently defective electronic network limits galvanic behavior and causes particle corrosion to occur more predominantly via basic chemical oxidation (Figure 15.18).

Thermal treatment, or annealing, is a process commonly used in metallurgy to relieve internal stress, refine grain structure and produce equilibrium conditions within a metal. It has therefore been considered a method for the modification of BNP to promote galvanic coupling between the Fe alloying metal and also in INP to promote electron transfer from the metallic core to the outer surface of the encapsulating oxide. By annealing under vacuum, further oxidation of the particles can be significantly limited and, at higher temperatures, reversed to cause oxide thinning (Scott et al., 2010 and references therein).

Vacuum annealing of INP at 500°C and pressures better than $10^{-5}$ mbar for periods of 24 h has been observed to result in numerous physiochemical effects on INP structure (Scott et al., 2010). Changes included a reduction in surface area by up to 75% (from 19.0 $m^2$ $g^{-1}$ to 4.8 $m^2$ $g^{-1}$) related to diffusion bonding of previously discreet particles but with an accompanying improvement in physical structure which included (1) reordering and recrystallization of the metallic cores, (2) concurrent thinning, dehydration, and stoichiometric refinement of the surface oxide, and (3) migration of impurities toward the particle surfaces and grain boundaries (Figure 15.17).

Comparative studies of annealed and nonannealed INP for contaminant remediation in environmental, industrial, and simple laboratory solutions (Crane et al., 2009; Dickinson and Scott, 2010) have since shown that vacuum-annealed INPs exhibit similar levels of contaminant removal, but with a marked decrease in corrosion. Taking into account the reduced surface area, annealing was demonstrated to significantly increase particle reactivity and has been ascribed to the formation of an effective electronic network within the annealed INP structures, related to the formation of a uniform magnetite ($Fe_3O_4$) oxide layer around the metal core (Figure 15.18).

Magnetite is a mixed Fe(II)-Fe(III) oxide and has the ability to chemically reduce contaminants through surface interactions with structural Fe(II). The oxide is also recognized to behave electrochemically due to its high conductivity ($10^2$–$10^3$ $\Omega^{-1}$ $cm^{-1}$) at temperatures above 122 K (the Verwey Transition of magnetite) (Verwey et al., 1947) when "electron hopping" is observed between adjacent Fe(II) sites. Resultantly, magnetite in itself is a possible candidate material for contaminant remediation (McCormick and Adriaens, 2004; He and Traina, 2005; Scott et al., 2005a) and will display catalytic surface reactivity, which is limited by the quantity of available electrons from within the bulk structure. Exhaustive redox reactivity will result in the oxidation of all Fe(II) to Fe(III), forming isostructural maghemite and ultimately haematite ($Fe_2O_3$).

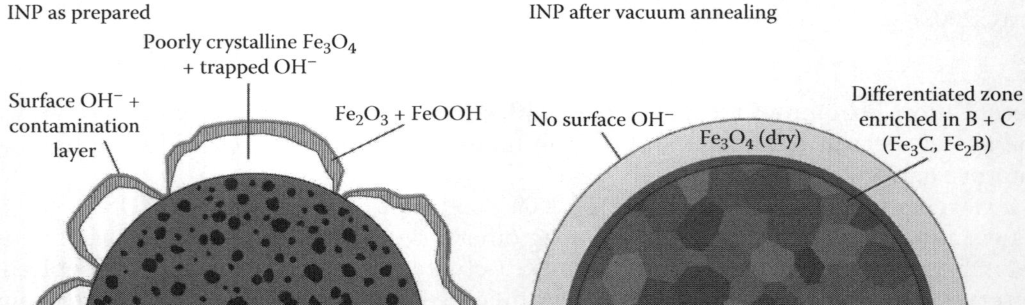


**FIGURE 15.17**
An illustration of the perceived structural transformation of an idealized INP (a) before and (b) after the vacuum heat treatment.

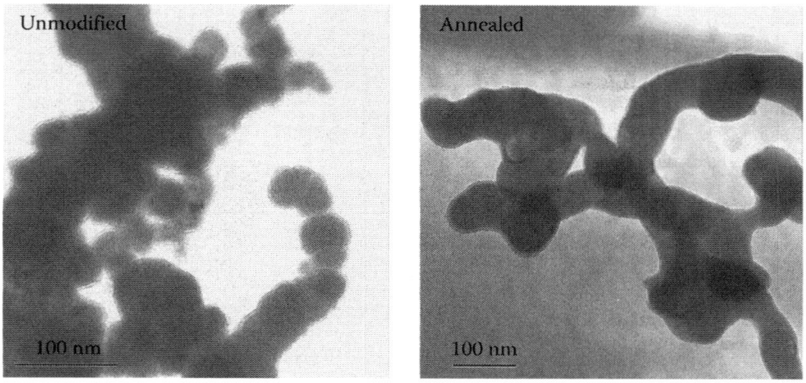


**FIGURE 15.18**
Transmission electron microscope images of (left) unmodified INPs and (right) annealed INPs.

Similarly, the magnetite present on the surface of INPs will behave in just the same way. The only difference is that the underlying metal can provide a pool of electrons that can be passed up through the magnetite, where surface reactions are catalyzed.

In unmodified INP, the surface magnetite is defective and poorly crystalline and may coexist with other nonconductive oxide phases. Consequently, the catalytic behavior of the surface oxide is limited and in the early stages of aqueous activity, direct electrochemical corrosion and dissolution of the metallic iron may be observed until a uniform oxide is formed. By comparison, the magnetite present on annealed INPs is determined to be uniform with refined stoichiometry and good electron conductivity such that catalytic behavior is greatly improved and direct contact of the zero-valent iron with water (or constituent contaminants) is prevented. Resultantly, particle reactivity is improved while also limiting initial corrosion rates. Results have also indicated that contaminants such as uranium are retained for longer periods on annealed INPs, perhaps indicating that in the early stages of reaction, a greater proportion of the sequestered U(VI) is converted to U(IV) oxide via surface-catalyzed reductive precipitation (Dickinson et al., 2009).

In relation to bimetallic particles, vacuum heat treatments are also observed to induce favorable changes including improved galvanic behavior. In recent years, numerous experimental studies of bimetallic nanoparticles for contaminant remediation have been made, including Fe/Pd (Grittini et al., 1995; Wang and Zhang, 1997; Zhang et al., 1998; Elliott and Zhang, 2001; Lien and Zhang, 2005, 2007), Fe/Pt, (Zhang et al., 1998), Fe/Ag (Xu and Zhang, 2000), and Fe/Ni (Zhang et al., 1998; Lien and Zhang, 1999; Schrick et al., 2002; Riba et al., 2010). Varying results have been yielded with Fe-Pd generally outperforming the other combinations. However, in many cases, corrosion of the noble metal was observed, indicating that galvanic processes were underperforming.

A recent comparative study of the aqueous behaviors of bimetallic FeNi and vacuum-annealed FeNi nanoparticles (hereafter termed ABNPs) has provided clear evidence that vacuum thermal treatment improves galvanic properties. The study concluded that unmodified BNPs failed to behave as galvanic substances, similar to INP, due to the aforementioned defects and imperfections arising during aqueous synthesis. In contrast, the annealed particles exhibited a significantly enhanced contaminant uptake, with only iron rather than nickel corrosion observed and indicating that classical galvanic corrosion was occurring (Wang and Zhang, 1997) (Figure 15.19).

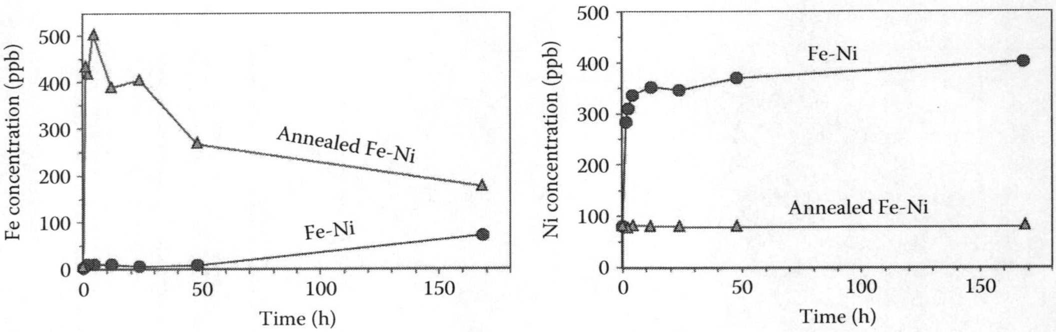


**FIGURE 15.19**
Corrosion profiles of annealed and nonannealed bimetallic $Fe_{78}$-$Ni_{22}$ nanoparticles during reaction with a waste effluent containing 0.42 ppm uranium (Dickinson and Scott, 2010). Both types of bimetallic nanoparticles were observed to remove 98% of the total U from the contaminated effluent. Corrosion of the annealed BNP is observed to be galvanic, while corrosion of the standard BNP was not.

While the nanoparticle alterations detailed in this section clearly result in an improved material, the commercial adoption of these processes will occur only if the increased performance is not heavily outweighed by the cost of primary materials and production. For example, noble metals such as Pd, Pt, and Ag are significantly more expensive than iron and, resultantly, BNPs will always be more expensive to produce than INPs. Additionally, while much emphasis has been placed on BNPs for enhancing particle reactivity for remediation, seemingly little attention has been paid to reducing production costs or improving environmental compatibility, both of which are important developmental drivers for nanoparticle technology.

#### 15.5.6.2 Enhancements in Nanoparticle Mobility for Subsurface Deployment

The majority of scientific attention for INP remediation research has been placed on determining the rates and levels of contaminant removal in sterile synthetic laboratory solutions and subsequently in more complex groundwater samples. From the scientific literature, it is clear that both INPs and BNPs are effective for the rapid and comprehensive cleanup of polluted water. Little consideration was initially placed on other criteria until early pilot-scale field studies indicated that particle aggregation and pore clogging were systemic issues for the environmental application of nanoparticles, with maximum practical transport distances of only a few reported for bare unsupported particles in saturated sediments (Schrick et al., 2004 and references therein). The very limited subsurface mobility of INPs can be explained by three primary mechanisms:

1. Particle aggregation and subsequent gelation (formation of a particle network), caused by poor colloidal stability, causing pore clogging and sedimentation.
2. Particle oxidation/corrosion which, in oxygenated waters, results from the formation of voluminous surface precipitates and "rust" phases that also cause particle sedimentation and pore clogging.
3. Particle removal from solution by interaction with subsurface components e.g., attachment to mineral surfaces and carbonaceous materials or via microbial removal (Elimelech et al., 1998).

Perhaps the most significant of these mechanisms is that of particle aggregation, with tests even at low aqueous INP concentrations (60–80 mg/L) showing that particle clusters caused by aggregation can grow large enough to sediment out under gravity within periods as short as 30 min (Phenrat et al., 2007a,b). At higher (g/L) particle concentrations, INP aggregation and sedimentation is observed to occur significantly faster. The reason for such rapid aggregation is ascribed to the relative imbalance between attractive forces which act to draw particles together and repulsive forces which act to push them apart and maintain them in suspension. Following classical colloidal dispersion theory, known as DLVO theory (Derjaguin and Landau, 1941; Verwey and Overbeek, 1948), the major attractive energy is the van der Waals energy which is influenced by particle size, chemistry, and aqueous concentration. Opposing this, the major repulsive force is electrostatic double layer interaction energy, which is influenced by the particle surface potential and ionic strength of the solution (in this case groundwater). For INP and BNP, a further attractive energy which is considered to tip the balance in favor of aggregation is provided by magnetic forces, due to the presence of magnetite ($Fe_3O_4$) and metallic Fe, which are both magnetic (McCurrie, 1994; Beke, 1998). Under an applied magnetic field, INPs are observed to rapidly form chain-like aggregates (Promislow et al., 1995; de Vicente et al., 2000) with similar aggregates frequently observed for dried particulates in the absence of such a field (Figure 15.20). In agreement with theoretical calculations (de Gennes and Pincus, 1970), the particles behave as nanoscale bar magnets, with particles aggregating and forming chains by linking pole to pole. Under aqueous conditions, Fe, with an intermediate ionic charge-to-radius ratio, also acts as a Lewis acid and readily hydroxylases water. This process is typically followed by further adsorption of water molecules that hydrogen-bond to the surface OH groups, thus promoting iron particle agglomeration.

In addition to rapid aggregation, the subsurface mobility of INPs is considered also to be limited by the formation of corrosion or "rust" phases, such as ferrihydrite ($Fe_2O_3 \cdot H_2O$), that arise from INP oxidation and occupy 5–6 times the volume of the precursor metal. Resultantly, these voluminous products can limit INP mobility by causing particle gelation, sedimentation, and pore clogging. As iron oxidation is the principal mechanism by which contaminants are remediated, it would be desirable that any chemical modification made to limit particle corrosion by dissolved oxygen and oxygen-containing compounds, would not adversely affect affinity and reactivity toward target contaminants.

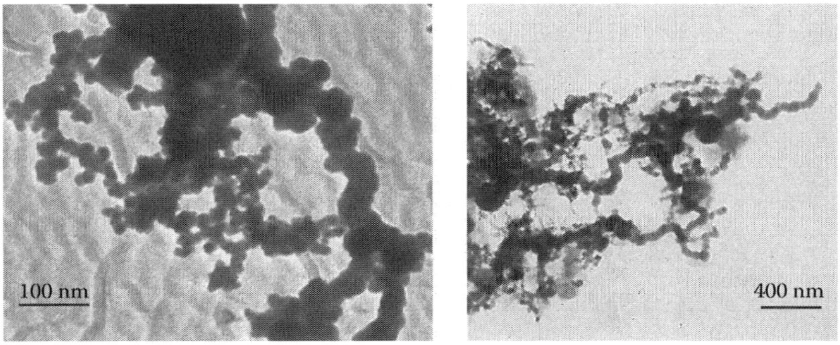


**FIGURE 15.20**
Transmission electron microscope images of fractal nanoparticle clusters of (left) bimetallic Fe-Ni, and (right) monometallic iron. On drying, the formation of chain-like aggregates is attributed to magnetic interactions between particles.

In groundwaters, the interaction of INP with mineral surfaces will be continual. Particle collisions with subsurface components (primarily mineral surfaces) are estimated to be on the order of hundreds per meter traveled (Logan et al., 1999). Thus, for particles to be effectively transported, nearly all of these collisions must be unsuccessful i.e., the *sticking probability* must be very low. In fine grained sediments such as silts and muds, pore spaces are typically around 500 nm in diameter (Nyer and Vance, 2001), meaning that particle sticking and aggregation combined with precipitation of secondary phases can quickly result in pore clogging. It therefore follows that sediments with higher porosity and pore size, e.g., coarse sands, will exhibit greatest INP mobility.

The key to improving particle mobility is found in modifying their surface properties such that the INP are resistant to the retardation mechanisms described. Many of the methods now available to improve INP mobility in the subsurface are derived from work carried out since the 1960s on ferrofluids (Papell, 1965) for applications in fields such as magnetism, optics, biophysics, medicine, rheology, and thermodynamics (Berkovsky et al., 1996). Ferrofluids are classified as stabilized colloidal mixtures of magnetic nanoparticles suspended in a carrier fluid, which become strongly polarized in the presence of a magnetic field. This reasonably describes the particulate slurries used to deliver INPs into the subsurface, except that basic INP suspensions are not well stabilized and precludes them from being true ferrofluids. Previous research of ferrofluids has shown that their colloidal stability can be controlled by using specific surfactants or polymeric surface coatings and by controlling the surface particle charge.

Where surfactants are used to promote colloidal stability, the stability of the suspension is due to the steric hindrance of the surfactant molecules that counteracts the electrical and dipolar attractions between particles. Surfactants associate themselves with the particle surfaces by forming hemimicelles (Figure 15.21). At low surfactant concentrations, hydrophobic groups are exposed to the aqueous phase while, under higher concentrations, the hydrophilic head is exposed to the aqueous media and inhibits flocculation via steric repulsion. This allows much longer suspension via Brownian motion in aqueous media. Indeed, it is also worth noting that iron is often used in industrial

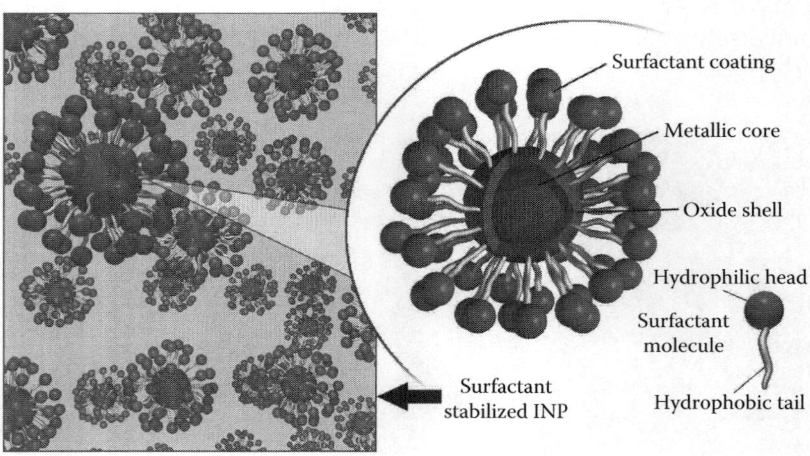


**FIGURE 15.21**
**(See color insert following page 302.)** An idealized representation of surfactant stabilized INPs in aqueous solution, showing (inset) the association of surfactants on INP surfaces to form hemimicelles. Constituent components are also labeled.

processes to remove surfactants from the environment via coagulation–flocculation processes which take advantage of the hydrophobic tail formed when using a high ratio of iron to surfactant (Aboulhassan et al., 2006). It is only when a high surfactant to iron ratio is used does the colloidal iron become more dispersed via the formation of a hydrophilic hemi-micelle, or micelle for fully dispersed particles. Consequently, for INP injection technology, the use of surfactants has limited environmental application: as in concentrated particle suspensions, typical of injection slurries, the surfactants may promote aggregation, and, during subsurface dispersion, the surfactant would readily desorb in surfactant-free water and therefore quickly cease to promote INP suspension in groundwater.

In contrast to the use of surfactants, the controlled coating of INP surfaces with high molecular weight polymers is an irreversible process and could, therefore, represent an appropriate INP modification to increase environmental mobility (Saleh, 2007; Saleh et al., 2007). Polymer coatings can either be fixed onto the INP surface via covalent bonding or physical adsorption, or, alternatively, polymer layers may be grown on the INP surfaces using techniques such as atom transfer radical polymerization (ATRP), where surface metal ions act as catalysts for carbon–carbon bond formation (Saleh et al., 2005). Polyelectrolyte (polymer) coatings, including polyaspartate, caboxymethyl cellulose, polystyrene sulfonate (Phenrat et al., 2007a,b, Phenrat et al., 2009), polyacrylic acid (Schrick et al., 2004), polymethacrylic acid, and polymethylmethacrylate or butyl methacrylate (Sirk et al., 2009) have been demonstrated as effective for stabilization of INP dispersions and are considered to work in a similar way to surfactants, based on steric hindrances of the surface coatings that counteract the attractions between particles. Adsorption of the polyelectrolyte is observed to increase the INPs' electrophoretic mobility over a broad pH range relative to unmodified INP (Hydutsky et al., 2007; Kanel and Choi, 2007; Kanel et al., 2008; Matyjaszewski et al., 2009; Phenrat et al., 2009; Sirk et al., 2009). Furthermore, under groundwater conditions, the coatings are observed to persist for periods up to 8 months, with polymer desorption occurring very gradually and most slowly for higher molecular weight polyelectrolytes (Phenrat et al., 2009). An additional benefit for remediation applications is that polymers may also fortuitously act as a material source to stimulate biological remediation of organic pollutant compounds, especially in carbon-limited environments (Phenrat et al., 2009).

Research has also concentrated on the use of INP carbon coatings to improve dispersion stability. The formation of anionic hydrophilic carbon coatings, typically by including carbon black during particle synthesis (Hoch et al., 2008) is observed to improve stability to a level comparable with that obtained by using polyelectrolytes.

Work by Schrick et al. (2004) used polyacrylic acid and anionic hydrophilic carbon-supported INP to remove chlorinated hydrocarbons from a selection of contaminated sands and soils. Their results demonstrated that both coatings significantly reduced the sticking coefficient of iron nanoparticles in Ottawa sand and in a high-clay soil; however, relatively poor transport was observed in the other soil types studied, implying a higher sticking coefficient with lower soil charge density.

One of the most popular polymeric materials for stabilizing INP dispersions is guar gum, a cheaply produced natural water-soluble biodegradable polysaccharide, formed from guar beans. This environmentally benign material is already extensively used in a variety of industries (cosmetics, mining, pharmaceuticals, textiles, paper, etc.) as a dispersing and stabilizing agent. In water, the gum is nonionic and hydrocolloidal, remaining neutrally charged and unaffected by ionic strength or pH across an environmentally relevant range (pH 5–6). As a coating on INPs, the gum is therefore ideal.

Tiraferri et al. (2008) examined the colloidal stability of INP modified by guar gum and compared them to pure INP. Electrophoretic mobility measurements demonstrated the ability of the gum to adsorb on the nanoparticles, forming a slightly negatively charged layer. Subsequently, the guar modified INPs were observed, using dynamic light scattering, to exhibit a significantly reduced hydrodynamic radius from 500 nm in bare INP to less than 200 nm. In aqueous solution, the sedimentation profiles recorded after 20 min demonstrated that the colloidal stability of the guar-modified INP was better than that of the pure INP with aggregation prevented even at very high salt concentrations (0.5 M NaCl and 3 mM $CaCl_2$).

In the future, it is possible that other "support" materials may be developed to enhance subsurface nanoparticle mobility. In addition to improved mobility, other commercial drivers including cost and environmental compatibility are likely to promote the examination of a variety of cheaply produced biopolymers. In comparison to guar gum, biopolymers such as alginate and potato starch have already been examined and have proven unsuccessful (Tiraferri et al., 2008). In this respect, other biomaterials are very likely to receive similar attention in the near future.

The developments outlined in the preceding text represent the maturation of INP remediation technology to the current status quo and primarily result from lessons learned in trying to successfully take INPs out from the laboratory and into the field and from trying to maximize remediation efficiency.

The final aspect of using nanoparticles for remediation of contaminated sites is still something of an unknown and relates to the ecological impact of these materials and the potential for endangering human health by deliberately deploying such materials in the environment.

## 15.6 Toxicological Studies of Iron Nanoparticles

Nanoparticles (NPs) are in many of the products that we use on a daily basis and are surreptitiously entering the environment in ever-increasing volumes. Although the current mass balance for NPs in the environment system is presently undefined, it is conceivable that a large amount of anthropogenic NPs, from an increasing variety of sources, could find their way into the soil, atmosphere, and aquatic systems (Navarro et al., 2008). Nanoparticles of $CeO_2$, $TiO_2$, ZnO, and Ag are likely to pose the greatest threat as they are considered to account for the most significant volumes of nanomaterials accumulating in the environment. Already, there is a growing body of research demonstrating the toxicological effects of these materials on bacteria, invertebrates, and higher level animals such as mice and fish (Nel et al., 2006; Oberdorster et al., 2007; Auffman et al., 2008; Baun et al., 2008; Baun, Hartmann, Grieger et al., 2008; Navarro et al., 2008; Santschi and Sigg 2008; Handy et al., 2008a,b; Nowack, 2009; Ramsden et al., 2009).

By association, the deliberate introduction of appreciable volumes of iron-based nanoparticles into the environment for counteracting pollution may well prove highly controversial (Masciangioli and Zhang, 2003; Zhang, 2003; Waychunas et al., 2005; Yue and Economy, 2005). The same properties that make INPs potentially useful for environmental remediation, specifically their small size and high redox reactivity, also make them potentially harmful to living things (Nel et al., 2006). To date, a limited number of studies have reported demonstrable toxicity to cells of different types (Waychunas et al., 2005; Pisanic et al., 2007; Lee et al., 2008; Keenan et al., 2009; Phenrat et al., 2009). A study by Keenan et al. (2009) demonstrated that INPs can rapidly react with oxygen and cause lung cells to die while Pisanic et al. (2007) observed limited growth and damage in nerve cells exposed to INPs. The primary mechanism for cellular damage is considered to be related to INP oxidation reactions, where redox cycling and the generation of reactive oxygen species (ROS) from reduced Fe within a cell can cause lipid peroxidation and damage to internal structures such as DNA (Stohs and Bagchi, 1995; Donaldson et al., 1997; Valko et al., 2005; Xia et al., 2006). Evidence suggests INPs to have greater toxicity than nanoscale magnetite (Phenrat et al., 2009), highlighting the significant role of reduced iron ($Fe^0$ or $Fe^{2+}$) in causing toxic effects. This result also suggests that iron oxides/hydroxides produced by INP corrosion present significantly less risk, because their inherent nanotoxicity is lower and their increased volume (into the macroscale) significantly reduces the likelihood of cellular uptake.

The study by Pisanic et al. (2007) also showed that INPs with stabilization coatings, such as polyacrylic acid, resulted in more cellular damage since the particles existed for a longer period without degradation. The shape and size of particles has also been linked to the uptake and toxicology of nanoparticles, with increasingly smaller particles displaying intensifying toxicity.

Laboratory tests have clearly demonstrated INP toxicity within isolated test systems, but more research has yet to be performed away from isolated cell-culture experiments in more complex systems representative of the natural environment. How the environment can influence the toxicity of iron and other nanoparticles is still poorly understood, due to the sheer complexity and variability of such systems. It is highly likely that NPs in natural environments are unlikely to be as toxic as demonstrated by laboratory experiments. For example, in a study by Tong et al. (2007), $C_{60}$ fullerenes, which in the laboratory have destroyed microbes, were not observed to seriously damage the microbial cultures present in soil after 30 days' exposure.

Although there are many complexities in understanding the fate of exotic ENPs such as $CeO_2$ in environmental systems, the reactions that determine the fate of metallic iron, and by proxy INPs, are well understood. Resultantly, the fate of INPs in subsurface and aqueous environments can be reasonably well predicted. The acute redox sensitivity of INP, which drives the high rates of contaminant reaction and corrosion observed, dictates that their persistence in subsurface environments will be limited. Studies have clearly shown that even the most effectively stabilized and engineered INP will be gradually immobilized within periods of 8–9 months in the subsurface via aggregation, mineral sorption, or oxidative degradation. Resultantly, INPs injected into the subsurface in aqueous suspension are highly unlikely to make it into the blood streams of humans or other mammals as they will have broken down long before contact. The greatest risk that INPs pose to humans is most obviously via inhalation and suitable care and precautions should quite obviously be taken when handling loose, dry INP. By using INPs in the way they are intended i.e., in solution, any immediate human danger is removed.

In 2008, the international organization for economic cooperation and development (OECD) proposed 14 classes of nanoparticles for testing toxicology and those of their break-down products in environmental systems. INPs were listed among the candidate materials and will receive rigorous testing in the near future (Stone et al., 2009). Until these experiments are performed, it is very hard to make any clear conclusions on the consequences of using nanoparticles for environmental remediation. While INP technology has already been used commercially in countries including the United States, Canada, the Czech Republic, Germany, Italy, and Slovakia (Zhang, 2003; Theron et al., 2008), other countries such as the UK have yet to establish a legislative framework for commercial application. It is likely that in future the findings of the OECD and other similar groups will significantly influence the legislation that governs the use of environmental nanotechnology.

## 15.7 Concluding Remarks

Worldwide, the number of sites in need of remedial attention is constantly increasing. In 2005, the U.K. Environment Agency estimated that there were 37,000 sites (78,000 ha) in need of remediation in England and Wales, with the number of contaminated sites increasing at 250 per year (EA, 2004). Resultantly, the U.K. contaminated land remediation market is expected to be worth £630M by 2010, rising from £404M in 2005 (Randall, 2007). On a larger scale, a similar situation exists in the United States and across Europe. In 2007, the European Environment Agency (EEA, 2007) reported an estimated 250,000 polluted sites in the EEA member countries that required cleanup. Similarly, the U.S. Environmental Protection Agency (EPA) has estimated that over 138,000 polluted sites in the United States require cleanup, with associated costs of up to $250 billion (EPA, 2004a,b, 2008), a significant cleanup bill.

The incentive for cost reduction applies globally, with many classical remediation methods regarded as being too costly for extensive deployment in the developing world and well beyond economic feasibility for most rural communities. Nanotechnology clearly provides a new answer for solving environmental pollution. By providing rapid and extensive remediation at a significantly reduced cost, the appeal of this technology is likely to be global. Only time will tell if environmental remediation using nanoparticles will receive universal acceptance and widespread application.

## References

Aboulhassan, M.A., Souabi, S., Yaacoubi, M. et al. 2006. Removal of surfactant from industrial wastewaters by coagulation flocculation process. *Int. J. Environ. Sci. Technol.* 3(4):327–332.

Abrams, B. and Wilcoxon, J. 2005. Nanosize semiconductors for photooxidation. *Crit. Rev. Solid State Mater.* 30:153.

Acar, Y.B. and Alshawabkeh, A.N. 1993. Principles of electrokinetic remediation. *Environ. Sci. Technol.* 27:2638–2647.

Adebajo, M.O., Frost, R.L., Kloprogge, J.T. et al. 2003. Porous materials for oil spill cleanup: A review of synthesis and absorbing properties. *J. Porous Mater.* 10:159–170.

Agency for Toxic Substances and Disease Registry. 1999. *Toxicological Profile for Uranium*. ATSDR, U.S. Department of Health and Human Services. Public Health Service. Atlanta, GA.

Ahmad, T., Kahlown, M.A., Tahir, A. et al. 2004. Arsenic an emerging issue: Experiences from Pakistan. *Proceedings of the 30th WEDC International Conference*, Vientiane, Lao PDR, pp. 459–466.

Aiken, J. 1884. On the formation of small clear spaces in dusty air. *Trans. R. Soc Edinb*. 30:337–368.
Allen, G.C., Scott, T.B., Lee, D.F. et al. 2004. The extraction of uranium from groundwaters on iron surfaces. *Philos. Mag*. 84(11):691–696.
Alowitz, M.J. and Scherer, M.M. 2002. Kinetics of nitrate, nitrite, and Cr(VI) reduction by iron metal. *Environ. Sci. Technol*. 36:299–306.
Auffan, M., Achouak, W., Rose, J.B. et al. 2008. Relation between the redox state of iron-based nanoparticles and their cytotoxicity toward *Escherichia coli*. *Environ. Sci. Technol*. 42:6730–6735.
Baer, D.R., Burrows, P.E., and El Azab, A.A. 2003. Enhancing coating functionality using nanoscience and nanotechnology. *Prog. Org. Coat*. 47:342–356.
Barnes, R.J. 2009. Dechlorinating bacterial strains and nano-scale iron particles for the remediation of CAH contaminated sites. D.Phil dissertation, Department of Earth Sciences, University of Oxford, Oxford, U.K.
Baun, A., Hartmann, N.B., Grieger, K. et al. 2008. Ecotoxicity of engineered nanoparticles to aquatic invertebrates: A brief review and recommendations for future toxicity testing. *Ecotoxicology* 17(5):387–395.
Beke, D.L. 1998. Intrinsic and domain magnetism in nanocrystalline materials. *Cryst. Res. Technol*. 33:1039–1059.
Berkovsky, B., Bashtovoy, V., and Berkovskii, B.M. 1996. *Magnetic Fluids and Applications Handbook*. Begell House Publishers, New York.
Bezbaruah, A.N., Krajangpan, S., Chisholm, B.J. et al. 2009. Entrapment of iron nanoparticles in calcium alginate beads for groundwater remediation applications. *J. Hazard. Mater*. 166(2–3):1339–1343.
Bigg, T. and Judd, S.J. 2000. Zero-valent iron for water treatment. *Environ. Technol*. 21(6):661–670.
Billon, G., Ouddane, B., Proix, N. et al. 2005. Distribution coefficient and redox behaviour of uranium in Authie Bay (northern France). *Int. J. Environ. Anal. Chem*. 85(14):1013–1024.
Blowes, D.W., Ptacek, C.J., Benner, S.G. et al. 2000. Treatment of inorganic contaminants using permeable reactive barriers. *J. Contam. Hydrol*. 45:123–137.
Bonin, P.M.L., Jedral, W., Odziemkowski, M.S. et al. 2000. Electrochemical and Raman spectroscopic studies of the influence of chlorinated solvents on the corrosion behaviour of iron in borate buffer and in simulated groundwater. *Corrosion Sci*. 42:1921–1939.
Brown Jr., G.E., Henrich, V.E., Casey, W.H. et al. 1999. Metal oxide surfaces and their interactions with aqueous solutions and microbial organisms. *Chem. Rev*. 99:77–174.
Brus, L. 1986. Electronic wave functions in semiconductor clusters: Experiment and theory. *J. Phys. Chem*. 90:2555–2560.
Burghardt, D., Simon, E., Knöller, K. et al. 2007. Immobilization of uranium and arsenic by injectable iron and hydrogen stimulated autotrophic sulphate reduction, *J. Contam. Hydrol*. 94:305–314.
Campbell, C.T. and Parker, S.C. 2002. The effect of size-dependent nanoparticle energetics on catalyst sintering. *Science* 298:811–814.
Cantrell, K.J., Kaplan, D.I., and Wietsman, T.W. 1995. Zero-valent iron for the in-situ remediation of selected metals in groundwater. *J. Hazard. Mater*. 42:201–212.
Cao, J., Elliott, D., and Zhang, W.J. 2005. Perchlorate reduction by nanoscale iron particles. *Nanopart. Res*. 7:499–506.
Çelebi, O., Üzüm, Ç., Shahwan, T. et al. 2007. A radiotracer study of the adsorption behavior of aqueous $Ba^{2+}$ ions on nanoparticles of zero-valent iron. *J. Hazard. Mater*. 148:761–767.
Charlet, L., Liger, E., and Gerasimo, P. 1998. Decontamination of TCE- and U-rich waters by granular iron: Role of sorbed Fe(II). *J. Environ. Eng*. 124:25–30.
Cheng, S.F. and Wu, S.C. 2000. The enhancement methods for the degradation of TCE by zero-valent metals. *Chemosphere* 41(8):1263–1270.
Cheng, I.F., Fernando, Q., and Korte, N. 1997. Electrochemical dechlorination of 4-chlorophenol to phenol. *Environ. Sci. Technol*. 31(4):1074–1078.
Cheng, R., Wang, J.-L., and Zhang, W.-X. 2007. Comparison of reductive dechlorination of p-chlorophenol using Fe-0 and nanosized Fe-0. *J. Hazard. Mater*. 144:334–339.

Choe, S., Chang, Y.Y., Hwang, K.Y. et al. 2000. Kinetics of reductive denitrification by nanoscale zero-valent iron. *Chemosphere* 41:1307–1311.

Choi, C.J., Dong, X.L., and Kim, B.K. 2001. Characterization of Fe and Co nanoparticles synthesized by chemical vapor condensation. *Scr. Materialia* 44:2225–2229.

Commission on Geosciences, Environment, and Resources. 2000. Natural attenuation for groundwater remediation. National Academy Press, Washington, DC.

Cornell, R.M. and Schwertmann, U. 1996. *The Iron Oxides 573*. VCH-Verlag, Weinheim, Germany.

Crane, R., Dickinson, M.D., Popescu, I.C., and Scott, T.B. 2009. Iron Nanoparticles for the Remediation of Uranium Contaminated Waters from Lişava, the Banat District, Romania. Session P, 39èmes journées des actinides, La Grande Motte, France.

Crawford, R.J., Harding, I.H., and Mainwaring, D.E. 1993. Adsorption and co-precipitation of multiple heavy metal ions onto the hydrated oxides of iron and chromium. *Langmuir* 9:3057–3062.

Cundy, A.B., Hopkinson, L., and Whitby, R.L.D. 2008. Use of iron-based technologies in contaminated land and groundwater remediation: A review. *Sci. Total Environ*. 400:42–51.

Grolimund, D. and Borkovec, M. 2005. Colloid-facilitated transport of strongly sorbing contaminants in natural porous media: Mathematical modeling and laboratory column experiments. *Environ. Sci. Technol*. 39(17):6378–6386.

Dabro, P., Cyr, A., Laplante, F. et al. 2000. Electrocatalytic dehydrochlorination of pentachlorophenol to phenol or cyclohexanol. *Environ. Sci. Technol*. 34:1265.

Darab, J.G., Amonette, A.B., Burke, D.S.D. et al. 2007. Removal of pertechnetate from simulated nuclear waste streams using supported zerovalent iron. *Chem. Mater*. 19:5703–5713.

De Carlo, E.H., Zeltiln, H., and Fernando, Q. 1981. Simultaneous separation of trace levels of germanium, antimony, arsenic, and selenium from an acid matrix by adsorbing colloid flotation. *Anal. Chem*. 53:1104–1107.

de Gennes, P.-G. and Pincus, P.A. 1970. Pair correlations in a ferromagnetic colloid. *Phys. Kondens. Mater*. 11:189–198.

de Vicente, J., Delgado, A.V., Plaza, R.C. et al. 2000. Stability of cobalt ferrite colloidal particles: Effect of pH and applied magnetic fields. *Langmuir* 16:7954–7961.

Derjaguin, B. and Landau, L. 1941. Theory of the stability of strongly charged lyophobic sols and of the adhesion of strongly charged particles in solutions of electrolytes. *Acta Phys.–Chem. URSS* 14:633.

Dickinson, M. and Scott, T.B. 2010. The application of zero-valent iron nanoparticles for the remediation of a uranium-contaminated waste effluent. *J. Hazard. Mater*. 178(1–3):171–179 (10.1016/j.jhazmat-2010-01-060).

Dickinson, M., Scott, T.B., Crane, R. et al. 2009. The effects of vacuum annealing on the structure and surface chemistry of iron:nickel alloy nanoparticles. *J. Nanoparticle Res*. 10.1007/s11051-009-9767-y.

Domenica, P. and Schwartz, F.W. 1990. *Physical and Chemical Hydrogeology*. John Wiley & Sons, New York.

Donaldson, K., Brown, D.A., Mitchell, C. et al. 1997. Free radical activity of PM10: Iron-mediated generation of hydroxyl radicals. *Environ. Health Perspect*. 105:1285–1289.

EEA. 2007 Progress in management of contaminated sites. Report CSI015. Copenhagen, Denmark: European Environment Agency.

Elihn, K., Otten, F., and Boman, M. 1999. Nanoparticle formation by laser-assisted photolysis of ferrocene. *Nanostruct. Mater*. 12:79–82.

Elihn, K., Otten, F., Boman, M. et al. 2001. Size distributions and synthesis of nanoparticles by photolytic dissociation of ferrocene. *Appl. Phys. A* 72:29.

Elimelech, M., Gregory, J., Jia, X., and Williams, R.A. 1998. *Particle Deposition and Aggregation: Measurement, Modelling, and Simulation*. Elsevier.

Elliott, S. 1998. *The Physics and Chemistry of Solids*. Wiley & Sons, New York.

Elliott, D.W. and Zhang, W. 2001. Field assessment of nanoscale bimetallic particles for groundwater treatment. *Environ. Sci. Technol*. 35:4922–4926.

Environment Agency. 2004. The State of Soils in England and Wales. Environment Agency, Bristol, 32.

EPA. 2004a. EPA 542-R-04-015. Technology Innovation and Field Services Division. Office of Solid Waste and Emergency Response. http://www.epa.gov/tio/download/market/2004market.pdf

EPA. 2004b. Cleaning Up the Nations Waste Sites: Markets and Technology Trends. http://www.clu-in.org/download/market/2004market.pdf (Accessed August 2009).

Farrell, J., Bostick, W.D., Jarabek, R.J. et al. 1999. Uranium removal from ground water using zero valet iron media. *Ground Water* 37:618–624.

Feitz, A.J., Joo, S.H., Guan, J. et al. 2005. Oxidative transformation of contaminants using colloidal zero-valent iron. *Colloids Surf. A* 265:88–94.

Fendorf, S.E. and Li, G. 1996. Kinetics of chromate reduction by ferrous iron. *Environ. Sci. Technol.* 30:1614–1617.

Fiedor, J.N., Bostick, W.D., Jarabek, R.J. et al. 1998. Understanding the mechanism of uranium removal from groundwater by zero-valent iron using x-ray photoelectron spectroscopy. *Environ. Sci. Technol.* 32:1466–1473.

Frimmel, F.H., von der Kammer, F., and Flemming, H.C. 2007. *Colloidal Transport in Porous Media*. Springer, Berlin, Germany.

Fryxell, G.E., Shane, A.R., Mattigod, S.V. et al. 2004. *Dekker Encyclopedia of Nanoscience and Nanotechnology*, Vol. 1, Marcel-Dekker, New York, pp. 1125–1135.

Gan, S., Liang, Y., Baer, D.R. et al. 2001. Effect of platinum nanocluster size and titania surface structure upon CO surface chemistry on platinum support on $TiO_2$(110). *J. Phys. Chem. B* 105:2412–2415.

Gavaskar, A., Gupta, N., Sass, B., Fox, T., Janosy, R., Cantrell, K., and Olfenbuttle, R. 1997. Design Guidance for Application of Permeable Barriers to Remediate Dissolved Chlorinated Solvents. U.S. Air Force Report, Battelle Press, OH.

Gavaskar, A.R., Sass, B.M., Drescher, E., Cumming, L., Giammar, D., and Guptam N. 1998. Enhancing the reactivity of permeable reactive barrier media. Paper presented at *First International Conference on Remediation of Chlorinated and Recalcitrant Compounds*, Columbus, OH, May 18–21, 1998, pp. 91–96.

Gavaskar, A., Tatar, L., and Condit, W. 2005. Cost and performance report, nanoscale zero-valent iron technologies for source remediation. NAVFAC, Naval Facilties Engineering Command, Port Hueneme, CA. Contract report CR-05-007-ENV.

Gerasimov, Y., Dreving, V., Eremin, E. et al. 1985. *Physical Chemistry* 2:624. MIR, Moscow.

Gillham, R.W. and O'Hannessin, S.F. 1994. Enhanced degradation of halogenated aliphatics by zero-valent iron. *Ground Water* 32(6):958–967.

Glazier, R., Venkatakrishnan, R., Gheorghiu, F. et al. 2003. Nanotechnology takes root. *Civil Eng.* 73:64–69.

Golab, A.N., Indraratna, B., and Peterson, M.A. 2006. Performance of alkaline reactive materials in the remediation of acidic groundwater. *Chin. J. Geochem.* 25:88–89.

Grittini, C., Malcomson, M., and Fernando, Q. 1995. Hydrodechlorination of chlorinated ethanes by nanoscale Pd/Fe bimetallic particles. *Environ. Sci. Technol.* 29:2898–2900.

Grolimund, D. and Borkovec, M. 2005. Colloid-facilitated transport of strongly sorbing contaminants in natural porous media: Mathematical modeling and laboratory column experiments. *Environ. Sci. Technol.* 39(17):6378–6386.

Grolimund, D., Borkovec, M., and Barmettler, K. 1996. Colloid facilitated transport of strongly sorbing contaminants in natural porous media: A laboratory column study, research communication. *Environ. Sci. Technol.* 30:3118–3123.

Gu, B., Liang, L., Dickey, M.J. et al. 1998. Reductive precipitation of uranium (VI) by zero-valent iron. *Environ. Sci.* 21:3366–3373.

Handy, R.D., von der Kammer, F., Lead, J. R. et al. 2008a. The ecotoxicology and chemistry of manufactured nanoparticles. *Ecotoxicology* 17(4):287–314.

Handy, R.D., Owen, R., and Valsami-Jones, E. 2008b. The ecotoxicology of nanoparticles and nanomaterials: Current status, knowledge gaps, challenges, and future needs. *Ecotoxicology* 17(5):315–325.

He, Y.T. and Traina, S.J. 2005. Cr(VI) reduction and immobilization by magnetite under alkaline pH conditions: The role of passivation. *Environ. Sci. Technol.* 39(4):4499–4504.

Henderson, A.D. and Demond, A.H. 2007. Long-term performance of zero-valent iron permeable reactive barriers: A critical review. *Environ. Eng. Sci.* 24:401–423.

Hoch, L.B., Mack, E.J., Hydutsky, B.W. et al. 2008. Carbothermal synthesis of carbon-supported nanoscale zero-valent iron particles for the remediation of hexavalent chromium. *Environ. Sci. Technol.* 42:2600–2605.

Hoffmann, M.R., Martin, S.T., Choi, W. et al. 1995. Environmental applications of semiconductor photocatalysis. *Chem. Rev.* 95:69–96.

Hristov, J. and Fachikov, L. 2007. Overview of separations performed by magnetically assisted fluidized beds with a special emphasis to waste stream cleaning. *China Particuology* 5(1):11–18.

Hsi, C.D. and Langmuir, D. 1985. Adsorption of uranyl onto ferric oxyhydroxides. Application of the surface complexation site-binding model. *Geochim. Cosmochim. Acta* 49:1931–1941.

Huang, Q., Xiangyang, S., Pinto, R.A. et al. 2008. Tunable synthesis and immobilization of zero-valent iron nanoparticles for environmental applications. *Environ. Sci. Technol.* 42:8884–8889.

Huber, D. 2005. Synthesis, properties and applications of iron nanoparticles. *Small* 1(5):482–501.

Hydutsky, B.W., Mack, E.J., Beckerman, B.B. et al. 2007. Optimization of nano- and microiron transport through sand columns using polyelectrolyte mixtures. *Environ. Sci. Technol.* 41(18):6418–6424.

Hyne, R.V., Rippon, G.D., and Ellender, G. 1992. PH-dependent uranium toxicity of fresh water hydra. *Sci. Total Environ.* 125:159–73.

Interstate Technology and Regulatory Council (ITRC). 2005. Permeable Reactive Barriers: Lessons Learned/New Directions, ITRC.

Johnson, R.L., Thoms, R.B., Johnson, R.O.B. et al. 2008. Field evidence for flow reduction through a zero-valent iron permeable reactive barrier. *Ground Water Monit. Remed.* 28:47–55.

Johnson, T.L., Scherer, M.M., and Tratnyek, P.G. 1996. Kinetics of halogenated organic compound degradation by iron metal. *Environ. Sci. Technol.* 30(8):2634–2640.

Johnson, S.A., Ollivier, P.J., and Mallouk, T.E. 1999. Ordered mesoporous polymers of tunable pore size from colloidal silica templates. *Science* 283:963–965.

Joo, S.H., Feitz, A.J., Sedlak, D.L. et al. 2005. Quantification of the oxidizing capacity of nanoparticulate zero-valent iron. *Environ. Sci. Technol.* 39:1263–1268.

Kamat, P.V. and Meisel, D. 2003. Nanoscience opportunities in environmental remediation. *C. R. Chimie* 6:999.

Kanel, S.R. and Choi, H. 2007. Transport characteristics of surface-modified nanoscale zero-valent iron in porous media. *Water Sci. Technol.* 55:157–162.

Kanel, S.R., Manning, B., Charlet, L. et al. 2005. Removal of arsenic(III) from groundwater by nanoscale zero-valent iron. *Environ. Sci. Technol.* 39:1291–1298.

Kanel, S., Greneche, J.M., and Choi, H. 2006. Arsenic(V) removal from groundwater using nano scale zero-valent iron as a colloidal reactive barrier material. *Environ. Sci. Technol.* 40:2045–2050.

Kanel, S.R., Nepal, D., Manning, B. et al. 2007. Transport of surface-modified iron nanoparticle in porous media and application to arsenic(III) remediation. *J. Nanopart. Res.* 9(5):725–735.

Kanel, S.R., Goswami, R.R., Clement, T.P. et al. 2008. Two dimensional transport characteristics of surface stabilized zero-valent iron nanoparticles in porous media. *Environ. Sci. Technol.* 42(3):896–900.

Karabelli, D., Uzum, C., Shahwan, T. et al. 2008. Batch removal of aqueous $Cu^{2+}$ ions using nanoparticles of zero-valent iron: A study of the capacity and mechanism of uptake. *Ind. Eng. Chem. Res.* 47(14):4758–4764.

Karlsson, M.N.A., Deppert, K., Wacaser, B.A. et al. 2005. Size-controlled nanoparticles by thermal cracking of iron pentacarbonyl. *Appl. Phys. A* 80(7):1579–1583.

Keenan, C.R., Goth-Goldstein, R., Lucas, D., and Sedlak, D.L. 2009. Oxidative stress induced by zero-valent iron nanoparticles and Fe(II) in Human Bronchial Epithelial Cells. *Environ. Sci. Technol.* 43:4555–4560.

Khudenko, B.M. 1991. Feasibility evaluation of a novel method for destruction of organics. *Water Sci. Technol.* 23:1873–1881.

Kim, Y.-H. and Carraway, E.R. 2000. Dechlorination of pentachlorophenol by zero valent iron and modified zero valent irons. *Environ. Sci. Technol.* 34(10):2014–2017.

Klabunde, K.J., Stark, J., Koper, O. et al. 1996. Nanocrystals as stoichiometric reagents with unique surface chemistry. *J. Phys. Chem.* 100(30):12142–12153.

Klimenkov, M., Nepijko, S., Kuhlenbeck, H. et al. 1997. The structure of Pt-aggregates on a supported thin aluminium oxide film in comparison with unsupported alumina: A transmission electron microscopy study. *Surf. Sci.* 391:27–36.

Kosmulski, M. 2001. *Chemical Properties of Material Surfaces* (Technical University of Lublin, Lublin, Poland). Surfactant Science Series, Vol. 102, Edited by Arthur T. Hubbard (Santa Barbara Science Project, Santa Barbara, CA. Marcel Dekker, Inc., New York.

Kovalic, W.W. 1995. In situ remediation technology status report: Electrokinetics. US Environmental Protection Agency. EPA 542-K-94-007. Washington, DC.

Kuhn, L.T., Bojesen, A., Timmermann, L. et al. 2002. Structural and magnetic properties of core–shell iron–iron oxide nanoparticles. *J. Phys: Condens. Matter* 14:13551.

Langmuir, D. 1997. *Aqueous Environmental Geochemistry*. Prentice Hall, Upper Saddle River, NJ.

Lee, C., Kim, J.Y., Lee, W.I. et al. 2008. Bactericidal effect of zero-valent iron nanoparticles on *Escherichia coli*. *Environ. Sci. Technol.* 42:4927–4933.

Lenhart, J.J. and Honeyman, B.D. 1999. Uranium (VI) sorption to haematite in the presence of humic acid. *Geochim. Cosmochim. Acta* 63:2891–2901.

Li, X.Q. and Zhang, W.X. 2007. Sequestration of metal cations with zerovalent iron nanoparticles—A study with high resolution X-ray photoelectron spectroscopy (HR-XPS). *J. Phys. Chem.* 111:6939–6946.

Li, D., Wang, Y., and Xia, Y. 2004. Electrospining nanofibers as uniaxially aligned arrays and layer-by-layer stacked films. *Adv. Mater.* 16(4):361–366.

Li, L., Fan, M., Brown, R.C. et al. 2006. Synthesis, properties, and environmental applications of nanoscale iron-based materials: A review. *Crit. Rev. Environ. Sci. Technol.* 36:405–431.

Liang, L., Korte, N., Goodlaxson, J.D. et al. 1997. Byproduct formation during the reduction of TCE by zero-valence iron and palladized iron. *Ground Water Monit. Rem.* 17:122–127.

Lien, H.L. and Zhang, W.X. 1999. Transformation of chlorinated methanes by nanoscale iron particles. *J. Environ. Eng.* 125:1042–1047.

Lien, H.L. and Zhang, W.X. 2001. Nanoscale iron particles for complete reduction of chlorinated ethenes. *Colloids Surf. A* 191:97–105.

Lien, H.L. and Zhang, W.X. 2005. Hydrodechlorination of chlorinated ethanes by nanoscale Pd/Fe bimetallic particles. *J. Environ. Eng.-ASCE* 131:4–10.

Lien, H.L. and Zhang, W.X. 2007. Nanoscale Pd/Fe bimetallic particles: Catalytic effects of palladium on hydrodechlorination. *Appl. Catal. B-Environ.* 77:110–116.

Liger, E., Charlet, L., and Van Cappellen, P. 1999. Surface catalysis of uranium(VI) reduction by iron(II). *Geochim. Cosmochim. Acta* 63:2939–2955.

Liu, Y., Majetich, S.A., Tilton, R.D. et al. 2005. TCE dechlorination rates, pathways, and efficiency of nanoscale iron particles with different properties. *Environ. Sci. Technol.* 39:1338–1345.

Logan, B.E., Camesano, T.A., DeSantis, A.A. et al. 1999. Comment on "A method for calculating bacterial deposition coefficients using the fraction of bacteria recovered from laboratory columns" by Bolster et al. *Environ. Sci. Technol.* 33:1316–1317.

Luther, G.W. and Rickard, D.T. 2005. Metal sulfide cluster complexes and their biogeochemical importance in the environment. *J. Nanopart. Res.* 7:389–407.

Margeat, O., Dumestre, F., Amiens, C. et al. 2005. Synthesis of iron nanoparticles: Size effects, shape control and organisation. *Prog. Solid State Chem.* 33:71–79.

Masciangioli, T. and Zhang, W.X. 2003. Environmental technologies at the nanoscale. *Environ. Sci. Technol.* 37(5):102A–108A.

Matheson, L.J. and Tratnyek, P.G. 1994. Reductive dehalogenation of chlorinated methanes by iron metal. *Environ. Sci. Technol.* 28:2045–2053.

Matyjaszewski, K., Lowry, G.V., and Tilton, R.D. 2009. Effect of adsorbed polyelectrolytes on nanoscale zero valent iron particle attachment to soil surface models. *Eviron. Sci. Technol.* 43(10):3803–3808.

McCarthy, J.F. and McKay, L.D. 2004. Colloid transport in the subsurface. *Vadose Zone J.* 3:326–337.

McCormick, M.L. and Adriaens, P. 2004. Carbon tetrachloride transformation on the surface of nanoscale biogenic magnetite particles. *Environ. Sci. Technol.* 38(12):1045–1053.

McCurrie, R.A. 1994. *Ferromagnetic Materials: Structure and Properties*. Academic Press, London, U.K.

Meyer, D.E., Wood, K., Bachas, L.G. et al. 2004. Degradation of chlorinated organics by membrane-immobilized nanosized metals. *Environ. Prog.* 23(3):232–242.

Miehr, R., Tratnyek, P.G., Bandstra, J.Z. et al. 2004. Diversity of contaminant reduction reactions by zerovalent iron: role of the reductate. *Environ. Sci. Technol.* 38:139–147.

Missana, T., Maffiotte, C., and García-Guttiérrez, M. 2003. Surface reaction kinetics between nano-crystalline magnetite and uranyl. *J. Colloid Interface Sci.* 261:154–160.

Mondal, K., Jegadeesan, G., and Lalvani, S.B. 2004. Removal of selenate by Fe and NiFe nanosized particles. *Ind. Eng. Chem. Res.* 43:4922–4934.

Morrison, S.J. and Spangler, R.R. 1993. Chemical barriers for controlling groundwater contamination. *Environ. Prog.* 12:175–181.

Morrison, S.J., Metzler, D.R., and Carpenter, C.E. 1998. Uranium precipitation in a permeable reactive barrier by progressive irreversible dissolution of zero-valent iron. *Environ. Sci. Technol.* 35:385–390.

Morrison, S.J., Mushovic, P.S., and Niesen, P.L. 2006. Early breakthrough of molybdenum and uranium in a permeable reactive barrier. *Environ. Sci. Technol.* 40:2018–2024.

Muftikian, R., Nebesny, K., Fernando, Q. et al. 1996. X-ray photoelectron spectra of the palladium-iron bimetallic surface used for the rapid dechlorination of chlorinated organic environmental contaminants. *Environ. Sci. Technol.* 30(12):3593–3596.

Mulvaney, P. 2001. Metal nanoparticles: Double layers, opticalproperties, and electrochemistry. In Klabunde, K. J. (ed.), *Nanoscale Materials in Chemistry*, Wiley, NewYork, pp. 121–167.

Naka, D., Kim, D., and Strathmann, T.J. 2006. Abiotic reduction of nitroaromatic compounds by aqueous iron(II)-catechol complexes. *Environ. Sci. Technol.* 40:3006–3012.

Navarro, E., Baun, A., Behra, R. et al. 2008. Environmental behaviour and ecotoxicity of engineered nanoparticles to algae, plants and fungi. *Ecotoxicology*. 17:372–386.

Nel, A., Xia, T., Madler, L. et al. 2006. Toxic potential of materials at the nanolevel. *Science* 311:622–627.

Nepijko, S.A., Klimenkow, M., Adelt, M. et al. 1999. Structural investigation of palladium clusters on ç-$AlO_3$(111)/NiAl(110) with transmission electron microscopy. *Langmuir* 15:5309–5313.

Nepijko, S.A., Klimenkow, M., Kuhlenbeck, H. et al. 1998. TEM study of tantalum clusters of $Al_2O_3$/NiAl(110). *Surf. Sci.* 412/413:192–201.

Noubactep, C.A. 2008. A critical review on the process of contaminant removal in $Fe^0$-$H_2O$ systems. *Environ. Technol.* 29:909–920.

Noubactep, C., Meinrath, G., Dietrich, P. et al. 2005. Testing the suitability of zerovalent iron materials for reactive walls. *Environ. Chem.* 2:71–76.

Nowack, B. 2009. The behavior and effects of nanoparticles in the environment. *Environ.Pollut.* 157(4):1063–1064.

Nurmi, J.T., Tratnyek, P.G., Sarathy, V. et al. 2005. Characterization and properties of metallic iron nanoparticles: Spectroscopy, electrochemistry, and kinetics. *Environ. Sci. Technol.* 39:1221–1230.

Nyer, E. K. and Vance, D. B. 2001. *Ground Water Monit. Rem.* Spring, 41.
Obare, S.O. and Meyer, G.J.J. 2004. Nanostructured materials for environmental remediation of organic contaminants in water. *Environ. Sci. Health* A39:2549.
Oberdorster, G., Stone, V., and Donaldson, K. 2007. Toxicology of nanoparticles: A historical perspective. *Nanotoxicology* 1(1):2–25.
Odziemkowski, M.S., Schuhmacher, T.T., Gillham, R.W. et al. 1998. Mechanism of oxide film formation on iron in simulating groundwater solutions—Raman spectroscopic studies. *Corrosion Sci.* 40(2–3):371–389.
O'Hannesin, S.F. and Gillham, R.W. 1998. Long-term performance of an in situ "iron wall" for remediation of VOCs. *Ground Water* 36:164–170.
O'Loughlin, E.J., Kelly, S.D., Cook, R.E. et al. 2003. Reduction of uranium (VI) by mixed iron(II)/iron(III) hydroxide (green rust): Formation of $UO_2$ nanoparticles. *Environ. Sci. Technol.* 37:721–727.
Pamuka, S., Hannum, L., and Wittle, J.K. 2008. Delivery and activation of nano-iron by DC electric field. *J. Environ. Sci. Health A* 43:934–944.
Papell, S.S. 1965. Low viscosity magnetic fluid obtained by the colloidal suspension of magnetic particles. U.S. Patent 3215572.
Park, S.H. and Xia, Y. 1998. Macroporous membranes with highly ordered and three-dimensionally interconnected spherical pores. *Adv. Mater.* 10:1045–1048.
PARS Environmental. http://www.parsenviro.com/index.html (accessed August 2009).
Pavlakis, N., Pollock, C.K., McLean, G. et al. 1996. Deliberate overdose of uranium: Toxicity and treatment. *Nephron* 72:313–317.
Penchev, I.P. and Hristov, J.Y. 1990. Behavior of fluidized beds of ferromagnetic particles in an axial magnetic field. *Powder Technol.* 61:103–118.
Phenrat, T., Kim, H.J., Fagerlund, F. et al. 2009. Particle size distribution, concentration, and magnetic attraction affect transport of polymer-modified Fe(0) nanoparticles in sand columns. *Environ. Sci. Technol.* 43(13):5079–5085.
Phenrat, T., Saleh, N., Sirk, K. et al. 2007a. Stabilization of aqueous zerovalent iron dispersions by anionic polyelectrolytes: Adsorbed anionic polyelectrolyte layer properties and their effect on aggregation and sedimentation. *J. Nanopart. Res.* 10:795–814.
Phenrat, T., Saleh, N., Sirk, K. et al. 2007b. Aggregation and sedimentation of aqueous nanoscale zerovalent iron dispersions. *Environ. Sci. Technol.* 41(1):284–290.
Phenrat, T., Long, T.C., Lowry, G.V. et al. 2009. Partial oxidation ("aging") and surface modification decrease the toxicity of nanosized zerovalent iron. *Environ. Sci. Technol.* 43:195–200.
Pisanic, T.R., Blackwell, J.D., Shubayev, V.I. et al. 2007. Nanotoxicity of iron oxide nanoparticle internalization in growing neurons. *Biomaterials* 28(16):2572–2581.
Ponder, S.M., Ford, J.R., Darab, J.G. et al. 1999. Ferragels: *A new* family of materials for remediation of aqueous metal ion solutions. Scientific basis for nuclear waste management XXII. *Mater. Res. Soc.* 556:1269–1276.
Ponder, S.M., Darab, J.G., and Mallouk, T.E. 2000. Remediation of Cr(VI) and Pb(II) aqueous solutions using supported, nanoscale zero-valent iron. *Environ. Sci. Technol.* 34:2564–2569.
Ponder, S.M., Darab, J.G., Bucher, J. et al. 2001 Surface chemistry and electrochemistry of supported zerovalent iron nanoparticles in the remediation of aqueous metal contaminants. *Chem. Mater.* 13(2):479–486.
Powell, R.M., Puls, R.W., Hightower, S.K. et al. 1995. Coupled iron corrosion and chromate reduction: Mechanisms for subsurface remediation. *Environ. Sci. Technol.* 29:1913–1922.
Probstein, R.F. and Hicks, R.E. 1993. Removal of contaminants from soils by electric field. *Science* 260:498–503.
Project on Emerging Nanotechnologies. 2009. Woodrow Wilson International Center for Scholars, Washington DC. www.nanotechproject.org (Accessed August 2009).
Promislow, J.H.E., Gast, A.P., and Fermigier, M. 1995. Aggregation kinetics of paramagnetic colloidal particles. *J. Chem. Phys.* 102:5492–5498.
Punoose, A., Magnone, H., Seehra, M.S. et al. 2001. Bulk to nanoscale magnetism and exchange bias in CuO nanoparticles. *Phys. Rev. B* 64:174420.

Ramsden, C.S., Smith, T.J., Shaw, B.J. and Handy, R.D. 2009. Dietary exposure to titanium dioxide nanoparticles in rainbow trout (Oncorhynchus mykiss): No effect on growth, but subtle biochemical disturbances in the brain. *Ecotoxicology* October 18(7):939–951.

Randall, J. 2007. The land remediation industry. The EIC guide to the UK Environmental Industry 2007. Environmental Industries Commission, 61–62.

Riba, O., Scott, T.B., Ragnarsdottir, K.V. et al. 2008. Reaction mechanism of uranyl in the presence of zero-valent iron nanoparticles. *Geochim. Cosmochim. Acta* 72:4047–4057.

Riba, O., Barnes, R. J., Scott, T.B. et al. 2010. Optimization of nanoscale nickel/iron particles for the reduction of high concentration chlorinated aliphatic hydrocarbon solutions. *Chemosphere* 79(4):448–454.

Roy, S.B. and Dzombak, D.A. 1998. Sorption nonequilibrium effects in colloid-facilitated transport of hydrophobic organic compounds in porous media. *J. Contam. Hydrol.* 30:179–200.

Sabatini, D.A. and Knox, R.C. 1992. *Transport and Remediation of Subsurface Contaminants: Colloidal, Interfacial, and Surfactant Phenomena* (ACS Symposium Series), American Chemical Society, Washington, DC.

Saiers, J.E. and Hornberger, G.M. 1996. The role of colloidal kaolinite in the transport of caesium through laboratory sand columns. *Water Resour. Res.* 32:33–41.

Saleh, N. 2007. An assessment of novel polymeric coatings to enhance transport and in situ targeting of nanoiron for remediation of non-aqueous phase liquids (NAPLs). *Civil and Environmental Engineering*. Carnegie Mellon University, Pittsburgh, PA, p. 274.

Saleh, N., Sarbu, T., Sirk, K. et al. 2005. Oil-in water emulsions stabilized by highly charged polyelectrolyte-grafted silica nanoparticles, *Langmuir* 21(22):9873–9878.

Saleh, N., Sirk, K., Liu, Y. et al. 2007 Surface modifications enhance nanoiron transport and NAPL targeting in saturated porous media. *Environ. Eng. Sci.* 24(1):45–57.

Santschi, P.H. and Sigg, L. 2008. Environmental behavior and ecotoxicity of engineered nanoparticles to algae, plants, and fungi. *Ecotoxicology* 17(5):372–386.

Scherer, M.M., Richter, S., Valentine, R.L. et al. 2000. Chemistry and microbiology of permeable reactive barriers for in situ groundwater clean up. *Rev. Environ. Sci. Technol.* 30:363–411.

Schmuki, P. 2002. From Bacon to barriers: a review on the passivity of metals and alloys. *J. Solid State Electrochem.* 6:145–164.

Schrick, B., Blough, J.L., Jones, A.D. et al. 2002. Hydrodechlorination of trichloroethylene to hydrocarbons using bimetallic nickel-iron nanoparticles. *Chem. Mater.* 14:5140–5147.

Schrick B., Hydutsky B.W., and Blough, J.L. 2004. Delivery vehicles for zerovalent metal nanoparticles in soil and groundwater. *Chem. Mater.* 16(11):2187–2193.

Scott, T.B. 2005. Sorption of uranium onto iron bearing minerals, PhD thesis, Interface Analysis Centre, University of Bristol, Bristol, U.K.

Scott, T.B., Allen, G.C., Heard, P.J. et al. 2005a. Reduction of U(VI) to U (IV) on the surface of magnetite. *Geochim. Cosmochim. Acta* 69:5639–5646.

Scott, T.B., Allen, G.C., Heard, P.J. et al. 2005b. The extraction of uranium from groundwaters on iron surfaces. *Proc. R. Soc. A* 461:1247–1259.

Scott, T.B., Riba, O., and Allen, G.C. 2007. Aqueous uptake of uranium onto pyrite surfaces, reactivity of fresh versus weathered material. *Geochim. Cosmochim. Acta* 71:5044–5053.

Scott, T. B., Dickinson, M., Crane, R. et al. 2010. The effects of vacuum annealing on the structure and surface chemistry of iron nanoparticles. *J. Nano. Res.* 12(5):1765–1775.

Shao, H., Lee, H., Huang, Y. et al. 2005. Control of iron nanoparticles size and shape by thermal decomposition method. *IEEE Trans. Magnetics* 41(10):3388–3390.

Sharma, R.K., Sharma, P., and Maitra, A. 2003. Size-dependent catalytic behavior of platinum nanoparticles on the hexacyanoferrate-(III)/thiosulfate redox reaction. *J. Colloid Interface Sci.* 265:134–140.

Sheppard, D. 2007. Emerging challenges—Nanotechnology and the environment. Global Environmental Outlook (GEO) Yearbook. United Nations Environmental Programme. Earthprint Ltd. England, pp. 61–70.

Shimotori, T., Nuxoll, E.E., Cussler, E.L. et al. 2004. A polymer membrane containing Fe0 as a contaminant barrier. *Environ. Sci. Technol.* 38(7):2264–2270.

Shimotori, T., Nuxoll, E.E., Cussler, E.L. et al. 2006. High-density polyethylene membrane containing Fe0 as a contaminant barrier. *J. Environ. Eng.* 132(7):803–809.

Shokes, T.E. and Moller, G. 1999. Removal of dissolved heavy metals from acid rock drainage using iron metal. *Environ. Sci. Technol.* 33:282–287.

Signorini, L., Pasquini, L., Savini, L. et al. 2003. Size-dependent oxidation in iron/iron oxide core-shell nanoparticles. *Phys. Rev. B* 68(19):5423–5431.

Simon, F.-G., Meggyes, T., and McDonald, C. 2002. *Advanced Groundwater Remediation, Active and Passive Technologies*. Thomas Telford Publishing, London, U.K.

Singh, I.B. and Singh, D.R. 2003. Effects of pH on Cr-Fe interaction during Cr(VI) removal by metallic iron. *Environ. Technol.* 24:1041–1048.

Sirk, K.M., Saleh, N.B., Phenrat, T. et al. 1995. Reductive dechlorination of chlorinated ethenes by iron metal. *209th American Chemical Society National Meeting*, Anaheim, CA, Division of Environmental Chemistry, Conference Proceedings, 53:695–698.

Sirk, K.M., Saleh, N.B., Phenrat, T. et al. 2009. Effect of adsorbed polyelectrolytes on nanoscale zero valent iron particle attachment to soil surface models. *Environ. Sci. Technol.* 43:3803–3808.

Sivavec, T. and Horney, D. 1995. Reductive dechlorination of chlorinated ethenes by iron metal, in *Proceedings of the 209th ACS National Meeting*, Anaheim, California, April 2–6.

Smyth, D.A., Shikaze, S.G., and Cherry, J.A. 1997. Hydraulic performance of permeable barriers for the in situ treatment of contaminated groundwater. *Land Contam. Reclamation*, 5(3):131–137.

Snoeyink, V.L. and Jenkins, D. 1980. *Water Chemistry*. Wiley, New York.

Sridharan, N. and Lee, G.F. 1972. Co-precipitation of organic compounds from lake water by iron salts. *Environ. Sci. Technol.* 6:1031–1033.

Stohs, S.J. and Bagchi, D. 1995. Oxidative mechanisms in the toxicity of metal ions. *Free Radic. Biol. Med.* 18:321–336.

Stone, V., Nowack, B., Baun, A. et al. 2009. Nanomaterials for environmental studies: Classification, reference material issues, and strategies for physico-chemical characterisation. *Sci. Total Environ.* 408(7):1745–1754.

Stucki, J.W. 1988. Structural iron in smectites. In Stucki, J.W., Goodman, B.A., and Schwertmann, U. (eds.), *Iron in Soils and Clay Minerals*, D. Reidel, Dordrecht, the Netherlands, pp. 305–428.

Stumn, W., and Sulzberger, B. 1992. The cycling of iron in natural environments: Considerations based on laboratory studies of heterogeneous redox processes. *Geochim. Cosmochim. Acta*. 56: 3233–3257.

Sun, Y.P., Li, X.Q., Cao, J. et al. 2006. Characterization of zero-valent iron nanoparticles. *Adv. Colloid Interface Sci.* 120:47–56.

Surdo, E.M., Cussler, E.L., and Arnold, W.A. 2009. Sorptive and reactive scavenger-containing sandwich membranes as contaminant barriers. *J. Environ. Eng.* 135(2):69–76.

Suthersan, S. 1996. *Remediation Engineering: Design Concepts* (Environmental Science and Engineering Series). Lewis, Boca Raton, FL.

Tarr, M.A. (ed.). 2003. *Methods for Wastes and Pollutants: Environmental and Industrial Applications*. Marcel Dekker, New York, pp. 371–421.

Tee, Y.-H., Bachas, L., and Bhattacharyya, D. 2009. Degradation of trichloroethylene by iron-based bimetallic nanoparticles. *J. Phys. Chem. C* 113(22):9454–9464.

Theron, J., Walker, J.A., and Cloete, T.E. 2008. Nanotechnology and water treatment: Applications and emerging opportunities. *Crit. Rev. Microbiol.* 34:43–69.

Thomson, E. 2008. MIT scientists develop a 'paper towel' for oil spills. *MIT Tech Talk*, 52(28):4. June 4. Available at: http://web.mit.edu/newsoffice/2008/techtalk52-28.pdf (accessed August 2009).

Tiraferri, A., Chen, K.L., Sethi, R. et al. 2008. Reduced aggregation and sedimentation of zero-valent iron nanoparticles in the presence of guar gum. *J. Colloid Interface Sci.* 324(1–2):71–79.

Tong, Z., Bischoff, M., Nies, L. et al. 2007. Impact of fullerene (C60) on a soil microbial community. *Environ. Sci. Technol.* 41(8):2985–2991.

Tratnyek, P.G. 1996. Putting corrosion to use: Remediating contaminated groundwater with zero-valent metals. *Chem. Ind. Lond.* 13:499–503.

Tratnyek, P.G., Scherer, M.M., Johnson, T.J. et al. 2003. Permeable reactive barriers of iron and other zero-valent metals. In Tarr M.A. (Ed.), *Chemical Degradation Methods for Wastes and Pollutants: Environmental and Industrial Applications* (Environmental Science and Pollution Control), Marcel Dekker, New York, pp. 371–42.

Uzum, C., Shahwan, T., Eroglu, A.E. et al. 2008. Application of zero-valent iron nanoparticles for the removal of aqueous $Co^{2+}$ ions under various experimental conditions. *Chem. Eng. J.* 144(2):213–220.

Valko, M., Morris, H., and Cronin, M.T.D. 2005. Metals, toxicity and oxidative stress. *Curr. Med. Chem.* 12:1161–1208.

Verwey, E.J.W. and Heilmann, E.L. 1947. Physical properties and cation arrangement of oxides with spinel structures I. Cation arrangement in spinels. *J. Chem. Phys.*, 15:174–180.

Verwey, E.J.W. and Overbeek, J. 1948. *Theory of the Stability of Lyophobic Colloids*. Elsevier, Amsterdam, the Netherlands.

Vinten, A.J.A., Yaron, B., and Nye, P.E. 1983. Vertical transport of pesticides into soil when adsorbed on suspended particles. *J. Agric. Food Chem.* 31:662–664.

Vogel, T.M., Criddle, C.S., and McCarty, P.L. 1987. Transformation of halogenated aliphatic compounds. *Environ. Sci. Technol.* 21:722–736.

Wang, Y. and Herron, N. 1991. Nanometer-sized semiconductor clusters: Material synthesis, quantumsize effects, and photophysical properties. *J. Phys. Chem.* 95:525–532.

Wang, C.-B. and Zhang, W.X. 1997. Synthesizing nanoscale iron particles for rapid and complete dechlorination of TCE and PCBs. *Environ. Sci. Technol.* 31:2154–2156.

Waybrant, K.R., Blowes, D.W., and Ptacek, C.J. 1998. Selection of reactive mixtures for use in permeable reactive walls for treatment of mine drainage. *Environ. Sci. Technol.* 32:1972–1979.

Waychunas, G.A., Kim, C.S., and Banfield, J.F. 2005. Nanoparticulate iron oxide minerals in soils and sediments: Unique properties and contaminant scavenging mechanisms. *J. Nano Res.* 7(4–5):409–433.

Weber, E.J. 1996. Iron-mediated reductive transformations: Investigation of reaction mechanism. *Environ. Sci. Technol.* 30:716–719.

Wersin, P., Hochella Jr., M.F., Persson, P. et al. 1994. Interaction between aqueous uranium (VI) and sulphide minerals: Spectroscopy evidence for sorption and reduction. *Geochim. Cosmochim. Acta* 58:2829–2843.

Whatlington, K. 2005. Emerging Nanotechnologies for Site Remediation and Wastewater Treatment. U.S. Environmental Protection Agency. Washington DC.

White, A.F. and Paterson, M.L. 1996. Reduction of aqueous transition metal species on the surface of Fe(II)-containing oxides. *Geochim. Cosmochim. Acta* 60:3799–3814.

World Health Organisation. 2009. http://www.who.int/en/ (accessed August 2009).

Wu, L. and Ritchie, S.M.C. 2008. Enhanced dechlorination of trichloroethylene by membrane supported Pd-coated iron nanoparticles. *Environ. Prog.* 27(2):218–224.

Wu, L., Shamsuzzoha, M., and Ritchie, S.M.C. 2005. Preparation of cellulose acetate supported zerovalent iron nanoparticles for the dechlorination of trichloroethylene in water. *J. Nanopart. Res.* 7:469–476.

Xia, T., Kovochich, M., Brant, J. et al. 2006. Comparison of the abilities of ambient and manufactured nanoparticles to induce cellular toxicity according to an oxidative stress paradigm. *Nano Lett.* 6, 1794–1807.

Xiao, S., Shen, M., Guo, R. et al. 2009. Immobilization of zerovalent iron nanoparticles into electrospun polymer nanofibers: Synthesis, characterization, and potential environmental applications. *J. Phys. Chem. C* 113:18062–18068.

Xu, Y. and Zhang, W. 2000. Subcolloidal Fe/Ag particles for reductive dehalogenation of chlorinated benzenes. *Ind. Eng. Chem. Res.* 39:2238–2244.

Yang, G.C.C., Tu, H.C., and Hung, C.H. 2007. Stability of nanoiron slurries and their transport in the subsurface environment. *Sep. Purif. Technol.* 58(1):166–172.

Yavuz, C.T., Mayo, J.T., Mayo, Suchecki, C. et al. 2010. Pollution magnet: Nano-magnetite for assenic removal from drinking water. *Environ. Geochem. Health.* 10.1007/s10653-010-9293-y.

Yong, R.N. and Mulligan, C.N. 2003. *Natural Attenuation of Contaminants in Soils.* CRC Press/Taylor & Francis Group, Boca Raton, FL.

Yuan, J., Liu, X., Akbulut, O. et al. 2008. Superwetting nanowire membranes for selective absorption. *Nat. Nanotechnol.* 3:332–336.

Yue, Z. R. and Economy, J. 2005. Nanoparticle and nanoporous carbon adsorbents for removal of trace organic contaminants from water. *J. Nanopart. Res.* 7(4–5):477–487.

Zaera, F. 1989. A thermal desorption and X-ray photoelectron spectroscopy study of the surface chemistry of iron pentacarbonyl. *J. Vac. Sci. Technol.* A7:640–645.

Zhang, W.X. 2003. Nanoscale iron particles for environmental remediation: An overview. *J. Nanopart. Res.* 5:323–332.

Zhang, W.X., Wang, C.B., and Lien, H.L. 1998. Treatment of chlorinated organic contaminants with nanoscale bimetallic particles. *Catalysis Today* 40:387–395.

Zhang, M., Fang, S., Zakhidov, A.A. et al. 2005. Strong, transparent, multifunctional, carbon nanotube sheets. *Science* 309:1215–1219.

# 16

## *Inorganic Nanotubes and Fullerene-Like Structures—From Synthesis to Applications*

**Maya Bar-Sadan and Reshef Tenne**

**CONTENTS**



### 16.1 Introduction

Inorganic fullerene (IF)-like nanoparticles and inorganic nanotubes (INT) form a relatively new class of nanomaterials. They are generically produced from layered (2D) materials, which enable formation of stable, closed, hollow structures in the nanodomain. Taking advantage of the structural analogy between graphite and inorganic 2D compounds, researchers demonstrated that nanoparticles of $WS_2$ (Tenne et al. 1992), $MoS_2$ (Feldman et al. 1995, Margulis et al. 1993), BN (Chopra et al. 1995, Golberg et al. 1998), $NiCl_2$ (Rosenfeld Hacohen et al. 1998), and the like become unstable in the platelet (bulk) form and spontaneously assemble into hollow seamless structures, such as multiwall quasi-spherical spheres, nanooctahedra (Parilla et al. 1999), or nanotubes. The research on these nanostructures brought about major advances in chemistry and nanotechnology, and pointed to many possible applications.

Since the works of the early 1990s, numerous new IF and INT have been prepared, and the list of reported nanostructures of this kind is ever expanding. More recently, inorganic nanotubes from compounds with quasi-isotropic (3D) structure, like GaN (Goldberger et al. 2003), MgO (Zhan et al. 2004), and spinel ($ZnAl_2O_4$) (Fan et al. 2007) nanotubes, have also been described. Another recent major development represents the synthesis of arrays of polycrystalline $TiO_2$ and similar nanotubes through the electrochemical anodization of Ti foils in fluoride and other acidic solutions (Beranek et al. 2003, Gong et al. 2001). These developments broadened the concept of hollow nanostructures beyond that of carbon, deep into the realm of inorganic chemistry.

Nanotubes prepared from layered (2D) and quasi-isotropic (3D) compounds are distinguishable in a number of ways. Layered materials produce bent structures, where each of the layers preserves its fundamental structure. Nanotubes of 3D compounds are usually obtained using a solid or liquid template and they grow along the easy growth axis in a highly faceted form. Nanotubes of 2D compounds maintain the coordination of the bulk material and, therefore, they exhibit little surface-related phenomena, like reconstruction or chemisorption. The outer layer, which is exposed to the exterior, is usually quite inert, as it is in the bulk structure. These layers are held together mainly by van der Waals (vdW) forces alone. Contrarily, nanoparticles of 3D compounds, and in particular nanotubes, thereof suffer from surface oxidation and surface reconstruction, in some cases extensive enough to determine their physical behavior. The latter phenomena are known for the luminescence of various quantum dots, magnetic behavior of spintronic devices, and the transport and superconducting properties of nanowires and nanotubes.

The driving force for the formation of fullerenes and nanotubes stems from the abundant reactive atoms on the periphery of the quasi–2D planar nanostructure (see Figure 16.1).

Thus, the planar topology of the layered nanoparticles is unstable with respect to the hollow and seamless fullerenes. Using this reasoning, it has been proposed (Feldman et al. 1995, Margulis et al. 1993) that the formation of fullerenes is not unique to carbon and is a genuine property of 2D (layered) compounds. However, in contrast to graphite, each inorganic molecular sheet consists of multiple layers of different atoms chemically bonded together. By folding the molecular sheet and stitching the rim atoms together, seamless and stable nanotubular (1D) and spherical (0D) fullerene-like structures with all bonds satisfied were realized. In contrast to the unstable bulk material, the seamless structure of the nanoparticles provides kinetic stabilization. Thus, the closed nanoparticles do not expose the prismatic (*hk*0) edges, and therefore are less reactive. This kinetic stabilization is manifested, e.g., by the slow exfoliation of IF-$Cs_2O$ nanoparticles through water intercalation into the vdW gaps.

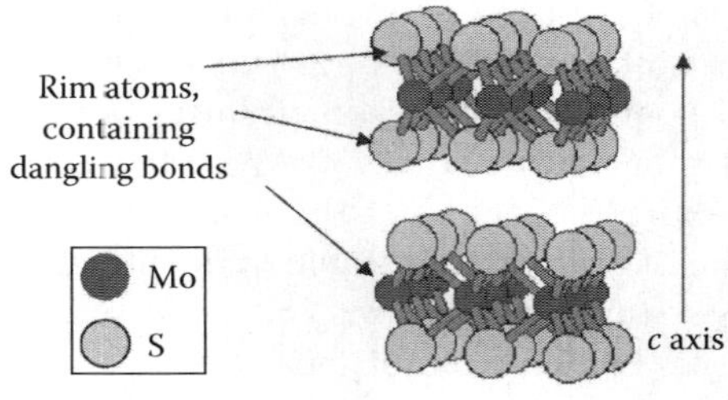


**FIGURE 16.1**
Ball and stick model of the layered (2D) crystal $MoS_2$. Mo is shown in dark gray, S in light gray. The covalently bonded layers are held together by vdW forces. In the edges of the structure, rim atoms contain reactive dangling bonds (shown by arrows).

In this chapter, we delineate a number of recent developments in the chemistry of IF and INT. New insights into the structure-properties relationship of such nanostructures have been gained through recent advances in calculations and new physical measurements. A concise description of this progress, and in particular the intimate relationship between theory and experiment, is emphasized. Alongside the progress in the synthesis and structural elucidation of such nanostructures, major progress has recently been achieved in scaling-up the production and applications of both IF and INT based on $WS_2$. The recent commercialization of the first series of products based on the superior tribological behavior of IF-$WS_2$ nanoparticles is indeed indicative of this trend. The large number of potential applications in the fields of lubrication, high performance nanocomposites, sensors, renewable energy, energy storage, and catalysis are briefly discussed.

## 16.2 Materials and Properties

Inorganic layered (2D) compounds are abundant, in particular among the transition metal chalcogenides (sulfides, selenides, and tellurides—transition metal sulfides TMS), halides (chlorides, bromides, and iodides—transition metal halides TMH), oxides, pnictides, and numerous ternary (quaternary) compounds. Therefore, there is a rich variety of materials, which can be potentially used for the fabrication of inorganic nanotubes and fullerene-like nanoparticles.

The first IF structures and nanotubes were reported in 1992 (Tenne et al. 1992). Since then, numerous synthetic routes have been applied to produce these structures. Large amounts of $WS_2$ and $MoS_2$ multiwall nanoparticles with fullerene-like structure (Feldman et al. 1995, 1996, Zak et al. 2000) and various inorganic nanotubes have been realized (Remskar et al. 1996, 1999, 2001, Rothschild et al. 2000, Spahr et al. 1998, Yin et al. 2005a,b, Zelenski and Dorhout 1998, Zhu et al. 2000). Today, 17 years later, it is possible to synthesize IF and to some extent also INT-$WS_2$ and $MoS_2$ in large quantities (see Figure 16.2).

This is a direct consequence of the relentless synthetic efforts during these years, meant to find new routes and in the understanding of the reaction mechanisms involved in the process.

The synthetic procedures will be divided mainly into two categories: synthetic routes in which a chemical reaction was applied, and synthetic routes in which the starting material is usually the bulk compound, which undergoes a structural change after an energy stroke and transforms into an assortment of closed hollow nanostructures.

In a complementary way, the synthesis of nanotubes from inorganic compounds has seen a burst of interest in recent years. Nanotubes of various inorganic compounds have been synthesized by variety of methods. The availability of some of these nanotubes in large amounts permitted a systematic study into their physical and chemical properties. Each of these techniques is very different from the others and produces nanotubes and fullerene-like material of somewhat different characteristics.

### 16.2.1 Chemical Routes to IF and INT

Metal chalcogenides are, by far, the most extensively researched group of layered materials in this connection. Among the 2D compounds, the transition metal chalcogenides nanotubes make the largest group and in addition to $WS_2$ it includes $MoS_2$ (Feldman et al.

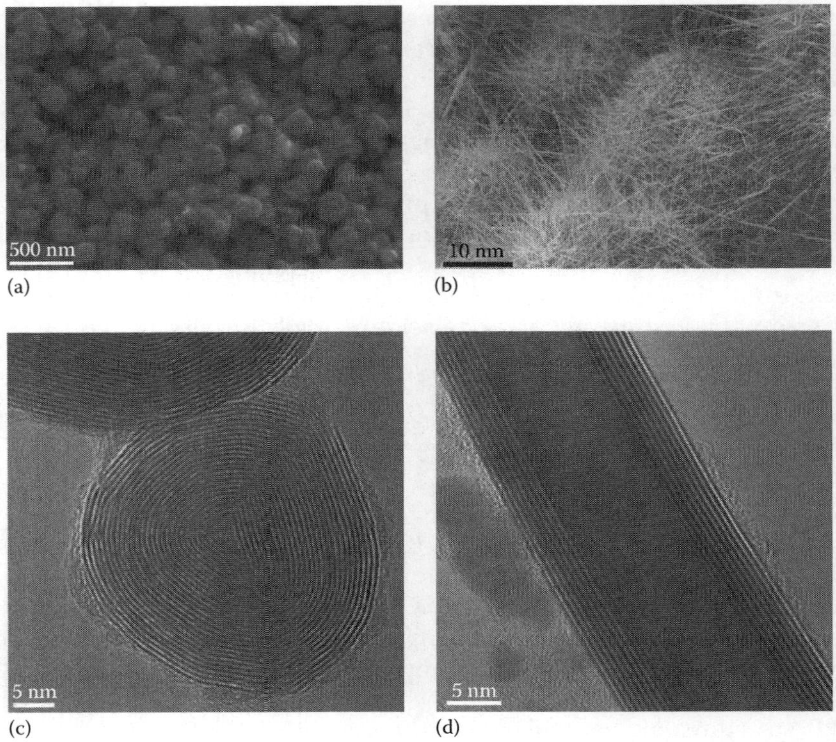


**FIGURE 16.2**
IF-like nanoparticles: SEM images of (a) IF-$MoS_2$ (Courtesy of R. Rosentsveig.), (b) INT-$WS_2$ (Courtesy of A. Zak.); TEM images of (c) IF-$MoS_2$ (Courtesy of R. Popovitz-Biro.), and of (d) INT-$WS_2$. (Courtesy of M. Bar-Sadan.)

1995), $NbS_2$ (Schuffenhauer et al. 2002), $TaS_2$ (Nath and Rao 2001b), $HfS_2$ (Nath et al. 2004), $ZrS_2$ (Nath and Rao 2002), $TiS_2$ (Margolin et al. 2005a, Nath and Rao 2002), some of the dislenides (Nath and Rao 2001a) and tellurides (Wu et al. 2007).

Here, the synthesis of IF and INT from $WS_2$ and $MoS_2$ will be explained in greater detail. The different synthetic routes result in different products, some of which are presented here, but new synthetic techniques are developed by the week. The first synthesis used was by sulfidizing ultrathin $WO_3$ ($MoO_3$) films in a reducing atmosphere at elevated temperatures (850°C) (Tenne et al. 1992). The reaction resulted in concentric polyhedral and cylindrical structures, 10–100 nm in size. To further increase the amount and improve the size control of the nanoparticles formation, trioxide powders were used as precursors (Feldman et al. 1995, Zak et al. 2002). Within the first few seconds of the reaction, the top surface of the oxide nanoparticles reacts with the $H_2S$ gas and a completely closed monomolecular $MS_2$ layer or two are formed. The inert surface-sulfide layer prohibits fusion of the nanoparticles into macroscopic entities, which would eventually lead to the formation of the bulk 2H-$MS_2$ phase. Subsequently, the oxide core is progressively converted into the respective sulfide (IF) through a slow diffusion-controlled reaction. Consequently, the size of the IF particle is determined by the size of the incipient oxide nanoparticle. Due to differences in the density of the oxide precursor and the sulfide product (5%), and the fact that the initial volume of the oxide nanoparticle is preserved during the chemical transformation, the resulting IF nanoparticle possesses a hollow core, occupying up to 20%–30% of the quasi-spherical particles' volume. Scaling-up the process and routine production of $WS_2$ nanotubes was established after the fluidized bed reactor (FBR) was developed (Zak et al. 2009). This technology became the basis for further scaling-up of the synthesis and commercialization of IF-$WS_2$, and more recently, the nanotubes thereof.

Substantial efforts have been devoted in recent years to alternative gas-phase reactions, which are kinetically controlled and can be ascribed to a nucleation and growth model. These kinds of reactions exhibit fast kinetics and better size control and allow for doping and alloying of the nanoparticles. Two kinds of metallic gas sources have been used: metal halides, which can be volatilized at mild temperatures (Lee et al. 2003, Margolin et al. 2005a, Schuffenhauer et al. 2002, 2005), and metal carbonyl vapors (Etzkorn et al. 2005, Li et al. 2004, Yella et al. 2008, Zink et al. 2007). This kind of reaction lends itself to a fast nucleation of the incipient $MX_2$ nuclei, which grow outward in a layer-by-layer fashion. Fullerene-like $NbS_2$ nanoparticles (Schuffenhauer et al. 2002) were the first to be synthesized according to this strategy. Subsequently, $TiS_2$ nanotubes (Chen et al. 2003c), IF-$TiS_2$ (Margolin et al. 2005b), and IF-$WS_2$ (Li et al. 2004) nanoparticles, were synthesized by the reaction of the respective metal halide with a sulfurization reagent. In the case of IF-$TiS_2$ (Margolin et al. 2005b), it is interesting to notice that the new reaction pathway affects the morphology of the product in the following way: The synthesized nanoparticles were found to have many layers and are quite perfectly spherical, displaying relatively smooth curvature. The mechanism of the IF-$TiS_2$ nanoparticles growth may be envisaged as a homogeneous nucleation of the fullerene-like structures from embryonic clusters formed in the vapor phase. Therefore, many of the IF nanoparticles contained cores made of a number of tiny spherical IF centers, having no appreciable voids (see Figure 16.3).

Interestingly enough, irrespective of the reaction time, the size of the nanoparticles (ca. 100 nm) seem to remain unchanged. This observation suggests that beyond this typical size the IF nanoparticles become unstable and consequently the number of the nanoparticles increases with the reaction time.

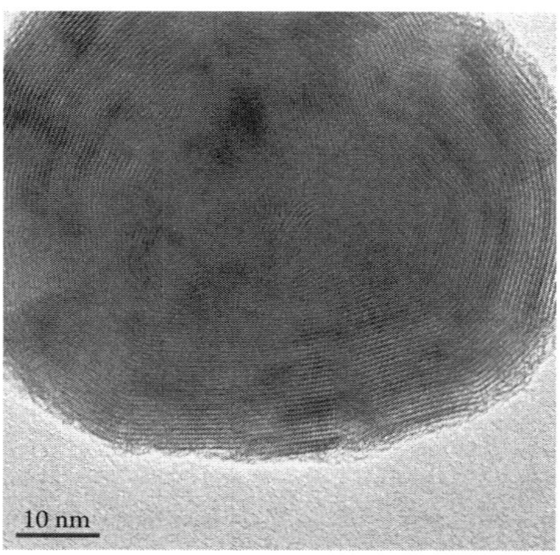


**FIGURE 16.3**
TEM image of the multinuclei IF-$TiS_2$ nanoparticle. (Reprinted from Margolin, A. et al., *Chem. Phys. Lett.*, 411, 162, 2005b. Copyright [2005], Reprinted with permission from Elsevier.)

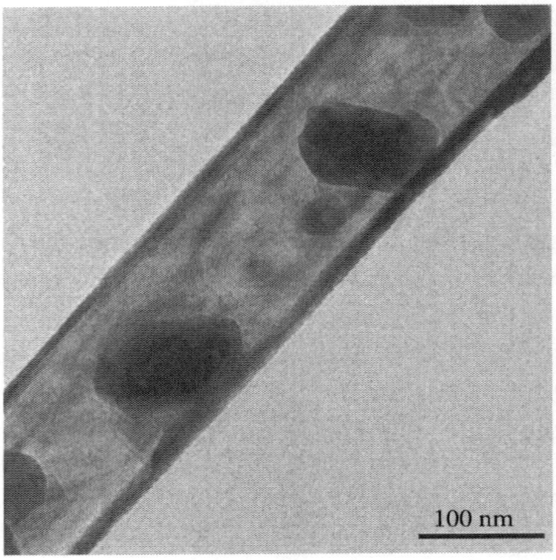


**FIGURE 16.4**
TEM image of a typical $MoS_2$ Mama tube obtained through the chemical vapor transport reaction. (Courtesy of M. Remskar.)

High-temperature sulfidation of $Mo_6S_2I_8$ nanowires resulted in the formation of the first $MoS_2$ nanotubes with fullerene-like $MoS_2$ nanoparticles occluded in their hollow core—the so-called Mama tubes (Remskar et al. 2007) (see Figure 16.4).

In contrast to freely grown IF nanoparticles, which exhibit a hexagonal 2H (*P63/mmc*) structure (Feldman et al. 1995), IF-$MoS_2$ nanoparticles growing in the confined space of the $MoS_2$ nanotube belong to the rhombohedral 3R polytype (*R3m*). This observation is reminiscent of the previous observation that inorganic salts, like KI, grow as a totally new crystallographic polytype in the confined space of carbon nanotubes core (Meyer et al. 2000). It is likely that the Mama tubes will exhibit very different physical behavior as compared with the truly hollow $MoS_2$ nanotubes. Fullerene-like nanoparticles hinging on the outer walls of $WS_2$ nanotubes (nanobuds) were synthesized by reacting $W_5O_{14}$ nanowires with $H_2S$ gas (Remskar et al. 2008). It appears that the fullerene-like $WS_2$ nanoparticles were produced on defective sites of the oxide nanowires, which served as the seed promoting their growth.

Hollow $WS_2$ nanocubes were obtained by spray pyrolysis of ethanolic $(NH_4)_2WS_4$ solutions at 900°C (Bastide et al. 2006). The synthesis of the hollow box-like morphology follows from a three-step process. First, the solvent evaporates at $<100°C$ from the droplet, leading to the rapid formation of $(NH_4)_2WS_4$ crystals with a parallelepiped shape. When the temperature rises to a few hundreds degrees, the ammonium thiotungstate decomposes to amorphous $WS_3$ (a-$WS_3$) nanoparticles, which maintain the nanobox form. In the third step (>700°C), the a-$WS_3$ loses 1S and crystallizes, forming the stable $WS_2$ hollow nanoboxes. These $WS_2$ nanoboxes are not perfectly crystalline, but they are reminiscent of the $MoS_2$ nanooctahedra (Parilla et al. 1999), which are much smaller in size and are perfectly crystalline.

The mechanistic aspects of the growth of IF-$MS_2$ (M = Mo, W) nanoparticles from amorphous $MS_3$ have been recently studied by in situ transmission electron microscopy (TEM) operating at high temperatures (Zink et al. 2008). Upon annealing, round amorphous particles in the pristine sample gradually transformed into hollow, onion-like nanoparticles. This study indicates that the driving force for the formation of seamless IF nanostructures is the chemical energy stored in the rim atoms of the nanoplatelets.

Another IF produced by an oxide precursor is $ReS_2$ (Coleman et al. 2002) (see Figure 16.5).

In contrast to other 2D $MX_2$ compounds (X = S, Se, Cl, I), which have a hexagonal symmetry and exhibit metal to nonmetal bonds only, the layered $ReS_2$ consists of repeating $Re_4$ parallelogram motifs with covalent bonds linking neighboring metal atoms, in addition to the ubiquitous metal to nonmetal bonds. $ReS_2$ is a semiconductor, which makes its IF nanostructures interesting for a host of potential applications in photocatalysis, nanotechnology, and more. In the case of IF-$WS_2$, the specific gravity of the starting material ($WO_3$) is lower than that of the sulfide product, thus leaving a hollow core of about 20% in the center of the nanoparticles. However, the specific gravity of the starting $ReO_2$ powder is appreciably higher than that of the end product (IF-$ReS_2$). Since the out-diffusion of rhenium during sulfurization is not likely, the core of the IF nanoparticles is not expected to be hollow at the end of the reaction, and residual $ReO_2$ is found in the core even at the end of the reaction. In fact, the reaction may cease as a result of a buildup of internal pressure, which results from the differences in the specific gravity between the oxide precursor and the sulfide compound. In a similar manner to the observations regarding IF-$WS_2$, when the precursors are $Re_2(CO)_{10}$ and elemental S, different morphology is observed (Yella et al. 2008), and the resulting IF are nonhollow.

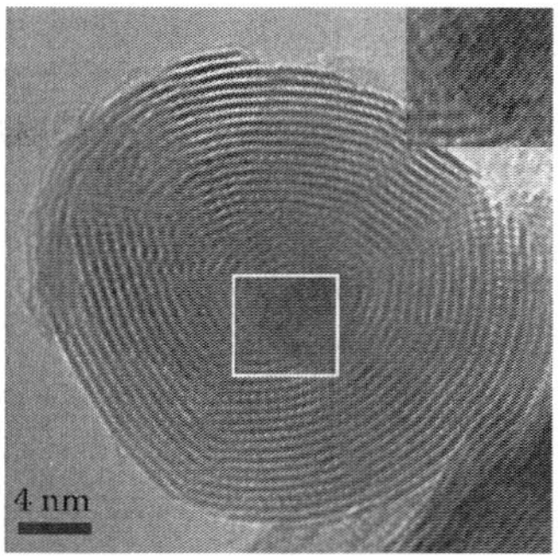


**FIGURE 16.5**
HRTEM micrograph showing a small $ReS_2$ particle encapsulating a $ReO_2$ nanoparticle at its core. The 0.29-nm lattice fringes (inset) correspond to the $ReO_2$ (111) planes. (Reprinted from Coleman, K.S. et al., *J. Am. Chem. Soc.*, 124, 11580, 2002. With permission. Copyright [2002], American Chemical Society.)

By using a sonochemical bath, hollow (IF) nanoparticles of the layered compound $Tl_2O$ with anti-$CdCl_2$ structure were synthesized in substantial amounts (Avivi et al. 2000). They were later isolated by selective heating of the reaction product to 300°C.

### 16.2.2 Physical Routes to IF

Physical methods are used for producing both carbon and IF structures, and they generally do not involve a net chemical reaction. In fact, the historic discovery of $C_{60}$ was accomplished by laser ablation of graphite (Kroto et al. 1985). The arc discharge technique was the method in which the first large quantities of $C_{60}$ as well as carbon nanotubes were produced (Iijima 1991). In most of these methods nanoparticles are produced from the vapor phase or plasma. These methods include laser ablation (Albu-Yaron et al. 2005, Bar-Sadan et al. 2006b, Hong et al. 2003, Kratschmer et al. 1990, Parilla et al. 1999, 2004, Rosenfeld Hacohen et al. 2002, 2003, Sen et al. 2001), arc discharge (Saito et al. 1993), resistive heating of graphite targets (Guo et al. 1991), e-beam irradiation (Banhart et al. 1994, Jose Yacaman et al. 1996, Popovitz-Biro et al. 2001, 2003, Stephan et al. 1998) as well as short electrical pulses from the tip of an STM (Homyonfer et al. 1996). In contrast to the high temperature chemical synthesis of IF and INT, the high power pulsed physical methods provide far from equilibrium conditions. Thus, while the chemical methods are tuned for the synthesis of nanoparticles with the lowest free energy, the physical methods are able to produce hollow nanoparticles of, e.g., lower radius of curvature or smaller number of layers (*vide infra*).

The main advantage of the energetic experimental systems, which do not use a specific chemical reaction, is their versatility. The ability to use different compounds as a precursor to the reaction allows for the synthesis of many different materials in the same experimental setup.

These include, e.g., IF and INT from BN (Stephan et al. 1998), TMS [$MoS_2$ (Bar-Sadan et al. 2006a, Enyashin et al. 2007a, Parilla et al. 1999, 2004), $SnS_2$ (Hong et al. 2003), $Hf_2S$ (Nath et al. 2004)], TMH [$NiCl_2$ (Rosenfeld Hacohen et al. 1998, 2002), $NiBr_2$ (Bar-Sadan et al. 2006b), $CdI_2$ (Sallacan et al. 2003), $CdCl_2$ (Popovitz-Biro et al. 2001)], and alkali metal oxides [$Cs_2O$ (Albu-Yaron et al. 2005, 2006)]. These compounds undergo structural changes and form closed-cage structures of different morphologies. It should be emphasized that fundamentally the lattice structure of the compounds is preserved in the IF nanoparticles. In other cases, like the "fullerenic oxides" (Hervieu et al. 2004), or the MET-CARS ($C_{20}M_{12}$, where M = Ti, Zr) (Guo et al. 1992a,b), which are actually huge cages arranged in a structure topologically similar to that of $C_{84}$, no parent compound with a layered structure is known in the bulk. Therefore, clusters of the same material, even if hollow or cage-like, cannot be considered as "inorganic fullerenes," unless their structure resemble one of the bulk polytypes (e.g., 2H- or 1T- in the $MoS_2$ system) (Bar-Sadan et al. 2006a, Bertram et al. 2006). An important advantage of the energetic methods, like laser ablation and arc discharge, is that they form a dense "soup" of the ablated atoms, which quench very rapidly. The fast cooling of the dense plasma leaves no time for a diffusion process of atoms and leads thereby to the formation of local minimum energy stable clusters, like $C_{60}$ and lately nanooctahedra of $MoS_2$ (Bar-Sadan et al. 2006a, Enyashin et al. 2007a,b), etc.

In the previous section, the formation of large multiwalled IFs by the sulfurization reaction of oxide nanoparticles was described. Since the reaction starts from the surface of the particle and progresses inward toward the core of the oxide particle, the size of the final product is predetermined by the reduction process. The mechanism for the production of the small closed polyhedral structures by energetic techniques is distinctly different from that leading to the synthesis of the large IF. The smaller structures are believed to represent a local minimum in the free energy of formation within a restricted configuration space. Specifically, they form under conditions where the number of atoms in a confined volume (a few nm wide) remains small and the energy of the constituents is large enough to reach a local energy minimum. Furthermore, once this minimum has been reached, the excess energy should be rapidly removed from the structure in order to prevent the coarsening of the nanoclusters.

The above considerations can be exemplified with the laser ablation setup. In this technique, a laser beam is aimed at a pressed pellet. Visible/IR laser light incident on a solid target is normally absorbed by the electronic modes, followed by thermalization process and conversion of the excess energy to vibrational excitations. Generally, laser ablation processes all occur in a very short timescale of a few nanoseconds and below. These processes include light-absorption, heat diffusion, melting, and evaporation. The ablation process depends on the target material, laser wavelength, and the power density. An important feature concerning the ablation is the recoil of the nanoparticles from the ablation spot through the hot vapors and the hot plasma. The recoil phenomenon is explained by the high pressure of the gases formed inside the cavities in the pellet during the ablation process. The recoiled particles may be carried by the gas turbulence back into the hot zone of the furnace. This annealing process provides additional energy to the nanoclusters allowing them to form stable clusters and nanoparticles. Subsequently, the nanoclusters are taken by the gas stream outside the reactor, where they rapidly cool down. The soot is collected from the reactor walls or from a substrate placed downstream of the reactor.

Unlike the chemical synthetic route, the mechanism of formation of the small and symmetric IFs shaped as nanooctahedra and the larger quasi-spherical (fullerene-like) structures in these high energy setups is not understood yet. Nonetheless, the reaction mechanism for the formation of the fullerenic cages probably avoids direct folding of the layered sheets. The inorganic closed-cage structures are quite complex, being built of a shell consisting of triatomic layers and containing several orders of magnitude more atoms than their carbon predecessors. Therefore, a full description of the formation mechanism of the IFs (like $MoS_2$ nanooctahedra) is still lacking.

The use of the energetic physical systems may also yield IF from materials that could not be produced in a different route, due to their high chemical reactivity. Alternatively, such methods can produce morphologies different from those obtained via the chemical routes. An extreme case, in which the laser ablation setup was used to prepare IF structures of the very reactive compound $Cs_2O$ (Albu-Yaron et al. 2005), is noteworthy. Here, the chemical methods failed to produce such nanostructures, so far. Films with a Cs/O ratio of approximately 2:1 are known to reduce the electron work function of semiconductor surfaces, increasing the sensitivity of optoelectronic detectors, used, e.g., in medical imaging. Unfortunately, these films are highly reactive and are damaged by even a short exposure to low vacuum conditions. Therefore, coating semiconductor surfaces with 3R-$Cs_2O$ (anticadmium-chloride structure) films involves strict vacuum conditions. Furthermore, the high reactivity of the product prevented the verification of its crystalline structure by TEM analysis, until recently. Nested closed nanoparticles of $Cs_2O$ could be observed in laser-ablated powder sealed in a closed quartz ampoule. Using a specially constructed environmental chamber attached to the TEM, the IF nanoparticles could be observed in tiny amounts, initially. Subsequently, they could be produced in larger quantities by ablation with a focused solar beam (Albu-Yaron et al. 2006). Remarkably though, these fullerene-like nanoparticles suffered only slow degradation in the ambient atmosphere, demonstrating thereby the kinetic stabilization brought about by the closed-cage structure.

The synthesis of small amounts of IF and INT from TMH compounds was also achieved via the physical synthetic routes and less so by the chemical routes. TMH differ from the TMS by having a higher degree of bond ionicity. Although the crystalline structure of TMH such as $NiCl_2$ and of TMS resemble each other to a certain extent, they substantially differ in their stacking fault, shear, and compression force constants (Rosenfeld Hacohen et al. 2003). In particular, the TMH are appreciably more ionic and, therefore, they are very hygroscopic up to being deliquescent (dissolve in their own water of absorption). These differences are attributed to their different chemistry and are clearly reflected in their tendency to form folded and closed nanostructures, which is a lot easier for $MoS_2$ ($WS_2$). Therefore, the IF structures of TMH produced by laser ablation are morphologically different from the large onion-like structures described before, which are much more relevant to the TMS. The closed structures of the TMH, which are produced by either laser ablation or e-beam irradiation, are smaller in size (<30 nm), as compared to the onion-like IF of the TMS, which may reach 200 nm in diameter (Feldman et al. 1996, Zak et al. 2000). See Figure 16.6, showing a variety of IF-TMH: $NiBr_2$ (Bar-Sadan et al. 2006b), $NiCl_2$ (Rosenfeld Hacohen et al. 2002), $CdI_2$ (Popovitz-Biro et al. 2003, Sallacan et al. 2003), $CdCl_2$ (Popovitz-Biro et al. 2001), and $MgCl_2$. The propensity of evenly spherical nanostructures to form faceted polyhedra while thickening is observed in TMH as well (Bar-Sadan et al. 2006b, Rosenfeld Hacohen et al. 2003).

Generally, the IF nanoparticles from TMH, which are one to two layers thick, are evenly folded (quasi-spherical). However, in general, the IF nanoparticles with more layers are

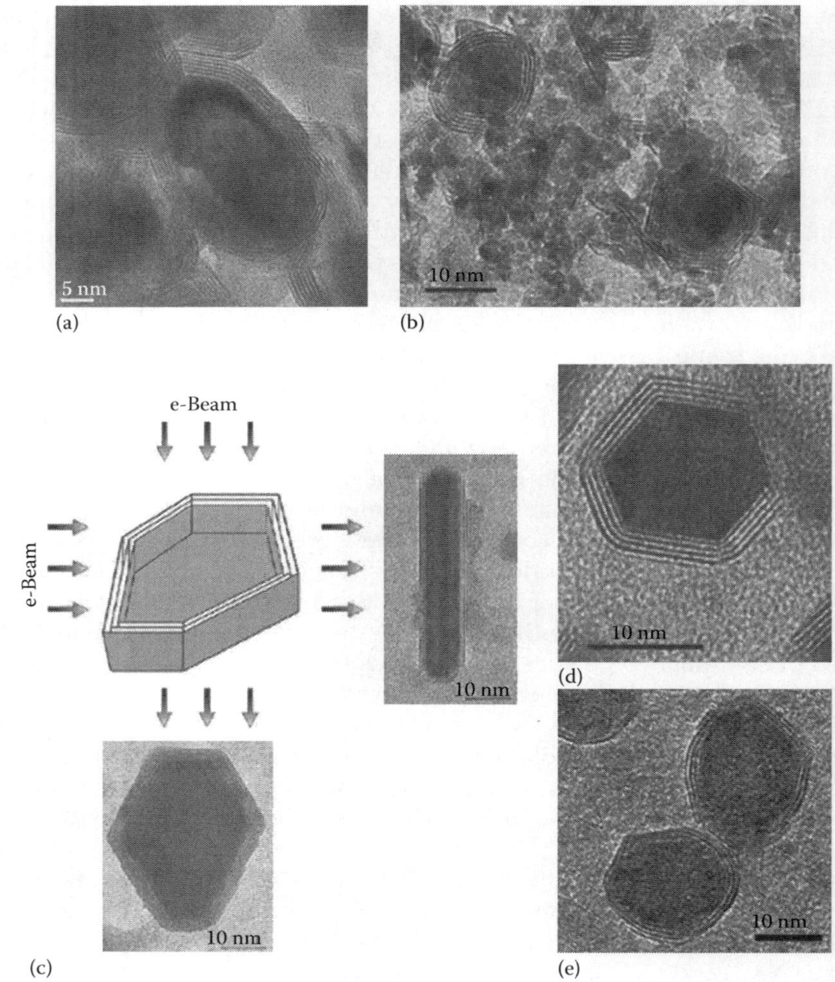


**FIGURE 16.6**
TMH IF nanoparticles: (a) $NiBr_2$. (Reprinted from Bar-Sadan, M. et al., *Mater. Res. Bull.*, 41, 2137, 2006b. With permission. Copyright [2006], American Chemical Society.) and (b) $NiCl_2$. (Reprinted from Rosenfeld Hacohen, Y. et al., *Adv. Mater.*, 14, 1075, 2002. With permission. Copyright Wiley-VCH Verlag GmbH & Co. KGaA.) (c) $CdI_2$: A schematic drawing showing the polyhedral nanoparticle viewed by TEM along the two main axes—one leading to a hexagonal shaped structure, while the other produces a rectangular silhouette. Included also are the respective IF-$CdI_2$ TEM images. (Reprinted from Popovitz-Biro, R. et al., *J. Mater. Chem.*, 13, 1631, 2003. With permission. Copyright the Royal Society of Chemistry.) (d) IF-$CdCl_2$ in a hexagonal projection. (Reprinted from Popovitz-Biro, R. et al., *Isr. J. Chem.*, 41, 7, 2001. With permission. Copyright Israel Science Journals.) (e) IF-$MgCl_2$ particles. (Courtesy of M. Bar Sadan and I. Pinkas.)

much more faceted. Figure 16.6e shows IF-$MgCl_2$, which were produced by a two-stage reaction—first laser ablation with a pulsed femtosecond laser (800 nm) and subsequent in situ irradiation in the electron microscope. Often there is outgassing of the light element during the reaction, leaving spherical droplets, probably consisting of the heavier metal. This inner core serves as a templating agent, providing a surface for recondensation and formation of closed-cage particles. In these cases it is difficult to draw a direct connection between the experimental data and the theory regarding the formation of facets as a strain-relief mechanism (Srolovitz et al. 1995).

Faceting may also serve as a stress-relief mechanism in some TMH IF nanoparticles: faceted $CdCl_2$ fullerene-like structures were produced in a high yield by e-beam irradiation of $CdCl_2 \cdot H_2O$ powder (Popovitz-Biro et al. 2001) and more recently facetted fullerene-like structures were seen in IF nanoparticles of $SnS_2$ (Hong et al. 2003), $CdI_2$ (Hong et al. 2003, Popovitz-Biro et al. 2003), and $NiBr_2$ (Bar-Sadan et al. 2006b) as well. TEM observations of the closed-cage $CdCl_2$ nanostructures showed that two main kinds of topologies were abundant—hexagonal and rectangular. A postulated $CdCl_2$ polyhedron with a faceted topology having essentially hexagonal cross section along one direction and rectangular cross section perpendicular to the previous one seem to comply well with the TEM observations (see Figure 16.6d). In this case, the rapid heating of the nanoparticles by e-beam irradiation differs from the slow heating of a nanopowder in a furnace. In the latter technique, the entire reactor is heated and consequently the gaseous phase is in equilibrium with the solid phase. In the former technique, the e-beam irradiation leads to a high local heating of the solid, which promotes the formation of dislocations and evaporation of the volatile constituents, like the halide atoms and the water molecules. Furthermore, the high energy of the electron beam flux can induce very fast diffusion of the atoms and rapid structural changes, leading, thereby, to a fast healing of the defects in the structure. A previous theoretical study by Bates and Scuseria (1998) tends to support this mechanism for e-beam induced damage–healing in (nested) carbon fullerenes. The vapor pressure of the volatile carbon residues in the e-beam column is sufficient to replenish the atoms, which are knocked out from the carbon structure. This is not the case, of course, for knocked-out sulfur or halide atoms of irradiated IF or INT. Obviously, the bending energy of 1D objects, such as nanotubes, is appreciably smaller than that of the fullerene-like nanoparticles, where folding along two axes is required in order to close the structure. Consequently, a polygonal cross section, which is so common in fullerene-like structures, is rarely observed in the nanotubes.

In analogy to $C_{60}$, the inorganic system has a recurring unit, which is the smallest symmetric closed-cage structure and is thermodynamically (meta-) stable. These structures are produced only by high energy methods, and are in the shape of an octahedron $<8$ nm in size. These structures reveal the basic symmetry prevalent for the inorganic system, being essentially the tendency to produce a square (rhombi)-like defect instead of the pentagonal or heptagonal defects, typical for the carbon system. Early indications for this idea were suggested already in 1993 (Margulis et al. 1993). The striking example of this divergence from the carbon chemistry is demonstrated by the boron nitride system. Boron nitride is isoelectronic and structurally analogous to carbon (Golberg et al. 1998). The bond length, the long-order parameters, and the lattice constants are very similar for BN and C. Furthermore, in analogy to the carbon system, boron nitride comes in two forms, the hexagonal (graphite-like) and the cubic (diamond-like) forms, with the first one being the stable form in ambient conditions. Thus, it has been expected that the similarities should transcend also to the nanoscale structures, i.e., the appearance of such structures as fullerenes and nanotubes.

Nevertheless, unlike the almost perfectly spherical carbon onions, medium size (app. 20 nm in diameter) IF-BN structures produced by laser ablation were found to be strongly facetted (Boulanger et al. 1995) sizes.

When produced in smaller sizes, by e-irradiation, the basic structure of the BN particles exhibit rectangular shapes, attributed to an octahedral 3D structure (see Figure 16.7, Golberg et al. 1998, Stephan et al. 1998). In accordance with Euler's theorem six squares (instead of 12 pentagons in carbon fullerenes) are needed to close the structure alongside the eight equilateral triangular faces.

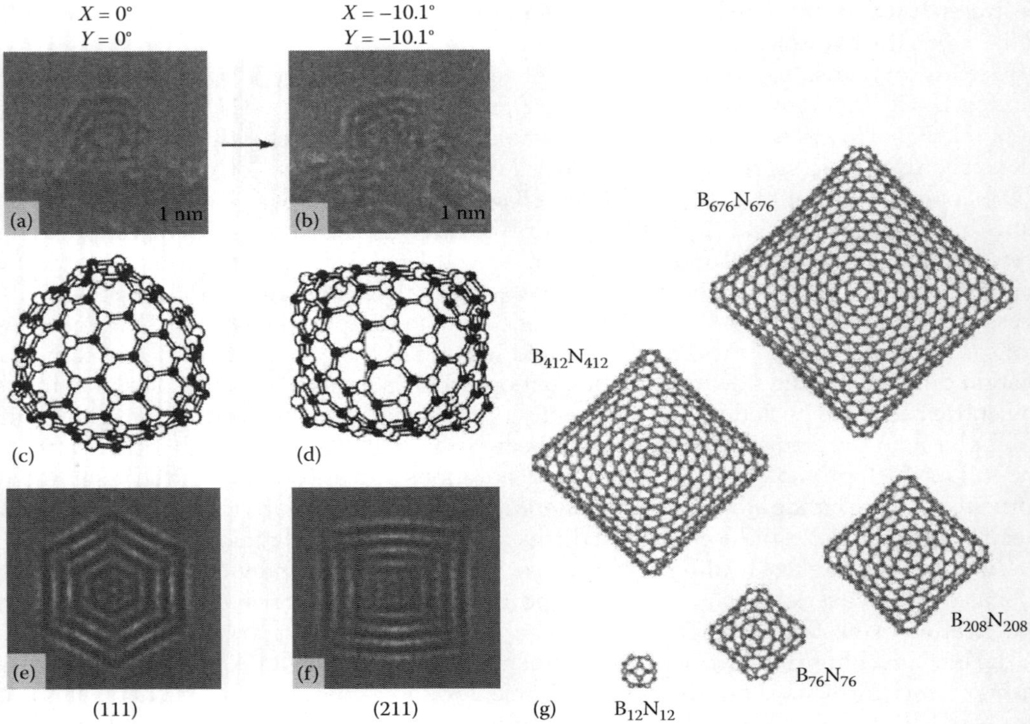


**FIGURE 16.7**
BN closed-cage structures: (a, b) A nested three-shelled irradiation-induced BN fullerene viewed at various angles; (c, d) the 3D structural octahedral models of the $B_{76}N_{76}$ molecule in [111] and [211] orientations; and (e, f) the computer simulated images of a five-layer BN onion ($B_{12}N_{12}$) in the corresponding projections, which are shown for comparison. (g) Series of octahedral cages building a 5-layer onion: $B_{12}N_{12}$, $B_{76}N_{76}$, $B_{208}N_{208}$, $B_{412}N_{412}$, $B_{676}N_{676}$, after structural relaxation by atomistic calculations. (Reprinted from Stephan, O. et al., *Appl. Phys. Mater. Sci. Process.*, 67, 107, 1998. With permission. Copyright Springer; Golberg, D. et al., *Appl. Phys. Lett.*, 73, 2441, 1998. With permission. Copyright the American Institute of Physics.)

Furthermore, small (<15 nm) structures of $MoS_2$ are also obtained by highly energetic processes like arc-discharge in water (Seifert et al. 1997). This method yields $MoS_2$ polyhedra of different shapes and 2–3 layers thick with sizes between 5 and 15 nm. The 2D TEM projections of these nanoparticles yielded mainly tetragons, triangles, and pentagons. Trying to avoid excessively close proximity of atoms at the vertices, the formation of four- or five-member Mo rings was hypothesized. Another technique used to produce polyhedral nanoparticles was the microwave plasma (Vollath and Szabo, 2000). The particles consisted typically of 3–4 layers with diameter of 10–15 nm. In one case, laser ablation of $MoS_2$ at 1050°C yielded closed-cage multiwall hollow nanoparticles along with metal-filled nanoparticles (Sen et al. 2001). The $WS_2$ nanoparticles were both quasi-spherical and faceted. The particles had five concentric layers and a diameter of about 10 nm. In another study, symmetric nanooctahedra of $MoS_2$ and $MoSe_2$ were produced by laser ablation for the first time in 1999 (Parilla et al. 1999, 2004). These new, fascinating structures were extensively studied, and their modeling and HRTEM characterization is discussed in the following section.

### 16.2.3 Intercalation and Doping of the Nanoparticles

Reversible alkali metal intercalation is one of the key technologies for energy storage in the form of rechargeable batteries. The large surface area of the IF nanopowder could be advantageous for electrode material, provided it would contain sufficient reactive sites for the intercalation/deintercalation process. Moreover, the intercalation may change the properties of the IF, stimulating the search for new possible applications.

It is likely that in the case of $VS_2$, Na intercalation facilitates the synthesis of IF nanoparticles, since it has been established previously that Na intercalation promotes 2H-$VS_2$ formation, which is otherwise not a stable phase at ambient conditions (Friend and Yoffe 1987). The intercalated moieties were found to be stable in air and even in water. The intercalated IF structures could be dispersed in alcoholic suspensions while the pristine IF nanoparticles did not form stable suspensions even after prolonged sonication (Homyonfer et al. 1997). These results indicate that the intercalation of alkali metal atoms in the vdW gap of the IF particles led to a partial charge transfer to the host lattice, which increased the polarizability of the nanostructures, enabling them to disperse in polar solvents. Suspensions prepared from IF powder and $WS_2$ nanotubes, which contained large amount of the sodium atoms as an intercalant (>5%) were found to be virtually indefinitely stable.

The optical absorption spectra of intercalated 2H-$MoS_2$ (bulk) did not show appreciable changes for alkali metal concentrations 30%. Since the concentration of the intercalating metal atoms in the IF did not exceed 10%, no changes in the optical transmission spectra of the IF phase were anticipated nor were they found to occur. Thin films of intercalated IF nanoparticles showed respectable and time-invariant photoeffects (Homyonfer et al. 1997). It is well known that the prevalence of dangling bonds on the prismatic faces of 2H-$WS_2$ platelets leads to rapid recombination of photoexcited carriers. Consequently, in contrast to the single crystal variants, the performance of photovoltaic devices, based on thin film layered compounds was found to be disappointing. The absence of dangling bonds in IF material was suggested as a means for alleviating this problem. However, low quantum efficiencies (number of collected charges/number of incident photons) and poor I-V performance was recorded for the IF-$WS_2$ films. The efficiencies were found to be even lower for films containing IF nanoparticles with substantial amounts of dislocations.

Intercalation of alkali metals into presynthesized IF nanoparticles was carried out by exposure of the powder to alkali metal (potassium and sodium) vapor at moderate temperatures and for long periods of time, up to 1 month, (Zak et al. 2000). The intercalation did not yield a pure phase. The large tensile strain imposed on the closed folded shells of the fullerene-like nanoparticles is believed to be responsible for the incomplete intercalation of these materials. The integrity of the intercalated IF nanostructures was confirmed by TEM imaging. Nonetheless partial distortions in the outermost layers of the nanoparticles were observed. The exposure of the intercalated IF nanoparticles to ambient conditions resulted in a significant expansion along the *c*-axis, but without any observable changes in the *a*- and *b*-lattice constants. Intercalation of one to two water molecules per intercalated atom was suggested as a possible explanation for this large expansion of the *c*-axis.

Following the intercalation, the transport and magnetic properties of the intercalated samples were significantly altered. Heavily (8%–18%) K intercalated 2H-$MoS_2$ exhibited a semiconductor to metal transition. Furthermore, a transition from diamagnetic to paramagnetic behavior as well as a decrease in room temperature resistivity and activation energy values were observed for all the intercalated phases. The structure and the electric/magnetic properties of the pristine host were recovered after a few weeks of exposure to the ambient suggesting slow deintercalation of the hydrated ions from the gallaries between the $MoS_2$ layers.

More recently, the intercalation of the IF nanoparticles was revisited (Kopnov et al. 2008); IF-$WS_2$ nanoparticles and 2H-$WS_2$ (bulk) were subjected to the intercalation process with sodium, potassium, and rubidium. In contrast to the above-mentioned study, special care was exerted to keep the intercalated phases in a moisture free environment during the study. Many of the nanoparticles were intercalated only at their outermost layers with the core remaining unchanged, probably because of the large tensile strain induced on the closed layers. A correlation was found between the vdW gap expansion of the intercalated particles and the radius of the alkali atom—the greatest interlayer expansion occurred for the rubidium intercalation, the least for the sodium intercalated IF powder. The *a*-axis of the intercalated phase was found to slightly expand as well. The fraction of the intercalated phases in the reacted (IF and 2H) powders depended on the specific alkali metal, with highest percentage of the intercalated phase achieved for rubidium intercalation in both kinds of powders.

By addition of $NbCl_4$ or $ReCl_4$ sources to the $MCl_n$ precursors, *p*- and *n*-type doped IF-Mo(W)$S_2$ and some doped nanotubes were obtained (Deepak et al. 2006, 2007). IF-$Mo_{1-x}Nb_xS_2$ nanoparticles have been synthesized by a vapor-phase reaction involving the respective metal halides with $H_2S$. The IF-$Mo_{1-x}Nb_xS_2$ nanoparticles, containing up to 25% Nb, were characterized by a variety of experimental techniques. Analysis of the x-ray powder diffraction, x-ray photoelectron spectroscopy, and different electron microscopy techniques showed that the majority of the Nb atoms are organized as nanosheets of $NbS_2$ within the $MoS_2$ host lattice. Most of the remaining Nb atoms (3%) are interspersed individually and randomly in the $MoS_2$ host lattice. Very few Nb atoms, if any, are intercalated between the $MoS_2$ layers. A subnanometer film of niobium oxide seems to engulf the majority of the nanoparticles. An x-ray photoelectron spectroscopy in the chemically resolved electrical measurement (CREM) mode (Cohen 2004, Doron-Mor et al. 2000) and scanning probe microscopy measurements of individual nanoparticles show that the mixed IF nanoparticles are metallic, independent of the substitution pattern of the Nb atoms in the lattice of $MoS_2$, whereas unsubstituted IF-$MoS_2$ nanoparticles are semiconducting. This is in accordance with the calculations of Seifert and coworkers (Ivanovskaya et al. 2006a,b) who showed that the IF-$Mo_{1-x}Nb_xS_2$ nanoparticles exhibit metallic character, irrespective of the substitution patterns of the Nb atoms. Furthermore, the IF-$Mo_{1-x}Nb_xS_2$ nanoparticles were found to exhibit interesting single electron tunneling effects at low temperatures.

### 16.2.4 Modifications of IF and INT

Surface functionalization of IF and INT was not explored until recently. The main goal of studies in this area is to make these nanoparticles compatible with different fluids or solid phases, i.e., to prevent their segregation into agglomerates or separate domains. Recent efforts by a few groups pointed out the rich chemistry and potential applications of this strategy. Noncovalent functionalization of BN nanotubes in aqueous solution was accomplished through πψ stacking of an anionic perylene-3,4,9,10-tetracarboxylic acid tetrapotassium salt (PTAS). PTAS functionalization resulted in a well-controlled modification of the BN nanotube surfaces with the carboxylate functional groups pointing outward and making the nanotubes dispersible in aqueous solutions (Wang et al. 2008).

Using nitrilotriacetic acid (NTA), Tremel and coworkers (Tahir et al. 2006, 2007) demonstrated complete solubilization of fullerene-like $MoS_2$ and $WS_2$ nanotubes in aqueous solutions. The NTA moiety served as a linker coordinating $Ni^{2+}$ ions that, in turn, used their vacant coordination sites for binding to the S surface atoms of the $WS_2$ nanotubes. Furthermore, 3-hydroxytyramine was used as an anchoring group for the attachment of $TiO_2$ nanoparticles to the nanotube surface. Moreover, fluorescing chromophore was attached to the nanotube surface, allowing its direct observation by confocal microscopy.

The synthesis of macroscopic amounts of multiwall $WS_2$ nanotubes with an inner hollow core of 10–15 nm suggests that they can be ideally suited as templates for the synthesis of core-shell nanotubular structures. Thus, core-shell $PbI_2$@$WS_2$ nanotubes were recently prepared by melting the lead salt in the presence of $WS_2$ nanotubes. The molten salt is "sucked" into the hollow core of the $WS_2$ nanotube by capillary forces and wets its inner walls (Kreizman et al. 2009). The crystallization of the $PbI_2$ upon cooling produces a core ($PbI_2$)–shell ($WS_2$) nanotubular structure (see Figure 16.8).

This scheme is likely to be used in the future for the synthesis of various other core-shell nanotubes bearing the possibility for observing new physical properties and realizing new

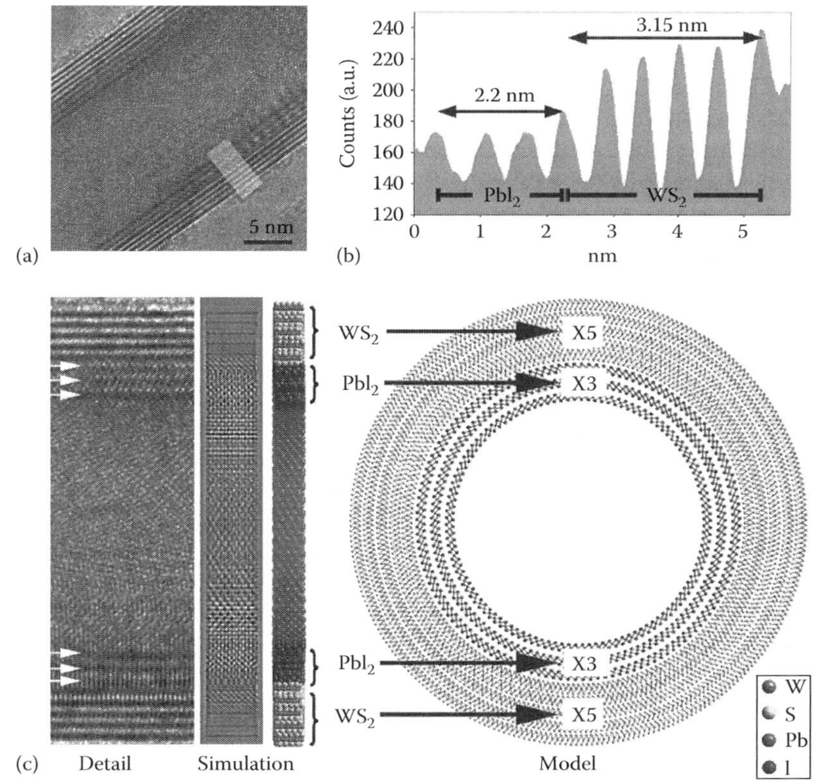


**FIGURE 16.8**
**(See color insert following page 302.)** (a) HRTEM micrograph showing a core-shell $PbI_2$@$WS_2$ composite nanotube. (b) Line profile obtained from the indicated region in (a) showing two types of nanotube layers; five "outer" $WS_2$ layers with sharper contrast and an average spacing of 0.63 nm and three "inner" layers with more complex contrast and an average spacing of 0.70 nm, corresponding to three concentric $PbI_2$ nanotubes. (c) Detail from (a) showing the complex contrast of the inner $PbI_2$ layers (arrowed) relative to the outer $WS_2$ layers. To the right of the detail is a simulation and a cutaway space filling model (left) and cross-sectional structure model (right) with both $WS_2$ (aba stacking) and $PbI_2$ layers (abc stacking) indicated. (Reprinted from Kreizman, R. et al., *Angew. Chem. Int. Ed.*, 48, 1230, 2009. With permission. Copyright Wiley-VCH Verlag GmbH & Co. KGaA.)

nanodevices. Interestingly, in contrast to the bulk material, the closed IF nanoparticles and INT of the metal halides are much less affected by humidity and avert water uptake. This observation suggests a kinetic stabilization of these seamless polyhedral nanostructures against chemical attack in a hostile environment.

## 16.3 Properties

### 16.3.1 Characterization and Modeling

It is obvious that the synthesis of the nanoparticles is merely the first step of the research methodology, which is followed by characterization. Due to the size scale involved and the limited amount of available nanostructures, this phase of the study is mainly focused on various microscopy techniques. These techniques should provide information regarding both the morphology (the 3D shape of the nanoparticle) and its physical properties. While scanning electron microscopy (SEM) and atomic force microscopy (AFM) provide information on the external topography and shape of the nanoparticles, TEM reveals the internal structure of the particles to the atomic level. In the past decade, TEM has gone through a revolution in terms of advancements in instrumentation as well as in image processing schemes. The result is a more closely related overlapping of first principle calculations and experimental data, facilitated by the construction of atomic structure models. This exciting development enables long-term collaborations between the different groups which perform the synthesis, the microscopy, and DFT calculations. It is, therefore, possible to follow the size evolution of a given system, from clusters through other small structures and onto the bulk regime. At the same time modern microscopy techniques can shed light on the electronic properties of the studied materials.

In this chapter, which is focused on IF/INT nanoparticles, the recent ultrahigh-resolution TEM studies of $MoS_2$ nanooctahedra and $WS_2$ tubes is discussed. In analogy to carbon fullerenes and the BN system, closed inorganic cages produced by energetic techniques, such as laser ablation, are often highly symmetric and occur in a specific size range (Parilla et al. 1999). Preliminary theoretical models for the closed-cage $MoS_2$ structures were suggested (Ascencio et al. 2003), presenting the three most stable shapes for a multi-layered structure. Thereafter, a detailed density functional tight binding model coupled with molecular dynamics (DFTB-MD) theoretical analysis for $MoS_2$ nanooctahedra was undertaken. This study permitted a direct comparison with experimental data obtained by conventional microscopes (Bar-Sadan et al. 2006a, Enyashin et al. 2007a,b). Furthermore, using a few other simplifications, a semiempirical model was conceived extending the calculations to IF of any size. These semiempirical calculations permitted establishing the phase stability diagram for different structures of $MoS_2$ nanoparticles. The great similarity between the experimental and calculated observations was compelling and gave further credibility to this kind of analysis. A few important conclusions regarding the $MoS_2$ octahedral closed-cage structures can be drawn from the combined theoretical–experimental study (Bar-Sadan et al. 2006a, Enyashin et al. 2007a,b). These particles manifested a strong correlation between shape and properties, even among structures of the same size and composition. In contrast to bulk and nanotubular $MoS_2$, which are semiconductors, DFTB-MD calculations indicate that the Fermi level of the nanooctahedra is situated within the Mo d-bands, endowing the nanoparticles a metallic-like character. According to these calculations, the occurrence of metallic-like d-band conductivity correlates sensitively with defect formation at the apices and the seaming of the triangular faces at their edges.

The theoretical calculations resulted in 15 suggested structural models of the nanooctahedra, which differed in their construction principles—the arrangement of facets, edges, and vertices. By estimating the energetics of these structure models, a stability curve was constructed from the model calculations, which could be compared with size statistics extracted from the experimental images. The favorable agreement between the calculations and the experimental findings establishes the validity of the model. In particular, nanooctahedra 3–6 nm in size, approx. $10^3$–$10^5$ atoms in total, consisting of 3–5 $MoS_2$ shells were found to be the most stable species in this size range.

A refinement of the model could be achieved, elucidating the nanooctahedra properties in greater detail, using aberration-corrected microscopy. This technique is particularly viable in order to enhance the sensitivity and imaging of the light atoms (S). The high resolution TEM images, taken in the negative Cs imaging (NCSI) mode, successfully revealed the details of the atomic structure, including the sulfur coordination. An example is shown in Figure 16.9, which presents a comparison between suggested structures calculated by the DFTB method and experimental images. Out of the 15 model structures, which were proposed by Parilla et al. (2004) and by Bar-Sadan et al. (2006a) and studied by the DFTB-MD analysis, two models were compared to the experimental data (Figure 16.9b and e compared with Figure 16.9d and f). The model structure shown in Figure 16.9b and e was calculated by the DFTB-MD method to be the most energetically stable structure. The structures presented in Figure 16.9b and e fit very well with the apex structure of the

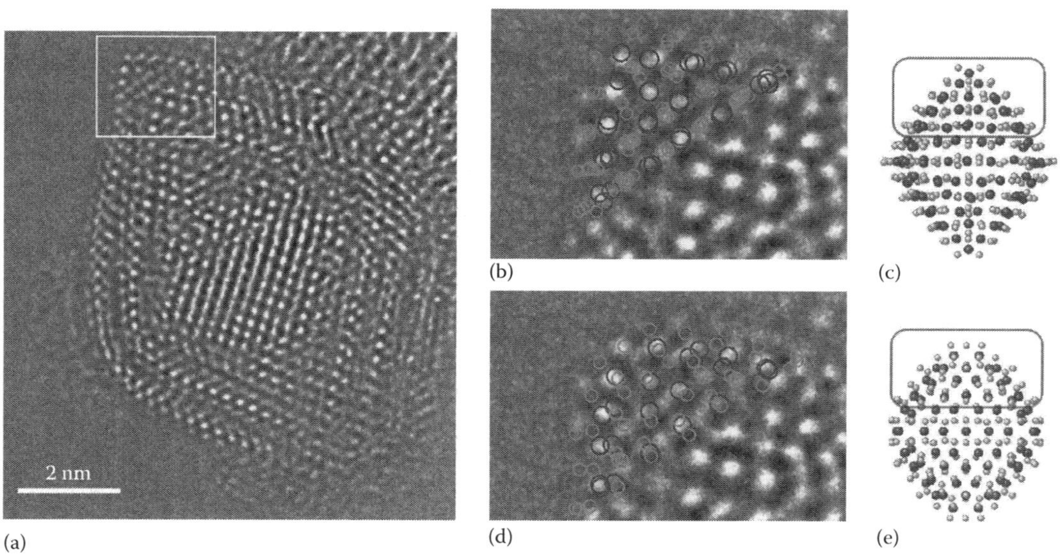


**FIGURE 16.9**
**(See color insert following page 302.)** (a) Atomic resolution image of a $MoS_2$ nanooctahedron taken in an image-side aberration-corrected FEI Titan 80–300 under NCSI conditions. (b, d) A magnified part of the tip of the octahedron revealing the sulfur atoms at the surfaces with (c, e) a superposition of models for two of the 15 hypothetical structures proposed by Parilla et al. (2004). Mo atoms are displayed in red, S atoms in yellow. (b) One of the most stable models (no. 5 of Bar-Sadan et al. 2006a) coincides, while (d) the less stable structure fails to match. (Reprinted from Bar Sadan, M. et al., *Proc. Natl. Acad. Sci.*, 105, 15643, 2008a. With permission. Copyright the National Academy of Sciences.)

experimental image, especially seen in the match of the positions of the light S atoms. The structures seen in Figure 16.9d and f do not show such correspondence.

By imaging nanooctahedra structures that lack mirror symmetry relative to the basal plane of the structure, the existence of less stable structures from the energetic point of view, could also be proven. Nonetheless, the reaction mechanism responsible to the production of these symmetric and rather complicated structures with $10^4$–$10^5$ atoms is a matter of great puzzle.

Recently, the IF and INT structures have been the focus of tomographic studies in ultra-high-resolution electron microscopy (Bar Sadan et al. 2008b). This progress was enabled by using the new aberration-corrected (TEM) instruments and by selecting IF nanoparticles with only a few contrastive shells. Simulations show that, in principle, atomic resolution tomography is feasible, and preliminary results already provide shell resolution (<0.6 nm). For such purpose, a structural model of a nanooctahedron, refined by DFTB, was generated first. In the next step, a tilt series of simulated images of the nanooctahedron was produced. Thereafter, the images were processed and used for the tomographic reconstruction, which is shown in Figure 16.10.

The reconstruction process was done in exactly the same way as performed for the experimentally acquired tilt series. Both Mo and S atoms are resolved in the simulated study. Figure 16.11 shows the experimental result. Due to a limited tilt range in the experiment, the tomogram reconstructs mainly the shell structure. Nonetheless, the Mo-Mo distance within the shell (0.28 nm) could be resolved within the xy-plane. This is the first step toward atomic resolution tomography in the bright field mode, where more examples are to follow.

For INT-$WS_2$, aberration-corrected microscopy combined with advanced modeling provided the opportunity to investigate the relationship between the chirality of the different shells of a tube and its handedness (Bar Sadan et al. 2008a). The main results are the understanding that in most of the investigated tubes there is a single chiral shell embedded within nonchiral ones. This shell probably serves the role of the continuous growth front (template) for the reaction. Thereafter the nonchiral, lower energy, shells are threaded, providing reinforcement and matching of the shell distance to the vdW distance in the bulk. Some tubes were nevertheless observed to have their shells with various small chiral angles, see Figure 16.12, where HAADF-STEM image reveals the W skeleton of the tube. In Figure 16.12, a simulated constructed image is presented, superimposed on top of the experimental one. There is an atom-to-atom match between the experimental result and the calculated structural model.

In another publication, aberration-corrected microscopy was used to measure (using EELS) the bandgap of a single tube (Bar Sadan et al. 2009). A good match was obtained between the observed bandgap (1.2 and 1.6 for the direct and indirect bandgaps, respectively) and the literature values. It is hoped that in the future, more works of the sort would provide better information regarding the structure–function relationship.

### 16.3.2 Electronic and Optical Properties

Optical absorption (Frey et al. 1998) and Raman spectra measurements (Frey et al. 1999) of the $MS_2$ (M = Mo, W) nanostructures were reported in the late 1990s. Shortly afterward, electronic band-structure calculations were carried out for these systems by the DFTB method (Enyashin et al. 2007a,b, Seifert et al. 2000). Milosevic and coworkers recently discussed line-symmetry–based calculations of the polarized optical absorption in single-wall $MoS_2$ and $WS_2$ nanotubes (Milosevic et al. 2007). The calculated optical spectra show highly anisotropic absorption. Calculations (Enyashin et al. 2007a,b, Milosevic et al. 2007),

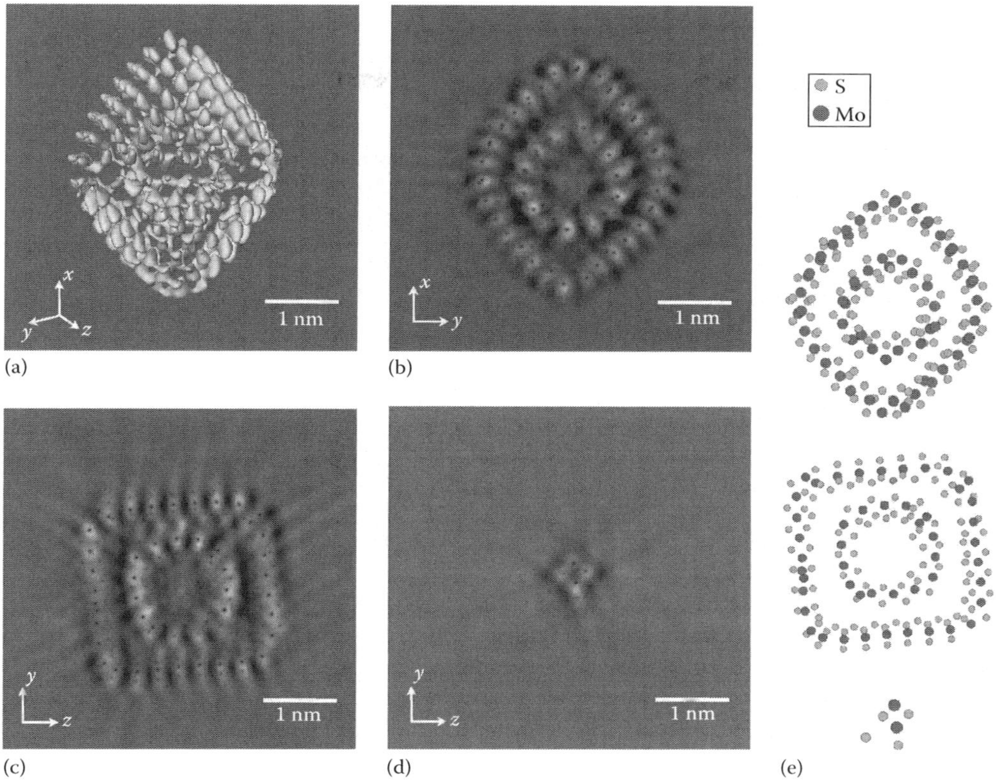


**FIGURE 16.10**
**(See color insert following page 302.)** Single-axis tomographic reconstruction of a $MoS_2$ nanooctahedron from a tilt series of simulated images. A tilt range of 120° and an angular increment of 1° were chosen. By convention, z refers to the axis of transmission and $x$ is parallel to the tilt axis in an orthogonal base system $(x,y,z)$. (a) Isosurface visualization of the reconstructed tomogram of the closed-cage structure model. (b–d) Slices taken in different planes and at different positions. Corresponding slices of the projected electrostatic potential of the model structure are superimposed in a red temperature scale. Strong dots represent Mo atoms; the weaker dots correspond to S atoms. (b) Tomogram slice in the $xy$-plane through the center of the octahedron. (c) Central tomogram slice in the $yz$ plane. (d) Tomogram slice in the $yz$ plane through the apex of the structure showing the square defect in the seaming of the triangular faces. (e) The corresponding structure model slices. (Reprinted from Bar Sadan, M. et al., *Nano Lett.*, 8, 891, 2008b. With permission. Copyright the American Chemical Society.)

scanning tunneling microscopy (STM) studies (Scheffer et al. 2002), and optical absorption spectra of $MoS_2$ nanotubes and IF structures (Frey et al. 1998) have shown that, in comparison with bulk material, there is a strong reduction of the gap size with decreasing radius of the tube or the IF nanoparticle. This reduction is caused by the compression of the inner S shell in the S-Mo-S triple layer in the tubular structure compared with the flat triple layer in layered $MoS_2$ (Scheffer et al. 2002).

Infrared (IR) investigations were performed on IF-$WS_2$ nanoparticles (Luttrell et al. 2006). The IR allowed modes in 2H-$WS_2$ are the $XY$-polarized (double degeneracy) $E_{1u}$ and the $Z$-polarized $A_{2u}$ modes at 356 and 437 $cm^{-1}$, respectively. The oscillator strength of the $E_{1u}$ transition is appreciably stronger in the bulk (2H) material as compared to the IF sample at room temperature and at 10 K. By analyzing the two modes in both the 2H (bulk) and IF-$WS_2$ nanoparticles, the researchers concluded that the interlayer charge polarization,

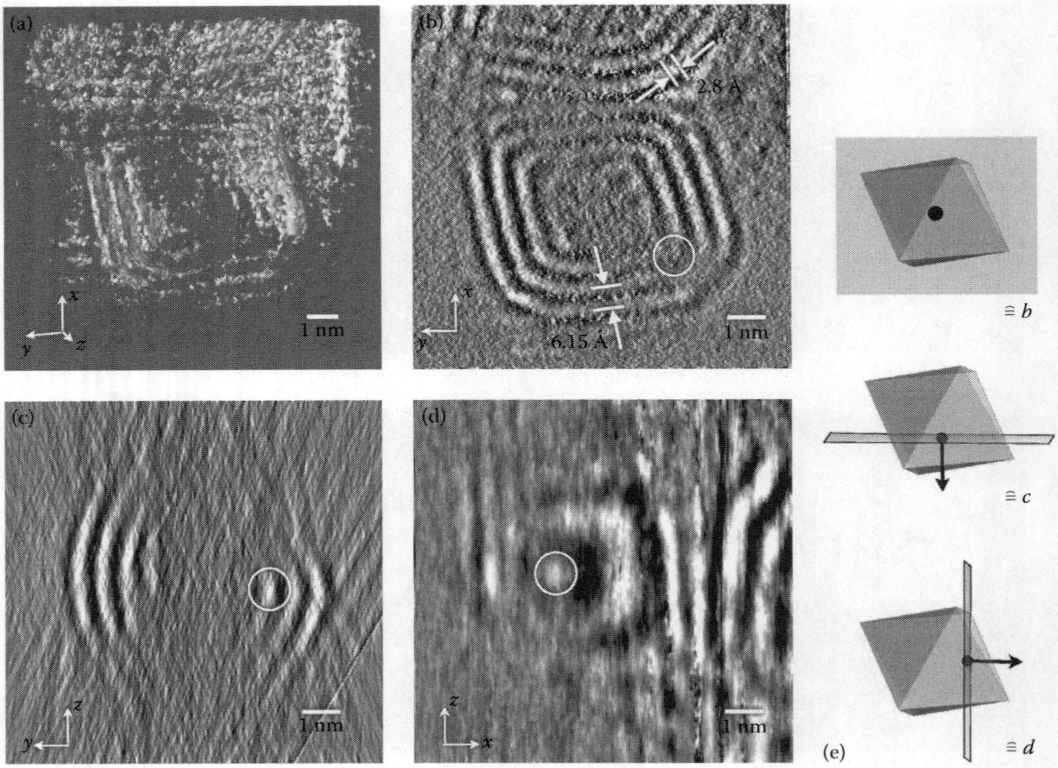


**FIGURE 16.11**
Experimental tomogram reconstructed from a single-axis tilt series of bright-field images. The images were taken over a tilt range of 63° in increments of 3°. *z* refers to the axis of transmission and *x* is parallel to the tilt axis in an orthogonal base system (*x*,*y*,*z*). (a) Isosurface representation showing the nested shell structure. (b–d) Tomogram slices in three orthogonal cut planes through the tip of the second shell (circular marker) of the lower octahedron. (e) Schematic drawings showing the orientation and position of the slice planes corresponding to (b–d). In all the slices the nested structure and the shell separation of 6.15 Å is clearly resolved. In the *xz*-slice in (c), the square coordination of the triangular $MoS_2$ faces is observed for the third and fourth shell. In the upper part of the *xy*-slice of the tomogram, the resolution of the projected distance between Mo units of ~2.8 Å is retained. (Reprinted from Bar Sadan, M. et al., *Nano Lett.*, 8, 891, 2008b. With permission. Copyright the American Chemical Society.)

i.e., e-transfer from the metal to the S atom, is appreciably smaller in the 2H compared with the IF nanoparticles. In contrast, the interlayer charge polarization between the layers is slightly larger in the IF compared with the 2H (bulk) particles. This observation suggests that the interlayer interaction in the IF nanoparticles is somewhat larger compared with the 2H-$WS_2$ particles. Furthermore, this result indicates that the hybridization of the W and S wavefunctions is perturbed in the folded structure and, consequently, the material possesses larger 3D character as compared with 2H (planar) bulk $WS_2$.

The so-called antenna effect, which has been predicted and confirmed for carbon nanotubes (Ajiki and Ando 1995), is also manifested in the multi-wall $MS_2$ nanotubes. Phonon dispersion calculations (Dobardzic et al. 2005, 2006, Virsek et al. 2007) and resonance Raman spectra of $MS_2$ tubes (Joly-Pottuz et al. 2006, Rafailov et al. 2005) have been reported quite recently. Resonance Raman ($\lambda_{exc}$ = 632.8 nm) spectra of individual $MoS_2$ and

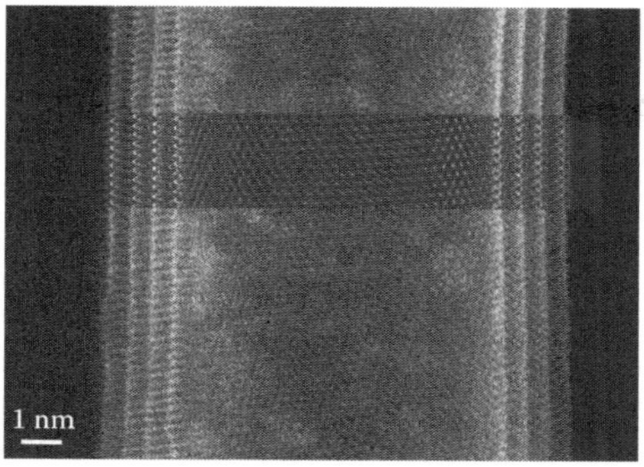

**FIGURE 16.12**
HAADF-STEM image of a$WS_2$ nanotube taken in a probe-side corrected FEI Titan 80–300 microscope operated at 300 kV. The projected potential of a four-shell tube is overlaid [roll-up vectors and chiral angles (4,92, 2.1°) (6,103, 2.8°) (5,115, 2.1°) (7,123, 2.7°)]. The correspondence between the two in the atomic scale is apparent and the helicity can be clearly seen. (Reprinted from Bar Sadan, M. et al., *Proc. Natl. Acad. Sci.*, 105, 15643, 2008a. With permission. Copyright the National Academy of Sciences.)

$WS_2$ nanotubes were measured and compared to the bulk materials (Virsek et al. 2007), showing an upshift of the peaks (3–10 $cm^{-1}$) of both the original $A_{1g}$ (408,421 $cm^{-1}$) and $E^1_{2g}$ (383,356 $cm^{-1}$) Raman modes. This shift was ascribed to the built-in strain in the nanotubes. Indeed, collapsed $MoS_2$ nanotubes (ribbons) showed no shift of the Raman peaks. Furthermore, electron diffraction analysis showed that locally, the nanotubes crystallized in the high pressure rhombohedral (3R) symmetry rather than the hexagonal symmetry, which is the stable polytype in ambient conditions. From the orientation dependence of the nanotube's resonant Raman intensity, the ratio of the perpendicular to the parallel polarizabilities $\alpha_{XX}/\alpha_{ZZ}$ is estimated at 0.16, showing the strong polarization along the nanotube axis (Rafailov et al. 2005). In another study (Yu et al. 2007), the Raman lines of W(core)/$WS_2$(shell) nanoparticles were studied as a function of the hydrostatic pressure in a diamond anvil cell. The $A_{1g}$ line at 420 $cm^{-1}$ was shown to contain a new low energy shoulder at 416 $cm^{-1}$, which is attributed to two-phonon coupling originating from longitudinal acoustic (LA) and transverse acoustic (TA) phonons at the K-point of the Brillouin zone. This low energy shoulder is enhanced by the curvature of the nanoparticles. Raman spectra of the W/$WS_2$ particles showed that the intensity ratio of the LA + TA mode and the $A_{1g}$ mode varies considerably with increasing pressure. This ratio first increases and then reaches a maximum and decreases with the application of further pressure. Upon releasing the pressure, the original lineshape is restored, indicating that the nanotubes are stable under hydrostatic pressure of up to 18 GPa and could serve in ultrahigh strength nanocomposites.

Optical absorption, photoluminescence (PL), and luminescence excitation studies of $TiO_2$ nanotubes showed that the bandgap of the nanotubes (3.87 eV) (Bavykin et al. 2005) is close to that of titania nanosheets (3.84 eV) (Sakai et al. 2004), but is appreciably higher than that of the anatase phase of titania (3.2 eV). These studies have also shown that changing the internal diameter of the $TiO_2$ nanotubes in the range of 2.5–5 nm did not lead to any changes in the positions of the absorption and emission bands. These observations indicate small quantum-size effects for titania nanotubes in this range. Thus, researchers conclude that the electronic structure of $TiO_2$ nanotubes is very close to that of $TiO_2$ nanosheets (Bavykin et al. 2005, Sakai et al. 2004). Another study observed strong and broad sub-bandgap PL with a peak at 570 nm in samples consisting of titanate nanotubes. The PL was associated with the Ti-OH complex within the tubular structure (Qian et al. 2005).

Although BN nanotubes are high-bandgap materials, metallic BC nanotubes with rectangular cross section are stable (Ponomarenko et al. 2006). These results emphasize the extraordinary richness of INT and IF in terms of optical and electrical properties. Spatially resolved electron energy loss spectroscopy experiments of several individual BN single-, double-, and triple-wall nanotubes were carefully measured (Arenal et al. 2005).

For all investigated nanotubes, with diameters ranging from 1.5 to 3 nm and the number of layers ranging from one to three, the value of the optical gap was found to be $5.8 \pm 0.2$ eV, which is very close to that of bulk h-BN. Ab initio calculations were also done for double-wall BN nanotubes of a diameter smaller than 1.5 nm (Jhi et al. 2005). The density of states (DOS) of the double-wall nanotubes is similar to those of the single-wall tubes, but the bandgap of the double-wall tube is smaller. The two sets of data do not necessarily contradict each other because the bandgap of BN nanotubes shrinks with decreasing diameter at 1.5 nm and below (Rubio et al. 1994). Bias-dependent STM and scanning tunneling spectroscopy were used to reveal a strong voltage dependence of the BN nanotube bandgap, i.e., a giant stark effect (Ishigami et al. 2005).

While much effort has been devoted to the study of the optical and the mechanical properties of IF and INT, the electronic and transport properties of these nanoparticles were studied rather scantly. The bulk electrical transport properties of IF-$WS_2$ nanopowders were characterized (Kopnov et al. 2006). The electrical conductivity dependence of the IF pellets on temperature is described well within the model of the electronically inhomogeneous (fluctuating potential barrier) grain boundaries. The free carrier type and its density at 300 K were found to be the same for 2H-$WS_2$ and IF-$WS_2$ pellets. However, the bulk resistivity of the compacted IF powder was found to be higher than that of the 2H (bulk) material by 2–8 orders of magnitudes. Furthermore, unlike most inorganic semiconductors, IF material exhibits unusually low mobility values. These discrepancies stem probably from the structural differences of the two materials and possibly also the loose contact between the small nanoparticles. IF pellets subjected to vacuum annealing at elevated temperatures demonstrated higher resistivity than the nonannealed ones, possibly due to releasing of trap sites by outgoing molecules like water and hydrogen. This assumption is strongly supported by the fact that the nanoparticles repel each other and disband the aggregates after annealing as if they acquired some surface charge.

A phototransistor based on individual $WS_2$ nanotubes, which is sensitive to the visible light was recently fabricated and tested (Unalan et al. 2008). The maximal sensitivity of the phototransistor was obtained when the visible light for a halogen lamp was polarized in parallel to the nanotube axis and the minimum sensitivity occurred when the light was polarized perpendicular to the nanotube axis. The carrier mobility and concentration increased from $4.1 \times 10^{-4}$ cm$^2$ V$^{-1}$ and $1.5 \times 10^7$ cm$^{-1}$ in the dark to $1.3 \times 10^{-3}$ cm$^2$ V$^{-1}$ and $2.57 \times 10^7$ cm$^{-1}$, respectively, under illumination. Transistors capable of detecting visible light would have a wide range of applications in consumer and medical electronics.

### 16.3.3 Thermal Properties

Very few systematic studies have been reported in the area of thermal properties. The thermal conductivity of a mat of multiwall BN nanotubes was studied as a function of the temperature and was similar to that of carbon nanotubes (Chang et al. 2005). The thermal conductivity of an individual multiwall BN nanotube was estimated to be roughly 1620 W mK$^{-1}$. This high thermal conductivity value, which is a factor of 3–4 higher than the thermal conductivity of bulk 2H-BN, is attributed to the ballistic-type thermal conductivity of the 1D nanostructure. The thermal conductivity at low temperatures is dominated by the heat capacity and reflects the size confinement of the phonons in the nanotubes. In another recent study, the low-temperature specific heats of bulk (platelets) and IF nanoparticles of $WS_2$ were measured (Brown et al. 2007). Below 9 K, the specific heat of the nanoparticles deviates from that of the bulk counterpart. It also deviates from the usual $T^3$ dependence below 4 K, which is attributed to finite size effects that eliminate long-wavelength acoustic phonons and interparticle-motion entropy.

### 16.3.4 Mechanical Properties

IF-$WS_2$ (IF-$MoS_2$) were found to exhibit, among other things, superior tribological behavior offering numerous applications (Rapoport et al. 1997, 2005). Their quasi-spherical shape and relatively high crystalline perfection endow them with high strength and elasticity (Joly-Pottuz et al. 2006, Zhu et al. 2003). Their small sizes (<120 nm) ensures their facile access into the rubbed surface, preventing on the one hand, the direct asperity contact of the reciprocating surfaces and at the same time allowing rolling and sliding, thereby, reducing both friction and wear. Furthermore, under extreme pressure (>1 GPa) and high shear stress, the IF nanoparticles gradually deform and peel their outermost layers (Tahir et al. 2006). These molecular layers are deposited on the asperities (third body) and prevent a direct metal to metal contact, thereby, slowing down the wear rate (Joly-Pottuz et al. 2005).

Various schemes have been proposed for the use of the nanoparticles for lubrication, e.g., as an additive to lubricating fluids. For this purpose, various mixtures of the solid powder and lubrication fluids were prepared and tested under standard conditions (Rapoport et al. 1997). More recently, a number of studies have indicated that the IF material can also serve very effectively as a dry solid lubricant either as a film or when impregnated to various matrices (Chhowalla and Amaratunga 2000, Leshchinsky et al. 2002, Rapoport et al. 2001). In one such study, IF-$MoS_2$ nanoparticles were produced by the arc-discharge technique and collected on a Ti foil, forming, thus, a thin film of this material (Chhowalla and Amaratunga 2000). The IF-based film exhibited very low friction coefficients (>0.01), even at 45% humidity. Under similar conditions, a sputtered $MoS_2$ film exhibited a friction coefficient >0.1 and rapid wear. Self-lubrication of mechanical parts can alleviate some of the technological complexities involved in the lubrication by fluids of mechanical systems, as well as the environmental impact of this technology. Long-term tests showed that the lifetime of the self-lubricating bearing impregnated with IF material can be extended by up to two orders of magnitudes. The remarkable effect of the IF material has been attributed to the slow release of the IF nanoparticles, which reside in the porous matrix of the metallic matrix. This work suggests numerous applications for the IF nanoparticles.

Self-lubricating metal coating containing fullerene-like $WS_2$ (IF) nanoparticles was demonstrated to significantly reduce friction with possible applications in orthodontics (Katz et al. 2006). Although these may be considered as preliminary report only, application of the self-lubricity behavior of such coatings in other medical devices, like needles, catheters, endoscopes, and coating surfaces of articulated joints are foreseen.

The mechanical properties of single $WS_2$ nanotubes were shown to be remarkably different from the bulk materials. The mechanical behavior of bulk materials is dictated by the nature of the chemical bond holding the atoms together as well as the structural and chemical defects. The occurrence of intrinsic defects, like vacancies, is dictated by thermodynamics of finite temperature systems. On the other hand, extrinsic defects, like grain boundaries and dislocations are induced by the growth and processing of the material. Therefore, most bulk materials are appreciably weaker than what would be predicted by considering the strength of their chemical bond. In contrast to this, a series of experiments with individual multiwall $WS_2$ nanotubes (Kaplan-Ashiri et al. 2006) have clearly demonstrated that their mechanical properties are predictable from first principle calculations, i.e., they can be referred to the strength of the chemical bond. The nanotubes showed elastic linear behavior almost to their failure. The Young's modulus calculated from the slope of the curve; strength and strain at yielding are 160 GPa, 16% and 10%, respectively. Around 50% of the measured nanotubes showed strength in excess of 13 GPa, which is indicative that they are essentially (critical) defect-free. DFTB-MD calculations of single-wall $MoS_2$ nanotubes were carried out and were compared, after normalization, to the experimental data of $WS_2$ nanotubes. The normalized (from $MoS_2$ to $WS_2$) theoretical data were about 30% higher than the experimentally observed strength and strain to failure. Furthermore, the theoretical calculations clearly show that under extreme strain one of the Mo-S bonds in the middle of the nanotube fails, which leads to stress concentration on the neighboring bonds and, subsequently, to their rapid failure, and so forth. This kind of failure does not involve any intrinsic or extrinsic defect. Future work may focus on mechanical testing of individual nanotubes within the TEM, which may shed clear picture on the atomic structure of the nanotube under strain and its failure.

In another series of experiments (Kaplan-Ashiri et al. 2007), individual $WS_2$ nanotubes were suspended on an empty channel and were pushed sideway in the middle (bending test). The sliding of one nanotube layer with respect to its neighbors could be calculated from the modified Timoshenko's bending equation. In bulk materials the ratio between the Young's ($E$) and shear ($G$) moduli is high (about 0.3). Consequently the contribution of the shear mode to the bending energy is rather small. The relatively free sliding, i.e., the low sliding modulus ($G/E \sim 0.01$) of neighboring walls with respect to each other in the nanotube allows this mode to take a relatively significant share of the bending energy. These studies together with the advent in the large-scale synthesis of fullerene-like $WS_2$ and its nanotubes (Zak et al. 2009) are nourishing a new wave of studies aimed at developing ultrahigh strength nanocomposites of various sorts.

## 16.4 Applications

### 16.4.1 Hydrogen Storage in Inorganic Nanotubes

Studies have been carried out to investigate inorganic nanotubes as host materials for hydrogen storage. The novel structure of 1D inorganic nanostructures makes them good candidates for energy storage. Their small size, which is correlated with high surface areas and large surface to volume ratio, favors physical or chemical interactions of the host materials and the guest molecules or ions. The large surface area and the short diffusion distance in the nanotubes results in a high capacity and fast kinetics. Furthermore, the stability of these materials allows them to withstand the impact of numerous intercalation/deintercalation cycles, while preserving their original structure (Cheng and Chen 2006).

Electrochemical hydrogen storage of $MoS_2$ nanotubes was studied by Chen et al. (2001). The motivation for the study was based on the earlier use of bulk $MoS_2$ as a cathode material for lithium-ion batteries (Whittingham 1978). Furthermore, the combination of sulfur, which is a strong hydride forming element, with molybdenum, which is a weak hydride forming element, similar to the well-known $LaNi_5$ intermetallic compound, makes the $MoS_2$ nanotubes a viable potential candidate for hydrogen storage. According to this study, the electrochemical charge–discharge mechanism occurring in $MoS_2$ nanotubes is somewhere in-between that of carbon nanotubes (physisorption) and metal hydride electrodes (chemical process), and consists of a charge-transfer reaction and a diffusion step. Electrochemical storage of 0.97 wt.% hydrogen was measured for the $MoS_2$ nanotubes. The energy of hydrogen adsorption on the surface (both outside and inside the tube) and interstitial sites can be reduced by inducing more defects.

The electrochemical hydrogen-storage measurement, and the hydrogen gas-adsorption properties of $MoS_2$ nanotubes were investigated as well (Chen et al. 2003a). Here, the amounts of hydrogen-adsorption in the open-ended $MoS_2$ nanotubes reached a maximum capacity of about 1.2 wt.%. Furthermore, this study revealed that most of the hydrogen could be desorbed at 298 K.

The hydrogen storage of $TiS_2$ nanotubes was studied as well; since Ti is a lighter element compared to Mo it might be a better candidate for hydrogen storage. High-purity $TiS_2$ nanotubes with open-ended tips were synthesized by a chemical transport reaction. They exhibited capacity to store 2.5 wt.% of hydrogen at a temperature of 298 K and under hydrogen pressure of 4 MPa (Chen et al. 2003b). It was found out that 1.0 wt.% of the hydrogen is chemisorbed. The hydrogen adsorption and desorption of $TiS_2$ nanotubes were found to depend on the number of cycles. HRTEM studies showed that different types of defects including point defects and plane defects were formed. This behavior was attributed to the irregular expansion of the interlayer spacing during the hydrogen adsorption–desorption and to the high chemical reactivity of the host compound, especially with respect to oxidation. On the one hand, the formation of some specific defects would be expected to enhance the catalytic reactivity toward reversible hydrogen storage and rapid adsorption/desorption kinetics. On the other hand, the collapse of the crystalline structure of the nanotube decreases their stability toward further oxidation and consequently influences their long-term hydrogen storage capacity. Mechanistic questions such as the possible formation of S–H bonds in the $TiS_2$ nanotubes have to be addressed in future studies.

Another candidate for hydrogen storage are the BN nanotubes. Taking into account the chemical and thermal stability, BN nanotubes may be stable lightweight hydrogen-gas accumulators. Ma et al. (2002) studied the hydrogen adsorption properties of two kinds of BN nanotubes. The measurements show that the multiwalled and bamboo-shaped BN nanotubes can adsorb hydrogen up to 1.8 and 2.6 wt.%, respectively, at about 10 MPa, in striking contrast to the negligible 0.2 wt.% in conventional BN. This study confirmed the hypothesis that BN nanotubes possess much higher hydrogen-adsorption capacity than bulk BN powder. The faster hydrogen uptake and higher capacity could be attributed to the higher specific surface area of the nanotubes as compared to the bulk phase. Note, however, that the capped feature of BN multiwalled nanotubes prevents the hollow cavity of the tubes from being easily accessed by the hydrogen molecules. As a comparison, the bamboo-shaped tubes, which have more defects in the structure and open-edge layers on the exterior surface, show a somewhat higher hydrogen uptake amount. This result indicates that the majority of the hydrogen is chemisorbed either on the surface of the tube or in between the layers. After complete desorption, the hydrogen uptake capacity remained almost unchanged in the following cycle, indicating the high reversibility of the adsorption–desorption processes. In another work (Chen et al. 2005), the electrochemical hydrogen storage of BN nanotubes was studied. The BN nanotubes were synthesized through a chemical vapor deposition method using $LaNi_5$ alloy as a catalyst precursor.

The nanotubes are straight with diameters of 30–50 nm and length of several microns. It was verified for the first time that BN nanotubes can store hydrogen by means of an electrochemical method, though the reported capacity is low at present (0.25 wt.%). It was tentatively concluded that by surface modification with metal or alloy the electrocatalytic activity could be improved, which would enhance the electrochemical hydrogen uptake of BN nanotubes.

### 16.4.2 Commercial Products Incorporating IF and INT

The advanced and innovative work regarding IF and INT has been incorporated into commercial products by ApNano Materials (www.apnano.com), which produces tonnage quantities of different grades of IF. Engine and transmission system oils formulated with IF-$WS_2$ nanoparticles as additives as well as greases formulated with IF-$WS_2$ nanoparticles, were launched in the market recently.

In recent years, much effort has also been dedicated to the development of self-lubricating coatings (see, e.g., Friedman et al. 2007, Katz et al. 2006) with numerous potential applications in the medical, machining, aerospace, and various other industries. The first products based on this technology were recently commercialized by ApNano Materials. In contrast to the case of fluid lubricants, for which the nanoparticles must be introduced into the contact area by flow, the metal-impregnated IF nanoparticles are released from the surface upon wear, leading to a reduction in the friction coefficient within a few run-in cycles. A typical reduction in the friction coefficient from 0.3 to 0.1 and below was observed for these metal coatings. These values are appreciably lower than the friction coefficient observed for the pure-metal substrate, which was 0.3–0.6. Another advantage of the IF nanoparticles is that they slow down the oxidation of the metal surface dramatically, thereby preventing decomposition of the brittle oxide and scratching of the metal surface. It is believed that the nanoparticles serve as a kind of cathodic protection for the rubbed surface, which otherwise undergoes relatively fast tribo-chemical oxidation. Black metallic coatings with IF nanoparticles incorporated, showing less than 1.7% reflectivity over the range of 300–1100 nm, were recently prepared at ApNano Materials. This development offers various applications ranging from solar-to-thermal converters to optical coatings. More work is needed, however, to optimize the optical properties of the coatings.

Similarly, IF-$WS_2$ nanoparticles were impregnated in polymer matrices, producing self-lubricating polymer coatings (Hou et al. 2008) with a large number of potential applications. The use of IF-$WS_2$ nanoparticles as the building blocks for polymer nanocomposites is another perspective application. Furthermore, in another study (Naffakh et al. 2007) it was found that the addition of small amounts of IF nanoparticles into isotactic polypropylene (iPP) matrix increased the thermal stability of the polymer and also had a remarkable effect on its crystallization rate. Furthermore, the storage moduli of the IF nanoparticles/iPP nanocomposites were appreciably higher than those of the pure iPP. In another work, the effect of IF closed-caged nanoparticles of IF-$WS_2$ on the mechanical properties and especially on the toughness of epoxy resins was studied (Shneider et al. 2010). The epoxy resin was a commonly used DGEBA (diglycidyl ether of bis-phenol A) cured with polyamido amine. Experimental results demonstrated enhanced shear strengths and shear modulus (see Figure 16.13) and a significant increase in the peel strengths at low concentrations of the IF-$WS_2$ nanoparticles. The simultaneous increase in stiffness and shear strength, as observed for the IF-$WS_2$ nanocomposites, indicate that stresses are efficiently transferred via the matrix/nanoparticle interface.

Similarly, polymer (polystyrene and polyaniline) nanocomposites with BN nanotubes have been prepared, and their mechanical properties have been evaluated (Golberg et al.

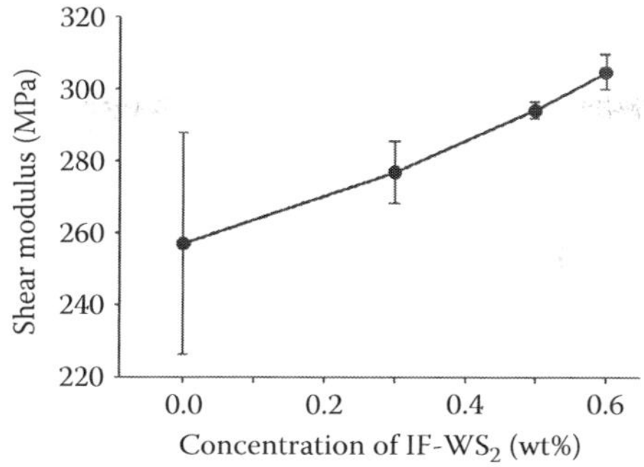


**FIGURE 16.13**
Dependence of the shear modulus of epoxy nanocomposites on the concentration of the IF-$WS_2$ nanoparticles. Mixing conditions were 18,000 rpm for 4 h. (Courtesy of M. Shneider.)

2007). Indeed, these polymer nanocomposites exhibited enhanced modulus (>20%) upon the addition of a small amount (1%) of the surface-modified BN nanotubes. Furthermore, they showed enhanced stability against oxidation as compared with the pristine polymer. Glasses loaded with 4% BN nanotubes showed substantial reinforcement with respect to the pristine glass (Golberg et al. 2007).

Arrays of polycrystalline $TiO_2$ nanotubes were prepared by anodization of Ti foils, and their structure and growth mechanism were investigated in quite a detail (Beranek et al. 2003, Gong et al. 2001). Such arrays have recently been studied as vectors for charge separation and transport in dye-sensitized solar cells (Shankar et al. 2008) with cell efficiencies exceeding 6%. Photocurrent modulation spectroscopy showed that the charge recombination of photoexcited carriers is slower by one order of magnitude in the nanotube based cell versus the nanoparticulate based cell (Zhu et al. 2006). The enhancement of the carrier transport behavior was attributed to the diminished effect of the grain boundaries resistance in the nanotubes as compared with a film consisting of $TiO_2$ nanoparticles.

Researchers have reported the application of various kinds of INT as materials for gas-sensing devices. The large surface area and the specificity of the nanotube-gas interaction may lead to very high sensitivities. Recently $VO_x$-alkyl amine nanotubes were tested as gas sensors for oxygen, nitrogen oxide, triethylamine (TEA), and dimethylmethylphosphonate (DMMP) (Grigorieva et al. 2008). In particular, such nanotubes demonstrate a detectable sensor signal with respect to TEA, even at 125°C and at gas concentrations of 0.6 ppm.

Various catalytic applications for INT and IF have been proposed in recent years. For example, $MoS_2$ nanotubes loaded with Ni nanoparticles may serve as a very efficient hydrodesulfurization catalyst (Cheng et al. 2006).

## 16.5 Conclusions and Perspectives

The burgeoning field of IF and INT provides large field for research and shows great potential for different applications. However the most daunting issue remains the synthesis of new nanoparticles belonging to this class of materials. A successful scale-up of the synthesis process provides the means to study the nanomaterial phase in great detail.

Based on such studies new applications arise and indeed some of which are already commercialized. However, in order to achieve that goal, the microscopic synthetic route should be revisited and the growth mechanism of the IF/INT nanophases should be better understood. The first breakthrough of the synthetic route already enabled major improvement of the produced quantities of $WS_2$ and $MoS_2$ nanoparticles. The recent progress regarding the large-scale synthesis of $WS_2$ nanotubes, IF-$MoS_2$, and to some extent also the $MoS_2$ nanooctahedra is very encouraging, suggesting that they will have a future impact analogous to that of the already commercialized IF-$WS_2$. As the main obstacle toward mass produced IF nanoparticles is gradually being removed in the last few years, various applications such as lubricating oils and greases, self-lubricating coatings, composite materials, etc. emerge making the research in this direction highly warranted. Noticeably, though, electronic, optical and magnetic applications for this class of materials are missing. This disturbing fact emphasizes the need to invest more heavily in the synthesis of IF/INT nanomaterials, which exhibit interesting properties in these regards.

## Acknowledgments

The authors thank the Minerva Fellowship program funded by the German Federal Ministry for Education and Research, the support of the G.M.J. Schmidt Minerva Center for Supramolecular Chemistry; the Harold Perlman Foundation and the ERC grant INTIF No. 226639 for funding. In addition, the authors thank the Deutsche Forschungsgemeinschaft (DFG) for their support of sub-angström microscopy at the Ernst Ruska Centre. RT holds the Drake family chair in nanotechnology and is the director of the Helen and Martin Kimmel Center for Nanoscale Science.

## References

Ajiki, H. and Ando, T. 1995. Carbon nanotubes: Optical absorption in Aharonov-Bohm flux. *Japanese Journal of Applied Physics Supplement* 34:107–109.

Albu-Yaron, A. et al. 2005. Preparation and structural characterization of stable $Cs_2O$ closed-cage structures. *Angewandte Chemie: International Edition* 44:4169–4172.

Albu-Yaron, A. et al. 2006. Synthesis of fullerene-like $Cs_2O$ nanoparticles by concentrated sunlight. *Advanced Materials (Weinheim, Germany)* 18:2993–2996.

Arenal, R. et al. 2005. Electron energy loss spectroscopy measurement of the optical gaps on individual boron nitride single-walled and multiwalled nanotubes. *Physical Review Letters* 95:127601.

Ascencio, J. A. et al. 2003. Structural models of inorganic fullerene-like structures. *Surface Science* 526:243–247.

Avivi, S. et al. 2000. A new fullerene-like inorganic compound fabricated by the sonolysis of an aqueous Solution of $TlCl_3$. *Journal of the American Chemical Society* 122:4331–4334.

Banhart, F. et al. 1994. The formation of curled concentric-shell clusters in boron-nitride under electron-irradiation. *Chemical Physics Letters* 231:98–104.

Bar-Sadan, M. et al. 2006a. Structure and stability of molybdenum sulfide fullerenes. *Journal of Physical Chemistry B* 110:25399–25410.

Bar-Sadan, M. et al. 2006b. Closed-cage (fullerene-like) structures of $NiBr_2$. *Materials Research Bulletin* 41:2137–2146.

Bar Sadan, M. et al. 2008a. Atom by atom: HRTEM insights into inorganic nanotubes and fullerene-like structures. *Proceedings of the National Academy of Sciences of The United State of America* 105:15643–15648.
Bar Sadan, M. et al. 2008b. Toward atomic-scale bright-field electron tomography for the study of fullerene-like nanostructures. *Nano Letters* 8:891–896.
Bar Sadan, M. et al. 2009. Inorganic $WS_2$ nanotubes revealed atom by atom using ultra-high-resolution transmission electron microscopy. *Applied Physics A: Materials Science & Processing* 96:343–348.
Bastide, S. et al. 2006. $WS_2$ closed nanoboxes synthesized by spray pyrolysis. *Advanced Materials (Weinheim, Germany)* 18:106–109.
Bates, K. R. and Scuseria, G. E. 1998. Why are buckyonions round? *Theoretical Chemistry Accounts* 99:29–33.
Bavykin, D. V. et al. 2005. Apparent two-dimensional behavior of $TiO_2$ nanotubes revealed by light absorption and luminescence. *Journal of Physical Chemistry B* 109:8565–8569.
Beranek, R. et al. 2003. Self-organized porous titanium oxide prepared in $H_2SO_4$/HF electrolytes. *Electrochemical and Solid State Letters* 6:B12–B14.
Bertram, N. et al. 2006. Nanoplatelets made from $MoS_2$ and $WS_2$. *Chemical Physics Letters* 418:36–39.
Boulanger, L. et al. 1995. Concentric shelled and plate-like graphitic boron-nitride nanoparticles produced by Co2-laser pyrolysis. *Chemical Physics Letters* 234:227–232.
Brown, S. et al. 2007. Bulk vs nanoscale $WS_2$: Finite size effects and solid-state lubrication. *Nano Letters* 7:2365–2369.
Chang, C. W. et al. 2005. Thermal conductivity of B-C-N and BN nanotubes. *Applied Physics Letters* 86:173102.
Chen, J. et al. 2001. Electrochemical hydrogen storage in $MoS_2$ nanotubes. *Journal of the American Chemical Society* 123:11813–11814.
Chen, J. et al. 2003a. Novel hydrogen storage properties of $MoS_2$ nanotubes. *Journal of Alloys and Compounds* 356:413–417.
Chen, J. et al. 2003b. Titanium disulfide nanotubes as hydrogen-storage materials. *Journal of the American Chemical Society* 125:5284–5285.
Chen, J. et al. 2003c. Lithium intercalation in open-ended $TiS_2$ nanotubes. *Angewandte Chemie-International Edition* 42:2147–2151.
Chen, X. et al. 2005. Preparation and electrochemical hydrogen storage of boron nitride nanotubes. *Journal of Physical Chemistry B* 109:11525–11529.
Cheng, F. Y. and Chen, J. 2006. Storage of hydrogen and lithium in inorganic nanotubes and nanowires. *Journal of Materials Research* 21:2744–2757.
Cheng, F. Y. et al. 2006. $MoS_2$-Ni nanocomposites as catalysts for hydrodesulfurization of thiophene and thiophene derivatives. *Advanced Materials* 18:2561–2564.
Chhowalla, M. and Amaratunga, G. A. J. 2000. Thin films of fullerene-like $MoS_2$ nanoparticles with ultra-low friction and wear. *Nature (London)* 407:164–167.
Chopra, N. G. et al. 1995. Boron-nitride nanotubes. *Science* 269:966–967.
Cohen, H. 2004. Chemically resolved electrical measurements using x-ray photoelectron spectroscopy. *Applied Physics Letters* 85:1271–1273.
Coleman, K. S. et al. 2002. The formation of $ReS_2$ inorganic fullerene-like structures containing $Re_4$ parallelogram units and metal-metal bonds. *Journal of the American Chemical Society* 124:11580–11581.
Deepak, F. L. et al. 2006. $MoS_2$ fullerene-like nanoparticles and nanotubes using gas-phase reaction with $MoCl_5$. *Nano* 1:167–180.
Deepak, F. L. et al. 2007. Fullerene-like (IF) $Nb_xMo_{1-x}S_2$ nanoparticles. *Journal of the American Chemical Society* 129:12549–12562.
Dobardzic, E. et al. 2005. Zero m phonons in $MoS_2$ nanotubes. *Physical Review B* 71:121405.
Dobardzic, E. et al. 2006. Raman and infrared-active modes in $MS_2$ nanotubes (M = Mo, W). *Physical Review B* 74:033403.
Doron-Mor, I. et al. 2000. Controlled surface charging as a depth-profiling probe for mesoscopic layers. *Nature* 406:382–385.

Enyashin, A. N. et al. 2007a. Structure and stability of molybdenum sulfide fullerenes. *Angewandte Chemie: International Edition* 46:623–627.

Enyashin, A. N. et al. 2007b. Simulation of inorganic nanotubes. In *Springer Series in Materials Science*, Sibylle Gemming, Michael Schreiber and Jens-Boie Suck, Eds., pp. 33–57. Berlin/Heidelberg, Germany: Springer.

Etzkorn, J. et al. 2005. Metal-organic chemical vapor deposition synthesis of hollow inorganic-fullerene-type $MoS_2$ and $MoSe_2$ nanoparticles. *Advanced Materials (Weinheim, Germany)* 17:2372–2375.

Fan, H. J. et al. 2007. Formation of nanotubes and hollow nanoparticles based on kirkendall and diffusion processes: A review. *Small* 3:1660–1671.

Feldman, Y. et al. 1995. High-rate, gas-phase growth of $MoS_2$ nested inorganic fullerenes and nanotubes. *Science* 267:222–225.

Feldman, Y. et al. 1996. Bulk synthesis of inorganic fullerene-like $MS_2$ (M = Mo, W) from the respective trioxides and the reaction mechanism. *Journal of the American Chemical Society* 118:5362–5367.

Frey, G. L. et al. 1998. Optical-absorption spectra of inorganic fullerenelike $MS_2$ (M = Mo, W). *Physical Review B: Condensed Matter and Materials Physics* 57:6666–6671.

Frey, G. L. et al. 1999. Raman and resonance Raman investigation of $MoS_2$ nanoparticles. *Physical Review B: Condensed Matter and Materials Physics* 60:2883–2892.

Friedman, H. et al. 2007. Fabrication of self-lubricating cobalt coatings on metal surfaces. *Nanotechnology* 18:115703.

Friend, R. H. and Yoffe, A. D. 1987. Electronic-properties of intercalation complexes of the transition-metal dichalcogenides. *Advances in Physics* 36:1–94.

Golberg, D. et al. 1998. Octahedral boron nitride fullerenes formed by electron beam irradiation. *Applied Physics Letters* 73:2441–2443.

Golberg, D. et al. 2007. Boron nitride nanotubes. *Advanced Materials* 19:2413–2432.

Goldberger, J. et al. 2003. Single-crystal gallium nitride nanotubes. *Nature* 422:599–602.

Gong, D. et al. 2001. Titanium oxide nanotube arrays prepared by anodic oxidation. *Journal of Materials Research* 16:3331–3334.

Grigorieva, A. V. et al. 2008. Sensor properties of vanadium oxide nanotubes. *Mendeleev Communications* 18:6–7.

Guo, T. et al. 1991. Doping bucky: Formation and properties of boron-doped buckminsterfullerene. *Journal of Physical Chemistry* 95:4948–4950.

Guo, B. C. et al. 1992a. Ti8c12+-metallo-carbohedrenes: A new class of molecular clusters. *Science* 255:1411–1413.

Guo, B. C. et al. 1992b. Metallo-carbohedrenes [$M_8C_{12}$ + (M = V, Zr, Hf, and Ti)]: A class of stable molecular cluster ions. *Science* 256:515–516.

Hervieu, M. et al. 2004. The route to fullerenoid oxides. *Nature Materials* 3:269–273.

Homyonfer, M. et al. 1996. Scanning tunneling microscope induced crystallization of fullerene-like $MoS_2$. *Journal of the American Chemical Society* 118:7804–7808.

Homyonfer, M. et al. 1997. Intercalation of inorganic fullerene-like structures yields photosensitive films and new lips for scanning probe microscopy. *Journal of the American Cemical Society* 119:2693–2698.

Hong, S. Y. et al. 2003. Synthesis of $SnS_2$/SnS fullerene-like nanoparticles: A superlattice with polyhedral shape. *Journal of the American Chemical Society* 125:10470–10474.

Hou, X. et al. 2008. Microstructures and tribological properties of PEEK-based nanocomposite coatings incorporating inorganic fullerene-like nanoparticles. *Surface and Coatings Technology* 202:2287–2291.

Iijima, S. 1991. Helical microtubules of graphitic carbon. *Nature* 354:56–58.

Ishigami, M. et al. 2005. Observation of the giant stark effect in boron-nitride nanotubes. *Physical Review Letters* 94:056804.

Ivanovskaya, V. V. et al. 2006a. Structure, stability and electronic properties of composite $Mo_{1-x}Nb_xS_2$ nanotubes. *Physica Status Solidi B: Basic Solid State Physics* 243:1757–1764.

Ivanovskaya, V. V. et al. 2006b. Electronic structure of niobium-doped molybdenum disulfide nanotubes. *Russian Journal of Inorganic Chemistry* 51:320–324.
Jhi, S. H. et al. 2005. Formation and electronic properties of double-walled boron nitride nanotubes. *Solid State Communications* 134:397–402.
Joly-Pottuz, L. et al. 2005. Ultralow-friction and wear properties of IF-$WS_2$ under boundary lubrication. *Tribology Letters* 18:477–485.
Joly-Pottuz, L. et al. 2006. Pressure-induced exfoliation of inorganic fullerene-like $WS_2$ particles in a Hertzian contact. *Journal of Applied Physics* 99:023524-1–023524-5.
JoseYacaman, M. et al. 1996. Studies of $MoS_2$ structures produced by electron irradiation. *Applied Physics Letters* 69:1065–1067.
Kaplan-Ashiri, I. et al. 2006. On the mechanical behavior of $WS_2$ nanotubes under axial tension and compression. *Proceedings of the National Academy of Sciences of the United States of America* 103:523–528.
Kaplan-Ashiri, I. et al. 2007. Microscopic investigation of shear in multiwalled nanotube deformation. *Journal of Physical Chemistry C* 111:8432–8436.
Katz, A. et al. 2006. Self-lubricating coatings containing fullerene-like $WS_2$ nanoparticles for orthodontic wires and other possible medical applications. *Tribology Letters* 21:135–139.
Kopnov, F. et al. 2006. Transport properties of fullerene-like $WS_2$ nanoparticles. *Physica Status Solidi B: Basic Solid State Physics* 243:1229–1240.
Kopnov, F. et al. 2008. Intercalation of alkali metal in $WS_2$ nanoparticles, revisited. *Chemistry of Materials* 20:4099–4105.
Kratschmer, W. et al. 1990. Solid C-60: A new form of carbon. *Nature* 347:354–358.
Kreizman, R. et al. 2009. Core-shell $PbI_2$@$WS_2$ inorganic nanotubes from capillary wetting. *Angewandte Chemie: International Edition* 48:1230–1233.
Kroto, H. W. et al. 1985. C-60: Buckminsterfullerene. *Nature* 318:162–163.
Lee, G. H. et al. 2003. A simple synthetic route to $MoS_2$ and $WS_2$ nanoparticles and thin films. *International Journal of Modern Physics B* 17:1134–1140.
Leshchinsky, V. et al. 2002. Inorganic nanoparticle impregnation of self-lubricated materials. *International Journal of Powder Metallurgy (Princeton, New Jersey)* 38:50–57.
Li, X.-L. et al. 2004. Atmospheric pressure chemical vapor deposition: An alternative route to large-scale $MoS_2$ and $WS_2$ inorganic fullerene-like nanostructures and nanoflowers. *Chemistry: A European Journal* 10:6163–6171.
Luttrell, R. D. et al. 2006. Dynamics of bulk versus nanoscale $WS_2$: Local strain and charging effects. *Physical Review B: Condensed Matter and Materials Physics* 73:035410/1–035410/6.
Ma, R. Z. et al. 2002. Hydrogen uptake in boron nitride nanotubes at room temperature. *Journal of the American Chemical Society* 124:7672–7673.
Margolin, A. et al. 2005a. Fullerene-like nanoparticles of titanium disulfide. *Current Nanoscience* 1:253–262.
Margolin, A. et al. 2005b. Inorganic fullerene-like nanoparticles of $TiS_2$. *Chemical Physics Letters* 411:162–166.
Margulis, L. et al. 1993. Nested fullerene-like structures. *Nature (London)* 365:113–114.
Meyer, R. R. et al. 2000. Discrete atom imaging of one-dimensional crystals formed within single-walled carbon nanotubes. *Science* 289:1324–1326.
Milosevic, I. et al. 2007. Electronic properties and optical spectra of $MoS_2$ and $WS_2$ nanotubes. *Physical Review B (Condensed Matter and Materials Physics)* 76:233414–4.
Naffakh, M. et al. 2007. Influence of inorganic fullerene-like $WS_2$ nanoparticles on the thermal behavior of isotactic polypropylene. *Journal of Polymer Science Part B: Polymer Physics* 45:2309–2321.
Nath, M. and Rao, C. N. R. 2001a. $MoSe_2$ and $WSe_2$ nanotubes and related structures. *Chemical Communications* 2001:2236–2237.
Nath, M. and Rao, C. N. R. 2001b. New metal disulfide nanotubes. *Journal of the American Chemical Society* 123:4841–4842.
Nath, M. and Rao, C. N. R. 2002. Nanotubes of group 4 metal disulfides. *Angewandte Chemie-International Edition* 41:3451–3454.

Nath, M. et al. 2004. Nanoparticles produced by laser ablation of $HfS_3$ in liquid medium: Inorganic fullerene-like structures of $Hf_2S$. *Chemistry of Materials* 16:2238–2243.
Parilla, P. A. et al. 1999. The first true inorganic fullerenes? *Nature* 397:114.
Parilla, P. A. et al. 2004. Formation of nanooctahedra in molybdenum disulfide and molybdenum diselenide using pulsed laser vaporization. *Journal of Physical Chemistry B* 108:6197–6207.
Ponomarenko, O. et al. 2006. Small-radius clean and metal-doped boron carbide nanotubes: A density functional study. *Physical Review B* 74:125421.
Popovitz-Biro, R. et al. 2001. Nanoparticles of $CdCl_2$ with closed cage structures. *Israel Journal of Chemistry* 41:7–14.
Popovitz-Biro, R. et al. 2003. $CdI_2$ nanoparticles with closed-cage (fullerene-like) structures. *Journal of Materials Chemistry* 13:1631–1634.
Qian, L. et al. 2005. Bright visible photoluminescence from nanotube titania grown by soft chemical process. *Chemistry of Materials* 17:5334–5338.
Rafailov, P. M. et al. 2005. Orientation dependence of the polarizability of an individual $WS_2$ nanotube by resonant Raman spectroscopy. *Physical Review B* 72:205436.
Rapoport, L. et al. 1997. Hollow nanoparticles of $WS_2$ as potential solid-state lubricants. *Nature (London)* 387:791–793.
Rapoport, L. et al. 2001. Slow release of fullerene-like $WS_2$ nanoparticles as a superior solid lubrication mechanism in composite matrices. *Advanced Engineering Materials* 3:71–75.
Rapoport, L. et al. 2005. Applications of $WS_2$ ($MoS_2$) inorganic nanotubes and fullerene-like nanoparticles for solid lubrication and for structural nanocomposites. *Journal of Materials Chemistry* 15:1782–1788.
Remskar, M. et al. 1996. $MoS_2$ as microtubes. *Applied Physics Letters* 69:351–353.
Remskar, M. et al. 1999. Stabilization of the rhombohedral polytype in $MoS_2$ and $WS_2$ microtubes: TEM and AFM study. *Surface Science* 435:637–641.
Remskar, M. et al. 2001. Self-assembly of subnanometer-diameter single-wall $MoS_2$ nanotubes. *Science* 292:479–481.
Remskar, M. et al. 2007. Inorganic nanotubes as nanoreactors: The first $MoS_2$ nanopods. *Advanced Materials (Weinheim, Germany)* 19:4276.
Remskar, M. et al. 2008. $WS_2$ nanobuds as a new hybrid nanomaterial. *Nano Letters* 8:76–80.
Rosenfeld Hacohen, Y. et al. 1998. Cage structures and nanotubes of $NiCl_2$. *Nature (London)* 395:336–337.
Rosenfeld Hacohen, Y. et al. 2002. Vapor-liquid-solid growth of $NiCl_2$ nanotubes via reactive gas laser ablation. *Advanced Materials (Weinheim, Germany)* 14:1075–1078.
Rosenfeld Hacohen, Y. et al. 2003. Synthesis of $NiCl_2$ nanotubes and fullerene-like structures by laser ablation: Theoretical considerations and comparison with $MoS_2$ nanotubes. *Physical Chemistry Chemical Physics* 5:1644–1651.
Rothschild, A. et al. 2000. Growth of $WS_2$ nanotubes phases. *Journal of the American Chemical Society* 122:5169–5179.
Rubio, A. et al. 1994. Theory of graphitic boron-nitride nanotubes. *Physical Review B* 49:5081–5084.
Saito, Y. et al. 1993. Growth and structure of graphitic tubules and polyhedral particles in arc-discharge. *Chemical Physics Letters* 204:277–282.
Sakai, N. et al. 2004. Electronic band structure of titania semiconductor nanosheets revealed by electrochemical and photoelectrochemical studies. *Journal of the American Chemical Society* 126:5851–5858.
Sallacan, N. et al. 2003. Nanoparticles of $CdI_2$ with closed cage structures obtained via electron-beam irradiation. *Solid State Sciences* 5:905–908.
Scheffer, L. et al. 2002. Scanning tunneling microscopy study of $WS_2$ nanotubes. *Physical Chemistry Chemical Physics* 4:2095–2098.
Schuffenhauer, C. et al. 2002. Synthesis of $NbS_2$ nanoparticles with (nested) fullerene-like structure (IF). *Journal of Materials Chemistry* 12:1587–1591.
Schuffenhauer, C. et al. 2005. Synthesis of fullerene-like tantalum disulfide nanoparticles by a gas-phase reaction and laser ablation. *Small* 1:1100–1109.

Seifert, G. et al. 1997. Boron-nitrogen analogs of the fullerenes: Electronic and structural properties. *Chemical Physics Letters* 268:352–358.
Seifert, G. et al. 2000. Novel $NbS_2$ metallic nanotubes. *Solid State Communications* 115:635–638.
Sen, R. et al. 2001. Encapsulated and hollow closed-cage structures of $WS_2$ and $MoS_2$ prepared by laser ablation at 450–1050°C. *Chemical Physics Letters* 340:242–248.
Shankar, K. et al. 2008. Highly efficient solar cells using $TiO_2$ nanotube arrays sensitized with a donor-antenna dye. *Nano Letters* 8:1654–1659.
Shneider, M. et al. 2010. The effect of tungsten sulfide fullerene-like nanoparticles on the toughness of epoxy adhesives. *Journal of Adhesion Science and Technology* 24:1083–1095.
Spahr, M. E. et al. 1998. Redox-active nanotubes of vanadium oxide. *Angewandte Chemie-International Edition* 37:1263–1265.
Srolovitz, D. J. et al. 1995. Morphology of nested fullerenes. *Physical Review Letters* 74:1779–1782.
Stephan, O. et al. 1998. Formation of small single-layer and nested BN cages under electron irradiation of nanotubes and bulk material. *Applied Physics A: Materials Science & Processing* 67:107–111.
Tahir, M. N. et al. 2006. Overcoming the insolubility of molybdenum disulfide nanoparticles through a high degree of sidewall functionalization using polymeric chelating ligands. *Angewandte Chemie: International Edition* 45:4809–4815.
Tahir, M. N. et al. 2007. Hierarchical assembly of $TiO_2$ nanoparticles on $WS_2$ nanotubes achieved through multifunctional polymeric ligands13. *Small* 3:829–834.
Tenne, R. et al. 1992. Polyhedral and cylindrical structures of tungsten disulfide. *Nature* 360:444–446.
Unalan, H. E. et al. 2008. ZnO nanowire and $WS_2$ nanotube electronics. *IEEE Transactions on Electron Devices* 55:2988–3000.
Virsek, M. et al. 2007. Raman scattering of the $MoS_2$ and $WS_2$ single nanotubes. *Surface Science* 601:2868–2872.
Vollath, D. and Szabo, D. V. 2000. Nanoparticles from compounds with layered structures. *Acta Materialia* 48:953–967.
Wang, W. et al. 2008. Aqueous noncovalent functionalization and controlled near-surface carbon doping of multiwalled boron nitride nanotubes. *Journal of the American Chemical Society* 130:8144–8145.
Whittingham, M. S. 1978. Chemistry of intercalation compounds: Metal guests in chalcogenide hosts. *Progress in Solid State Chemistry* 12:41–99.
Wu, X. et al. 2007. Single-walled $MoTe_2$ nanotubes. *Nano Letters* 7:2987–2992.
Yella, A. et al. 2008. Large scale MOCVD synthesis of hollow $ReS_2$ nanoparticles with nested fullerene-like structure. *Chemistry of Materials* 20:3587–3593.
Yin, L. W. et al. 2005a. Self-assembled highly faceted wurtzite-type ZnS single-crystalline nanotubes with hexagonal cross-sections. *Advanced Materials (Weinheim, Germany)* 17:1972.
Yin, L. W. et al. 2005b. Single-crystalline AIN (meaning aluminium nitride) nanotubes with carbon-layer coatings on the outer and inner surfaces via a multiwalled-carbon-nanotube-template-induced route. *Advanced Materials (Weinheim, Germany)* 17:213.
Yu, S. D. et al. 2007. Study of the hydrostatic pressure dependence of the Raman spectrum of W/$WS_2$ fullerene-like nanosphere with core-shell structure. *Journal of Physics: Condensed Matter* 19:425228.
Zak, A. et al. 2000. Growth mechanism of $MoS_2$ fullerene-like nanoparticles by gas-phase synthesis. *Journal of the American Chemical Society* 122:11108–11116.
Zak, A. et al. 2002. Alkali metal intercalated fullerene-like $MS_2$ (M = W, Mo) nanoparticles and their properties. *Journal of the American Chemical Society* 124:4747–4758.
Zak, A. et al. 2009. Insight into the growth mechanism of $WS_2$ nanotubes in the scaled-up fluidized-bed reactor. *Nano* 4:91–98.
Zelenski, C. M. and Dorhout, P. K. 1998. Template synthesis of near-monodisperse microscale nanofibers and nanotubules of $MoS_2$. *Journal of the American Chemical Society* 120:734–742.
Zhan, J. et al. 2004. Bulk synthesis of single-crystalline magnesium oxide nanotubes. *Inorganic Chemistry* 43:2462–2464.

Zhu, Y. Q. et al. 2000. Production of $WS_2$ nanotubes. *Chemistry of Materials* 12:1190–1194.

Zhu, Y. Q. et al. 2003. Shock-wave resistance of $WS_2$ nanotubes. *Journal of the American Chemical Society* 125:1329–1333.

Zhu, K. et al. 2006. Enhanced charge-collection efficiencies and light scattering in dye-sensitized solar cells using oriented $TiO_2$ nanotubes arrays. *Nano Letters* 7:69–74.

Zink, N. et al. 2007. Selective synthesis of hollow and filled fullerene-like (IF) $WS_2$ nanoparticles via metal-organic chemical vapor deposition. *Chemistry of Materials* 19:6391–6400.

Zink, N. et al. 2008. In situ heating TEM study of onion-like $WS_2$ and $MoS_2$ nanostructures obtained via MOCVD. *Chemistry of Materials* 20:65–71.

# 17

# *Inorganic Nanoparticles for Catalysis*

**Naoki Toshima**

**CONTENTS**



## 17.1 Introduction

Metal nanoparticles have attracted a great interest as a subject for scientific research and industrial application under the internationally big research boom of nanoscience and nanotechnology starting from the Clinton's statement in 2000, because metal nanoparticles are easily prepared even in single-nanometer size and their applications to catalysis successfully give the productive results as a fundamental research.

Metal nanoparticles already have a long history as nanometer-scale materials before the Clinton's statement. In the history of colloid chemistry, metal nanoparticles, which are called colloidal metal or metal fine particles, have been recognized as a research subject by scientists more than 150 years ago. For example, Michael Faraday reported the preparation of gold sols (colloidal dispersion of gold nanoparticles in water) in 1857 (Faraday 1857). Of course, Faraday could not measure the size of gold nanoparticles. However, the gold nanoparticles he prepared were reproduced by his successor at Royal Institution in London, J. M. Thomas, who demonstrated that the particle size is 3–30 nm in diameter (Thomas 1988).

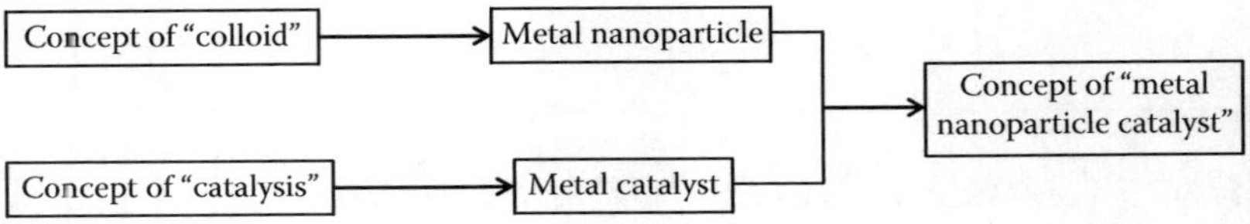


**FIGURE 17.1**
Schematic illustration of historical background to provide the concept of "metal nanoparticle catalyst."

On the other hand, metal nanoparticles in catalysts have a long history, too. Human beings had used some materials as a special additive to accelerate practical reactions in history. However, J. J. Berzelius recognized the catalysis as a special phenomenon in 1835 and Wo. Ostwald gave the definition of catalysis in 1901, when the chemistry of catalysis started practically. Wo. Ostwald also defined "colloid" as a dispersion state of materials. Thus, both concepts of "colloid" and "catalysis" have a strong relationship even from the beginning. Now, a new concept of metal nanoparticle catalysts becomes very popular in science and technology and is going to provide a new frontier in science of material conversion (cf. Figure 17.1).

Now let us go back to the original subject of catalysis. A catalyst can be defined as a reactant that is essential to the pathway or mechanism of a reaction and often appears in the rate law, but not is the overall stoichiometry of the reaction. The materials used as a catalyst can be classified into five categories, i.e., (1) metals, (2) metal oxides, (3) metal complexes, (4) other chemicals, and (5) biocatalysts. Metals are one of the most important materials for catalysis. The catalytic function (activity and selectivity) of catalysts depends on their composition and structure. The particle size is one of the factors affecting the catalytic properties. Since the catalysis occurs only on the surface of metal, the small metal particle is favored as the catalyst.

Figure 17.2 shows the relationship between the diameter (*d*/nm) of fcc (face-centered-cubic) structured Pt nanoparticle and the percentage (*r*/%) of surface atoms in total atoms constructing a nanoparticle. Actually, the percentage of surface atoms rapidly increases with decreasing the particle size, especially less than 5 nm. If the diameter is less than 2.5 nm, more than half of the atoms are located at the outermost surface layer of the particle.

The small particle has not only a large percentage of surface atoms, but also a large percentage of the active atoms. The surface atoms in fcc-structured metal nanoparticles are

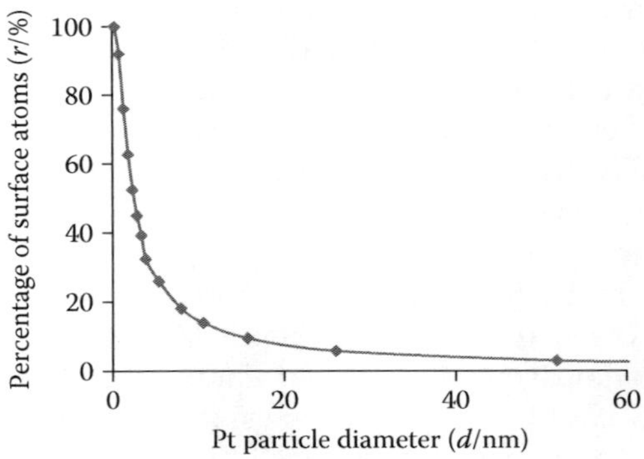


**FIGURE 17.2**
The relationship between Pt particle diameter (*d*/nm) and the percentage of surface atoms in total atoms (*r*/%).

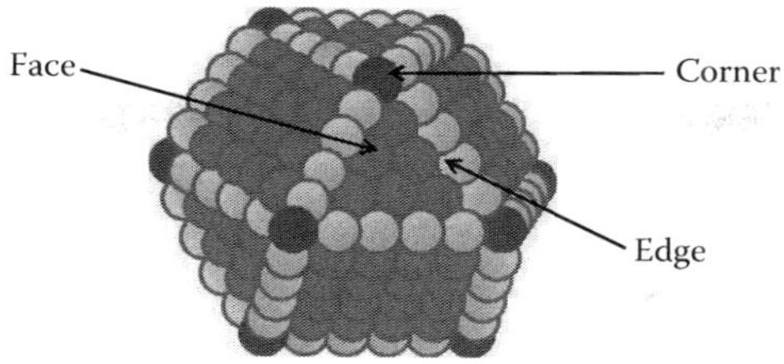


**FIGURE 17.3**
A model of fcc-structured $Pt_{561}$ cluster. The surface atoms at corner, edge, and face are shown by black, white, and gray balls, respectively.

composed of atoms at corner, edge, and face. Figure 17.3 illustrates these atoms in an fcc-structured Pt nanoparticle constructed by 561 atoms and having the diameter of 2.8 nm. As illustrated in Figure 17.3, the numbers of total surface atoms, and those at corner, edge, and face in the $Pt_{561}$ cluster ($d = 2.8$ nm) are 252, 12, 96, and 144, respectively. Whereas in the $Pt_{55}$ cluster ($d = 1.3$ nm), those numbers are 42, 12, 24, and 6, respectively. As mentioned in the previous paragraph, the percentage of surface atoms increases with decreasing the size of the metal nanoparticle. Thus, the catalytic activity per mol-metal increases with decreasing the size of metal nanoparticles. In addition, the catalytic activity normalized by surface area sometimes increases with decreasing the size of metal nanoparticle. This phenomenon can be explained by the variation of the different kinds of surface atoms, i.e., atoms at corner (vertex, or top), edge, and face.

As shown in Table 17.1, the percentage of corner and edge atoms to all surface atoms, especially the percentage of corner atoms, increases with decreasing the particle size. In addition, the corner atoms commonly have the highest number of vacant coordination position among surface atoms. If the activity of the surface atom increases with increasing the number of vacant coordination position, then the catalytic activity of the surface atoms increases in the order of face < edge < corner. Thus, the increase of the corner atom with decreasing the size could provide the very effective catalyst for such reactions, which we call structure-sensitive reactions.

**TABLE 17.1**
Number of Atoms Constructing Fcc-Structured Pt Nanoparticle

| Shell Number ($n$) | Number of Atoms | | | | | | Diameter ($d$/nm) |
|---|---|---|---|---|---|---|---|
| | Core | Shell | | | | Total | |
| | | Corner (%) | Edge (%) | Face (%) | Surface (Subtotal%) | | |
| 0 | 1 | 0 | 0 | 0 | 0 | 1 | 0.26 |
| 1 | 1 | 12 (100) | 0 (0) | 0 (0) | 12 (100) | 13 | 0.77 |
| 2 | 13 | 12 (29) | 24 (57) | 6 (14) | 42 (100) | 55 | 1.29 |
| 3 | 55 | 12 (13) | 48 (52) | 32 (35) | 92 (100) | 147 | 1.81 |
| 4 | 147 | 12 (7) | 72 (44) | 78 (48) | 162 (100) | 309 | 2.32 |
| 5 | 309 | 12 (5) | 96 (38) | 144 (57) | 252 (100) | 561 | 2.84 |
| 6 | 561 | 12 (3) | 120 (33) | 230 (64) | 362 (100) | 923 | 3.35 |
| 7 | 923 | 12 (2) | 144 (28) | 336 (68) | 492 (100) | 1415 | 3.87 |
| 8 | 1415 | 12 (2) | 168 (26) | 462 (72) | 642 (100) | 2057 | 4.39 |

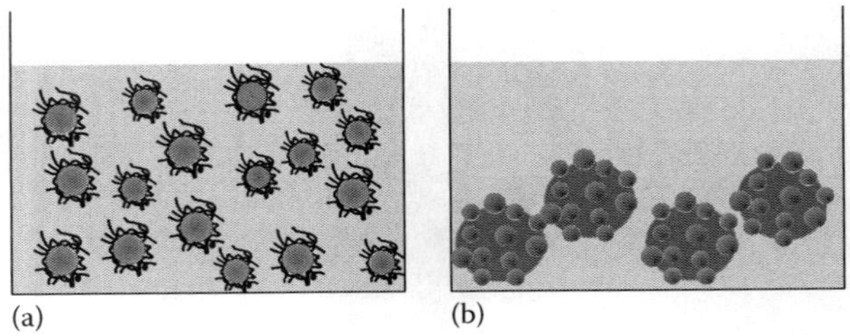


**FIGURE 17.4**
Schematic illustration of (a) homogeneous and (b) heterogeneous catalysts of metal nanoparticles.

Catalysts can be classified into two categories according to the applied reaction phase, i.e., homogeneous catalysts and heterogeneous catalysts, which can be used in a homogeneous and a heterogeneous phase, respectively. For example, metal complex catalysts and biocatalysts are usually used in solution, into which catalysts and reactants can be dissolved. In contrast, metal catalysts are often used in heterogeneous phases such as gas/solid and liquid/solid phases. For this purpose, small metal particles are immobilized on the inorganic oxide supports such as alumina ($Al_2O_3$) and silica ($SiO_2$). Solid supports have advantages not only in the separation of catalysts from reactants and products in a heterogeneous phase but also in the enhancement of catalytic functions, including life time by the dispersion and dilution of metal particles, tuning of the electronic state of catalytic sites by interactions between metal particles and supports, and so on.

In the case of metal nanoparticles, they can be used like the homogeneous catalyst as colloidal dispersion in solution as well as the heterogeneous catalyst supported on inorganic supports, as shown in Figure 17.4. When metal nanoparticles are used in a colloidal state in solution, the stability of the colloid is a key factor to get active catalysts. Colloidal dispersion can be stabilized by electric or steric repulsion between particles. In order to get the stable dispersion even under the severe reaction conditions, the electric stabilization is not enough but the steric stabilization using polymer stabilizers are effective. In the history of colloid chemistry, several researchers tried to use metal colloid as a catalyst. However, the activity could not be kept for enough long period for the reaction. Hirai et al. developed a new method to prepare stable polymer-protected metal colloids by the reduction of metal ions with alcohol in the presence of water-soluble polymers (Hirai et al. 1979, Hirai and Toshima 1986). The polymers can coordinate to the metal ions before reduction, and the coordination to metal atoms and the metal nanoparticles are kept after the reduction, as shown in Figure 17.5, which results in stable colloidal dispersions of metal.

The benefit to use polymer, especially poly(*N*-vinyl-2-pyrrolidone) (PVP), as the stabilizer should be emphasized. In the case of PVP, the coordination bond of PVP to metal nanoparticles is very weak, but PVP can coordinate to metal nanoparticles at many points. This polydentate coordination can provide a strong bond of PVP to a metal nanoparticle as a whole, which results in a stable dispersion of metal nanoparticles. On the other hand, the weak bond of individual coordination can have another benefit. When a reaction substrate approaches to a polymer-protected metal nanoparticle catalyst, the PVP-metal nanoparticle coordination bond will be broken at one site because of weak coordination bond at an individual site. However, the PVP molecule cannot go apart from the particle because the PVP molecule can coordinate to the nanoparticle at many sites (Toshima 2003). The catalysis process of the PVP-protected metal nanoparticle in colloid is illustrated in Figure 17.6 (Toshima 2004).

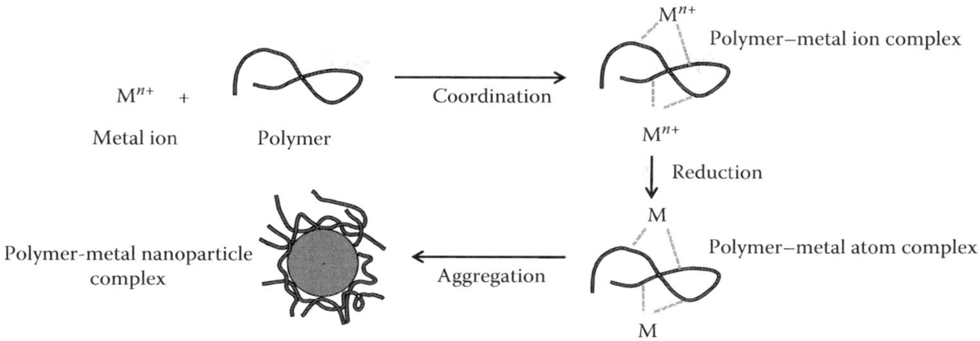


**FIGURE 17.5**
Preparation of stable dispersions of polymer-protected metal nanoparticles due to polydentate coordination starting from polymer–metal ion complexes.

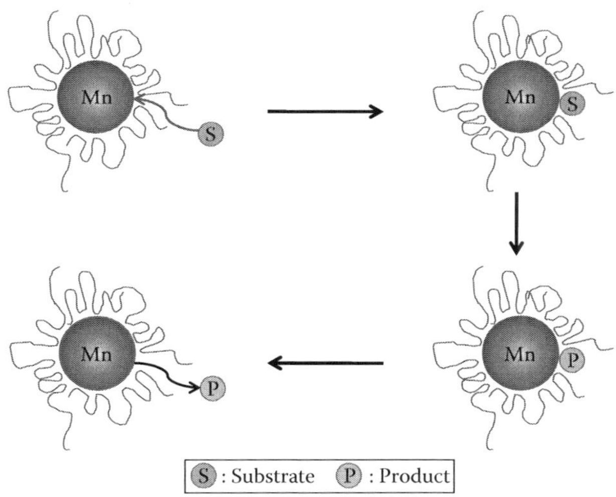


**FIGURE 17.6**
**(See color insert following page 302.)** Catalysis process on the surface of polymer-protected metal nanoparticles.

The practical metal catalysts, i.e., the metal nanoparticles supported on inorganic supports, are usually prepared by very complicated processes, so-called "impregnation method," in which metal ions in solution are impregnated into the pores of inorganic supports and then the metal ions on the supports are treated by heat and/or hydrogen to produce the supported metal catalysts. In this method, it is difficult to design the structure of the metal nanoparticles produced on the supports. In contrast, a new preparation method via metal nanoparticles is proposed recently, i.e., well-designed metal nanoparticles can be prepared as colloidal dispersion in advance, and then the prepared nanoparticles will be immobilized on inorganic supports. Although a suitable adsorption technique of colloids on supports is required in this method, it has a benefit to design the structure of metal nanoparticles in advance. In addition, the inorganic supports in nanometer scale receive an attention recently, which gives a new concept on catalysis, too.

## 17.2 General Methods for Preparation and Characterization of Metal Nanoparticles

The inorganic nanoparticle catalysts may involve various elements. However, the most popular catalytic site is given by metal nanoparticles. Thus, the preparation and characterization of metal nanoparticles will be focused and mentioned in general in this section.

### 17.2.1 Preparation of Metal Nanoparticles

In principle, the preparations of metal nanoparticles could be classified into two categories: physical and chemical methods (cf. Figure 17.7). In physical methods, evaporation or laser ablation from bulk metal is utilized to generate nanoparticles, while the reduction of metal ions to neutral atoms, followed by the aggregation of atoms to form particles, is a common strategy in chemical methods. The reduction can be conducted by conventional chemical (one- or two-phase systems), photochemical, sonochemical, electrochemical, or radiolytic energy.

In general, chemical methods have a significant advantage of being easily able to control primary structures of nanoparticles, such as size, shape, and composition, as well as to achieve mass production. The chemical method generally involves two processes in solution, i.e., (1) formation of metal atoms by the reduction of metal ions or the decomposition of metal complexes, and (2) growth of metal atoms to produce metal nanoparticles by controlling aggregation of metal atoms. In order to control the particle size and structure, the tuning of reaction conditions, especially the selection of stabilizer, is very important. At the beginning of preparation of metal colloid in history, Faraday prepared a colloidal dispersion of gold nanoparticles by reduction of gold(III) ions with white phosphorus. Recently, Turkevich prepared the gold and palladium sol by reduction with citrate (Turkevich and Kim 1970). Other reductants involve formaldehyde, hydrazine, hydrogen, hydrogen peroxide, carbon monoxide, etc. To prepare colloidal dispersions of metal nanoparticle in this method, very careful experiments are required. For example, the vessel should be completely cleaned by using steam after washing with concentrated nitric acid, and at least twice-distilled water should be used at any time as well. In this method, the stabilizers are not used during the preparation. Probably, the counter anions used for the synthesis may work to give a negative charge to the particles, which can prevent the aggregation of

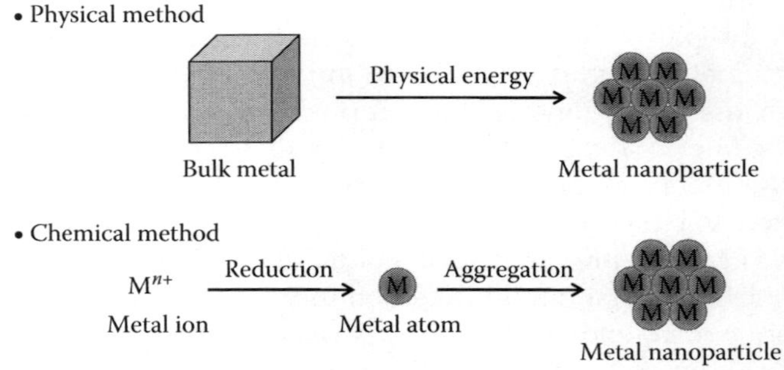


**FIGURE 17.7**
Brief concept of preparation of metal nanoparticles by physical and chemical method.

1. Alcohol reduction

$$PdCl_2 + C_2H_5OH \xrightarrow{h\nu} Pd + 2HCl + CH_3CHO$$

2. Hydrogen reduction

$$K_2PtCl_4 + 2NaOH + H_2 \longrightarrow Pt + 2NaCl + 2KCl + 2H_2O$$

3. Citrate reduction

$$2KAuCl_4 + 3C_6H_5Na_3O_7 \longrightarrow 2Au + 2KCl + 3NaCl + 3HCl + 3C_5H_4Na_2O_5 + 3CO_2$$

(Trisodium citrate) (Disodium 1,3-acetone-dicarborylate)

4. Photo-reduction

$$H_2PtCl_6 + 2H_2O \xrightarrow{h\nu} Pt + 4HCl + 2HClO$$

**FIGURE 17.8**
Typical reactions to produce metal atoms from metal salts. The reaction formulae depend on the reaction conditions.

nanoparticles forming precipitates by repulsive interaction among the negatively charged particles. Thus, the colloidal dispersions prepared by this method are not enough stable to be kept under ambient conditions. To stabilize these unstable colloids, the remaining ions are required to be removed from the system by using osmotic membranes and/or ion exchange resins. The latter are often used in recent years for their easiness. The other method to stabilize the unstable sols is the addition of stabilizers after the synthesis of the sol. Faraday used gelatin for this purpose (Faraday 1857). Synthetic water-soluble polymers like poly(vinylalcohol) and PVP are more recently used as well. This process is essential to use metal nanoparticles as homogeneous catalysts.

In contrast, metal ions are recently reduced in the presence of stabilizers to form metal nanoparticles (Hirai et al. 1979). This method provides rather stable dispersions of nanoparticles. In this process, various kinds of reduction methods can be utilized. The interaction, possibly the coordination bond formation between metal precursors and stabilizers in the reduction system, is a very important factor to produce stable colloidal dispersions of metal nanoparticles (Hirai and Toshima 1986). In this case, any kind of reductants can be applied for the reduction of metal ions, such as alcohol, ascorbic acid, hydroborate, hydrogen, etc. In addition, any kinds of extra energy such as ultra-violet (UV) and visible light, $\gamma$-ray, ultrasonic wave, etc. can be used for the reduction of metal ions or the decomposition of metal complexes. Figure 17.8 summarizes typical reactions to form metal atoms, which easily aggregate to produce colloidal dispersions of metal nanoparticles in the presence of stabilizers.

### 17.2.2 Preparation of Bimetallic Nanoparticles

A bimetallic nanoparticle is defined as a nanoparticle that contains two kinds of elements in a particle. If the particle contains three elements, it is called a trimetallic nanoparticle. The synthesis of bimetallic nanoparticles by chemical methods involves (1) simultaneous or co-reduction of two kinds of metal ions, (2) successive or two-step reduction of two kinds of metal ions, and (3) physical fusion or "self-organization" of two kinds of monometallic nanoparticles prepared in advance with or without after-treatments (Toshima 2008). In general, bimetallic alloy nanoparticles are prepared by the simultaneous reduction, while bimetallic nanoparticles with a core/shell structure are by the successive reduction. In the preparation of bimetallic nanoparticles, one of the most interesting aspects is the control of a core/shell structure, since the surface element plays an important role, especially in the catalysis of metal nanoparticles.

The simultaneous reduction of two kinds of metal ions can result in mixtures of two kinds of monometallic nanoparticles or a single product of bimetallic nanoparticles. The answer to the question which is the result depends not only on the combination of elements but also on the reaction conditions. In addition, even if the bimetallic nanoparticles are obtained, their structure, i.e., an alloy or a core/shell structure, depends on both the combination of elements and reaction conditions, too.

Miner et al. prepared monodispersed AuPt and PdPt alloys by the simultaneous reduction of the corresponding salt mixtures at various molar ratios by citrate (Miner et al. 1981). They confirmed that AuPt alloys are formed at any atomic ratio, even if the two metals show a broad miscibility gap between 2 and 85 wt% Au. The homogeneous character of the various AuPt alloys was proven by means of optical spectra, sedimentation measurements, and electron microscopy.

In contrast, the simultaneous reduction of a combination of Au and Pt ions or Pd and Pt ions with refluxing ethanol/water in the presence of poly(*N*-2-pyrrolidone)(PVP) can produce a core/shell-structured Au/Pt or Pt/Pd bimetallic nanoparticles (Toshima et al. 1989, Toshima and Yonezawa 1992). The core/shell structures were confirmed by EXAFS analyses (Toshima et al. 1991).

The formation mechanism of the core/shell structure by the simultaneous alcohol reduction of the two components of noble metal ions $M_1^+$ and $M_2^+$ (superscript "+" means cationic species) in the presence of PVP has been proposed as follows (Figure 17.9) (Toshima et al. 2008). At first, both metal ions $M_1^+$ and $M_2^+$ can coordinate to PVP (Step a). Then, one of the metal ions $M_1^+$ having a higher redox potential can be reduced at first (Step b). At this stage, another metal ion $M_2^+$ having a lower redox potential than $M_1^+$ still remains as an ion. The next step can be divided into two ways. In one path, $M_2^+$ ion can also be reduced to form atom $M_2$ (Step c). At this stage, both $M_1$ and $M_2$ exist as atoms. Then $M_1$ atoms aggregate to form a $M_1$ cluster (Step d), probably because the coordination bond between $M_1$ and PVP is weaker than that between $M_2$ and PVP. In another path, $M_1$ atoms coagulate to form an $M_1$ cluster while $M_2^+$ ions exist as ions (Step e). Then $M_2^+$ ions, which coordinate to PVP protecting $M_1$ clusters, can be reduced to form $M_2$ atoms (Step f). Thus, the $M_1$ clusters protected by PVP having $M_2$ atoms can be produced via two separate paths. At the last step, $M_2$ atoms can deposit on seed $M_1$ clusters to form $M_1$-core/$M_2$-shell bimetallic nanoparticles (Step g). Thus, to summarize, the core/shell structure can be controlled by the difference in both redox potentials of $M_1^+$ and $M_2^+$, and coordination ability of $M_1$ and $M_2$ atoms onto PVP.

The Brust method (Brust et al. 1994), proposed to prepare alkanethiol-protected Au nanoparticles using two-phase reaction with a phase transfer catalyst, can be applied to synthesize Au-based Ag, Pt, Pd, and Cu alloy bimetallic nanoparticles (Hostetler et al. 1998). This is a good method to prepare small metal nanoparticles, but the prepared nanoparticles can seldom be used as catalysts since thiols usually work as poisons for catalyses.

Late transition metal or 3d-transition metal, such as Co, Ni, Cu, is important for catalysis. The reduction of such metal ions to zero-valent metals is quite difficult because of their low redox potentials, but a production of bimetallic nanoparticles of 3d-transition metal and noble metal is not so difficult. In 1993, we successfully established a new preparation method of PVP-protected CuPt bimetallic nanoparticles (Toshima and Wang 1993,

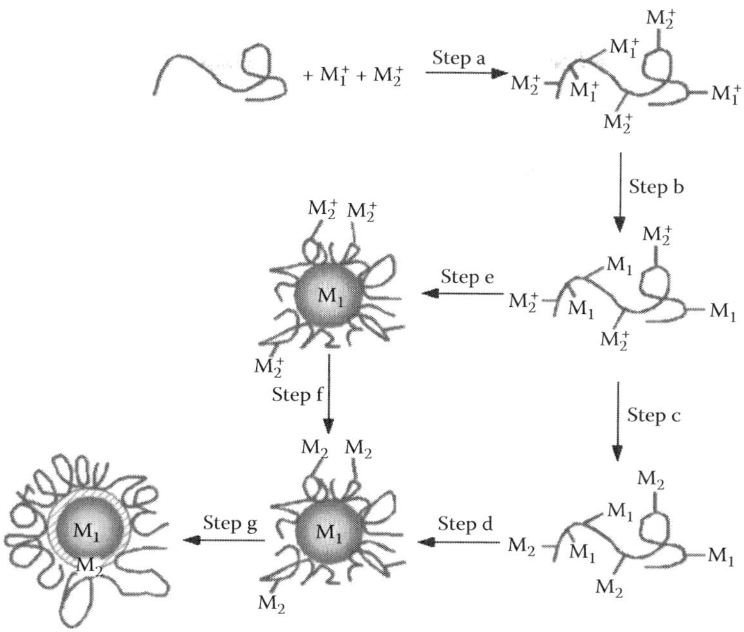


**FIGURE 17.9**
Formation mechanism of $M_1$-core/$M_2$-shell structured bimetallic nanoparticles. (Reprinted from Toshima N. et al., *Metal Nanoclusters in Catalysis and Materials Science: The Issue of Size Control*, Corain, B. et al. (eds), Elsevier, Amsterdam, the Netherlands, 2008. With permission from Elsevier.)

1994a,b). In this method, noble metal clusters, produced at first, are considered to catalyze the reduction of 3d-transition metal ions, which results in the production of bimetallic alloy nanoparticles.

There are many reports on the simultaneous reduction of two kinds of metal ions by various reductants and under various conditions in the presence of various protective reagents. Recent results were summarized in a table (cf. (Toshima et al. 2008), p. 51, Table 1).

Usually, the simultaneous reduction method provides alloyed bimetallic nanoparticles or mixtures of two kinds of monometallic nanoparticles. The bimetallic nanoparticles with a core/shell structure can form in the simultaneous reduction if the reduction is carried out under mild conditions, and there is difference in redox potentials between the two kinds of metals. As mentioned before, the metal ions with higher redox potential is first reduced to form a core part of the bimetallic nanoparticles, and then the metal with lower redox. The coordination ability may play a role in some extent to form a core/shell structure. Therefore, the simultaneous reduction method cannot provide bimetallic nanoparticles with so-called "inverted" core/shell structure, in which the metal of the core has lower redox potential.

For control of the core/shell structure, the successive reduction (or two-step reduction) method is more acceptable than the simultaneous reduction. Successive reduction involves the reduction of first metal ions, followed by the reduction of second metal ions. Second metals are usually deposited on the surface of the first metals due to the formation of the strong metallic bond, resulting in the core/shell structured bimetallic nanoparticles. However, this method is not always successful.

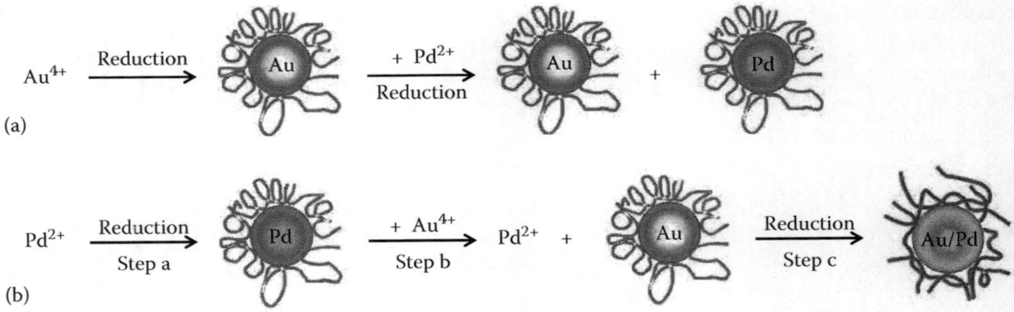


**FIGURE 17.10**
(**See color insert following page 302.**) Formation of (a) mixture of distinct Au and Pd monometallic NPs and (b) Au-Pd bimetallic nanoparticles by the successive reduction of Pd and Au ions with ethanol in the presence of PVP.

Our first attempt of a successive reduction method was utilized to prepare PVP-protected Au/Pd bimetallic nanoparticles (Harada et al. 1993a). An alcohol reduction of Pd ions in the presence of Au nanoparticles did not provide the bimetallic nanoparticles but the mixtures of distinct Au and Pd monometallic nanoparticles, while an alcohol reduction of Au ions in the presence of Pd nanoparticles can provide AuPd bimetallic nanoparticles, as shown in Figure 17.10. Unexpectedly, these bimetallic nanoparticles did not have a core/shell structure. This difference in the structure may be derived from the redox potentials of Pd and Au ions. When Au ions are added in the solution of enough small Pd nanoparticles, some Pd atoms on the particles reduce the Au ions to Au atoms, and the Pd atoms themselves are oxidized to Pd ions. The oxidized Pd ions are then reduced again by alcohol to deposit on the particles. This process may form the particles with a cluster-in-cluster structure and does not produce Pd-core/Au-shell bimetallic nanoparticles.

On the other hand, Schmid et al. successfully prepared ligand-stabilized Au-core/Pt- or Pd-shell bimetallic nanoparticles by a successive reduction (Schmid et al. 1991, 1996, Lee et al. 1995). Core of Au nanoparticles with a diameter of 18nm can be covered by Pt or Pd shells, when an aqueous solution of $H_2PtCl_6$ or $H_2PdCl_4$ was reduced with $H_3NOHCl$ in the presence of Au nanoparticles. The core/shell nanoparticles were stabilized by water-soluble $p$-$H_2NC_6H_4SO_3Na$. The original red color of the Au nanoparticles then changes to brown-black, which indicates the formation of Pt- or Pd-shell on the surface of Au nanoparticles. In the case of Au/Pt nanoparticles, a Au core was surrounded by a Pt shell of about 5nm in thickness, while a Au core was covered by the shell of well-ordered Pd atom in the case of Au/Pd nanoparticles. These core/shell structures were characterized by the energy dispersive x-ray (EDX) microanalysis. By this method, the inverted Pd-core/Au-shell bimetallic nanoparticles were also produced, probably because the nanoparticles are too large to adopt an alloy structure. The formation of PVP-protected Pd-core/Ni-shell bimetallic nanoparticles also proceeded by a successive alcohol reduction (Teranishi and Miyake 1999). Thus, the successive reduction is a successful method to prepare the core/shell structure if the bimetallic nanoparticles are enough large.

In order to realize the precise control of core/shell structures of small bimetallic nanoparticles, some problems have to be overcome. For example, the problems involve that the oxidation of the preformed metal core often takes place by the metal ions for making the shell when the metal ions have a high-redox potential, and that large islands of shell metal are produced on the preformed metal core. Therefore, we previously developed a so-called "hydrogen-sacrificial protective" strategy to prepare the bimetallic nanoparticles in the size range 1.5–5.5 nm with controllable core/shell structures (Wang and Toshima 1997). The strategy can be extended to other systems of bi- or multimetallic nanoparticles.

In general, it is not so easy to prepare bimetallic nanoparticles with an inverted core/shell structure like Pd-core/Au-shell. The sacrificial hydrogen strategy was used to construct the inverted core/shell structure, where the colloidal dispersions of Pd-cores are treated with hydrogen and then the ion solution of the second element, Au ions, is slowly added to the dispersions. This novel method, developed by us, gave the inverted core/shell structured bimetallic nanoparticles. The Pd-core/Au-shell structure was confirmed by FT-IR spectra of adsorbed CO.

Lee et al. succeeded in the preparation of Co-based bimetallic nanoparticles with core/shell structure via transmetalation reaction. The Co-core/Au-shell nanoparticles, for example, were confirmed to be almost the same in particle size with the seeded Co nanoparticle (Lee et al. 2005).

Thus, we concluded that the successive reduction method easily provides the bimetallic nanoparticles with a core/shell structure according to versatile design. For example, different reducing agents may be used for the first reduction and the second one, respectively, depending on the property of metals. In such cases, as two kinds of metals having much different redox potentials, however, "inverted" core/shell-structured nanoparticles are difficult to form even in the successive reduction. The "inverted" core/shell structure can be realized by an assistance of other technique such as the hydrogen-sacrificial protective strategy or transmetalation reaction.

Recently, we found that core/shell-structured nanoparticles with low entropy form spontaneously from the physical mixture of a dispersion of Ag nanoparticles and that of another noble metal (Rh Pd, or Pt) at room temperature. This discovery initiated from the disappearance of plasmon absorption, attributed to Ag nanoparticles, by the mixture of Ag nanoparticles with Rh nanoparticles within an hour (Hirakawa and Toshima 2003). Change of transmission electron micrography (TEM) image of the mixtures suggested the formation of pseudo-core/shell structures for the bimetallic products. We used isothermal titration calorimetry (ITC) for investigation on the forming process of the bimetallic nanoparticles (Toshima et al. 2005). The experimental results showed that the initial step of such a spontaneous process is strongly exothermic. When the alcohol dispersion of PVP-protected Rh nanoparticles with an average diameter of 2.3 nm was titrated into the alcoholic dispersion of PVP-protected Ag nanoparticles, a strong exothermic enthalpy change $\Delta H$ was observed, i.e., $\Delta H = -908$ kJ/mol for Ag (small) nanoparticle with an average diameter 10.8 nm and—963 kJ/mol for Ag (large) nanoparticles with an average diameter 22.5 nm. The strength of interaction was in the order of Rh/Ag > Pd/Ag > Pt/Ag. The strong exothermic interaction was considered as a driving force to from low entropy bimetallic nanoparticles by the physical mixture of two kinds of monometallic nanoparticles. We also revealed that exothermic interactions occur between a pair of noble metal nanoparticles themselves by using ITC. Although the detailed mechanism to form smaller core/shell particles from pseudo-core/shell aggregated particles is not clear yet, the exothermic interaction may play an important role for this realignment.

In summary, bimetallic nanoparticles, especially core/shell-structured bimetallic nanoparticles, are important candidates for effective catalysts. The core/shell structures can be created by (1) simultaneous reduction, (2) successive reduction, and (3) physical mixture. Thus, the characterization of the core/shell structure of purified bimetallic nanoparticles, especially small nanoparticles, is now a new target for the analysis.

### 17.2.3 Purification

The purification of metal nanoparticles dispersed in solution is not so easy. In traditional colloid chemistry, contamination is carefully avoided during preparation. For example, people used pure water distilled at least three times and glass vessels cleaned by steam for the preparation of colloidal dispersion. In addition, the reagents, which could not by-produce contaminants, were used for the preparation. Recently, however, various kinds of reagents were used for the reaction and protection. Thus, the special purification is often required, especially when the nanoparticles are prepared by chemical methods.

The purification methods for metal nanoparticles involve (1) evaporation, (2) centrifugalization, (3) extraction, (4) filtration, and (5) other methods.

The evaporation of volatile by-products and/or solvents is often used to obtain the solid metal nanoparticles. The residue may contain metal nanoparticles and protective reagents. When the nanoparticles are well protected by ligands or polymers, then the solid residues can be dispersed in solvents again without the coagulation of particles. When the nanoparticles are not well protected, however, the evaporation often results in the aggregation of nanoparticles.

The centrifugation is often used to separate metal nanoparticles from contaminants. If the size of nanoparticles is too small, the usual centrifuge is not sufficient for the separation. The centrifuge with super-high speed is required to get precipitates from nanoparticles. This method is also applied to get ultrafine nanoparticles by separation from the larger nanoparticles.

Extraction by an organic solvent or water can be used to separate metal nanoparticles soluble in an organic solvent or water. This technique can be applied only to the separation of the nanoparticles protected by organic ligands or polymers. The solubility of protecting reagents in the solvent is crucially important in this technique.

Conventional filtration cannot be applied to the separation of metal nanoparticles for purification. If the metal nanoparticles are protected by polymer, however, the ultrafilter, which can cut off the polymer over certain molecular weight, can be used to separate the polymer-protected metal nanoparticles. Free metal nanoparticles that are not protected by polymer can pass through the ultrafilter. Ion filter like cellulose can be used to separate ionic species from the reaction mixtures.

Other purification methods include a liquid phase chromatography, electrophoretic separation with mass spectroscopy, separation using magnetic properties, and so on. These separation methods are limited only for the metal nanoparticles having a special property useful for these purification methods.

### 17.2.4 Characterization

After purification, the metal nanoparticles are offered to characterization. The characterization techniques were well reviewed previously in literatures (Toshima and Yonezawa 1998, Toshima et al. 2008). Typical characterization methods are summarized in Table 17.2.

The most important information about nanoparticles is the size, shape, and their distributions, which crucially influence physical and chemical properties of nanoparticles. TEM is a powerful tool for the characterization of nanoparticles. TEM specimen is easily prepared by placing a drop of the dispersion of nanoparticles onto a carbon-coated copper microgrid, followed by the natural evaporation of the solvent. Even with low-magnification TEM, one can distinguish the difference in contrast derived from the atomic weight and

**TABLE 17.2**

Typical Characterization Methods of Metal Nanoparticles

| Method | Characteristic Property |
|---|---|
| TEM (HRTEM) | Size, shape, crystal structure |
| AFM, STM | Size, shape, surface structure |
| EDX, EELS | Elemental composition, valence state of element |
| UV-Vis (Tailing reflection and plasmon absorption) | Suggestion of particle size and coagulation |
| XRD | Lattice constant, average size |
| XPS | Valence state of element, elemental composition |
| EXAFS | Structure of bimetallic nanoparticles, size |
| CO-IR | Elemental composition of surface |
| Small angle x-ray scattering | Size, superstructure |
| Light scattering | Size |

the lattice direction. Furthermore, selective area electron diffraction can provide information on the crystal structure of nanoparticles.

High-resolution TEM (HRTEM) can provide the atomic-resolution real-space imaging of the nanoparticles (Wang 1998, 2000a). Although crystal structures can be surely determined by x-ray, electron, and neutron diffraction, the HRTEM is indispensable for the characterization of nanoparticles, particularly when the particle shape and composition are concerned. Scanning tunneling microscopy (STM) and atomic force microscopy (AFM) also can provide atomic-resolution images of large crystal surfaces, but they are almost impossible to clearly resolve the atomic lattices of nanoparticles because of the surface coating and the wobbling of nanoparticles under the scanning tip. HRTEM is a powerful and versatile tool that provides not only atomic-resolution lattice images but also chemical information at a spatial resolution of 1 nm or better, allowing direct identification of chemistry of a single nanoparticle (Wang and Kang 1988, Egerton 1996, Williams and Carter 1996, Wang 2000b). EDX analysis as well as electron energy-loss spectroscopy (EELS) analysis of nanoparticles are even more attractive for assessing the compositions and the valence states of constructing metal elements. If the EDX or EELS analysis of the parts of a nanoparticle can be carried out precisely, in other words, if the element distribution map can be obtained for one particle, then the structure of bimetallic nanoparticles can be estimated clearly (Matsushita et al. 2007).

In order to obtain the average size and size distribution, a size distribution histogram must be drawn by counting the diameters of at least 100, possibly more than 200 particles on an enlarged TEM photograph. If the particle has not a round shape, the average of long and short lengths should be measured. In this case, the distribution of aspect ratios should be reported, too.

The dispersion of metal nanoparticles usually has no characteristic peaks in ultraviolet and visible (UV-Vis) absorption, providing only the tailing reflection. The strength of reflection possibly depends on the size of nanoparticles and the extent of coagulation of nanoparticles in dispersion. In the case of nanoparticles of coinage metal like Au, Ag, and Cu, the dispersions have the respective plasmon absorption in a visible region. The plasmon absorption peak depends on not only the extent of coagulation (separation length of two adjacent particles) but also the kind of metal. Thus, the UV-Vis absorption spectra give an exact evidence for characterizing Au, Ag, Cu, and Hg nanoparticles.

Another important feature of UV-Vis measurement is to provide us the useful information about formation processes from metal ions to metal nanoparticles. During reduction of metal ions to produce the corresponding zero-valent nanoparticles, the color of the solution is drastically changed, i.e., plasmon absorption appears by the formation of zero-valent nanoparticles while the absorption of metal ions disappears by the reduction. Therefore, UV-Vis spectroscopy is useful to confirm both the degree of consumption of precursors by monitoring their ligand-to-metal or metal-to-ligand charge transfer transitions and the formation of band structures of nanoparticles by monitoring the plasmon band or the broad tailing absorption in the range from UV to visible region derived from the inter- and intra-band charge transfer transitions.

In 1993, we examined the formation processes of PVP-protected AuPt bimetallic nanoparticles by in-situ UV-Vis spectroscopy during the reduction (Yonezawa and Toshima 1993). In the case of PVP-protected AuPt bimetallic system, Au(III) ions are reduced first accompanying a decrease of peak at ca. 320 nm, followed by the reduction of Pt(IV) ions, decreasing in intensity of the peak at ca. 265 nm. The order of the reduction is consistent with the difference of standard redox potentials, i.e., 1.002 V for $[AuCl_3^-]$/Au and 0.68 V for $[PtCl_4^{2-}]$/Pt. After complete reduction of all Au(III) or Pt(IV) ions, the Au atoms aggregate first, followed by the deposition of the Pt atoms, indicated by the UV-Vis spectrum where the plasmon band at ca. 540 nm due to Au nanoparticles increases first, and then decreases accompanying increase of the plasmon band at ca. 370 nm due to Pt nanoparticles. According to these results in UV-Vis absorption spectra, the formation processes are proposed as schematically illustrated in Figure 17.11.

X-ray diffraction (XRD) provides useful information on crystal phase and lattice constant as well as average particle size of nanoparticles. Usually, the lattice constants of metal nanoparticles are the same as those of bulk metal. However, if the nanoparticles are so small, e.g., less than 2 nm in diameter, then the lattice constant or interatomic distance has

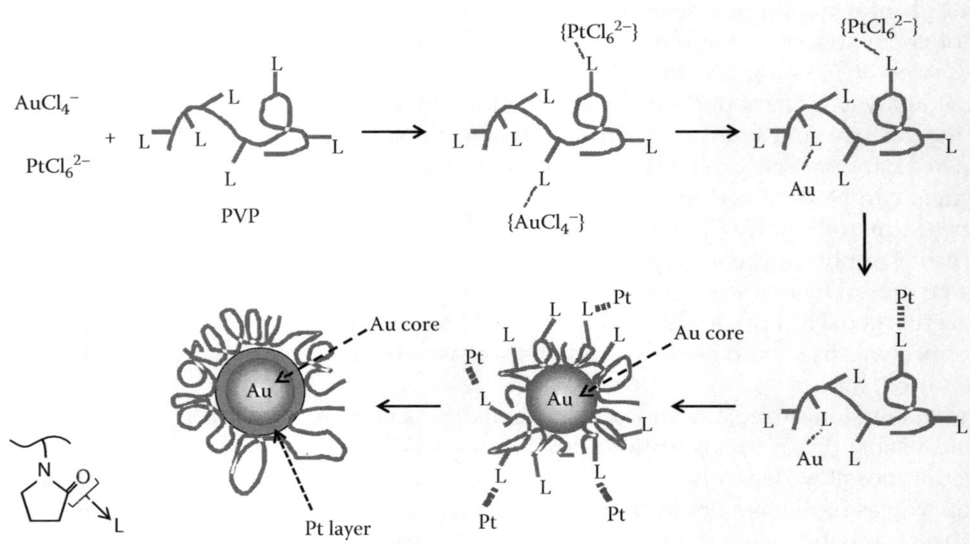


**FIGURE 17.11**
**(See color insert following page 302.)** Proposed formation process of PVP-protected AuPt bimetallic nanoparticle. (Reprinted from Yonezawa, T. and Toshima, N., *J. Mol. Catal.*, 83, 167, 1993. With permission from Rightslink.)

a trend to be a little shortened compared with that of bulk metal. This is considered due to that the number of neighboring atoms interacting with a particular atom is smaller in tiny nanoparticles than in the bulk. In the case of bimetallic nanoparticles, XRD is important to confirm whether the bimetallic nanoparticles adopt an alloy structure or not. Generally, an alloy consisting of two kinds of metals shows diffraction peaks between those of two pure metals.

In the case of CuPd (2:1) bimetallic nanoparticles, the XRD peaks of PVP-protected CuPd nanoparticles appeared between the corresponding diffraction lines of Cu and Pt nanoparticles (Toshima and Wang 1993). Thus, the bimetallic alloy phase was clearly found to be formed in CuPd (2:1) bimetallic nanoparticles. Ag-core/Rh-shell bimetallic nanoparticles, which formed by simple physical mixture of the corresponding monometallic ones, can be also characterized by XRD coupled with TEM (Hirakawa and Toshima 2003). The XRD and TEM showed that the bimetallic nanoparticles with Ag-core/Rh-shell structure spontaneously form by the physical mixture of Ag and Rh nanoparticles. It has to be mentioned that, in the case of enough small nanoparticles, the lattice constant or interatomic distance of core/shell-structured bimetallic nanoparticles is nearly the same as that of alloy-structured bimetallic ones. It is because the heteroepitaxial growth occurs in the interface between the core and the shell within one or two atomic layers from the interface. If the particle is enough large, different respective lattice images can be observed in the core and shell part of a particle by HRTEM.

The sharpness of XRD peaks is corresponding with the size of metal nanoparticles. Sherrers's equation is used to estimate the crystalline size of metal nanoparticles. Note that the size estimated from XRD peak width is sometimes larger than the size measured by TEM, especially when the size is very small. If the size estimated from XRD peak width is smaller than that directly measured by TEM, the particles could be polycrystalline.

Generally, one can obtain the surface (<10 nm) information such as the kind of metals and their valences by means of x-ray photoelectron spectroscopy (XPS). In the case of metal nanoparticles, the information on valences of metals and the ratio of metals with the respective valences can be obtained by the XPS measurement. Especially, for bimetallic nanoparticles, XPS is a good tool to get the information on the ratio of two metals near the surface of nanoparticles. Since the size of nanoparticles is enough small, not only the shell elements but also core ones can be observed by XPS in the case of core/shell-structured bimetallic nanoparticles. Since the shell element is considered to be more sensitive than the core one, however, the ratio of two elements observed by XPS can give some suggestion on a core/shell structure.

On the other hand, the XPS data near the Fermi level provide us the valuable information about the band structures of nanoparticles. XPS spectra near the Fermi level of the PVP-protected Pd nanoparticles, Pd-core/Ni-shell (Ni/Pd = 15/51, 38/561) bimetallic nanoparticles, and bulk Ni powder were investigated by T. Teranishi et al. (Teranishi and Miyake 1999). The XPS spectra of the nanoparticles become close to the spectral profile of bulk Ni as the amount of the deposited Ni increases. The change of the XPS spectrum near the Fermi level, i.e., the density of states, may be related to the variation of the band or molecular orbit structure. Therefore, the band structures of the Pd/Ni nanoparticles at Ni/Pd > 38/561 are close to that of the bulk Ni, which greatly influence the magnetic property of the Pd/Ni nanoparticles. PtRu bimetallic nanoparticles, prepared by w/o reverse microemulsions of water/Triton X-100/propanol-2/cyclohexane (Zhang and Chan 2003), were characterized by XPS and other techniques. The XPS analysis revealed the presence of Pt and Ru metal as well as some oxide of ruthenium.

It should be mentioned that XPS measurement of metal nanoparticles can be inhibited by organic materials, if the metal nanoparticles are covered by a lot of organic materials. In this case, the organic materials have to be removed without damaging the metal nanoparticles. For this purpose, Ar plasma or heat treatment is applied with care.

Extended x-ray absorption fine structure (EXAFS) analysis is a powerful spectroscopic method for structural analysis, which has been extensively applied to the problem of structure determination in nanoparticles, especially bimetallic nanoparticles (Sinfelt et al. 1984). The x-ray absorption spectrum of an element contains absorption edges corresponding to the excitation of electrons from various electronic states at energies characteristic of element. The oscillations (fine structure) observed at x-ray energy above an edge can be extracted from the x-ray absorption spectrum, which, after data manipulation, is best fit to a computed EXAFS spectrum for a model structural environment for the absorbing atom. The EXAFS is element specific and structure sensitive, and gives information on the number and identity of neighboring atoms and their distances from the absorbing atom. The information usually sought in an EXAFS measurement comprises the number of scattering atoms of each type and their distances from the absorbing atom. When multiple elements are present, they can be analyzed both as the absorbing atom and as the scattering atoms. If synchrotron radiation is used as x-ray source, EXAFS data acquisition is enormously shortened, and under favorable circumstances, an absorption spectrum can be obtained in less than an hour. Higher concentration of sample is favorable for EXAFS analysis.

Previously, the structural determination of bimetallic nanoparticles was carried out by the EXAFS measurements. The PVP-protected Pd/Pt(4/1) and Pd/Pt(1/1) bimetallic nanoparticles prepared by means of a simultaneous reduction of $PdCl_2$ and $H_2PtCl_6$ have a mean diameter of ~1.5 nm with a quite narrow size distribution, indicating that each nanoparticle is composed of 55 metal atoms (magic number) (Toshima et al. 1991). In the case of Pd/Pt(4/1) bimetallic nanoparticles, the coordination number of Pt atoms around the Pt atom suggests that the Pt atom coordinates predominantly to the other Pt atoms. Moreover, the coordination numbers are quite different from those calculated for the random model, where 42 Pd atoms and 13 Pt atoms are located completely at random. If 42 Pd atoms are located on the surface and the other 13 Pt atoms are at the core of the fcc-structured nanoparticles, then the Pd/Pt ratio is almost 4/1 and the coordination numbers calculated on the basis of the Pt-core model are quite consistent with the values observed from EXAFS. We succeeded in proposing a model structure by EXAFS analysis because our target bimetallic nanoparticles were homogeneous in size and structure. Other structural analyses by means of EXAFS were carried out for Pd/Rh (Harada et al. 1993b), Au/Pd (Toshima et al. 1992), NiPd (Lu et al. 1999) nanoparticles, and so on. The NiPd bimetallic nanoparticles were first proposed to have an alloy structure, but later proved to have heterobond-phillic structure (Bion et al. 2002). Recently, the characterization of bimetallic nanoparticles by EXAFS were extensively reported (Garcie-Gutierrez et al. 2005, Chen et al. 2006a,b).

The surface composition and structure of bimetallic nanoparticles are crucially important for their catalytic property as well as their optical property. IR measurement of carbon monoxide (CO) adsorbed on surface metals (CO-IR) is utilized for this purpose. CO is adsorbed on metals not only on-top sites but also in twofold or threefold sites, depending on the kinds of metals and their surface structures. The dramatical changes of wave number of adsorbed CO occur depending on the binding structure (Bradley et al. 1991, 1992, 1995, 1996, de Caro and Bradley 1997).

CO-IR measurement was performed on Pt/Pd bimetallic nanoparticles with core/shell structures and characterized their structures as Pd-core/Pt-shell (Wang and Tohima 1997). Au/Pd bimetallic nanoparticles were also characterized by the CO-IR (Shiraishi et al. 2003).

Solla-Gullon et al. (2003) carried out FT-IR experiments of adsorbed CO for PdPt nanoparticles prepared by the reduction of $H_2PtCl_6$ and $K_2PdCl_4$ with hydrazine in a w/o microemulsion of water/poly(ethyleneglycol) dodecylether (BRIJ6R)30)/*n*-heptane. The experiments gave information on the relative amount of linear- and bridge-bonded CO, which is known to depend on the surface distribution of the two elements.

## 17.3 Characteristic Properties of Metal Nanoparticle Catalysts

As mentioned in the previous section, metal nanoparticles can be utilized as a homogeneous and a heterogeneous catalyst. The catalytic properties of the nanoparticle catalysts depend on (1) surface area, (2) particle size and shape, (3) metal and composition, (4) protective organic material, and/or (5) inorganic support. These characteristic properties are reviewed in this section.

### 17.3.1 Surface Area

Ostwald has defined colloidal dispersions in terms of the size of the particles dispersed in the medium. A colloidal dispersion should have particle sizes within the range from 1 nm to 100 nm. The number of atoms involved and the surface area of a particle in this size range can be calculated. In the case of Pt, if spherical colloidal particles of atomic radius 0.138 nm are close-packed (fcc-structured), then particles of diameter 1 nm and 10 nm would contain 48 and 48,000 atoms, respectively, as mentioned in the previous section. In other words, the number of particles obtained from 1 mol of Pt is proportional to $r^{-3}$, where $r$ is the radius of the particle. The surface area of a particle is proportional to $r^2$. Thus, the total surface area of the particles contained in 1 mol of Pt is inversely proportional to $r$, as illustrated in Figure 17.12.

In the case of homogeneous catalysts, all the surface of metal nanoparticles can be used directly as active sites in catalytic reactions. In the case of heterogeneous catalyst, however, the parts of surface of metal nanoparticles are covered by inorganic supports or inhibited from the approach of reaction substances by the wall of inorganic supports. Thus, the total surface area is not proportional to $r$ but depends on the supporting conditions. The real surface area can be measured directly by the adsorption of gaseous molecules according to Brunauer, Emmet, Teller (BET) method.

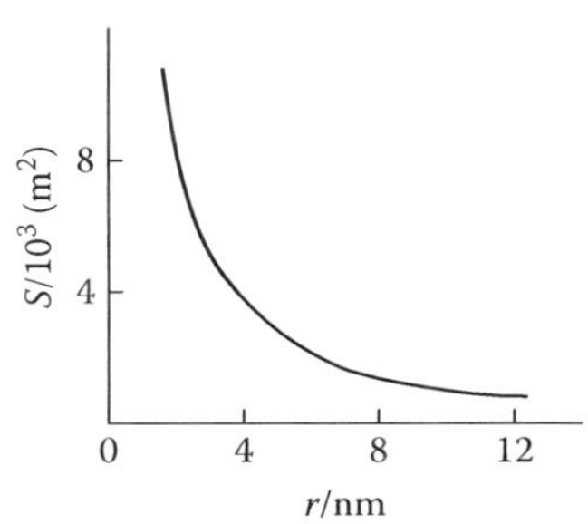


**FIGURE 17.12**
The relationship between the radium ($r$) of an individual platinum nanoparticle and total surface area ($S$) of the particles contained in 1 mol of platinum.

### 17.3.2 Particle Size and Shape

One of the most important characteristics of metal nanoparticle catalyst is its small particle size. Generally speaking,

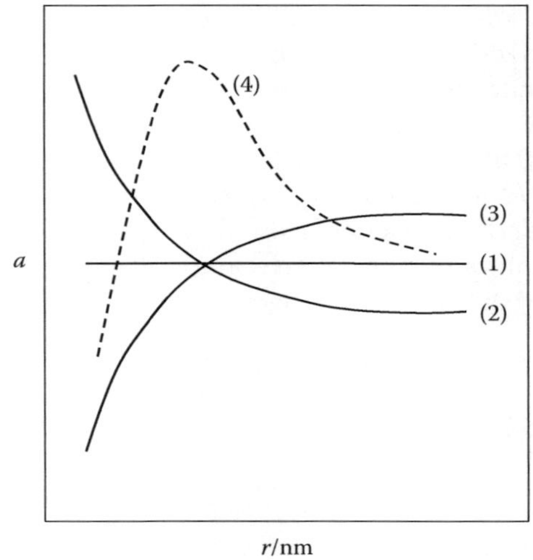


**FIGURE 17.13**
Effect of particle radium ($r$) on catalyst activity ($a$) per atom. In the curves illustrated, the catalytic activity is independent of the particle size in curve 1, it increases (curve 2) or decreases (curve 3) with a decrease in particle size, or exhibits a maximum at a definite size of particle (curve 4).

the particle size of a metal catalyst can influence both the catalytic activity and selectivity in chemical reactions. Particle size effects can be classified into four groups, as illustrated in Figure 17.13. The catalytic activity per atom (1) is independent of particle size, (2) increases with a decrease of particle size, (3) decreases with a decrease of particle size, and (4) exhibits a maximum at a definite size of particle. These effects are also observed in reaction selectivity.

Because the catalysis occurs only on the surface of metal particle, and total surface area increases with a decrease of particle size, the catalytic activity per atom increases with a decrease of particle size. If the particle size is too small, however, the character of the surface atoms of the particle may change. For example, the ratio of atoms at corner, edge, and face in the surface (shell) atoms can vary with the size as precisely discussed in Table 17.1. These ratios may vary also by altering the crystal structure even when the average particle size is kept constant.

Colloidal dispersions of PVP-protected Rh nanoparticles, prepared by refluxing solutions of $RhCl_3$ in alcohols (Hirai et al. 1976), work as catalysts for the hydrogenation of olefins. The catalytic activity depends on the particle size, which can be altered by altering the preparation conditions (Hirai et al. 1978). The most interesting observation arising from these results is that the rate of hydrogenations of an internal olefin is faster than that of the corresponding terminal olefins when a 0.9 nm size PVP-protected Rh nanoparticles are employed as the catalyst. The typical results are show in Figure 17.14 for mesityl oxide (4-methyl-3-penten-2-one, an internal olefin) and methylvinylketone (3-buten-2-one, a terminal olefin).

More drastic size effect was observed in the aerobic oxidation of $p$-hydroxybenzyl alcohol with PVP-protected Au clusters (Tsunoyama et al. 2006). Au in the bulk state is known not to have a catalytic activity, but Au clusters in small size have been discovered as a very active catalyst. The seed clusters with a diameter of $1.3 \pm 0.3$ nm was prepared by reducing $AuCl_4^-$ with $NaBH_4$ in a low-temperature aqueous solution of PVP. Subsequent reduction of more $AuCl_4^-$ by $Na_2SO_3$ in the presence of the seed clusters yielded a series of larger Au clusters. The size effect is shown in Figure 17.15.

The colloidal dispersion of PVP-protected Pt nanoparticles catalyzed the photochemical generation of hydrogen from $Na_2EDTA$ solutions in the presence of Ru $[(bpy)_3]^{2+}$ and methyl viologen, where the colloidal dispersion is advantageous to photoreaction because of its transparency. The rate of hydrogen generation was dependent on the particle size of Pt, as illustrated in Figure 17.16, a maximum rate being observed with a catalyst of ca. 3 nm particle size (Toshima et al. 1981).

Selectivity of the reaction is also influenced by the particle size. For example, the selectivity of partial hydrogenation of cyclopentadiene to produce cyclopentene catalyzed by PVP-protected Pd nanoparticles increases with decrease in Pd particle size (Hirai et al. 1985). The shape or crystal structure of particles also influences the catalytic activity and selectivity. For example, rods or wires, obtained by growth of (111) face of a particle, usually have a large area of (100) face and small (111) face. Thus, they could have a high catalytic activity if (100) face can provide an active site. Thus, the studies on the effect of particle shape or crystal structure on the catalytic activity and selectivity are still in progress. This kind of researches may increase in future.

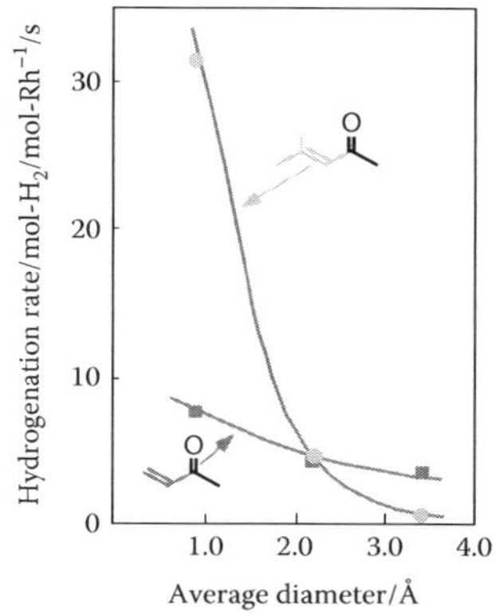


**FIGURE 17.14**
Size effect of catalytic activity of PVP-protected Rh nanoclusters for hydrogenation of 4-methyl-3-penten-2-one (circle) and 3-buten-2-one (square). (Reprinted from Toshima, N., *Shokubai*, 27, 488, 1985. With permission from Catalysis Society of Japan.)

### 17.3.3 Metal Composition

The catalytic activity and selectivity of metal certainly depends on the kind of the metal and additives. So, a trial-and-error has been repeated to discover the best catalyst for

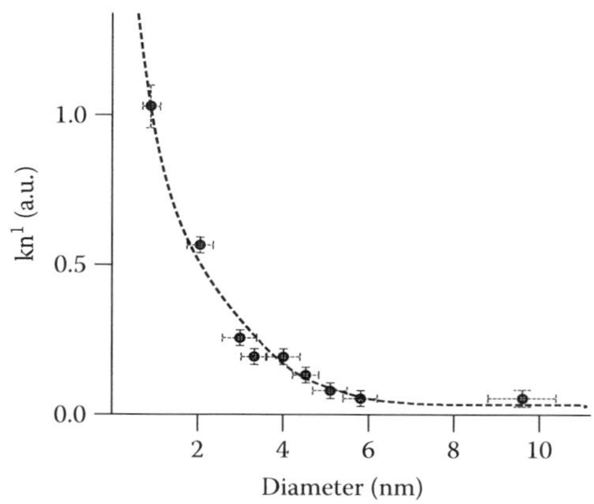


**FIGURE 17.15**
Rate constants per unit surface area as a function of cluster size for oxidation of *p*-hydroxybenzyl alcohol. Error bars in the horizontal and longitudinal axes of panel represent standard deviations of the core size and rate constant obtained from three independent batches of samples. The curve is a guide for the eye. (Reprinted from Tsunoyama, H. et al., *Chem. Phys. Lett.*, 429, 528, 2006. With permission from Rightslink.)

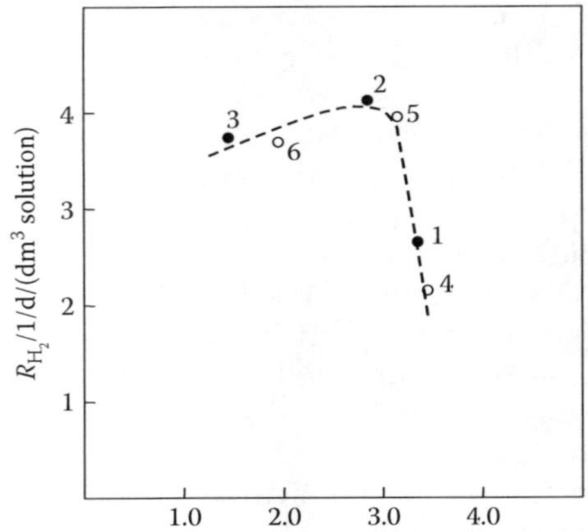


**FIGURE 17.16**
The relationship between the rate of photochemical evolution of hydrogen ($R_{H_2}$) and the average particle size (*r*) of poly(vinylpyrrolidone) (PVP)- or poly(vinyl alcohol) (PVA)-protected colloidal platinum catalysts. Reaction conditions: [Ru(bpy)$_3^{2+}$] = 5 × 10$^{-5}$ mol/dm$^3$; [MV$^{2+}$] = 5 × 10$^{-2}$ mol/dm$^3$; [Na$_2$ EDTA] = 5 × 10$^{-2}$ mol/dm$^3$; [Pt] = 6.6 × 10$^{-5}$ mol/dm$^3$; 20 mL, λ > 390 nm. 1, Pt-PVP-MeOH/$H_2O$; 2, Pt-PVP-EtOH/$H_2O$; 3, Pt-PVP-EtOH/NaOH; 4, Pt-PVA-EtOH/$H_2O$; 5, Pt-PVA-EtOH/$H_2O$, NaOH; 6, Pt-PVA-MeOH/$H_2O$. (Reprinted from Toshima, N. et al., *Chem. Lett.*, 793, 1981. With permission from The Chemical Society of Japan.)

practical reactions. There are more than 100 elements in the periodical table. If two elements are used to create a catalyst, more than 10,000 combinations are available. In practical catalysts, after the main metal element is chosen by trial-and-error, the second and the third elements are used for additives. The second and the third elements may form an alloy with the main element or sometimes distribute locally in the main element catalyst. The catalytic property may vary depending on the complete distribution or maldistribution. Some combinations of elements cannot form an alloy, which results in the maldistribution of the second element.

In the case of bimetallic nanoparticles, even the combination of two elements, which cannot form an alloy in bulk, can provide alloy nanoparticles. Thus, novel catalytic properties may be achieved by using bimetallic nanoparticles. In addition, more precise structures may be constructed by using nanoparticles. Figure 17.17 shows the cartoon to produce bimetallic catalysts of element A and B in bulk and in nanoparticle. Not only the increase in total surface area but also the construction of a designed structure in catalysts can be achieved by using nanoparticles. For example, a core/shell structure with enough thin shell layer could be constructed in nanoparticles. In the core/shell structure, only the shell metal is located on the surface and works as an active site of the catalyst, and the core metal can influence the electronic property of the shell layer.

It is generally said that the additives have an effect on the activity and selectivity of metal catalysts in two ways, i.e., electronic (a ligand effect) and geometrical (an ensemble effect) ways. In the core/shell-structured bimetallic nanoparticles, the core metal can have a ligand effect on the shell metal. In other words, the electronic property of surface atoms in the shell can be varied by charge transfer between core and shell metals. If two elements are located on the surface in the bimetallic nanoparticles, a substrate can interact with both elements, which may result in a new catalytic effect on the reaction of the substrate. This ensemble effect is geometrically controlled. Thus, both elements should be located in neighbor on the surface of catalysts. When both elements are

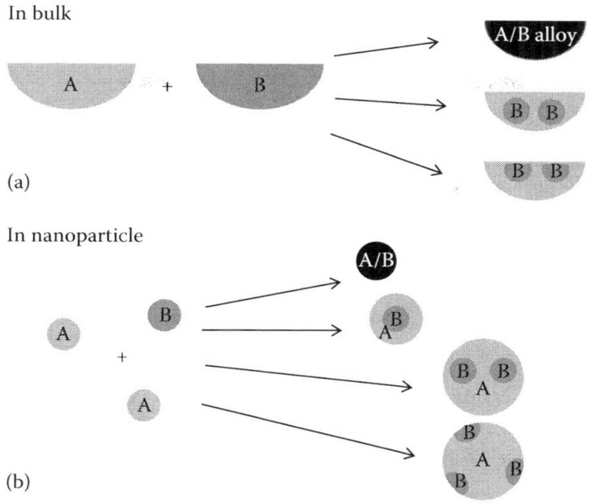


**FIGURE 17.17**
Schematic illustration of (a) bulk and (b) nanoparticle catalysts produced by combination of element A and B.

located in neighbor on the surface, however, the main metal can be influenced not only by a geometrical ensemble effect but also by an electronical ligand effect. In contrast, only the ligand effect can be realized in the case of core/shell-structured bimetallic nanoparticles.

Now let us show practical examples. We successfully prepared colloidal dispersions of Pt/Pd bimetallic nanoparticles by the simultaneous reduction of the corresponding metal ions in a refluxing solution of water/ethanol (Toshima et al. 1989). The prepared dispersions were used as a good catalyst for the hydrogenation of olefins. The catalytic activity for hydrogenation per metal is plotted against the composition (Pd ratio in %) in Figure 17.18 (Toshima et al. 1991). Although the Pd nanoparticle is known as an active catalyst and Pt as poor, Figure 17.18 reveals that the bimetallic nanoparticle catalyst at 80% Pd mole ratio or at mole ratio of Pd/Pt = 4/1 has the highest activity. Exactly at this ratio, the bimetallic nanoparticles have a complete core/shell structure. In other words, 13 atoms of Pt form a core and 42 atoms of Pd cover the core to form a one-atomic layer of shell, since the bimetallic nanoparticles have an average size of 1.4 nm, which is consistent with a particle composed of 55 (magic number) atoms. Thus, although Pt has poor activity for olefin hydrogenation, the existence of Pt core can improve the catalytic activity of Pd on the surface by decreasing the electronic density with electron transfer from Pd to Pt.

Another example is illustrated in Figure 17.19 for Cu/Pd bimetallic nanoparticle catalysts (Toshima and Wang 1994b). In this case, both Cu and Pd atoms are located on the surface of bimetallic nanoparticles. Since Cu and Pd are known to work as a catalyst for the hydrolysis of acrylonitrile to acrylamide and partial hydrogenation of cyclooctadiene to cyclooctene, respectively, the bimetallic nanoparticles work as catalysts for both reactions depending on the mole ratio of Pd/Cu. In fact, at high mole ratio of Pd, the bimetallic nanoparticles work as a catalyst for the partial hydrogenation of diolefin, but at high mole ratio of Cu, they work as a catalyst for the hydrolysis of acrylonitrile. In addition, the bimetallic nanoparticles containing more Cu than Pd are much more active than pure Cu nanoparticles as catalysts. This high catalytic activity is attributed to an ensemble effect of

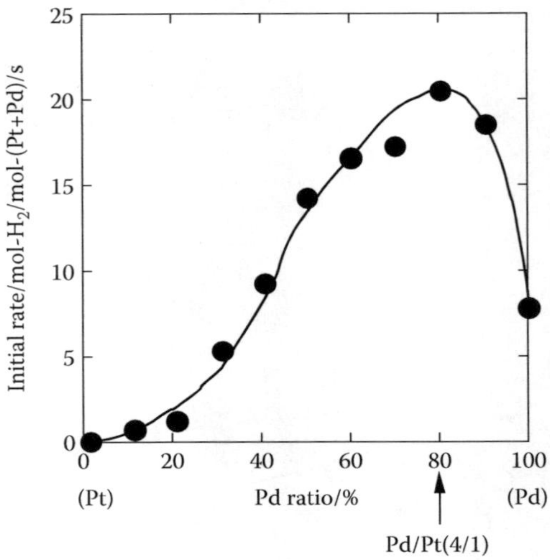


**FIGURE 17.18**
Relationship between the Pd composition and the catalytic activity of core/shell-structured Pd/Pt bimetallic nanoparticles. (Reprinted from Toshima, N. et al., *J. Chem. Soc. Faraday Trans.* 89, 2537, 1993. With permission from Royal Society of Chemistry.)

Cu and Pd atom adjacent to each other. This ensemble effect is illustrated in Figure 17.19b, where the Pd atom may coordinate to the olefinic part of acrylonitrile and the Cu atom to the nitrile part via water molecule (Toshima and Wang 1994a). This kind of geometrical effect can be achieved only when Cu and Pd atoms are adjacent to each other on the surface of a bimetallic nanoparticle.

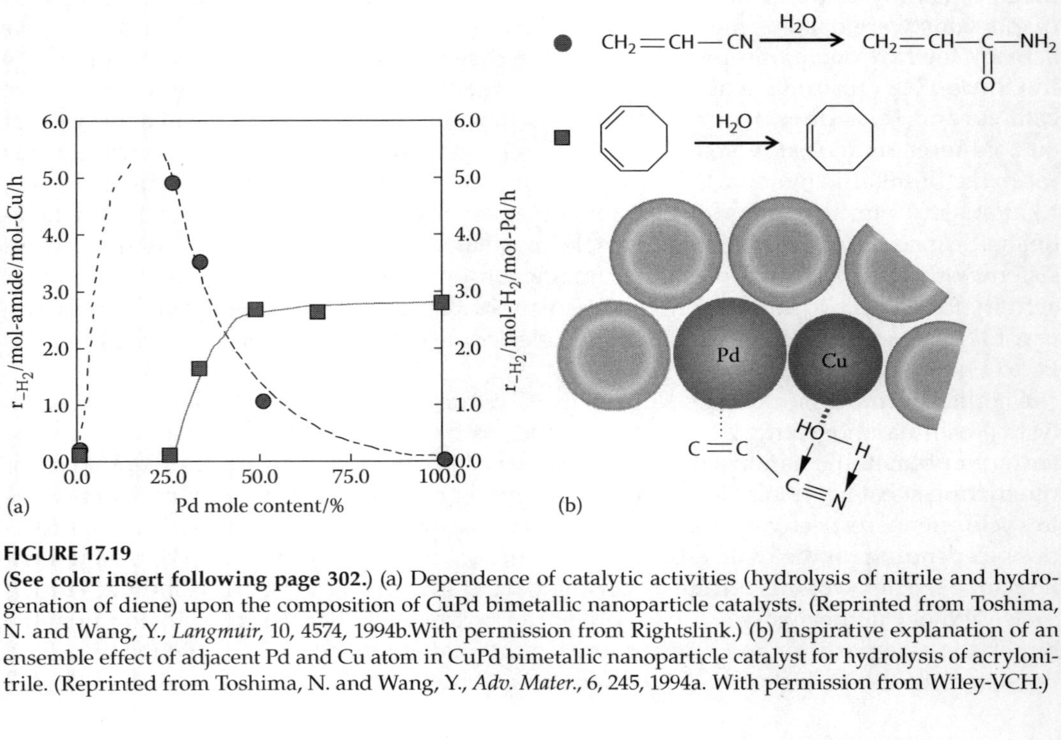


**FIGURE 17.19**
**(See color insert following page 302.)** (a) Dependence of catalytic activities (hydrolysis of nitrile and hydrogenation of diene) upon the composition of CuPd bimetallic nanoparticle catalysts. (Reprinted from Toshima, N. and Wang, Y., *Langmuir*, 10, 4574, 1994b.With permission from Rightslink.) (b) Inspirative explanation of an ensemble effect of adjacent Pd and Cu atom in CuPd bimetallic nanoparticle catalyst for hydrolysis of acrylonitrile. (Reprinted from Toshima, N. and Wang, Y., *Adv. Mater.*, 6, 245, 1994a. With permission from Wiley-VCH.)

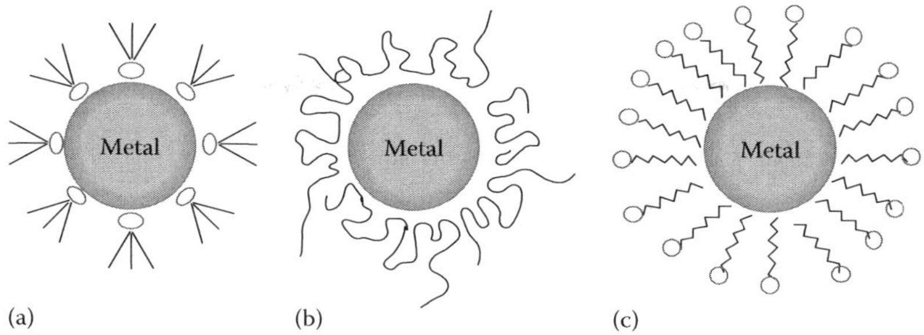


**FIGURE 17.20**
Metal nanoparticles stabilized by ligand molecules with low molecular weight (a), polymers (b), and surfactants (c).

### 17.3.4 Protective Organic Material

Protective organic materials like organic polymers, organic ligands, and organic surfactants play an important role to stabilize the dispersions of metal nanoparticles. The protected metal nanoparticles can be illustrated as shown in Figure 17.20. Without these organic protectants, the metal nanoparticles cannot work as homogeneous catalysts. If the dispersions are not stable enough, the nanoparticles can form precipitates, which results in poor catalytic activity. Thus, the stability of nanoparticles in solution is a very important factor if they are used as a homogeneous catalyst.

For this purpose, protective polymers are often used. The advantage of polymers as protective reagents was already mentioned previously using Figure 17.6, when the metal nanoparticles were used as homogeneous catalyst in dispersions. The protective function of polymers has been expressed quantitatively in terms of its "gold number" or "protective value," when nanoparticles are dispersed in water. The gold number is the amount of the protective colloid in milligrams, which just prevents 10 $cm^3$ of a red gold sol from changing color to violet on addition of 1 $cm^3$ of a 10% aqueous solution of NaCl. The smaller the gold number, the stronger is the protective function of the polymer (Zsigmondy 1901, Zsigmondy and Thiessen 1925). The protective value is the weight of a red gold sol in grams, which can just be protected from aggregation by 1 g of the protective colloid on addition of a 1% NaCl solution. Thus, the larger the protective value, the greater is the protective function. The gold number is inversely proportional to the protective value (Williams and Chang 1951, Thiele and von Levern 1965) The gold numbers and protective values of typical protective colloids are summarized in Table 17.3 (Hirai and Toshima 1986).

The protective value or gold number could be useful as a measure of the stable formation of colloidal dispersions in water. When Pt colloids were prepared from $H_2PtCl_6$ by reduction through refluxing in ethanol/water, polymers with a large protective value functioned well as protective colloids and produced polymer-protected Pt nanoparticles. However, when polymers with a small protective value were employed, precipitates or complexes were formed. Note that these values were measured in an aqueous solution. So, they cannot be applied to the dispersions of metal nanoparticles in a hydrophilic organic solvent. In this case, hydrophobic materials could be more useful for dispersions than the hydrophilic materials shown in Table 17.3 (Toshima and Liu 1992).

**TABLE 17.3**

Gold Numbers and Protective Values of Typical Protective Colloids (Hirari and Toshima 1986)

| Protective Colloid | Protective Value | Gold Number |
|---|---|---|
| *Natural materials* | | |
| Gelatine | 90 | 0.005, 0.03 |
| Gum arabic | | 0.2 |
| Agar agar | | 1.1 |
| Algin acid amid | 0.2 | 2 |
| Sodium oleate | | 4.0 |
| Sodium alginate | 0.04 | 10 |
| Pepsin, trypsin | 0.04 | 10 |
| Starch | | 16 |
| *Synthetic materials* | | |
| Poly(acrylic acid hydrazide) | 400 | 0.001 |
| Poly(*N*-vinyl-5-methoxazolidone) | 70 | 0.006 |
| Poly(vinylpyrrolidone) | 50 | 0.009 |
| Poly(vinyl alcohol) | 5 | 0.09 |
| Poly(acrylamide) | 1.3 | 0.3 |
| Poly(L-lysin hydrobromide) | 1 | 0.4 |
| Poly(acrylic acid) | 0.07 | 6 |
| Poly(ethyleneimine) | 0.04 | 10 |

### 17.3.5 Inorganic Support

In traditional catalysts, inorganic supports are generally used to control the functions of catalytic active components like metals. The supports are usually inactive as a catalyst but necessary to functionalize the active components. The roles of supports can be classified into five categories:

1. Increase in specific surface area of active component: To control the deposition of the active component on the surface of inorganic supports can increase dispersion or a ratio of active sites on the surface to whole active components, which can result in the increase of catalytic activity.
2. Increase in thermal stability of active component: Since the melting point of metal nanoparticles is less than the corresponding bulk metal, the metal nanoparticles easily melt and sinter, which results in thermal instability. The immobilization of metal nanoparticles on an inorganic support can depress their mobility, leading to their thermal stability.
3. Dilution effect: Multifunctional catalysts can be provided by immobilizing different active sites on the surface of inorganic support(s). If the support has a high thermal conductivity, then the local heat at active sites driven by heat of reaction can be removed through the supports, which is important especially in a heterogeneous gas/solid phase reaction.

4. Molding characteristic and mechanical strength: In practical catalysts, it is important to design the shape of catalysts as desired, e.g., porous structures. In this case, the mechanical strength is also an important factor to form the designed structure and keep the structure for a long period.
5. Interaction between metal and inorganic support: Contact between metal nanoparticles and inorganic supports induces the electron transfer, which can vary the electronic state of metal nanoparticles. Especially, the smaller the nanoparticle, the stronger is the interaction between the metal nanoparticles and the inorganic supports, and the higher is the area of the interface between metal particles and inorganic supports, which results in increasing the effect of supports.

Here, only the inorganic supports with large size are mentioned, which is useful for practical reactions in heterogeneous phases. However, the recent development of nanotechnology can make the preparation of small inorganic nanomaterials possible. These inorganic nanomaterials have other benefits, which will be described later (in Section 17.4.2). In addition, not only inorganic supports but also organic polymer supports have received the attention. Although organic polymer supports have weak mechanical strength, they are easily obtained and easily designed. So they have benefits different from inorganic supports. Since some polymers work as protective reagents as people know, so organic polymer supports have an advantage to prepare small metal nanoparticles on the supports. In contrast, it is not easy to prepare small metal nanoparticles with controlled structures on inorganic support. So, metal nanoparticles on inorganic supports can be prepared by the deposition of the metal nanoparticles, prepared in advance, on the inorganic supports. The organic protective reagents used for the preparation of dispersions of metal nanoparticles can be easily removed by heat treatment. Other deposition methods have been reported too. The concept of immobilization of metal nanoparticles on inorganic supports is illustrated in Figure 17.21.

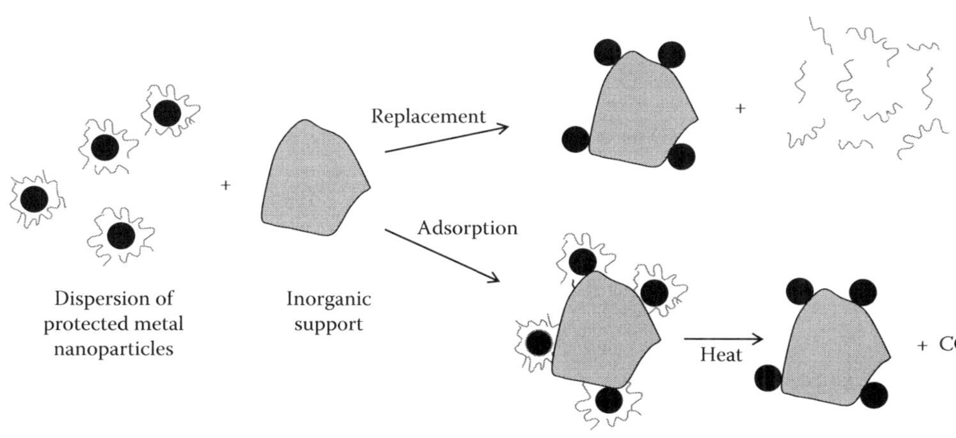


**FIGURE 17.21**
Schematic illustration for immobilization of metal nanoparticles on inorganic supports from dispersion of protected metal nanoparticles and inorganic supports.

## 17.4 Recent Progress of Metal Nanoparticle Catalysts

Much progress has been achieved on metal nanoparticle catalysts. Number of publications on this subject is increasing. Here, the author would like to focus on two important concepts developed recently concerning metal nanoparticle catalysts: Trimetallic nanoparticle catalysts and hybrid nanocatalysts.

### 17.4.1 Trimetallic Nanoparticle Catalysts

In the previous section (Section 17.3.3), composition and structure of bimetallic nanoparticles provide an important concept to design metal catalysts. Furthermore, the addition of the second and the third element on the main metal element can so often improve the catalytic activity and selectivity. From the viewpoint of these concepts, the trimetallic nanoparticle catalysts have received much attention.

In the case of metal nanoparticles, the contact of two different metals causes charge transfer between the two metals or localization of electron. Thus, the core/shell structure in bimetallic nanoparticles is of great importance, which can bring about charge transfer between core and shell metals and change the electronic density of the shell metal by an electronic ligand effect. Based on this concept, the triple core/shell structure in trimetallic nanoparticles should bring about stronger electronic effect than the bimetallic nanoparticles because of sequential electron transfer (Toshima 1991). Thus, the triple core/shell structure is of great importance in the catalytic properties of trimetallic nanoparticles, as shown in Figure 17.22.

In order to achieve the sequential electron transfer in the trimetallic nanoparticles, the order of the layered metals A, B, and C in Figure 17.22b is of great importance. We have succeeded in constructing such triple-layered core/shell-structured trimetallic nanoparticles by combination of two methods to prepare core/shell-structured bimetallic nanoparticles, i.e., sacrificial hydrogen reduction or homogeneous reduction and self-organization (physical mixing) (Toshima 2008). In the case of Pd/Ag/Rh trimetallic nanoparticles, mixture of the dispersion of Pd-core/Ag-shell bimetallic nanoparticles and those of Rh nanoparticles at room temperature results in the formation of trimetallic nanoparticles by self-organization, as shown in Figure 17.23 (Toshima et al. 2003, Matsushita et al. 2007). The trimetallic nanoparticles having an atomic composition of Pd/Ag/Rh = 1/2/13.5 and an average diameter of 2.2 nm show the highest catalytic activity among the metal

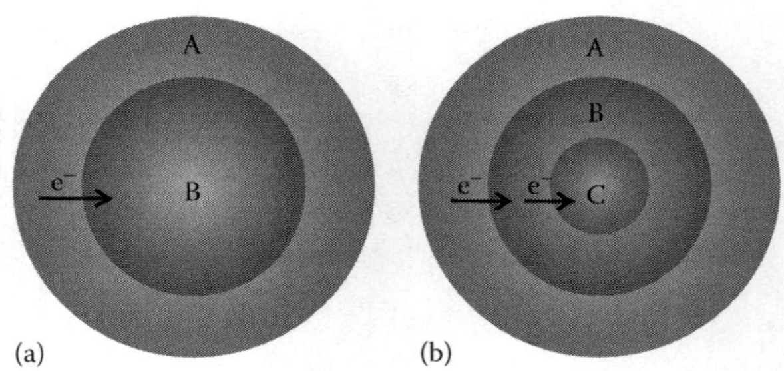


**FIGURE 17.22**
**(See color insert following page 302.)** Schematic illustration for cross sections of charge transfers between different metals A, B, and C in core/shell-structured (a) bimetallic and (b) trimetallic nanoparticles.

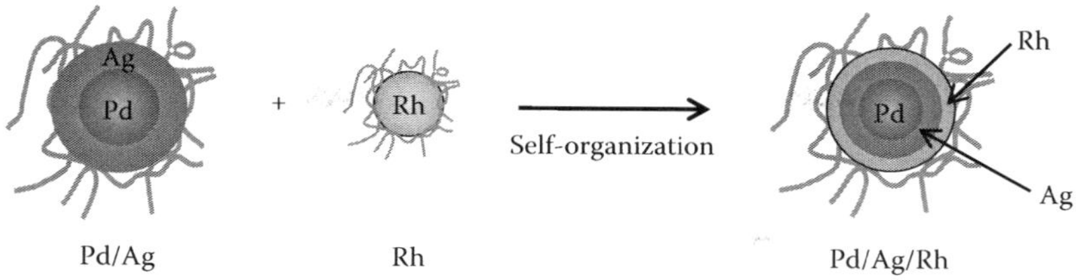


**FIGURE 17.23**
**(See color insert following page 302.)** Schematic illustration of spontaneous formation of Pd/Ag/Rh triple core/shell-structured trimetallic nanoparticles by self-organization at room temperature. (Reprinted from Matsushita, T. et al., *Bull. Chem. Soc. Jpn.*, 80, 1217, 2007. With permission from The Chemical Society of Japan.)

nanoparticle catalysts tested for the hydrogenation of methyl acrylate at 30°C under 1 atm. of hydrogen (Matsushita et al. 2007).

In the case of Au/Pt/Rh trimetallic nanoparticles, the triple core/shell structure was constructed by using a self-organization method in physical mixture at room temperature between Rh nanoparticles and Au/Pt bimetallic nanoparticles having a Au-core/Pt-shell structure, prepared in advance by the homogeneous reduction of Au and Pt ions by alcohol (Toshima et al. 2007). The highest catalytic activity for the hydrogenation of olefin was observed for the trimetallic nanoparticles having the atomic ration of Au/Pt/Rh = 1/4/20 and an average diameter of 2.9 nm. They could have a thermodynamically stable structure because the catalytic activity did not decrease after heat treatment. The Au-core/Pt-interlayer/Rh-shell structure was supported by the binding energy shifts in XPS.

In both cases of Pd/Ag/Rh and Au/Pt/Rh trimetallic nanoparticles, it is difficult to show the evidence for triple core/shell structure because they have very small sizes less than 3 nm and the ratios of the core element are very small, 6% and 4%, respectively. However, the following results can support the triple core/shell structure: (1) The trimetallic nanoparticles at these ratios had extraordinary high activities. (2) EF-TEM images showed that at least the interlayer element was located at an inner part of a particle and the shell element at an external part of a particle. (3) In the case of Au/Pt/Rh nanoparticles, binding energy shifts in XPS spectra were consistent with those expected from the triple core/shell structure.

In contrast, large trimetallic nanoparticles with a triple core/shell structure could be constructed by successive reduction recently. Zhon et al. prepared onion-like Pd-Bi-Au/C catalyst with average diameter of 13 nm by successive chemical reduction of precursor Au, Bi, and Pd salts in aqueous solution and immobilization on active carbon (Zhou et al. 2008). HR-TEM, XRD, XPS, and Auger electron spectroscopy were used to analyze the Au-core/Bi-interlayer/Pd-shell structure. The trimetallic nanoparticle catalyst was more active as a catalyst for aerobic liquid phase oxidation than Pd-Au/C bimetallic catalyst. Au/Pb/Pt trimetallic nanoparticles were prepared by the deposition of sequentially reduced Pb and Pt onto Au seed nanoparticles (Patra and Yang 2009). The Au-core/Pb-interlayer/Pt-shell structure was confirmed by UV-Vis, TEM, EDS, and cyclic voltammetry. The nanoparticles show high electrocatalytic activity for formic acid and methanol electrooxidation.

Of course, the synergetic effect in trimetallic nanoparticles is highly efficient in the three-layered core/shell structure. However, even trimetallic nanoparticles with random alloy or homogeneous alloy structure can show the synergetic effect. From this point of view, several works have been reported.

AuAgPd trimetallic alloy nanoparticles with average diameters of 44 nm were prepared by laser irradiation of a mixture containing Au, Ag, and Pb colloids and applied to the

Heck reaction (Tsai et al. 2003). PtRuCo trimetallic nanoparticles with an average diameter of 2.7 nm and a homogeneous alloy structure were prepared by microemulsions procession and used as more active catalysts for methanol oxidation than PtRu bimetallic nanoparticles (Zhang et al. 2004).

$Ru_5PtSn$ nanoparticle cluster supported on mesoporous silica was prepared from the carbonyl cluster $[PtRu_5(CO)_{15}(\mu\text{-}SnPh_2)(\mu_6\text{-}C)]$ and used as an excellent catalyst in the single-step hydrogenation of dimethyl terephthalate to cyclohexane dimethanol under mild conditions (100°C, 20 bar $H_2$) (Hungria et al. 2006). The structure and electronic properties of the trimetallic catalysts were reported (Groenbeck and Thomas 2007) and the details were discussed (Thomas et al. 2008). High-throughput pulsed laser ablation (HT-PLA) was used to prepare single- and multimetallic nanoparticles (Senkan et al. 2006). The supported nanoparticles created by HT-PLA were screened for catalytic activity and selectivity for the partial oxidation of propylene.

### 17.4.2 Hybrid Nanocatalysts

In the previous section (Section 17.3.5), the roles of inorganic supports were emphasized. Traditionally, the inorganic supports were composed of bulk metal oxides like $Al_2O_3$, $SiO_2$, $TiO_2$, $CeO_2$, etc. However, nanoparticles of metal oxides have been developed recently, and are going to be used for practical purposes. In addition, some effects of inorganic supports are attributed to the contact with active sites, i.e., metal nanoparticles. Thus, the combination of metal nanoparticles and nanoparticles of inorganic supports like metal oxides should enhance the catalytic functions of metal nanoparticles. Such nanoparticles combined in nanometer-scale are called "hybrid nanocatalysts" here. Since not only inorganic supports like metal oxides but also other materials like organic polymers can have an electronic ligand effect (electronic effect) on metal nanoparticle catalysts, such systems including organic materials will be discussed too in this section.

In principle, the preparation methods of hybrid nanocatalysts composed of metal nanoparticles and inorganic nanomaterials like metal oxide nanoparticles can be classified into four categories as illustrated in Figure 17.24:

1. Combination of metal nanoparticles and metal oxide nanoparticles: Metal and metal oxide nanoparticles are prepared in advance. The mixture of two dispersions of nanoparticles may produce the hybrid nanocatalysts. If the mixture does not form the hybrid, then a kind of technique should be designed such as to use bidentate ligands, to select the charged protective reagents, and so on.
2. Reactions starting from precursors of metal and metal oxide: Both reactions from metal precursor to metal (e.g., reduction) and from metal oxide precursors to metal oxide (e.g., hydrolysis) should be carried out in one pot. It is sometimes very difficult to control both reaction conditions in one vessel.
3. Reactions between metal oxide nanoparticles and metal precursors: The reduction of metal precursors can be carried out in the presence of dispersed metal oxide nanoparticles. In practice, metal precursors adsorbed on metal oxide nanoparticles may be reduced using chemical reductants.
4. Reactions between metal nanoparticles and metal oxide precursors: So-called "sol-gel" method is applied to the preparation of metal oxide nanoparticles from the corresponding precursors in the presence of dispersed metal nanoparticles. By this method, the catalysts of metal nanoparticles occluded in metal oxide were prepared.

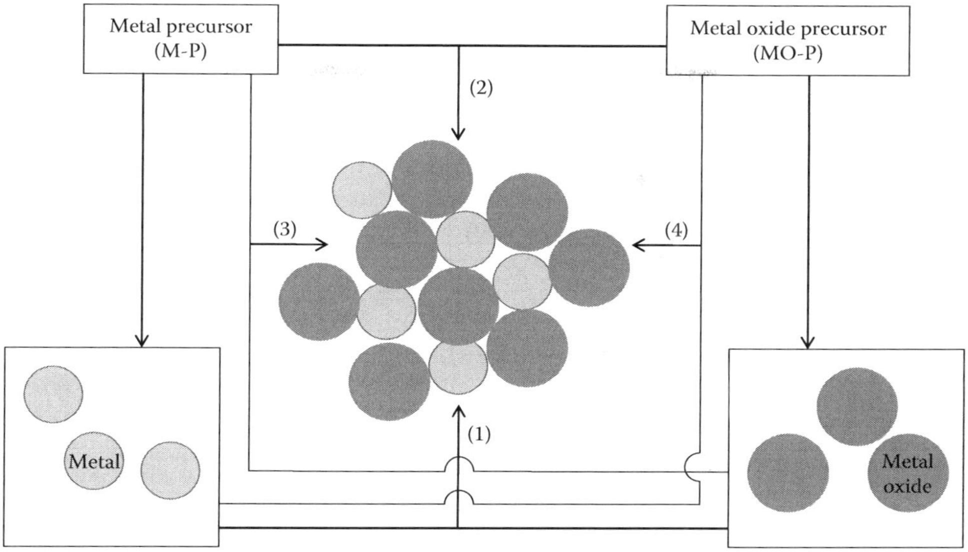


**FIGURE 17.24**
Concept of preparation of composite nanocatalysts composed of metal nanoparticles (M-NP) and metal oxide nanoparticles (MO-NP); (1) combination of M-NP and MO-NP, (2) reactions starting from metal precursors (M-P) and metal oxide precursor (MO-P), (3) reactions between MO-NP and M-P, and (4) reactions between MNP and MO-P.

The hybrid nanocatalysts composed of metal nanoparticles and nanometer-sized inorganic supports have received much attention, especially in the case of Au catalysts, because Au catalysts are strongly influenced by the inorganic supports and small Au nanocatalysts are effective as a catalyst (Haruta 2008). In addition, organic–inorganic hybrid composites can be also considered as supports, which have a possibility to be applied to various new fields such as electrocatalysts for fuel cells. Thus, the reports on hybrid nanocatalysts of Au nanoparticles and those of organic–inorganic hybrid will be referred here.

The hybrid nanocatalysts composed of Au and Pd nanoparticles, and $CeO_2$, CuO, and ZnO nanoparticle supports were prepared by a microwave method and applied to catalyst for CO oxidation (Glaspell et al. 2005). The supported Au/$CeO_2$ nanocatalysts exhibit excellent activity for low-temperature CO oxidation. The report on the importance of the presence of mixture in supported Au nanoparticle catalyst for CO oxidation (Date et al. 2004) as well as visible light illumination for the oxidation of formaldehyde and methanol in air (Chen et al. 2008) may suggest the importance of the hybrid nanocatalysts. Novel methods to prepare the supported Au nanoparticle catalysts were reported, e.g., the laser vaporization-controlled condensation technique (Yang et al. 2006), the liquid-phase reductive deposition method through the adsorption of specific metal complexes (Sunagawa et al. 2008), and so on.

The mixed-oxide nanoparticle supports were developed to enhance the activity of Au nanoparticle catalysts (Haider and Baiker 2008). The mixed-oxide supports were prepared by flame spray pyrolysis, resulting in agglomerated primary nanoparticles in the 10–15 nm range, onto which 6–9 nm Au particles were deposited by means of deposition-precipitation. The mixed-oxide-supported Au catalysts with noble metal loading of $0.6 \pm 0.17$ wt% were tested in the aerobic liquid-phase oxidation of 1-phenylethanol to phenyl methyl ketone, which showed that the activity depends strongly on the composition of the support, with

Cu and Mg being crucial components. Recently, a new Au catalytic system was prepared on ceria-modified meso-/macroporous binary metal oxide support ($CeO_2/TiO_2$-$ZrO_2$) and used as water-gas shift reaction catalyst (Idakien et al. 2009). Organic–inorganic hybrid can be used as supports. Pd nanoparticle catalysts stabilized by organic–inorganic hybrid Gn-PAMAM-SBA-15 ($n=1$–4) were prepared and used as heterogeneous hydrogenation of allyl alcohol. The activity over G4 was 1.5 times that over the fourth-generation PAMAM-encapsulated Pd homogeneous catalyst (Jieng and Gao 2006). A method for coating the inside of a glass tube like micro reactors by polyionic polymers like a Merrifield resin, and depositing and reducing Pd ions on it to provide Pd nanoparticle catalyst supported on glass by the polymer-bound ammonium groups was reported (Mennechke and Kischning 2009). This catalyst was employed in Suzuki-Miyamura and Heck cross couplings under flow condition. Metal-organic framework (MOF) was also used as a precursor for the synthesis of Pt supporting ZnO nanoparticle catalysts (Liu et al. 2009). The introduction of inorganic Pt salts into the MOF-5 pores ($[Zn_4O(bdc)]_3$, bdc = 1, 4-benzendicarboxylate), and heating the composites in air at 60°C gave the catalysts the activity of which was higher than conventional Pt/ZnO catalyst for CO oxidation. Polyionic polymers like a perfluorinated sulfonic acid copolymer (PFSA) are usually difficult to use as a protective reagent to stabilize metal nanoparticles in aqueous dispersions. We successfully prepared Pt nanoparticles protected by PFSA, which has good proton conductivity, gas permeability, and chemical stability (Naohara et al. 2010). PFSA-protected Pt nanoparticles formed the nano-network on dried films. The porous structures might improve the diffusion of reactants and products in the catalyst layer of polymer electrolyte fuel cells. In fact, PFSA-protected Pt nanoparticles showed a good electrocatalytic activity for oxygen reduction reaction as a conversional Pt/C catalyst.

## 17.5 Concluding Remarks

In general, research in nanoscience and nanotechnology is expected to have a great impact on the development of new catalysts, because the detailed understanding of chemistry of catalytic materials in the nanometer-scale and the ability to control their preparation will lead to rational and cost-efficient catalyst design. In fact, metal nanoparticles exhibit unique properties that differ from the bulk substance, e.g., different heat capacity, vapor pressure, and melting point. Moreover, when decreasing the metal particle size sufficiently enough, there occurs the transfer of the electronic state from metallic to a nonmetallic one. Metal nanoparticles also exhibit large surface-to-volume ratio and increased number of edges, corners, and faces leading to altered catalytic activity and selectivity. In addition, the structure of metal nanoparticles, especially bimetallic and trimetallic nanoparticles, is now controlled as designed, which is one of the topics in this chapter. Not only the recent topics but also the traditional preparation, purification, and characterization methods are reviewed briefly. The author expects that this fundamental knowledge summarized here could be useful for not only the newcomers in this field but also the specialists who are expert in this field to overview and summarize the knowledge.

Metal nanoparticles supported on inorganic and organic matrixes have shown promising features like higher catalytic activity and/or selectivity than conventional catalyst in many catalytic reactions. Especially hybrid nanocatalysts, in which metal nanoparticles and inorganic metal oxide nanoparticles and/or organic materials keep contact in nanometer-scale, are now going to develop new fields such as catalysts of Au nanoparticles, electrocatalysts for fuel cell, and so on. The problems associated with nanocatalysts are interdisciplinary ones and require the understanding of several mutually related sciences, chemical kinetics and catalysis, quantum chemistry, chemical engineering, and so on. The author expects this chapter may be useful for the development of a new field in nanoscience and nanotechnology as well as catalysis chemistry.

## Acknowledgments

The author expresses his thanks to all the coworkers for their help to develop a new field with him. The works on trimetallic nanoparticles were supported by a Grant-in-Aid for Scientific Research (B) (No. 15310078) from the Ministry of Education, Culture, Sports, Science and Technology (MEXT), Japan, and those on gold catalysts by a Core Research for Evolutional Science and Technology (CREST) program from Japan Science and Technology Agency (JST), Japan.

## References

Bion, C.-R., Suzuki, S., Asakura, K. et al. 2002. Extended x-ray absorption fine structure studies on the structure of the poly(vinylpyrrolidone)-stabilized Cu/Pd nanoclusters colloidally dispersed in solution. *J. Phys. Chem. B* 106: 8587–8598.

Bradley, J. S., Millar, J. M., Hill, E. W. et al. 1991. Surface chemistry on colloidal metals: Spectroscopic study of adsorption of small molecules. *Faraday Discuss*. 92: 255–268.

Bradley, J. S., Hill, E. W., Chaudret, B., and Duteil, A. 1992. Preparation and characterization of organosols of monodispersed nanoscale palladium. Particle size effects in the binding geometry of adsorbed carbon monoxide. *Chem. Mater*. 4: 1234–1239.

Bradley, J. S., Hill, E. W., Chaudret, B., and Duteil, A. 1995. Surface-chemistry on colloidal metals—Reversible adsorbate-induced surface-composition changes in colloidal palladium-copper alloys. *Langmuir* 11: 693–695.

Bradley, J. S., Via, G. H., Bonneviot, L. et al. 1996. Infrared and EXAFS study of compositional effects in nanoscale colloidal palladium-copper alloys. *Chem. Mater*. 8: 1895–1903.

Brust, M., Walker, M., Bethell, D., and Schiffrin, D. J. 1994. Synthesis of thiol derivatised gold nanoparticles in a two phase liquid/liquid system. *J. Chem. Soc. Chem. Commun*. 1994: 801–802.

de Caro, D. and Bradley, J. 1997. Surface spectroscopic study of carbon monoxide adsorption on nanoscale nickel colloids prepared from a zerovalent organometallic precursor. *Langmuir* 13: 3067.

Chen, H. M., Liu, R. S., Jiang, L. Y., Lee, J. F., and Hu, S. F. 2006a. Characterization of core-shell type and alloy Au/Ag bimetallic clusters by using extended x-ray absorption fine structure spectroscopy. *Chem. Rev. Lett*. 421: 118.

Chen, H. M., Peng, H. C., Liu, R. S., Hu, S. F., and Jieng, L. Y. 2006b. Local structural characterization of Au/Pt bimetallic nanoparticles. *Chem. Rev. Lett*. 420: 484.

Chen, X., Zhu, H.-Y., Zhao, J.-C. et al. 2008. Visible-light-driven oxidation of organic contaminants in air with gold nanoparticle catalysts on oxide supports. *Angew. Chem. Int. Ed*. 47: 5353–5356.

Date, M., Okumura, M., Tsubota, S. et al. 2004. Vital role of moisture in the catalytic activity of supported gold nanoparticles. *Angew. Chem. Int. Ed*. 43: 2129–2132.

Egerton, R. F. 1996. *Electron Energy-Loss Spectroscopy in the Electron Microscope*, 2nd ed., Plenum Press, New York.

Faraday, M. 1857. The Bakerian Lecture: Experimental relations of gold (and other metals) to light. *Philos. Trans. R. Soc.* 147: 145–181.

Garcie-Gutierrez, D. I., Gutierez-Wing, C. E., Giovanetti, L. et al. 2005. Temperature effect on the synthesis of Au–Pt bimetallic nanoparticles. *J. Phys. Chem. B* 109: 3813–3821.

Glaspell, G., Fuoco, L., and EL-Shell, M. S. 2005. Microwave synthesis of supported Au and Pd nanoparticle catalysts for CO oxidation. *J. Phys. Chem. B* 109: 17350–17355.

Groenbeck, H. and Thomas, J. M. 2007. Structural and electronic properties of a trimetallic nanoparticle catalyst: Ru5PtSn. *Chem. Phys. Lett.* 443: 337–341.

Haider, P. and Baiker, A. 2008. Gold supported on Cu–Mg–Al-mixed oxides: Strong enhancement of activity in aerobic alcohol oxidation by concerted effect of copper and magnesium. *J. Catal.* 248: 175–187.

Harada, M., Asakura, K., and Toshima, N. 1993a. Catalytic activity and structural analysis of polymer-protected gold/palladium bimetallic clusters prepared by the successive reduction of hydrogen tetrachloroaurate(III) and palladium dichloride. *J. Phys. Chem.* 97: 5103–5114.

Harada, M., Asakura, K., Ueki, Y., and Toshima, N. 1993b. Structural analysis of polymer-protected palladium/rhodium bimetallic clusters using EXAFS spectroscopy. *J. Phys. Chem.* 97: 10742–10749.

Haruta, M. 2008. Polymer attached catalysts. *Metal Nanoclusters in Catalysis and Materials Science: The Issue of Size Control*, ed. B. Corain, G. Schmid, and N. Toshima, Elsevier, Amsterdam, the Netherlands, pp. 183–199.

Hirai, H. and Toshima, N. 1986. Relevance of metal nanoclusters size control in gold (0) catalytic chemistry. *Tailored Metal Catalysts*, ed. Y. Iwasawa, R. Reidel, Pub. Co., Dordrecht, the Netherlands, pp. 87–140.

Hirai, H., Chawanya, H., and Toshima, N. 1985. Colloidal palladium protected with poly(N-vinyl-2-pyrrolidone) for selective hydrogenation of cyclopentadiene. *Reactive Polym.* 3: 127–141.

Hirai, H., Nakao, Y., and Toshima, N. 1978. Colloidal rhodium in poly(vinylpyrrolidone) as hydrogenation catalyst for internal olefin. *Chem. Lett.*, 1978: 545–546.

Hirai, H., Nakao, Y., and Toshima, N. 1979. Preparation of colloidal transition metals in polymers by reduction with alcohols or ethers. *J. Macromol. Soc. Chem.* A13: 727–750.

Hirai, H., Nakao, Y., Adachi, K., and Toshima, N. 1976. Colloidal rhodium in polyvinyl alcohol as hydrogenation catalyst for internal olefin. *Chem. Lett.*, 1976: 905–906.

Hirakawa, K. and Toshima, N. 2003. Ag/Rh bimetallic nanoparticles formed by self-assembly from Ag and Rh monometallic nanoparticles in solution. *Chem. Lett.* 32: 78–79.

Hostetler, M. J., Zong, C.-J., Yen, B. K. H. et al. 1998. Stable, monolayer-protected metal alloy cluster. *J. Am. Chem. Soc.* 20: 9396–9397.

Hungria, A. B., Raja, R., Adams, R. D. et al. 2006. Single-step conversion of dimethyl terephthalate into cyclohexanedimethanol with Ru5PtSn, a trimetallic nanoparticle catalyst. *Angew. Chem. Int. Ed.* 45: 4782–4785.

Idakien, V., Tabakova, T., Tenchen, K. et al. 2009. Gold nanoparticles supported on ceria-modified mesoporous–macroporous binary metal oxides as highly active catalysts for low-temperature water–gas shift reaction. *J. Mater. Sci.* 44: 6637–6643.

Iwasawa, Y., 1996. *X-Ray Absorption Fine Structures for Catalysts and Surfaces, World Scientific Series on Synchrotron Radiation Techniques and Applications*, Vol. 2, World Scientific, Singapore.

Jieng, Y. and Gao, Q. 2006. Heterogeneous hydrogenation catalyses over recyclable Pd(0) nanoparticle catalysts stabilized by PAMAM-SBA-15 organic–inorganic hybrid composites. *J. Am. Chem. Soc.* 128: 716–717.

Lee, A. F., Baddeley, C. J., Hardacre, C., Ormerod, R. M., Lambert, M., Schmid, G. et al. 1995. Structural and catalytic properties of novel Au/Pd bimetallic colloid particles: EXAFS, XRD, and acetylene coupling. *J. Phys. Chem.* 99: 6096–9102.

Lee, W. R., Kim, G., Choi, J. R. et al. 2005. Redox–transmetalation process as a generalized synthetic strategy for core–shell magnetic nanoparticles. *J. Am. Chem. Soc.* 127: 16090.

Liu, B., Hau, S., Tanoka, K. et al. 2009. Metal–organic framework (MOF) as a precursor for synthesis of platinum supporting zinc oxide nanoparticles. *Bull. Chem. Soc. Jpn*. 82: 1052–1054.

Lu, P., Teranishi, T., Asakura, K., Miyake, M. and Toshima, N. 1999. Polymer-protected Ni/Pd bimetallic nano-clusters: Preparation, characterization and catalysis for hydrogenation of nitrobenzene. *J. Phys. Chem. B* 103: 9673–9682.

Matsushita, T., Shiraishi, Y., Horiuchi, S. et al. 2007. Synthesis and catalysis of polymer-protected Pd/Ag/Rh trimetallic nanoparticles with a core–shell structure. *Bull. Chem. Soc. Jpn*. 80: 1217–1225.

Mennechke, K. and Kischning, A. 2009. Polyionic polymers—Heterogeneous media for metal nanoparticles as catalyst in Suzuki–Miyaura and Heck–Mizoroki reactions under flow conditions. *Beilstein J. Org. Chem*. 5 (21).

Miner, R. S., Namba, S., and Turkevich, J. 1981. Synthesis, characteristics and catalytic activity of monodisperse platinum alloys with gold and palladium. *Stud. Surf. Sci. Catal*. 7(Pt. A): 160–172.

Naohara, H., Yohimoto, T., and Toshima, N. 2010. Platinum nanoparticles protected by a perfluorinated sulfonic acid copolymer: Preparation, nano-network formation and electrocatalytic activity. *J. Power Sources* 195: 1051–1053.

Patra, S. and Yang, H. 2009. Synthesis of trimetallic Au@Pb@Pt core-shell nanoparticles and their electrocatalytic activity toward formic acid and methanol. *Bull. Korean Chem. Soc*. 30: 1485–1488.

Schmid, G., Lehnert, A., Malm, J.-O. et al. 1991. Ligand-stabilized bimetallic colloids identified by HRTEM and EDX. *Angew. Chem. Int. Ed. Engl*. 30: 874–876.

Schmid, G., West, H., Malm, J.-O. et al. 1996. Catalytic properties of layered gold-palladium colloids. *Chem. Eur. J*. 2: 1099–1103.

Senkan, S., Khn, M., Duan, S. et al. 2006. High-throughput metal nanoparticle catalysis by pulsed laser ablation. *Catal. Today* 117: 291–296.

Shiraishi, Y., Ikenaga, D., and Toshima, N. 2003. Preparation and catalysis of inverted core/shell structured Pd/Au bimetallic nanoparticles. *Aust. J. Chem*. 56: 1025–1029.

Sinfelt, J. H., Via, B. H., and Lytle, F. W. 1984. Application of EXAFS in catalysis. Structure of bimetallic cluster catalysts. *Catal. Rev. Sci. Eng*. 26: 81–140.

Solla-Gullon, J., Rodes, A., Montiel, V. et al. 2003. Electrochemical characterization of platinum-palladium nanoparticles prepared in a water-in-oil microemulsion. *J. Electroanal. Chem*. 554–555: 273–284.

Sunagawa, Y., Yamamoto, K., Takahashi, H., and Muramatsu, A. 2008. Liquid-phase reductive deposition as a novel nanoparticle synthesis method and its application to supported noble metal catalyst preparation. *Catal. Today* 132: 81–87.

Teranishi, T. and Miyake, M. 1999. Novel synthesis of monodispersed Pd/Ni nanoparticles. *Chem. Mater*. 11: 3414–3416.

Thiele, H. and von Levern, H. S. 1965. Synthetic protective colloids. *J. Colloid. Sci*. 20: 679–694.

Thomas, J. M. 1988. Colloidal metals: past, present and future. *Pure Appl. Chem*. 60: 1517–1527.

Thomas, M. J., Adams, R. D., Boswell, E. M., Captain, B., Gronbeck, H., and Raja, R. 2008. Synthesis, characterization, electronic structure and catalytic performance of bimetallic and trimetallic nanoparticles containing tin. *Faraday Discuss*. 138: 301–315.

Toshima, N. 1985. Metal colloid catalysts carried by polymers. *Shokubai* 27: 488–494.

Toshima, N. 1991. Sequential potential field and electron transfer in metal complex clusters. *Macromolecular Complexes: Dynamic Interactions and Electronic Processes*, ed. E. Tsuchida, VCH Pub. Inc., New York, pp. 331–352.

Toshima, N. 2003. Metal nanoparticles for catalysis. *Nanoscale Materials*, ed. L. M. Liz-Marzán and P. V. Kamat, Kluwer Academic Publishers, Boston, MA, pp. 79–96.

Toshima, N. 2004. Polymer-capped bimetallic nanoclusters as active and selective catalysts. *Macromolecular Nanostructured Materials*, ed. N. Ueyama and A. Harada, Kodansha/Springer, Tokyo, Japan/Berlin, Germany, pp. 182–196.

Toshima, N. 2008. Capped bimetallic and trimetallic nanoparticles for catalysis and information technology. *Macromol. Symp*. 270: 27–39.

Toshima, N. and Liu, H.-F. 1992. Preparation of organosol of noble metal clusters with novel method. *Chem. Lett*. 21: 1925–1928.

Toshima, N. and Yonezawa, T. 1992. Preparation of polymer-protected gold/platinum bimetallic clusters and their application to visible light-induced hydrogen evolution. *Macromol. Chem., Macromol. Symp.* 59: 281–295.

Toshima, N. and Wang, Y. 1993. Novel preparation, characterization and catalytic properties of polymer-protected Cu/Pd bimetallic colloid. *Chem. Lett.* 1993: 1611–1614.

Toshima, N. and Wang, Y. 1994a. Polymer-protected Cu/Pd bimetallic clusters. *Adv. Mater.* 6: 245–247.

Toshima, N. and Wang, Y. 1994b. Preparation and catalysis of novel colloidal dispersions of copper/noble metal bimetallic clusters. *Langmuir* 10: 4574–4580.

Toshima, N. and Yonezawa, T. 1998. Recent progress in bimetallic nanoclusters: Their preparation, structures and functions. Bimetallic nanoparticles—Novel materials for chemical and physical applications. *N. J. Chem.* 22: 1179–1201.

Toshima, N., Yan, H., and Shiraishi, Y. 2008. *Metal Nanoclusters in Catalysis and Materials Science: The Issue of Size Control*, ed. B. Corain, G. Schmid, and N. Toshima, Elsevier, Amsterdam, pp. 49–75.

Toshima, N., Yonezawa, T., and Kushihashi, K. 1993. Polymer-protected palladium–platinum bimetallic clusters: Preparation, catalytic properties and structural considerations. *J. Chem. Soc. Faraday Trans.* 89: 2537–2543.

Toshima, N., Kuriyama, M., Yamada, Y., and Hirai, H. 1981. Colloidal platinum catalyst for light induced hydrogen evolution from water. A particle size effect. *Chem. Lett.*, 1981: 793–796.

Toshima, N., Shiraishi, Y., Matsushita, T., Mukai, H., and Hirakawa, K. 2003. Self-organization of metal nanoparticles and its application to syntheses of Pd/Ag/Rh trimetallic nanoparticles catalysts with triple core/shell structures. *Int. J. Nanosci.* 1: 397–401.

Toshima, N., Kushihashi, K., Yonezawa, T. et al. 1989. Colloidal dispersions of palladium–platinum bimetallic clusters protected by polymers. Preparation and application to catalysis. *Chem. Lett.* 18: 1769–1772.

Toshima, N., Harada, M., Yonezawa, T. et al. 1991. Structural analysis of polymer-protected palladium/platinum bimetallic clusters as dispersed catalysts by using extended x-ray absorption fine structure spectroscopy. *J. Phys. Chem.* 95: 7448–7453.

Toshima, N., Harada, M., Yamazai, Y. et al. 1992. Catalytic activity and structural analysis of polymer-protected gold-palladium bimetallic clusters prepared by the simultaneous reduction of hydrogen tetrachloroaurate and palladium dichloride. *J. Phys. Chem.* 56: 9927–9933.

Toshima, N., Kanemaru, M., Shiraishi, Y. et al. 2005. Spontaneous formation of core/shell bimetallic nanoparticles: A calorimetric study. *J. Phys. Chem. B* 109: 16326–16331.

Toshima, N., Ito, R., Matsushita, T., and Shiraishi, Y. 2007. Trimetallic nanoparticles having a Au-core structure. *Catal. Today* 122: 239–244.

Tsai, S.H., Liu, Y.-H., Wu, P.-L. et al. 2003. Preparation of Au–Ag–Pd trimetallic nanoparticles and their application as catalysts. *J. Mater. Chem.* 13: 978–980.

Tsunoyama, H., Sakurai, H., and Tsukuda, T. 2006. Size effect on the catalysis of gold clusters dispersed in water for aerobic oxidation of alcohol. *Chem. Phys. Lett.* 429: 528–532.

Turkevich, J. and Kim, G. 1970. Palladium: Preparation and catalytic properties of particles of uniform size. *Science* 169: 873–879.

Wang, Y. and Toshima, N. 1997. Preparation of Pd–Pt bimetallic colloids with controllable core/shell structures. *J. Phys. Chem. B* 101: 5301–5306.

Wang, Z. L. 1998. Structural analysis of self-assembling nanocrystal superlattices. *Adv. Mater.* 10: 13–30.

Wang, Z. L. 2000a. Transmission electron microscopy of shape-controlled nanocrystals and their assemblies. *J. Phys. Chem. B* 104: 1153–1175.

Wang, Z. L., ed. 2000b. *Characterization of Nanophase Materials*, Wiley-VCH, Weinheim, Germany, p. 406.

Wang, Z. L. and Kang, Z. C. 1988. *Functional and Smart Materials-Structural Evolution and Structure Analysis*, Chapter 6, Plenum Press, New York.

Williams, H. B. and Chang, L. T. 1951. Gold number studies, with some observations on sensitization and protection. *J. Phys. Colloid Chem.* 55: 719–725.

Williams, D. B. and Carter, C. B. 1996. *Transmission Electron Microscopy*, Plenum Press, New York, NY.
Yang, Y., Soud, K. M., Abdelsayed, V. et al. 2006. Vapor phase synthesis of supported Pd, Au, and unsupported bimetallic nanoparticle catalysts for CO oxidation. *Catal. Commun.* 7: 281–284.
Yonezawa, T. and Toshima, N. 1993. Polymer- and micelle-protected gold/platinum bimetallic systems. Preparation, application to catalysis for visible-light-induced hydrogen evolution, and analysis of formation process with optical methods. *J. Mol. Catal.* 83: 167–181.
Zhang, X. and Chan, K. Y. 2003. Water-in-oil microemulsion synthesis of platinum–ruthenium nanoparticles, their characterization and electrocatalytic properties. *Chem. Mater.* 15: 451–459.
Zhang, X., Zhang, F., and Chan, K.-Y. 2004. Preparation of Pt–Ru–Co trimetallic nanoparticles and their electrocatalytic properties. *Catal. Commun.* 5: 749–753.
Zhou, Y., Wang, S., Ding, B. et al. 2008. Preparation of onion-like Pd–Bi–Au/C trimetallic catalyst and their application. *J. Sol-Gel Sci. Technol.* 47: 182–186.
Zsigmondy, R. 1901. Electrochemical characterization of platinum-palladium nanoparticles prepared in a water-in-oil microemulsion. *Z. Anal. Chem.* 40: 697–719.
Zsigmondy, R. and Thiessen, P. A. 1925. *Das Kolloide Gold*. Akad. Verlagsges., Leipzig, Germany.

# 18

# *Nanocatalysts: A New "Dimension" for Nanoparticles?*

**Paolo Ciambelli, Diana Sannino, and Maria Sarno**

**CONTENTS**



## 18.1 Introduction

During the initial years of the twenty-first century, a widely debated scientific topic, the potential impact of the growing interest in nanoscience, drew the attention of the catalysis community (see, as examples, Somorjai and Borodko 2001, Bell 2003, Kung and Kung 2003, Pernicone 2003). The discussion focused on a basic issue, well synthesized by the title of a short review (Schlögl and Hamid 2004): "Nanocatalysis: mature science revisited or something really new?" The review reminded that the use of nanosize materials in catalysis is not an innovation of recent years, since industrial catalysis has involved nanoparticles and performed their chemical transformations since the beginning of the twentieth century, effecting the oldest commercial application of nanotechnology. Moreover, erionite-based catalyst is the first example of selective cracking catalyst exploiting the nanosize dimension of the zeolite pores. Apart from these historical clarifications, it was recognized that only the expected progresses in nanoscience could have led to significant improvement of the design and synthesis of catalysts, since at that time the capability of controlling the uniformity of catalyst size and composition was absent. Therefore, as a conclusion of that debate, the strict relationship between research in catalysis and nanoscience was clearly recognized. More specifically, it was highlighted that both the advances in characterization techniques allowing a molecular-level understanding of the effect of nanoparticle size on catalytic performance, and the definition of novel approaches to nanoparticle synthesis for the construction of nanostructured catalysts are the keys for designing and developing novel catalysts capable of enhanced performances profiting from their nano nature.

## 18.2 Nanocatalyst Market

Concerning market and technological trends, several analyses frequently appear. The value of nanocatalysts in chemical industry is due to properties that allow enhancement of the rate of chemical reactions and that provide total selectivity to the desired product, which helps the chemical industry control the production of toxic waste.

Zeolites still constitute the largest segment in the international market for nanoporous materials. The global market for nanocatalysts or nanoscaled materials that have at least one dimension in the order of nanometers or are subject to a structural change at the nanoscale during the development of their catalytic activity was worth nearly 3.5 billion in 2003, about 40% in the petrochemical industry, 20% in the food industry, and 15% in the environmental sector (NANOTECH IT 2004).

In the highlights of a recent report (BCC Report NANO17E, 2007), it is estimated that the global market for nanoparticles used in energy, catalytic, and structural applications will increase by 400% from 2006 to 2012. As a result of this increase, even if catalytic applications are expected to drop to 26.6% of the total market by 2012, they will increase two times their market with respect to 2006. Energy applications are expected to grow from 15.1% of the total market in 2006 to 45% by 2012.

The increased use of nanocatalysts for refinery and petrochemical industry is mostly due to the superior selectivity performance obtained by a better control at the molecular level. The analysts agree that environmental applications will give the fastest growing to the nanocatalyst market over the period 2006–2015, since the growing environmental concerns such as air pollution, depleting energy sources, etc., are proving to be the factors influencing the rapid adoption of nanocatalysts for such applications (BizAcumen 2009).

## 18.3 Requirements for Nanocatalysts

To be used as a catalyst, a nanomaterial is required to fulfill several requirements, such as operation at very low or very high temperatures, in the presence of poisons, at high space velocity, with very low reactant concentrations, or operation under different or rapid changes of feed, and, more importantly, to be stable and not aggregate during the operation process.

This leads not only to seek more active or selective catalysts, but also new catalytic materials with high hydrothermal tolerance, duration, deactivation resistance, etc. Further improvements in the use of nanomaterials could minimize restrictions of heat or mass-transfer kinetics. Therefore, on these catalytic materials, in-depth studies of molecular aspects (understanding of reaction mechanisms, relationship between structure and activity or selectivity, modulation of active sites to change the diffusion characteristics and adsorption of reactants, products, or poisons for the reaction, well-controlled particle and pore size distribution, and optimized morphology) are a critical issue for the development of nanocatalysts. And the final acceptance will be always based on their cost.

All processes regarding the development and implementation of a nanomaterial require the study of production techniques and advanced analysis. The production of nanocatalysts first requires the development of synthetic methods that allow proper adjustment of the size and shape of nanoparticles, because these parameters are fundamental in controlling their physicochemical properties. The determination of the parameters that characterize the nanostructure is done not with the classical procedures adopted when working on a macroscopic scale (e.g., mechanical testing), but largely using tools such as electronic microscopy (SEM, TEM, EDS chemical analysis), atomic force microscopy (AFM), Raman spectroscopy, x-ray diffraction (XRD), x-ray photoelectron spectroscopy (XPS), etc.

This chapter will focus on nanoparticles applied in catalysis for green processes such as photocatalysis and for the synthesis of nanomaterials.

## 18.4 Nanophotocatalysts

### 18.4.1 Photocatalysis Features and Requirements

Before discussing the role of nanoparticles in photocatalysis, it seems useful to synthetically remind some relevant basic concepts. Photocatalysis is applied to a great extent in both environmental treatment (emission cleaning and water purification) and renewable energy. Photocatalysis is also regarded as an emerging technology for chemical transformations, allowing the accomplishment of cleaner productions in industry, as a response to growing environmental regulations (Hjeresen et al. 2002). From the late 1970s, the initial stages of this development were boosted by very attractive potential in water splitting using UV energy (Fujishima and Honda 1972).

Photocatalysis takes advantage of the properties of semiconductor materials to use energy, potentially at very limited cost, from the absorption of photons from solar or artificial light (Hoffmann et al. 1995).

Light absorption by a semiconductor catalyst promotes oxidation and reduction reactions, removing the need for expensive and dangerous solvents and chemicals (Vidal and Martin Luengo 2001). A further advantage is that photocatalytic processes in general do not require much severe operating conditions, resulting in reduced cost and limited safety precautions with respect to high temperature and pressure processes. Photocatalysis is also intrinsically very selective and therefore the production of by-products is reduced.

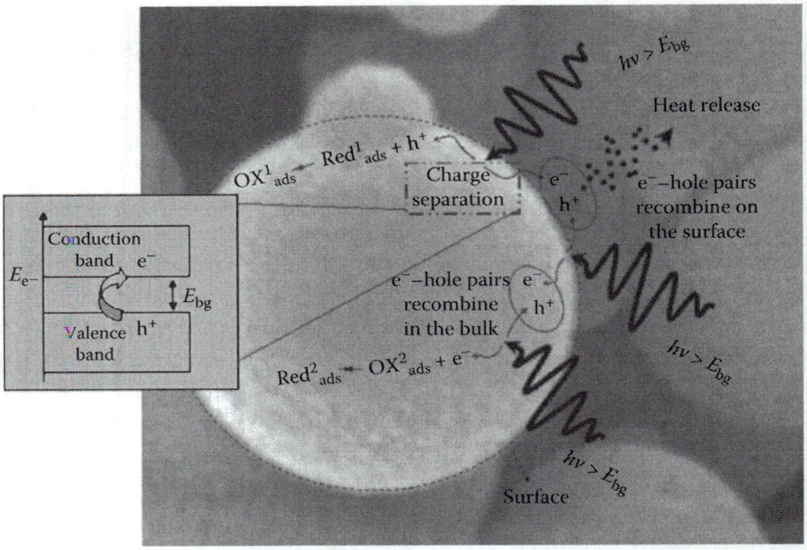


**FIGURE 18.1**
Photocatalysis scheme ($E_{bg}$ = band gap energy).

In a semiconductor at the ground state all electrons exist in the valence band. The conduction band is empty because it is separated from the valence band by an energy gap (band gap energy). Some electrons in the valence band could be excited to the conduction band by photons of suitable energy that has to be equal or greater than the band gap energy. The lack of these electrons in the valence band generates positive "holes." Under irradiation, there is charge separation and generation of electrons and hole pairs. The photogenerated pairs may recombine in the bulk or on the surface of the semiconductor, releasing heat, at a slower rate with respect to their formation; otherwise, electrons and holes on the surface of the semiconductor cause reduction and oxidation reactions, respectively, by interaction with adsorbed surface species. The photoelectron can be easily trapped by electronic acceptors like adsorbed $O_2$ to further produce a superoxide radical anion ($O^{2-}$), while the photo-induced holes are captured by electronic donors such as organic pollutants, which are oxidized (Palmisano et al. 1997).

Very small semiconductor particles have shown peculiar photophysical and photocatalytic properties. Nanosized particles, with diameters ranging between 1 and 10 nm, possess properties that fall into the region of transition between the molecular and the bulk phases (Beydoun et al. 1999 and references therein). Nanosized semiconductor particles, which exhibit size-dependent optical and electronic properties, are called quantized particles (Q-particles) or quantum dots (Kamat, 1995); their valence and conduction bands split into discrete electronic states (quantized levels) (Figure 18.1) (Beydoun et al. 1999 and references therein).

### 18.4.2 Effect of Particle Size, Crystal Structure, and Crystallinity on Titania Photoactivity

The ideal semiconductor photocatalyst should possess suitable band edges, chemical stability, corrosion resistance, and light harvesting ability. Several compounds such as metal oxides, metal sulfides, oxysulfides, oxynitrides, and composites thereof are potential photocatalysts (Beyduon et al. 1999). Many photocatalysts such as ZnO, $Nb_2O_5$, and $TiO_2$ have been used to degrade organic pollutants (Prado et al. 2009). $TiO_2$ in general being the preferred one. It is an *n*-type semiconductor having three major allotropic forms: anatase, rutile, and brookite, but only anatase and rutile seem to be photocatalytically active and therefore of commercial importance (Diebold 2003).

Rutile has a density of 4.2 g/cm$^3$, while anatase has a density of 3.9 g/cm$^3$ because the crystal structure of rutile is more tightly packed. Due to chemical stability and non toxicity, it is used as a pigment in pharmaceuticals and food colorings. Moreover, it does not suffer from photodecomposition or degradation.

The photoactivity of titania is strongly influenced by particle size, crystal structure, and crystallinity (Jung et al. 2002). The larger titania particles are less performing than the smaller ones, since the recombination of the photoexcited electron/hole pairs in the bulk is dominant due to the lower surface/volume ratio. However, it must be noticed that any thermal treatment necessary to control the crystal structure typically causes the aggregation of particles and, hence, the reduction of surface area. Therefore, it is a critical step in catalyst preparation.

The morphology is another important factor for the performance of photoreactors and devices. As an example, titania nanotubes provide channels for enhanced electron transfer, thereby helping to increase the efficiency in photocatalysis, as well as in solar cells and for electrolysis.

In a photoreactor the catalyst is usually placed in the form of a thin film or as particles in slurry and fluidized-bed photoreactors in order to favor light transfer. In the case of a film elongated and/or flat particles offer a higher capture surface to a perpendicular incident radiation; otherwise, round particles can catch the radiation in any direction when they are mixed to fluid streams and can prove useful to avoid the attrition phenomena that erode irregular shaped particles, facilitating elutriation phenomena or complicating the catalyst separation.

Nanostructured photocatalysts allow improvement of photoactivity, since they enhance both the adsorption of reactants and the desorption of products, due to the high surface area offered by the nanostructures, and reduce the electron–hole recombination, due to the short charge-transfer distance toward adsorbed species.

The modification of a surface with metal nanoparticles of Pt and Fe is frequently employed as a way of enhancing the photoactivity, by effectively reducing or retarding the surface recombination (Jung et al. 2002 and references therein). The presence of metal nanoparticles on the surface of nanosized semiconductor metal oxides induces the capture of photo-promoted electrons, if the Fermi level has a lower energy than the conduction band potential (Chiarello et al. 2008). As electrons accumulate into the noble metal particle, their Fermi level shifts to more negative values, closer to the conduction band (CB) level of $TiO_2$; this upward shift is more negative the smaller the metal particle size (Subramanian et al. 2004). Anyway, the metal doping over large-sized titania particles is less efficient due to the large bulk volume of recombination.

For environmental applications water suspensions of titania are often used (Herrmann 1999, Herrmann et al. 2002). The suspension is really a very high stability hydrocolloid, which makes the catalyst separation from water difficult. It was found that structuring titania into nanotube shapes facilitates the recovery (Prado et al. 2009).

Nanostructured titania in the form of nanotubes (Sreekantan et al. 2009), often in arrayed configurations, gives a quantum-confinement effect that varies the band gap of the material, leads to a larger surface area, and permits filling of the interior free space with active materials (chemical compounds, enzymes, noble metals, etc.) to assess multifunctionality. In addition, 1D nanoscale titania offers easy handling and simple preparation.

### 18.4.3 Preparation Methods of Photocatalysts

Several nanosized titania are commercially available, Titania P25 by Degussa being the most investigated and used benchmark among Ti-based photocatalysts. It contains both anatase and rutile phases, and possesses a primary particle size of about 20 nm and a specific surface area (SSA) of about 50 $m^2/g$ (Porter et al. 1999). Pure anatase titania with crystallite size ranging from 5 (PC 500) to 85 (PC10) nm are produced by Millennium Inorganic Chemicals (Ciambelli et al. 2008). HombiCat UV is an anatase-modified titanium dioxide (claimed crystal size less than 10 nm) developed for photocatalytic processes by Sachtleben Chemie GmbH. NaBond offers <100 nm anatase and rutile nanopowders.

SSA, phase composition, and average crystallite size of several commercial products are reported in Table 18.1.

The principal current technologies that use titania dense nanoparticles are the sol–gel technique, the gas-phase decomposition process, and the spray flame pyrolysis (FP).

In the sol–gel technique titania precursors in organic solvent such as titanium ethoxide (TEOT) (Jung et al. 2002), titanium tetraisopropoxide (TPT), (Jung et al. 2002, Hafizah and Sopyan 2009), $TiCl_4$ (Tseng et al. 2006), tetrabutyl titanate (TBT) (Mao et al. 2005), $TiOCl_2$, and similar precursors are added by distilled water drops, resulting in a slow process of hydrolysis-polymerization. The obtained white solution is then filtered to get a precipitate of $TiO_2$ gel powder and dried at 200°C for a short duration with time. The $TiO_2$ gel is then calcined at a particular with selected temperature to achieve the desired crystalline titania phase. It is well known that the morphology and the crystallization of particles depend on sol preparation conditions.

In the case of the TPT/water/ethanol system (Hafizah and Sopyan 2009), the anatase phase, typically as pseudospherical-shaped particles, is obtained at around 400°C. The degree of hydrolysis plays a significant role in the formation of powders. A lesser quantity of added water slows down the hydrolysis rate and, as a consequence, the polymerization

**TABLE 18.1**

Specific Surface Area, Crystalline Phase, and Crystallite Average Size for Several Commercial Nanotitania

| Sample | Specific Surface Area (BET), $m^2/g$ | Anatase Phase, wt. % | Rutile Phase, wt. % | Crystallite Average Size[a], nm |
|---|---|---|---|---|
| PC10 | 11 | >99 | — | 85 |
| PC50 | 50 | >99 | — | 25 |
| PC100 | 87 | >99 | — | 20 |
| PC105 | 86 | >99 | — | 23 |
| PC500 | 345 | >99 | — | 5 |
| $TiO_2$ CISE | 120 | >99 | — | 12 |
| P25 | 54 | 80 | 20 | 25 |
| UV100 | 337 | >99 | — | 5 |
| Na bond anatase-$TiO_2$ | 120 | 99.5+ | — | 20 |
| Na bond rutile-$TiO_2$ | 30 | — | 99+ | 80 |

[a] By Scherrer equation.

process, yielding nanosized particles that are more uniformly sized. When the water/TPT molar ratio increases, hydrolysis and polymerization accelerate, and rapid hydrolysis could result in the formation of large inhomogeneous nonspherical particles.

A two-step modification of sol–gel method has also been performed. $TiO_2$ sol is prepared by a chemical coprecipitation–peptization method (Tseng et al. 2006). An aqueous $NH_3$ solution is added to a $TiCl_4$/DI water solution at 48°C to ensure complete hydrolysis, producing a white precipitate. Subsequently, a yellow transparent $TiO_2$ sol is obtained after 2h of peptization with hydrogen peroxide (10%) and 24h of heating at 95°C. The resulting $TiO_2$ sol contains arrowhead-like crystals less than 30nm in size, with a degree of crystallization in anatase phase strongly dependent on the drying temperature.

Another approach in the sol–gel method is based on microemulsions, in which an aqueous phase, a surfactant, and an oil phase are stably and isotropically dispersed in an oil phase. Dispersed water phase droplets (typical size 10–50nm) are used for nanoconfined synthesis of particles. The formation of stable and nanosized $TiO_2$ nanoparticles via hydrolysis of titanium isopropoxide in microemulsion has been also reported (Zhang and Gao 2002). A solution of TPT in isopropanol was added to water, Span-Tween80, and toluene microemulsion, yielding $TiO_2$ precipitate. The water/surfactant ratio controlled the diameter of nanosized particles, allowing a narrow distribution of spherical nanoparticles to be obtained.

High-temperature techniques to produce stable nanophotocatalysts have also been investigated.

In the gas-phase decomposition, titanium alkoxide is evaporated at a chosen temperature, usually below 100°C, and the vapor is carried by a high purity inert gas (typically nitrogen), in the absence of water vapor and oxygen, to a reactor kept at 500°C–900°C to induce the organic precursor pyrolysis. Thimble filters are used to collect the particles produced. In the case of nanosized titania prepared via gas-phase decomposition, it is possible to control the crystallinity by changing the reaction temperature with a relatively low influence on the particle size (Jung et al. 2002).

Despite the gas-phase decomposition producing nanosized particles, the method suffers from low productivity and is limited to only volatile precursors as raw materials. In order to overcome this limit, the development of aerosol or liquid-feed (FP) has been proposed (Chiarello et al. 2005, 2008).

Ultrasound-assisted spray pyrolysis is well known as a technique to prepare ceramic powders of submicrometer size (Kang et al. 1996). In the spray pyrolysis method, which is similar in principle to gas-phase decomposition where a precursor is decomposed at high temperature, a nebulized solution containing the precursor, usually an aqueous and acidic phase where the alkoxide precursors are hydrolyzed, is fed to the decomposition section. The droplets generated by a nebulizer are air carried into the furnace at a temperature of 500°C–900°C and the formed particles are collected by a thimble filter (Figure 18.2). In this range of temperatures, submicronized particles are obtained.

FP is a very high temperature, and hence effective, modification of the spray pyrolysis technique.

A specifically designed burner (Chiarello 2005) allows the feeding of a mixture of oxygen and an organic solution of the titanium precursors through a nozzle (Figure 18.3), where the solvent acts as fuel for the flame. The mixture is ignited by a surrounding ring of $O_2$ and $CH_4$ or other fuel. The short residence time in the flame and the high temperature assure the decomposition; moreover, optimizing the main operating parameters such as liquid feeding rate, the flow rate of $O_2/CH_4$ mixture, the linear velocity of the dispersing-oxidizing oxygen, the required crystal phase of the product, and its structural homogeneity

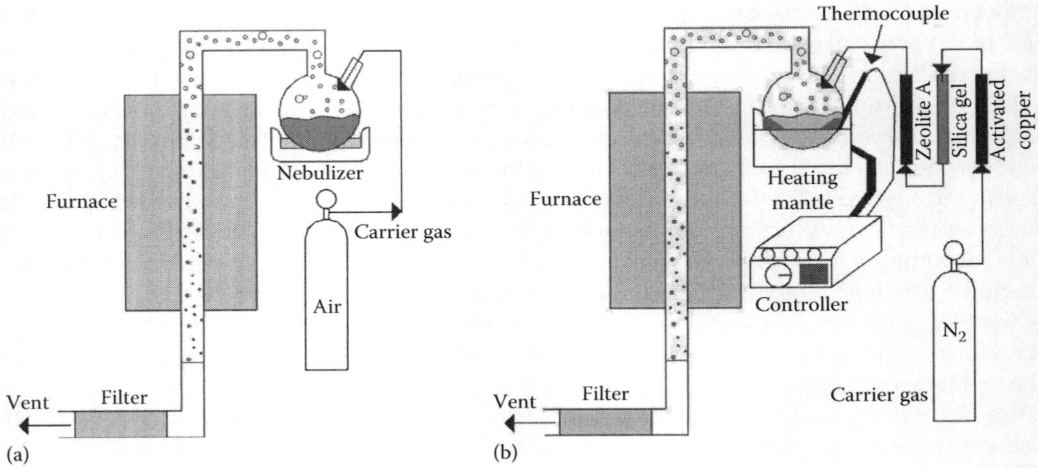


**FIGURE 18.2**
Sketched lab plants for the preparation of titania particles by (a) spray pyrolysis and (b) gas-phase decomposition of TPT. (Reprinted from Jung, K.Y. et al., *Appl. Catal. A: Gen.*, 22, 229, 2002. With permission of Elsevier.)

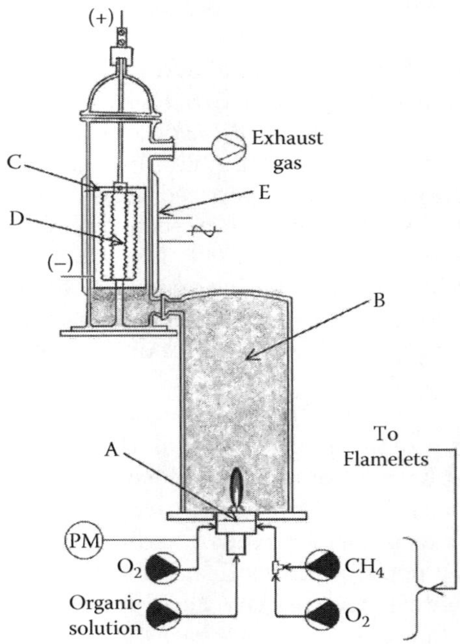


**FIGURE 18.3**
Scheme of the FP apparatus. A, burner; B, Pyrex glass conveyor; C, collector; D, multipin effluviator; E, heating mantle. (Reprinted from Chiarello, G.L. et al., *Appl. Catal. B: Environ.*, 84, 332, 2008. With permission of Elsevier.)

can be achieved. Nanometer-sized particles with high surface area (>100 $m^2/g$) and high purity can be obtained with high productivity, phase purity, and improved thermal resistance. In the synthesis of titania nanoparticles, the phase transformation of metastable anatase to rutile can be retarded and reduced by achieving fast crystal growth with a short residence time at a high reaction temperature.

Several organic solvents can be used such as xylene, propionic acid, pyridine, methanol, etc. It was found that the surface area of the FP-synthesized nanocatalysts linearly decrease with an increase in the combustion heat of the organic solvent/fuel, as a result of a higher flame temperature with a consequent increase in the rate of particle growth (see below). Thus, larger particles, possessing lower SSA are obtained.

It is worthwhile underlining that both the surface area and the titania crystallite dimensions of FP materials depend on the geometry of the burner employed for the synthesis, which allows a more efficient dispersion of the liquid solution by a narrower nozzle.

Other high heat sources have been used, as in the case of laser-pyrolysis with titanium alkoxide. It is worthwhile to note that this technique was developed as early as 1992. The titania nanoparticles obtained by $CO_2$ IR laser pyrolysis at high power were quite monodimensional in size (around 50 nm) and very stable against aggregation (Ciambelli et al. 1992, Musci et al. 1992).

Titania nanotubes are produced by a variety of techniques, including hydrothermal hydrolysis, template synthesis, and anodization. The latter method enables the production of well-ordered titania nanotube arrays.

An example of hydrothermal synthesis of titania nanotubes is reported by Kasuga et al. (Kasuga et al. 1999). In a typical procedure, 2 g of rutile $TiO_2$ powder in 85 mL of 10 M NaOH aqueous solution was into a Teflon-lined autoclave 130°C for 72 h. After filtration, water washing, and drying, in order to improve the crystallization of titanate nanotubes, calcination at 400°C for 1 h in air was performed, yielding $TiO_2$ nanotubes with high crystalline structure.

Template synthesis can shape the form and dimensions of a material through a matrix that constitutes the "negative" of the desired architecture. Template synthesis could be conducted as replicas of a porous membrane, typically with cylindrical pores of uniform diameter, and a nanocylinder or a nanofibril could be tailored in each pore in dependence of the properties of the material and the chemistry of the pore wall. On the other hand, a nanocylinder could be shaped around needle-like crystals, such as aragonite calcium carbonate. The tailoring was carried out (Qian et al. 2010) by using needle-like calcium carbonate and octadecylamine as double templates at room temperature in a nonaqueous system with tetrabutoxytitanium titania nanotubes with regular tubular morphology. Titania nanotubes, up to 15 μm long, with inner diameter of 400 nm, and a wall thickness of 40 nm, resulted in a high SSA (112 $m^2/g$).

Vertically oriented nanotube arrays were obtained by electrochemical anodization (Ghicov and Schmuki 2009). By adjusting the anodization parameters (temperature, potential rate, applied potential, electrolyte species, electrolyte pH, viscosity, aqueous or organic electrolyte, etc.), well-defined, self-organized, orthogonal titania nanotubular layers can be obtained. The morphological characteristics can be finely controlled, resulting in uniform titania nanotube arrays of various pore sizes (22–110 nm), length (200 nm–1000 μm), and wall thickness (7–34 nm).

### 18.4.4 Characterization of Nanoparticles for Photocatalysis

Scanning electron microscopy (SEM) allows examination of nanoparticle topographies at very high magnifications (up to 300,000×). SEM inspection is often used for the analysis of pores, cracks, and fractures of surfaces as well as morphology of samples.

Transmission electron microscopy (TEM) permits higher magnifications and spatial resolution than SEM, in the range of a few nanometers, and gives evidence also for the crystallographic structure, morphology, and of the composition of a nanoparticle.

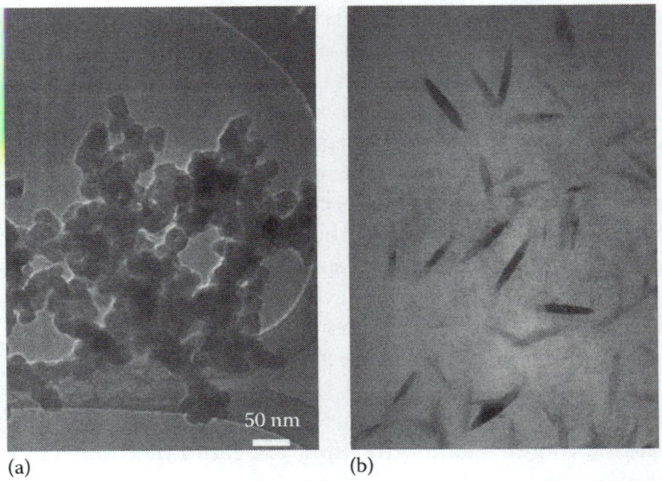


**FIGURE 18.4**
TEM images of (a) pseudospherical (P25) and (b) elongated titania nanocrystals by coprecipitation–peptization method. (Reprinted from Tseng, Y.-H. et al., *Micro Nano Lett.*, online no. 20065035, doi:10.1049/mnl:20065035 ristretto, 2006. With permission of Elsevier.)

Two examples regarding titania nanoparticles are presented in Figure 18.4.

Characterization in terms of SSA gives another significant contribution to evaluate and compare the nanocatalyst moieties. The measurement is typically based on the physical adsorption (or van der Waals adsorption) of suitable molecules (adsorbate), which can enrich the interfacial layer of a solid (adsorbent) upon exposure to an adsorbing solid to a gas or vapor, without chemical reaction occurring. Energy of interaction is low, less than 15 kJ/mol, and thus the adsorption is a reversible phenomenon.

The adsorption extent (denoted $n$) on a solid surface can be described at constant temperature and within the limits of vacuum and the saturation vapor pressure at which condensation takes place, i.e.,

$$n = f\left(\frac{p}{p_0}\right)T, \text{adsorbent, adsorbate}$$

where $p/p_0$ is the relative gas/vapor pressure. The amount of gas adsorbed when the mono-layer is saturated is proportional to the entire surface area of the sample. From the $N_2$ equilibrium adsorption isotherm as function of $N_2$ partial pressure, typically performed at the boiling point of pure liquid nitrogen ($T = 77.2$ K), the number of $N_2$ molecules necessary to have a uniform monolayer coverage of solid surface adsorption can be evaluated; it is multiplied by the area projected for a single molecule (6.2 Å$^2$ for $N_2$), to get the SSA value.

The low temperature is necessary to guarantee that no dissociation will occur or transformation of nitrogen, which can change the $N_2$ projected area. The measurement has to be carried out on a pretreated sample at suitable temperature and under vacuum to remove the surface contaminants beforehand. Both static volumetric or dynamic apparatus can be useful for the measurement.

According to IUPAC classification, six complete adsorption/desorption isotherm types take place in dependence of the different gas–solid interactions, and their trend represents a standardized way to classify the solid and to select the appropriate model to get the value of SSA and porosity characteristics.

The Type I isotherm (Figure 18.5) is characteristic of microporous solids (pores below 2 nm). Type II and IV arise from non-porous solids representing unrestricted monolayer–multilayer adsorption (Figure 18.5). Point B, at the beginning of the almost linear rise of adsorbate amount, is often taken to individuate the completion of monolayer coverage and the beginning of multilayer adsorption. Type III and type V are typical of low interaction between adsorbent and adsorbate, for example, water vapor on hydrophobic materials, while type VI is a quite rare step-like isotherm. The hysteresis loop is a feature of Type VI and V isotherms, generated by the capillary condensation of the adsorbate within the mesopores (pores in the ranges 2–50 nm) of the solid.

In particular, for type II and IV, the Brunnamer, Emmett, and Teller (BET) model is appropriate to evaluate the SSA, while Dubinin or Langmuir models are used for the evaluation of micropore volume from types I and III isotherms. The micro- and mesopore volume and size distributions can be obtained by the desorption branches of isotherm by different methods such as Dollimore-Heal (DH), Barrett Joiner Halenda (BJH), S&F or by adsorption branch with Horwath and Kawazoe (H&K) theory. For a deeper description, see the book by Gregg and Sing (1982).

The pores of the solid could be formed by the aggregation of primary particles (interparticle porosity) or be present in the primary particles (intraparticle porosity). In the absence

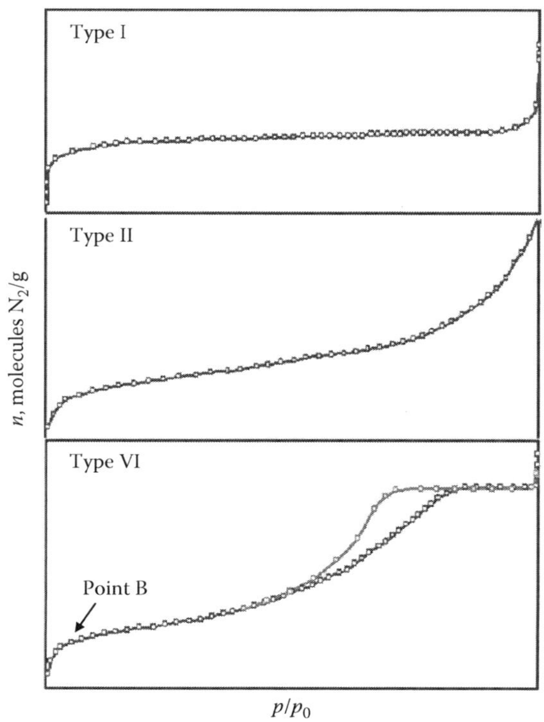


**FIGURE 18.5**
Type I, II, and IV standard adsorption isotherms according to IUPAC classification.

of intraparticle porosity, simple geometrical models could be employed to evaluate the particle size. For dense particles, the surface-to-volume ratio, $S_p/V_p$, divided by the particle density is the specific surface area, SSA:

$$\text{SSA} = \frac{S_p}{V_p \times \rho_p} \equiv \text{BET surface area}$$

For a sphere, the particle diameter is

$$D_p = \frac{6}{\text{SSA} \times \rho_p}$$

where
$V_p$ is the particle volume
$S_p$ is the surface area of a single particle
$\rho_p$ is the particle density
$D_p$ is the particle size

Since typically there is a particle size distribution, this simple calculation yields the average particle size. Specific techniques such as laser diffraction are used to evaluate size distribution, apart from direct electron microscopy observation.

This calculated size is in good agreement with that measured by SEM for pseudospherical titania particles (Jung et al. 2002), and the comparison permits to confirm of the absence of intraparticle pores. When the latter are present, they contribute to the total specific surface area, giving higher BET values.

For the titania particles prepared by the spray pyrolysis of TEOT and gas-phase decomposition of TPT (Jung et al. 2002), the comparison indicated that dense particles were obtained both in submicrometer and nanometer sizes. At similar crystallite sizes, evaluated by XRD (see Figure 18.8), the surface area of titania particles increased by reducing the particle size in the nanometer range.

SSA of FP-synthesized samples vs. combustion enthalpy of the solvent/fuel is shown in Figure 18.6. It was observed that both the surface area and the titania crystallite dimensions of FP materials depend on the geometry of the burner. Indeed, FP titania nanopowders prepared with a different burner could possess a higher surface area ($106\,m^2/g$) as a consequence of the more efficient dispersion of the liquid solution by a narrower nozzle. This would suppress the formation of the bigger particles, with the consequent increase of surface area.

As it is well known, the identification of the crystalline phase and the relevant size is obtained by the XRD patterns. The crystallite size (Figure 18.7) is evaluated by the Scherrer formula

$$t = \frac{K \times \lambda}{B \times \cos\theta_B}$$

where
$t$ is the thickness of crystallite in the direction individuated by Miller indices
$K$ is the constant dependent on crystallite shape (0.89 for Cu K$\alpha$)

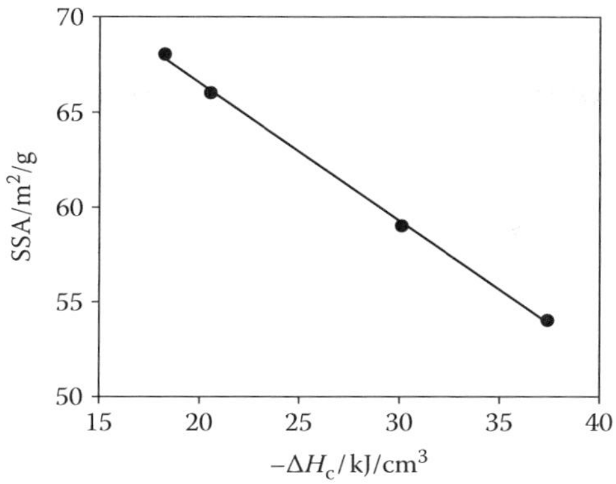


**FIGURE 18.6**
SSA of titania particles prepared by spray pyrolysis at several temperatures vs. combustion enthalpy of the fuel. (Reprinted from Chiarello, G.L. et al., *Appl. Catal. B: Environ.*, 84, 332, 2008. With permission of Elsevier.)

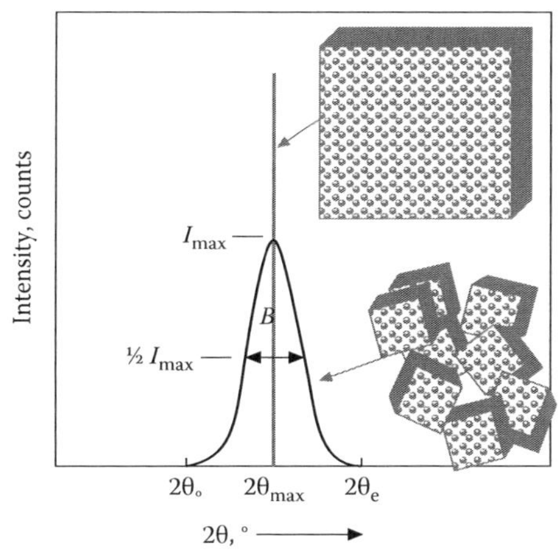


**FIGURE 18.7**
XRD peak broadening as function of crystallite dimension and relevant terms of Scherrer equation.

$\lambda$ is the x-ray wavelength
$B$ is the FWHM (full width at half max) or integral breadth
$\theta_B$ is the Bragg angle

The Scherrer formula can be applied when the crystallite size is <1000 Å, after excluding the peak broadening due to other factors, for example, the instrumental ones. The accuracy of estimation is at best within 20%–30%.

In Figure 18.8, XRD patterns of titania samples prepared by gas-phase TPT pyrolysis are reported and compared to $TiO_2$ Degussa (50 $m^2$/g). Pure anatase without rutile or brookite is formed up to 600°C. A small amount of rutile phase is formed at 700°C. The crystallite size of titania particles evaluated by the Scherrer equation increased from 28 to 32 nm with increasing process temperature.

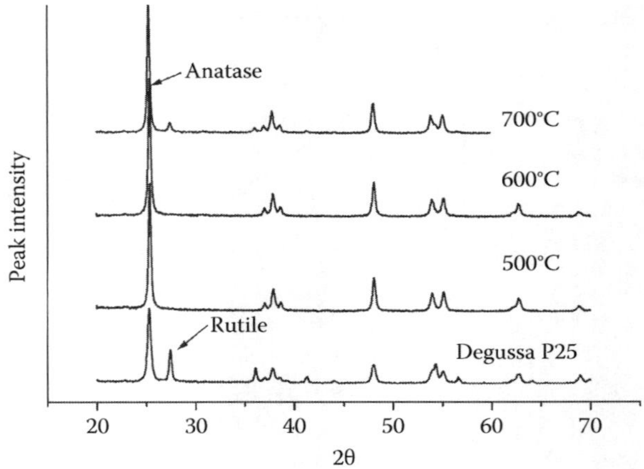


**FIGURE 18.8**
XRD patterns of nanosize titania prepared by the gas-phase pyrolysis of TPT. (Reprinted from Jung, K.Y. et al., *Appl. Catal. A: Gen.*, 22, 229, 2002. With permission of Elsevier.)

Degussa P25, widely used as the benchmark photocatalyst for its outstanding photocatalytic activity in both oxidative and reductive paths, is synthesized through the Aerosil process, consisting of the hydrolysis of $TiCl_4$ in an oxy-hydrogen flame.

By applying the Scherrer formula, the diameter of crystallites of the different phases can be evaluated as a function of the operating conditions of preparation, as reported by Chiarello et al. (2008). Different FP-synthesized photocatalysts exhibited smaller anatase crystallite size with respect to rutile crystallites when compared to P25 (Teoh et al. 2007). FP nanophotocatalysts with a high amount of anatase phase (90 wt.%) and smaller crystallite size result in powders with higher surface area (Chiarello et al. 2008).

Gold-loaded titania samples were also prepared by FP, and the presence of very small and well-dispersed gold nanoparticles on the $TiO_2$ surface, with a mean diameter around 1 nm was evidenced by HRTEM, appearing as bright dots (Chiarello et al. 2008). Moreover, the average gold particle size (2 nm) of the FP-made $Au/TiO_2$ was smaller than that of the gold-modified sample obtained by deposition of gold on Degussa P25 indicating the effectiveness of the FP technique to achieve high metal dispersion on $TiO_2$.

### 18.4.5 Size Effects on the Photocatalytic Activity

The main operating conditions in photocatalytic tests are reaction temperature, catalyst loading, incident light intensity, initial pH, feed, and concentration of reactants.

The use of nanosized metal oxides, typically semiconductors, as photocatalysts was focused on the degradation of pollutants in water (Alfano et al. 2000), self-cleaning windows and buildings, and self-sterilization (Jensen et al. 2004). Many applications are in fact devoted to the purification of air and water, both in outdoor and indoor environments.

The photodecoloration of methylene blue was largely adopted as test reaction to compare the activity of nanophotocatalysts. A well-dispersed $TiO_2$ aquatic sol of anatase, with high photoactivity and small particle size, was prepared by chemical coprecipitation–peptization method (Tseng et al. 2006). In the sol state, $TiO_2$ exhibited a peculiar visible-light-responsive photoactivity in the photodecoloration of methylene blue. The plausible cause was the N impurity on $TiO_2$ particles in the sol state.

In a study on the effect of the hydrolysis degree on the properties of nanosized $TiO_2$ powder photocatalyst obtained via sol–gel method (Hafizah and Sopyan 2009), the optimization of both crystallinity and particle size induced the efficient removal of phenol in water, the former responsible for the higher turnout of charge carriers at the photocatalyst's surface to enhance the photocatalytic activity. By this method, very small particles could create a "screening effect" in the solution that can limit the penetration of UV light and slow down the degradation rate.

By spray pyrolysis, (Jung et al. 2002) submicrometer titania particles can be obtained in the range 500°C–900°C, but their dimensions remain quite large—about 600 nm. A nanosized titania was successfully prepared by the gas-phase decomposition of TPT, showing higher photoactivity in trichloroethylene decomposition with respect to spray pyrolyzed particles, because of the increased surface area and the larger anatase crystallite size obtained.

Kormann et al. (1988) reported quantization effects with 20–40 Å titania particles (Kormann et al. 1988). The synthesized anatase $TiO_2$ particles exhibited a band gap shift of 0.15 eV with respect to bulk anatase $TiO_2$, while the synthesized rutile $TiO_2$ particles (2–5 nm diameter) were blue shifted by 0.1 eV.

When considering semiconductors other than $TiO_2$, many reports have observed quantum size effects with many different particles. For example, CdS and PbS, a few nanometers in size, exhibit quantum mechanical effects (Beydoun et al. 1999). The Q-particles can have a different color depending on the particle size. For example, cadmium sulfide is typically yellow, but it becomes colorless when the particle size is lower than 22 Å. Size-quantized (20 Å) CdS nanocrystals have been applied as photocatalysts for reduction of nitrate to ammonia and for $CO_2$ photoreduction (Beydoun et al. 1999 and references therein).

### 18.4.6 Influence of Morphology on the Photocatalytic Activity

A recent example of morphological effects on photocatalytic properties concerns nanowires. $TiO_2$ nanowires (NWs) were synthesized through a one-step hydrothermal process in 10 M NaOH (aq.) at 150°C for 72 h and post–heat treatment at 300°C–1000°C for 2 h. As effect of the temperature increase the $TiO_2$ NWs were first transformed into $TiO_2$ (B), a metastable polymorph of titanium dioxide derived from natural anatase, then into anatase and rutile gradually, preserving the 1D morphology (Figure 18.9). The final nanowires exhibited a high photocatalytic $H_2$ evolution rate (Jitputti et al. 2008).

Nanocatalysts are highly suitable for the formation of thin films of photoactive materials. Nanostructured $TiO_2$ films with controlled morphology and thickness were synthesized (Zhu and Zäch 2009) by single-step flame aerosol reactor at atmospheric pressure for use in water-splitting photocells and dye-sensitized solar cells (Thimsen et al. 2008). Two different morphologies were compared: a granular and a highly crystalline columnar morphology. In particular, the latter morphology consisted of single-crystal structures, approximately 85 nm in width and oriented normally to the substrate. Due to differences in electron transport and lifetime in the $TiO_2$ film, the columnar morphology outperformed the granular morphology in UV water splitting with a conversion efficiency of 11% for an optimum thickness of 1.5 μm as a swap between light absorption and electron transport losses. In the same way, superior performances with respect to granular morphology were shown by columnar titania nanostructures, with a visible-light-to-electricity-conversion efficiency of 6.0% in the dye-sensitized solar cell.

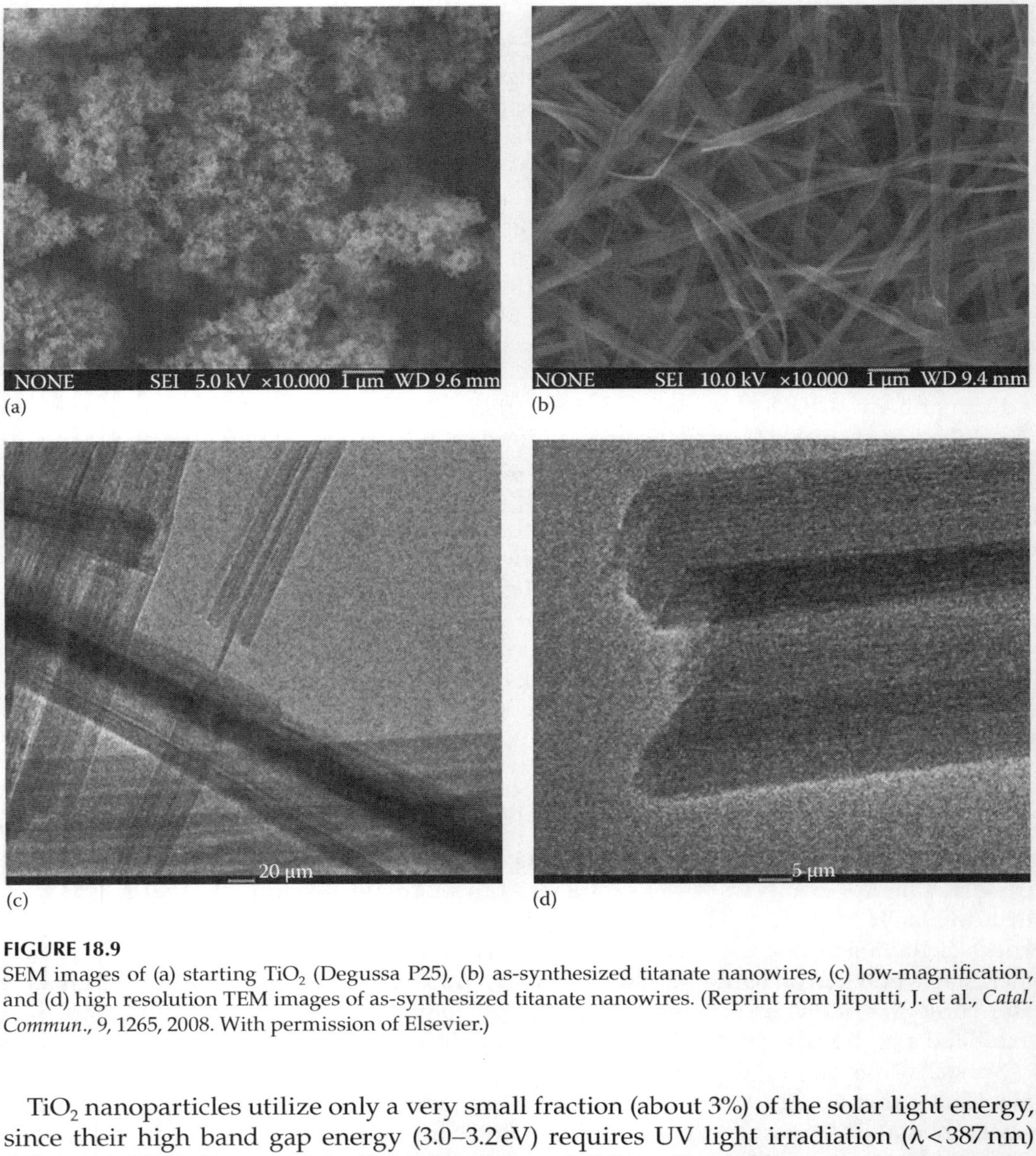


**FIGURE 18.9**
SEM images of (a) starting $TiO_2$ (Degussa P25), (b) as-synthesized titanate nanowires, (c) low-magnification, and (d) high resolution TEM images of as-synthesized titanate nanowires. (Reprint from Jitputti, J. et al., *Catal. Commun.*, 9, 1265, 2008. With permission of Elsevier.)

$TiO_2$ nanoparticles utilize only a very small fraction (about 3%) of the solar light energy, since their high band gap energy (3.0–3.2 eV) requires UV light irradiation ($\lambda < 387$ nm) (Zhu et al. 2008). To extend the light absorption of $TiO_2$ to the visible region, nanocomposites of $TiO_2$ with other semiconductors or sensitizing agents are often investigated.

CdS, a narrow band semiconductor with a higher conduction band than $TiO_2$, can photosensitize $TiO_2$ but also induce an efficient and longer charge separation, minimizing the electron–hole recombination (Linsebigler et al. 1995). Therefore, in order to efficiently use the visible light in the photocatalytic reaction, bamboo-like CdS/$TiO_2$ nanotubes were prepared. These CdS/$TiO_2$ nanotube composites have shown a much higher visible light photocatalytic activity for the degradation of methylene blue compared to $TiO_2$ nanotubes and CdS nanoparticles. Maximum photodegradation efficiency after 6 h irradiation can reach 84.5%. Framework-embedded, CdS quantum-dot-sensitized, ordered mesoporous $TiO_2$ (Figure 18.10) was fabricated by planting CdO as a seed

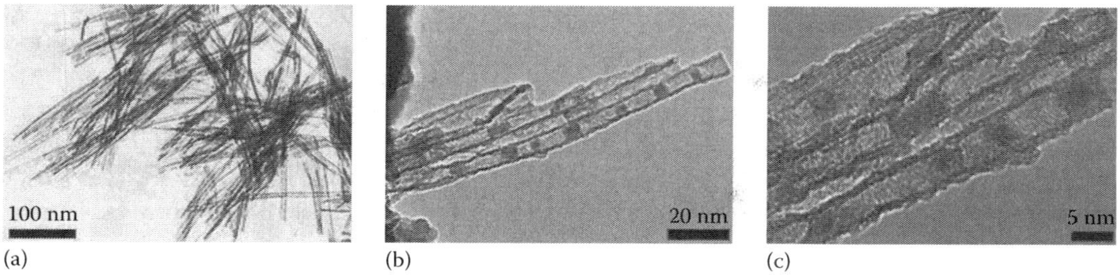


**FIGURE 18.10**
TEM and HRTEM images of (a) $TiO_2$ nanotubes and (b,c) $CdS/TiO_2$ nanotubes at different magnifications. (Reprinted from Zhu, J. et al., *J. Nanopart. Res.*, 10, 729, 2008. With permission of Springer.)

into the $TiO_2$ network then converted to CdS by ion-exchange at room temperature. The resulting ordered mesoporous $CdS/TiO_2$ composites possess a well-crystallized anatase phase, large specific surface area, low band gap energy, and a tight contact between CdS and $TiO_2$, resulting in an excellent visible-light photocatalyst (Zhu et al. 2008).

### 18.4.7 Some Examples of Photocatalytic Reactions Affected by the Catalyst Nanostructure

Since 2004, the number of publications on nanophotocatalytic $H_2$ production has increased by a factor of about 1.5 times every year. Many papers studied the impact of different nanostructures and nanomaterials on the performance of photocatalysts, since their energy conversion efficiency is principally determined by the nanoscale properties. A good example of the effect of such nanoscale properties on the photocatalytic performance is given by an investigation on $H_2$ production by water photosplitting or methanol photoreforming on $TiO_2$ or $Au/TiO_2$ produced by FP synthesis (Chiarello et al. 2008).

The hydrogen production rate by water photosplitting increased with increasing anatase content, surface area, and crystallinity of the nanophotocatalyst, which are related to the decrease of crystal defects, brought about by the high temperature experienced by the particles being formed during their passage along the flame. The highest rate of hydrogen production was attained with Au-modified nanotitania samples. Further addition of methanol increased the hydrogen production rate by 3 orders of magnitude with respect to that obtained with the corresponding bare $TiO_2$ and in the absence of methanol. Moreover, the addition of gold entrained a 30-fold increase of $H_2$ formation rate in methanol photoreforming compared with that obtained using bare $TiO_2$.

Nanosized titania nanoparticles offer a high surface area useful for the dispersion of active and/or selective species for catalytic reactions. A typical example of selectivity requirement is the partial oxidation of hydrocarbons to obtain organic compounds.

Selective photocatalytic oxidative dehydrogenation of cyclohexane to cyclohexene or benzene (Ciambelli et al. 2008), and of ethanol to acetaldehyde (Ciambelli et al. 2009a), was studied using a gas–solid fluidized-bed photoreactor (Ciambelli et al. 2009b).

In particular, the effect of sulfate doping of titania in promoting activity and selectivity of $MO_x/TiO_2$ catalysts for the cyclohexane photooxidative dehydrogenation was investigated on nanosized titania (Ciambelli et al. 2008 and references therein). Titania modification with sulfate and/or molybdate was carried out by incipient wet impregnation of suitable precursor on anatase nanocrystallites 5–10 nm in size. Due to the high specific area, at 60% of titania surface coverage by $MoO_x$, sulfate surface density up to 19 μmol/m² was obtained without the segregation of molybdenum species as $MoO_3$. Photocatalytic selective oxidation of cyclohexane to benzene on $MoO_x$/nanosized titania was enhanced by the sulphation. For gas–solid systems, the direct application in the packed fixed-bed photoreactors involves a high pressure drop inside the bed, limiting the amount of loadable nanocatalyst. In contrast, in fluidized-bed photoreactors, nanocatalysts have been successfully fluidized by using a Geldart class A diluent without the occurrence of elutriation phenomena. On Mo-based catalysts, polymolybdate species enabled 2.4 wt.%-sulfated titania to convert cyclohexane to benzene (99% selectivity) and cyclohexene, without any $CO_2$ formation. Selective cyclohexane conversion grows almost linearly with sulfate surface load up to 2.4 wt.%, enhancing the benzene yield. Improvements in both photoxidative dehydrogenation activity and benzene yield were attributed to the increased surface acidity that promotes the cyclohexane adsorption.

Ciambelli et al. dispersed $VO_x$ species at different vanadium loading on titania PC105 nanoparticles (Ciambelli et al. 2009) by wet impregnation and studied these catalysts in ethanol photocatalytic partial oxidation with a gas–solid photocatalytic fluidized-bed reactor. Modification of the titania surface with V species allows high photoactivity and high selectivity to acetaldehyde.

Figure 18.11a shows the V/Ti atomic ratio by XPS and SSA values on $V_2O_5/TiO_2$ catalysts surface as function of coverage degree with respect to the $V_2O_5$ theoretical monolayer, evidencing a very good dispersion of vanadium species on catalyst surface up to $V_2O_5$ nominal load 9%. Parallelly, band gap energies and photoreactivity show the same trend as function of coverage degree (Figure 18.11b).

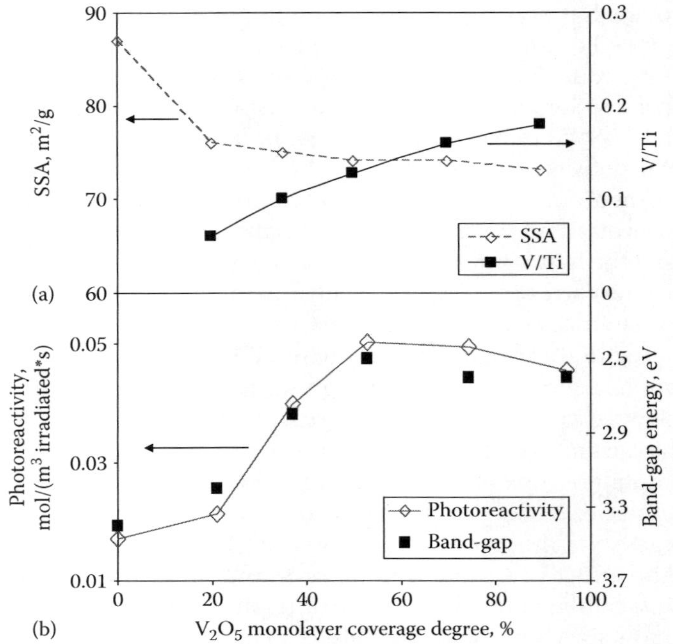


**FIGURE 18.11**
SSA and V/Ti atomic surface ratio (from XPS) on $V_2O_5/TiO_2$ catalysts (a) and surface and (b) ethanol reaction rates and band gap energy values as function of $V_2O_5$ coverage degree.

A quite linear correlation between photoreactivity and band gap energy values has been found (Ciambelli et al. 2009) for all sub-monolayer V-based catalysts because the decrease in band gap energies promoted the light absorption by the photocatalyst, yielding larger photoactivity. The latter increases since, in the presence of V-species, the recombination of the e$^-$–h$^+$ pair is slow and high photo-quantum efficiency of $V_2O_5/TiO_2$ catalysts is achieved (Chen et al. 2008).

## 18.5 Nanoparticles as Catalyst: The Case of Carbon Nanotube Growth

The increasing use of electron microscopy allowed to first observe carbon filaments in tubular form around 1950 (Tibbetts 1984), but only in the early 1990s were the so-called carbon nanotubes (CNT) having diameters values in the order of the nanometers evidenced (Iijima 1991). Since their discovery, carbon nanotubes became increasingly "popular" materials and generated a lot of research, due to their unique physical properties and potential applications such as in catalyst support, microsensors, field emission displays, supercapacitors, nanotube heterojunctions, nanoprobes, and electrical contacts (Chico et al. 1996, Niu et al. 1997, Hafner et al. 1999, Kong et al. 2000).

Carbon nanotubes are concentric graphitic cylinders, they can be multiwalled (MWNT) with a central tube of nanometric diameter surrounded by graphitic layers separated by about 0.34 nm, while single-walled nanotubes (SWNT) are constituted of only one graphitic layer. The first observed MWNTs were grown in an arc-discharge process, and two years later SWNTs were produced by the laser-ablation technique. During that time, the catalytic chemical vapor deposition (CCVD) method was first used to grow CNTs. CCVD immediately appeared as an effective way to the large-scale production of carbon nanotubes, at lower cost, for some specific applications. As an example, the capability to grow CNTs directly on a substrate at a desired position (a great challenge from the technological point of view) and at lower temperatures than arc discharge or laser ablation allowed the CNT growth by CCVD to be integrated in the fabrication processes of microelectronic circuits (Cheng et al. 1998).

It has been found that transition metals such as iron, cobalt, and nickel are catalysts for growing carbon nanotubes. However, bulk iron, as itself for example, is not able to catalyze the decomposition of methane to form carbon filaments; it must be first dispersed on some support (Ermakova et al. 2001). The CCVD process on a supported catalyst consists of several steps. Starting with the preparation of supported metal nanoparticles, different techniques can be used: sol–gel, coreduction of precursors, impregnation, incubation, ion exchange–precipitation, ion adsorption–precipitation, reverse micelle method, thermal decomposition of carbonyl complexes, metallo-organic chemical vapor deposition, and physical deposition (by evaporation, sputtering, etc.). After catalyst preparation, the CNT growing process requires that it be put in contact with a carbon precursor as vapor or gas phase inside a furnace at a given temperature. Carbon deposition occurs by decomposition of the precursor on the catalyst nanoparticles at temperatures generally ranging from 500°C to 1200°C.

The CCVD process for carbon nanotube growth is a very nice system for studying the eventual correlation between carbon nanotube geometry and nanoparticle property and size. Before discussing the state of knowledge on this issue, it would be helpful to start with a discussion on the growth mechanism of carbon nanotubes.

### 18.5.1 Growth Mechanism of Carbon Nanotubes

Two alternative basic models have been suggested in the literature: *tip growth* and *root growth*, which depend on the strength of interaction between catalyst nanoparticle and support (Figure 18.12). In the first case, the particle is detached from the support and is found inside the nanotube, typically at the tip. This occurs quite frequently when the metal–support interaction is not strong, or at least when this interaction is weaker than the metal–carbon interaction.

In the case of root growth, the metal particle always sticks to the support surface and the growth occurs as an extrusion of carbon from the metal particle. The advantage of the latter method becomes evident at the purification step, where both the support and the catalyst particle must be removed.

The mechanism of tip growth for filamentous carbon, as proposed by Baker et al. (Baker et al. 1972, 1973), is presented schematically in Figure 18.13. The deposition of amorphous carbon takes place at the site of contact of the metal particle with the support.

The formation of amorphous carbon is explained by the gas-phase polymerization of the hydrocarbon.

At the same time, the hydrocarbon is decomposed on the metal particle surface and the carbon atoms formed are dissolved in the bulk of the metal. Decomposition of the hydrocarbon and solubilization of carbon are accompanied by heat release, and this creates a temperature gradient along the particle. Owing to this gradient, carbon moves to a colder part of the particle in contact with the support. The deposited carbon raises the metal particle and removes it from the support. The initial filament growth causes curvature of the metal particle which becomes pear-shaped, and this in turn leads to the formation of

(a) (b)

**FIGURE 18.12**
Mechanism of tip and basal growth: (a) tip growth and (b) basal growth.

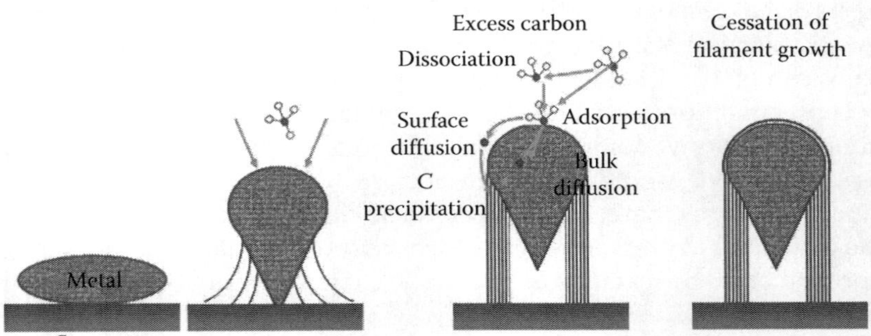


**FIGURE 18.13**
**(See color insert following page 302.)** Schematic representation of the tip mechanism of the filamentous carbon growth.

a channel. The increase in the filament growth rate is explained by the insulation of the particle from the support and the corresponding increase in the temperature gradient.

The excess carbon formed on the outer surface of the particle migrates from the surface and builds the external wall of the filament. It is presumed that the external wall of the filament has a different structure from that of its core. When the "hot" side of the particle is covered by carbon, the filament growth stops.

This hypothesis is supported by the fact that the activation energy of carbon filament growth on transition metals is similar to the activation energy of carbon diffusion in the bulk of the corresponding metal.

Therefore, if the formation of filaments is governed by the diffusion of carbon atoms in the corresponding metal, then under the same experimental conditions (temperature, time, carbon source, and catalytic particles), the growth rate will be proportional to the diffusion coefficient of carbon atoms in the corresponding metal.

For this reason, because the coefficient of carbon diffusion in metallic iron is 2–3 orders of magnitude larger than that in cobalt or nickel, the growth rate of carbon filaments on iron is higher than on other metal particles.

The mechanism of root growth differs from that proposed for the tip growth, essentially for the path of carbon atoms, that can diffuse to the colder part of the metal particles, as in the case of tip growth, on the surface of metal, as shown in Figure 18.14, determining a supersaturation and the formation of the cap. Finally, the growth of filaments, either in an isolated form or in bundles, takes place.

It was assumed that the driving force for diffusion is the temperature gradient arising between the sites where the decomposition of hydrocarbons occurs and the sites where the graphite phase grows. The higher the metal temperature, the larger the solubility of carbon in it; for this reason, a concentration gradient also appears.

The mechanism of growth of filamentous carbon brought about by carbon diffusion in a crystal has been accepted by most investigators; whereas the appearance of mass gradient determined by the temperature gradient has become the subject of a lively discussion (Robertson 1969, Evans et al. 1973, Keep 1977).

In order to assess the magnitude of the temperature gradient, Tibbetts et al. (1987) have simulated the process of the growth of carbon filaments from acetylene on Fe. To simplify the calculations, they assumed that the filament has no hollow channel and its diameter is equal to the diameter of the metal crystal initiating the filament growth. The

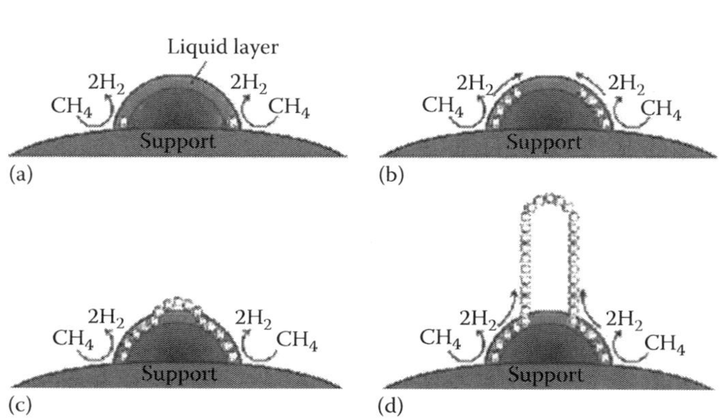


**FIGURE 18.14**
Schematic representation of the basal growth.

thermal balance of the growing filament end was calculated on the assumption that all the heat released in the decomposition of hydrocarbons is absorbed by the metal particle. Calculations showed that the temperature gradient is smaller than $2 \times 10^{-3}$ K, and such a small value cannot ensure the really observed growth rates of carbon filaments. Unsuccessful attempts were undertaken to prove a substantial influence of the temperature gradient (Yang and Yang 1985, Yang and Chen 1989).

An alternative explanation of the nature of the driving force of diffusion is based on the appearance of concentration gradient. In the model proposed by Nielsen and Trimm (1977), it was suggested that the solubility of carbon at the hydrocarbon–metal boundary is higher than that at the metal–graphite boundary. However, the data on solubility used in this model and those based on the results reported in Wada et al. (1971) are questionable.

The hypothesis of concentration gradient was also supported by investigating the growth of filamentous carbon upon disproportionation of CO on Co/Fe alloys (Audier et al. 1983a,b,c). It was assumed that the activity of carbon at the gas–metal interface is determined by the carbon activity in the gas phase. The activity of carbon at the metal–filament interface was assumed to be equal to the activity of graphite. On the basis of this simple assumption, a kinetic equation was deduced according to which the rate of carbon deposition depends linearly on the carbon activity in the gas phase. However, the explanation of the reasons for the appearance of the concentration gradient (Wada et al. 1971) seems to be supported by insufficient argumentations.

Another explanation is based on the carbide cycle mechanism. The carbide cycle mechanism includes two basic steps: (1) the "chemical" step, in which the catalytic decomposition of hydrocarbons occurs on the surface of a metal particle with the formation of carbon atoms, whose concentration increases to definite limiting values; (2) the "physical" step, in which the crystallization centers (nuclei) of the graphite phase are formed on definite faces of a metal particle. In this step, migration (diffusion in the bulk of the metal particle) of carbon atoms toward these centers and growth of a definite variety of graphite particles, predominantly in the form of filaments, begin. Despite the fact that the formation of carbon comprises two steps, the mechanism, on the whole, was called the mechanism of the carbide cycle since the carbon atoms involved in the graphite formation "originate" from intermediate carbide compounds.

The mass transfer of carbon atoms occurs owing to their diffusion in the bulk of metal particles from the formation site to the crystallization centers. Depending on the ratios of the rates of formation and diffusion of carbon atoms, the limiting steps of the process may be different. Let us consider the growth of carbon filaments under the conditions where the limiting step is the diffusion of carbon atoms in the metal particle.

In a saturated solution, the solid and dissolved phases possessing equal chemical potentials are in equilibrium. In our case, however, carbon on the front and rear sides of the metal particle is in different states that are characterized by different chemical potentials. The pool of carbon on the front side of the particle is continually replenished due to the decomposition of hydrocarbons. On the rear side, carbon is formed as graphite phase.

The concentrations of saturation (solubility) of carbon at these sites are substantially different; therefore, a concentration gradient arises, which ensures continuous dissolution of carbon on the front side and its diffusive mass transfer to the rear side of the metal particle. This gradient is rather high. The concentration of carbon atoms on the front side of the nickel particle formed by decomposition of intermediate carbide-like compounds is close to the concentration of carbon in $Ni_3C$ carbide, i.e., it reaches 25 wt.%. This value is confirmed by the calculation of the linear rate of carbon filament growth ($V$) (Chesnokov et al. 1982), as determined by the amount of carbon $Q$ that has diffused through an area section unit of a metal particle in unit time:

$$Q = \frac{D(c_1 - c_2)}{L} \quad V = \frac{D(c_1 - c_2)}{Ld}$$

where
- $D$ is the coefficient of carbon diffusion in the metal
- $c_1$ is the carbon concentration in the surface carbide-like compound
- $c_2$ is the carbon concentration in the saturated solution of carbon in nickel on the rear side of the metal particle
- $L$ is the diameter of the metal particle at the end of the filament
- $d$ is the density of graphite

These formulae were used to calculate the rates of the growth of carbon filaments, 60 nm in diameter, on metallic nickel at 873 K.

The decomposition of an intermediate carbide-like compound of the $Ni_3C$ type gives the concentration $c_1 = 0.514\,g/cm^3$; the values of $c_2 = 3.8 \times 10^{-10}\,g/cm^3$, $D = 3.89 \times 10^{-10}\,cm^2/s$, and $d = 2\,g/cm^3$ were taken from Buyanov (1983). The growth rate of carbon filaments found (165 nm/s) is close to the values measured experimentally (80–160 nm/s).

An analogous explanation of the reasons for the diffusion of carbon atoms in the metal particle is given in Sacco et al. (1984), De Bokx et al. (1985), Alstrup (1988), and Kuvshinov et al. (1998).

Until the beginning of the twenty-first century, the concept of the formation of carbon filaments by carbide cycle mechanisms was commonly accepted for the iron subgroup metals and their alloys (Kuvshinov et al. 1998). X-ray photoelectron spectroscopy and XRD measurements acquired in situ (de Heer et al. 1997) show the presence of iron carbide during the synthesis.

### 18.5.2 The Chemical Nature of Nanoparticles during CNT Growth

In general, the chemical nature of the catalyst is derived by TEM or diffraction measurements of the lattice constant of catalyst after growth. But two problems must be considered: (1) during growth, the catalyst contains excess carbon, which, on cooling, can precipitate as a carbide phase (Ducati et al. 2004) as the carbon solubility limit drops; (2) the metal and carbide phases can have rather similar lattice constants, so that they are not so easily distinguishable by diffraction, especially if the particles are small, and the lattice constant could be distorted by their small size. If TEM is used, care must be taken that the projection is correctly oriented against the lattice planes. For these reasons, more recently, the use of a combination of in situ photoemission and in situ TEM measurements (Hofmann et al. 2007, Mattevi et al. 2008, Hofmann et al. 2009) has been explored for characterization. The combination of these measurements suggests that the Ni catalyst during growth was in the form of metal, not as an oxide or a carbide.

In particular, the Ni and C spectra during CNTs growth result in their being dominated by the peak at 852.6 eV, which corresponds to Ni metal, and by a 284.3 eV peak, which corresponds to the $sp^2$ carbon of nanotube walls. In the same way, for Fe catalyst, the peak spectrum was dominated during and after growth by the peak at 707.3 eV due to Fe metal. There are weak features in the 711–709 eV range, which would be due to Fe oxides, but only before growth starts.

The carbon spectrum is dominated by the peak at 284.6–284.5 eV due to $sp^2$ carbon. This suggests that Fe exhibits initially a surface oxide, but during growth it is in the metallic state, and not as a carbide.

The conclusion of these studies is that for the main nanotube growth catalysts, the active state is the metal. This is consistent with most recent work. On the other hand, it can be noted that in situ photoemission measurements of de los Arcos et al. (2004, 2007) favored the oxide form. Yoshida et al. (2008) also show TEM evidence in favor of a carbide catalyst.

The results help to explain the behavior of Fe, Co, and Ni catalysts. Ni is active over a wide temperature range, Fe gives rise to the highest density nanotube forests. However, Fe is less active at low temperatures than Ni and, in surface growth, it is less active at highest temperatures; the decline in activity below 450°C could be due to it still being as an oxide. To reduce $FeO_x$, formed during sample transfer or preparation, can be used ammonia. This is not the case of NiO that has a lower heat of formation and consequently is more easily reduced. On the other hand, Fe also has a metastable carbide, while Ni has none. Thus, Fe catalyst can become slightly deactivated by the side reaction of carbide formation at higher temperatures.

Consideration of the other transition metals, for example, Ru, Re, Rh, Pd, etc., has been found to work as catalysts. The catalyst (see the schematic representation of the CNT growth, Figures 18.13 and 18.14) must perform two key functions: hydrocarbon dissociation and carbon solubility. Catalytic activity is often linked to carbon solubility. However, the highest solubilities occur for those metals that form carbides. A catalyst works by lowering the energy barrier to the overall reaction. This often occurs by forming intermediate species. On the other hand, any intermediate species such as a carbide should not be too stable, or it will not dissociate to give the product.

Figure 18.15 plots the formation energy of the carbides of the transition metals; the metals on the left form stable carbides, and this lowers their suitability as growth catalysts. The transition metals Fe, Co, and Ni form metastable or unstable carbides and show a moderate C solubility, which is one reason for their catalytic activity.

Hofmann et al. (2005) studied the diffusion mechanism of carbon on Ni nanoparticles, showing that carbon atoms diffuse on top of the closely packed (111) surface and subsurface for the less closely packed (100) surface, especially across the open ridges. The Ni–C bond is strong, C can enter an octahedral interstitial site in the Ni lattice, requiring a volume expansion. A surface site has fewer bonds than a bulk interstitial site, but it requires volume expansion; these lead to the optimum site being a subsurface interstitial, which

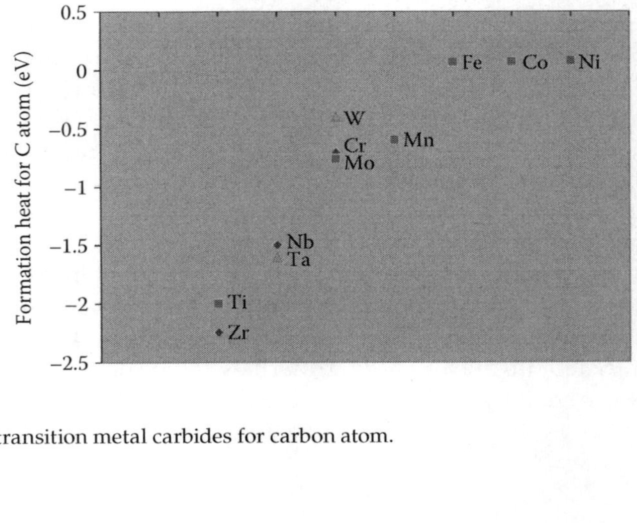


**FIGURE 18.15**
Heat of formation of transition metal carbides for carbon atom.

is fully bonded, but which allows some relaxation against the layer above it. This agrees with recent depth-resolved photoemission spectra of hydrocarbon reactions on Pd surfaces, which showed that the carbon actually lay 1–2 layers below the surface (Gracia et al. 2005, Teschner et al. 2006, Vass et al. 2008). In conclusion, the presence of a carbide signal in XPS does not necessarily prove the presence of a bulk carbide during growth, but just a subsurface carbide as in previous cases.

Hofmann et al. (2005) argued that surface (and subsurface) diffusion of C was the most probable diffusion path for SWNTs growth by chemical vapour deposition (CVD), as well as in plasma enhanced CCVD (Hofmann et al. 2003). A similar conclusion was reached by Raty et al. (2005).

Gold, which is a noble metal, was expected to be a poor catalyst for nanotube growth, with carbon being rather insoluble in bulk gold. This is consistent with calculations that show that C atoms are unstable in the center of a gold cluster (Raty et al. 2005). Ding et al. (2008) have observed that C–Au bonds are weaker than on most transition metals, so the nucleation cap would have difficulty to form. Recently, it has been found that gold nanoparticles are active for some oxidation reactions. This is due to two basic explanations: (1) the nanosize of the nanoparticles that exhibit a much larger surface area and much more of the atoms are active surface atoms with low coordination (Lopez et al. 2004), which might also alter their band structure; (2) electronegative metals can interact with defects in the oxide support and lead to a charge transfer that catalyzes reactions (Pacchioni et al. 2008).

On the experimental point of view, a number of researchers have been able to grow carbon nanotubes or carbon nanostructures using gold as catalyst, but in any case with low yield (Takagi et al. 2006); in some cases, SWNTs have been formed as indicated by the presence of radial breathing modes in the Raman spectra. This requires that gold catalyzes the hydrocarbon dissociation contemporaneously showing a C solubility. Teschner et al. (2006) studied the subsurface diffusion of C in Pd and suggested that the mechanism might involve subsurface diffusion of C in gold. In general, it must be considered that gold has a low solubility for carbon, and it is not so good in the dissociation of hydrocarbon precursors, requiring a higher temperature to overcome the reaction rate limit.

### 18.5.3 The Influence of Nanoparticle Electronic Structure on CNT Growth

The electronic structure also regulates the catalytic activity of nanoparticles. It is clear that the first action of the catalyst is to bond the hydrocarbon molecules to its surface, after that the hydrocarbon, in an adsorbate state, can interact with the catalyst by transferring an amount of its electron density to the catalyst. Generally, electron transfer from the catalyst to the non-occupied, antibonding orbitals of the adsorbate molecule takes place, resulting in a simultaneous back-donation. The dissociation of the molecule can occur as a consequence of the change in the electronic structure of the adsorbate. The transition metals with their non-filled *d* shells are able to interact with hydrocarbons showing catalytic activity. In particular, the properties that contribute to the ability to make and break adsorbate bonds are (1) the center of the *d*-bands, (2) the degree of filling of the *d*-bands, and (3) the coupling matrix element between the adsorbate states and the metal *d*-states (Ruban et al. 1997). We can conclude that the ability of a metal to catalyze the dissociation of a hydrocarbon molecule is correlated to its electronic structure. This can explain why iron is found to be more efficient than nickel and cobalt in hydrocarbon decomposition (Fonseca et al. 1996, Klinke et al. 2001). Some papers also reported that an added component that changes the electronic structure of the catalyst can lower the activation energy for dissociation, and thus the growth temperature (Harutyunyan et al. 2002). However, it is not clear whether differences in electronic structure between different catalyst materials can account for the observed differences in the quality of MWNT in terms of graphitization. This can explain why copper is able to yield only amorphous carbon, it is a non-transition metal with the 3d shell completely filled (Ivanov et al. 1994).

### 18.5.4 The Action of Stabilizing Components on the Catalytic Activity of Nanoparticles

Another important action during the catalytic activity of the nanoparticles is exercised by a stabilizing component. Molybdenum has been found completely inactive to catalyze CNT growth, while cobalt is unselective (i.e., yields formation of both SWNTs and MWNTs and eventually other carbonaceous deposits), on the other hand molybdenum–cobalt alloys have been found able to grow SWNT (Alvarez et al. 2001). Liao et al. (2003) also have studied the composition of cobalt–molybdenum nanoparticles and its influence on the carbon morphology. They found that particles with 5–15 wt.% cobalt tend to produce long CNTs, those with higher Co content, 40–45 wt.%, tend to produce short CNTs, and particles with 85–98 wt.% cobalt tend to produce onionated morphology.

In the cobalt–molybdenum alloys, cobalt is the real catalyst and molybdenum stabilizes $Co^{2+}$ ions. Indeed, in the absence of molybdenum, cobalt sinters in the reduced state and forms large metal aggregates, which generate defective MWNTs, filaments, and graphite nanofibers.

Not only molybdenum has been added as a non-transition metal additive to a transition metal catalyst (iron, cobalt, nickel), but also, for example, magnesium. The few authors who studied magnesium-, cobalt-, or nickel-based catalysts described formation of a solid solution like Ni(or Co)$_x$Mg$_{1-x}$O in the case of Ni, the $Ni^{2+}$ in the solid solution is found to be highly dispersed (host-dopant type) and thus difficult to reduce completely (valence stabilization by MgO crystal field) to $Ni^0$, which constitute the active sites. Deep reduction of nickel is therefore inhibited in $Ni_xMg_{1-x}O$, and thus the tendency to aggregate $Ni^0$ to form large particles (Chen et al. 1997).

Concerning the mixture of two catalysts, Pinheiro and Gadelle (2001) reported a thermodynamic study of the chemical state of a supported iron–cobalt catalyst during CO disproportionation. They showed clearly the dependence of the stability of the catalyst, and thus the progress of the disproportionation, on the composition of the catalyst alloy. The most stable alloy is found to be that with about 50 wt.% Co. More iron leads to the formation of $Fe_3C$ or $Fe_3O_4$ and more cobalt to $Co_2C$, in contact with CO-$CO_2$ mixtures.

### 18.5.5 The Effect of Supports on the Catalytic Activity of Nanoparticles

Since Co nanoparticles exhibit catalytic activity (Ciambelli et al. 2005) that changes when dispersed on different supports, and bi-metal catalysts exhibit better performances, we have studied a mixture of cobalt and iron catalyst for the synthesis of MWNTs at 700°C by CCVD of ethylene as carbon feedstock gas (Ciambelli et al. 2007). To explore the effect of the support, different aluminum hydroxides such as gibbsite, boehmite, and bayerite were used, which resulted in finding very high carbon and reaction yields with gibbsite and bayerite. Reaction time, gas volume/catalyst mass ratio and catalyst preparation influence the carbon nanotubes (CNT) growth. We have found, also, that both the drying time before the synthesis and the aging time affect the catalyst activity. CNT selectivity evaluated by TEM images (Figure 18.16) and thermogravimetric and Raman analysis depends both on support and operating conditions, reaching very high values with bayerite and gibbsite. CNT bundles grown on gibbsite show coiled ends (Figure 18.17), suggesting a different catalyst deactivation mechanism.

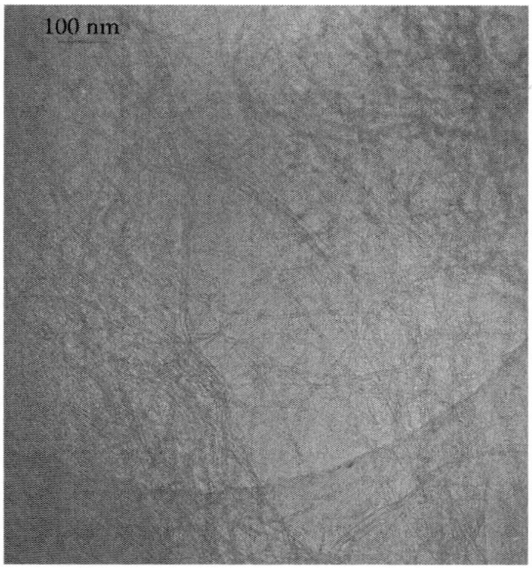


**FIGURE 18.16**
TEM image of carbon nanotubes produced by Co/Fe supported gibbsite.

**FIGURE 18.17**
SEM image of CNT produced on Co/Fe supported gibbsite.

### 18.5.6 The Nanoparticle Phase during CNT Growth

Another important question regarding the catalyst phase is whether the catalyst is in the solid or liquid form during the synthesis. It is necessary, first of all, to consider that the melting temperature of the catalyst is lowered by two effects: (1) The melting temperature is depressed by the Gibbs–Thomson effect for small diameter particles (Buffat et al. 1976):

$$\Delta T_{\mathrm{m}} = \frac{2T_{\mathrm{m}}\gamma}{L\rho r}$$

where

$T_m$ is the melting temperature in K
$\gamma$ is the surface tension
$L$ is the latent heat
$\rho$ is the density
$r$ is the particle radius

(2) The melting point is depressed by forming an eutectic with carbon. For CCVD, a single nanotube tends to grow from each catalyst nanoparticle (Choi et al. 2002), so the diameter of the catalyst tends to be equal to that of the resulting nanotubes (see below); hence the typical nanoparticle diameters are of the order of a few nanometers. This can reduce melting temperatures from 700°C–800°C down to 550°C, a significant reduction. This trend has been confirmed by molecular dynamics simulations (Ding et al. 2004).

For bulk CVD, such as in the injection methods for growing SWNTs, which use temperatures of about 1000°C (Li et al. 2006), the catalyst is likely to be in the liquid state. For low temperature CVD (480°C) on surfaces, recent in situ transmission electron microscopy studies of growth (Hofmann et al. 2007) show the presence of lattice planes in the catalyst nanoparticles, indicating that the catalyst has remained in the solid state. However, the severe mechanical re-shaping, especially during the tip growth of CNTs, can give the impression that the catalyst is liquid. On the other hand, this re-shaping occurs by creep in the solid state because of the large forces exerted by the surrounding carbon tube, and it is compatible with a solid form. The catalyst is solid in the remaining data.

### 18.5.7 The Relation between Carbon Nanotube Inner Diameter and Nanoparticle Size

As mentioned previously, the catalyst is able to catalyze the formation of CNTs only if it is in the form of particles. In the following text, the effect of the size and crystallographic orientation of the nanoparticles on the growth process will be discussed.

There is a consensus in the literature concerning the correlation between the size of the catalyst nanoparticles and the CNT diameter (Ivanov et al. 1994, Cheung et al. 2002, Zhang et al. 2002a,b). When the nanoparticles are prepared in holes or pores, the diameter of the former is subjected to the size of the latter and thus the resulting CNTs have a diameter roughly equal to the diameter of hole or pore (Duesberg et al. 2003). Nikolaev et al. said that the particle size after CNT growth is larger than the CNT diameter, suggesting that particles continue to grow even after nucleating a tube (Nikolaev et al. 1999). Thus, the relevant size of the nanoparticles, for the resulting diameter of the CNTs, is their size at the time of nucleation.

Dai et al. suggested that larger particles always appear to be onionated and so then are inactive for catalysis of CNTs (Dai et al. 1996). The formation of non-nanotube carbon was also reported by Perez-Cabero et al. with iron rich iron-silica catalysts (Perez-Cabero et al. 2003), they suggest that the degree of metallic iron aggregation upon reduction into metallic particles depends on the iron content. Thus, catalysts with higher iron content exhibit larger particles, and result inactive to form carbon nanotubes.

Therefore, there is experimental evidence that large particles are unable to catalyze the selective growth of CNT. It is probably due to their nanometer range that there may be a "size effect" contributing to the catalytic properties. For a particle size smaller than 5 nm, the number of atoms in low-coordinated positions is greater than 10% of the total

(Lambert et al. 1997). This may modify the electron density of the nanoparticle material or stabilize unusual faces or sites on the surface of the aggregates. Both will change the surface electronic structure, which can in turn affect the catalysis process, as discussed previously.

Beyond the size of the catalyst nanoparticles, another morphologic parameter might play a role in the catalytic growth process: the crystallographic orientation of the supported nanoparticles. Audier et al. (1981) studied the crystallographic characteristics of the catalyst particles at the end of carbon filaments (tip-growth). They found a correlation between the particle crystallographic orientation and the position of the tube axis. More recently, Ermakova et al. (2001) stated that hydrocarbon decomposition on nickel occurs on different edges of the nanoparticle due to anisotropy of nickel and that the filament axis is found to be parallel to the nickel (111) planes. The theoretical work of Shibuta and Maruyama (2003) indicates a strong interaction between the hexagonal carbon network formed in the first stage of the growth process and the crystallographic structure of catalytic metal atoms. The crystal orientation may thus play a role in the determination of chirality. Similarly, Vinciguerra et al. (2003) showed that the (110) plane of iron (bbc) and the (111) planes of cobalt and nickel (fcc) have the symmetry and distances to overlap with the lattice of graphene sheet. All discussions about the effect of crystallographic orientation of catalyst nanoparticles on growth of CNTs make sense only if the nanoparticle remains solid during the growth process.

As seen below, it is essential, in order to get CNTs of a given diameter, to have the ability to produce nanoparticles strictly controlling their size. The chosen method to get catalyst nanoparticles, therefore, has a critical influence on the catalytic growth process via the morphology of the obtained nanoparticles. Various preparation methods of nanoparticles have been listed previously; they may yield different catalytic properties.

Chhowalla et al. (2001) prepared iron-molybdenum nanoparticles by thermal decomposition of their carbonyl complexes in octyl ether solution under a nitrogen atmosphere. They investigated the dependence of the particle size on different parameters such as the reactant concentration, reaction time, and molar ratio of metal carbonyl and protective agents. In such a complex system, the size of the produced nanoparticles depends on a great number of factors, including the number of nuclei created, the total concentration of reactants, and the effect of protective agents. All these factors influence the size of the obtained products.

Vander Wal et al. (2001) compared the decomposition of a metal precursor on a metal oxide surface with ferrofluid, i.e., magnetite particles stabilized as a water-based suspension, finding that the morphology and size of the resulting nanotubes or nanofibers is much more uniform with the ferrofluid. Keeping in mind that the size of the nanotubes is directly dependent on the size of the catalyst particle, the authors explained this result by considering that the morphology of the preformed catalyst particle does not depend on the support. The size distribution of the particles will thus roughly stay the same on the support as in the solution. In contrast, the size of in situ–synthesized catalyst particles, produced by decomposition of a precursor on a support, is determined by the physical structure of the support, in particular by its void volume. Therefore, with in situ synthesized catalyst particles, a further difficulty still remains, since the growth process implies high temperatures (500°C–1200°C) and therefore can yield sintering of the nanoparticles. Ago et al. (2000) indeed prepared well-separated cobalt nanoparticles using the reverse micelle method, finding a mean diameter of 4 nm. Nevertheless, the inner diameters of the obtained MWNTs ranged from 8 to 20 nm and the outer ones from 20 to 40 nm, indicating that the as-prepared cobalt particles aggregate to form bigger clusters.

Ivanov et al. (1994) obtained a better metal dispersion on $SiO_2$ with precipitation-ion-exchange than with impregnation. In our previous work (Ciambelli et al. 2007) we have found that, starting from the same active phase prepared in the same way, carbon nanotubes with different diameters and length, as a consequence of different deactivation phenomena, can be obtained. The results also suggest that not only the support used, with its specific porosity determines the geometrical characteristics of the produced materials, but also the preparation step and the pretreatment conditions.

Recently (Altavilla et al. 2009), we have demonstrated the possibility of using monodispersed ferrite nanoparticles (oleic acid-capped $CoFe_2O_4$), obtained by a wet chemistry synthesis (Figure 18.18), as catalyst for the growth of CNT (Figure 18.19) by CCVD on spin coating covered silicon substrate.

**FIGURE 18.18**
TEM image of $CoFe_2O_4$ nanoparticles; the insert shows the electron diffraction pattern.

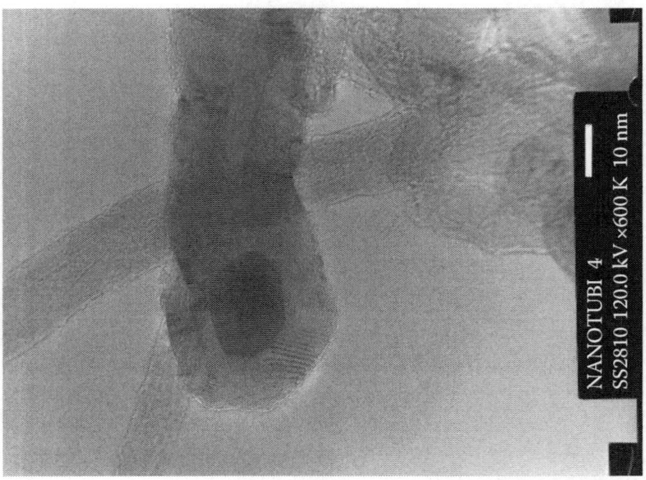


**FIGURE 18.19**
TEM image of a catalytic nanoparticle enclosed in the nanotube.

The $CoFe_2O_4$ nanoparticles on silicon catalyzed the growth of carbon nanotubes. The morphology of MWNT seems to be strongly dependent on the thickness and density of catalytic film. A more concentrated solution gives rise to the formation, at the end of the pretreatment, of larger particles, which leads to the formation of MWNTs with larger diameters. A close correlation of nanotube diameter with nanoparticle size has been verified. Additionally, CNT density and length can also be controlled by NPs concentration.

## Acknowledgments

Contributions by Vincenzo Vaiano, PhD, for SSA, band gap, and photoreactivity data and by Pierre Eloy and Eric Gaigneaux for XPS data of Figure 18.11 are acknowledged.

## References

Ago, H., Komatsu, T., Ohshima, S. et al. 2000. Dispersion of metal nanoparticles for aligned carbon nanotube arrays. *Applied Physics Letters* 77: 79–81.

Alfano, O. M., Bahnemann, D., Cassano, A. E., Dillert, R., and Goslich, R. 2000. Photocatalysis in water environment using artificial and solar light. *Catalysis Today* 58: 199–230.

Alstrup, I. 1988. New model explaining carbon filament growth on nickel, iron, and Ni-Cu alloy catalysts. *Journal of Catalysis* 109: 241–251.

Altavilla, C., Sarno, M., and Ciambelli, P. 2009. Synthesis of ordered layers of monodisperse $CoFe_2O_4$ nanoparticles for catalyzed growth of carbon nanotubes on silicon substrate. *Chemistry of Materials* 21: 4851–4858.

Alvarez, W., Kitiyana, N. B., Borgna, A. et al. 2001. Synergism of Co and Mo in the catalytic production of single-wall carbon nanotubes by decomposition of CO. *Carbon* 39: 547–558.

Audier, M., Coulon, M., and Bonnetain, L. 1983a. Disproportionation of CO on iron-cobalt alloys—I: Thermodynamic study. *Carbon* 21: 93–97.

Audier, M., Coulon, M., and Bonnetain, L. 1983b. Disproportionation of CO on iron-cobalt alloys—II: Kinetic study on iron-cobalt alloys of different compositions. *Carbon* 21: 99–103.

Audier, M., Coulon, M., and Bonnetain, L. 1983c. Disproportionation of CO on iron-cobalt alloys—III: Kinetic laws of the carbon growth and catalyst fragmentation. *Carbon* 21: 105–110.

Audier, M., Oberlin, A., and Coulon, M. 1981. Crystallographic orientations of catalytic particles in filamentous carbon: Case of simple conical particles. *Journal of Crystal Growth* 55: 549–556.

Baker, R. T. K., Barber, M. A., Harris, P. S. et al. 1972. Nucleation and growth of carbon deposits from the nickel catalyzed decomposition of acetylene. *Journal of Catalysis* 26: 51–62.

Baker, R. T. K., Harris, P. S., Thomas, R. B. et al. 1973. Formation of filamentous carbon from iron, cobalt and chromium catalyzed decomposition of acetylene. *Journal of Catalysis* 30: 86–95.

BCC Research. 2007. Nanostructured materials for energy, catalysis & structural applications, Report NANO17E.

Bell, A. T. 2003. The impact of nanoscience on heterogeneous catalysis. *Science* 299: 1688–1691.

Beydoun, D., Amal, R., Low, G., and McEvoy, S. 1999. Role of nanoparticles in photocatalysis. *Journal of Nanoparticle Research* 1: 439–458.

BizAcumen. 2009. Nanocatalysts—International technology and market trends, Report BIZ00011.

Buffat, P. and Borel, J. P. 1976. Size effect on the melting temperature of gold particles. *Physical Review A* 13: 2287–2298.

Buyanov, R. A. 1983. *Zakoksovanie Katalizatorov* [Coking of catalysts], Novosibirsk, Nauka.

Chen, F., Wang, J., Xu, J. Q., and Zhou, X. P. 2008. Visible light photodegradation of organic compounds over $V_2O_5/MgF_2$ catalyst. *Applied Catalysis* A 348: 54–59.

Chen, P., Zhang, H. B., Lin, G. D. et al. 1997. Growth of carbon nanotubes by catalytic decomposition of $CH_4$ or CO on a Ni–MgO catalyst. *Carbon* 35: 1495–1501.

Cheng, H. M., Li, S. F., Pan, G. et al. 1998. Large-scale and low-cost synthesis of single-walled carbon nanotubes by the catalytic pyrolysis of hydrocarbons. *Applied Physics Letters* 72: 3282–3284.

Chesnokov, V. V., Buyanov, R. A., and Afanasev, A. D. 1982. Izv. Sib. Otd. Akad. Nauk SSSR. Ser. Khim. 60

Cheung, C. L., Kurtz, A., Park, H. et al. 2002. Diameter-controlled synthesis of carbon nanotubes. *Journal of Physical Chemistry B* 106: 2429–2433.

Chhowalla, M., Teo, K. B. K., Ducati, C. et al. 2001. Growth process conditions of vertically aligned carbon nanotubes using plasma enhanced chemical vapour deposition. *Journal of Applied Physics* 90: 5308–5312.

Chiarello, G. L., Rossetti, I., and Forni, L. 2005. Flame-spray pyrolysis preparation of perovskites for methane catalytic combustion. *Journal of Catalysis* 236: 251–261.

Chiarello, G. L., Selli, E., and Forni, L. 2008. Photocatalytic hydrogen production over flame spray pyrolysis-synthesized $TiO_2$ and Au/$TiO_2$. *Applied Catalysis B: Environmental* 84: 332–339.

Chico, L., Crespi, V. H., Benedict, L. X. et al. 1996. Pure carbon nanoscale devices: Nanotube heterojunctions. *Physical Review Letters* 76: 971–974.

Choi, H. C., Kim, W., Wang, D. et al. 2002. Delivery of catalytic metal species onto surfaces with dendrimer carriers for the synthesis of carbon nanotubes with narrow diameter distribution. *Journal of Physical Chemistry B* 106: 12361–12365.

Ciambelli, P., Bagnasco, G., Lisi, L. et al. 1992. Vanadium oxide catalysts supported on laser-synthesized titania powders: Characterization and catalytic activity in the selective reduction of nitric oxide. *Applied Catalysis B: Environmental* 1: 61–77.

Ciambelli, P., Sannino, D., Sarno, M. et al. 2005. Selective formation of carbon nanotubes over comodified beta zeolite by CCVD. *Carbon* 43: 631–640.

Ciambelli, P., Sannino, D., Sarno, M. et al. 2007. Effects of alumina phases and process parameters on the multiwalled carbon nanotubes growth. *Diamond and Related Materials* 16: 1144–1149.

Ciambelli, P., Sannino, D., Palma, V., and Vaiano, V. 2008. The effect of sulphate doping on nanosized $TiO_2$ and MoOx/$TiO_2$ catalysts in cyclohexane photooxidative dehydrogenation. *International Journal of Photoenergy*. Article ID 258631, 8 pages doi:10.1155/2008/258631.

Ciambelli, P., Sannino, D., Palma, V., Vaiano, V., and Mazzei, R. S. 2009a. A step forwards in ethanol selective photo-oxidation. *Photochemical and Photobiological Sciences* 8: 699–704.

Ciambelli, P., Sannino, D., Palma, V., and Vaiano, V. 2009b. Photocatalytic fluidized bed reactor with high illumination efficiency for photocatalytic oxidation processes. PCT application WO2009IT00239. Università di Salerno.

Dai, H., Rinzler, A. G., Nikolaev, P. et al. 1996. Single-wall nanotubes produced by metal-catalyzed disproportionation of carbon monoxide. *Chemical Physics Letters* 260: 471–475.

De Bokx, P. K., Kock, A. J. H. M., Boellaard, E. et al. 1985. The formation of filamentous carbon on iron and nickel catalysts. *Journal of Catalysis* 96: 454–467.

de Heer, W. A., Bonard, J. M., Fauth, K. et al. 1997. Electron field emitters based on carbon nanotube films. *Advanced Materials* 9: 87–89.

de los Arcos, T., Garnier, M. G., Seo, J. W. et al. 2004. The influence of catalyst chemical state and morphology on carbon nanotube growth. *Journal of physical Chemistry B* 108: 7728–7734.

de los Arcos, T., Oelhafen, P., Thommen, V. et al. 2007. The influence of catalyst's oxidation degree on carbon nanotube growth as a substrate-independent parameter. *Journal of Physical Chemistry C* 111: 16392–16396.

Diebold, U. 2003. The surface science of $TiO_2$. *Surface Science Reports* 48: 53–229.

Ding, F., Bolton, K., and Rosen, A. 2004. Iron-carbide cluster thermal dynamics for catalysed carbon nanotube growth. *The Journal of Vacuum Science and Technology A* 22: 1471–1476.

Ding, F., Larsson, P., Larsson, A. et al. 2008. The importance of strong carbon–metal adhesion for catalytic nucleation of single-walled carbon nanotubes. *Nano Letters* 8: 463–468.

Ducati, C. I., Alexandrou, M., Chhowalla, J. et al. 2004. The role of the catalytic particle in the growth of carbon nanotubes by plasma enhanced chemical vapor deposition. *Journal of Applied Physics* 95: 6387–6391.

Duesberg, G. S., Graham, A. P., Liebau, M. et al. 2003. Growth of isolated carbon nanotubes with lithographically defined diameter and location. *Nano Letters* 3: 257–259.

Ermakova, M. A., Ermakov, D. Y., Chuvilin, A. L. et al. 2001. Decomposition of methane over iron catalysts at the range of moderate temperatures: The influence of structure of the catalytic systems and the reaction conditions on the yield of carbon and morphology of carbon filaments. *Journal of Catalysis* 201: 183–197.

Evans, E. L., Thomas, J. M., Thrower, P. A. et al. 1973. Growth of filamentary carbon on metallic surfaces during the pyrolysis of methane and acetone. *Carbon* 11: 441–445.

Fonseca, A., Hernadi, K., Nagy, J. B. et al. 1996. Optimization of catalytic production and purification of buckytubes. *Journal of Molecular Catalysis A: Chemistry* 107: 159–168.

Fujishima, A. and Honda, K. 1972. Electrochemical photolysis of water at a semiconductor electrode. *Nature* 238: 37–38.

Ghicov, A. and Schmuki, P. 2009. Self-ordering electrochemistry: A review on growth and functionality of $TiO_2$ nanotubes and other self-aligned $MO_x$ structures. *Chemical Communications* 20: 2791–2808.

Gracia, L., Catatayud, M., Minot, J. A. C. et al. 2005. Migration of the subsurface C impurity in Pd(111). *Physical Review B* 71: 033407 (4 pages).

Gregg, S. J. and Sing, K. S. W. 1982. *Adsorption, Surface Area, and Porosity*, Academic Press, London, U.K.

Hafizah, N. and Sopyan, I. 2009. Nanosized $TiO_2$ photocatalyst powder via sol-gel method: Effect of hydrolysis degree on powder properties. *International Journal of Photoenergy* Article ID 962783, 8 pages doi:10.1155/2009/962783.

Hafner, J. H., Cheung, C. L., and Lieber, C. M. 1999. Growth of nanotubes as probe microscopy tips. *Nature* 398: 761.

Harutyunyan, A. R., Pradhan, B. K., Kim, U. J. et al. 2002. CVD synthesis of single wall carbon nanotubes under "soft" conditions. *Nano Letters* 2: 525–530.

Herrmann, J.-M. 1999. Heterogeneous photocatalysis: Fundamentals and applications to the removal of various types of aqueous pollutants. *Catalysis Today* 53: 115–129.

Herrmann, J. M., Guillard, C., and Disdier, J. 2002. New industrial titania photocatalysts for the solar detoxification of water containing various pollutants. *Applied Catalysis B: Environmental* 35: 281–294.

Hjeresen, D. L., Kirchhoff, M. M. E., and Lankey, R. L. 2002. Green chemistry: Environment, economics, and competitiveness. *Corporate Environmental Strategy* 9: 259–266.

Hoffmann, M. R., Martin, S. T., Choi, W., and Bahnemann, D. 1995. Environmental application of semiconductor photocatalysis. *Chemical Review*. 95: 69–96.

Hofmann, S., Ducati, C., Kleinsorge, B. et al. 2003. Low-temperature growth of carbon nanotubes by plasma-enhanced chemical vapor deposition. *Applied Physics Letters* 83: 135–137.

Hofmann, S., Csanyi, G., Ferrari, A. C. et al. 2005. Electron transport and hot photons in carbon nanotubes. *Physical Review Letters* 95: 036101 (4 pages).

Hofmann, S., Sharma, R., Ducati, C. et al. 2007. In situ observations of catalyst dynamics during surface-bound carbon nanotube nucleation. *Nano Letters* 7: 602–608.

Hofmann, S., Blume, R., Wirth, C. T. et al. 2009. State of transition metal catalysts during carbon nanotube growth. *Journal of Physical Chemistry C* 113: 1648–1656.

Iijima, S. 1991. Helical microtubules of graphitic carbon. *Nature* 354: 56.

Ivanov, V., Nagy, J. B., Lambin, P. et al. 1994. The study of carbon nanotubules produced by catalytic method. *Chemical Physics Letters* 223: 329–335.

Jensen, H., Joensen, K. D., Jørgensen, J.-E., Pedersen, J. S., and Søgaard, E. G. 2004. Characterization of nanosized partly crystalline photocatalysts. *Journal of Nanoparticle Research* 6: 519–526.

Jitputti, J., Suzuki, Y., and Yoshikawa, S. 2008. Synthesis of $TiO_2$ nanowires and their photocatalytic activity for hydrogen evolution. *Catalysis Communications* 9: 1265–1271.

Jung, K. Y., Park, S. B., and Ihm, S.-K. 2002. Linear relationship between the crystallite size and the photoactivity of non-porous titania ranging from nanometer to micrometer size. *Applied Catalysis A: General* 22: 229–237.

Kamat, P. V. 1995. Tailoring nanostructured thin films. *Chemtech* 25: 22–28.

Kang, Y. C., Park, S. B., and Kwon, S. W. 1996. Preparation of submicron size gamma lithium aluminate particles from the mixture of alumina sol and lithium salt by ultrasonic spray pyrolysis. *Journal of Colloid Interface Science* 182: 59–62.

Keep, C. W., Baker, R. T. K., and France, J. A. 1977. Origin of filamentous carbon formation from the reaction of propane over nickel. *Journal of Catalysis* 47: 232–238.

Kasuga, T., Hiramatsu, M., Hoson, A., Sekino, T., and Niihara, K. 1999. Titania nanotubes prepared by chemical processing. *Advanced Materials* 11: 307–311.

Klinke, C., Bonard, J. M., and Kern, K. 2001. Comparative study of the catalytic growth of patterned carbon nanotube films. *Surface Science* 492: 195–201.

Kong, J., Franklin, N., Zhou, C. et al. 2000. Nanotube molecular wires as chemical sensors. *Science* 287: 622–625.

Kormann, C., Bahnemann, D. W., and Hoffmann, M. 1988. Preparation and characterisation of quantum-size titanium dioxide. *Journal of Physical Chemistry* 92: 5196–5201.

Kung, H. H. and Kung, M. C. 2003. Heterogeneous catalysis: What lies ahead in nanotechnology. *Applied Catalysis A* 246: 193–196.

Kuvshinov, G. G., Zavarukhin, S., Yu Mogilnykh, I. et al. 1998. Implementation of production of granular catalytic filamentous carbon on the scale of a pilot reactor. *Khim. Promst* 5: 48.

Lambert, R. M. and Pacchioni, G. 1997. *Chemisorption and Reactivity on Supported Clusters and Thin Films*. NATO ASI Series E,331. Kluwer, Dordrecht, the Netherlands.

Li, Y. L., Kinloch, I. A., and Windle, A. H. 2006. Direct spinning of carbon nanotube fibers from chemical vapour deposition synthesis. *Science* 304: 276–278.

Liao, X. Z., Serquis, A., Jia, Q. X. et al. 2003. Effect of catalyst composition on carbon nanotube growth. *Applied Physics Letters* 82: 2694.

Linsebigler, A. L., Lu, G., and Yates, J. T. Jr. 1995. Photocatalysis on $TiO_2$ surfaces: Principles, mechanisms, and selected results. *Chemical Review* 95: 735–758.

Lopez, N., Janssens, T. V. W., Clausen, B. S. et al. 2004. On the origin of the catalytic activity of gold nanoparticles for low-temperature CO oxidation. *Journal of Catalysis* 223: 232–235.

Mao, L., Li, Q., Dang, H., and Zhang, Z. 2005. Synthesis of nanocrystalline $TiO_2$ with high photoactivity and large specific surface area by sol-gel method. *Materials Research Bulletin* 40: 201–208.

Mattevi, C., Wirth, C. T., Hofmann, S. et al. 2008. In-situ x-ray photoelectron spectroscopy study of catalyst–support interactions and growth of carbon nanotube forests. *Journal of Physical Chemistry C* 112: 12207–12213.

Musci, M., Notaro, M., Curcio, F., Casale, C., and De Michele, G. 1992. Laser synthesis of vanadium-titanium oxide catalysts. *Journal of Materials Research* 7: 2846–2852.

Frost & Sullivant. 2008. Nanocatalysts—Application impact analysis, Report FRSO 1073, December 2008.

NANOTECH IT. 2004. Newsletter. 1, April 2004.

Ni, M., Leung, M. K. H., Leung, D. Y. C., and Sumathy, K. 2007. A review and recent developments in photocatalytic water-splitting using $TiO_2$ for hydrogen production. *Renewable and Sustainable Energy Reviews* 11: 401–425.

Nielsen, J. R. and Trimm, D. L. 1977. Mechanism of carbon formation on Ni-containing catalysts. *Journal of Catalysis* 48: 155–165.

Nikolaev, P., Bronikowski, M., Bradley, R. et al. 1999. Gas-phase catalytic growth of single-walled carbon nanotubes from carbon monoxide. *Chemical Physics Letters* 313: 91–97.

Niu, C., Sichel, E. K., Hoch, R. et al. 1997. High power electrochemical capacitors based on carbon nanotube electrodes. *Applied Physics Letters* 70: 1480–1482.

Pacchioni, G., Sicolo, S., Di Valentino, C. 2008. A route toward the generation of thermally stable Au cluster anions supported on the MgO surface. *Journal of the American Chemical Society* 130: 8690–8695.
Palmisano, L. and Sclafani, A. 1997. Thermodynamics and kinetics for heterogeneous photocatalytic processes, In *Heterogeneous Photocatalysis* ed. M. Schiavello. John Wiley & Sons, London, U.K., p. 109.
Perez-Cabero, M., Rodrıguez-Ramos, I., and Guerrero-Ruız, A. 2003. Characterization of carbon nanotubes and carbon nanofibers prepared by catalytic decomposition of acetylene in a fluidized bed reactor. *Journal of Catalysis* 215: 305–316.
Pernicone, N. 2003. Catalysis at the nanoscale level. *Cattech* 7: 196–204.
Pinheiro, J. P. and Gadelle, P. 2001. Chemical state of a supported iron–cobalt catalyst during CO disproportionation. I. Thermodynamic study. *Journal of Physical Chemistry Solids* 62: 1015–1021.
Porter, J. F., Li, Y. -G. and Chan, C. K. 1999. The effect of calcination on the microstructural characteristics and photoreactivity of degussa P-25 $TiO_2$. *Journal of Materials Science* 34: 1523–1531.
Prado, A., Leonardo, G. S., and Costa, L. 2009. Photocatalytic decoloration of malachite green dye by application of $TiO_2$ nanotubes. *Journal of Hazardous Materials* 169: 297–301.
Qian, H., Xia, T., Yun-Jie, Y., and Yi, M. 2010. Optimal synthesis of mesostructured hollow titania nanotubes templated on $CaCO_3$ nanoparticles. *Powder Technology* 198: 429–434.
Raty, J. V., Gygi, F., and Galli, G. 2005. Growth of carbon nanotubes on metal nanoparticles: Dynamics simulations. *Physical Review Letters* 95: 096103.
Robertson, S. D. 1969. Carbon formation from methane pyrolysis over some transition metal surfaces—I. Nature and properties of the carbons formed. *Nature* 221: 1044–1046.
Ruban, A., Hammer, B., Stoltze, P. et al. 1997. Surface electronic structure and reactivity of transition and noble metals. *Journal of Molecular Catalysis A Chemistry* 115: 421–429.
Sacco, Jr., A., Thacker, P., Chang, T. N. et al. 1984. The initiation and growth of filamentous carbon from α-iron in $H_2$, $CH_4$, $H_2O$, $CO_2$, and CO gas mixtures. *Journal of Catalysis* 85: 224–236.
Schlögl, R. and Abd Hamid, S. B. 2004. Nanocatalysis: Mature science revisited or something really new? *Angewandte Chemie International Edition* 43: 1628–1637.
Shibuta, Y. and Maruyama, S. 2003. Molecular dynamics simulation formation process of single-walled carbon nanotubes by CCVD method. *Chemical Physics Letters* 382: 381–386.
Somorjai, G. A. and Borodko, Y. G. 2001. Research in nanosciences—Great opportunity for catalysis science. *Catalysis Letters* 76: 1–5.
Sreekantan, S. and Hung, L. M. 2009. Effects of electrolyte nature on the formation of well-ordered Titania nanotubes produced via anodisation. *Journal of Physical Science* 20: 61–71.
Subramanian, V., Wolf, E. E., and Kamat, P. 2004. Catalysis with $TiO_2$/gold nanocomposites. Effect of metal particle size on the fermi level equilibration. *Journal of the American Chemical Society* 126: 4943–4950.
Takagi, D. Y., Homma, H., Hibino, S. et al. 2006. Single-walled carbon nanotube growth from highly activated metal nanoparticles. *Nano Letters* 6: 2642–2645.
Teoh, W. Y., Denny, F., Friedmann, D. et al. 2007. Photocatalytic mineralisation of organic compounds: A comparison of flame-made $TiO_2$. catalysts. *Topics in Catalysis* 44: 489–497.
Teschner, D., Vass, A., Havecker, M. et al. 2006. Alkyne hydrogenation over Pd catalysts: A new paradigm. *Journal of Catalysis* 242: 26–37.
Thimsen, E., Rastgar, N., and Biswas, P. 2008. Nanostructured $TiO_2$ films with controlled morphology synthesized in a single step process: Performance of dye-sensitized solar cells and photo water splitting. *Journal of Physical Chemistry C* 112: 4134–4140.
Tibbetts, G. G., Devour, M. G., and Rodda, E. J. et al. 1987. An adsorption-diffusion isotherm and its application to the growth of carbon filaments on iron catalyst particles. *Carbon* 25: 367–375.
Tibbetts, G. G. 1984. Why are carbon filaments tubular? *Journal of Crystal Growth* 66: 632–638.
Tseng, Y.-H., Liu, S.-L., and Kuo, C.-S. 2006. Preparation of nano-sized $TiO_2$ sol and its visible-light-responsive photocatalysis in aquatic state. *Micro & Nano Letters* online no. 20065035 doi:10.1049/mnl:20065035 ristretto

Vander Wal, R. L., Ticich, T. M., Curtis, V. E. et al. 2001. Substrate–support interactions in metal-catalyzed carbon nanofiber growth. *Carbon* 39: 2277–2289.

Vass, E. M., Havecker, M., Zafeiratos, Z. et al. 2008. The role of carbon species in heterogeneous catalytic processes: An in situ soft x-ray photoelectron spectroscopy study. *Journal of Physical Condensed Matter* 20: 184016 (13 pages).

Vidal, A. and Martín Luengo, M. A. 2001. Inactivation of titanium dioxide by sulphur: Photocatalytic degradation of Vapam®. *Applied Catalysis B: Environmental* 32: 1–9.

Vinciguerra, V., Buonocore, F., Panzera, G. et al. 2003. Growth mechanisms in chemical vapour deposited carbon nanotubes. *Nanotechnology* 14: 655–660.

Wada, T., Wada, H., Elliott, J. F. et al. 1971. Thermodynamics of the fcc Fe-Ni-C and Ni-C alloys. *Metallic Transaction* 2: 2199–2208.

Yang, R. L. and Yang, K. T. 1985. Evidence for temperature driven carbon diffusion mechanism of coke deposition on catalysts. *Journal of Catalysis* 93: 192–198.

Yang, R. T. and Chen, J. P. 1989. Mechanism of carbon filament growth on metal catalysts. *Journal of Catalysis* 115: 52–64.

Yoshida, H., Takeda, S., Uchiyama, T. et al. 2008. Atomic-scale in-situ observation of carbon nanotube growth from solid state iron carbide nanoparticles. *Nano Letters* 8: 2082–2086.

Zhang, R. and Gao, L. 2002. Preparation of nanosized titania by hydrolysis of titanium alkoxide in micelles. *Materials Research Bulletin* 37: 1659–1666.

Zhang, W. D., Wen, Y., Tjiu, W. C. et al. 2002a. Growth of vertically aligned carbon-nanotube array on large area of quartz plates by chemical vapor deposition. *Applied Physics A* 74: 419–422.

Zhang, Y., Li, Y., Kim, W. et al. 2002b. Imaging as-grown single-walled carbon nanotubes originated from isolated catalytic nanoparticles. *Applied Physics A* 74: 325–328.

Zhu, J., Yang, D., Geng, J., Chen, D., and Jiang, Z. 2008. Synthesis and characterization of bamboo-like $CdS/TiO_2$ nanotubes composites with enhanced visible-light photocatalytic activity. *Journal of Nanoparticles Research* 10: 729–736.

Zhu, J. and Zäch, M. 2009. Nanostructured materials for photocatalytic hydrogen production. *Current Opinion in Colloid and Interface Science* 14: 260–269.

# Index

## D



## E



## F

## Q



## R